OCCUPATIONAL SAFETY AND HYGIENE VI

SELECTED CONTRIBUTIONS FROM THE INTERNATIONAL SYMPOSIUM OCCUPATIONAL SAFETY AND HYGIENE (SHO 2018), GUIMARÃES, PORTUGAL, 26–27 MARCH 2018

Occupational Safety and Hygiene VI

Editors

Pedro M. Arezes
University of Minho, Guimarães, Portugal

João Santos Baptista
University of Porto, Porto, Portugal

Mónica P. Barroso, Paula Carneiro, Patrício Cordeiro & Nélson Costa
University of Minho, Guimarães, Portugal

Rui B. Melo
Technical University of Lisbon, Cruz Quebrada—Dafundo, Portugal

A.S. Miguel & Gonçalo Perestrelo
University of Minho, Guimarães, Portugal

CRC Press
Taylor & Francis Group
Boca Raton London New York Leiden

CRC Press is an imprint of the
Taylor & Francis Group, an **informa** business

A BALKEMA BOOK

CRC Press/Balkema is an imprint of the Taylor & Francis Group, an informa business

© 2018 Taylor & Francis Group, London, UK

Typeset by V Publishing Solutions Pvt Ltd., Chennai, India
Printed and bound in Great Britain by CPI Group (UK) Ltd, Croydon, CR0 4YY

All rights reserved. No part of this publication or the information contained herein may be reproduced, stored in a retrieval system, or transmitted in any form or by any means, electronic, mechanical, by photocopying, recording or otherwise, without written prior permission from the publisher.

Although all care is taken to ensure integrity and the quality of this publication and the information herein, no responsibility is assumed by the publishers nor the author for any damage to the property or persons as a result of operation or use of this publication and/or the information contained herein.

Published by: CRC Press/Balkema
   Schipholweg 107C, 2316 XC Leiden, The Netherlands
   e-mail: Pub.NL@taylorandfrancis.com
   www.crcpress.com – www.taylorandfrancis.com

ISBN: 978-1-138-54203-7 (Hardback)
ISBN: 978-1-351-00888-4 (eBook)

Table of contents

Foreword	xi
List of reviewers	xiii
A new approach—use of effectiveness and efficiency concepts in the identification of activities related to occupational safety management on motorways *J.E.M.R. Silva & C. Rodrigues*	1
Sampling methods for an accurate mycobiota occupational exposure assessment—overview of several ongoing projects *C. Viegas*	7
A systematic literature review on Thermal Response Votes (TSV) and Predicted Mean Vote (PMV) *E.E. Broday & A.A. de P. Xavier*	13
Participatory ergonomics training for a warehouse environment—a process solution *A. Mocan & A. Draghici*	19
Wheelchair users' anthropometric data: Analysis of existent available information *S. Bragança, I. Castellucci & P. Arezes*	23
A review on occupational risk in gasification plants processing residues of sewage sludge and refuse-derived fuel *O. Alves, M. Gonçalves, P. Brito, E. Monteiro & C. Jacinto*	29
Safety assessment of Ammonium Nitrate Fuel Oil (ANFO) manufactory *D.S. Bonifácio, E.B.F. Galante & A.N. Haddad*	35
The relevance of surfaces contamination assessment in occupational hygiene interventions in case of exposure to chemicals *S. Viegas*	41
Optimization of the CoPsoQ questionnaire in a population of Archipelago of Azores *F.O. Nunes & A. Gaspar da Silva*	47
Exposure to occupational noise in police—a systematic review *A.C. Reis & M. Vaz*	53
Geostatistics applied to noise exposure—a systematic review *A.C. Reis & J.M. Carvalho*	61
Muscle fatigue assessment in manual handling of loads using motion analysis and accelerometers: A short review *F.G. Bernardo, R.P. Martins & J.C. Guedes*	67
Evaluation of physical fatigue based on motion analysis *F.G. Bernardo, R.P. Martins & J.C. Guedes*	73
Maintenance and safety according to the Brazilian performance standard: NBR 15.575 *M.A.S. Hippert & O.C. Longo*	79
Sensation seeking and risk perception as predictors of physical and psychosocial safety behavior in risk and non-risk professions *G. Gonçalves, C. Sousa, M. Pereira, E. Pinto & A. Sousa*	85

Work ability, ageing and musculoskeletal symptoms among Portuguese municipal workers 91
C.A. Ribeiro, T.P. Cotrim, V. Reis, M.J. Guerreiro, S.M. Candeias, A.S. Janicas & M. Costa

Psychological health and wellbeing in intralogistics workplaces—an empirical analysis 97
M. Hartwig & E. Mamrot

Detailed evaluation of human body cooling techniques through thermo-physiological modelling 103
A.M. Raimundo, D.A. Quintela, A. Virgílio & M. Oliveira

Air quality assessment in copy centers 109
M. Cunha, A. Ferreira, F. Silva & J.P. Figueiredo

The contribution of digital technologies to construction safety 115
D. Pinto, F. Rodrigues & J. Santos Baptista

Tower crane safety: Organizational preventive measures vs. technical conditions 121
J.A. Carrillo-Castrillo, J.C. Rubio-Romero & M.C. Pardo-Ferreira

Occupational and public exposure to radionuclides in smoke from forest fires—a warning 125
F.P. Carvalho, J.M. Oliveira & M. Malta

Annual school safety activity calendar to promote safety in VET 131
S. Tappura & J. Kivistö-Rahnasto

Accidents prevention at home with elderly 137
E.M.G. Lago, B. Barkokébas Jr., F.M. da Cruz, A.R.B. Martins, B.M. Vasconcelos, T. Zlatar, R. Manta & S. Porto

Occupational safety and health for workers managing wild animals 143
A.R.B. Martins, B. Barkokébas Jr., E.M.G. Lago, B.M. Vasconcelos, F.M. da Cruz, T. Zlatar, T.C.M. de França & L.R. Pedrosa

Pulmonary disease due to exposure to nanoparticles 149
B. Calvo-Cerrada, A. López-Guillén, P. Sanz-Gallen, G. Martí-Armengual & A. López-Guillen

Assessment of air quality in professional kitchens of the city of Coimbra 155
A. Ferreira, A. Lança, D. Barreira, F. Moreira & J.P. Figueiredo

Analyse of occupational noise and whole-body vibration exposure at bus drivers—case study 161
A.F. Teixeira, M.L. Matos, L. Pedrosa & P. Costa

Evaluation of the infrastructural conditions and practices of the manipulators—case study: Nursing home and day center 165
I.N. Roxo, C. Santos, J.P. Figueiredo & A. Ferreira

Neuromuscular fatigue assessment in operators performing occupational-transportation/ transshipment tasks: A short systematic review 171
T. Sa-ngiamsak, N. Phatrabuddha & T. Yingratanasuk

BIM (Building Information Modelling) as a prevention tool in the design and construction phases 177
M. Tender, J.P. Couto, C. Lopes, T. Cunha & R. Reis

Recent trends in safety science and their importance for operations practice 183
V. Rigatou, K. Fotopoulou, E. Sgourou, P. Katsakiori & Em. Adamides

Latest efforts aimed at upgrading the IMS-MM 189
C. Saraiva, P. Domingues, P. Sampaio & P. Arezes

Exposure to particles in the poultry sector 195
F. Capitão, A. Ferreira, F. Moreira & J.P. Figueiredo

Scale psychometrics of the Portuguese short and middle length variants of the Copenhagen Psychosocial Questionnaire III pilot study 201
D.A. Coelho & M.L. Lourenço

Evaluation of occupational safety and health conditions in a Brazilian public hospital 207
C.F. Machado & J.S. Nóbrega

Levels of non-ionizing radiation in vertical residences in João Pessoa (PB) 213
A.A.R. Silva, L.B. Silva, R.M. Silva & R.B.B. Dias

Human error prevention by training design: Suppressing the Dunning-Kruger effect in its infancy 219
E. Stojiljković, A. Bozilov, T. Golubovic, S. Glisovic & M. Cvetkovic

Workload and work capacity of the public agent at a department of a federal institution of higher education 225
L.L. de F. Cavalcanti & L.B. Martins

Exposure to psychosocial risks in footwear industry workers: An exploratory analysis 231
D. Sousa, M.M. Sá, R. Azevedo, O. Machado, J. Tavares & R. Monteiro

Inclusive, interdisciplinary and digital-based OSH resources: A booklet for students 235
F. Rodrigues, F. Antunes, A.M. Pisco Almeida, J. Beja, L. Pedro, M. Clemente, R. Neves & R. Vieira

When the working hours become a risk factor: The debate about $3 \times 8h$ and $2 \times 12h$ shifts 239
L. Cunha, D. Silva, I. Ferreira, C. Pereira & M. Santos

Occupational risks in hospital mortuary 245
A. Ferreira, A. Lança, C. Mendes, M. Sousa & S. Paixão

Occupational exposure in fitness clubs to indoor particles 251
K. Slezakova, M.C. Pereira, C. Peixoto, C. Delerue-Matos & S. Morais

Occupational risks in hospital sterilization centers 257
A. Ferreira, A. Lança, C. Mendes, M. Sousa & S. Paixão

Chemical risks associated with the application of plant protection products 261
T. Neves, A. Ferreira, A. Lança, A. Baltazar & J.P. Figueiredo

Levels of urinary biomarkers of exposure and potential genotoxic risks in firefighters 267
M. Oliveira, C. Delerue-Matos, S. Morais, K. Slezakova, M.C. Pereira, A. Fernandes, S. Costa & J.P. Teixeira

Whole body vibration and acoustic exposure in construction and demolition waste management 273
M.L. de la Hoz-Torres, M.D. Martínez-Aires, M. Martín-Morales & D.P. Ruiz Padillo

Combined effects of exposure to physical risk factors: Impact on blood oxygen saturation (SpO_2) in construction workers 279
F.M. Cruz, P. Arezes, B. Barkokébas Jr. & E.B. Martins

Evaluation of professional exposure to nanoparticles in automotive repair activities 285
G. Oliveira, H. Simões, A. Ferreira, F. Silva & J.P. Figueiredo

Occupational exposure to bioaerosols in the waste sorting industry 291
V. Santos, J.P. Figueiredo, M. Vasconcelos Pinto & J. Santos

Nanomaterials *versus* safety of workers—a short review 297
D. da Conceição Peixoto Teixeira & P.A.A. Oliveira

Use of the 5S methodology for improving work environment 301
A. Górny

Investigation and analysis of work accidents at the Urban Hygiene Department in a Portuguese Municipality 307
P.G. da Fonseca, R.B. Melo & F. Carvalho

An investigation of forearm EMG normalization procedures in a field study 313
S. Aia & M. Reinvee

The impact of pesticides on the cholinesterase-activity in serum samples A. Teixeira, G. Oostingh, A. Valado, N. Osório, A. Caseiro, A. Gabriel, A. Ferreira & J.P. Figueiredo	319
Comparative assessment of work-related musculoskeletal disorders in an industrial kitchen D.M.B. Costa, R.V. Ferreira, E.B.F. Galante, J.S.W. Nóbrega, L.A. Alves & C.V. Morgado	325
Work-related musculoskeletal disorders in the transportation of dangerous goods M.V.T. Rabello, D.M.B. Costa & C.V. Morgado	331
Scaffold use risk assessment model: SURAM M. Rebelo, P. Laranjeira, F. Silveira, K. Czarnocki, E. Blazik-Borowa, E. Czarnocka, J. Szer, B. Hola & K. Czarnocka	335
The response of phagocytes to indoor air toxicity L. Vilén, J. Atosuo, E. Suominen & E.-M. Lilius	341
Modelling the effect of human presence in a single room with computational fluid dynamics simulation—a short review P.H. Moreira, J.C. Guedes & J. Santos Baptista	347
Bacteria bioburden assessment and MRSA colonization of workers and animals from a Portuguese swine production: A case report E. Ribeiro, A. Pereira, C. Vieira, I. Paulos, M. Marques, T. Swart & A. Monteiro	351
Assessment of fatigue through physiological indicators: A short review J.A. Pérez & J.C. Guedes	355
Workload mental analysis of the workers in a collective road transport company T.G. Lima, F.S.M. Gonçalves & R.M.F. Marinho	361
Ergonomic analysis of ballistic vests used by police officers in Paraíba State—Brazil G.A.M. Falcão, T.F. Chaves, M.B.F.V. Melo & L.V. Alves	367
Facilitators for safety of visually impaired on the displacements in external environments: A systematic review E.R. Araújo, C. Rodrigues & L.B. Martins	371
Impacts of accidents with workers in rock processing industries: Short review A. do Couto, J. Santos Baptista, B. Barkokébas Jr. & I. da Silva	377
Whole-body vibration in mining equipment—a short review J. Duarte, M.L. Matos, J. Castelo Branco & J. Santos Baptista	383
The effect of two training methods on workers' risk perception: A comparative study with metalworking small firms B.L. Barros, M.A. Rodrigues & A.R. Dores	389
Diagnoses of the acoustic perceptions of workers for auditory signal design G. Dahlke & T. Ptak	395
Evacuation study of a high-rise building in fire safety, prescriptive *versus* performance analysis A.C. Matos & M. Chichorro Gonçalves	401
Radiological characterization of the occupational exposure in hydrotherapy spa treatments M.L. Dinis & A.S. Silva	407
Labor claims and certification in occupational health and safety management F.D. Dionísio, N. Costa & C.P. Leão	413
Ergonomic risk measurement in prioritizing corrective actions at workstations W. Czernecka & A. Górny	419
Main mitigation measures—occupational exposure to radon in thermal spas A.S. Silva & M.L. Dinis	425
Safety performance and its measurement: An empirical study concerning leading and lagging indicators A. Job & I.S. Silva	431

Comparison between methods of assessing the risk of musculoskeletal disorders on upper limb extremities: A study in manual assembly work *S. Costa, P. Carneiro & N. Costa*	437
Stress misconduct and reduced ability to express dissent: A study on a sample of students at the University of Siena *O. Parlangeli, P. Palmitesta, M. Bracci, E. Marchigiani & P.M. Liston*	443
Wearable technology for occupational risk assessment: Potential avenues for applications *I. Pavón, L. Sigcha, P. Arezes, N. Costa, G. Arcas & J.M. López*	447
Maturity level of hearing conservation programs in metallurgical companies and the workers noise perception *I.C. Wictor & A.A. de P. Xavier*	453
Pitfalls of measuring illuminance with smartphones *R.B. Melo, F. Carvalho & D. Cerqueira*	459
Effects of thermal (dis)comfort on student's performance: Some evidences based on a study performed in a secondary school *M. Talaia, M. Silva & L. Teixeira*	465
Usability of an open ERP System in a manufacturing company: An ergonomic perspective *K. Hankiewicz & K.R. Kumara Jayathilaka*	471
Postural adjustment for balance in asymmetric work. A practical example *G. Dahlke & K. Turkiewicz*	477
Working conditions and occupational risks in the physiotherapist's activity *L.S. Costa & M. Santos*	483
Influencing factors in sustainable production planning and controlling from an ergonomic perspective *M. Zarte, A. Pechmann & I.L. Nunes*	489
Proof of concept—work accident costs analysis tool *C. Correia, J. Santos Baptista & P.A.A. Oliveira*	495
Wrist-hand work-related musculoskeletal disorders in a dairy factory: Incidence, prevalence and comparison between different methods for disease validation *A. Raposo, R. Pinho, J. Santos Baptista & J. Torres Costa*	501
Work ability of teachers in Brazilian private and public universities *A.J.A. Silva & T.G. Lima*	507
Maintenance manual for equipment on construction site *C. Reis, C. Oliveira, M.A. Araújo Mieiro, P. Braga, J.F. Silva & J.A.F.O. Correia*	513
Ergonomic risk management of pruning with chainsaw in the olive sector *M.C. Pardo-Ferreira, A. Zambrana-Ruíz, J.A. Carrillo-Castrillo & J.C. Rubio-Romero*	517
What kind of lower limb musculoskeletal disorders can be associated with bus driving? *M. Cvetković, M.L. Dinis, E. Stojiljković & A.M. Fiuza*	523
Lean thinking practices in ergonomics in industrial sector *M. Gibson & B. Mrugalska*	529
Occupational activity on blood pressure profile *J. Pereira, R. Souza & H. Simões*	535
Reusing single-use medical devices, where are we standing now? *L.B. Naves & M.J. Abreu*	541
Active workstations to improve job performance: A short review *S. Maheronnaghsh, M. Vaz & J. Santos*	547
Case study of risk assessment in single family housing *C. Reis, V. Gomes, P. Braga, J.F. Silva, J.A.F.O. Correia & C. Oliveira*	551

Evaluation of the luminous performance of a public university library in the Northeast of Brazil *A.R.M.V. Silva, M.N. Almeida, J.I.F. Machado, B.C.C.B. Carvalho & H.C. Albuquerque Neto*	555
Thermal comfort assessment of orthopaedic health professionals in an operating room *N. Rodrigues, A.S. Miguel, S. Teixeira, C. Fernandes & J. Santos Baptista*	561
Study of the impact of investments in occupational safety and accident prevention in the construction sector *R. Pais & P.A.A. Oliveira*	567
Lean construction and safety *A.M. Reis, B.V. Zeglin & L.L.G. Vergara*	573
Analysis of the noise level in a university restaurant: A case study *A.D.S. Oliveira, H.C. Albuquerque Neto, M.N. Almeida, A.R.M.V. Silva & B.C.C.B. Carvalho*	577
The influence of noise on the perceptions of discomfort, stress, and annoyance *R. Monteiro, M.A. Rodrigues & D. Tomé*	583
Assessment and characterization of WMSDs risk in nurses who perform their activity in surgical hospitalization *M. Torres, P. Carneiro & P. Arezes*	589
Risk analysis in the execution of the Aguas Santas tunnel *E. Carpinteiro, C. Reis, P. Braga, J.A.F.O. Correia, J.F. Silva & C. Oliveira*	595
Characterization of school furniture in a basic education school *A. Fernandes, P. Carneiro, N. Costa & A.C. Braga*	601
Gas distribution companies: How can knowledge management promote occupational health and safety? *C.M. Dufour & A. Draghici*	607
Cytostatic-drugs handling in hospitals: Impact of contamination at occupational environments *J. Silva, P. Arezes, N. Costa, A.C. Braga & R. Schierl*	613
Author index	617

Foreword

Occupational Safety and Hygiene VI collects recent papers of selected authors from 21 countries in the domain of Occupational Safety and Hygiene (OSH). The contributions cover a wide range of topics, including:

– Occupational safety
– Risk assessment
– Safety management
– Ergonomics
– Management systems
– Environmental ergonomics
– Physical environment
– Construction safety, and
– Human factors

Occupational Safety and Hygiene VI represents the state-of-the-art on the above mentioned domains, and is based on research carried out at universities and other research institutions. Some contributions focus more on practical case studies developed by OSH practitioners within their own companies. Hence, the book provides practical tools and approaches currently used by OHS practitioners in a global context.

Each manuscript was peer reviewed by at least 2 of the 126 members from 16 different countries of the International Scientific Committee of the International Symposium on Occupational Safety and Hygiene, annually organised by the Portuguese Society of Occupational Safety and Hygiene (SPOSHO). These international experts covered all academic fields of the book.

As in all the previous editions, the editors would like to take this opportunity to thank the academic partners, namely, the School of Engineering of the University of Minho, the Faculty of Engineering of the University of Porto, the Faculty of Human Kinetics of the University of Lisbon, the Polytechnic University of Catalonia and the Technical University of Delft. The editors also would like to thank the scientific sponsorship of 24 other academic and professional institutions, the official support of the Portuguese Authority for Working Conditions (ACT), as well as the valuable support of several companies and institutions, including media partners, which have contributed to the broad dissemination. Finally, the editors wish also to thank all the reviewers, listed below, which were involved in the process of reviewing and editing the included papers. Without them, this book would not be possible.

The editors hope that the articles included in this book will be a valuable contribution to the dissemination of research conducted by OSH-researchers. Their work shows the latest research and approaches, giving visibility to emerging issues and presenting new solutions in the field of occupational safety and hygiene.

The Editors,
Pedro M. Arezes, J. Santos Baptista, Mónica P. Barroso,
Paula Carneiro, Patrício Cordeiro, Nélson Costa,
Rui B. Melo, A.S. Miguel & Gonçalo Perestrelo

List of reviewers

A.S. Miguel
Alberto Villarroya López
Álvaro Cunha
Ana C. Meira Castro
Ana Ferreira
Anabela Simões
Angélica Acioly
Anil Kumar
Anna Sophia Moraes
Antonio Lopez Arquillos
António Oliveira e Sousa
Beata Mrugalska
Béda Barkokébas Júnior
Camilo Valverde
Carla Barros
Carla Viegas
Catarina Silva
Celeste Jacinto
Celina P. Leão
Cristina Reis
Delfina Ramos
Denis Coelho
Divo Quintela
Eliane Lago
Ema Leite
Emília Duarte
Emília Rabbani
Enda Fallon
Fernanda Rodrigues
Fernando Amaral
Filipa Carvalho
Filomena Carnide
Florentino Serranheira
Francisco Fraga
Francisco Másculo

Francisco Silva
Hélio Cavalcanti Neto
Hernâni Neto
Ignacio Castellucci
Ignacio Pavón
Isabel Loureiro
Isabel Nunes
Isabel Silva
J. Santos Baptista
Jack Dennerlein
Jesús Carrillo-Castrillo
Joana C.C. Guedes
Joana Santos
João Ventura
Jorge Gaspar
José C. Torres Da Costa
José Carvalhais
Jose L. Meliá
José Miguel Cabeças
José Pedro Domingues
Juan C. Rubio-Romero
Laura Martins
Liliana Cunha
Luis Franz
Luiz Bueno Da Silva
M.D. Martínez-Aires
Mahmut Ekşioğlu
Mahrus K. Umami
Manuela Silva
Marcelo Pereira da Silva
Maria Antónia Gonçalves
Maria José Abreu
Maria Luísa Matos
Marino Menozzi
Marta Santos

Martin Lavallière
Martina Kelly
Matilde Rodrigues
Maurília Bastos
Mohammad Shahriari
Mónica Barroso
Nélson Costa
Nelson Rodrigues
Olga Mayan
Paula Carneiro
Paulo A.A. Oliveira
Paulo Flores
Paulo Sampaio
Paulo de Carvalho
Pedro Arezes
Pedro Ferreira
Pere Sanz Gallen
Rui Azevedo
Rui B. Melo
Rui Garganta
Salman Nazir
Sara Bragança
Sérgio Sousa
Sílvia da Silva
Susana Costa
Susana Sousa
Susana Viegas
Szabó Gyula
Tânia Miranda Lima
Teerayut Sa-ngiamsak
Teresa Patrone Cotrim
Tomi Zlatar
Walter Correia

A new approach—use of effectiveness and efficiency concepts in the identification of activities related to occupational safety management on motorways

J.E.M.R. Silva & C. Rodrigues
Research Centre for Territory, Transport and Environment, Faculty of Engineering, University of Porto, Porto, Portugal

ABSTRACT: The need to improve the safety management in the maintenance and operational works on a motorway is a request that all the organizations demand to achieve better goals on safety and in the use of resources. The use of instruments linked to efficiency and effectiveness is one possible solution to this problem. The objective of the present investigation is to prove that it is possible to identify the main activities contained in the A area of an ABC curve and linked to the Pareto's concepts, defining this way the 20% most dangerous maintenance and operational activities on motorway works. The methodology used was a DELPHI panel formed by experts from different areas linked to safety at work on motorways. The statistic treatment was made through the informatic program IBM SPSS Statistics 24. The results show that it is possible to identify 36 main activities in maintenance and operational works, being relevant the activities related to emergency response, safety guards, expansion joints, provisional signaling and the access to working stations in toll areas. The conclusions of the study reveal that these 36 activities represent the 20% most dangerous activities in maintenance and operational works on a motorway, and show that besides the risks of the work itself, there are four crucial traffic factors to develop a correct risk analysis: weather conditions, traffic characteristics, pavement characteristics and motorway morphologic type.

1 INTRODUCTION

The complexity produced by multiple variables of causes in the working accidents on operational motorways is one most the defying problems that health and safety professionals must deal with in order to ensure the safety of all the users and workers involved in maintenance activities in these infrastructures. To ensure the outcome of zero serious or mortal accidents, different factors are crucial to the risk analysis that must be conducted. Besides the inherent risks associated with the construction work or the activities that must be developed, conditions related to external factors cannot be forgotten. Climatic conditions, morphologic motorway factors, traffic type and conductors' behavior are examples of some problems that should be considered when analyzing risks. Nevertheless, another type of challenges has arrived for all the organizations particularly in the Health and Safety area. The allocation decrease of human and material resources is a reality affecting organizations and specially the areas that are not considered to be direct wealth generators. In this context, it is essential that both health and safety professionals and the scientific community can give an adequate response to all these problems. New instruments can be adopted to all these questions. For example, some researches have conducted a systematic review on the relationship between effectiveness and efficiency with the management of OSH on motorways and with the main critical factors on H&S risk analysis (Silva & Baptista & Rodrigues, 2014). Another research is focused in the discovery of the most dangerous places to work in these infrastructures (Hallowell et al., 2011). However, this second article only deals with construction activities on construction phase, without the interference of other external critical factors to Health and Safety. A third article is related with the incompatibility and interference of construction activities on motorways, but once more without referring other constrains (Esmaieli et al., 2013). In the case of motorways in operation, new risks are arising from the need to maintain their operation combined with the traditional risks in the construction sector, exponentially increasing the hazards for the workers and for the users of these infrastructures. Due to the lack of studies that explore a relation between motorways in operation and the maintenance and operational activities, a new approach is needed to identify the main risks that affect workers and motorway users. The use of effectiveness and efficiency tools can be one possible answer.

This context needs also to be explored by the scientific community as a research stimulus on management tools related to the effectiveness and efficiency of the human and material resources, amenable to being incorporated in the traditional instruments of Occupational Safety and Health (OSH).

In fact, this means the need to identify the main activities through multicriteria concepts as well as rationalization and optimization tools. This can help the researchers to identify what to do and how to do it, in terms of Health and Safety. Basically, this also proves that besides the identification of the most relevant activities the researchers must identify all the main internal and external factors that can make a difference at a risk level.

All in all, this research was carried out bearing these goals in mind.

2 METHODOLOGY

The research methodology is based on one concept, one validated method and one informatic tool. The fundamental concept utilized is based on the Pareto's Principle, whose basic principle states that 80% of events come from 20% of the possible causes (Randson, Darrell & Boyd, Amy). This Principle adapted to our thematic states that 80% of the accidents are caused by 20% of all the activities. The first step to achieve this goal is to select the validated method that will be used in the research. Among different hypotheses, the selected method was the Delphi methodology. Besides being a validated one, developed by de RAND Corporation after the second world war, this methodology has the objective of achieving a feasible consensus among experts. Some authors (C. Okoli, S.D. Pawlowski), consider this methodology as a valuable instrument to avoid conflicts among experts and to make predictions, to identify a hierarchy of variables with relevance to the research. Other authors (Skulmoski, Hartman, & Krahn) consider this methodology as the most suitable to "investigate what does not yet exist" and suitable to PHD theses. This method requires the production of inquiries that must be answered by the experts. The research carried out (C. Okoli, S.D. Pawlowski) suggests the adoption of the Kendall's coefficient as the most suitable to be adopted by experts and of 0.7 as the minimum value to achieve consensus. When this value is not achieved, the experts receive for each question, information about the following statistics elements: median, average, mode and standard deviation. After this step, they are asked if they want to review their score. Another issue that must be carefully handled concerns the experts' choice. According to some authors (Skulmoski, Hartman, & Krahn), the experts must accomplish four requirements: knowledge and experience in the area, will to participate in the research, communication skills and available time. Other requirements are needed to ensure a methodological success. The first step is to elaborate an identification matrix of experts; the second step has to do with the selection of experts; The third step refers to the with identification of additional experts and the fourth step is related the expert's hierarchy level. One last step is required to make a formal invitation to the experts. According to all these requirements, thirteen experts were identified, all of them with complementary experiences in the same area of these research. Their areas are shown in Fig. 1.

The informatic toll used to achieve the response values consensus among experts was the informatic program IBM SPSS Statistics 24.

3 RESULTS

To reach a consensus among experts, it became necessary to produce and invoice two types of inquiries.

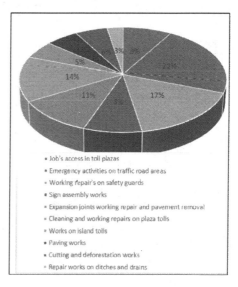

Figure 2. Most dangerous activities.

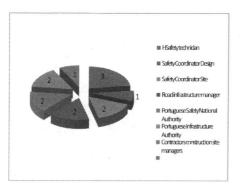

Figure 1. Experts' areas.

A first inquire with close questions and numeric score, which each expert could suggest for each activity and a second one with open questions, where each expert could suggest new activities or factors. In the second inquire, each expert was questioned not only to review his score for each case where the consensus was not achieved but also to analyze the other experts' proposals.

The number of activities identified and analyzed by the experts' panel were 187 activities. Besides the

Table 1. Relevant activities and factors.

Hazard level	Activities description	Relevant factors	Experts panel points
1	Access to working place through electronic toll "green way" on toll plazas	Light traffic, bad weather conditions and hard pavement	128
2	Motorway emergency activities in car accidents (motorway with two lanes on the same direction)	Less than 300 m curve, congested traffic, bad weather conditions and hard pavement	125
3	Cleaning works in the middle of the motorway	Less than 300 m curve, normal traffic, bad weather conditions and hard pavement	122
4	Safety rails repair on the median strip, with rails on both sides and without inner vegetal band	Less than 300 m curve, normal traffic, bad weather conditions and hard pavement	121
5	Signaling assembly and disassembly in car accidents support, by emergency team	Less than 300 m curve, congested traffic, bad weather conditions and hard pavement	120
6	Safety rails repair on the median strip, with rails on both sides and with inner vegetal band	Less than 300 m curve	120
7	Motorway emergency activities in car accidents (motorway with two lanes on the same direction)	Less than 300 m curve	119
8	Motorway emergency activities in car accidents (motorway with tree lanes on the same direction)	Less than 300 m curve	119
9	Motorway emergency activities in car accidents (motorway with tree lines on the same direction)	Less than 300 m curve, congested traffic, bad weather conditions and hard pavement	119
10	Safety rails repair on the median strip, with rails on both sides and without inner vegetal band	Less than 300 m curve	117
11	Access to working place through electronic toll "green way" in toll plazas	none	116
12	Signaling assembly and disassembly in working site construction, on the new jersey median strip and without safety rails	Less than 300 m curve, normal traffic, bad weather conditions and hard pavement	116
13	Safety rails repair on the motorway narrow right hard shoulder, with excavation slopes over 2 m and 45 degrees/inclination and narrow hard shoulder	Less than 300 m curve, normal traffic, bad weather conditions and hard pavement	116
14	Expansion joints works, in motorways with tree lanes and narrow hard shoulder	Less than 300 m curve, normal traffic, bad weather conditions and hard pavement	115
15	Cleaning works near the new jersey median strip	Less than 300 m curve, normal traffic, bad weather conditions and hard pavement	115
16	Cleaning works before the toll plaza	Normal traffic, bad weather conditions and hard pavement	114
17	Signaling assembly and disassembly in car accidents support, by emergency team	Less than 300 m curve	114
18	Cleaning works near the new jersey median strip	Less than 300 m curve	114
19	Maintenance working repairs in island toll plaza with traffic interruption	Light traffic, bad weather conditions and hard pavement	113
20	Equipment assembly, disassembly and repair after an island toll plaza with traffic	Light traffic, bad weather conditions and hard pavement	113

inherent risks of the activities, the expert's panel considered that there are four fundamental factors that affect the risk analysis and the safety level of the selected activities. These factors are: motorway infrastructure morphology, type of pavement, climatic conditions and traffic type. After the second interaction, the experts' panel agreed in the selection of the 36 most dangerous activities and the factors that can influence their hazard level. It must be emphasized that between the first and the second interaction, only two new activities were integrated in the selected group. A simple note to explain why 36 activities were chosen instead of 37 activities. In fact, 37 activities are the real percentage of 20% of the global number of activities, but the number of draws in the 37th position, was substantial and it was decided to assume 36 activities as the percentage of 20% of the global activities. Fig. 2 shows the areas and the type of activities that the Delphi panel consider as the most dangerous among that selection.

Table 1 shows in detail, the influence between activities and factors and the identification of the twenty most dangerous activities.

4 DISCUSSION

One of the fundamental issues addressed by this research is the relationship between the activities hazards induced by the maintenance construction works and the traffic circulation. Apart from exceptional situations, motorways as public infrastructures must always be open to traffic. This problem has been addressed in some research (Esmaieli et al., 2013), as one of the main hazards to workers and motorways users when traffic and maintenance works exist on operational motorways. The relevant question is to find out the main traffic factors and the main maintenance construction works that can cause damage to the both groups. The increase of the effectiveness and efficiency prevention measures to this problem is related to the assertive activities planning (Esmaieli et al., 2013) and these assumptions can only be achieved with a clear definition of the main hazardous construction activities or the relevant traffic factors. It has a fundamental importance for this research, the discovery of the several factors that can induce the loss of control of vehicles, in the presence of maintenance works on the motorway. The Delphi experts panel have identified 34 of the 36 most dangerous maintenance work activities as affected by the following motorway factors: Motorway infrastructure morphology, type of pavement, traffic climatic conditions and traffic type.

For the present research and according with the identification of the most relevant activities, the subdivision of above the referred factors are as follows: motorway infrastructure morphology with or without a less 300 meters curve; three types of pavement—hard (bigger adherence and less visibility in the case of rain), rough (medium adherence and medium visibility in case of rain) and draining (less adherence and bigger visibility in case of rain); traffic climatic conditions—favorable (dry weather and good visibility) and unfavorable (rainy weather and poor visibility); traffic type—light traffic, normal traffic and congested traffic. The consensus achieved about the motorway infrastructure morphology factors stated that 8 activities are only influenced by a less 300 m curve, 6 activities by a conjunction of pavement type, traffic and climatic conditions and 19 activities by a conjunction of a less 300 m curve, pavement type, traffic climatic conditions and traffic type. As for the type of pavement, the achieved consensus showed that a hard pavement is the most unfavorable one for all applicable conditions. Concerning the traffic climatic conditions, the experts' panel considered that the most dangerous situations for workers and drivers are produced by unfavorable climatic conditions (poor visibility and rainy weather). As for the traffic type, the experts' panel considered that this factor is relevant in the same 25 possible referred combinations, and stated that the most unfavorable situations founded are related to the following situations: light traffic for activities in toll plazas and signage frames areas, congested traffic for emergency activities and normal traffic for all the activities.

These assumptions are all consistent with some research (Wu, Jianjun & Abdel-Aty, Mohamed & Yu, Rongjie & Gao, Zyou) and (Bergel-Hayat, Ruth & Debbarh, Mohammed & Antoniou, Constantinos & Yannis, George), which stated that factors related to bad visibility, high velocity, low temperature, type of car occupation and type of traffic flow are relevant factors to the loss control of vehicles. However, another factor could be relevant for an optimized result that is not consider in this research. In effect, the number of repetition of the same working operations in the same activities on a motorway can change in some way the level of hazard exposition that the workers and drivers are subject to. This problem can only be solved by studying specific parts of a single motorway, situation that is not in the scope of this research. The types of factors versus activities can be seen in Fig. 3.

Analyzing the remaining data, it is possible to see that there is a catalyzing effect that risks can produce namely between the factors associated with the loss of car control and the working maintenance risks. This effect had already been noted in a research by Hadad (2007). Another fact that is also possible to verify is the need of an intervention

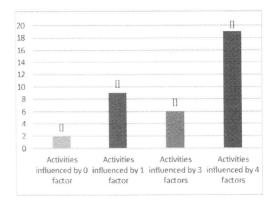

Figure 3. The influence of factors versus activities.

on the design phase, as Atkinson (2010) suggests in his proposals, where he states there is an interaction between designers and contractors that must be followed with the intention of minimizing risks during the construction and maintenance phase of a motorway. In fact, some hazard level that has been detected can only be completely solved with another kind of options on design phase. The most paradigmatic example of this statement is the case of the toll plaza, where it's only possible to eliminate the running over risk with the construction of access tunnels or superior passages.

5 CONCLUSIONS

The research results show that almost all the scrutinized activities have at least one option factors included in the final group of the most dangerous activities. Only those included in the group of structural repairs on superior passages or support devices are not included in the final selected group. This outcome is quite conditioned by factors that are not only related to hazards from the civil works and traffic in operational motorways. As a clear example of this statement, it is possible to show the case of the emergency activities. It is a fact already stated in former researches (Prati, Gabriele & Pietrantoni, Luca) that the traffic impact in emergency workers is one of the main risks for this kind of activity. This conclusion is in line with the conclusions from the present research that shows that 22% of the most dangerous identified activities are connected with emergency teams' response. This group of activities is identified as the most numerous activities group in all the research. Another important conclusion that can be drawn from the analyzed data regards the type of road, traffic, weather and pavement on the motorway. More than 50% of the final selected activities are influenced by the conjunction of 4 factors and about 70% by three or more factors. This evidence proves that the global hazard level of the maintenance activities can only be properly analyzed with the incorporation of the hazards from the activities, as well as simultaneously with the hazards from the traffic and the motorway infrastructure. Another complementary conclusion came from the score level that the experts give to the same activities without the motorway and traffic factors. Only in two cases, access to working places on toll plazas and expansion joints repairs, the experts panel considered that the referred factors don't increase the dangerous level of those activities and concluded that whatever the kind of traffic, weather, pavement and/or motorway, all the risks level are the same. In all other selected activities and in 90% of the cases, every time that a factor is incorporated in the equation, the risk level increased substantially. As corollary point of the research, all the panel members agreed that the clear identification of the most dangerous activities are vital to make a correct integration of general prevention principles into the design and construction phase. This research leads us to new challenges associated with this subject. Next steps are needed to increase effectiveness and efficiency in the level of H&S on motorways. After the identification of the 20% most dangerous activities, the next logical move is the identification of the 20% most dangerous risks, followed in the future by identification of the corresponding 20% most efficient protection measures. This contribution aims at giving the first step on that direction.

REFERENCES

Atkinson, A & Westwall, R. 2010. The relationship between inte-grated design and construction and safety on construction projects. In Constr. Manag. and Econ. 28: 1007–1017.

Bergel-Hayat, Ruth & Debbarh, Mohammed & Antoniou, Constantinos & Yannis, George. 2013. Explaining the road accident risk: Weather effects. In Accident Analysis and Prevention 60 (2013) 456–465.

Esmaieli, B. & Hallowell, M., 2013. Integration of safety risk data with highway construction schedules. In Construction Management and Economics 31(6): 528–541.

Hadad, Y. & Laslo, Z & Ben-Yair, A. 2007. Safety improvements by eliminating hazardous combinations. In Ukio Technologinis Ir Ekonominis Vystymas 13(2): 114–119.

Hallowell, M. & Esmaieli, B. & Chinowsky, P., 2011. Safety risk interactions among highway construction work tasks. In Construction Management and Economics 29:417–429.

Okoli, C., S.D. Pawlowski. 2004. The Delphi method as a research tool: an example, design considerations and applications.Information & Management 42 (2004) 15–29. In Science Direct.

Prati, Gabriele & Pietrantoni, Luca. 2012. Predictors of safety behaviour among emergency responders on the. In Journal of Risk Research Vol. 15, No. 4, April 2012.

Randson, Darrell & Boyd, Amy. 1997. The Pareto Principle and Rate Analysis. In Quality Engineering, 10(2), 223–229 (1997–98).

Silva, J. & Baptista, J. & Rodrigues, C. 2014. Use of effectiveness and efficiency concepts in occupational management on motorways: A systematic review. In Occupational Safety and Hygiene IV—Taylor & Francis: 427–433.

Skulmoski, Hartman, & Krahn. 2007. The Delphi Method for Graduate Research. In Journal of Information Technology Education Volume 6.

Wu, Jianjun & Abdel-Aty, Mohamed & Yu, Rongjie & Gao, Zyou. 2013. A novel visible network approach for freeway crash analysis. In Transportation Research Part C 36 (2013): 72–82.

Sampling methods for an accurate mycobiota occupational exposure assessment—overview of several ongoing projects

C. Viegas
GIAS, ESTeSL—Escola Superior de Tecnologia da Saúde de Lisboa, Instituto Politécnico de Lisboa, Lisbon, Portugal
Centro de Investigação e Estudos em Saúde Pública, Escola Nacional de Saúde Pública, ENSP, Universidade Nova de Lisboa, Lisbon, Portugal

ABSTRACT: Exposure to bioaerosols is a critical occupational issue that requires close attention. Workers in several occupational environments, such as health care, agriculture, animal production, waste, fishery, forestry, mining, construction, and day care are exposed to higher risks of biological hazards due to work characteristics. This review intends to provide information on what is currently known about sampling methods to achieve mycobiota exposure assessment, since fungal burden characterization continuous to be a challenging task for every industrial hygienist. A brief description about Research Group Environment & Health (GIAS) contribution on this topic is also given with an overview of the developed and ongoing projects. Passive and active methods should be applied in parallel to ensure a more precise occupational exposure assessment to the fungal contamination. Increasing the number of different sampling methods will enrich data findings enabling industrial hygienists to perform risk characterization.

1 INTRODUCTION

Airborne microorganisms might pose an occupational hazard when present in high concentrations in occupational environments resulting in health problems (Stetzenbach, Buttner & Cruz 2004). The presence of high levels of bioaerosols, and more specifically the mycobiota content, is frequently the result of the natural colonization of an organic substrate present in the workstation but may also be intentionally added (the case of food industry) (Oplliger 2014). Therefore, exposure to bioaerosols is a critical occupational issue that requires close attention (Wang et al. 2015). The workers in different settings, such as health care, agriculture, animal production, waste, fishery, forestry, mining, construction, and day care are exposed to higher risks of biological hazards because of the work characteristics (Wang et al. 2015; Viegas et al. 2015a). Numerous studies have indicated that these workers have higher prevalence rates of respiratory diseases and airway inflammation (Heldal et al. 2003; Bang et al. 2005; Heederik et al. 2007; Cox-Ganser et al. 2009).

Of note, is the uniquely of each bioaerosol sample as its composition varies in time and space (abundance and diversity of species) (Oplliger 2014). Thus, exposure assessment to bioaerosols and more specifically to the mycological content, remains to be a challenging task for every industrial hygienist. Occupational exposure to microbiological risks can be estimated using a variety of different methods and each situation is unique and requires specific methodology (Oppliger 2014; Viegas et al. 2015a).

Information about on what is currently known concerning sampling methods to achieve mycobiota exposure assessment will be provided. In addition, a brief description about Research Group Environment & Health (GIAS) contribution on this topic is also given with an overview of the developed and ongoing projects.

2 SAMPLING METHODS

2.1 *Active methods*

The active methods devices used to sample airborne fungi mainly rely on three different principles namely, impaction, impingement and filtration (Mandal & Brandl 2011; Viegas et al. 2015a). Impactors use solid media such as agar to collect bioaerosols by impactation. Their easiness to handle and the cheap cost are major advantages (Zollinger, Krebs & Brandl, 2006; Viegas et al. 2015a). The colonies number can be counted by visual inspection after incubation resulting in a direct quantitative estimate of the number of culturable fungi in the volume of air sampled

(Zollinger, Krebs & Brandl 2006; Viegas et al. 2015a). This method is the chosen by ACGIH (Verhoeff, Van Wijnen & Brunekreef 1992) and recommended by the Canadian Health Organization (Health Canada 1993).

In impingers particle collection is based on liquid media. Normally, sampled air is drawn by suction through a narrow inlet tube into a small flask containing the collection medium. Once the sampling is complete, aliquots of the collection liquid can be cultivated in appropriate growth media to enumerate viable microorganisms allowing quantitative determinations, since the sample volumes and sampling times can be previous defined (Zollinger, Krebs & Brandl 2006; Viegas et al. 2015a). Formerly known as scrubbers, they are often necessary in occupational settings with higher fungal loads. This approach, besides allowing dilution of the sample prior to plate incubation, also easier the application of molecular tools since a liquid air sample is expected after the sampling. Both these features are not generally possible with samplers that employ impaction on solid media (Thorne & Heederik 1999; Viegas et al. 2015a). On the downside, fungi present in small numbers and as single units may be less represented (Macher 2001) and impingers cannot operate for long periods since liquid evaporation can hamper the fungi viability (De Nuntiis et al. 2003).

Filtration samplers collect particles air through suction filters. Air is drawn by a vacuum line through a membrane filter that can be made of glass fibre, Polyvinylchloride (PVC), polycarbonate or cellulose acetate or gelatin (Mandal & Brandl 2011). The filter membrane can be placed on a culture media and incubated to allow fungal growth or even digested with a tampon solution such as sterile Phosphate Buffered Saline (PBS) and then inoculated in the selected media (Sudharsanam et al. 2012; Viegas et al. 2015a). Filter samples can also be dispersed in a liquid prior to cultivation enabling higher colony counts (Viegas et al. 2015) and a more straightforward bench work in case of applying molecular tools (Viegas et al. 2015a). Due to the risk of dehydration—since the surface where the particles are collected is completely dry—this method is only suited for resistant microorganisms, such as fungal spores (De Nuntiis et al. 2003).

Personal sampling is the ideal method for assessing personal bioaerosol exposure (Wang et al. 2015). An perfect personal bioaerosol sampler used in occupational environments should be light and robust, noninterfering with the work tasks, able to collect selected bioaerosols, and with suitable biological sampling efficiencies (Agranovski et al. 2002). The filtration method is the one generally applied for personal bioaerosol exposure assessment.

2.2 *Passive methods*

Passive sampling provides a valid risk assessment as it measures the harmful part of the airborne population which falls onto a critical surface (French et al. 1980). In less contaminated occupational environments, such as hospital facilities, passive monitoring uses mainly "settle plates", which are standard Petri dishes. These plates contain culture media exposed to the air for a given time in order to collect biological particles which "sediment" out and are then incubated. Results are expressed in CFU/plate/time or in $CFU/m^2/hour$ (Pasquarella, Pitzurra & Savino 2000). This method was already reported as the only method applied to ensure the bioburden exposure assessment (Hayleeyesus & Manaye 2014).

In settings with higher mycobiota burden the trend is to complement active methods with different passive methods such as surface swab (Viegas et al. 2016c) and, more recently, Electrostatic Dust Cloth (EDC) (Normand et al. 2009; Viegas et al. 2017c).

Surface swabs complement microbiological characterization of the air and are used in order to identify contamination sources and to evaluate efficacy of surface cleaning and disinfection procedures (Klánová & Hollerová 2003; Stetzenbach, Buttner & Cruz 2004; Viegas et al. 2016c). Surface samples are collect by swabbing the surfaces using a 10×10 cm^2 stencil disinfected with 70% alcohol solution between samples according to the International Standard ISO 18593 (2004). Specifically for the mycobiota assessment surface swabs already showed higher diversity in terms of the number of fungal species detected, as well as a higher fungal load, when compared with the air samples proving the relevance of analysis of the latter samples in complementing the results obtained by air sampling (Viegas et al. 2016c).

The EDC is an easy-to-use passive device that consists on a polypropylene cloth (Kilburg-Basnyat et al. 2016) and is increasingly being used because it is electrostatic, inexpensive, easy to obtain, and effective at collecting dust (Cozen et al. 2008). EDCs employ electric fibers which have revealed to increase particle retention (Kilburg-Basnyat et al. 2016). Despite not widely used in occupational exposure assessments the main advantage from EDC is that they can collect contamination from a larger period of time (weeks to several months) as in the analyzed papers, whereas air samples can only reflect the load from a shorter period of time (mostly minutes) (Viegas et al. 2015a; Viegas et al. 2017c).

The importance to employ, in parallel, active methods and passive methods was already reported (Brenier-Pinchart et al. 2009; Viegas et al. 2016c;

Table 1. Sampling methods applied from the different projects developed and ongoing on GIAS.

Occupational setting	Active methods	Passive methods	Sampling and analyses methods	Reference
Poultries	Impaction Impinger	Surface swabs	Impaction and surface swabs: Culture based methods	Viegas et al. 2014c
WWTP	Impaction Impinger	Surface swabs		Viegas et al. 2014a
WTP	Impaction Impinger	Surface swabs	Impinger: Molecular tools	Viegas et al. 2014b
Cork Industry	Impaction Impinger	Surface swabs		Viegas et al. 2015b
Feed Industry	Impaction Impinger	Surface swabs		Viegas et al. 2016a
Slaughterhouses	Impaction Impinger	Surface swabs		Viegas et al. 2016b
Swine (on going)	Impaction Impinger	Surface swabs	Impaction: Culture based methods	–
			Impinger and surface swabs: Molecular tools	
Bakeries (ongoing)	Impaction Impinger Filtration	Surface swabs EDC	Impaction: Culture based methods	–
Hospital facilities (starting)	Impaction Impinger Filtration	Surface swabs EDC	All the other sampling methods applied culture based-methods and molecular tools	–

Viegas et al. 2017c) since active methods provide information about the contamination load, while us-ing passive methods such as surface swabs and EDCs gives us a more detailed scenario regarding occupational exposure to mycobiota (Viegas et al. 2017c).

Of note, depending on each occupational environment the assessment can be complemented with different environmental matrices, such as litter and feed in animal production (Sabino et al. 2012; Viegas et al. 2012; Viegas et al. 2013), air conditioning filters from fork lifters managing waste (Viegas et al. 2017b), raw materials on feed industry (Viegas et al. 2016a) settled dust in bakeries (data not available). The obtained results besides providing information about the contamination sources can also enrich exposure assessment with data and, consequently, the risk characterization.

3 PROJECTS OVERVIEW

Since 2010 the Research Group Environment & Health (GIAS) from Escola Superior de Tecnologia da Saúde de Lisboa, Instituto Politécnico de Lisboa, has been focusing on the mycobiota occupational exposure assessment in different occupational environments (Poultries, Waste Water Treatment Plants—WWTP, Waste Treatment Plants—WTP, Cork industry, Feed industry, Slaughterhouses, Swine, Bakeries and Hospital facilities) (Table 1). Active and passive methods have being applied to pursuit the most suitable and accurate fungal contamination exposure assessment with an increase of sampling methods, and also analyses, performed through the years.

No significant correlation was found on fungal loads obtained through both active and passive samplings (Viegas et al. 2014–2017) as in other study (Petti, Iannazzo & Tarsitani 2003). The lack of correlation can be justified by a wide range of environmental variables, such as workers who may carry the mycobiota indoors (Scheff et al. 2000), as well as the developed activities and work practices that may affect fungal load (Jürgensen and Madsen 2016). Furthermore, when correlation is tested only with culture base methods results, we cannot disregard also the fact that viable fungi constitute a small percentage of the total concentration of the mycobiota (Huang et al. 2013; Viegas et al. 2015a) and, therefore, a bias about the fungal load is expected.

4 CONCLUSIONS

Passive and active methods should be applied in parallel to ensure a more precise occupational exposure assessment to the fungal contamination. Increasing the number of different sampling methods will enrich data findings, enabling industrial hygienists to perform risk characterization.

ACKNOWLEDGEMENTS

This overview would not have been possible to develop without the institutional support given by Escola Superior de Tecnologia da Saúde de Lisboa (ESTeSL) and Research Group Environment & Health. The author is grateful to Instituto Politécnico de Lisboa, Lisbon, Portugal for funding the

Project "Waste Workers' Exposure to Bioburden in the Truck Cab during Waste Management—W2E Bioburden" (IPL/2016/W2E_ESTeSL), to Portuguese Authority for Working Conditions for funding the Project "Occupational exposure assessment to particulate matter and fungi and health effects of workers from Portuguese Bakeries " (005DBB/12) and also to Occupational Health Services from the industries engaged in the projects covered in this overview.

REFERENCES

Agranovski, I.E., Agranovski, V., Reponen, T., Willeke, K. & Grinshpun, S.A. 2002 Development and evaluation of a new personal sampler for culturable airborne microorganisms. Atmos Environ 36: 889–898.

Bang, B., Aasmoe, L., Aamodt, B.H., Aardal, L., Andorsen, G.S., Bolle, R., et al. 2005. Exposure and airway effects of seafood industry workers in Northern Norway. J Occup Environ Med 47: 482–492.

Cox-Ganser, J.M., Rao, C.Y., Park, J.H., Schumpert, J.C. & Kreiss, K. 2009. Asthma and respiratory symptoms in hospital workers related to dampness and biological contaminants. Indoor Air 19: 280–290.

Cozen, W., Avol, E., Diaz-Sanchez, D., McConnell, R., Gauderman, W.J., Cockburn, M.G. & Mack, T.M. 2008. Use of an electrostatic dust cloth for self-administered home allergen collection. Twin Res Hum Genet 11(2), 150–155.

De Nuntiis, P., Maggi, O, Mandrioli, P., Ranalli, G. & Sorlini, C. 2003. Monitoring the Biological Aerossol. In Mandrioli P, Caneva G, Sabbioni C. (Eds.), Cultural Heritage and Aerobiology—Methods and Measurement Techniques for Biodeterioration Monitoring. Dordrecht. Kluwer Academic Publishers: 107–144.

French, M.L.V., Eitzen, H.E., Ritter, M.A. & Leland, D.S. 1980. Environmental control of microbial contamination in the operating room. In Wound Healing and Wound Infection. Edited by Hunt TK. New York: Appleton-Century Crofis: 254–261.

Hayleeyesus, S.F. & Manaye, A.M. 2014. Microbiological Quality of Indoor Air in University Libraries Asian Pac J Trop Biomed 4(Suppl 1): S312-S317.

Health Canada—Indoor air quality in office buildings: a technical guide. 1993 Vancouver: Health Canada. Available in: http://www.hc-sc.gc.ca/ewh-semt/alt_formats/hecs-sesc/pdf/pubs/air/office_buildingimmeubles_bureaux/93ehd-dhm166-eng.pdf.

Heederik, D., Sigsgaard, T., Thorne, P.S., Kline, J.N., Avery, R., Bønløkke, J.H., et al. 2007. Health effects of airborne exposures from concentrated animal feeding operations. Environ Health Perspect 115: 298–302.

Heldal, K.K., Halstensen, A.S., Thorn, J., Djupesland, P., Wouters, I., Eduard, W., et al. 2003. Upper airway inflammation in waste handlers exposed to bioaerosols. Occup Environ Med 60: 444–450.

Huang, P.Y., Shi, Z.Y., Chen, C.H., Den,W., Huang, W.M. & Tsai, J.J. 2013. Airborne and Surface-Bound Microbial Contamination in Two Intensive Care Units of a Medical Center in Central Taiwan. Aerosol Air Qual. Res 13: 1060–1069.

ISO 18593:2004(E). 2004. Microbiology of food and animal feeding stuffs—Horizontal methods for sampling techniques from surfaces using contact plates and swabs. Geneva: International Organization for Standardization.

Jürgensen, C.W. & Madsen, A.M. 2016. Influence of everyday activities and presence of people in common indoor environments on exposure to airborne fungi. AIMS Environmental Science 3(1): 77–95.

Kilburg-Basnyat, B., Metwali, N., & Thorne, P.S. 2016. Performance of electrostatic dust collectors (EDCs) for endotoxin assessment in homes: effect of mailing, placement, heating and electrostatic charge. J Occup Environ Hyg 13(2): 85–93.

Klánová, K. & Hollerová, J. 2003 Hospital indoor environment: Screening for micro-organisms and particulate matter. Indoor Built Environ. 12: 61–7.

Macher, J.M. 2001. Evaluation of a Procedure to Isolate Culturable Microorganisms from Carpet Dust. Indoor Air 11: 134–140.

Mandal, J. & Brandl, H. 2011. Bioaerosols in Indoor Environment—A Review with Special Reference to Residential and Occupational Locations. The Open Environmental & Biological Monitoring Journal 4: 83–96.

Normand, A.C., Vacheyrou, M., Sudre, B., Heederik, D.J.J., & Piarroux, R. 2009. Assessment of dust sampling methods for the study of cultivable—microorganism exposure in stables. Applied and Environmental Microbiology 75(24), 7617–7623.

Oppliger, A. 2014. Advancing the Science of Bioaerosol Exposure Assessment. Ann. Occup. Hyg. 58 (6): 661–663.

Pasquarella, C., Pitzurra, O. & Savino, A. 2000. The index of microbial air contamination. J. Hosp. Infect. 46: 241–256.

Petti, S., Iannazzo, S. & Tarsitani, G. 2003. Comparison between different methods to monitor the microbial level of indoor air contamination in the dental office. Ann. Ig. 15: 725–733.

Sabino, R., Faísca, V.M., Carolino, E., Veríssimo, C. & Viegas, C. 2012. Occupational Exposure to Aspergillus by Swine and Poultry Farm Workers in Portugal. Journal of Toxicology and Environmental Health, Part A 75: 1381–1391.

Scheff, P., Pulius, V., Curtis, L. & Conroy, L. 2000. Indoor air quality in a middle school, Part II: Development of emission factors for particulate matter and bioaerosols. Appl. Occup. Environ. Hyg. 15, 835–842.

Stetzenbach, L.D., Buttner, M.P. & Cruz, P. 2004. Detection and enumeration of airborne contaminants. Curr Biotechnol. 15: 170–4.

Sudharsanam, S., Swaminathan, S., Ramalingam, A., Thangavel, G., Annamalai, R., Steinberg, R., Balakrishnan, K. & Srikanth, P. 2012. Characterization of indoor bioaerosols from a hospital ward in a tropical setting. African Health Sciences 12(2): 217–225.

Verhoeff, A.P., Van Wijnen, J.H. & Brunekreef, B. 1992. Presence of viable mould propagules in indoor air in relation to house damp and outdoor air. Allergy 47; 2: 83–91.

Viegas, C, Malta-Vacas, J., Sabino, R., Viegas, S. & Veríssimo, C. 2014c. Accessing indoor fungal contamination using

conventional and molecular methods in Portuguese poultries. Environmental Monitoring and Assessment. 186, 3: 1951–1959.
Viegas, C., Carolino, E., Malta-Vacas, J., Sabino, R., Viegas, S. & Veríssimo, C. 2012. Fungal contamination of poultry litter: a public health problem. Journal of Toxicology and Environmental Health, Part A, 75: 1341–1350.
Viegas, C., Carolino, E., Sabino, R., Viegas, S. & Veríssimo, C. 2013. Fungal Contamination in Swine: A Potential Occupational Health Threat. Journal of Toxicology and Environmental Health, Part A 76: 4–5: 272–280.
Viegas, C., Faria, T., Aranha Caetano, L., Carolino, E., Quintal Gomes, A. & Viegas, S. 2017a. Aspergillus spp. prevalence in different occupational settings. Journal of Occupational and Environmental Hygiene. DOI: 10.1080/15459624.2017.1334901.
Viegas, C., Faria, T., Carolino, E., Sabino, R., Quintal Gomes, A. & Viegas, S. 2016a. Occupational Exposure to Fungi and Particles in Animal Feed Industry. Medycyna Pracy 67(2).
Viegas, C., Faria, T., Cebola de Oliveira, A., Aranha Caetano, L., Carolino, E., Quintal-Gomes, A., Twarużek, M.; Kosicki, R., Soszczyńska, E. & Viegas, S. 2017b. A new approach to assess fungal contamination and mycotoxins occupational exposure in forklifts drivers from waste sorting. Mycotoxin Research. DOI: 10.1007/s12550-017-0288-8.
Viegas, C., Faria, T., dos Santos, M., Carolino, E., Sabino, R., Quintal Gomes, A. & Viegas, S. 2016b. Slaughterhouses Fungal Burden Assessment: A Contribution for the Pursuit of a Better Assessment Strategy Int. J. Environ. Res. Public Health 13, 297; doi:10.3390/ijerph13030297.
Viegas, C., Faria, T., Gomes, A.Q., Sabino, R., Seco, A. & Viegas, S. 2014a. Fungal Contamination in Two Portuguese Wastewater Treatment Plants, Journal of Toxicology and Environmental Health, Part A: Current Issues 77:1–3: 90–102.
Viegas, C., Faria, T., Meneses, M., Carolino, E., Viegas, S., Quintal Gomes, A. & Sabino, R. 2016c. Analysis of Surfaces for Characterization of Fungal Burden—Does it Matter? International Journal of Occupational Medicine and Environmental Health 29(4).
Viegas, C., Gomes, A.Q., Abegão, J., Sabino, R., Graça, T. & Viegas, S. 2014b. Assessment of Fungal Contamination in Waste Sorting and Incineration—Case Study in Portugal. Journal of Toxicology and Environmental Health, Part A: Current Issues, 77:1–3: 57–68.
Viegas, C., Pinheiro, C., Sabino, R., Viegas, S., Brandão, J. & Veríssimo, C., editors. 2015a. Environmental mycology in public health: fungi and mycotoxins risk assessment and management. Academic Press.
Viegas, C., Ramalho, I., Alves, M., Faria, T., Aranha Caetano, L. & Viegas, S. 2017c Electrostatic dust cloth—A new sampling method for occupational exposure to bioaerosols. International Symposium on Occupational Safety and Hygiene SHO2017, Arezes, P. et al. Portuguese Society of Occupational Safety and Hygiene: 40–41.
Viegas, C., Sabino, R., Botelho, D., dos Santos, M., Quintal Gomes, A. 2015b. Assessment of exposure to the Penicillium glabrum complex in cork industry using complementing methods. Archives of Industrial Hygiene and Toxicology, 66; 3: DOI: 10.1515/aiht-2015-66-2614.
Wang, C-H., Chen, B.T., Han, B-C., Liu, AC-Y., Hung, P-C., Chen, C-Y., et al. 2015. Field Evaluation of Personal Sampling Methods for Multiple Bioaerosols. PLoS ONE 10(3): e0120308.
Whyte, W. 1986. Sterility assurance and models for assessing airborne bacterial contamination. J. Parenter. Sci. Technol. 40:188–197.
Zollinger, M., Krebs, W. & Brandl H. 2006. Bioaerosols formation during grape stemming and crushing. Sci Total Environ 363: 253–9.

A systematic literature review on Thermal Response Votes (TSV) and Predicted Mean Vote (PMV)

E.E. Broday & A.A. de P. Xavier
UTFPR—Federal University of Technology of Paraná, Ponta Grossa, Brazil

ABSTRACT: The Fanger PMV model is widely used to evaluate Thermal Comfort. However, this model is limited and does not represent reality when compared to people's real thermal sensations, collected in a field study, in different regions of the world. The aim of this article is to verify, which studies were made and where discrepancies occur between thermal sensation votes and PMV. The main roles of this search consisted of searching for information, definition of inclusion criteria, selection of studies that met these criteria, confirmation of the chosen studies and critical analysis of selected works. The research was conducted in 48 sources of scientific information, resulting in 6718 articles. After applicable exclusion and inclusion criteria, 22 articles were left to be analyzed. It was verified that the discrepancies found did not correspond to just one range of environmental conditions but to several. Due to this problem, some authors suggest procedures to decrease discrepancies and others have developed adaptations and new models to better predict the response of a given group.

1 INTRODUCTION

In the last decades, the growth of industrialization has resulted in a large number of people spending a great amount of time indoors doing their activities, whether these indoor areas be air-conditioned or not. This has caused a growing interest in studies on the conditions of the environment because it's important to create a comfortable environment where people find themselves in.

A comfortable environment, from a thermal point of view, improves the productivity, satisfaction and well-being of the occupants of a building. The evaluation of thermal comfort is an important point which must be known by the engineers who plan an HVAC system (heating, ventilation and air conditioning), the aim of which is to create an environment which can be used by a maximum of users with the maximum comfort—therefore, knowing thermal responses from people in the environments is fundamental.

P.O. Fanger, a pioneer in studies of thermal comfort, began his studies in air-conditioned chambers in Denmark in the 1970s. The studies in air-conditioned chambers have been the advantage of having total control over all the environmental variables. However, with the passing of time, studies in air-conditioned chambers it has been complemented with field studies.

In spite of the differences between the studies in air-conditioned chambers and field studies, the final goal of the two approaches is always the same: thermal comfort, a state that seems to occur when body temperature is kept with minimal physiological regulatory effort (Kilic et al., 2006).

These studies in air-conditioned chambers served as the basis from which Fanger created the index for the *Predicted Mean Vote* (PMV). This index shows the average thermal sensation of a large group of people exposed to the same environment and this index is the most used for thermal comfort to evaluate the thermal sensation in a moderate environment, that is, the environment which does not provide an extreme feeling of heat or cold.

However, when the PMV index is compared to real thermal sensations collected in field studies, it presents significant discrepancies. The discrepancy found between the PMV index and real thermal sensations collected in field studies is recounted by Mors (2011), Maiti (2014), Broday (2014) and Yun (2014).

In order to better know the studies that present these discrepancies, this article presents a systematic literature review, using the PRISMA statement and verifies where and under what conditions these studies occurred, as well as verifying which are the most important results found.

2 MATERIALS AND METHOD

2.1 Search strategy

This article used the PRISMA statement (*Preferred Reporting Items for Systematic Reviews and Meta-Analyses*) (Liberati et al. 2009). Following authorship order, authors developed the following tasks in the review process: research and treatment information; design criteria and search; verification of information collected and validation of the research.

This research was done on 48 sources of scientific information: ACM Digital Library, ACS Journals, ASME Digital Library, BioMed Central Journals, Cambridge Journals Online, CE Database (ASCE), Directory of Open Access Journals (DOAJ), Highwire Press, IEEE Xplore, Informaworld (Taylor and Francis), IOP Journals, MetaPress, nature.com, Oxford Journals, Royal Society of Chemistry, Sage Journals Online, SciELO, Science Magazine, ScienceDirect, SIAM, SpringerLink, The Chronicle of Higher Education and Wiley Online Library, Academic Search Compelte, AGRICOLA Articles, AGRICOLA Books, Arts & Humanities Citation Index, Beilstein via SCIRUS (ChemWEB), Business Source Complete, CitiSeerX, Compendex, Current Contents, Datamonitor, Energy Citations (DOE), ERIC, Inspec, Library, Information Science & Technology Abstracts (LISTA), PsycArticles, PubMed, Science Citation Index, Science & Technology Proceedings, SCOPUS, Social Sciences Citation Index, Social Sciences & Humanities Proceedings, SourceOECD, TRIS Online, Web of Science and Zentralblatt Math.

To perform the research, a combination of keywords was used. The keywords chosen for the research were: Thermal Comfort Model, Predicted Mean Vote, Modeling, Thermal Response, Discrepancies and PMV. Figure 1 shows the five sets of combinations for the key words.

As can be seen in the Figure 1, the key words were combined in twos, in all data banks and scientific magazines. The order in which the key words are researched does not alter the meaning of the research.

2.2 Screening and eligibility criteria

After the collection, systematically, all the information from all the above sources, based on the same set of keywords, it has proceeded to the first screening of the relevant work for the ongoing study. This selection was performed using the following steps and rejection criteria in the following order:

1. Sorting items alphabetically by author, to facilitate the removal of duplicates;
2. Removal of articles prior to 2004 (over 10 years). Any relevant articles with over 10 years are obtained from the reference lists of articles found in the first selection. This step allows remove a significant number of minor publications, thus reducing the search time;
3. Removal of items that do not provide the complete basic information (author, title, year of publication or source revised);
4. Removal of duplicate articles;
5. Removal of articles whose title does not show a relationship with the subject.

After an initial screening, where only are analyzed basic references of articles (author, title, year of publication), goes to the analysis of abstracts:

6. Removal of articles whose stated goals in the summary do not match the objectives of this study;
7. Removal of articles whose methodology are inadequate or does not fit with the present study;
8. Removal of articles without full text available.

The next step is a preliminary analysis of the items selected with complete and accessible texts, being accepted to fulfill the following eligibility criteria:

– Articles which are not part of the theme, a characteristic which can be easily detected from the title or abstract;
– Articles prior to 2004;
– Articles do not have the name of the authors who wrote them;
– Articles that don't present date;

In the other hand, articles were considered eligible if they meet the following criteria:

– Studies which whose objectives are involved with Thermal Comfort area;
– Studies which provide information about new models or adaptation for PMV index;
– Studies which present discrepancies between the real thermal sensation collected in field studies and PMV.

After obtaining the final list of selected articles, the analysis of the references was performed, in order to find other works that might be relevant to the subject. Through this methodology it is possible to know the latest studies on Thermal

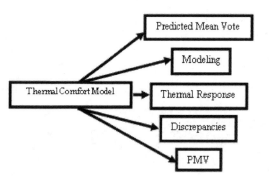

Figure 1. Combinations of key words; Source: The Authors.

Comfort, particularly in the relationship between PMV Fanger index and field studies.

3 RESULTS

3.1 Study selection

The details of the selection of all relevant articles for this paper were performed based on several criteria, based on the PRISMA statement established for reporting systematic reviews and meta-analyses of studies (Plos Medicine, 2009). All these details are available on Figure 2.

As can be seen in Figure 2, based on the defined search method it was found 6718 articles, which four articles were selected from references to compose the analysis discussed in this article. Of all articles found, after elimination of 1097 repeated and applied the exclusion and inclusion criteria, 22 studies were selected for a more detailed analysis.

3.2 Selected studies

Nineteen articles were selected because they dealt specifically with the discrepancies between the PMV models and the real thermal sensations collected in field studies. A brief summary of these articles can be seen in Table 1.

It is found that there is a great diversity of field studies, which are applied in many different areas. All studies cited above show that PMV index

Table 1. Main characteristics found in the studies.

Author	Year	Country	Building	Type of ventilation
Al-Ajmi	2007	Kwait	Mosque	AC
Albadra	2017	Jordan	Refugee Camp	NV
Andreasi	2009	Brazil	Army Headquarter Bank	NV AC
Broday	2014	Brazil	Mettalurgical Branch	NV
Broday	2017	Brazil	Mettalurgical Branch	NV
Buratti	2009	Italy	University Classroom	NV/AC
Feriadi	2004	Indonesia	Residential Buildings	NV
Han	2006	China	Rural and Urban Residences	NV
Huang	2014	China	Climate Chamber	Environment controlled
Indraganti	2010	India	Residential Buildings	NV
Indraganti	2013	Japan	University Buildings	NV/AC
Ji	2006	China	Office	NV
Kajtár	2016	Hungary	Office Building	AC
Kim	2014	South Korea	Office Building	AC
Maiti	2014	India	University	Environment controlled
Mors	2014	The Netherlands	Children Classrooms	NV
Shen	2010	China	Residential Buildings	NV
Simone	2013	Italy	Hypermarket	NV/AC
Teli	2012	England	Primary School Children	NV
Yau	2009	Malaysia	Hospital	AC
Yun	2014	South Korea	Kindergarten children	NV
Zhang	2007	China	Classroom	NV

*AC—Air-conditioned/NV—Naturally Ventilated.

does not fully meet the thermal sensation of votes obtained in field studies.

It can be verified that China is the country where more researchers study on the subject, followed by Italy, India and Brazil. There are many studies and they have come up with proposals in order to reduce the discrepancies between the thermal sensation votes and the PMV, either by minimizing errors in data collection in field studies or the adoption and development of new models instead the use of PMV.

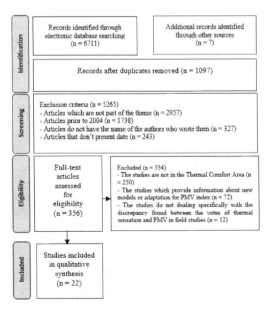

Figure 2. PRISMA flow diagram; Source: Plos Medicine (2009).

Table 2. Main characteristics found in the studies.

Author	Year	Equation*
Al-Ajmi	2007	AMV = 0.1334*T_o − 3.48
Andreasi	2017	TSV = 0.0277 + 0.5949*PMV
Albadra	2009	TSV = 0.1446*T_o − 3.2762
Broday	2014	TSV = 0.6591 + 0.5133*PMV
Broday	2017	PMV = 0.5155*TSV + 0.6441
Buratti	2009	–
Feriadi	2004	TSV = −1.61 + 1.33*PMV
Han	2006	MTSV = −0.11 + 0.78*PMV
Huang	2014	–
Indraganti	2010	TSV = 0.248*Tg + 0.159*Tw − 10.959
Indraganti	2013	TSV = 0.311*Tg − 7.749 (NV)
		TSV = 0.299*Tg − 8.109 (AC)
Ji	2006	–
Kajtár	2016	AMV = PMV + 0.275 (−1.7 ≤ PMV ≤ 0.5)
Kim	2014	–
Maiti	2014	TSV = 0.31978*Ta − 7.9388
Mors	2014	–
Shen	2010	TSV = 0.1049*Ta − 1.7126 (winter)
		TSV = 0.1612*Ta − 2.79 (summer)
Simone	2013	–
Teli	2012	TSV = 0.26*T_o − 5.68 (NE classroom)
		TSV = 0.18*T_o − 3.09 (SE classroom)
Yau	2009	–
Yun	2014	TSV = 0.293*T_o − 6.471
Zhang	2007	TSV = 0.448*T_o − 0.9628

*Please note that TSV (Thermal Sensation Vote), MTSV (Mean Thermal Sensation Vote) and AMV (Actual Mean Vote) are the same thing.

4 DISCUSSION

After the selection of the papers, this topic will show the main characteristics of the papers found. In these studies, is very common relate the variable TSV with another variable, like PMV, T_o (operative temperature), T_a (air temperature), T_g (globe temperature), T_w (wet bulb temperature) in an equation of first degree. Table 2 shows these relations found by the studies analysed.

With regard to Table 2 above, the objective of performing these regressions is to check how this equation can get closer to reality with the collected data. When TSV relates PMV and for example, it can be seen that the index Fanger underestimates or overestimates the actual feeling of the people. These comparisons become relevant in order to know the characteristics of the environment and its occupants.

5 CONCLUSIONS

Studies in thermal comfort are becoming more widespread because there is a need to create a thermically comfortable environment, because it is in these environments that people spend most of their time every day. For the engineers who dimension the environments, thermal comfort is important, since in environments where there is no thermal comfort, people are obliged to use alternative methods of electricity to reach a level of comfort.

The main aim of this article was to do a bibliographic survey of studies on thermal comfort which show discrepancies between thermal sensation votes and PMV. According to the studies that were analyzed here, these discrepancies occur in schools, hospitals, markets, residences, universities and industries.

In this research, a type of field study caught our attention for not being very common until that moment: studies on thermal comfort with children. Due to the difficulties that this kind of study imposes, it is believed that there is still a lot to explore in this specific area, since Yun (2012) showed that children repond differently to thermal sensation.

The fact remains that studies in thermal comfort are becoming more widespread because of the necessity to create a comfortable indoor environment since it is in these environments that people spend most of their time. For the engineers who design the environments, thermal comfort is important, because in environments where there is no thermal comfort, people are forced to use alternative ways of electricity to reach a level of comfort.

As people spend a great deal of their time inside closed environments, the indoor quality of this environment either at an air quality level, luminosity or temperature become a principal factor. A thermally comfortable environment bolsters its users' productivity, satisfaction and well-being (IIDA, 2005). Whenever it is possible to determine the PMV and when this result is closer to people's responses to thermal sensation, one can provide a more appropriate environment to users in order to have Thermal Comfort and, principally, to avoid Thermal Stress.

The importance of this article lies in the fact that, through the literature review, one can easily see which are the most recent studies published on the subject and it helps researchers and readers to make a more precise search on the topic and verify more precisely where there are discrepancies between the PMV index and the actual thermal sensation reported in field studies.

REFERENCES

Al-Ajmi, F.F. 2010. Thermal comfort in air-conditioned mosques in the dry desert climate. *Building and Environment*, 45, 2407–2413.

Albadra D, Vellei M, Coley D, Hart J, Thermal comfort in desert refugee camps, Building and Environment (2017), doi: 10.1016/j.buildenv.2017.08.016.

Broday, E.E., Xavier, A.A.P., Oliveira, R. 2014. Comparative analysis of methods for determining the metabolic rate in order to provide a balance between man and the environment. *International Journal of Industrial Ergonomics*, 44, 570–580.

Broday, E.E., Xavier, A.A.P., Oliveira, R. 2017. Comparative analysis of methods for determining the clothing surface temperature (t_{cl}) in order to provide a balance between man and the environment. *International Journal of Industrial Ergonomics*, 57, 80–87.

Buratti, C., Ricciardi, P. 2009. Adaptive analysis of thermal comfort in university classrooms: correlation between experimental data and mathematical models. *Building and Environment*, 44, 674–687.

Djongyang, N., Tchinda, R., Njomo, D. 2010. Thermal Comfort: a review paper. *Renewable and Sustainable Energy Reviews*, 14, 2626–2640.

Fanger, P.O. 1970. *Thermal comfort, analysis and application in environmental engineering*. Copenhagen: Danish Technical Press.

Feriadi, H., Wong, N.H. 2004. Thermal comfort for naturally ventilated houses in Indonesia. *Energy and Buildings*, 36, 614–626.

Han, J., Yang, W., Zhou, J., Zhang, G., Zhang, Q., Moschandreas, D.J. 2009. A comparative analysis of urban and rural residential thermal comfort under natural ventilation environment. *Energy and Buildings*, 41, 139–145.

Huang, L., Arens, E., Zhang, H., Zhu, Y. 2014. Application of whole-body heat balance models for evaluating thermal sensation under non-uniform air movement in warm environments. *Building and Environment*, 75, 108–113.

Iida, I. 2005. *Ergonomia: projeto e produção* (2nd ed.) São Paulo: Blucher.

Indraganti, M. 2010. Thermal comfort in naturally ventilated apartments in summer: Findings from a field study in Hyderabad, India. *Applied Energy*, 87, 866–883.

Indraganti, M., Ooka, R., Rijai, H.B. 2013. Thermal comfort in offices in summer: Findings from a field study under the "setsuden" conditions in Tokyo, Japan. *Building and Environment*, 61, 114–132.

Ji, X.L., Lou, W.Z., Dai, Z.Z., Wang, B.G., Liu, S.Y. 2006. Predicting thermal comfort in Shangai's non-air-conditioned buildings. *Building Research & Information*, 34, 507–514.

Kajtár L, Nyers J, Szabó J, Ketskeméty L, Herczeg L, Leitner A, Bokor B. 2016. Objective and subjective thermal comfort evaluation in Hungary. *Thermal Science*.

Kilic, M., Kaynakli, R., Yamankaredeniz, R. 2006. Determination of required core temperature for thermal comfort with steady-state energy balance method. *International Communications in Heat and Mass Transfer*, 33, 199–210.

Kim, J.T., Lim, J.H., Cho, S.H., Yun, G.Y. 2014. Development of the adaptive PMV model for improving prediction performances. *Energy and Buildings*, http://dx.doi.org/10.1016/j.enbuild.2014.08.051.

Liberati, A., Altman, D. G., Tetzlaff, J., Mulrow, C., Ioannidis, J. P. A., Clarke, M., Moher, D. 2009. Academia and Clinic The PRISMA Statement for Reporting Systematic Reviews and Meta-Analyses of Studies that Evaluate Health Care Interventions: *Annals of Internal Medicine*, 151(4).

Maiti, R. 2014. PMV model is insufficient to capture subjective thermal response from Indians. *International Journal of Industrial Ergonomics*, 44, 349–361.

Mors, S. ter, Hensen, J.L.M, Loomans, G.L.C., Boerstra, A.C. 2011. Adaptive thermal comfort in primary school classrooms: creating and validating PMV-based comfort charts. *Building and Environment*, 46, 2454–2461.

Plos Medicine. 2009. The PRISMA Statement for Reporting Systematic Reviews and Meta-Analyses of Studies that evaluate health care interventions: explanation and elaboration. Guidelines and Guidance.

Shen, C., Yu, N. 2010. Study of Thermal Comfort in Free-Running Buildings Based on Adaptive Predicted Mean Vote. E-Product E-Service and E-Entertainment (ICEEE), 2010 International Conference on, 1–4.

Simone, A., Crociata, S.D., Martellotta, F. 2013. The influence of clothing distribution and local discomfort on the assessment of global thermal comfort. *Building and Environment*, 59, 644–653.

Teli, D., Jentsch, M.F., James, P.A.B. 2009. Naturally ventilated classrooms: an assessment of existing comfort models for predicting the thermal sensation and preference of primary school children. *Energy and Buildings*, 53, 166–182.

Yao, R., Li, B., Liu, J. 2009. A theoretical adaptive model of thermal comfort—Adaptive Predicted Mean Vote (aPMV). *Building and Environment*, 44, 2089–2096.

Yau, Y.H., Chew, B.T. 2009. Thermal comfort study of hospital workers in Malaysia. *Indoor Air*, 19, 500–510.

Yun, H., Nam, I., Kim, J., Yang, J., Lee, K., Sohn, J. 2014. A field study of thermal comfort for kindergarten children in Korea: An assessment of existing models and preferences of children. *Building and Environment*, 75, 182–189.

Zhang, G., Zheng, C., Yang, W., Zhang, Q., Moschandreas, D.J. 2007. Thermal Comfort Investigation of Naturally Ventilated Classrooms in a Subtropical Region. *Indoor and Built Environment*, 16, 148–158.

Participatory ergonomics training for a warehouse environment—a process solution

A. Mocan
PhD Student, Politehnica University Timisoara, Romania

A. Draghici
Prof. Dr. Ing., Politehnica University Timisoara, Romania

ABSTRACT: Within a non-mechanized warehouse environment the development of musculoskeletal disorders is an ever present risk. While the introduction of new technology to replace excessive manual manipulation can help workers reduce their strain, a lack of ergonomics understanding and clear goals not just from workers, but from managers as well as industrial engineers, can lead to inadequate solutions. In a business environment that cuts budgets for training programs the present paper presents a low cost easy to use participative ergonomics process that cen be implemented in the warehouse environment of a medium sized manufacturing facility. This will allow the collaborative creation and implementation of the most suitable ergonomic solutions for all workers problems and aid in optimizing the warehouse working environment by creating occupational wellbeing. The presented approach is aligned with an organizations' needs to face the new challenges and goals for sustainable development.

1 INTRODUCTION

Within a low mechanized warehouse environment, execution of tasks by warehouse employees requires the twisting of body parts such as back, knees and shoulders. The repetitiveness of these tasks lead on the short term to injuries and on a long term to permanent musculo-skeletal deformation. The term work related musculoskeletal disorders, as Bernard et al. (1997) defines it refers to conditions that involve the nerves, tendons, muscles, and supporting structures of the body. Several factors have been associated with work related musculoskeletal disorders, such as repetitive motion, excessive force, awkward and/or sustained postures, prolonged sitting and standing. In a warehouse environment, the basic process steps such as receiving, storing, retrieving and shipping goods create an environment that is ripe for incorrect ergonomics practices.

As Mocan et al. (2017) mentioned in previous research, there are three areas in which ergonomic principles can be used to improve the quality of the work environment and help reduce negative incidents such as injuries, one of them being the creating of structured processes and training implemented to reduce the number of errors caused by lack of knowledge. Safety at work along with ergonomics are areas that are extensively covered within the EU legislation. Besides the minimal requirements these offers there are firms, such as Siemens, that dedicate significant amount of resources to create and maintain their own ergonomics tools and assesments, as Labuttis analized (2015). This is a best case scenario, in which firms strive for a better ergonomic environement for their workers.

On the other end of the spectrum some firms, such as the manufacturing facility analyzed in the current paper, do not show to be interested in advancing their ergonomic knowlede, due to a set of high management decisions that severely limited the amount of money to be invested in trainings.

The purpose of the currrent paper is to present an easy to use, low cost iterative process that leverages the concept of participative ergonomics, in order to improve working conditions for the firm's warehouse employees.

2 HOW DO WE TRAIN?

According to PWC's Global Industry 4.0 Survey (2016) companies are investing in training employees and driving required organizational change with an expected payback within two years. In the era of Industry 4.0 a one solution fits all is no longer feasible, especially in the case of ergonomics, which has to deal with a wide range of body types and movements. Because of this, participatory ergonomics along with individual tracking has the possibility of attracting the best results.

Wilson & Haines (1997) defined participatory ergonomics (PE) as 'the involvement of people in planning and controlling a significant amount of their own work activities, with sufficient knowledge and power to influence both processes and outcomes in order to achieve desirable goals'. As Laing et al. (2004) explain, the main characteristic of PE, which is a subset of macro ergonomics, is the creation of companywide cross functional teams that first undergo ergonomic training. After building up a foundation of ergonomic knowledge their purpose is to begin making improvements to their workplace.

Loisel et al. (2001) have shown that PE has been used to improve work organizational climate, reduce mental workload and rehabilitate workers with back pain. Now it is the internationally recommended approach to reducing musculoskeletal disorders associated with manual tasks, as Stubbs discusses (2002). PE is also the reference strategy to be implemented in order to develop ergonomic measures from the bottom up. Because of the heavy level of manual labor still performed in warehouse picking, storing and shipping operations, as well as the bottom up approach required in an individualized ergonomics solution, PE is the recommended ergonomics implementation strategy for warehouse environments.

According to a survey done by the learning service provider KnowledgePool, 61% of Learning and Development Managers agree that the training budget is the first to be cut when times are tough due to an inability to establish a return on investment (2012). This is why a grass roots solution such as participatory ergonomics can help support the workers without requiring a large amount of investment.

3 WHAT IS THE CURRENT SITUATION?

In order to understand what needs to be changed an analysis of the exiting situation is necessary.

The analyzed location is a West European manufacturing facility of a multinational aerospace and transportation company. The 173.006 m^2 site can be split into a warehouse area, a manufacturing area and an office area. Due to recent cuts in the facility's budget the warehouse area has changed both in size as well as in location. Instead of its own separate building with product flows and organization that have been established for decades the warehouse is now forced to operate from half of a production hall. The incoming goods area is comprised of 150 m^2 of covered space from which the goods are sorted and sent to their respective racks.

The warehouse, such as the rest of the site, operates health and safety procedures following their OHSAS 18001:2007 certification. Weekly KPIs track the rate of accidents (split into small and serious) and the number of risk alerts raised. While the number of risk alerts in 2017 is below the 1/month accepted threshold and their serious accident rate (accident that lead to an absence from work) has been 0 for more than 3 years, there is no mechanism that captures suboptimal work practices or repetitive strain problems.

There are 2 health and safety responsibles for the 500 people working on site. They are involved in managerial discussion in order to provide legislation input, and their internal auditing is based on national and EU standards and legislation. Within 2017 they championed the company's "0 accidents" strategy and policies, but did not participate in any extra ergonomics trainings or monitorings.

At the moment the "0 accidents" strategy documentation and awareness posters are presented in the manufacturing facility, the warehouse and the offices in specially designated areas. They can also be accessed by the company employees on an open access folder. There is no tracking in place for how often they are accessed/read and no process to check if their teachings have been absorbed by the employees.

In summary, the company operates at a minimum effort with regards to ergonomics, health and safety, following the legislation in place and doing little more extra to promote workplace wellbeing.

4 HOW TO ENGAGE ERGONOMICS PRACTICE IN THE WAREHOUSE?

In order to properly implement PE there are a series of tools that should be implemented depending on the organizational structure and industrial context that the firm is working in. As Pandey and Vinay (2016) discuss in their paper, the tools implemented need to provide optimal progression from data collection to analysis to conceptualizing within the firm's context.

A lot of methods have already been developed to help figure out the incorrect postures related to activities in a warehouse environment. Battini et al. (2014) describe how these tools range from simple self-reporting like interviews, questionnaires or diaries to more complex observational or simulation tools.

The abundance and easy access to quantitative data however will not help in creating efficient and effective ergonomics policies and trainings if not applied in the proper manner. As mentioned before, PE relies on working together with a group of people in order to achieve a full understanding of ergonomic aspects within the company. That means that training has to be given to more than just the workers.

The facility under analysis has experienced a training budget cut for the past few years and has not been able to offer any official trainings to their staff in that period. One way they have so far managed to circumvent that decision is by allocating a large percentage of the budget to consultancies.

These consultancies however are used to help the production and engineering area and are not at all involved in warehouse analyses of any kind. This is the baseline of decisions that the current approach has to work with. There is no possiblity to redirect budget towards training activities and the addition of a consultancy team for warehouse specific improvements would prove very difficult to approve.

Designing a workspace is not a one man job. Information about the job's tasks, the operational and strategic requirements of management, the usage and throughput of the warehouse, etc. all have to be analyzed at the same time in order to provide an output that is optimal for all who use it. Because of this, as shown in Figure 1, training shouldn't just be a punctual employee activity, but rather an integrated learning structure, where all involved are taught how their activities influence and are influenced by others.

The following sections will present the process through which workspace design can be iteratively improved by the workspace participants themselves, with the guidance of an ergonomic expert.

1. Observation week
2. Initial process mapping
3. Improvement process mapping

4.1 Observation week

The goal of the observation week is to create a baseline observation of the warehouse processes.

The ergonomics expert or consultant begins by spending a week in the warehouse area silently participating and observing operational activities, mapping the flow of goods and noting the managerial controls in place for these processes. Based on his analysis he begins the process mapping iteration by creating a workplace snapshot to be used during the discussions as well as the theoretical background needed to argument his position in the upcoming discussions. Given the above mentioned trifecta of participants the arguments need to be prepared based on the area of interest of each of the participants, so that they are effective in convincing them.

The cost of the observation week is equivalent to the consultancy cost of the ergonomics expert.

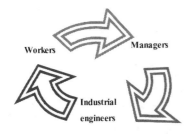

Figure 1. Warehouse training trifecta.

4.2 Initial process mapping

The goal of the first meeting workshop is to create a rough draft of warehouse processes as they currently exist and are perceived by the users. In a brown paper session, the ergonomics expert along with the key users define the general warehouse processes. Each of the key users is then instructed to draw the processes as they are currently happening in a standard process flow diagram. The first set of drawings is analyzed to discuss the differences in viewpoint of the key users. The discussion will help set the baseline understanding of how the processes are really carried out in the warehouse and provide a clear as-is picture.

Then the same process is repeated, this time the key users being asked to draw the agreed upon process on a spaghetti diagram of the warehouse locations. Even if the warehouse processes were clarified in the previous step it is expected that the users will once again provide different spaghetti charts with regards to the flow. This is because people that are not directly involved in the flow are not aware of how workers actually complete their tasks.

The ergonomics specialist joins the discussion by providing his own spaghetti chart, based on the visual observations he has made in his observation week. This outsider point of view serves to show the people involved in the process what is actually going on in terms of flows and the means by which they are done. It is expected that this presentation varies from all the previous ones, as real life processes suffer from interference, task switching, worker preference, etc. This is used to start a discussion on the means to reduce interferences and optimizing interprocess flows, while also aligning the business needs of all key users. Based on these discussions a new process flow is designed and common KPIs are put in place to measure its performance.

The first session ends by optimizing the existing processes without changing the location of the goods or the understandings of the flow. This is the "quick win" round, where understanding the process and the varying viewpoints is more important than the amount of ergonomic benefits reached. It is left to run for three months.

The cost of the initial process mapping is equivalent to the consultancy cost of the ergonomics expert and the time of those participating in the workshop meeting.

4.3 Improvement process mapping

The initial process redefinition is monitored throughout its 3 months and changes in process speed and accuracy are monitored weekly via the commonly agreed KPIs. It has the purpose of introducing a mentality change in the workers by aknowledging that change is possible and desirable. At the end of the 3 months the ergonomics expert is once again present on site for a week.

During this week he checks the warehouse status, the monitoring of the new KPIs and the way in which the new process is upheld by the 3 types of workshop participants. At the end of the week a new workshop meeting is set up by the ergonomics expert in which the participants are once agains asked to draw the current process map of the warehouse activities. This time around the ergonomics expert plays the devil's advocate and facilitates discussions regarding process and location reorganization with the goal of changing them for the better. By challenging the existent status quo and providing technical expertize as to why certain changes would be beneficial he guides the team towards improving ergonomic soutions.

A new process map and spaghetti chart is created, this time changing the processes and locations of goods within the warehouse. Even if the solution is suboptimal the ergonomic expert agrees with it and its implementation. Again, the iterative process allows those involved in the process to learn from their decisions. The ergonomic hints passed on along the way in the discussion will be the basis of the following iteration's improvement. This process repeats every 3 to 6 months, bringing together around the table the people involved.

The cost of the improvement process mapping is once again equivalent to the consultancy cost of the ergonomics expert for the check-up week and the time of those participating in the workshop meeting.

4.4 *Discussion*

The purpose of the iterative process is to create a mentality of change where new ideas are discusses and accepted. It leverages the knowledge and experience of people involved in the warehouse processes, as they are ultimately the ones that know the most about their job. As Zalk (2001) discusses, PE creates a culture of participation and learning where collaboration between different functions is seen as the way to work, and a sense of a community, of a team is reinforced.

5 CONCLUSIONS

Warehouse ergonomics implies a wide range of movement tracking and analyzing based on the type of industry the warehouse operates in. One sided analysis and transformation of the data into ergonomics trainings and procedures is useless. It has been shown that imposing rules on workers does not guarantee the following of those rules, if they are seen as being imposed.

Participatory ergonomics can be the answer to this standstill. Because of the wide range of activities, their outspread location, as well as worker differences in behavior and perception, a one size fits all solution not sustainable. Thus by involving multiple departments and experts in the analysis process the chances of coming up with a better solution increase. In the process of finding out a better solution along with other colleagues community is formed, goals are aligned and discussions from various perspective foster a better understanding of the entire warehousing process.

With the help of a few punctual directions by an ergonomics specialist the understanding and training process can result in significant benefits for the company's ergonomic understanding with minimal financial investment.

REFERENCES

2016 Global Industry 4.0 Survey—Industry key findings www.pwc.com/industry40, Accessed 10.09.2017.

Battini, D., Persona, A., Sgarbossa, F. 2014. Innovative real-time system to integrate ergonomic evaluations into warehouse design and management. *Computers & Industrial Engineering* 77: 1–10.

Bernard BP, Putz-Anderson V, Burt SE, et al. 1997. 2nd edition. *Muskuloskeletal disorders and workplace factors: a critical review of epidemiologic evidence for work-related musculoskeletal disorders of the neck, upper extremity, and lower back*, Cincinnati: DHHS (NIOSH).

Haines, H., Wilson, J.R., Vink, P., and Koningsveld, E. 2002. Validating a framework for participatory ergonomics (the PEF). *Journal of Ergonomics*, 45(4): 309–327.

Knowledgepool, Training Budgets fall foul of cuts, http://spec.knowledgepool.com/knowledge-centre/news/20120214-training-budgets.html, Accessed 08.12.2017.

Labuttis, J. 2015. Ergonomics as Element of Process and Production Optimization, *Procedia Manufacturing*, 3:4168–4172.

Laing, A.C.; Frazer, M.B.; Cole, D.C.; Kerr, M.S.; Wells, R.P. and Norman, R.W. 2005. Study of the effectiveness of a participatory ergonomics intervention in reducing worker pain severity through physical exposure pathways. *Ergonomics*, 48: 150–170.

Loisel, P., Gosselin, L., Durand, P., Lemaire, J., Poitras, S. and Adenheim, L. 2001. Implementation of a participatory ergonomics pogrom in the rehabilitation of workers suffering from subacute back pain. *Journal of Applied Ergonomics*, 32: 53–60.

Mocan, A., Draghici, A., Mocan, M. 2017. A way of gaining competitive advantage through ergonomics improvements in warehouse logistics. *Proceedings of the 15th Management and Innovative Technologies (MIT) Conference*.

Pandey, K. and Vinay, D. 2016. Participatory ergonomics, its benefits and implications: a systematic review, *India International Journal of Humanities and Social Sciences*. 5(3):25–32.

Stubbs, D.A. 2002. Ergonomics and occupational medicine: future challenges. *Journal of Occupational Medicine*. 50: 277–282.

Zalk, D.M., T.W. Biggs, C.M. Perry, R. Tageson, P. Tittiranonda, S. Burastero and Barsnick. 2000. Participatory ergonomics approach to waste container handling utilizing a multidisciplinary team, *14th Triennial congress of the international ergonomics association, 44th Annual Meeting of the Human Factors and Ergonomics society San Diego, CA July 3–August 4*.

Wheelchair users' anthropometric data: Analysis of existent available information

S. Bragança
Research and Innovation, Southampton Solent University, Southampton, UK

I. Castellucci
Escuela de Kinesiología, Facultad de Medicina, Universidad de Valparaíso, Valparaíso, Chile

P. Arezes
Centro ALGORITMI, School of Engineering, University of Minho, Guimarães, Portugal

ABSTRACT: Anthropometric data represents the characteristics of a given population and, as such, it varies accordingly. Wheelchair users have, most times, different body characteristics than able-bodied people. These differences can be noticed also in terms of anthropometric data. This paper demonstrates the differences between the results obtained in several studies where anthropometric data of wheelchair users was collected. The results of the comparisons between studies demonstrated that there are some dissimilarities on the data collected. The several studies use different data collection protocols, which compromises the results and the comparability of the results obtained. Hence, it become unclear if the data collected in these studies is accurate and valid enough to be used in other future studies.

1 INTRODUCTION

Anthropometric data is still limited for the able-bodied population, but it is even scarcer in terms of the structural and functional anthropometric dimensions of wheelchair users (Kozey & Das 2004).

Research showed that the anthropometric characteristics of wheelchair users are different from the anthropometric characteristics of the able-bodied (Lucero-Duarte et al. 2012).

In fact, anthropometric data varies according to the type of disability—some are characterized by atypical distributions of muscle bulk, bone mass or body stature (Hobson & Molenbroek 1990). Frequent use of a wheelchair also promotes an overly developed upper body and a more atrophied and weakened lower body (Dingley et al. 2015).

As such, designing workplaces and products for wheelchair users based on the able-bodied population anthropometric is not the most correct procedure to be adopted. According to Das and Kozey (1999), the design guidelines used to develop products for wheelchair users are usually based on timeworn information (Floyd et al. 1966) or extrapolated from the able-bodied population (Pheasant & Haslegrave 2006). Kozey and Das (2004) and reinforce this by stating that it is not appropriate, nor possible, to properly design a workstation for the wheelchair users' population using information about seated able-bodied workers. Consequently, these authors suggested that there is the need to further investigate the differences between these populations and to also generate more reliable anthropometric data of wheelchair users, which will certainly play an important role in the design of products and spaces. Hobson and Molenbroek (1990) even argue that anthropometric data for the disabled population will need to differentiate between disabilities, and in some cases within disabilities, in order to achieve the desired usefulness. Gonzalez et al. (2012) also referred to the importance of having a European wheelchair users' database to facilitate the development of products, such as clothes.

None of these situations represents the best-case scenario, as this specific population should have its representative anthropometric database. In fact, some previous studies tried to investigate and define the anthropometric characteristics of the disabled population (Urrutia et al. 2015), even trying to compare them with the anthropometric characteristics of the able-bodied population (Goswami et al. 1987). However, it seems to be a difficult task as many of these studies have limited and small sample sizes and do not use the more accurate data collection methods available (such as 3D body scanners).

The main objective of this paper is to compare the data collected in a variety of studies where anthropometric measurements of wheelchair users was gathered. The purpose of this comparison is to understand what are the measurements collected in each study and if there are similarities between the studies, but also to understand if the numerical values obtained are similar and reliable so that they can be used in other future studies.

2 METHODOLOGY

This work is based on the results of a Systematic Literature Review (SLR) that was previously conducted by the authors. This SLR followed a five-step approach and tried to answer the research question: *"How can the available literature on anthropometric data collection of wheelchair users be characterized and compared, to allow for a better understanding of the current scenario and to promote the use of data for future studies?"*

Three bibliographic databases were selected (*ISI Web of Science*, *Scopus* and *PubMed*) and, using variations of a search string (wheelchair OR "mobility impairment" OR "physical disability" OR "assistive mobility technology" OR "mobility aid") AND ("anthropometr* database" OR anthropometr* OR "body characteristics" OR "physical characteristics" OR "body measurements"), the answer to that research question was found. The SLR resulted in the identification of 40 articles.

For the present study only a selection of these 40 articles has been considered. As the main goal of this study is to compare the data from the different studies, it was extremely important to ensure that the studies were comparable and that there were the largest number of similarities between them as possible. As such, to select the papers to be included, some exclusion criteria were defined:

- papers that did not present the results of the measurements in numbers, but instead in graphs or figures;
- the data collected was regarding children or elderly people;
- the results were clustered in classes or groups, rather than representative of the entire population.

This selection lead to the inclusion of 15 papers in this study.

3 RESULTS

The 15 papers selected to this study have several similarities but they are also very different from one another. This difference is especially notorious in the area each paper focused on.

Eight papers were specifically related to the collection of anthropometric data; four focused on sports issues; one on medical issues; and two on product design. As expected this fact impacted not only on the number and type of data collected, but also on the type of statistical analysis performed. Table 1 summarizes this information.

As can be seen, the papers that had the main objective of collecting anthropometric data were the ones where more measurements were gathered and where the percentiles were calculated. A considerable amount of measurements was also collected in the studies that focused on product design. However, and surprisingly, percentiles were not considered.

The other papers, with less focus on ergonomics concerns, collected fewer measurements and analyzed the data only in terms of mean and standard deviation.

3.1 *Type of anthropometric measurement*

Regarding the measurements collected across the several papers there are also some discrepancies. The 15 papers collected a total of 80 different anthropometric measurements. Forty-six of these

Table 1. Characterization of the papers selected for this study.

T	[ID] Reference	N	Stat. A
A	[1] Floyd et al., 1966	9	M; SD; P(5); P(95)
	[2] Hosler et al., 1982	10	M; SD; Rg
	[3] Goswami et al., 1987	12	M; SD; P(5); P(95)
	[4] Jarosz, 1996	18	M; SD; P(5); P(95)
	[5] Das and Kozey, 1999	16	M; SD; P(5); P(95); Rg
	[6] Paquet and Feathers, 2004	27	M; SD; P(5); P(95)
	[7] Barros and Soares, 2012	15	M; SD; Rg; Md
	[8] Lucero-Duarte et al., 2012	14	M; SD; P(5); P(95)
S	[9] Gass and Camp, 1979	7	M; SD; Rg
	[10] Cooper, 1992	5	M; SD
	[11] Gil et al., 2015	4	M; SD
	[12] Granados et al., 2015	6	M; SD
M	[13] Barreto et al., 2009	6	M; SD
D	[14] Nitz and Bullock, 1983	8	M; SD
	[15] Urrutia et al., 2015	10	I

T: Type; [ID]: Study ID number; N Number of Measurements Collected; Stat. A: Statistical Analysis; Anthropometry; S: Sports issues; M: Medical issues; D: Product Design; M: Mean; SD: Standard Deviation; P(5): 5th percentile; P(95): 95th percentile; Rg: Range; Md: Median; I: Individual data.

measurements were only collected in a single study, leaving only 34 measurements common to two or more studies.

This indicates that there is an inconsistency in terms of what are the most appropriate measurements to be collected. From this information, it is possible to conclude that there is clearly no specific protocol that studies should use. Instead, the measurements collected are the ones that seem more relevant to the specific application of each study.

Nonetheless, there are a few measurements that seem to be commonly collected by several studies – 13 of these measurements were common to at least four studies (Table 2).

3.2 Numerical value of the anthropometric measurements

Regarding the numerical value of the measurements, they are somewhat consistent and similar throughout the different studies. Figure 1 shows examples of specific anthropometric measurements (buttock-popliteal length and sitting height) that were collected in different studies and that present similar values.

Despite not being expected that the values of the measurements are exactly the same, as they represent different populations, it would be

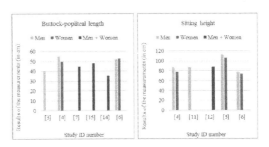

Figure 1. Examples of similarities between the mean values of the measurements collected in different studies.

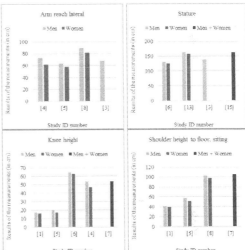

Figure 2. Examples of dissimilarities between the mean values of the measurements collected in different studies.

expected that the variations were small. However, this was not the case in some of the measurements collected. As can be seen in Figure 2, four of the measurements analyzed presented large differences according to the study where they were collected, namely: arm reach lateral; stature; knee height; shoulder height to floor, sitting.

4 DISCUSSION

There are some speculative reasons that can lead to the inconsistencies found in the several studies analyzed. Even though the names given to the measurements collected in each study are more or less similar, it is unclear if they represent the same body parts. Moreover, this is aggravated by the fact that most studies do not give a definition of the measurements to be collected, which makes the process of understanding exactly what is being measured even more difficult. Table 3 demonstrates

Table 2. Measurements collected in the selected papers.

Measurements collected	
1. Abdominal depth, sitting	2. Abdominal skinfold
3. Arm girth*	4. Arm girth (contracted)
5. Arm reach forward*	6. Arm reach lateral*
7. Arm reach overhead from floor, sitting	8. Buttock-knee length
9. Buttock-popliteal length*	10. Chest depth, sitting*
11. Chest girth, sitting	12. Elbow breadth
13. Elbow height to chair, sitting	14. Elbow height to the floor, sitting*
15. Eye height to floor, sitting	16. Forearm-fingertip length
17. Hand breadth	18. Height to floor, sitting*
19. Hip breadth, sitting*	20. Knee height*
21. Popliteal height*	22. Shoulder (biacromial) breadth
23. Shoulder (bideltoid) breadth	24. Shoulder breadth
25. Shoulder height to chair sitting	26. Shoulder height to floor, sitting*
27. Shoulder-elbow length	28. Sitting height*
29. Stature*	30. Subscapular skinfold
31. Sum of skinfolds	32. Supra iliac skinfold
33. Triceps skinfold	34. Waist girth

*Measurements common to at least four papers.

Table 3. Measurements collected in the selected papers.

ID	Original name	Definition
Arm reach lateral		
[3]	Arm grasp (Max.)	NA
[4]	Lateral reach	NA
[5]	Radial arm reach	Sagittal Shoulder marker to the proximal interphalangeal joint
Stature		
[8]	Lateral arm reach to middle finger	NA
[13]	Stature	NA
[15]	Height	NA
[6]	Overall height	Vertical distance from the floor plane to the vertex
[3]	Stature	NA
Knee height		
[7]	Height of the knee	Anterior side of the knee—floor
[5]	Knee height	Sagittal Floor to anterior surface of the thigh (distal end)
[6]	Knee height	Vertical distance from floor plane to suprapatella landmark
[1]	Lower leg length	NA
[4]	Knee height	NA
Shoulder height to floor, sitting		
[1]	Floor to shoulder	NA
[7]	Height of the shoulder	Acromion—floor
[8]	Shoulder height	NA
[5]	Shoulder height	Sagittal Floor to acromion

ID: Study ID number; NA: Not Available.

Table 4. Mean values obtained in different studies for the *Arm Reach Lateral* measurement.

		Mean value in cm	
ID	Measuring devices	Men	Women
[3]	Measuring tape + Anthropometer + Callipers	67.90	N/A
[4]	N/A	72.75	61.72
[5]	Photogrammetry	63.10	58.10
[8]	Measuring tape + Callipers	89.30	81.12

ID: Study ID number.

Table 5. Mean values obtained in different studies for the *Knee Height* measurement.

		Mean value in cm		
ID	Measuring devices	Men	Women	Men + Women
[1]	Fixed anthropometer	17.30	16.20	N/A
[4]	N/A	53.65	46.83	N/A
[5]	Photogrammetry	19.90	17.20	N/A
[6]	Photogrammetry	64.50	62.80	N/A
[7]	Photogrammetry	N/A	N/A	53.45

ID: Study ID number.

the differences between the names given to the same measurement and the definitions provided by the studies.

As can be noticed, all the studies could be measuring the same body part and giving it different name, but, without a clear definition it is extremely difficult to be certain.

Another possible cause for the discrepancies might be the use of different measuring equipment. Table 4 shows the example of the arm reach lateral measurement, collected by four different studies. As can be observed, not all of the studies used the same data collection technique/equipment and as a result, the measurements obtained are somewhat different.

Furthermore, it was also found that even when the equipment and technology used is the similar, the results vary considerably. Examples of this were found not only for the traditional data collection technique, with measuring tapes and callipers but also for more advanced techniques, as photogrammetry.

In Table 4, it can be noticed that studies [3] and [8] allegedly used the same equipment but the values obtained are very different: a mean of 67.90 cm in study [3] versus a mean of 89.30 cm in study [8], which is a difference of 21.40 cm.

In Table 5, three of the studies use photogrammetry but the results are very different: a mean of 19.90 cm in study [5] versus a mean of 64.50 cm in study [6], which is a difference of 44.60 cm.

Moreover, by comparing all the measurements collected by all the studies it became clear that some of them tended to present measurements with numerical values always smaller than the other studies.

All the measurements presented in study [1] had smaller values that every other study that collected the same measurements. In this particular study, the data was collected with a fixed anthropometer. Any of the other studies used this technique, which might indicate that this is the cause for the discrepancies and that these techniques might not be the most accurate and appropriate one.

On the other hand, study [8] presented, for many of the measurements collected, value higher than the other studies that collected the same measurements. In this case, similar techniques were used by several comparable studies (measuring tapes and callipers) but the results obtained were different.

All of these issues compromise the validity of the results presented in the several studies. The cause of these differences remains unclear. It is recognizable that re-using the data collected in any of these studies might be risky as it is not possible to know which one is the most accurate or representative one.

5 CONCLUSIONS

The findings of the present study were very important to clearly understand the current scenario in terms of anthropometric data of wheelchair users.

There are very few measurements that are collected in several studies, which makes it difficult to understand which of the measurements are the most important one to consider.

It can be concluded that the adoption of different measurement protocols result in large differences in the results obtained. Hence, comparing the results obtained in the several studies in extremely difficult as the characteristics are not always the same—comparisons would only be reliable if similar protocols were used.

Moreover, this study allowed to understand that using the data collected in previous studies in future studies should be done with caution, if even done at all. It was found that the same measurement collected by different people are sometimes considerably different and, as such, it would be complicated to ascertain what would the most reliable data to be used.

Only by comparing the studies side by side is it possible to verify the great dissimilarities between them and the impact that this has on the application of the data.

REFERENCES

Barreto, F.S. et al. 2009. Nutritional assessment of disabled subjects practitioners of swimming. *Revista Brasileira de Medicina do Esporte* 15(3): 214–218.

Barros, H.O. & Soares, M.M. 2012. Using digital photogrammetry to conduct an anthropometric analysis of wheelchair users. *Work* 41: 4053–4060.

Cooper, R.A. 1992. The contribution of selected anthropometric and physiological variables to 10K performance of wheelchair racers: A preliminary study. *Journal of Rehabilitation Research and Development* 29(3): 29–34.

Das, B. & Kozey, J.W. 1999. Structural anthropometric measurements for wheelchair mobile adults. *Applied Ergonomics* 30(5): 385–390.

Dingley, A.A., Pyne, D.B. & Burkett, B. 2015. Relationships Between Propulsion and Anthropometry in Paralympic Swimmers. *International journal of sports physiology and performance* 10(8): 978–985.

Floyd, W.F. et al. 1966. A study of the space requirements of wheelchair users. *Paraplegia* 4(1): 24–37.

Gass, G.C. & Camp, E.M. 1979. Physiological characteristics of trained australian paraplegic and tetraplegic subjects. *Medicine and science in sports and exercise* 11(3): 256–259.

Gil, S.M. et al. 2015. The Functional Classification and Field Test Performance in Wheelchair Basketball Players. *Journal of Human Kinetics* 46(1): 219–230.

Gonzalez, J.C. et al. 2012. FASHION-ABLE: Needs and requirements for clothing, footwear and orthotics of consumers groups with highly individualised needs. *Proceedings of the 18th International Conference on Engineering, Technology and Innovation*: 1–10.

Goswami, A., Ganguli, S. & Chatterjee, B.B. 1987. Anthropometric characteristics of disabled and normal Indian men. *Ergonomics* 30(5): 817–823.

Granados, C. et al. 2015. Anthropometry and Performance in Wheelchair Basketball. *Journal of Strength and Conditioning Research* 29(7): 1812–1820.

Hobson, D.A. & Molenbroek, J.F.M. 1990. Anthropometry and design for the disabled: Experiences with seating design for the cerebral palsy population. *Applied Ergonomics* 21(1): 43–54.

Hosler, W.W., Morrow, J.R. & Boelter, J. 1982. Isokinetic Strength Characteristics of Men Confined to Wheelchairs. *Proceedings of the Human Factors and Ergonomics Society*: 181–183.

Jarosz, E. 1996. Determination of the workspace of wheelchair users. *International Journal of Industrial Ergonomics* 17(2): 123–133.

Kozey, J.W. & Das, B. 2004. Determination of the normal and maximum reach measures of adult wheelchair users. *International Journal of Industrial Ergonomics* 33(3), pp. 205–213.

Lucero-Duarte, K. et al. 2012. Anthropometric data of adult wheelchair users for Mexican population. *Work* 41: 5408–10.

Nitz, J.C. & Bullock, M.I. 1983. Wheelchair design for people with neuromuscular disability. *Australian Journal of Physiotherapy* 29(2): 43–47.

Paquet, V. & Feathers, D. 2004. An anthropometric study of manual and powered wheelchair users. *International Journal of Industrial Ergonomics* 33(3): 191–204.

Pheasant, S. & Haslegrave, C.M. 2006. *Bodyspace: Anthropometry, ergonomics and the design of work*, London: CRC Press.

Urrutia, F. et al. 2015. User centered design of a wheelchair based in an anthropometric study. *Proceedings of the 2015 IEEE Chilean Conference on Electrical, Electronics Engineering, Information and Communication Technologies*: 235–243.

A review on occupational risk in gasification plants processing residues of sewage sludge and refuse-derived fuel

O. Alves & M. Gonçalves
MEtRICs, Department of Science and Technology of Biomass, Universidade NOVA de Lisboa, Portugal

P. Brito & E. Monteiro
Polytechnic Institute of Portalegre, Portugal

C. Jacinto
UNIDEMI, Mechanical and Industrial Engineering, Universidade NOVA de Lisboa, Portugal

ABSTRACT: This paper presents a review on Occupational Safety and Health (OSH) issues of gasification plants using residues of sewage sludge and refuse-derived fuel. The survey (2006–2016) consisted of a systematic review of literature retrieved from scientific databases. Despite abundant literature on environmental impacts of gasification plants, few publications focus on OSH aspects. Of these, 16 were considered relevant for the purpose. The results are summarized in a Table that provides a short description and the main findings of each study. It can be concluded that explosion and inhalation of toxic gases are the most common OSH risks reported in gasification processes; apparently, both are considered negligible if safety measures are adopted. Other risks identified comprise the release (and potential inhalation) of heavy metals and harmful tar production, but the main concern is on the environmental impact. Prevention measures mostly include adjusting granulometry to the type of biomass, workers' training, good ventilation and reliable equipment.

1 INTRODUCTION

Due to problems stemming from an ever-increasing energy demand, exhaustion of fossil fuels and the problem of global warming caused by their use, other alternatives for energy generation from renewable resources such as gasification are currently receiving more attention.

Generally speaking, a gasification process means the total or partial transformation of solid biomass into gases (Figure 1). It consists in a thermochemical process where biomass is subjected to high temperature (>700°C) in a medium with deficit of oxygen, generating solid by-products (ashes and chars), a liquid rich in hydrocarbons (tars), and a synthesis gas (syngas) with a good heating value that can be burned to obtain energy (Kumar, 2015). Different materials have been used in the process (e.g.: coal, or animal residues), including urban and industrial residues, e.g.: Sewage Sludge (SS) or Refuse-Derived Fuel (RDF), because global costs and the negative impact for the environment are both attenuated as compared with conventional waste treatments of landfilling, incineration and valorization in soil fertilization (Azapagic, 2007; Furness et al., 2000; Khoo, 2009).

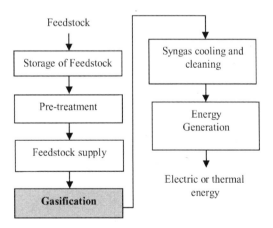

Figure 1. Main operating blocks of gasification installations (adapted European Commission, n.d., p. 12).

The discipline of Occupational Safety and Health (OSH) deals with the identification of risks responsible for accidents and illnesses at work and the prediction of their frequency and gravity, in order to define the appropriate safety measures to mitigate their occurrence (Vasilescu et al., 2008).

Inside a gasification plant, the flammable nature of the syngas, the presence of toxic compounds in all products (e.g.: heavy metals, acid gases and phenols) and the demanding working conditions may pose OSH problems for humans and environmental damage (Brisolara & Qi, 2013; Mishra et al., 2015).

The systematic prevention of harmful effects requires careful attention from industrial producers, preferably supported by scientific studies and/ or guidance from regulating authorities; this is why a literature survey on the topic was felt necessary and opportune. Moreover, OSH effects are less studied as compared with environmental impacts; thus, the objective of this work it to provide a review of relevant literature related to OSH studies in gasification plants.

The focus on risk prevention was primarily given to gasification of SS and RDF. It should be noted that standards and legislation are not in the scope of this work, unless they are explicitly referred to, or analyzed in some publication.

2 METHODOLOGY

The methodology used consisted of a "systematic review of literature" illustrated in Figure 2. In this approach, data are analyzed and synthesized and relations between published materials are pinpointed, from which conclusions are drawn about what is known and unknown on the subject under consideration (Torreglosa et al., 2016).

Step 1 coincides with the search objective. It can be translated into the question: "what relevant studies, and how many, can be traced in the literature related to OSH risks in gasification plants?" This research question was restricted to a time frame of around 11 years (2006–2016), and it focused, as much as possible, on gasification of SS and RDF.

The literature survey (step 2) was performed by searching online databases of scientific publications, thesis and European studies and recommendations such as Elsevier, Taylor and Francis, Springer and the European Commission portal. Generic search engines were used to support the quest for publications as well (e.g.: B-on, Google Scholar, Scielo and Open Science Directory) and other papers were added by cross-referencing.

The following combinations of keywords were used: "gasification sewage sludge safety", "gasification sewage sludge risk assessment", "gasification RDF safety" and "gasification RDF risk assessment". This search returned several hundreds of records, for which a refinement was carried out by adding four additional keywords "review", "occupational", "plant" and "facility", one at a time. This action helped to restrict the scope; at the same time, all papers using any language other than English were excluded at this stage. Since only few documents included SS and RDF as raw materials in studies about safety and risk assessment, the review was extended to others that occasionally appeared and which covered traditional materials like coal and vegetable specimens. The reasoning was that matters regarding workers safety and risks associated with the gasification of these other materials probably are not much different and may give an interesting starting point for the situation of the residues in cause.

After examination of titles and abstracts of the records retrieved (Step 3) from the databases and the elimination of duplicates, 55 documents were downloaded and read. More than half focused solely on environmental impacts and were disregarded.

From this first analysis (Step 4), a final list of 16 documents was considered relevant for the present work, and these were therefore subjected to further scrutiny. Each publication was then classified according to its nature (e.g.: experimental study, review, thesis, modelling framework, or practical guidance).

The present review (Step 5) focused on the selected publications, from which the respective aims, identified risks and main findings were extracted and reported here.

3 RESULTS

Table 1 synthetizes 16 publications reporting relevant information primarily concerned with safety

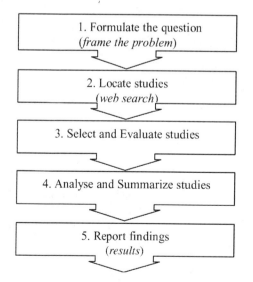

Figure 2. Steps for performing a systematic literature review (adapted from Torreglosa et al., 2016, p. 320).

Table 1. Summary of published work on occupational (OSH) risks in gasification plants (time span: 2006–2016).

Article type	Reference, Origin	Type of risk	Relevant findings/comments
Experimental studies—occupational safety and health	(Tian et al., 2009), China	Gas explosion	Compares the risk between a conventional and a two-heat source gasifier. The second is safer since high temperature is more homogeneously distributed, providing that some precautions are adopted (e.g.: reactor cooling or injection of inert gas, in case of eruption).
	(Di Sarli et al., 2014), Italy	Gas explosion	Evaluates the explosion of a simulated syngas composed by CO, CO_2, H_2, CH_4 and N_2. For lower amounts of CO (<27% v/v), the syngas showed a lower reactivity due to the smaller values of maximum pressure rise and deflagration index. The worst case occurred at the lowest temperature (10°C) and the highest quantity of CO (60% v/v), where maximum pressure, maximum pressure rise and deflagration index achieved 5.7 bar, 89.1 bar/s and 35.6 bar.m/s, respectively, being the last one lower than the value of CH_4 (55 bar.m/s).
	(Werle & Dudziak, 2014), Poland	Absorption of toxic compounds (heavy metals and organic contaminants)	Higher content of contaminants were found in by-products of gasification of SS than in initial residue. Heavy metals Zn, Cr and Cu moved mainly to ash and char (total concentration ≈7750 mg/kg): resultant tars contained phenols, polychlorinated biphenyls (PCB's) and traces of heavy metals. Some compounds are above limits defined by Polish regulations, so further treatments are required.
	(Rong et al., 2015), Singapore	Absorption of toxic compounds (heavy metals and organic contaminants)	Analyses the toxicity effect on human cells of ashes produced in the gasification of SS and wood. Ashes show a high toxicity level mainly because of high pH (>9) and ionic strength originated by alkali and alkaline earth metals, so a careful management and disposal are essential.
	(Torrent et al., 2015), Spain	Fire	Examines factors influencing self-ignition of vegetable biomasses like its composition, treatment and flammability. Higher contents of lignin and smaller granulometry increase the risk of self-ignition of biomass. Wastes of olive oil and grape seeds may self-ignite easily.
Case study—occupational safety	(Abidin et al., 2011), Malaysia	Gas explosion, CO inhalation, fire	Uses FTA (fault-tree analysis) and HAZOP (hazard and operability) methods. Shows how failure of monitoring equipment and rupture by over pressurization are unlikely to occur (<10^{-5}/year), and the consequences (increase of flammable gas concentration and release of CO) cause little injury to workers.
	(Arena et al., 2008), Italy	Gas explosion, toxic gas inhalation	Assesses risks of inadequate feeding. Probability of risk is very low and limited to mechanical damages (less than 2×10^{-9}), increasing to 3.7×10^{-5} if the plant works continuously. Training of operators and use of reliable devices may reduce risks to more than 97%.

(Continued)

Table 1. (Continued).

Article type	Reference, Origin	Type of risk	Relevant findings/comments
	(Hirano, 2006), (Gao & Hirano, 2006), Japan	Gas explosion	Identifies causes and processes behind gas explosions at three different sites. In the RDF storage plant, the natural oxidation of the organic fraction and the high height of the pile caused a rise in temperature that promoted a spontaneous ignition of the combustible and the explosion. Cooling of space and evacuation of produced gases are suggested to prevent these phenomena.
Guidance—occupational safety	(European Commission, n.d.)	Gas explosion, fire, toxic gas inhalation, mechanical injuries	Gives recommendations for the correct design of a gasification plant, choice of equipment, operation of installation and procedures in case of emergency. Incidents are more probable to happen during start-up and shut down of the plant. Proposes measures like the duplication of sensors, anti-backfiring valves in ducts and positive pressure in ambient-air of the rooms to avoid accidents.
	(Lettner et al., 2007) Austria	Gas explosion, fire, toxic gas inhalation, mechanical injuries	Applies the HAZOP to evaluate risks. Plant was segmented in four modules: fuel supply, gasifier, gas treatment and gas combustion. Common hazards and risks are gas leakages, inappropriate air intake, inhalation and contamination by toxic products and fire. Apparently, the most frequent consequence is equipment damage.
MSc thesis—occupational safety	(Huuskonen, 2012) Finland	Toxic gas inhalation	Evaluates the impact of occupational safety and environmental issues for human health in all processes of a gasification plant, by calculating the disability adjusted life year for several operations. The greatest impacts are caused during the reactor operation. Occupational accidents constitute a lesser problem than environmental emissions, and improvement of the second may degrade the first. The approach helped to identify the worst hazards, but other risk assessment tools may be required to study effects in detail.
Review—occupational safety	(Mishra et al., 2015) India	Gas explosion, toxic gas inhalation	Risks of fire, gas poisoning and noise issues are present in most parts of gasification plants. Toxic substances may be released as gases (CO, S, Cl, NO_x, SO_x), liquids (phenols, benzene and polycyclic aromatic hydrocarbons (PAH)) or solids (dust), causing pathologies like headaches, anemia, irritation and cancer. Treatments are required to clean all products, including effluents from gas cleaning.
Risk modelling—occupational safety and health	(Molino et al., 2012) Italy	Gas explosion	Compares two mathematical methodologies (based on Italian and International Electrotechnical Commission's (IEC's) norms, and on a computer dynamic model) to determine the volume of syngas released by leakage. It was seen that norms gave conservative results for the potentially explosive area and distance as compared with the computer model, which in turn is more feasible. Dangerous distance is greater for flanges than for valves (10 cm vs. 5 cm).
	(Ragazzi & Rada, 2012) Italy	Absorption and inhalation of toxic substances (gas pollutants)	Simulates and compares the emissions of gaseous contaminants by combustion, pyrolysis and gasification of waste. In all cases negligible concentrations were released, but gasification and two-step pyrolysis-gasification are slightly less pollutant than combustion. Emissions of Cd (a carcinogenic element) may be higher in some cases and more prevention measures must be adopted.
	(Lonati & Zanoni, 2013) Italy	Absorption of Hg	Uses a Monte-Carlo probabilistic model to determine the contamination by mercury emissions in RDF gasification. Concentrations calculated for air and soil (<0.022 ng/m^3 and <0.017 mg/kg) were found to be two and one order of magnitude lesser than original values, respectively, which showed a negligible influence for human health. Food ingestion was the main pathway of contamination.

and health of workers in gasification facilities, despite some covering environmental issues as well. They range from experimental studies, European guidance, academic thesis and other reviews, to the mathematical modelling of gas release and explosion risk.

These 16 papers originate from 10 different countries, giving evidence that preoccupations are world-wide. Table 1 shows that studies related to occupational safety and risk assessment are scarce when dealing with the use (or processing) of waste residues, and many are centered in the risks of gas *explosion* and *inhalation of toxic gases* (e.g.: Lettner et al., 2007; Arena et al., 2008; Abidin et al., 2011; Huuskonen, 2012). Apparently these two modalities are the most frequent risks in this field.

This search revealed that there is lack of quantitative data about the occurrence of accidents in gasification facilities, but the existing few studies seem to agree that accident risk level is considered negligible when the appropriate safety measures are implemented (Abidin et al., 2011; Arena et al., 2008).

For assessing OSH risk in gasification processes, different methods have been applied. Basic methods identified in the literature include: (a) HAZOP; (b) FTA; (c) disability adjusted life year (DALY); (d) methods defined by Italian standards CEI 31–35 and 35/A and CEI EN 60079-10 (see Abidin et al., 2011; Huuskonen, 2012; Lettner et al., 2008; Molino et al., 2012). Most assessments are merely "qualitative" for exploring causes and consequences of abnormal events (e.g.: HAZOP), whilst others give some "quantitative" data for the severity and frequency of risk (e.g.: FTA, DALY and the approaches recommended by Italian norms).

The studies reviewed also highlighted that preventive measures must be implemented not only during the gasification process in the reactor and treatment of gases, but also in the storage of biomass, namely the volume and height of piles and a suitable granulometry for the materials, since these may constitute a fire hazard (self-ignition) if not correctly accommodated (Torrent et al. 2015; Hirano 2006). According to most studies, important measures to implement in all facilities include, for instance, the storage of feedstocks with a granulometry not so fine and in piles not so high (<8 m), training of workers, good ventilation, cooling of the stored materials, reliable equipment, generation of positive pressure inside working rooms, duplication of sensors (e.g. temperature, pressure and CO levels), back-firing valves inside gas ducts, protective casing involving the reactor and replacement of air by an inert gas in the gas collecting section. In case of eruption or energy fault, additional actions like injection of an inert gas to cool the reactor and installation of a safety valve located at the exit of the reactor to close it may be also implemented.

It must also be pointed out that appropriate decontamination treatments of gases and a careful management of the toxic by-products that are generated (ashes, chars and liquid effluents) must also be implemented as preventive measures. These may include physical separation of volatile particles using cyclones or filters, catalytic cracking and wet scrubbing for the treatment of gases; regarding liquid effluents, wet oxidation, adsorption on activated carbon and biological treatment may be eligible. Finally, valorization of chars for other purposes (e.g. adsorption of aqueous contaminants) is seen as a way to minimize the health impact caused by their deposition in landfills.

4 CONCLUDING REMARKS

This review covered OSH risks in gasification plants. It was found that gas explosion and inhalation of toxic substances are the most studied cases, although there is agreement that risk levels are generally low in most cases. However, this situation may alter if a newer technology is employed. HAZOP and FTA were the preferred methods for assessing such risks.

When the publication covered both occupational and environmental issues simultaneously, it was found that occupational accidents constitute a lesser problem than environmental emissions (and impacts); moreover, it appears that improvement of the second may degrade the first.

Overall, the review shows that there are insufficient studies covering OSH risks in gasification plants and even fewer on installations dealing with residues of SS and RDF. The findings suggest that further work must be developed to extend the existing scarce data and to evaluate the potential for incidents and/or ill-health cases in this industry, in terms of their frequency and/or harmful potential.

Since gasification of residues is still considered to be in a developing stage, any future change in the technology must be followed by a detailed risk analysis, preferably embracing the whole life cycle, to evaluate conveniently the effects on OSH issues. Specifically, and as suggested by some authors, future work in this area include more investigations focused on the effect of gases in human health, analysis of accidents that occurred in existing plants to understand the phenomena and to define more precise preventive measures, as well as adoption of life cycle analysis tools in OSH risk assessment, and creation of databases containing OSH information about gasification facilities for an effective evaluation of risks inside new plants.

ACKNOWLEDGEMENTS

The authors acknowledge financial support (PhD grant) from the Foundation for Science and Technology of the Portuguese Ministry of Science, Technology and Higher Education (ref. SFRH/BD/111956/2015), and also from Faculdade de Ciências e Tecnologia—Universidade NOVA de Lisboa.

REFERENCES

Abidin, N.A.Z., Ariffin, M.A., Rusli, R. 2011. *Preliminary risk assessment for the bench-scale of biomass gasification system*, Proc. National Postgraduate Conference—Energy and Sustainability: Exploring the Innovative Minds, September 19–20, 2011, Perak.

Arena, U., Romeo, E., Mastellone, M.L. 2008. Recursive Operability Analysis of a pilot plant gasifier. *Journal of Loss Prevention in the Process Industries*, **21**:50–65.

Azapagic, A. 2007. Energy from municipal solid waste: large-scale incineration or small-scale pyrolysis? *Environmental Engineering and Management Journal*, **6**:337–346.

Brisolara, K.F. & Qi, Y. 2013. Biosolids and Sludge Management. *Water Environment Research*, 85:1283–1297.

European Commission, n.d., Deliverable D18 - Final Guideline for Safe and Eco-friendly Biomass Gasification, On-line at: https://ec.europa.eu/energy/intelligent/projects/sites/iee-projects/files/projects/documents/gasification_guide_gasification_guideline.pdf.

Furness, D.T., Hoggett, L.A., Judd, S.J. 2000. Thermochemical Treatment. *Chartered Institution of Water and Environmental Management*, **14**:57–65.

Gao, L. & Hirano, T. 2006. Process of accidental explosions at a refuse derived fuel storage. *Journal of Loss Prevention in the Process Industries*, **19**:288–291.

Di Sarli, V., Cammarota, F., Salzano, E. 2014. Explosion parameters of wood chip-derived syngas in air. *Journal of Loss Prevention in the Process Industries*, **32**:399–403.

Hirano, T. 2006. Gas explosions caused by gasification of condensed phase combustibles. *Journal of Loss Prevention in the Process Industries*, **19**:245–249.

Huuskonen, A. 2012. *Inclusion of Occupational Safety in Life Cycle Assessment*, MSc Thesis, Tampere University of Technology, Finland.

Khoo, H.H. 2009. Life cycle impact assessment of various waste conversion technologies. *Waste Management*, **29**:1892–1900.

Kumar, Y. 2015. Biomass Gasification—A Review. *International Journal of Engineering Studies and Technical Approach*, **1**:12–28.

Lettner, F., Timmerer, H., Haselbacher, P. 2007. Deliverable 9 : Report on possible Health, Safety and Environmental (HSE) hazards from biomass gasification plants, European Commission, On-line at: https://ec.europa.eu/energy/intelligent/projects/en/projects/gasification-guide.

Lonati, G. & Zanoni, F. 2012. Probabilistic health risk assessment of carcinogenic emissions from a MSW gasification plant. *Environment International*, **44**:80–91.

Mishra, A., Singh, R., Mishra, P. 2015. Effect of Biomass Gasification on Environment. *Mesopotamia Environmental Journal*, **1**:39–49.

Molino, A., Braccio, G., Fiorenza, G., Marraffa, F.A., Lamonaca, S., Giordano, G., Rotondo, G., Stecchi, U., La Scala, M. 2012. Classification procedure of the explosion risk areas in presence of hydrogen-rich syngas: Biomass gasifier and molten carbonate fuel cell integrated plant. *Fuel*, **99**:245–253.

Ragazzi, M. & Rada, E.C. 2012. Multi-step approach for comparing the local air pollution contributions of conventional and innovative MSW thermo-chemical treatments. *Chemosphere*, **89**:694–701.

Rong, L., Maneerung, T., Ng, J.C., Neoh, K.G., Bay, B.H., Tong, Y.W., Dai, Y., Wang, C.H. 2015. Co-gasification of sewage sludge and woody biomass in a fixed-bed downdraft gasifier: Toxicity assessment of solid residues. *Waste Management*, **36**:241–255.

Tian, Y.X., Wu, M.H., Wang, X.G., Zhang, Y.P., Qiang, J.F., Tian, X.W., Wang, X.L. 2009. Safety research in the gasification process of novel multi-thermal-source coal gasifier. *Mining Science and Technology*, **19**:210–215.

Torreglosa, J.P., García-Triviño, P., Fernández-Ramirez, L., Jurado, F. 2016. Control strategies for DC networks: A systematic literature review. *Renewable and Sustainable Energy Reviews*, **58**:319–330.

Torrent, J.G., Anez, N.F., Pejic, L.M., Mateos, L.M. 2015. Assessment of self-ignition risks of solid biofuels by thermal analysis. *Fuel*, **143**:484–491.

Vasilescu, G., Draghici, A., Baciu, C. 2008. Methods for analysis and evaluation of occupational accidents and diseases risks. *Environmental Engineering and Management Journal*, 7:443–446.

Werle, S. & Dudziak, M. 2014. Analysis of organic and inorganic contaminants in dried sewage sludge and by-products of dried sewage sludge gasification. *Energies*, **7**:462–476.

Safety assessment of Ammonium Nitrate Fuel Oil (ANFO) manufactory

D.S. Bonifácio
PEA, UFRJ, Rio de Janeiro, Brazil

E.B.F. Galante
IME, Rio de Janeiro, Brazil

A.N. Haddad
PEA, UFRJ, Rio de Janeiro, Brazil

ABSTRACT: The work presents the final results from a safety analysis of a typical manufactory of the commercial explosive known as ANFO—ammonium nitrate and fuel oil. ANFO is a commercial explosive whose fundamental characteristics in terms of industrial production are low cost and ease to produce. As a general rule, ANFO is a mechanical mixture of an oxidizer (role played by the ammonium nitrate) and a source of carbon/fuel. The later role is where the composition usually varies across manufacturers, thus these are not a single composition for the ANFO. In our case study, we selected the most common process of producing ANFO, as well as considered the most effective (in terms of explosive performance) and environmentally friendly formulation. Then we applied a preliminary hazard analysis to the case study, whose results are consolidated in Table 4. The PHA indicated that the most critical risk scenarios are those related to possibility of explosion, and the cause attached to the highest risk relates to metal contamination within the Ammonium nitrate, reason why we strongly advise any manufacturer to implement metal detection screening prior to dose the Ammonium Nitrate into the mixer.

1 INTRODUCTION

Explosives have already been used in pyrotechnic displays in ancient china. As time moved forward, explosives have had many other uses, from military applications to open tunnels and mining operations. The later takes advantage of the shock wave and high pressure generated during detonation of explosives to cutting in rock boulders. Explosives whose main application are in mining are classified as commercial explosives, being Dynamite, emulsion and ANFO (ammonium Nitrate Fuel Oil) important examples (Akhavan, 2011; Cooper, 1996; Meyer, Kohler, Homburg, Köhler, & Homburg, 2007).

ANFO stands out from the others commercial explosives for its fundamental characteristics in terms of industrial production: economic factor and ease to produce. In addition, this explosive has a simplified production route and a low manufacturing cost. Still, several studies on the research of the ideal composition for ANFO, modifying the types of fuels used, were made. All of them focused on the determination of the ideal technical parameters that lead to the best work during rock removal, and which could provide adequate fragmentation of the rock mass at the lowest possible cost.

Due to costs and availability, A high percentage mining operations uses ANFO in its more traditional version, a mixture of only ammonium nitrate and fuel oil. However, international conflicts and in the Middle East region might limit access to petrol and, as consequence, to diesel oil (Munaretti, 2002), which booster a need for a replacement fuel in the ANFO formulation. Alongside this, another driver of change is the growing increase in legislation on the use of chemicals and their consequences on the environment. Concern about environmental issues has become a reality in Brazil, given the vast number of control bodies set up at the national and state levels for direct inspection and expected damages in the event of non-compliance with laws created for the protection and preservation of the environment.

Replacing diesel using biofuels might not be a straight forward decision, since its production and end use can have serious environmental impacts, such as too much use of water, destruction of forests, reduction of production increase in soil degradation. Furthermore, there is risk associated with the handling, manufacture and use of the compound because it is an energetic and explosive material.

Replacing diesel in ANFO has been addressed by (Resende, 2011), when the performances of ammonium nitrated mixed with diesel, biodiesel, sugarcane bagasse, rice husk and corn cob were tested for velocity of detonation and fragmentation power. Later, Bonifácio (Bonifácio, 2017) tested the environmental impact of these same compositions, carrying out the life cycle analysis (International Organization for Standardization (ISO) & ISO—International Organization for Standardization, 1998; Iso, 2000).

Both analysis converge with the same conclusions. Whilst this convergence confirms the completeness of the ANFO study, it remains to have its manufactory process assessed for safety. Hence this work provides a risk analysis of the manufacture of ANFO based on the most efficient (Resende, 2011) and environmentally (Bonifácio, 2017) friendly composition: The typical ammonium nitrate and diesel oil.

2 METHODOLOGY

The chosen risk study (ISO, 2009) was the Preliminary Hazards Analysis (PHA) as proposed by the MIL-STD-882-E (Galante, 2015; USDoD (US Department of Defence), 2012). The PHA tool aims to guide in establishing risk and hazard scenarios and, for each scenario, determine causes and effects. After this, the causes are parameterized and a frequency is stabilized using Table 1.

The same procedure is than repeated for the effects, which are parameterized using the categories from Table 2.

Once the frequency and severity parameters are determined, the level of risks is determined using a multi-criteria matrix (Table 3). Once the level of risk is determined, the scenarios can be placed in

Table 1. Frequency categories. (USDoD (US Department of Defence), 2012).

Cat	Name	Description
A	Frequent	Likely to occur often in the life of an item.
B	Probable	Will occur several times in the life of an item.
C	Occasional	Likely to occur sometime in the life of an item
D	Remote	Unlikely, but possible to occur in the life of an item.
E	Improbable	So unlikely, it can be assumed occurrence may not be experienced in the life of an item.
F	Eliminated	Incapable of occurrence. This level is used when potential hazards are identified and later eliminated.

Table 2. Severity categories (USDoD (US Department of Defence), 2012).

Cat	Name	Description
1	Catastrophic	Could result in one or more of the following: death, permanent total disability, irreversible significant environmental impact, or monetary loss equal to or exceeding $10M.
2	Critical	Could result in one or more of the following: permanent partial disability, injuries or occupational illness that may result in hospitalization of at least three personnel, reversible significant environmental impact, or monetary loss equal to or exceeding $1M but less than $10M.
3	Marginal	Could result in one or more of the following: injury or occupational illness resulting in one or more lost work day(s), reversible moderate environmental impact, or monetary loss equal to or exceeding $100 K but less than $1M.
4	Negligible	Could result in one or more of the following: injury or occupational illness not resulting in a lost work day, minimal environmental impact, or monetary loss less than $100 K.

Table 3. Risk matrix (USDoD (US Department of Defence), 2012).

		Severity			
		1	2	3	4
Frequency	A	High	High	Serious	Medium
	B	High	High	Serious	Medium
	C	High	Serious	Medium	Low
	D	Serious	Medium	Medium	Low
	E	Medium	Medium	Medium	Low
	F	Eliminated			

a hierarchy and the proposed mitigating measures should feed into a plan of action.

3 CASE STUDY

ANFO is one of the most easily explosives to be produced mainly due to its composition consisting of a mechanical mixture of one oxidant and one fuel. Figure 1 presents a schematic model of the production process. The most common production process involves a doing silo pouring ammonium nitrate onto a screw mixer, where the diesel in dropped on the ammonium nitrate.

For this work, the composition of interest is one in which Ammonium nitrate is the oxidizing agent (94.5%) and diesel acts as fuel (5.5%).

3.1 Ammonium Nitrate (AN)

Ammonium nitrate has its main use in agriculture as a nitrogen-rich (35%) nutrient. The preparation of ammonium nitrate requires natural gas, which can be found in petroleum deposits, coal and decomposition of organic waste (Munaretti, 2002). Coal, in Europe, for example, is included in Annex XIV of the European Regulation REACH (UK HSE Executive, 2016) due to carcinogenic effects and will need an authorization to be used in future, which might affect Ammonium Nitrate production. Ammonium nitrate is also used as an oxygen source in explosive formulations, such as ANFO, dynamites, emulsions and acquagel (Akhavan, 2011; Cooper, 1996; Valença, Reis, Palazzo, Rocha, & Athayde, 2013).

3.2 Diesel

Diesel is one of the several hydrocarbon compounds distilled from Petroleum. Its physical-chemical properties vary across oilfields. In Brazil, according to the Brazilian Petroleum Agency (ANP, 2017), about 20% of the processed Petroleum is converted into automotive gasoline and 36% into diesel oil. The chemical structure of diesel contains larger size of chains molecules, which gives this fuel greater specific mass, lower volatility and lower solubility in water.

4 RESULTS

The work developed by Bonifácio (Bonifácio, 2017) discussed environmental impacts of different raw materials in the ANFO formulation and concluded that diesel oi list the better option, hence this work considers the ANFO made from Ammonium Nitrate and Diesel.

The result of the risk study of each scenario was consolidated in Table 4. Each risk scenario was studied, determining causes and effects. Using the PHA methodology, the causes and effects were parameterized as frequency ("F" column) and severities ("S" column).

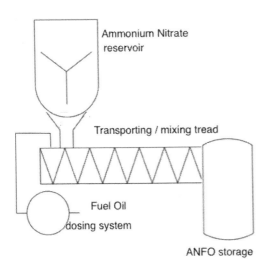

Figure 1. ANFO production process.

Table 4. Hazard analysis.

Hazard	Causes	Effects	F	S	Risk
Metal contamination within the Ammonium Nitrate packages	• Uneffective sieving/ lack of scanning with metal detector	• Damage to people • Damage to equipment • Damage to installations • Accidental detonation within the mixing screw.	B	1	High
Metal contamination within the Ammonium Nitrate dosing silo	• Uneffective maintenance/ decontamination	• Damage to people • Damage to equipment • Damage to installations • Accidental detonation within the mixing screw.	B	1	High
Lack of fuel being added into the mixture	• Dosing pump failure • Low level of fuel in its reservoir	• Loss of production batch • Need to disposal of unknown quantities of AN/ANFO/ low quality ANFO	E	2	Medium

(*Continued*)

Table 4. (Continued).

Hazard	Causes	Effects	F	S	Risk
Dosing system (pump and hove) failure	• Failure in Operation procedures • Blockades within the hove • Failure in any seals and connections	• Lack of fuel being added into the mixture • fires	E	2	Medium
Low level of fuel in its reservoir	• Leakage • Failure in Operation procedures	• Lack of fuel being added into the mixture	E	2	Medium
Trip and falling	• Use steps to transport materials • Fuel leakages/floor slippery	• Damage to people • Loss of material	D	4	Low
Manually handling heavy materials	• Need to transport raw materials • Remove final products from the workshop • Decontamination • Maintenance of equipment	• Damage to people • Damage to equipment • Damage to installations • Loss of material	E	4	Low
Ammonium Nitrate spreads	• Failure in dosing • Overflow dosing silo	• Loss of material • Loss of time	E	4	Low
ANFO spreads	• Overflow into the final reservoir • Issues with dosing screw	• Loss of material • Loss of time • Risk of fires and explosion • Need to disposal/decontamination	E	1	Medium
Fires	• ANFO spreads • Metal contamination within the Ammonium Nitrate dosing silo • Dosing pump failure • Electrical failure	• Damage to people • Damage to equipment • Damage to installations • Loss of material • Explosion	D	1	Serious
Explosion	• ANFO spreads • Metal contamination within the Ammonium Nitrate dosing silo • Dosing pump failure • Electrical failure • Fires	• Loss of life • Loss of equipment • Loss of installation • Loss of capability	D	1	Serious

By using the risk matrix, it was possible to establish a priority for each risk. It should be emphasized that in a risk study to be carried out in a professional way, there would be one last column in Table 4, called "recommendations", where the team would propose measures to be taken to reduce the risk level.

5 CONCLUSIONS

We presented the case study as the most common process of producing ANFO and developed a preliminary hazard analysis to the case study, whose results are consolidated in Table 4. The PHA indicated that the most critical risk scenarios are those related to possibility of explosion, and the cause attached to the highest risk relates to metal contamination within the Ammonium nitrate, reason why we strongly advise any manufacturer to implement metal detection screening prior to dose the Ammonium Nitrate into the mixer.

REFERENCES

ANP. (2017). Agência Nacional de Petróleo. Retrieved January 1, 2017, from http://www.anp.gov.br/wwwanp/.

Akhavan, J. (2011). *The Chemistry of Explosives* (Third Edit). Norfolk: Biddles Ltd., Kings Lynn, Norfolk.

Bonifácio, D. de S. (2017). *ANÁLISE DO CICLO DE VIDA DE MISTURAS EXPLOSIVAS COM NITRATO DE AMÔNIO*. Universidade Federal do Rio de Janeiro (UFRJ).

Cooper, P. (1996). *Explosives Engineering.* Wiley-VCH, Inc. United States of America.

Galante, E.B.F. (2015). *Princípios de Gestão de Riscos* (1st ed.). Curitiba: Appris.

International Organization for Standardization (ISO), & ISO—International Organization for Standardization. ISO 14041 Environmental Management—Life cycle Assessment—Goal and Scope Definition and Inventory Analysis., Brussels: International Organisation for Standardisation (1998). Geneva, Switzerland, Switzerland.

Iso. (2000). ISO 14042. Environmental management—Life cycle assessment—Life cycle impact assessment. Management environnemental—Analyse du cycle de vie—Évaluation de l'impact du cycle de vie. *International Standard*, *2000*, 24.

ISO. ISO 31.000 - Risk Management—Principles and Guidelines (2009). Geneva, Switzerland: International Organisation for Standardization.

Meyer, R.R., Kohler, J., Homburg, A., Köhler, J., & Homburg, A. (2007). *Explosives* (6th, Compl ed.). Weinheim: Wiley-VCH Verlag GmbH, Weinheim.

Munaretti, E. (2002). *Desenvolvimento e avaliação de desempenho de misturas explosivas a base de nitrato de amônio e óleo combustível.* UNIVERSIDADE FEDERAL DO RIO GRANDE DO SUL Escola.

Resende, S.A. (2011). *DESENVOLVIMENTO DE EXPLOSIVOS UTILIZANDO COMBUSTÍVEIS NÃO-CONVENCIONAIS.* Universidade Federal de Ouro Preto.

UK HSE Executive. (2016). Registration, Evaluation, Authorisation & restriction of CHemicals (REACH). Retrieved February 8, 2016, from http://www.hse.gov.uk/reach/.

USDoD (US Department of Defence). MIL STD 882-E—Standard Practice for System Safety (2012). Washington, USA: USA.

Valença, U. da S., Reis, S.S., Palazzo, M., Rocha, J.F., & Athayde, A.A.C. (2013). *Engenharia dos Explosivos: um enfoque dual* (1st ed.). Rio de Janeiro, Brazil: FRF—Fundação Ricardo Franco. Retrieved from http://disseminar.ime.eb.br/explo.html.

The relevance of surfaces contamination assessment in occupational hygiene interventions in case of exposure to chemicals

S. Viegas
GIAS, ESTeSL—Escola Superior de Tecnologia da Saúde de Lisboa, Instituto Politécnico de Lisboa, Lisbon, Portugal
Centro de Investigação em Saúde Pública, Escola Nacional de Saúde Pública, Universidade NOVA de Lisboa, Lisbon, Portugal

ABSTRACT: Dermal exposure assessment is still not a priority in the intervention of occupational hygiene when assessing exposure to chemical substances. However, for some substances in specific occupational settings, dermal intake can be the most important exposure route. Skin contamination can occur from contact with contaminated surfaces. Thus, knowing the levels of the substance present on the surfaces of the workplace is an important tool to indirectly measure dermal exposure. This paper aimed to claim attention for the relevance of surfaces contamination assessment in the occupational hygiene interventions. An extensive search was performed to identify scientific papers published after 2010, reporting data of surfaces contamination in the scope of occupational hygiene interventions. Twenty seven papers were considered and, from those, 63% were devoted to antineoplastic drugs occupational exposure assessment. This short review allowed concluding that surfaces contamination assessment is a very useful tool that, besides giving an indirect measure of dermal exposure, can also give relevant information to guide interventions to prevent exposure.

1 INTRODUCTION

Nowadays, dermal exposure assessment is still not a priority in the intervention of occupational hygiene and methods to study exposure by inhalation are much more studied and validated. However, for some chemical substances in specific occupational settings, dermal route can be a very important exposure route.

Skin contamination can occur from the deposition of aerosols, via direct immersion into a chemical substance (liquid or solid), as a result of spills and splashes, through vapour deposition and penetration, or from contact with contaminated surfaces (Fenske, 1993). Considering this, Fenske in 1993 suggested that dermal exposure levels could be assessed from data resulting from: 1) biomonitoring values, 2) levels of contamination on clothing, or 3) levels of deposition on skin or surfaces of the workplace.

Surface contamination assessment, which can be done by wipe sampling or vacuuming of surfaces, may serve as a predictor of dermal exposure to chemicals. Wipe sampling provides information about the mass of the contaminant on a surface. This sampling approach can be used as an indirect method to measure exposure (Fenske, 1993; Schneider et al., 1999), since provide information about contamination on surfaces which may lead to dermal uptake by workers (Connor and Smith, 2016).

The objective of this paper was to claim attention for the relevance of surfaces contamination assessment in the occupational hygiene interventions by conducting a review of studies where surfaces contamination assessment has been performed.

2 MATERIALS AND METHODS

An extensive search was performed to identify scientific papers, available in different scientific databases (PubMed and Web of Science) published after 2010, reporting data of surfaces contamination in the scope of occupational hygiene interventions, particularly related with the assessment of occupational exposure to chemical substances. Only the articles written in English were considered. The search was done using the following key-words in different combinations: surfaces contamination assessment, occupational exposure and exposure assessment.

3 RESULTS AND DISCUSSION

Twenty seven papers were selected and included in a more detailed analysis concerning objectives of the study and the main findings (Table 1). Seventeen papers (63%) were devoted to antineoplastic

Table 1. Papers considered for further analysis (n = 29).

Reference	Substance/group of substances analized	Objective of the study	Main conclusions
Siderov et al., 2010	Antineoplastic drugs	Determine the impact of a CSTD on cytotoxic surface contamination.	When used inside a CDSC, the CSTD further reduces surface contamination, in some instances to undetectable levels.
Sottani et al., 2010	Antineoplastic drugs	Analyze trends of surfaces contamination.	A strong reduction of contamination was observed since 2003, when recommended safe handling procedures started to be followed by workers.
Moretti et al., 2010	Antineoplastic drugs	Study protocol: Assess occupational exposure in nurses involved in preparation and administration of drugs.	Findings will be used to prevent exposure.
Villarini et al., 2011	Antineoplastic drugs	Evaluate work environment contamination: preparation and administration of drugs.	Highest contamination found in preparation. Procedures adopted are not sufficient to prevent exposure. Important to use protective equipment and implement correct work practices.
Konate et al., 2011	Antineoplastic drugs	Assess platinum contamination in the operating room and exposure of health workers.	Contamination on surfaces and nearby area was found. Based on the level of floor contamination, there is a high risk of exposure for cleaning and maintenance staff.
Hedmer and Wohlfart, 2012	Antineoplastic drugs	Propose hygienic guidance values for surface monitoring of CP and IFO.	Surface monitoring combined with hygienic guidance values is a useful tool for health care workers to regularly benchmark their own surface loads.
Naito et al., 2012	Antineoplastic drugs	Examine platinum contamination on the exterior surface of vials containing cisplatin or carboplatin.	External contamination was confirmed in all cisplatin and carboplatin vials tested.
Chu et al., 2012	Antineoplastic drugs	Determine if antineoplastic drug contamination of surfaces exists despite being cleaned.	Results suggest that drug contamination is common in hospital pharmacies and that current cleaning practices may not be effective in removing residual drug from the surfaces.
Hon et al., 2013	Antineoplastic drugs	Describe the contamination of frequently contacted surfaces and identify factors that may be associated with surface contamination.	Frequently contacted surfaces at every stage of the hospital medication system had measureable levels of antineoplastic drug contamination. Drug preparation stage had the highest average contamination.
Odraska et al., 2013	Antineoplastic drugs	Compare the surface contamination level of the conventional preparation room and outpatient clinic before and after the implementation of measures.	Measures implemented in the outpatient clinic were shown to reduce workplace contamination effectively. However, measures implemented in the preparation room, where relatively strict regulations had already been adopted before the study, were less effective.
Luo et al., 2013	DDTs and dicofol	Evaluate the potential contamination of dicofol manufacturing equipment.	The surfaces were contaminated with DDT and dicofol.
Kopp et al., 2013	Antineoplastic drugs	Evaluate working practices and safety measures during drug administration and to assess workplace contamination in outpatient oncology health care.	Workplace contamination with antineoplastic drugs is still present. As patients have to be considered as potential source of contamination. The study revealed that it is possible to administer a large number of preparations without causing high workplace contamination.
Gribovich et al., 2013	Arsenic	Assess potential arsenic contamination of work surfaces in an anthropology department in a museum.	Workplace observations and wipe sampling data enabled the development of recommendations to help to further reduce potential occupational exposure to arsenic.

Reference	Substance	Objective	Findings
Viegas et al., 2014	Antineoplastic drugs	Characterize occupational exposure to antineoplastic drugs in pharmacy and administration units.	Workplace contamination with antineoplastic drugs was observed. Specific measures should be taken, particularly those related with the promotion of good practices and safety procedures.
Shepard and Brenner, 2014	Nanoparticles	Identify potential exposure scenarios and evaluate the presence of these materials on surfaces.	Nanoparticles were identified in surface samples from work areas where engineered nanoparticles were used or handled. Precautionary measures should be applied.
Fleury-Souverain et al., 2014	Antineoplastic drugs	Evaluate the contamination of surfaces during preparation of injectable chemotherapies in hospital pharmacies.	Most of the hospital pharmacies had some contamination of surfaces by different cytotoxic agents. Higher levels of contamination were mainly detected inside biosafety cabinets, but contamination was also found in logistical and storage areas.
Ceballos et al., 2015	Pesticides	Assess occupational exposures to sea lamprey pesticides.	Found surfaces and worker's skin contaminated with pesticides. Recommended minimizing exposures by implementing engineering controls and use of personal protective equipment.
Hedmer et al., 2015	Nanomaterials	Know the presence of carbon-based nanomaterials in surfaces at a small-scale producer.	Nanomaterials were identified in all parts of the workplace, thus, increasing the risk for secondary inhalation and dermal exposure of the workers.
Anastasi et al., 2015	Antineoplastic drugs	Evaluate two cleaning solutions for the chemical decontamination of the surfaces.	Higher contamination levels were distributed on areas frequently touched by the pharmacy technicians. Both cleaning solutions were able to reduce contamination.
Vyas et al., 2016	Antineoplastic drugs	Investigate the surface contamination from the preparation of five anticancer drug infusions with two different types of equipment.	It was possible to conclude the equipment that implicates less contamination.
Ceballos et al., 2016	Metals	Characterize employee exposure to metals and recommend control strategies to reduce exposure.	Found metals on non-production surfaces. Provided recommendations for improving local exhaust ventilation, using respirators until exposures are controlled, and reducing the migration of contaminants from production to non-production areas.
Schenk et al., 2016	Antineoplastic drugs	Evaluate surface contamination by platinum drugs in the environment of patients in ICUs and wards treated by hyperthermic intraperitoneal chemotherapy.	High platinum-drug concentrations in urine and drainage liquids are the main source of contamination. Therefore, safe handling of these liquids is the best way to avoid cross-contamination on surfaces in wards and ICUs.
Poupeau et al., 2016	Antineoplastic drugs	To monitor environmental contamination in pharmacy and patient care areas in Canadian hospitals.	In comparison with other multicenter studies that were conducted in Canada, the concentration of antineoplastic drugs measured on surfaces is decreasing.
Paik et al., 2017	Beryllium	Assessment of current conditions in the facility and a create baseline for future impacts.	All the data have shown that beryllium has been effectively managed to prevent exposures to workers during routine and non-routine work.
Beaucham et al., 2017	Metals (lead)	Discuss wipe sampling for measuring lead on surfaces in three facilities.	Wiping sampling demonstrated lead in non-production surfaces in all three workplaces and that the potential that employees were taking lead home to their families existed.
Beattie et al., 2017	Metals (Cr)	Investigate whether repeat biological monitoring over time could drive sustainable improvements in exposure control in the industry.	This study has shown that exposures to chromium VI and nickel in the electroplating industry occur via a combination of inhalation, dermal and ingestion routes. Surface contamination found in areas such as canteens highlights the potential for transferal from work areas, and the importance of a regular cleaning regime.
Brouwer et al., 2017	4,4'- methylene dianiline (MDA)	Evaluate two different techniques for assessing dermal exposure to MDA.	Results indicated a general contamination of the workplace and equipment.

drugs occupational exposure assessment. The others papers were dedicated to occupational exposure assessment to other substances, such as metals (chromium and lead), beryllium, arsenic, nanomaterials, 4,4'- methylene di-aniline (MDA) and pesticides. However only after 2013 it was observed other substances being considered for this approach, besides antineoplastic drugs.

The assessment of the contamination present on work surfaces is commonly reported in the case of antineoplastic occupational exposure assessment, since skin absorption is a more important route of exposure than inhalation, and also because these drugs are non-volatile and remains on work surfaces for long periods of time (European Parliament, 2016). These characteristics are common to almost all the substances reported on the selected studies but in some cases the deposition on surfaces means that exposure by inhalation can also occur, since resuspension of the substances from the surfaces can happen. This is a reality particularly in the case of nanomaterials (Hedmer et al., 2015).

Considering the main conclusions in each study analysed (Table 1), it is possible to realize that, more than an indirect measure of exposure, the surfaces contamination assessment allows understanding if the workplace contamination can promote exposure by dermal absorption since the information obtained provides an indication of the potential for dermal exposure (Sottani et al., 2010; Villarini et al., 2011). This is particular relevant if the criteria to choose the surfaces to sample were based on tasks observation to identify the surfaces that workers handle or touch more frequently (Hon et al., 2013; Viegas et al., 2014).

The information obtained can also help to investigate mechanisms of release and spread and thus help to identify possible sources and routes of exposure (Naito et al., 2012; Hon et al., 2013; Kopp et al., 2013; Schenk et al., 2016). Several studies also use the information obtained to recognize surface contamination concentrations and trends (Sottani et al., 2010; Villarini et al., 2011; Konate et al. 2011; Naito et al., 2012; Anastasi et al., 2015). The information allows for instance to identify what are the areas with higher contamination and with more surfaces contaminated (Viegas et al., 2014; Fleury-Souverain et al., 2014). Those areas are the ones where the risk management measures should be applied first.

Frequently, the information was also used to identify the most relevant risk management measures and to evaluate effectiveness of those measures. This was one of the main objectives in several studies (Sottani et al., 2010; Siderov et al., 2010; Chu et al., 2012; Odraska et al., 2013; Viegas et al., 2014; Ceballos et al., 2016; Beaucham et al., 2017).

The need of cleaning and the assessment of the effectiveness of the cleaning process were also reported in numerous studies since the results obtained allowed having a detailed picture of the surfaces contamination and to understand where the cleaning process is failing (Chu et al., 2012; Anastasi et al., 2015; Beattie et al., 2017; Brouwer et al., 2017).

4 CONCLUSION

This short review allowed to demonstrated that surfaces contamination assessment is a very useful tool that, besides giving an indirect measure of exposure by the dermal route, can also give relevant information to guide interventions to prevent exposure by dermal route, inhalation route and, even, digestive route that can occur due to contaminated hands after touching a contaminated surface. Occupational hygienists should keep in mind this tool as a complementary resource to a more accurate exposure assessment and to identify the most suitable risk management measures to apply. This will also guarantee more data available related with dermal and surface contamination in different settings promoting that the dermal exposure assessment models that are being developed are more robust and reliable.

REFERENCES

Anastasi, M., Rudaz, S., Lamerie, T.Q., Odou, P., Bonnabry, P., Fleury-Souverain, S. 2015. Efficacy of Two Cleaning Solutions for the Decontamination of 10 Antineoplastic Agents in the Biosafety Cabinets of a Hospital Pharmacy. *The Annals of Occupational Hygiene* 59 (7): 895–908.

Beattie, H., Keen, C., Coldwell, M., Tan, E., Morton, J., McAlinden, J., Smith, P. 2015. The use of biomonitoring to assess exposure in the electroplating industry. *Journal of Exposure Science and Environmental Epidemiology* 1–9.

Beaucham, C., Ceballos, D., King, B. 2017. Lessons learned from surface wipe sampling for lead in three workplaces. *J Occup Environ Hyg* 14(8): 611–619.

Ceballos, D., Beaucham, C., Page, E. 2017. Metal Exposures at three U.S. electronic scrap recycling facilities. *Journal of Occupational and Environmental Hygiene* 14:6, 401–408.

Ceballos, D., Beaucham, C.C., Kurtz, K., Musolin, K. 2015. Assessing occupational exposure to sea lamprey pesticides. *International Journal of Occupational and Environmental Health* 21(2): 152–160.

Chu, W.C., Hon, C.Y., Danyluk, Q., Chua, P.P., Astrakianakis, G. 2012. Pilot assessment of the antineoplastic drug contamination levels in British Columbian hospitals pre- and post-cleaning. *J Oncol Pharm Pract.* 18(1): 46–51.

Connor, T.H., Smith, J.P. 2016. New Approaches to Wipe Sampling Methods for Antineoplastic and Other Hazardous Drugs in Healthcare Settings. *Pharm Technol Hosp Pharm* 1(3): 107–114.

European Parliament. 2016. Preventing occupational exposure to cytotoxic and other hazardous drugs: European policy recommendations. Strasbourg: Author. Retrieved from: *http://www.europeanbiosafetynetwork.eu/Exposure%20to%20Cytotoxic%20Drugs_Recommendation_DINA4_10–03–16.pdf.*

Fenske, R.A. 1993. Dermal exposure assessment techniques. *Annals of Occupational Hygiene* 37:687–706.

Fenske, R.A. 2000. Dermal exposure: a decade of real progress. *Annals of Occupational Hygiene.* 44: 489–491.

Fleury-Souverain, S., Mattiuzzo, M., Mehl, F. 2015. Evaluation of chemical contamination of surfaces during the preparation of chemotherapies in 24 hospital pharmacies. *Eur J Hosp Pharm* 22: 333–341.

Gribovich A, Lacey S, Franke J, Hinkamp D. 2013. Assessment of arsenic surface contamination in a museum anthropology department. *J Occup Environ Med.* 55(2): 164–7.

Hedmer, M., Ludvigsson, L., Isaxon, C., Nilsson, P.T., Skaug, V., Bohgard, M., Pagels, J.H., Messing, M.E., Tinnerberg, H. 2015. Detection of Multi-walled Carbon Nanotubes and Carbon Nanodiscs on Workplace Surfaces at a Small-Scale Producer. *Ann. Occup. Hyg.* 59 (7): 836–852.

Hedmer, M., Wohlfart, G. 2012. Hygienic guidance values for wipe sampling of antineoplastic drugs in Swedish hospitals. *J. Environ. Monit.* 14, 1968.

Hon, C.Y., Teschke, K., Chu, W., Demers, P., Venners, S. 2013. Antineoplastic drug contamination of surfaces throughout the hospital medication system in Canadian hospitals. *J Occup Environ Hyg* 10(7): 374–83.

Konate, A., Poupon, J., Villa, A., Garnier, R., Hasni-Pichard, H., Mezzaroba, D., Fernandez, G., Pocard, M. 2011. Evaluation of environmental contamination by platinum and exposure risks for healthcare workers during a heated intraperitoneal perioperative chemotherapy (HIPEC) procedure. *J Surg Oncol.* 103(1): 6–9.

Kopp, B., Schierl, R., Nowak, D. 2013. Evaluation of working practices and surface contamination with antineoplastic drugs in outpatient oncology health care settings. *Int Arch Occup Environ Health* 86(1): 47–55.

Luo, F., Song, J., Chena, M., Wei, J., Pana, Y., Yu, H. 2013 Risk assessment of manufacturing equipment surfaces contaminated with DDTs and dicofol. *Science of the Total Environment* 468–469: 176–185.

Moretti, M., Bonfiglioli, R., Feretti, D., Pavanello, S., Mussi, F., Grollino, M.G., Villarini, M., Barbieri, A., Ceretti, E., Carrieri, M., Buschini, A., Appolloni, M., Dominici, L., Sabatini, L., Gelatti, U., Bartolucci, G.B., Poli, P., Stronati, L., Mastrangelo, G., Monarca, S. 2011. A study protocol for the evaluation of occupational mutagenic/carcinogenic risks in subjects exposed to antineoplastic drugs: a multicentric project. BMC Public Health 11:195

Naito, T., Osawa, T, Suzuki N, Goto T, Takada A, Nakamichi H, Onuki Y, Imai K, Nakanishi K, Kawakami J. Comparison of contamination levels on the exterior surfaces of vials containing platinum anticancer drugs in Japan. Biol Pharm Bull. 2012;35(11): 2043–9.

Odraska, P., Dolezalova, L., Kuta, J., Oravec, M., Piler, P., Blaha, L. 2013. Evaluation of the efficacy of additional measures introduced for the protection of healthcare personnel handling antineoplastic drugs. *Ann Occup Hyg.* 57(2): 240–50.

Paik, S.Y., Epperson, P.M., Kasper, K.M. 2017. Assessment of personal airborne exposures and surface contamination from x-ray vaporization of beryllium targets at the National Ignition Facility. *J Occup Environ Hyg.* 14(6): 438–447.

Poupeau, C., Tanguay, C., Caron, N.J., Bussières, J.F. 2016. Multicenter study of environmental contamination with cyclophosphamide, ifosfamide, and methotrexate in 48 Canadian hospitals. *J Oncol Pharm Pract. Pii:* 1078155216676632.

Schenk, K.E., Schierl, R., Angele, M., Burkhart-Reichl, A., Glockzin, G., Novotny, A., Nowak, D. 2016. Cisplatin and oxaliplatin surface contamination in intensive care units (ICUs) and hospital wards during attendance of HIPEC patients. *Int Arch Occup Environ Health* 89(6): 991–996.

Schneider, T., Vermeulen, R., Brouwer, D.H., Cherrie, J.W., Kromhout, H., Fogh, C.L. 1999. Conceptual model for assessment of dermal exposure. *Occup Environ Med:* 56: 765–773.

Shepard, M., Brenner, S. 2014. Cutaneous exposure scenarios for engineered nanoparticles used in semiconductor fabrication: a preliminary investigation of workplace surface contamination. *Int J Occup Environ Health* 20(3): 247–57.

Siderov, J., Kirsa, S., McLauchlan, R. Reducing workplace cytotoxic surface contamination using a closed-system drug transfer device. 2010. *J Oncol Pharm Pract.* 16(1): 19–25.

Sottani, C., Porro, B., Comelli, M., Imbriani, M., Minoia, C. 2010. An analysis to study trends in occupational exposure to antineoplastic drugs among health care workers. *Journal of Chromatography B* 878: 2593–2605.

Viegas, S., Pádua, M. Veiga, A.C., Carolino, E., Gomes, M. 2014. Antineoplastic drugs contamination of workplace surfaces in two Portuguese hospitals. *Environ Monit Assess* 186: 7807–7818.

Villarini, M., Dominici, L., Piccinini, R., Fatigoni, C., Ambrogi, M., Curti, G., Morucci, P., Muzi, G., Monarca, S., Moretti, M. 2011. Assessment of primary, oxidative and excision repaired DNA damage in hospital personnel handling antineoplastic drugs. *Mutagenesis* 26(3): 359–69.

Vyas, N., Turner, A., Clark, J.M., Sewell, G.J. 2016. Evaluation of a closed-system cytotoxic transfer device in a pharmaceutical isolator. *J Oncol Pharm Pract.* 22(1): 10–9.

Waldron, H.A. 1983. A brief history of scrotal cancer. British *Journal of Industrial Medicine* 40:390–401.

Optimization of the CoPsoQ questionnaire in a population of Archipelago of Azores

F.O. Nunes
Instituto Superior de Engenharia de Lisboa, Instituto Politécnico de Lisboa, Lisbon, Portugal

A. Gaspar da Silva
Instituto Superior de Educação e Ciências, Lisbon, Portugal

ABSTRACT: The present study analyzes the internal consistency and reliability of an already widely disseminated and nationally and internationally validated questionnaire applied to a selected population of the Autonomous Region of Azores. The questionnaire consists of the 119 items in the long-length version of the Portuguese adaptation of the Copenhagen Psychosocial Questionnaire (CoPsoQ). Applying factor analysis techniques to the results of the 234 questionnaires collected, an optimized questionnaire was constructed with 54 items that, besides its ease of application, also presents better cost-benefit and reliability in the majority of item's scales/dimensions.

1 INTRODUCTION

Health and well-being at work are currently addressed by a set of disciplines without a tradition of close interdisciplinary cooperation (Schabracq et al., 2003). Clinical psychology addresses issues such as trauma and post-traumatic stress, burnout, and therapeutic interventions in individuals and groups. Social psychology deals with group dynamics, social support, and person-environment adaptation. Developmental psychology is concerned with the different stages of life and issues related to the professional career. Finally, industrial, organizational and occupational psychology, deals with issues such as stress and fatigue at work, job characteristics and organizational issues.

The causal relationship between work-related psychosocial stressors and the incidence and prevalence of occupational morbidity and mortality leads to the need for a better understanding of the work environment, e.g., stress-health relationships, in modern work life in a rapidly changing world (Lundberg & Cooper, 2011).

The goal of the present study is to analyze the internal consistency and reliability of the Copenhagen Psychosocial Questionnaire (CoPsoQ), a widely disseminated questionnaire which has been validated both nationally and internationally (Kristensen et al., 2005; Nübling et al. 2006; Moncada et al., 2008 and 2014), applied to a selected population in the Autonomous Region of Azores.

The questionnaire consisted of the 119 items in the long version of the Portuguese adaptation of the Copenhagen Psychosocial Questionnaire, developed for research purposes. The Portuguese version also offers a medium and a short version developed empirically from the long version (Silva et al., 2006).

The use of factor analysis tools and weighting of calculated reliability coefficients allowed for a non-empirical reduction in size of the applied questionnaire increasing the ease of its future application.

2 METHODOLOGY

2.1 *Psychosocial risk assessment instruments*

Over the past few decades workplaces have undergone major changes involving the introduction of downsizing and outsourcing, lean production and just-in-time, more hours of work, temporary work and part-time employment.

The attitude of the workers as a psychological state, acquired and organized through their own experience that urges them to react in a characteristic way to certain people, objects or situations, is not subject to direct observation; this can only be inferred, indirectly, through verbal expressions that are uttered or of one's own conduct.

A measuring instrument is a mechanism for measuring phenomena, which is used to collect and record information for evaluation, decision-making and, in the final stages, understanding the phenomenon (Colton & Covert, 2007).

The models established for psychosocial risks assume that the relationship between stress and pressure is not deterministic, but rather mediated and moderated by variables such as personal resources (Cox et al., 2000).

There are three models that deal with the relationship between stressors and the consequences that follow:

- The "demand-control" model (Karasek, 1979), which was later expanded to "demands-controls-supports" (Johnson, 1989);
- The "person-environment adjustment" model (French et al., 1982);
- The "effort-reward imbalance" model (Siegrist, 1996).

ISO 10075-3: 2004 (E) differentiates the aspects of mental stress, and mental tension and its effects on the individual. According to this standard, the assessment and measurement should refer to these different steps in the exposure process that can affect health and safety, wellness, performance and productivity.

CoPsoQ was originally developed in 1997 by the Danish National Institute for Occupational Health in Copenhagen and validated in 2000 by Kirstensen and Borg and seeks a standard and multidimensional approach to the spectrum of psychosocial factors. The original instrument is intended to cover the general needs involved in the concept of "stress at work", based on the demand-control model (Cox et al., 2000; Daniels et al., 2002), which attempts to explain stress as a consequence of high demands on work and low social support.

Despite most of the scales that integrate the various dimensions being based on instruments already known and previously validated, others were built specifically for this application. CoPsoQ has been adopted as a standard for the evaluation of psychosocial factors in the workplace; notwithstanding, some scales have been modified and adapted to the realities of each country, thus not allowing a direct comparison of the results in all dimensions.

2.2 Population and sample

The population covered in the study involved 1,246 individuals, from six of the nine islands of Azores, carrying out security and emergency activities (75.7%) and 401 involved in trading and distribution of goods and services (24.3%). The global response was of 14.2% (13.1% and 17.7%, respectively). The sample distribution by age group was mainly 21–30 (26.5%), 31–40 (45.3%), and 41–50 (22.2%) years old, where a total of 63.8% were male.

2.3 Validation of psychosocial risk assessment instruments

The validity of a questionnaire describes the success obtained in measuring the intended concept and is a characteristic of the answers obtained (Colton & Covert, 2007).

Internal consistency is a form of validity measurement based on the correlation of responses between different items of the same questionnaire (or between the same subscales in a longer questionnaire); It assesses whether the various items that are proposed to measure the same general construct produce similar results.

The internal consistency of a questionnaire (or factors of the questionnaire) is usually measured with Cronbach's alpha coefficient (Cronbach, 1951), a statistic calculated from the correlations between items. Although the range of interest lies in values between 0 and 1, the coefficient will be negative whenever there is greater variability within the subject than the variability between subjects (Knapp, 1991).

The Cronbach's alpha coefficient, which in sum is the lower limit of the reliability coefficient of the responses to a questionnaire, is calculated from the sum of the variances of the individual items and the covariance between items by equation (1),

$$\alpha = \frac{k}{k-1}\left(1 - \frac{\sum_{i=1}^{k}\sigma_i^2}{\sum_{i=1}^{k}\sigma_i^2 + 2\sum_{i>}^{k}\sum_{j}^{k}\sigma_{ij}}\right) = \frac{k}{k-1}\left(1 - \frac{\sum_{i=1}^{k}s_i^2}{S_{sum}^2}\right) \quad (1)$$

where

- k is the number of items in the questionnaire,
- S_i^2 is the variance of the N responses to item i,
- S_{sum}^2 is the total variance of the responses to the questionnaire (from the N responses to the k items).

A very high value of Cronbach's alpha (e.g., 0.95 or higher) is not necessarily the most desirable since it is indicative of redundancy in the items (Streiner, 2003). The goal in designing a reliable measuring instrument is for the score obtained on similar items to be comparable (internal consistency), but for each one to contribute with unique information. Very high alpha values can be an indication of excessively large scales, parallel items, redundant items, or a sub-representation of the dimension (Kline, 1979).

Cronbach's alpha is necessarily higher for tests (scales) that measure more directed constructs and lower when measuring more generic constructs. This phenomenon, along with a number of other reasons, argues against the use of objective limit values for internal consistency measurements (Peters, 2014). In addition, Cronbach's alpha also increases with the number of items (Cortina, 1993), with shorter scales often having lower reliability estimates (reduced Cronbach's alphas), but will be preferable in many situations because they are lighter and practical to apply.

Thus, while the measurement ideal is that all items in a test measure the same latent variable, the alpha coefficient often reaches very high values, even when the set of items measures several unrelated latent variables (Cortina, 1993).

Nevertheless, using reference values generically accepted to describe internal consistency, the Cronbach alpha coefficients obtained for the scales/dimensions of the questionnaire used were 2 excellent ($\alpha \geq 0,9$), 12 good ($0,8 \leq \alpha < 0,9$), 9 acceptable ($0,7 \leq \alpha < 0,8$), 6 questionable ($0,6 \leq \alpha < 0,7$), 1 poor ($0,5 \leq \alpha < 0,6$), 1 non-interpretable (negative alpha), and 1 (health) for which the alpha cannot be calculated because it has only one variable.

2.4 Size reduction of the questionnaire used

Factor analysis is an interdependence technique whose main objective is to define the underlying structure among the variables under analysis, since the variables are the building blocks of the relationships (Hair et al., 2014).

The general objective of the techniques of factor analysis is to find a way to condense (summarize) the information contained in several original variables into a smaller set of dimensions or new variables (factors) with a minimal loss of information.

The total variance of any variable can be divided into three types of variance: common, single and error variances. When a variable is more correlated with one or more variables, the common variance (commonality) increases.

The common variances (commonalities) obtained for the 119 original variables under analysis were between 0.629 and 0.860.

Principal component analysis is the most appropriate technique when data reduction is the main concern, focusing on the minimum number of factors required to explain the maximum share of the total variance represented in the original set of variables, and when prior knowledge suggests that the specific and error variances represent a relatively small proportion of the total variance.

The extraction of factors from the data collected for the 119 variables in the 234 responses obtained was performed according to three steps: correlation matrix calculation (containing correlation coefficients) for all variables and application of the Bartlett test; extraction of initial factors (by analysis of the main components) and rotation of the extracted factors for a final solution.

The technique used to establish a cut-off point in the number of factors to be extracted is based on the criterion that any individual factor must explain the variance of at least one variable if it is retained for interpretation.

In the first extraction the number of factors to be extracted according to the previously defined criterion was fixed at 30, with the last factor considered, an eigen value of 1.010.

The criterion of the percentage of variance explained is an approach based on obtaining a cumulative percentage of the total variance extracted by successive factors. Despite the lack of an absolute limit adopted for all applications, in the social sciences, where information is relatively inaccurate, it is normal to consider satisfactory a solution that accounts for 60% of the total variance (Hair et al., 2014).

The 30 factors retained in the first extraction explain 74.0% of the total variance, which can be considered quite satisfactory.

The matrix of correlation coefficients (matrix of components) was obtained in SPSS (Statistical Package for the Social Sciences), version 21, with the following configurations:

- Application of the KMO and Bartlett test (statistical test for the general meaning of all correlations within the correlation matrix);
- Matrix rotation method—VARIMAX (one of the most commonly used orthogonal factor rotation methods focused on simplifying columns in a matrix of factors);
- Concealment of correlation coefficients lower than 0.33.

Table 1 identifies the correlations obtained between the extracted factors and the dimensions that group the totality of the 119 items involved in the first extraction.

Dimensions "23. Health", "24. Work-family conflict", "27. Vitality" and "28. Behavioral stress" appears to correlate with the extracted factor #1, indicating a correlation with a factor that measures an overall sense of health and well-being. Dimension "23. Health" also has a correlation with factor

#28 and "28. Behavioral stress" a correlation with factor #29, so their items are not retained for the second extraction.

Table 1. Correlations identified between the factors extracted and the dimensions of the applied questionnaire.

Extracted factor	Dimensions (Number of items not involved)
1	23. Health; 24. Work-family conflict (−1); 27. Vitality; 28. Behavioral stress (−3)
2	9. Recognition; 16. Vertical trust (−1); 17. Justice and respect
3	28. Behavioral stress (−4); 29. Somatic stress (−1); 30. Cognitive stress
4	14. Quality of leadership
5	31. Self-efficacy
6	10. Role conflict
7	26. Sleep problems
8	32. Offensive behavior (−2)
9	6. Possibilities for development
10	13. Sense of community
11	19. Meaning of work
12	25. Family-work conflict
13	4. Demands to hide emotions
14	5. Influence at work
15	12. Social support
16	1. Quantitative demands
17	22. Job insecurity
18	21. Satisfaction with work (−1)
19	18. Social responsibility (−1)
20	11. Social support from colleagues
21	8. Role clarity
22	3. Emotional demands
23	2. Cognitive demands
24	32. Offensive behavior (−4)
25	?
26	20. Commitment to work (−1)
27	29. Somatic stress (−1)
28	?
29	28. Behavioral stress (−6)
30	28. Behavioral stress (−5)

Dimensions "9. Recognition", 16. Vertical trust" and "17. Justice and respect" appear to be correlated with factor #2, indicating a correlation with a factor measuring a general feeling about leadership being fair and trustworthy. Dimensions "28. Behavioral stress", "29. Somatic stress" and "30. Cognitive stress" appear to be correlated with factor #3, indicating that they are correlated with a measure of a general sensation of the stress experienced by the respondents. Dimension "28. Behavioral stress" (minority represented) also shows a correlation with factor #30, leading to its items not being retained for the second extraction.

The remaining dimensions appear related to only one factor each. The exception occurs in dimensions "7. Predictability of work" and "8. Role clarity", indicating that these are correlated with a factor measuring a general sensation about the organization of work.

3 RESULTS

With the manipulation of the rows of the components matrix, to order within the factors extracted according to the order of the variables (numbering of the items), it was possible to identify and associate the factors extracted with the factors/constructs of the questionnaire. The final optimized questionnaire was obtained on the 6th and last extraction.

Table 2 summarizes the results obtained in the extractions carried out with the methodology described above.

Table 3 identifies retained items (*underlined italics*) from the long version of the applied CoPsoQ (32 dimensions with 119 items) to the short (optimized) version (19 dimensions with 54 items).

Table 2. Results obtained in the six factor extractions carried out.

Extraction	Number of items used	Bartlett test (χ^2)	Number of factors extracted	Number of dimensions identified	Percentage variance explained (%)
1.[a]	119	12764.846	30	32	73.992
2.[a]	93	11286.611	25	26	72.952
3.[a]	69	7930.238	19	22	72.497
4.[a]	63	7107.858	18	21	73.410
5.[a]	59	6408.678	17	22	72.942
6.[a]	57	6111.642	17	20	73.524
Final	54	–	17	19	–

Table 3. CoPsoQ applied *versus* the resulting optimized version (*underlined italics*): contexts, dimensions and items.

Workplace	Demands and pressures at work	1. Quantitative demands	*1, 2, 3*, 4
		2. Cognitive demands	5, 6, 7
		3. Emotional demands	8, 9, 10
		4. *Demands to hide emotions*	11, *12, 13, 14*
	Organization and work content	5. Influence at work	15, 16, 17, 18
		6. *Possibilities for development*	19, *20, 21*, 22
		7. Predictability of work	23, 24
		8. Role clarity	25, 26, 27
		9. *Recognition*	28, *29, 30, 31*
	Interpersonal relations and leadership	10. *Role conflicts*	32, *33, 34, 35*
		11. *Social support*	36, *37, 38*
		12. Head social support	39, 40, 41
		13. Social community	42, 43, 44
		14. *Leadership quality*	45, 46, *47, 48*
		15. Horizontal trust	49, 50, 51
		16. *Vertical trust*	52, 53, *54*
		17. Justice and respect	55, 56, 57, 58
		18. *Social responsibility*	59, *60, 61*, 62
Workplace-individual interface		19. *Meaning of work*	*63, 64*, 65
		20. Commitment to work	66, 67, 68
		21. *Job satisfaction*	69, 70, *71, 72*
		22. *Insecurity in employment*	73, *74, 75, 76*
Individual	Health and well-being	23. Health	77
		24. Work-family conflict	78, 79, 80
		25. *Family-work conflict*	*81, 82*
		26. Sleep problems	83, 84, 85, 86
		27. *Vitality*	*87, 88, 89, 90*
		28. *Behavioral stress*	*91, 92, 93*, 94, 95, 96, 97, 98
		29. *Somatic stress*	99, *100*, 101, 102, 103
		30. *Cognitive stress*	*104, 105, 106, 107*
	Personality	31. *Self-efficacy*	108, *109, 110, 111, 112*, 113
		32. *Offensive behavior*	114, 115, 116, *117, 118, 119*

4 CONCLUSIONS

In terms of reliability of the scales, with the respective items retained in successive factor extractions, the optimized questionnaire presents several cases in which the dimension of the optimized version with fewer items presents a superior Cronbach alpha. In the optimized version only two dimensions (10.5%) have an alpha lower than 0.70 against seven (21.9%) of the applied version.

According to the most demanding criterion of ISO 10075-3: 2004 (E) the optimized version has eight dimensions (42.1%) with an alpha equal to or greater than 0.80 against fourteen dimensions (43.8%) of the applied version.

The analysis of the dimensions/scales of the applied version and the optimized version shows a good correspondence in the five theoretically defined contexts: demands and pressures at work, work organization and content, interpersonal relations and leadership, workplace-individual interface, health and well-being and personality.

Where possible, in the comparison between the questionnaire used and the optimized version, several situations are identified where the cost-benefit ratio has been improved, e.g., less or equal number of items in the scale with higher or equal alpha coefficient, such as the concrete cases of the eight scales: demands to hide emotions, possibilities of development, role conflicts, social support of colleagues, social responsibility, meaning of work, behavioral stress and offensive behavior.

Despite the specificities of the sample analyzed in this study, the comparison with the available results in tertile distributions of each of the dimensions presented in the Spanish version (Moncada et al., 2008) and in the Portuguese version applied in continental Portugal (Silva et al., 2006), both of which of medium-length, allowed for the identification of similar results in several comparable dimensions.

Of the 54 items in the optimized questionnaire only 38.2% and 39,0% overlap with those in the medium and short-length Portuguese versions of the questionnaire, respectively.

REFERENCES

Colton, D. and Covert, R.W. (2007). Designing and constructing instruments for social research and evaluation. John Wiley & Sons, Inc.

Cortina, J.M. (1993). What is coefficient alpha? An examination of theory and applications. Journal of Applied Psychology, 78, 98–104.

Cox, T., Griffiths, A., Barlowe, C., Randall, R., Thomson, L. and Rial-Gonzalez, E.. (2000). Organisational interventions for work stress: A risk management approach. Contract Research Report 286/2000. Health and Safety Executive.

Cronbach, L.J. (1951). Coefficient alpha and the internal structure of tests. Psychometrika, 16(3), 297–334.

Daniels, K., Harris, C. and Briner, R. (2002). Understanding the risks of stress: A cognitive approach. Contract Research Report 427/2002. Health and Safety Executive.

French, J.R.P., Jr., Caplan, R.D., & Harrison, R.V. (1982). The mechanisms of job stress and strain. Jonh Wiley & Sons.

Hair Jr., J.F., Black, W.C., Babin, B.J. and Anderson, R.E. (2014). Multivariate Data Analysis. 7th Ed. Pearson Education Limited.

ISO 10075-3:2004(E). Ergonomic principles related to mental workload—Part 3: Principles and requirements concerning methods for measuring and assessing mental workload.

Johnson, J.V. (1989). Collective control: Strategies for survival in the workplace. International Journal of Health Services, 19, 469–480.

Karasek, R.A. (1979). Job demands, job decision latitude and mental strain: Implications for job redesign. Administrative Science Quarterly, 24, 285–308.

Kline, P. (1979). Psychometrics and psychology. Academic Press. London.

Kristensen, T.S., Hannerz, H, Høgh, A., Borg, V. (2005). The Copenhagen Psychosocial Questionnaire-a tool for the assessment and improvement of the psychosocial work environment. Scandinavian Journal of Work, Environment & Health, 31(6), 438–449.

Lundberg, U. and Cooper, C.L. (2011). The Science of Occupational Health: Stress, Psychobiology and the New World of Work. John Wiley & Sons.

Moncada, S., Serrano C., Corominas, A., Camps, A., y Giné, A. (2008). Exposición a riesgos psicosociales entre la población asalariada en España (2004-05): valores de referencia de las 21 dimensiones del cuestionario COPSOQ ISTAS21. Rev. Esp. Salud Pública, N.º 6, 667–675.

Moncada, S., Utzet, M, Molinero, E., Llorens, C., Moreno, N., Galtés, A. and Navarro, A. (2014). The Copenhagen Psychosocial Questionnaire II (COPSOQ II) in Spain-A Tool for Psychosocial Risk Assessment at the Workplace. American Journal of Industrial Medicine, 57, 97–107.

Nübling, M., Stößel, U., Hasselhorn, H-M., Michaelis, M. and Hofmann, F. (2006). Measuring psychological stress and strain at work: Evaluation of the COPSOQ Questionnaire in Germany. GMS Psycho-Social-Medicine, Vol. 3.

Schabracq, M.J., Winnubst, J.A.M. and Cooper, C.L. (2003). The Handbook of Work and Health Psychology. 2nd ed. John Wiley & Sons.

Siegrist, J. (1996). Adverse Health Effects of High-Effort/Low-Reward Conditions. Journal of Occupational Health Psychology, 1, 27–41.

Silva, C., Amaral, V., Pereira, A., Bem-haja, P., Pereira, A., Rodrigues, V., Cotrim, T., Silvério, J., Nossa, P. (2006). Copenhagen Psychosocial Questionnaire COPSOQ—Versão Portuguesa. Fundação para a Ciência e Tecnologia.

Streiner, D.L. (2003). Starting at the beginning: An introduction to coefficient alpha and internal consistency. Journal of Personality Assessment, 80(1), 99–103.

Exposure to occupational noise in police—a systematic review

A.C. Reis & M. Vaz
PROA/LABIOMEP/INEGI, Faculty of Engineering, University of Porto, Porto, Portugal

ABSTRACT: The exposure to occupational noise can cause hearing problems, induce accidents and decrease the performance of professional activities. The aim of this study was to synthesize the state-of-the-art in what concerns Occupational Noise Exposure in Police forces. For this purpose, PRISMA Statement methodologies were applied. The bibliographic survey was carried out using six keywords, in combinations, on articles indexed to databases. The procedure led to the identification of 135 papers, of which 114 were discarded and 21 included in this review. The survey revealed that in police activity, the influence of noise is directly related to a variety of sources that can cause hearing loss.

1 INTRODUCTION

Police work is considered one of the most dangerous and stressful, resulting in disturbances and health problems (Hartley, et al., 2012). However, worldwide studies in order to assess the associated risk factors in this group are scarce (Win, Balalla, Lwin, & Lai, 2015).

Noise is an unwanted sound with characteristics (in terms of dislike and intensity) such that it can cause discomfort and danger to health. Harmful health effects are primarily auditory; in fact, prolonged exposure has been shown to cause Noise-Induced Hearing Loss (NIHL) (Chiovenda, et al., 2007).

The development of NIHL is affected by many factors such as: individual susceptibility, age, noise level, noise characteristics, duration of exposure, associated risk factors such as smoking, and use of hearing protectors (Shrestha, Shrestha, Pokhare, Amatya, & Karki, 2011).

Gilbertson & Vosburgh, (2015), claim that police are a group of individuals with multiple sources of noise at work, which have had little focus on the literature.

2 SOURCES AND METHODOLOGY

2.1 Research strategy

The literature review was carried out according to the (Preferred Reporting Items for Systematic Reviews and Meta-Analysis) PRISMA® recommendations.

The present review was based on bibliographical research carried out over databases and scientific journals, using the search engines available at the Faculty of Engineering of the University of Porto. The research was initially done on all databases and scientific journals available, and subsequently gradually restricted to the relevant resources and papers of interest.

The papers found to be relevant for evaluation were published in journals linked to the following databases: Scopus; ScienceDirect; PubMed, Academic Search Complete, Medline, Web of Science (Inspec), Eric and in the journal: AIP Journals.

The research was conducted by combining a set of pre-defined keywords. The AND boolean operator was used between them, and publications were searched by title, abstract and keywords: "police officers AND noise exposure"; "police officers AND occupational noise"; "police officers AND noise measurement"; "hearing noise AND induced noise AND police".

2.2 Exclusion criteria

These were:

- Publication date: articles published previously to 1990 were excluded.
- Document type: conference articles, editorials and review articles were excluded.
- Language: the review was restricted to articles in English.

2.3 Eligibility criteria

To be included, articles needed to deal with both occupational noise and police professionals. Those that did not, were excluded.

Not included, with justification: articles that did not directly relate exposure to noise with police professionals were excluded from the research; for example, articles with the themes: stress, optoacoustic emissions, noise with dogs, pregnant police women, noise affecting other professionals, etc.

3 RESULTS AND DISCUSSION

The research strategy identified a total of 135 papers. After the removal of the repeated ones (66) and papers falling under exclusion criteria (20), finally 49 full text papers were analysed. After applying the eligibility criteria, 28 papers were not included with justification and only 21 papers were included in the final study.

The Table 1 below summarizes the characteristics of the study groups on those 21 papers which were selected.

Table 2 presents the study data collected by the different authors, namely: the activity of each police professional at the time of study, the study method and the period of the study, the type of result measured, norms and other references used, such as patterns and the type of data analysis or software that each author used to obtain results.

The noise exposure data columns summarize key findings with respect to noise exposure.

Some notes on how the above studies were conducted are now given.

According to Sanju & Kumar, (2016), participants were randomly selected based on willingness to participate in the study, but those with a family history of ear diseases and exposure to noise (unrelated to traffic), were excluded. Similarly, Lesage, Jovenin, Deschamps, & Vincent, (2009), Nagodawithana, Pathmeswaran, Pannila, Wickramasinghe, & Sathiakumar, (2015) and Gupta, Khajuria, Manhas, Gupta, & Singh, (2015) and Singh & Mehta, (1999), excluded from their studies professionals with documented external, internal or middle ear diseases, and policemen with congenital hearing problems.

In the majority of the studies carried out, authors concluded that the occupational health division performed medical tests annually. Medical examinations included obtaining detailed occupational history, past and current noise exposure history, as well as compliance with the use of hearing protection devices, and physical examination, which included body mass index and otoscopic examination. Other tests, such as blood tests, blood sugar levels, cholesterol, liver function tests, kidney function tests and urine microscopy, as well as audiometry test, were performed to complete the medical examination.

The audiometry tests were also performed at least 16 hours after the last exposure to noise in most studies, to exclude the temporary displacement of the threshold, a condition in which there is temporary hearing loss after exposure to noise. This closes the notes on studies conduction.

The study of Shrestha, Shrestha, Pokhara, Amatya, & Karki, (2011), showed that NIHL was common in traffic police professionals because of the nature of their work. This study has shown that increasing age, length of service, alcohol consumption, and smoking are significant risk factors that cause NIHL.

Several authors have concluded that NIHL is an incurable but preventable occupational condition. Other authors found that increasing age (Gupta, Khajuria, Manhas, Gupta, & Singh, 2015) and the presence of high blood pressure (Shrestha, Shrestha, Pokhare, Amatya, & Karki, 2011) are significant associated factors for NIHL. The development of NIHL is affected by many factors such as individual susceptibility, age, noise level, noise characteristic, duration of exposure, risk factors such as associated smoking and (non) use of hearing protectors (Shrestha, Shrestha, Pokhare, Amatya, & Karki, 2011), (Win, Balalla, Lwin, & Lai, 2015).

According to Shrestha, Shrestha, Pokhare, Amatya, & Karki, (2011), bilateral hearing loss was commonly observed in their study. Among some patients with unilateral hearing loss, the left ear was more involved than the right ear. This is comparable to the study among operational engineers (Hong, 2005). Similarly, in the study of (Nair & Kashyap, 2009), right hearing loss was predominant, showing that hearing losses can be unilateral, depending on noise context.

Police officers are potentially exposed to various sources of noise, including sirens, horns, shots, police dogs barking and traffic noise. Specifically, for motorcyclists. (Win, Balalla, Lwin, & Lai, 2015).

In the same study, Win et al. registered NIHL as being the most prevalent among male police officers (37.7%), as compared to female police officers (23.9%). This is similar to the results of other studies, which also show a prevalence of NIHL.

According to the study by Sharif, Taous, Siddique, & Dutta, (2009), 24% of Dhaka traffic police had mild to moderate hearing loss due to noise exposure and this was related to duration of exposure. Analogously, Lesage, Jovenin, Deschamps, & Vincent, (2009) also investigated the NIHL in French police and found that they had a 28% incidence rate, as compared to the administrative ones who totalled a 16% rate. The study by Gupta, Khajuria, Manhas, Gupta, & Singh, (2015) revealed that 22% of traffic police had NIHL and most had mild to moderate impairment. Significant association was observed between NIHL and duration of exposure to noise. A high prevalence was found by Shrestha, Shrestha, Pokhare, Amatya, & Karki, (2011), with 66.4% of Katmanda Metropolitan Transit police professionals having NIHL, of which 40.9% had bilateral involvement. However, the highest prevalence of NIHL was seen by Singh & Mehta, (1999), Ingle, Pachpande, Wagh, & Attarde, (2005), Guida, Diniz, & Kinoshita, (2011), who observed that police (shooting and transit) professionals had NIHL values above of 80% in their study.

Table 1. Characteristics of the study groups.

Reference (Authors, year, country)	Sample size	Average age	Years in duty	Duration of daily service	Smoking	Alcoholic habits	Diabetes	Hypertension
Singh & Mehta, (1999), India	421	40.05	> 2	–	–	–	–	–
Pawlaczyk-Łuszczyńska, et al., (2004), Poland	18 C 28 P	25.2	–	–	–	–	–	–
Ingle, et al., (2005), India	50	39.50	–	10–12h	–	–	–	–
Chiovenda, et al., (2007), Italy	39 P 42 C	42.20 P 43.50 C	97%: 83%: 5	8h P	–	–	–	–
Lesage, Jovenin, Deschamps, & Vincent, (2009), França	920 P 805 C	37.60 P 41.80 C	> 0,5	–	–	516 Y 1166 N	–	–
Omidvari & Nouri, (2009), Iran	79	–	–	–	–	–	–	–
Wu & Young, (2009), Taiwan	20 SI 12 SII	31.00 SI 42.00 SII	–	–	–	–	–	–
Sharif, Taet et al., (2009), Bangladesh	100	–	6–20	8h	–	–	–	–
Guida, et al., (2011), Brazil	30	41.90	$\bar{X}$: 19.40	–	–	–	–	–
Heupa, Gonçalves, & Coifman, (2011), Brazil	65 P 50 C	32.20 P 33.00 C	$\bar{X}$: 9.10 P $\bar{X}$: 11.10 C	–	–	–	–	–
Shrestha, Shrestha, Pokhare, Amatya, & Karki, (2011), Nepal	110	29.82	$\bar{X}$: 11.86	< 8h: 30 > 8h: 80	42 Y 68 N	70 Y 40 N	–	–
Caciari, et al., (2013), Italy	357 P 357 C	38.1 P 38.9 C	≥ 4	7 h	–	–	–	–
Venkatappa, et al., (2013), India	60	42.20	–	> 8h	–	–	–	–
Guida, et al., (2014), Brazil	12	–	–	–	–	–	–	–
Patel, et al., (2014), India	110	–	–	–	–	–	–	–
Gilbertson & Vosburgh, (2015), USA	16	31.00	–	–	–	–	–	–
Gupta, et al., (2015), India	150	36.65	1 to 6	–	–	–	–	–
Jazani, Saremi, Rezapour, Kavousi, & Shirzad, (2015), Iran	246	26.37	≥ 3: 173 ≤ 3: 73	–	31 Y 215 N	–	–	–
Nagodawithana, et al., (2015), USA	287	> 30: 94 ≤ 30: 193	> 4: 165 ≤ 4: 122	> 8h: 122 ≤ 8h: 165	–	–	–	–
Win, Balalla, Lwin, & Lai, (2015), Myanmar	543	35.55	$\bar{X}$: 14.75	–	162 Y 203 N	28 Y 337 N	33 Y 332 N	58 Y 307 N
Sanju & Kumar, (2016), India	60 P 80 BD	40.50 39,31	> 8 > 8	8h 8–15h	60 N 80 N	–	–	–

Legend: P – Policemen, C – Control, BD – Bus Driver, M – Male, F – Female; S – Study, Y – Yes, N – No; U – Uses, NU – No Use.

Concerning susceptibility factors, the above studies found uncertainty as to whether smoking or alcoholic habits increase the risk of NIHL (negative for one study, postive for two, all cross-sectional). In respect to diabetes and hypertension, Win, Balalla, Lwin, & Lai, (2015) and

Table 2: Study data.

Reference (Authors, year, country)	Activity	Study type	Methodologies and testing	Quantities measured	Noise Exposure Data - Registered noise levels	NIHL
Singh, et al., (1999), India	Traffic police	Cross-sectional	Tonal audiometric test	—	65–75 dB for light vehicles 70–90 dB for heavy vehicles	81.2% moderate 37.7% severe
Pawlaczyk-Łuszczyńska, et al., (2004), Poland	Shooting	Between-group design	Audiometric test Otoacoustic emissions	$L_{peak,C}$ [1]; $L_{Aeq,Tc}$ [2]; $L_{A\,F\,max}$ [3]; $L_{F\,max}$ [4]	148.3–160.9 dB	—
Ingle, Pachpande, Wagh, & Attarde, (2005), India	Traffic police	Cross-sectional	Tonal audiometric test Questionnaires	L_{eq} [2]; $MaxL$ [5]; $MaxP$ [6]	87.5 dB	—
Chiovenda, et al., (2007), Italy	Traffic police	Case-control	Audiometric test	—	—	—
Lesage, Jovenin, Deschamps, & Vincent, (2009), França	Policemen vs Policemen on motorcycle vs Civil servants	Cross-sectional	Medical examination Audiometric and otological Interviews	L_{eq}	63 to 90 dB (A) driving 105 dB (A) open air	28% P 16% C
Omidvari & Nouri, (2009), Iran	Traffic police	Cross-sectional	Dosimeter	L_{eq}; $L10$; Lp_{max}; Lp_{min}	64.5 e 77.2 dB	61%
Wu & Young, (2009), Taiwan	Shooting	Longitudinal	Dosimeter Otological examination Audiometric test Otoscopy Tuning fork test	dB(HL)	—	4000 and 6000 Hz of the left ear showed significant deterioration
Sharif, Taous, Siddique, & Dutta, (2009), Bangladesh	Shooting	Cross-sectional	Audiometric test	—	500 Hz a 4000 Hz	75%
Guida, Diniz, & Kinoshita, (2011), Brazil	Shooting (38 revolver, 40 pistol, both from Taurus Co)	Cohort, cross-sectional	Decibelimeter Tonal audiometry Acoustic spectrum: DAT recorder coupled to stereo microphone	LC_{peak}; L_{max}	113,1 dB Lc_{peak} for 40 pistol 16,8 dB Lc_{peak} for 38 revolver	Hearing loss at 4,000Hz in 86.7% of the cases
Heupa, Gonçalves, & Coifman, (2011), Brazil	Shooting	Case-control retrospective	Sound level meter Tonal audiometric test Transient otoacoustic emissions test Questionnaires	—	119 to 133 dB(C)	25% of exposed

Authors, year, country	Population	Study design	Measurement method	Noise metrics	Noise levels	Results
Shrestha, Shrestha, Pokhare, Amatya, & Karki, (2011), Nepal	Traffic police	Cross sectional, descriptive	Audiometric impedance and tonal test; Clinical examination of hearing	–	–	66.4%
Caciari, et al., (2013), Italy	Traffic police	Case-control	Otoscopic examination Tonal audiometric test Sound level meter	L_{Aeq}; L_{Ceq}; LC_{peak};	L_{EX}: 74.3 dB(A)	NIHL low, is very disabling
Venkatappa, et al., (2013), India	Traffic police	Cross-sectional	Sound level meter Questionnaires	L_{eq}; LC_{peak}; L_{max}; $L_{Ex,8h}$	71.2 to 91 dB	3.33% hearing ability was below average
Guida, Taxini, Gonçalves, & Valenti, (2014), Brazil	Shooting	Cross-sectional	Dosimeter Two microphones with accoustic calibrator	LC_{peak}; L_{max}; L_{eq}	146 dB(C) (peak) 129.4 dB(C) (L_{max})	27.5% and 25% of the evaluated military police personnel
Patel, et al. (2014), India	Traffic police	Cross-sectional	Otorhinolaryngological examination Audiological tests	–	–	68.2%
Gilbertson & Vosburgh, (2015), USA	Patrol officers	Between-group design	Dosimeter (right shoulder)	L_{Aeq}, $L_{eq,8h}$ $L_{OSHA,8h}$	61 to 79 dB(A)	28% of police officers had NIHL
Gupta, Khajuria, Manhas, Gupta, & Singh, (2015), India	Policemen	Cross-sectional	Otoscopy Tests of Rinne Weber and Schwabach Audiometric test Medical examination	–	69.2 to 80.1 dB(A)	22% of sample
Jazani, Saremi, Rezapour, Kavousi, & Shirzad, (2015), Iran	Traffic police	Cross-sectional	Sound level meter Questionnaires Periodic medical examinations	$L_{Aeq,8h}$	71.63 e 88.51 dB(A)	–
Nagodawithana, et al., (2015), USA	Traffic police	Cross-sectional	Tonal audiometric test Questionnaires	–	–	41% found in 118 policemen
Win, Balalla, Lwin, & Lai, (2015), Myanmar	Shooting and policemen	Cross-sectional	Audiometric test Medical examinations	–	–	34.2%
Sanju & Kumar, (2016), India	Traffic police vs bus drivers	Between-group design	Questionnaires	–	–	60% Normal hearing

Legend: [1]$Lc_{peak.}$ – Peak sound pressure level, C-weighted; [2]L_{Aeq} = L_{eq} – Equivalent continuous noise level; [3]$L_{A F max}$ – Pressure level with fast time constants; [4]$L_{T F max}$ – Maximum sound pressure levels in bands of 1/3 of the frequency range 40-20,000 Hz; [5]MaxL – Maximum noise level; [6]MaxP – Maximum sound pressure level; [7]$L_{EX,8h}$ – Level of exposure to noise, A-weighted, normalized to daily exposure of 8 hours of work.

Nagodawithana, et al., (2015) found a likelihood of increased risk for both, especially for hypertension. Summarising further on susceptibility factors, two (cross-sectional) studies yielded positive for age and two (case-control, group design) did not. Years in duty and duration of daily service were always found to increase susceptibility to NIHL (three cross-sectional and one longitudinal study).

4 CONCLUSION

Exposure to occupational noise can cause hearing problems as well as cause accidents and decrease performance of professional activities. In police activity, the influence of noise is directly related to the various sources that can cause hearing loss. All reviewed articles agree that, except for policemen performing purely administrative duties, noise exposure leads to some sort of hearing loss. This is the main conclusion of this review.

The scientific literature has thus validated the fact that occupational NIHL is a condition among police professionals. This creates a necessity to train police professionals to deal with the hazardous effects of noise exposure, and motivates further studies in this occupational group, in order to assess the associated risk factors and point possible preventive actions.

REFERENCES

Caciari, T., Rosati, M.V., Casale, T., Loreti, B., Sancini, A., Riservato, R., ... Tomei, G. (2013). Noise-induced hearing loss in workers exposed to urban stressors. *Science of the Total Environment*, 302–308. doi:10.1016/j.scitotenv.2013.06.009.

Chiovenda, P., Pasqualetti, P., Zappasodi, F., Ercolani, M., Milazzo, D., Tomei, G., ... Tecchio, F. (2007). Environmental noise-exposed workers: Event-related potentials, neuropsychological and mood assessment. *International Journal of Psychophysiology*, 228–237. doi:10.1016/j.ijpsycho.2007.04.009.

Gilbertson, L.R., & Vosburgh, D.H. (2015). Patrol officer daily noise exposure. *Journal of Occupational and Environmental Hygiene*, 686–691. doi:10.1080/15459624.2015.1043051.

Guida, H.L., Diniz, T.H., & Kinoshita, S.K. (2011). Acoustic and psychoacoustic analysis of the noise produced by the police force firearms. *Brazilian Journal of Otorhinolaryngology*, 163–170.

Guida, H.L., Taxini, C.L., Gonçalves, C.O., & Valenti, V.E. (2014). Evaluation of hearing protection used by police officers in the shooting range. *Source of the DocumentBrazilian Journal of Otorhinolaryngology*, 515–521. doi:10.1016/j.bjorl.2014.08.003.

Gupta, M., Khajuria, V., Manhas, M., Gupta, K.L., & Singh, O. (2015). Pattern of noise induced hearing loss and its relation with duration of exposure in traffic police personnel. *Indian Journal of Community Health*, 276–280.

Heupa, A.B., Gonçalves, C.G., & Coifman, H. (2011). Effects of impact noise on the hearing of military personnel. *Brazilian Journal of Otorhinolaryngology*, 747–53.

Hong, O. (2005). Hearing loss among operating engineers in American construction industry. *Int Arch Occup Environ Health*, 565–574. doi:10.1007/s00420-005-0623-9.

Ingle, S.T., Pachpande, B.G., Wagh, N.D., & Attarde, S.B. (2005). Noise exposure and hearing loss among the traffic policemen working at busy streets of Jalgaon urban centre. *Transportation Research Part D: Transport and Environment*, 69–75. doi:10.1016/j.trd.2004.09.004.

Jazani, R.K., Saremi, M., Rezapour, T., Kavousi, A., & Shirzad, H. (2015). Influence of traffic-related noise and air pollution on self-reported fatigue. *International Journal of Occupational Safety and Ergonomics*, 193–200. doi:10.1080/10803548.2015.1029288.

Kamal, A.-A.M., Eldamati, S.E., & Faris, R. (1989). Hearing Threshold of cairo traffic policemen. *International Archives of Occupational and Environmental Health*, 543–545. doi:10.1007/BF00683124.

Lesage, F.-X., Jovenin, N., Deschamps, F., & Vincent, S. (2009). Noise-induced hearing loss in French police officers. *Occupational Medicine*, 483–486. doi:10.1093/occmed/kqp091.

Nagodawithana, N.S., Pathmeswaran, A., Pannila, A.S., Wickramasinghe, A.R., & Sathiakumar, N. (2015). Noise-induced hearing loss among traffic policemen in the City of Colombo, Sri Lanka. *Asian Journal of Water, Environment and Pollution*, 9–14. doi:10.3233/AJW-150002.

Nair, C.S., & Kashyap, R.C. (2009). Prevalence of Noise Induced Hearing Loss in Indian Air Force Personnel. *Medical Journal Armed Forces India*, 247–251. doi:10.1016/S0377-1237(09)80015-4.

Omidvari, M., & Nouri, J. (2009). Effects of noise pollution on traffic policemen. *International Journal of Environmental Research*, 645–652.

Patel, R.B., Patel, J.A., Contractor, J.A., Chaudhari, A.V., Patel, D.C., & Parmar, H.M. (2014). Prevalence of hearing loss of people working in heavy traffic areas of Surat city. *National Journal of Otorhinolaryngology and Head and Neck Surgery*, 34–36.

Pawlaczyk-Łuszczyńska, M., Dudarewicz, A., Bąk, A., Fiszer, M., Kotyło, P., & Śliwińska-Kowalska, M. (2004). Temporary changes in hearing after exposure to shooting noise. *International Journal of Occupational Medicine and Environmental Health*, 285–293.

Sanju, H., & Kumar, P. (2016). Self-assessment of noise-induced hearing impairment in traffic police and bus drivers: Questionnaire-based study. *Indian Journal of Otology*, 162–167. doi:10.4103/0971-7749.187971.

Sharif, A., Taous, A., Siddique, B.H., & Dutta, P.G. (2009). Prevalence of noise induced hearing loss among traffic police in Dhaka Metropolitan City. *Mymensingh Medical Journal*, S24-28.

Shrestha, I., Shrestha, B.L., Pokhare, M., Amatya, R.M., & Karki, D.R. (2011). Prevalence of noise induced hearing loss among traffic police personnel of Kathmandu Metropolitan City. *Kathmandu Univ Med J*, 9(36):274-8. doi:10.3126/kumj.v9i4.6343.

Singh, V.K., & Mehta, A.K. (1999). Prevalence of occupational noise induced hearing loss amongst traffic police personnel. *Indian Journal of Otolaryngology and Head and Neck Surgery*, 23–26.

Thorne, P.R., Ameratunga, S.N., Stewart, J., Reid, N., Williams, W., Purdy, S.C., ... Wallaart, J. (2008). Epidemiology of noise-induced hearing loss in New Zealand. *Journal of the New Zealand Medical Association*, 33–44.

Venkatappa, K.G., Vinutha Shankar, M.S., & Annamalai, N. (2013). Assessment of knowledge, attitude and practices of traffic policemen regarding the auditory effects of noise. *Indian Journal of Physiology and Pharmacology*, 69–73.

Win, K.N., Balalla, N.B., Lwin, M.Z., & Lai, A. (2015). Noise-Induced Hearing Loss in the Police Force. *Safety and Health at Work*, 134–138. doi:10.1016/j.shaw.2015.01.002.

Wu, C.-C., & Young, Y.-H. (2009). Ten-year longitudinal study of the effect of impulse noise exposure from shooting on inner ear function. *Source of the DocumentInternational Journal of Audiology*, 655–660. doi:10.1080/14992020903012481.

Geostatistics applied to noise exposure—a systematic review

A.C. Reis
PROA/LABIOMEP/INEGI, Faculty of Engineering, University of Porto, Porto, Portugal

J.M. Carvalho
CERENA, Faculty of Engineering, University of Porto, Portugal

ABSTRACT: The aim of this study was to synthesize the state-of-the-art in what concerns geostatistics applied to occupational noise. For this purpose, PRISMA Statement methodologies were applied. Articles published before year 1996 were excluded. The search was carried out using four keywords indexed to databases. The procedures led to the identification of 231 articles, of which 204 were discarded and 27 included in this review. After the study of the various works, it was concluded that geostatistical analysis was used mainly for occupational and environmental noise analysis, namely road traffic noise.

1 INTRODUCTION

Geostatistics is an invaluable tool that can be used to characterize spatial or temporal phenomena. Geostatistics originated from the mining and petroleum industries, starting with the work by Danie Krige in the 1950's and was further developed by Georges Matheron in the 1960's. In both industries, geostatistics was successfully applied to solve cases where decisions concerning expensive operations were based on interpretations from sparse data located in space. Geostatistics has since been extended to many other fields in or related to the earth sciences, e.g., hydrogeology, hydrology, meteorology, oceanography, geochemistry, geography, soil sciences, forestry, landscape ecology (Zhang, 2011).

Geostatistics offers a way of describing the spatial continuity of natural phenomena and provides adaptations of classical regression techniques to take advantage of this continuity (Srivastava & Isaaks, 1989).

Physically, there is no difference between sound and noise. Sound is a sensory perception and noise corresponds to undesired sound. By extension, noise is any unwarranted disturbance within a useful frequency band (NIOSH, 1991). Noise is present in every human activity, and when assessing its impact on human well-being, it is usually classified either as occupational noise (i.e. noise at the workplace), or as environmental noise, which includes noise in all other settings, whether at the community, residential, or domestic level (e.g. traffic, playgrounds, sports, music) (WHO, 2004). The maximum noise value recommended by the legislation for 8 hours daily exposure is 85 dB (A) (WHO, 2004).

Several elements affect workers' health and noise is a major factor causing stress and harm to operators which are exposed for many hours to loud noises.

The main objective of this study was to synthesize current scientific knowledge regarding geostatistics applied to noise exposure.

2 SOURCES AND METHODOLOGY

2.1 Research strategy

The literature review was performed according to the PRISMA® recommendations (Preferred Reporting Items for Systematic Reviews and Meta-Analysis) and was based on bibliographical research, using the databases available at the Faculty of Engineering of the University of Porto. The research was initially done in all databases available and gradually restricted to the relevant resources and articles of interest. The main databases scanned were: *Scopus*; *ScienceDirect*, *PubMed* and *Medline*.

The search was carried out combining a set of keywords that were pre-defined. The Boolean operator used between them was AND and sources were searched by title, abstract and keywords: "Geostatistics And Noise"; "Geostatistics And Occupational Health"; "Geostatistics And Noise Level".

2.2 Exclusion criteria

These were:

– Publication date: articles published previously to 1996 were excluded;

- Document type: conference articles, editorials and review articles were excluded in a first stage;
- Language: the review was restricted to articles in English and Portuguese.

2.3 Eligibility criteria

To be included in this review, articles should relate to the application of geostatistics to occupational noise. Thus:

- Articles that did not relate to geostatistics and occupational noise were excluded, with reason;
- Articles relating to geostatistics and environmental noise were included whenever environmental noise was judged to relate to occupational noise (e.g. environmental noise affecting workers in the open space).

3 RESULTS AND DISCUSSION

The search strategy identified a total of 231 articles. After the removal of the repeated articles (1) and the articles by the exclusion criteria (81), 149 full text articles were analyzed. After applying the eligibility criteria, 122 articles were excluded, with justification, and 27 articles were included in the final study.

Figure 1 shows the flowchart of the research strategy.

Table 1 below presents the different geostatistical models and modeling approaches applied to noise that were found in the 27 included articles.

The most relevant findings of the above articles are now discussed.

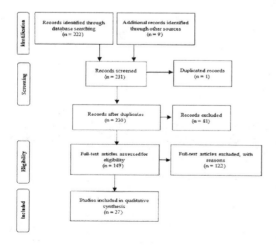

Figure 1. Flowchart of the research strategy.

Hamed & Effat, (2007), claim that spatial modeling is a valuable tool for quantifying the potential level of environmental consequences within the context of an environmental impact assessment (EIA) study. Their article presents a GIS-based tool for the assessment of airborne-noise and ground-borne vibration from public transit systems, and its application to an actual project.

Vehicle traffic is the major source of noise in urban environments, which in turn has multiple impacts on health. Evaluating road traffic noise and its likely impacts on the local community was the major objective of the studies of Seto, Holt, Rivard, & Bhatia, (2007); Banerjee, Chakraborty, Bhattacharyya, & Gangopadhyay, (2009); Birk, Ivina, Klot, Babisch, & Heinrich, (2011); Ko, Chang, Kim, Holt, & Seong, (2011); Mehdi, Kim, Seong, & Arsalan, (2011); Seong, et al., (2011); Renterghem, Botteldooren, & Dekoninck, (2012); Méline, Hulst, Thomas, Karusisi, & Chaix, (2013); Lee, et al., (2014); Zuo, et al., (2014) and Barceló, et al., (2016). In (Seto, Holt, Rivard, & Bhatia, 2007) the spatial distribution of community noise exposures and annoyance was also investigated.

The noise generated by machines and equipment may cause discomfort to workers, possibly even jeopardizing their health. The spatial analyzes of noise data, using geostatistics, allows for the determination of the level of harm that operators are subject to, fostering the definition of a work regime and the need of personal protection equipment, (Yanagi Junior, Schiassi, Rossoni, Ponciano, & Lima, 2012). The study of Ferraz, Silva, Nunes, & Ponciano, (2012), Yanagi Junior, Schiassi, Rossoni, Ponciano, & Lima, (2012), Gonçalves, Ferraz, Silva, Oliveira, & Ferraz, (2014), Silva, et al., (2014) and Spadim, Marasca, Batistuzzi, Denadai, & Guerra, (2015), was aimed at characterizing the spatial variability of the noise level generated by agricultural machines, using geostatistics, and to verify whether the values were within the limits of human comfort. All of machines presented noise levels above than 85 dB (A) near to the operator, demanding the use of hearing protection.

Urban environmental noise pollution has impact on the quality of life and it is a serious health and social problem (Oloruntoba, et al., 2012). Noise mapping has been used as an evaluation tool, not only for occupational noise, but also for environmental noise, helping to support decision making in urban planning. The studies of Ventura, (2008); Costa & Lourenço, (2011); Oloruntoba et al., (2012); and Scariot, Paranhos Filho, Torres, & Victório, (2012), were based on geotechnologies that consisted of noise mapping. This is a powerful tool to visualize and understand the sound distribution in a specific area. The noise

Table 1. Geostatistical models applied to noise.

Author	Activity/Situation	Sample
Thiéry, Wackernagel, & Lajaunie, (1996)	Automatic presses	2 workers
Hamed & Effat, (2007)	Public transportation	Middle East traffic
Seto, Holt, Rivard, & Bhatia, (2007)	Road traffic noise	235 streets of S. Francisco, USA, 709 measurements
Ventura, (2008)	Environmental noise	Campus of the School of Technology and Management of Leiria, Portugal
Banerjee, Chakraborty, Bhattacharyya, & Gangopadhyay, (2009)	Road traffic noise	25 locations in Asansol, India, 869 participants
Birk, Ivina, Klot, Babisch, & Heinrich, (2011)	Road traffic noise	1154 children aged 10 years
Costa & Lourenço, (2011)	Environmental noise	Urban area of Sorocaba, Brasil
Ko, Chang, Kim, Holt, & Seong, (2011)	Road traffic noise	48 measurement locations in Youngdeungpo-gu, S. Korea
Mehdi, Kim, Seong, & Arsalan, (2011)	Road traffic noise	308 locations of major intersections with congestion, in Karachi, Pakistan
Seong, et al., (2011)	Road traffic noise	Average daily traffic data set for 2008
Ferraz, Silva, Nunes, & Ponciano, (2012)	Agricultural machinery (coffee growing)	Portable coffee harvesting machine-coffee field
Yanagi Junior, Schiassi, Rossoni, Ponciano, & Lima, (2012)	Agricultural machinery	2 Agricultural machines: hedge trimmer and backpack blower
Oloruntoba, et al., (2012)	Noise in residential neighborhoods	11 localities, 341 housing
Renterghem, Botteldooren, & Dekoninck, (2012)	Road traffic noise	250 locations in Flanders, Belgium
Santos, Nascimento, & Silva, (2012)	Thermal and acoustic environment in an aviary (chicken cut)	Aviary
Santos M.B., et al., (2012)	Thermal and acoustic environment in an aviary (chicken cut)	Aviary
Scariot, Paranhos Filho, Torres, & Victório, (2012)	Environmental noise	Municipality of Campo Grande, Brazil
Méline, Hulst, Thomas, Karusisi, & Chaix, (2013)	Noise from road, rail and air traffic	7290 participants, aged 30 to 79 years
Gonçalves, Ferraz, Silva, Oliveira, & Ferraz, (2014)	Agricultural machinery	Tobatta tawny tractor coupled to a mower
Lee, et al., (2014)	Road traffic noise	25 districts of Seoul, S. Korea
Silva, et al., (2014)	Agricultural machinery	4 Agricultural machines: harvester, chainsaw, brushcutter and tractor
Zuo, et al., (2014)	Road traffic noise	554 locations in the city of Toronto, Canada
Pena, Lourençoni, Carvalho, & Yanagi Junior, (2015)	Agricultural machinery	Cleaning and destoning coffee machine ("beneficiadora de café"), 22 readings of sound intensity
Spadim, Marasca, Batistuzzi, Denadai, & Guerra, (2015)	Agricultural engines (engine revs)	2 Tractors of 78 horsepower
Barceló, et al., (2016)	Road traffic noise	Population of Barcelona, by gender and different ages
Lippiello, Degan, & Pinzari, (2016)	Open Quarry	50 workers
Oliveira, Damasceno, Ferraz, Nascimento, & Abreu, (2016)	Swine farrowing unit	3 first weeks of piglet growth, measured at 36 equidistant points

maps allow the identification of the most critical regions, noise sources and their influence areas, as well as verifying where the normative limit values for sound pressure are exceeded (Scariot, Paranhos Filho, Torres, & Victório, 2012).

The study of Lippiello, Degan, & Pinzari, (2016), did a comparison of stochastic and deterministic methods for mapping environmental noise from opencast quarries. Authors compared performances obtained by stochastic and deterministic methods to determine the acoustic climate of the areas surrounding opencast quarries.

The case study of Oliveira, Damasceno, Ferraz, Nascimento, & Abreu, (2016), evaluated the noise levels in a pig farrowing house during the first three weeks of life of piglets and Santos, et al., (2012) and Scariot, Paranhos Filho, Torres, & Victório, (2012) studied the use of geostatistics for the spatial characterization of the thermal and acoustic environment of aviaries. This tool has been used in several studies aimed at the characterization and spatialization of sound pressure levels emitted by animals in facilities.

4 CONCLUSION

This review shows that geostatistical analysis is often used for occupational noise analysis.

Geostatistics methods can be used to study multiple issues relating to occupational noise, in many facets. A vast array of techniques exist that are relevant for this purpose, helping decision-makers gain a better understanding of how noise affects workers and subsequently plan for ways to mitigate the impact of this important public health problem.

REFERENCES

Banerjee, D., Chakraborty, S.K., Bhattacharyya, S., & Gangopadhyay, A. (2009). Attitudinal response towards road traffic noise in the industrial town of Asansol, India. *Environmental Monitoring and Assessment*, 37–44. doi:10.1007/s10661-008-0247-0.

Barceló, M.A., Varga, D., Tobias, A., Diaz, J., Linares, C., & Saez, M. (2016). Long term effects of traffic noise on mortality in the city of Barcelona, 2004–2007. *Environmental Research*, 193–206. doi:10.1016/j.envres.2016.02.010.

Birk, M., Ivina, O., Klot, S.V., Babisch, W., & Heinrich, J. (2011). Road traffic noise: self-reported noise annoyance versus GIS modelled road traffic noise exposure. *Journal Environmental Monitoring*, 3237–3245. doi:10.1039/c1em10347d.

Costa, S.B., & Lourenço, R.W. (2011). Geoprocessing applied to the assessment of environmental noise: a case study in the city of Sorocaba, São Paulo, Brazil. *Environmental Monitoring and Assessment*, pp. 329–37. doi:10.1007/s10661-010-1337-3.

Ferraz, G.A., Silva, F.C., Nunes, R.A., & Ponciano, P.F. (2012). Variabilidade espacial do ruído gerado por uma derriçadora portátil em lavoura cafeeira. *Coffee Science, Lavras*, 276–283.

Gonçalves, L.M., Ferraz, G.A., Silva, C.J., Oliveira, M.M., & Ferraz, P.F. (2014). Mapeamento da distribuição espacial por ruído emitido por um trator de rabiça por meio da geoestatística. *XLII Congresso Brasileiro de Engenharia Agrícola*. Brasil.

Hamed, M., & Effat, W. (2007). A GIS-based approach for the screening assessment of noise and vibration impacts from transit projects. *Journal of Environmental Management*, pp. 305–313. doi:10.1016/j.jenvman.2006.06.010.

Ko, J.H., Chang, S.I., Kim, M., Holt, J.B., & Seong, J.C. (2011). Transportation noise and exposed population of an urban area in the Republic of Korea. *Environment International*, 328–334. doi:10.1016/j.envint.2010.10.001.

Lee, J., Gu, J., Park, H., Yun, H., Kim, S., Lee, W.,… Cha, J.-S. (2014). Estimation of populations exposed to road traffic noise in districts of Seoul metropolitan area of Korea. *International Journal of Environmental Research and Public Health*, 2729–2740. doi:10.3390/ijerph110302729.

Lippiello, D., Degan, G.A., & Pinzari, M. (2016). Comparison of stochastic and deterministic methods for mapping environmental noise from opencast quarries. *American Journal of Environmental Sciences*, pp. 68–76. doi:10.3844/ajessp.2016.68.76.

Mehdi, M.R., Kim, M., Seong, J.C., & Arsalan, M.H. (2011). Spatio-temporal patterns of road traffic noise pollution in Karachi, Pakistan. *Environmental International*, 97–104. doi:10.1016/j.envint.2010.08.003.

Méline, J., Hulst, A.V., Thomas, F., Karusisi, N., & Chaix, B. (2013). Transportation noise and annoyance related to road traffic in the French RECORD study. *International Journal of Health Geographics*, 12–44. doi:10.1186/1476-072X-12-44.

NIOSH. (1991). *Noise-induced loss of hearing*. Cincinnati: National Institute for Occupational Safety and Health.

Oliveira, C.E., Damasceno, F.A., Ferraz, G.A., Nascimento, J.C., & Abreu, L.P. (2016). Distribuição espacial dos níveis de ruído em uma maternidade de suínos utilizando a geoestatística. *XLV Congresso Brasileiro de Engenharia Agrícola*. Brasil.

Oloruntoba, E.O., Ademola, R.A., Sridhar, M.K., Agbola, S.A., Omokhodion, F.O., Ana, G.E., & Alabi, R.T. (2012). Urban Environmental Noise Pollution and Perceived Health Effects in Ibadan, Nigeria. *African Journal of Biomedical Research*, 77–84.

Pena, M.S., Lourençoni, D., Carvalho, A., & Yanagi Junior, T. (2015). Variabilidade espacial do nível de ruído emitido por uma benificiadora de café. *Congresso Técnico Científico da Engenharia e da Agronomia*. Brasil.

Renterghem, T.V., Botteldooren, D., & Dekoninck, L. (2012). Evolution of building façade road traffic noise levels in Flanders. *Journal Environmental Monitoring*, 677–686. doi:10.1039/C2EM10705H.

Santos, M.B., Nascimento, J.W., Furtado, D.A., Monteiro, L.F., Borba, J.T., & Farias, R. (2012). Utilização da geoestatística para caracterização espacial do ambiente térmico e acústico em galpões para criação de frangos de corte. *International Symposium on Occupational Safety and Hygiene*, 397–399.

Scariot, E.M., Paranhos Filho, A.C., Torres, T.G., & Victório, A.B. (2012). O uso de geotecnologias na elaboração de mapas de ruído. *Engenharia Sanitaria e Ambiental*, 51–60.

Seong, J.C., Park, T.H., Ko, J.H., Chang, S.I., Kim, M., Holt, J.B., & Mehdi, M.R. (2011). Modeling of road traffic noise and estimated human exposure in Fulton County, Georgia, USA. *Environment International*, 1336–1341. doi:10.1016/j.envint.2011.05.019.

Seto, E.Y., Holt, A., Rivard, T., & Bhatia, R. (2007). Spatial distribution of traffic induced noise exposures in a US city: an analytic tool for assessing the health impacts of urban planning decisions. *International Journal of Health Geographics*, 6–24. doi:10.1186/1476-072X-6-24.

Silva, M.O., Yanagi Junior, T., Schiassi, L., Rossoni, D.F., Barbosa, J.A., & Yanagi, S.N. (2014). Spatial variability of the noise level of a hedge trimmer and backpack blower. *Revista Brasileira de Ciências Agrárias*, pp. 454–458. doi:10.5039/agraria.v9i3a3770.

Spadim, E.R., Marasca, I., Batistuzzi, M.M., Denadai, M.S., & Guerra, S.F. (2015). Dependência espacial do ruído de tratores agrícolas em diferentes. *Revista de Agricultura Neotripical*, 29–33.

Srivastava, R.M., & Isaaks, E.H. (1989). *An Introduction to Applied Geostatistics*. Oxford: Oxford University Press.

Thiéry, L., Wackernagel, H., & Lajaunie, C. (1996). Geostatistical Analysis of time series of Short Laeq Values. *Proceedings of Inter-noise*, 2065–2068.

Ventura, J. (2008). *Modelação do ruído ambiente no campus da ESTG. Comparação entre a utilização de uma técnica geoestatística (Kriging) e software de previsão acústica*. Portugal.

WHO. (2004). *Occupational noise*. Geneva: Protection of the Human Environment.

Yanagi Junior, T., Schiassi, L., Rossoni, D.F., Ponciano, P.F., & Lima, R.R. (2012). Spatial variability of noise level in agricultural machines. *Eng. Agric., Jaboticabal*, pp. 217–225. doi:10.1590/S0100-69162012000200002.

Zhang, Y. (2011). *Introduction to Geostatistics*. Wyoming: University of Wyoming.

Zuo, F., Li, Y., Johnson, S., Johnson, J., Varughese, S., Copes, R., . . . Chen, H. (2014). Temporal and spatial variability of traffic-related noise in the City of Toronto, Canada. *Science of the Total Environment*, 1100–1107. doi:10.1016/j.scitotenv.2013.11.138.

Muscle fatigue assessment in manual handling of loads using motion analysis and accelerometers: A short review

F.G. Bernardo, R.P. Martins & J.C. Guedes
FEUP, Faculty of Engineering, University of Porto, Portugal

ABSTRACT: Muscle fatigue can cause productivity loss, human errors, unsafe actions, injuries and Work Related Musculoskeletal Disorders (WMSDs). To compensate muscle fatigue, people adapt their working strategy, changing movement patterns, recruiting different muscles and changing kinetic or kinematic components of the movement (like joint angles and velocities). This review, according to PRISMA Statement, was performed to summarize and analyze studies concerning muscle fatigue assessment using accelerometers and motion analysis. It was based on relevant articles published in 6 databases, namely Academic Search Complete, Inspec, PubMed, Science Direct, Scopus and Web of Science. A total of 15 articles were included in the systematic review. The following topics were analyzed: muscle groups evaluated, the tasks performed by the volunteer subjects, the assessment methods applied and the equipment and software used. Similar conclusions were obtained, regarding movement variability, muscular adaptations and changes in movement strategies, due to fatigue.

1 INTRODUCTION

The term muscle fatigue refers to the transient decrease in the capacity to perform physical actions and the decrease in the maximal force or power (velocity of muscle contraction) that the involved muscles can produce (Enoka et al., 2008).

Muscle Fatigue can be categorized as one of the symptoms of blood occlusion, because limited blood flow delivers insufficient oxygen and nutrients, alongside inadequate removal of metabolic waste products, causing lactate concentrations to rise (Oyewole, 2014).

Muscle fatigue can be divided in 3 stages: Non-Fatigue (the fresh muscle is able to exert its maximum force), Transition-to-Fatigue (once the fresh muscle starts to fatigue, new recruitment of muscle fibers occurs) and Fatigue (the onset of fatigue during a muscle contraction) (Al-Mulla et al., 2011). The Transition-to-Fatigue stage can be extended by practicing proper work/rest time ratio and controlling the lifting variables when performing manual handling of loads, that consist in any of the following activities: lifting, holding, putting down, pushing, pulling, carrying or moving of a load (animate or inanimate) (Halim et al., 2014). A significant level of muscle fatigue can cause productivity loss, human errors, unsafe actions, injuries and work related musculoskeletal disorders (WMSDs) (Sluiter et al. 2003; Toole 2005; Huang and Hinze 2006; Hallowell 2010).

During repetitive and fatiguing work, the musculoskeletal system adapts and uses momentary muscle substitution patterns, which result in more variable and less coordinated movements (Mehta et al., 2015).

The workers experience also plays a role, as demonstrated by Authier et al. (1996), were distinct lifting techniques were used by experienced workers in transferring boxes and these techniques were tested and reported to reduce back loading. A study of dynamic lifting conducted by Chen (1999), revealed that the lifting range differed significantly when participants felt fatigue in the upper limbs and that they used increasingly stooped and accelerated techniques at the beginning of the lift, followed by stiffening of the arms at the end of the lift.

The selection of volunteer subjects has to comply with certain criteria. It is important that the subjects are healthy, because any diseases or injuries may influence muscle performance and lead to inconsistent findings. Smoking and alcohol consumption also influence the results, alongside age and gender differences (Al-Mulla et al., 2011).

The assessment methods can be classified into invasive and non-invasive. Some examples of invasive methods are: blood lactate level, blood oxygen level, pH of muscle and needle Electromyography (EMG). Some examples of non-invasive methods are: surface Electromyography (sEMG), Near-Infrared Spectroscopy (NIRS), Mechanomyography (MMG), Acoustic Myography (AMG) and Sonomyography (SMG) (Bhat et al., 2016).

A recent review, performed by Srinivasan and Mathiassen (2012) concluded that motor variability is a relevant issue in an occupational context and that there is a great need for studies of motor variability. They suggest future research in creating methods to assess motor variability and study the relationship with occupational tasks and outcomes like fatigue and performance.

The main objective of this study involves the definition of a muscle fatigue assessment method based on movement measuring instruments, specifically accelerometers and motion analysis, in order to objectify a measurement technique susceptible to being applied in a real work environment. A systematic review was performed, to review the literature concerning muscle fatigue assessment using accelerometers and motion analysis, in order to analyze what muscle groups were evaluated, the tasks performed by the volunteer subjects, the assessment methods used, the equipment and software used, and collect the most important conclusions, relevant to the present study.

2 METHODS

Figure 1 displays the flowchart of the systematic review stages, according to PRISMA statement. The applied criteria are presented in Table 1.

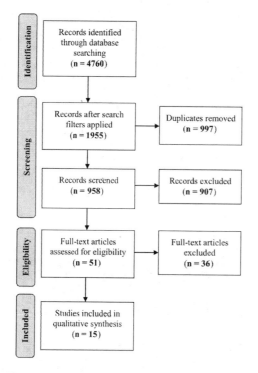

Figure 1. Systematic review stages.

3 RESULTS AND DISCUSSION

Analyzing the countries of each study, 8 were performed in the United States of America, 2 in Canada, 2 in France, 1 in Germany, 1 in Greece and 1 in New Zealand.

The 15 articles included a total of 235 participants, more precisely 162 men (68,9%) and 73 women (31,1%).

The most studied muscle groups were the Anterior Deltoid (10 studies), Biceps Brachii (7 studies), Posterior Deltoids (6 studies), Triceps (5 studies) and Erector Spinae (4 studies).

The most performed task consisted in repetitive lifting of a box/load (5 studies), followed by a task similar to sawing (3 studies), push/pull weights (2 studies), arm flexion/extension (2 studies), biceps curl (2 studies) and automobile assembly task (1 study).

Analyzing the fatigue measurement methods, 9 studies used surface Electromyography (sEMG), 2 studies used Near-Infrared Spectroscopy (NIRS), 2 studies used a combination of sEMG and accelerometry, 1 study combined sEMG and NIRS. 1 study evaluated the effects of fatigue through torso kinematics/kinetics, linear/angular momentum and Lyapunov exponents.

Comparing the equipment used, only 2 studies used accelerometers (triaxial accelerometers), 10 used motion analysis system with cameras and 2 studies used magnetic motion capture system. In order to maintain a certain pace in the task performance, 5 studies used a metronome. The dynamometer was included in 6 studies.

The software used included Matlab in 3 studies, Visual-3D in 3 studies and LabView in 1 study. A total of 7 studies applied the Borg scale (ratings of perceived exertion) or Borg CR-10 scale.

The articles had similar conclusions regarding movement variability, muscular adaptations and changes in movement strategies, due to fatigue.

The 2 studies that included accelerometry, (Brown et al., 2016; Dong et al., 2014), used sEMG as the fatigue measurement method and performed biceps curls. According to Brown et al. (2016), it is possible to distinguish between fatigued and non-fatigued sets of strength training based on acceleration data with a single parameter (calculated by a linear combination of duration and range of motion of each repetition). As a limitation, the fact that the volunteers could voluntarily slow down movement and manipulate kinematic data is mentioned. In the study performed by Dong et al. (2014), accelerometers were used to capture dynamic movements and impact simultaneously with sEMG data measurements. They tracked localized muscular fatigue levels by constantly updating the measured parameters, considering the fatigue process as a dynamic process.

Table 1. Criteria applied in the systematic review.

Identification and Search Strategy	The search was conducted through 6 electronic databases, namely: Academic Search Complete, Inspec, PubMed, Science Direct, Scopus and Web of Science. The systematic review focused on literature that addresses muscle fatigue assessment, using accelerometers and motion analysis, especially applied to upper limbs. Thus, the selected keywords used in the research had to be related to accelerometers, upper limbs, movement and kinetics. In order to assess other possible methods that have been applied, a couple of more generic keywords were used, regarding muscle fatigue determination and the evaluation over time. Finally, a relation between fatigue and musculoskeletal disorders had to be analyzed, more specifically in a workplace environment. The search terms were as follows: "Muscle Fatigue" combined with "Accelerometer", "Repetitive Work", "Upper Limbs", "Force Sensor", "Postural Control", "Lifting", "Tracking", "Movement Variability", "Forearm Muscles", "Repetitive Lifting", "Determination", "Manual Handling", "Reflex", "Biomechanic", "Arm Movement", "Kinetic", "Workplace" and "Musculoskeletal disorders". A total of 4760 items were found, using these keywords only (371 from Academic Search Complete, 210 from Inspec, 483 from PubMed, 317 from Science Direct, 2344 from Scopus and 1035 from Web of Science). The search was initially conducted by inserting the keywords and selecting, when possible, "Article title, Abstract, Keywords". The selected years were from 2010 to 2016 and the selected document type were articles, reviews and journal articles. Only the English language was considered. A total amount of 1955 items were gathered, using this search filters, meaning 2805 were excluded.
Screening Criteria	Since a lot of these items were duplicates, these had to be removed and counted, before continuing the screening criteria. A total of 997 duplicates were removed, resulting in a final value of 958 records screened. The next step in the screening had the following procedure: a) Title and Abstract were analyzed; Studies were automatically excluded if one of these conditions were met: 1) studies related to muscle rehabilitation; 2) studies applied to people with injuries/disorders/diseases; 3) studies applied to elders; 4) studies related to prosthetics; 5) studies related to sports like running, football, cycling, where lower limbs are predominantly used. b) the full text article was retrieved and considered. Whenever the title and abstract weren't enough to determine if the selection criteria were met, the article proceeded to the next step. If the article couldn't be retrieved, it was also excluded. This procedure led to a total of 51 articles, meaning 907 articles were excluded.
Eligibility Criteria	Studies were included if the following conditions were met: a) Muscle fatigue development in upper limbs (forearm, arm and shoulder muscles) and torso (2 excluded); b) Experimental evaluation based on movement (7 excluded); c) Well defined sample/participants (age, body mass and height) (7 excluded); d) Sample without any injuries/chronic pain (1 excluded); e) Studies not comparing different tools to assess muscle fatigue (4 excluded); f) Studies not validating methods or related to virtual simulation (3 excluded); g) Studies not assessing fall risk (2 excluded); h) Studies not related to assembly/pegboard or similar activity (6 excluded); i) Studies not evaluating task rotation effect on fatigue (2 excluded); j) Studies performed in a workplace/occupational environment (2 excluded); This eligibility criteria led to a total of 15 articles, meaning 36 articles were excluded. All selected articles have ethical considerations: informed consent obtained on human subjects and the protocol was approved by an ethical committee.

In the study performed by McDonald et al. (2016), they concluded that even tough kinematic and muscular changes allow workers to recover, they may not perceive existing fatigue, which may contribute to overuse injuries in the workplace. In order to maintain task performance, participants employed small compensations in several joint angles.

It was demonstrated in the study conducted by Mehta et al. (2015) that the behavioral changes observed when performing a prolonged repetitive asymmetric lifting activity likely increase the risk of back injury. When precise placement is required, larger sustained twisting postures and lateral bending moments on the spine occur.

The age factor was analyzed in Boocock et al. (2015) study, showing differences in lifting postures were found between the two age groups during the repetitive handling task. Lumbar mobility and dynamic range increased more in the younger group, which expose them to increased risk of injury. Peak lumbosacral, trunk, hip and knee flexion angles differed between age groups over the duration of the task, as did lumbosacral and trunk angular velocities.

The effect of local and widespread fatigue was analyzed in the study conducted by Cowley et al. (2014). The conclusion was that after localized fatigue, subjects made shorter, slower movements, while after widespread fatigue subjects exerted less control over non-goal-relevant variability and did not change movement patterns. Although they altered their control strategies, they continued to achieve the timing goal after both fatigue tasks.

Physiological and biomechanical responses to prolonged repetitive asymmetric lifting activity was assessed by Mehta et al. (2014). Participants repetitively lifted the box in a stooped posture they bent more over time.

The study conducted by Spyropoulosa et al. (2013) evaluated two different lifting ranges and concluded that volunteers used the stoop posture and achieved a better blood flow to the upper limbs, which helped to recover and delay fatigue accumulation.

The effect of different heights was evaluated in Gates et al. (2011), showing that subjects fatigued more quickly when movements were performed at the high height, they altered their kinematic patterns and exhibited increased kinematic variability of their movements post-fatigue.

4 CONCLUSIONS

The 15 articles included in this systematic review demonstrated that muscle fatigue leads people to change their movement patterns and changing kinetic or kinematic components of the movement.

Only 2 studies included accelerometers, in a combination with sEMG. Studies using only accelerometers to assess fatigue were not found, showing that there is a future research possibility in using this methodology, combined with a motion analysis system. This way, by monitoring adaptations to muscle fatigue and evaluating kinetic and kinematic changes, muscle fatigue can be assessed.

REFERENCES

Al-Mulla, M.R, Sepulveda, F. & Colley, M. (2011) A review of non-invasive techniques to detect and predict localized muscle fatigue. Sensors, 11, pp. 3545–3594. http://dx.doi.org/10.3390/s110403545.

Authier, M., M. Lortie, & M. Gagnon. (1996). Manual Handling Techniques: Comparing Novices and Experts. International Journal of Industrial Ergonomics, 17, pp. 419–429.

Bhat, Praveen & Gupta, Ajay. (2016). A Novel Approach to Detect Localized Muscle Fatigue during Isometric Exercises. pp. 224–229. htttp://dx.doi.org/10.1109/BSN.2016.7516264.

Boocock, M., Mawston, G., & Taylor, S. (2015). Age-related differences do affect postural kinematics and joint kinetics during repetitive lifting. Clinical Biomechanics, 30, pp. 136–143. http://dx.doi.org/10.1016/j.clinbiomech.2014.12.010.

Brown, N., Bichler, S., Fiedler, M., & Alt, W. (2016). Fatigue detection in strength training using three-dimensional accelerometry and principal component analysis. Sports Biomechanics, 15:2, pp. 139–150. http://dx.doi.org/10.1080/14763141.2016.1159321.

Chen, Y. (2000). Changes in lifting dynamics after localized arm fatigue. International Journal of Industrial Ergonomics. 25 (6), pp. 611–9. http://dx.doi.org/10.1016/S0169-8141(99)00048-7.

Cowley, J., Dingwell, J., & Gates, D. (2014). Effects of local and widespread muscle fatigue on movement timing. Experimental Brain Research, 232:12, pp. 3939–3948. http://dx.doi.org/10.1007/s00221-014-4020-z.

Dong, H., Ugalde, I., Figueroa, N., & Saddik, A. (2014). Towards Whole Body Fatigue Assessment of Human Movement: A Fatigue-Tracking System Based on Combined sEMG and Accelerometer Signals. Sensors, 14, pp. 2052–2070. http://dx.doi.org/10.3390/s140202052.

Enoka, R.M. & Duchateau, J., 2008. Muscle fatigue: what, why and how it influences muscle function. J. Physiol. 586 (1), 11–23. http://dx.doi.org/10.1113/jphysiol.2007.139477.

Gates, D., & Dingwell, J. (2011). The effects of muscle fatigue and movement height on movement stability and variability. Experimental Brain Research, 209:4, pp. 525–536. http://dx.doi.org/10.1007/s00221-011-2580-8.

Halim, I., Rawaida, R, K., A, R., Saptari, A., & Shahrizan, M. (2014). Analysis of muscle activity using surface electromyography for muscle performance in manual lifting task. Applied Mechanics and Materials, 564, pp. 644–649. http://dx.doi.org/10.4028/www.scientific.net/AMM.564.644.

Hallowell, M.R. (2010). Worker fatigue: Managing concerns in rapid renewal highway construction projects. Professional Safety, 55 (12), pp. 18–26.

Huang, X., & Hinze, J. (2006). Owner's role in construction safety. J. Constr. Eng. Manage., pp. 164–173. http://dx.doi.org/10.1061/(ASCE)0733-9364(2006)132:2(164).

McDonald, A., Tse, C., & Keir, P. (2016). Adaptations to isolated shoulder fatigue during simulated repetitive work. Part II: Recovery. Journal of Electromyography and Kinesiology, 29, pp. 42–49. http://dx.doi.org/10.1016/j.jelekin.2015.05.005.

Mehta, J., Lavender, S., & Jagacinski, R. (2014). Physiological and biomechanical responses to a prolonged repetitive asymmetric lifting activity. Ergonomics, 57:4, pp. 575–588. http://dx.doi.org/10.1080/00140139.2014.887788.

Mehta, J., Lavender, S., Jagacinski, R., & Sommerich, C. (2015). Effects of Task Precision Demands on Behavioral and Physiological Changes During a Repetitive Asymmetric Lifting Activity. Human Factors and Ergonomics Society, 57:3, pp. 435–446. http://dx.doi.org/10.1177/0018720814551556.

Oyewole, S. (2014). Enhancing Ergonomic Safety Effectiveness of Repetitive Job Activities: Prediction of

Muscle Fatigue in Dominant and Nondominant Arms of Industrial Workers. Human Factors and Ergonomics in Manufacturing & Service Industries, 24:6, pp. 585–600. http://dx.doi.org/10.1002/hfm.20590.

Sluiter, J.K., De Croon, E.M., Meijman, T.F., & Frings-Dresen, M.H.W. (2003). Need for recovery from work related fatigue and its role in the development and prediction of subjective health complaints. Occup. Environ. Med., 60 (Suppl. 1), pp. 62–70. http://dx.doi.org/10.1136/oem.60.suppl_1.i62.

Spyropoulosa, E., Chroni, E., Katsakiori, P., & Athanassiou, G. (2013). A quantitative approach to assess upper limb fatigue in the work field. Occupational Ergonomics, 11, pp. 45–57. http://dx.doi.org/10.3233/OER-130206.

Srinivasan, D., Mathiassen, SE. (2012). Motor variability in occupational health and performance. Clinical Biomechanics. 27, pp. 979–993. http://dx.doi.org/10.1016/j.clinbiomech.2012.08.007.

Toole, T.M. (2005). A project management causal loop diagram. Proc., 21st Conf. of Association of Researchers in Construction Management (ARCOM 2005), Association of Researchers in Construction Management (ARCOM), U.K., pp. 763–772.

Evaluation of physical fatigue based on motion analysis

F.G. Bernardo, R.P. Martins & J.C. Guedes
FEUP, Faculty of Engineering, University of Porto, Portugal

ABSTRACT: Muscle fatigue refers to the transient decrease in the capacity to perform physical actions and can cause productivity loss, human errors, unsafe actions, injuries and Work Related Musculoskeletal Disorders (WMSDs). A total of 13 participants repetitively lifted a 2.5 kg load at a total elevation of 0.5 m, until voluntary exhaustion or intensive pain. Several indicators of muscle fatigue were found, including increased forward bending, micro expressions, changing load support strategy, holding load closer to the chest during elevation and increase in eccentric movement speed. Volunteers that practiced sports regularly lasted longer in the experiment and it was found that smoking and sedentarism limited the exercise capacity of some subjects. It was concluded that to compensate muscle fatigue, people adapt their working strategy, changing movement patterns, recruiting different muscles and changing kinetic or kinematic components of the movement (like joint angles and velocities).

1 INTRODUCTION

According to Enoka et al. (2008), muscle fatigue refers to the transient decrease in the capacity to perform physical actions and the decrease in the maximal force or power (velocity of muscle contraction) that the involved muscles can produce. It can be categorized as one of the symptoms of blood occlusion, because limited blood flow delivers insufficient oxygen and nutrients, alongside the inadequate removal of metabolic waste products, causing lactate concentrations to rise (Oyewole, 2014).

Manual handling of loads consist in any of the following activities: lifting, holding, putting down, pushing, pulling, carrying or moving of a load. A significant level of muscle fatigue can cause productivity loss, human errors, unsafe actions, injuries and Work Related Musculoskeletal Disorders (WMSDs) (Sluiter et al., 2003; Toole, 2005; Huang and Hinze, 2006; Hallowell, 2010).

The main goals of this study were:

a. Identification of muscle fatigue indicators that could be listed and checked in a real work environment;
b. Definition of a muscle fatigue assessment method based on motion analysis, in order to objectify a measurement technique susceptible of being applied in a real work environment.

2 MATERIAL AND METHODS

2.1 *Participant selection*

Participants must be healthy, because any diseases or injuries may influence muscle performance and

Table 1. Subjects characteristics.

	Age	Height (m)	Weight (kg)	BMI (kg/m^2)
Males	24 ± 0.8	1.83 ± 0.08	76.8 ± 7.3	22.9 ± 2.2
Females	24 ± 1.2	1.65 ± 0.05	62.7 ± 3.8	23.0 ± 1.6

lead to inconsistent findings. Variables like lifestyle, eating habits and medical history are important.

In order to analyze these variables, a questionnaire divided in 4 topics (personal data, lifestyle, eating habits and medical history) was created.

2.2 *Sample characteristics*

A total of 13 participants were recruited, namely 4 men and 9 women. The same number of women and men were intended to be in the study, but more than double females were included, because it was a convenience sample. Their characteristics are presented in Table 1. All participants provided informed written consent.

2.3 *Apparatus/Equipment*

The software used consisted in a motion analysis software available from Video4Coach, namely SkillCapture version 2.0.6 and SkillSpector Version 1.2.4. The first one was used to record the experiment, alongside two Logitech C920 HD Pro Webcams, each one placed in the left and right side of the person, at around 45 degrees. The recordings were then studied in SkillSpector.

Volunteers had markers placed in strategic parts of the upper body, as shown in Figure 1, in order

Figure 1. Body segments and markers.

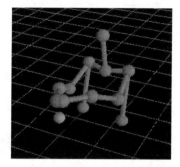

Figure 2. 3D model created in SkillSpector.

to improve the video analysis and creation of the body segments in the 3D model, as presented in Figure 2.

The 3D model was built by marking each reference point placed on the volunteer body in each frame (1 frame = 0.03 s) of the video. Since two cameras were used, this had to be done twice, one for each recorded video. Because of this limitation, only the differences between the first and last movements of lifting and lowering were studied, considering that the last movement before voluntary exhaustion is the most representative of muscle fatigue.

The body position was calculated using a DLT algorithm (Direct Linear Transformation) which was found based on the calibration object. This is the mathematical methods used to transform image data to real world coordinates.

2.4 Experimental procedure

The study started with the completion of a short questionnaire that covered 4 topics, namely: personal data, lifestyle, eating habits and medical history. The experiment was conducted inside a climatic chamber, where specific conditions were set, namely $25 \pm 2°C$ temperature and $49 \pm 5\%$ relative humidity.

At the beginning of the study, a cubic calibration structure with 50 cm edge was placed where the volunteers would perform the exercise, as shown is Figure 3, with the 3 adopted axis that were used.

The volunteers repetitively lifted a 2.5 kg load (dumbbell disk) from a 0.72 m height to 1.22 m, meaning a total elevation of 0.5 m, until voluntary exhaustion or intensive pain, that is, until they no longer could perform the task. While performing the exercise, two video cameras recorded the movements performed and these images were subsequently analyzed. A metronome (100 bpm) marked the pace that volunteers needed to perform the lifting and lowering movements of the load so that all volunteers performed the task under equal conditions. Figure 4 represents the lifting and lowering movement executed by the participants.

The experiments were conducted in the morning between 8 a.m. and 12 p.m., in order to have rested volunteers.

Ratings of Perceived Exertion (RPE) were gathered using the modified Borg scale. Subjects rated their level of fatigue on a scale from 0 (none at all) to 10 (maximal exertion) at the end of the experiment. Subjects were also asked about the physical

Figure 3. Calibration object.

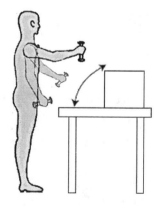

Figure 4. Lifting and lowering movement.

symptoms they experienced throughout the experiment and these were noted down.

2.5 Statistical analysis

Normality tests were performed in order to determine the normality of the data, namely Shapiro-Wilk, Anderson-Darling, Lillefors and Jarque-Bera tests. Outliers were detected with Grubbs test.

Student's paired t-test compared the means of two series of measurements performed on the same statistical units, namely the information of the first and final movements.

3 RESULTS AND DISCUSSION

3.1 Questionnaire results

The questionnaire covered 4 topics, namely: personal data, lifestyle, eating habits and medical history. Table 2 sums the main results of the questionnaire.

In terms of lifestyle, 9 in 13 participants played sports regularly (3 out of 4 males and 6 out of 9 females), only 1 female smoked and 4 in 13 (2 out of 4 males and 2 out of 9 females) admitted they had a sedentary lifestyle.

A total of 10 people drink coffee regularly (only 3 females do not drink). None of the participants consume alcoholic drinks regularly or was vegetarian.

Analyzing the data, there is a slight tendency for people that practice sports to hold on longer in the experiment. The only smoker was female 1, the one that only did the experiment for 7 minutes, which shows that smoking may harm physical condition. In terms of sedentarism, female 1 and 2 are the only females that admit they have a sedentary life, showing that the lack of exercise might have an influence on the results.

Female 1 combined the negative effects of smoking with sedentary life, which might explain her lower performance. She was also the person that consumed more coffees per day (4 coffees), the double than other volunteers.

3.2 Posture and behavioral changes recorded on camera and symptoms described by the volunteers

According to literature, higher movement variability occurs with the development of muscle fatigue (Brown et al., 2016). Muscular and kinematic adaptations occur to reduce the load on the fatigued muscles (McDonald et al., 2016). Other studies concluded that fatigue induced changes in movement strategies (Lee et al., 2014) and subjects altered their kinematic patterns significantly in response to muscle fatigue (Gates et al., 2011).

The posture and behavioral alterations observed in the recordings included:

a. Micro expressions (involuntary facial expressions) including biting or narrowing lips during elevation, eyebrows down and together;
b. Increased breathing pattern;
c. Changing load support strategy (hand and wrist posture); Tightening and releasing fingers on the load, trying to get better grip;
d. Increase in eccentric movement speed, following the conclusion of Brown et al. (2016).
e. Increased forward bending, also described by Mehta et al. (2014);
f. Head slightly bending forward;
g. Neck bending sideways;
h. Decrease in body stability;
i. Changing support leg (some volunteers started to feel numb legs);
j. Stretching body and spine;
k. Holding load closer to the chest during elevation in the final stages of the experiment, resulting in a less fluid movement. This shows volunteers adapted to compensate fatigued muscles, which was also concluded by McDonald et al. (2016), Lee et al. (2014) and Gates et al. (2011);
l. Increased difficulty to follow metronome rhythm.

3.3 Total execution time, Borg Scale, initial and final movements time

The total execution time, Borg Scale and initial and final movements time are presented in Table 3. The total time in the experiment of each participant, both males (M) and females (F), is presented in Figure 6, while Figure 6 presents the Borg

Table 2. Questionnaire results.

Name	Plays sports	smokes	Sedentary lifestyle	Coffe regularly
Male 1	Yes	No	No	Yes
Male 2	Yes	No	Yes	Yes
Male 3	No	No	Yes	Yes
Male 4	Yes	No	No	Yes
Female 1	No	Yes	Yes	Yes
Female 2	No	No	Yes	No
Female 3	Yes	No	No	Yes
Female 4	Yes	No	No	No
Female 5	No	No	No	Yes
Female 6	Yes	No	No	Yes
Female 7	Yes	No	No	No
Female 8	Yes	No	No	Yes
Female 9	Yes	No	No	Yes

Table 3. Total time, Borg Scale, initial and final movements time.

	t_{total} (s)	Borg Scale	t_i (s)	t_f (s)
Males	40.5 ± 15.8	4.3 ± 0.9	4.8 ± 0.4	5.0 ± 0.1
Females	18.3 ± 9.9	5.3 ± 1.9	5.0 ± 0.3	4.6 ± 0.7

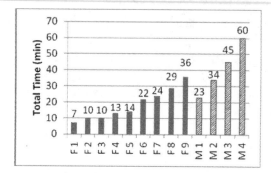

Figure 5. Total time in the experiment.

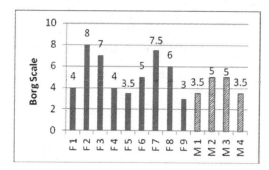

Figure 6. Borg Scale values.

Scale values, from 0 (none at all) to 10 (maximal exertion).

Analyzing Figure 5, it seems that the males could handle more time performing the experiment. Male 3 and 4 had the biggest times, with 45 and 60 minutes respectively. Female 1 had the lowest time, with only 7 minutes of exercise. The highest Borg Scale scores were given by female 2, 3 and 7. Female 2 and 3 did the same time (10 minutes) and gave similar Borg scale results (8 and 7 respectively), corresponding to a really hard exercise. Female 1 only gave a value of 4, half of female 2, which shows that it might not be self-aware of her fatigue level. In a work environment, this can lead to excessive effort and cause injuries.

3.4 Statistical results

The statistical analysis showed that individuals increased their range on the X axis (increased forward bending) and both wrists increased their velocity in the lifting and lowering stage. In terms of acceleration, half the participants had less fluid movement during lifting, with higher acceleration values in some parts and lower values in others. Relatively to the elbows, subjects closed the elbows and hold them near the body, in order to increase stability and hold the load better. Volunteers performed the lifting with higher velocity in the final movement of the experiment.

4 CONCLUSIONS

This study intended to identify muscle fatigue indicators that could be listed and checked in a real work environment and define a muscle fatigue assessment method based on motion analysis.

The questionnaires results showed that there is a slight tendency for people that practice sports to have more endurance. On the other side, the lack of exercise and smoking might harm physical condition. In order to compensate muscle fatigue, people adapted their working strategy and exhibited several muscle fatigue indicators. Males could handle more time performing the experiment.

4.1 Strengths of the methodology

This methodology is a new approach that allows the assessment of common symptoms and behaviors that people demonstrate when starting to feel muscle fatigue, which could be listed and used as indicators in the work environment. Motion analysis can be combined with accelerometry in order to assess muscle fatigue, by noticing changes in accelerations and movement patterns, as well as angles between wrist, shoulder and elbow.

4.2 Limitations of the methodology

The main limitation of this study is the fact that only the first and last movements of lifting and lowering were studied. The analyzed sample of 13 people is not the ideal sample size to generalize the results and future work should include a larger and representative sample. The uneven number of males and females is also a limitation (convenience sample).

REFERENCES

Brown, N., Bichler, S., Fiedler, M. & Alt, W. (2016). Fatigue detection in strength training using three-dimensional accelerometry and principal component analysis. Sports Biomechanics, 15:2, pp. 139–150. http://dx.doi.org/10.1080/14763141.2016.1159321.

Enoka, R.M. & Duchateau, J., (2008). Muscle fatigue: what, why and how it influences muscle function. J. Physiol. 586 (1), 11–23. http://dx.doi.org/10.1113/jphysiol.2007.139477.

Gates, D. & Dingwell, J. (2011). The effects of muscle fatigue and movement height on movement stability and variability. Experimental Brain Research, 209:4, pp. 525–536. http://dx.doi.org/10.1007/s00221-011-2580-8.

Hallowell, M.R. (2010). Worker fatigue: Managing concerns in rapid renewal highway construction projects. Professional Safety, 55 (12), pp. 18–26.

Huang, X. & Hinze, J. (2006). Owner's role in construction safety. J. Constr. Eng. Manage., pp. 164–173. http://dx.doi.org/10.1061/(ASCE)0733-9364(2006)132:2(164).

Lee, J., Nussbaum, M. & Kyung, G. (2014). Effects of work experience on fatigue-induced biomechanical changes during repetitive asymmetric lifts/lowers. Ergonomics, 57:12, pp. 1875–1885. http://dx.doi.org/10.1080/00140139.2014.957733.

McDonald, A., Tse, C. & Keir, P. (2016). Adaptations to isolated shoulder fatigue during simulated repetitive work. Part II: Recovery. Journal of Electromyography and Kinesiology, 29, pp. 42–49. http://dx.doi.org/10.1016/j.jelekin.2015.05.005.

Mehta, J., Lavender, S. & Jagacinski, R. (2014). Physiological and biomechanical responses to a prolonged repetitive asymmetric lifting activity. Ergonomics, 57:4, pp. 575–588. http://dx.doi.org/10.1080/00140139.2014.887788.

Oyewole, S. (2014). Enhancing Ergonomic Safety Effectiveness of Repetitive Job Activities: Prediction of Muscle Fatigue in Dominant and Nondominant Arms of Industrial Workers. Human Factors and Ergonomics in Manufacturing & Service Industries, 24:6, pp. 585–600. http://dx.doi.org/10.1002/hfm.20590.

Sluiter, J.K., De Croon, E.M., Meijman, T.F. & Frings-Dresen, M.H.W. (2003). Need for recovery from work related fatigue and its role in the development and prediction of subjective health complaints. Occup. Environ. Med., 60 (Suppl. 1), pp. 62–70. http://dx.doi.org/10.1136/oem.60.suppl_1.i62.

Toole, T.M. (2005). A project management causal loop diagram. Proc., 21st Conf. of Association of Researchers in Construction Management (ARCOM 2005), Association of Researchers in Construction Management (ARCOM), U.K., pp. 763–772.

Treloar, H.R., Piasecki, T.M., McCarthy, D.E. & Baker, T.B. (2014). Relations Among Caffeine Consumption, Smoking, Smoking Urge, and Subjective Smoking Reinforcement in Daily Life. pp. 93–99. http://dx.doi.org/10.1089/jcr.2014.0007.

Maintenance and safety according to the Brazilian performance standard: NBR 15.575

M.A.S. Hippert
Federal University of Juiz de Fora, Juiz de Fora, Brazil

O.C. Longo
Federal University Fluminense, Niterói, Brazil

ABSTRACT: The topic of building maintenance has grown in importance and overcome the culture that a construction process ends upon delivery of the building. Performance Standard NBR 15.575 has driven this discussion in Brazil. This Standard establishes a series of user needs that must be met, including maintainability. Based on a literature review, this article discusses Brazilian maintenance regulations with a focus on work safety. The purpose is to identify the safety requirements that must be considered in order to meet the NBR 15575 maintenance requirements. As a result, safety concerns in the Standard generally refer to access guarantee and the anchoring system. These should be considered at the beginning of a design process, as stated in the literature, so that they can contribute to work safety in a building's maintenance activities.

1 INTRODUCTION

Maintenance is aimed at preserving and restoring a building by taking user safety into account. In a broader sense, workers carrying out maintenance activities can also be considered as users. According to Melo Filho et al. (2012), the high risk potential for accidents in maintenance activities leads to the need for greater attention to risk analysis and its control so as to ensure worker safety and health in maintenance activities.

On the other hand, the Brazilian Performance Standard NBR 15575 (ABNT, 2013) includes in its scope a list of user needs that must be met, including maintenance-related requirements. Although technical standards are not laws, they are created by ABNT—the Brazilian official entity responsible for their publication—and therefore need to be addressed.

Based on these considerations, the purpose of this article is to identify the safety requirements that must be met, so as to comply with the maintenance requirements found in the Brazilian Performance Standard NBR 15.575. To this end, an international and national literature review is done in order to establish the necessary analytical basis.

2 BUILDING MAINTENANCE

Materials and buildings deteriorate over time. This is why planning and performing proper maintenance is required. Building maintenance can be defined as a "set of activities that must be carried out in order to preserve or restore the functional capacity of a building and its parts so as to meet the needs and safety of its users" (ABNT, 2011). However, according to (Gasparoli & Scaltritti, 2011), the topic of planned maintenance, as it appears in conferences and seminars, has not yet found an effective range of contributions in terms of application with the definition of its content and areas of intervention.

Maintenance should be considered at the beginning of a building project (Antunes & Calmon, 2005 and Melo Filho et al., 2012), where procedures needed for its preservation and restoration are set out. Sanches & Fabrício (2008) go even further and suggest the development of a maintenance project that corresponds to a "set of measures taken to contribute to increased maintainability, providing guidelines and subsidies for the next stage." Failure to consider maintenance aspects in the design stage can cause performance problems in a building (Sanches & Fabrício, 2008).

According to the cost evolution principle, the later maintenance is carried out in a building, the greater its cost will be, and the greater its performance will be impaired.

2.1 *Maintenance issues*

Failure to consider the maintenance aspects while still in the design stage results in problems.

Melo Filho et al. (2012) present a survey with a small company in Recife (PE) that conducted the maintenance of plumbing systems on a construction site. The authors found that access to the main water tank downpipe installation in the building was achieved through a poorly illuminated 40-cm high span, which made it difficult to carry out the maintenance activities. For Gnipper (2010), this is a frequent situation, especially in old vertical buildings where water distribution systems under elevated tanks have a clearance of only 0.50 m or 0.60 m, which makes access for maintenance services difficult.

Another example of failure to consider maintenance in a design project is shown by Sanches & Fabrício (2009). In a survey of a housing project of social interest in São Carlos (SP), containing 104 housing units, the authors found a number of issues:

– a sewage catchment box located within the private areas of the housing units, but serving three other units;
– humidity on the ground floor walls which, due to a lack of gutters on the 2nd floor, causes the water to fall from the top floor onto the ground and then climb up the non-waterproofed walls
– water meters installed in circulation areas, with no protection, causing frequent damage.

On the other hand, a lack of maintenance has been reported in the literature as responsible for a number of problems found in buildings. For Lessa & Souza (2010), the lack of maintenance causes inconveniences, loss and serious problems that could be avoided or even minimized if an efficient maintenance management program were adopted. This could prevent wear, equipment deterioration and even accidents.

Regarding accidents, the Instituto Brasileiro de Avaliações e Perícias de Engenharia (IBAPE) conducted a survey on accidents occurring in buildings aged more than 30 years in the city of Sao Paulo in 2009. These building accidents occurred exclusively during the use stage of the building. It was found that 66% of the probable causes and origins of the accidents are related to maintenance deficits, premature loss of performance and marked deterioration. The remaining 34% refer to the so-called constructive defects or endogenous anomalies (IBAPE, 2012).

Similarly, Medeiros & Grochoski (2007) surveyed 12 cases of collapsed ceiling overhangs and similar structures in Brazil from 1990 to 2007. For the authors, the most frequent causes of accidents are: iron corrosion, structure overload, design error, building misuse, construction error and water infiltration. However, most of them would be preventable if an inspection and periodic ceiling overhang maintenance program were adopted.

Inadequate maintenance can also cause health problems for building users. van der Zee et al. (2016) investigated whether a fine filter mechanical ventilation system could improve indoor air quality at a school located near a highway in Amsterdam. In the study they identified a number of maintenance-related problems such as the presence of old and overloaded filters. The authors concluded that the proposed application could reduce classroom exposure to traffic exhaust. However, filtration is only effective if filters are frequently replaced and the ventilation system is properly maintained.

In air-conditioned environments, ventilation system maintenance is mandatory. For Oliver & Shackleton (1998) apud Graus & Dantas (2007), both degradation and lack of maintenance in this system can lead to improper air quality due to its contamination by biological agents, inadequate air exchange and filtration rates, as well as the loss of control of thermal parameters.

2.2 Maintenance standards

Standards are established in order to regulate maintenance activities. At the international level, Gonçalves (2014) presents the main Portuguese and European references defining maintenance rules, requirements and procedures. These include EN 15331 – *Critérios para a conceção, gestão e controle de serviços de manutenção para edifícios* and NP 4483 – *Guia para a implementação do sistema de gestão da manutenção*.

Similarly, Brazil has Standards NBR 5674 – *Manutenção de edificações – Requisitos para o sistema de gestão da manutenção* (ABNT, 2012) and NBR 14037 – *Manual de operação, uso e manutenção das edificações – Conteúdo e recomendações para elaboração e apresentação* (ABNT, 2011).

NBR 5674 (ABNT, 2012) provides that maintenance services must be performed by a local maintenance team, a qualified company and/or a specialized company. Local maintenance personnel should perform various services for which they have received guidance. In addition, they need to be aware of risk and accident prevention. A qualified company is an organization or individual who has received training and works under the responsibility of a qualified professional. Finally, a specialized company is an organization or a self-employed professional who carries out duties for which specific qualifications are required.

NBR 14037 (ABNT, 2011), in turn, establishes the content that must be included in the Operation, Use and Maintenance Manual, including the applicable warranty periods. This document must be delivered to the owner by the person responsible for the production of the building, after completion of the construction work and beginning of the use and occupation stage. The need for information is also emphasized by Wu et al (2007) when stating that good access to information is needed regarding the performance of maintenance activities so that maintenance can be adequately performed.

3 BRAZILIAN PERFORMANCE STANDARD – NBR 15.575

Building Performance can be defined as the behavior of a building and its systems in use (ABNT, 2013). According to Szigeti & Davis (2005), "performance-based building is an encompassing approach related to the design, operation and maintenance of a building during its entire life cycle; essentially its general performance".

Lately, the need to consider performance in buildings has led several international organizations concerned with the issue to create regulatory systems based on this concept, such as in the United Kingdom, New Zealand, Australia, Canada, the Netherlands, Sweden, Norway and the United States. Spain, for example, created the Technical Building Code (CTE) in 2006, which is an official assessment program where performance targets and ways to achieve them are defined (CTE, 2006).

According to Oliveira & Mitidieri Filho (2012), in countries such as France, Canada and Japan, the performance of a building as a product and its subsystems is initially determined and only then are future constructive technologies defined. However, this is not the case found in Brazil, since the development of projects generally does not take performance into account, especially for housing projects. The architectural aspects and defined technologies are initially established, and only then are performance requirements considered (Melhado, 2001). But, the national reality started to change when Performance Standard NBR 15575 – Residential Buildings—Performance became effective. This Standard was first published in 2008, and after its review it became a requirement as of July 2013.

As a Performance Standard, NBR 15575 focuses on user requirements for residential buildings and their systems regarding their behavior when in use, not on the prescription of how systems are built. All of its provisions apply to systems that are found in "residential buildings, designed, constructed, operated and subject to maintenance interventions that comply with the specific instructions of the relevant Operation, Use and Maintenance Manual" (ABNT, 2013). This Standard defines (qualitative) requirements and performance criteria (quantitative or assumptions) for each user need and condition of exposure followed by assessment methods, based on which their compliance can be measured.

This Standard consists of six parts concerning the systems that make up residential buildings: General requirements; Structural systems requirements; Floor systems requirements; Internal and external vertical wall systems requirements; Roofing systems requirements; and Water/Sewage systems requirements. Although other systems, such as air conditioning, gas, and telecommunications, are considered by home users, these systems were not considered in the current standard version. Each of these parts was arranged according to a list of user requirements categorized as follows: safety (structural safety, fire safety, and safe use and operation), inhabitability (watertightness, thermal performance, acoustic performance, lighting performance, health, hygiene and air quality, functionality and accessibility, tactile and anthropodynamic comfort) and sustainability (durability, maintainability and environmental impact).

4 NBR 15.575: MAINTENANCE AND SAFETY

NBR 15575 maintenance requirements are described as follows:

– structural performance
– safe use and operation
– functionality and accessibility
– maintainability

Structural performance refers to maintaining a satisfactory level of safety for the structure and users during the project life cycle and exposure conditions considered in the project; safe use and operation should be considered in design, especially regarding aggressive agents; functionality and accessibility refer to compliance with system functions and access while maintenance refers to easy access or the means of access, considering that a project should favor access conditions for site inspection.

As to the structural performance requirement, only Part 5 of the Standard, which refers to the roofing system (RS), has a maintenance requirement for the item Assembly or Maintenance Requests which refers to the need for the structure to withstand person and object loads during the assembly or maintenance stages.

Likewise, for safe use and operation, only Part 5 of the Standard includes a maintenance and operation requirement. The criteria are as follows:

– parapets: must be provided to support suspended scaffolds or light platforms

- work safety on sloped roof systems: for sloped roofs with > 30% slope, the project should consider the use of devices anchored to the main structure in order to enable the attachment of ropes, safety belts and other personal protective equipment and the means of access for maintenance
- possibility of people walking on the roofing system: roofs and ceilings must be provided so that people can walk on them during assembly, maintenance or installation operations

Functionality and accessibility also include the requirement for maintenance of equipment and devices or RS components and members only in Part 5 of the Standard. The criteria are as follows:
- installation, maintenance and uninstallation of roof equipment and devices: the roofing system must provide means to easily and technically meet inspection, maintenance and installation needs as planned in the project

Maintainability is a requirement that is present in all systems addressed by the Standard (the Standard does not address floor maintenance). Performance assessment criteria and methods for each of the systems are shown in Table 1.

Table 1 shows that the maintainability requirement establishes as an assessment criterion the presence of easy access for the inspection of buildings and compliance with provisions in the Operation, Use and Maintenance Manual. The Performance Standard also refers to complying with the other two technical maintenance standards: NBR 14037 and NBR 5674.

Based on the joint analysis of Standards NBR 15575, NBR 5674 and NBR 14037, it is possible to see that, although the preparation of an Occupation, Use and Maintenance Manual is mandatory, it is not explicitly clear that it serves as a guide for maintenance activities, with a focus on worker safety. In general, the information in the manual refers to the safety and wholesomeness of buildings.

Consideration of work safety in maintenance is found in NBR 5674, which, in its Section 6.2, when addressing the means of maintenance control, provides for the need for the contractor to provide "means that ensure the necessary conditions for safe maintenance services" and "means to protect building users from any damage or losses arising from the performance of maintenance services" before starting the services.

It should be noted from the foregoing that, according to NBR 15.575 provisions, we must consider the following as to maintenance and safety together:

- easy access to conduct inspections
- adequate structure to support people and object loads during the maintenance stage
- anchoring system for roof and façade maintenance
- conditions for safe maintenance services
- protection of users from possible damage resulting from maintenance services.

Table 1. Maintainability.

Part of the standard	Criterion	Assessment method
Generalities	Easy access or means of access: a building and its systems must provide access conditions for inspection by installing scaffolding fixtures, platforms or other means that allow maintenance to be carried out.	- Project analysis: the project must be properly planned, so as to enable building inspection and maintenance - The developer or construction company must provide the user with a manual that complies with ABNT NBR 14037:2011 - Maintenance management must comply with ABNT NBR 5674:2012.
Structure	Structural System Operation, Use and Maintenance Manual.	Verification of maintenance processes as specified in the Manual.
Vertical walls	Wall System Operation, Use and Maintenance Manual.	Operation, Use and Maintenance Manual Review.
Roof	Roof Operation, Use and Maintenance Manual.	Operation, Use and Maintenance Manual Review.
Water/sewage Installations	Inspection devices shall be provided in sewage and rainwater pipes under the conditions in ABNT NBR 8160 and 10844.	Project verification or prototype inspection.
	Water/Sewage Installation Operation, Use and Maintenance Manual: specify the conditions for operation, use and maintenance, including the "As Built".	Operation, Use and Maintenance Manual Review.

Source: NBR 15575 (2013).

All these actions should be considered in the design stage of a building, in a simultaneous and integrated design process, with the participation of the various specialties involved in order to meet the specificities of each project. The inclusion of safety issues in the Performance Standard can be considered as a breakthrough for the construction industry. This is because it induces designers to think in advance, i.e., still during the development of a project, of which security actions should to be taken for the maintenance services in buildings.

On the other hand, decisions made in a project should be used to prepare the user manual which will be delivered to the user in order to guide the work to be done. In the case of common areas this manual must be delivered to the building administrator or superintendent. As we have seen, although the safety issue is not explicit in the manual, it would be advisable for the manual to also include guidelines for the safety of users and workers performing building maintenance activities in addition to building safety.

It should be noted that the safety issues listed here refer only to those set out in NBR 15.575 (which refers to other maintenance standards) and that compliance with them does not exempt an organization from complying with the existing work safety and health regulations and standards.

5 CONCLUSIONS

The increase in maintenance activities in the civil construction industry has been pointed out as a worldwide trend in the literature. This paper contributes to the discussion of the topic by suggesting the incorporation of the maintenance-related worker's health and safety requirements in the initial stage of a project as provided in the Brazilian legislation. These are extremely important topics for all companies in the construction industry. Thus, it is believed that this study can help meet the maintenance requirements included in the Performance Standard, in addition to preventing and predicting work-related incidents.

The inclusion of safety requirements in the user manual and the verification of their effective contribution to the improvement of worker's safety and health management as related to maintenance management activities are seen as a continuity of this study.

ACKNOWLEDGEMENTS

CAPES—Coordenação de Aperfeiçoamento de Pessoal de Nível Superior

REFERENCES

Associação Brasileira de Normas Técnicas—ABNT. 2011. NBR 14037: Diretrizes para elaboração de manuais de uso, operação e manutenção de edificações—Requisitos para elaboração e apresentação dos conteúdos. Rio de Janeiro.

Associação Brasileira de Normas Técnicas—ABNT. 2012. NBR 5674: Manutenção de Edificações—Requisitos para o sistema de gestão de manutenção. Rio de Janeiro.

Associação Brasileira de Normas Técnicas—ABNT. 2013. NBR 15575: Edificações Habitacionais—Desempenho. Rio de Janeiro.

Antunes, G.B.S. & Calmon, J.L. 2005. Manutenção de Edifícios: Importância no Projeto e Influência no Desempenho Segundo a Visão dos Projetistas. In *Anais* CONPAT 2005. Assunção: Paraguai.

Código Técnico de La Edificacion—CTE. 2006. España. Disponível em https://www.codigotecnico.org/. Acesso em: 10 Ago 2015.

Gnipper, S.F. 2010. Diretrizes para formulação de método hierarquizado para investigação de patologias em sistemas prediais hidráulicos e sanitários. Dissertação de mestrado. Universidade Estadual de Campinas, São Paulo.

Gasparoli P & Scaltritti M. 2011. Inspections and Maintenance Programmes for the Conservation of Historic Architecture—some Critical Issues Related to Degradation and Durability. In: *Proceedings International Conference Durability of Building Materials and Components.* Porto: FEUP. Portugal.

Gonçalves, C.D.F. 2014. Gestão da Manutenção em Edifícios: Modelos para uma abordagem LARG (Lean, Agile, Resilient e Green). Dissertação de Doutorado em Engenharia Industrial. Faculdade de Ciências e Tecnologia. Faculdade Nova de Lisboa.

Graudenz, G.S. & Dantas, E. 2007. Poluição dos ambientes interiores: doenças e sintomas relacionados às edificações. *Revista RBM ORL* Fev 07:23–31.

Instituto Brasileiro de Avaliações e Perícias de Engenharia de São Paulo. 2012. Inspeção predial: a saúde dos edifícios. Disponível em: http://www.ibape-sp.org.br/arquivos/CARTILHA-Inspecao-predial-a-saude-dosedif%C3%ADcios.pdf. Acesso em: 01 dez. 2017.

Lessa, A.K.M. C & Souza, H.L. 2010. Gestão da manutenção predial: uma aplicação prática. Rio de Janeiro: Qualitymark.

Medeiros, M.H.F. & Grochoski, M. 2007. Marquises: por que algumas caem? *Revista Concreto e Construções* 46: 95–102.

Melhado, S.B. 2001. Gestão, cooperação e integração para um novo modelo voltado a qualidade do processo de projeto na construção de edifícios. Tese de Livre Docência. Escola Politécnica da Universidade de São Paulo, São Paulo.

Melo Filho, E.C., Rabbanib, E.R.K. & Barkokebas Júnior, B. 2012. Avaliação da segurança do trabalho em obras de manutenção de edificações verticais. *Revista Produção*, v. 22, n. 4, p. 817–830.

Oliveira, A.L. & Mitidieri Filho, C.V. 2012. O projeto de edifícios habitacionais considerando a Norma Brasileira de Desempenho: análise aplicada para as vedações verticais. *Gestão e Tecnologia de Projetos,* São Carlos, v.7, n.1, p 90–100.

Sanches I.D. & Fabrício M.M. 2008. Projeto para manutenção. In: *Anais VII Workshop Brasileiro Gestão do Processo de Projetos na Construção de Edifícios*. São Paulo: ANTAC.

Sanches, I.D. & Fabricio, M.M. 2009. A importância do projeto na manutenção de HIS. In: *Anais Simpósio Brasileiro de Gestão e Economia da Construção - SIBRAGEC 2009*. João Pessoa: ANTAC.

Szigeti, F. & Davis, G. 2005. Performance Based Building: conceptual framework performance based building thematic network: 2001–2005. Rotterdam: CIB, 2005. (PeBBu Final Report, EC 5th Framework).

van der Zee, S.C., Strak, M., Dijkema, M.B.A., Brunekreef, B. & Janssen, N.A.H. 2016. The impact of particle filtration on indoor air quality in a classroom near a highway. *Indoor Air* 27: 291–302.

Wu, S., Lee, A., Tah, J.H.M. & Aouad, G. 2007. The use of a multi-attribute tool for evaluating accessibility in buildings: the AHP approach. *Facilities* 25 (9/10): 1849–59.

Sensation seeking and risk perception as predictors of physical and psychosocial safety behavior in risk and non-risk professions

G. Gonçalves & C. Sousa
Universidade do Algarve, Faro, Portugal
CIEO—Research Centre for Spatial and Organizational Dynamics, University of Algarve, Portugal

M. Pereira
Universidade do Algarve, Faro, Portugal

E. Pinto
Universidade do Algarve, Faro, Portugal
CESUAlg—Centre for Research and Development in Health, University of Algarve, Portugal

A. Sousa
Universidade do Algarve, Faro, Portugal
CIMA—Centre for Marine and Environmental Research, University of Algarve, Portugal

ABSTRACT: This study has as main objective to verify the predictive effect of sensation seeking and risk perception in the safety behaviors in individuals with risk and non-risk professions. Using a sample of 183 participants (N_{female} = 64; N_{male} = 119), aged between 17 and 63 years (M = 36.25; SD = 11.27), the results point the sensation seeking personality trait and risk perception as predictors of safety behaviors, only in individuals with non-risk professions. The individuals with higher risk professions are more likely to adopt physical and psychosocial safety behaviors. Gender differences were also found, and it is men who have a higher score on sensation seeking trait, as well as in safety behaviors and their participation.

1 INTRODUCTION

In recent years, research has sought to understand the predictive role of personality attributes and risk perception in workplace safety behaviors. Professional risk activities (e.g., firefighters, medical emergency) are extremely time and risk intensive activities (Adler-Tapia, 2014; Kunadharaju, Smith, & DeJoy, 2011; US Fire Administration, 2009). Despite the constant training that these professionals receive, there is a high rate of accidents, both in the training situation and in the exercise of their activity (US Fire Administration, 2014). According to Sabey and Taylor (1980), human factors contribute about 95% to the occurrence of accidents. Among the various attributes, the sensation seeking is one of the most referenced and analyzed attributes as a predictor of risk behaviors. The personality trait sensation seeking was originally defined by Zuckerman (1994, p. 27) as "a trait defined by the seeking of varied, novel, complex, and intense sensations and experiences, and the willingness to take physical, social, legal, and financial risks for the sake of such experience".

Thus, individuals with a high score in this personality trait tend to look for new and intense situations and experiences, neglecting aspects such as uncertainty and/or risk, predicting a greater degree of satisfaction in this type of situations (Zuckerman, 1994, 2008). According to several studies, it is adolescents and individuals of the masculine gender, those that present higher scores in the trait sensation seeking (e.g., Burri, 2017; Buss, 2004; Drane, Modecki, & Barber, 2017; Zakletskaia, Mundt, Balousek, Wilson, & Fleming, 2009). Associated with this trait, then arise risk behavior as practicing extreme sports, alcohol and drugs use, sexual experiences of *risk*, dangerous driving, antisocial behavior, crime and delinquency (e.g., Bratko & Butkovic, 2003; Zuckerman, 2008; Whissell & Bigelow, 2003) and a greater propensity to have accidents (e.g., Rahemi, Ajorpa, Esfahani, & Aghajani, 2017; Whishart, Somoray, & Rowland, 2017).

In addition to personality variables, the social cognition variables play a relevant role in occupational safety. Both theoretical frameworks contribute to the explanation of individual and social behavior. In order to integrate the perspective of social cognition we consider the variable risk perception. This variable is defined by the probability perception of an unpleasant event as an accident and concern with the eventuality of occurrence (LeRoy, 2005; Salminen, Klen, & Ojanen, 1999). In this regard, the Health Belief Model (e.g., Glanz, Rimer, & Lewis, 2002) considers that the greater the risk perception, the more individuals develop safety behaviors. Although risk perception is widely studied in certain contexts such as occupational health, social psychology, health psychology, and environmental psychology, the relationship between risk perception and occupational safety behavior remains unclear (Taylor & Snyder, 2017).

The output variable considered is the physical and psychosocial safety behavior, which encompasses compliance and participation with safety (Griffin & Neal, 2000; Neal & Griffin, 2006). Safety compliance describes the main activities that employees must perform to maintain safety in the workplace. Safety participation refers to behaviors that do not directly contribute to an individual's personal safety, but which help develop an environment that supports safety. In Portugal, studies are scarce, in particular the integration of personality and social cognition in the analysis of determinants of risk and safety behaviors. Thus, supported by the perspectives of personality and social cognition, it is the objective of the present study to evaluate the predictive role of the sensation seeking and risk perception, in the safety behaviors. Considering that the pressure and risk of certain professions affect risk behaviors, the objective of this study was to compare workers from a high-pressure and wear-resistant professional activity with professionals from lower-risk professions.

2 METHOD

2.1 Sample and procedure

A convenience sample of 183 participants (64 women and 119 men) is aged between 17 and 63 years (M = 36.25; SD = 11.27). About 44.8% are single and 45.4% have a university degree. Most respondents are employees (83.1%) and the majority are from the Alentejo and Algarve regions (63.9%).

The sample was divided according to the type of the individuals' professions. Thus, a sample of 92 participants with non-risk professions and a sample of 91 participants with risk professions were included. The risk professions were: firefighters (N = 33), nurses, medical and emergency medical technicians (N = 26) and police and military personnel (N = 32).

Participants were asked to answer a self-report questionnaire with an average completion time of 10 minutes. Data collection was performed in several places, collectively and individually, namely in university classes, public and private companies, public libraries, and other public places. Only the questionnaires completed correctly were included in the analysis.

2.2 Measures

A self-report questionnaire composed of sociodemographic items and the following instruments:

Brief Sensation Seeking Scale: Developed by Hoyle, Stephenson, Palmgreen, Puzzles, & Donohew (2002), and adapted to the Portuguese population by Sousa, Gonçalves, Sousa, & Pinto (2017). This scale consists of 8 items and the answers are given through a 5-point Likert scale (1 – strongly disagree to 5 – strongly agree). This scale contemplates items such as "I would like to take off on a trip with no pre-planned routes or timetables" (item 5) that presented the highest mean (M = 3.88; SD = 1.20) and "I would love to have new and exciting experiences, even if they are illegal" (item 8), that presented the lowest mean (M = 2.28; SD = 1.30). The scale has good reliability and validity ($\alpha = 0.88$).

Risk Perception: Developed by Moen (2007) is a scale that assesses the concern and fear of having accidents. It was translated into Portuguese for the present study. It is a one-dimensional scale composed of 4 items. The item that presented a higher mean (M = 4.44; SD = 1.75) was the item 2 ("When there are a lot of accidents in work, I become worried") and the one that presented a lower mean (M = 3.13; SD = 1.70) was the item 1 ("I become nauseated when I think about accidents"). Three more questions were raised about the probability of having accidents soon, the next year, and the probability of getting injured in an accident. These questions were adapted from the study of Clay, Trebarne, Hay-Smith and Milosavljevic (2014). All responses are given on a 7-point Likert scale, where a minimum score corresponds to less fear, concern, or probability of having accidents. The scale presented a good reliability ($\alpha = 0.84$).

Physical and Psychosocial Safety Behavior Scale: Developed by Bronkhorst (2015) and based on the work of Neal and Griffin (2006), has been translated into the Portuguese language through a translation-retranslation process. It is a two-dimensional scale, composed of 12 items that evaluate the behavior of employees, regarding physical and psychosocial safety, through a scale of 5 points (1 – totally disagree to 5 – totally agree). Each dimension is divided into two sub-dimensions: compliance and participation in relation to physical and psychosocial safety. Examples of items:

item 1 "I use all the necessary safety regulations and equipment to minimize physical/psychological strain in my job" and item 6 "I voluntarily carry out tasks or activities that help to improve workplace physical/psychological safety". Regarding the physical safety behavior, the item that presented a highest mean (M = 3.86; SD = 0.86) was the item 2 ("I use the correct procedures and regulations for physical safety when carrying out my job") of the compliance sub-dimension, and the lower mean (M = 3.33; SD = 1.09), was the item 4 ("I promote the physical safety program within the organization") belonging to the participation sub-dimension. In relation to the psychological safety behavior, the item with a highest mean (M = 3.70; SD = 0.93) was the item 7 ("I use measures to prevent or minimize psychological strain in my job") of the compliance sub-dimension, and the item with the lower mean (M = 3.14; SD = 1.10) was the item 10 ("I promote the psychological safety program within the organization") which belongs to the participation sub-dimension. The observed α ranged between 0.89 and 0.94.

Histograms and measures of skewness and kurtosis showed that the distributions of the all items were normal (skewness from −0.83 to 0.79 and kurtosis from −1.40 to 0.53), since they are below 2 and 7 respectively (Bentler & Wu, 2002; Curran, West, & Finch, 1996).

Once all scales used were completed by one person and used a similar response format, Harman's one-factor test (Podsakoff, MacKenzie, Lee, & Podsakoff, 2003) was conducted to detect potential bias caused by Common Method Variance (CMV). We entered all scales items into a principal component analysis and examined the unrotated factor solution. This analysis did not produce a single or assigned factor, since the main factor only explains 27.65% of the total variance. In addition, the analysis produced 18 factors with eigenvalue greater than 1, which are necessary to explain 93.8% of the variance and indicates the absence of the CVM.

3 RESULTS

In this subsection are shown the results obtained for each of the variables described, which are presented in tabular form. The presentation begins with means (M) and standard deviations (ST) values (Table 1), follows with the hierarchical regressions (Tables 2 and 3) and professional and gender differences.

3.1 Descriptive analysis

With the exception of the sensation seeking variable, participants with risk professions have higher

Table 1. Mean and standard deviation of the variables.

	Non-risk profession		Risk profession	
	M	SD	M	SD
Sensation seeking	2.98	0.92	2.93	1.02
Physical safety behaviour	3.59	0.88	3.74	0.69
– Compliance	3.78	0.85	3.93	0.76
– Participation	3.40	1.38	3.55	0.77
Psychosocial safety behaviour	3.35	0.84	3.49	0.85
– Compliance	3.49	0.82	3.64	0.88
– Participation	3.19	0.99	3.34	0.95
Risk perception	3.78	1.44	3.66	1.34

Table 2. Physical safety behaviour—non-risk profession.

	Physical safety behaviour—participation		
	β	t	p
Sensation seeking	0.274	2.693	0.008
Risk perception	0.071	0.683	0.496
	$r^2 = 0.28$; $p = 0.025$		

Table 3. Psychosocial safety behaviour—non-risk profession.

	Psychosocial safety behaviour—participation		
	β	t	p
Sensation seeking	0.279	2.785	0.007
Risk perception	0.057	0.551	0.583
	$r^2 = 0.29$; $p = 0.024$		

means values than those with non-risk professions. Table 1 shows that the highest mean dimension is physical safety behavior compliance (M = 3.93; SD = 0.76) and the smaller mean dimension is the sensation seeking (M = 2.93; SD = 1.02). It can also be observed that all dimensions have relatively low means.

3.2 Inferential analysis

To analyze the predictor effect of sensation seeking and risk perception on safety behaviors, multiple regression analyzes were performed on both samples. Statistically significant results were only found in the sample of participants with non-risk professions and in the participation dimension (vide Tables 2 and 3).

Table 2 shows that sensation seeking and risk perception explain about 28% of the participation dimension of physical safety behaviors ($p = 0.025$).

With regard to the predictive effect of the sensation seeking and risk perception variables on psychosocial safety behaviors (Table 3), they explain about 29% of the participation dimension ($p = 0.024$).

3.3 Profession differences

To analyze the difference between professions, an independent sample t test was performed. There were significant differences in the variables: 1) physical safety behavior ($t_{(181)} = -1.283$; $p = 0.019$); 2) physical safety behavior compliance ($t_{(181)} = -1.126$; $p = 0.043$); and 3) physical safety behavior participation ($t_{(181)} = -0.902$; $p = 0.031$).

3.4 Gender differences

To analyze the difference between genders, an independent sample t test was performed, considering the entire sample (N = 183). There were significant differences in the variables: 1) sensation seeking ($t_{(181)} = -1.338$; $p = 0.04$), with men (M = 3.03; SD = 1.01) having a higher mean than women (M = 2.83; SD = 0.88); 2) physical safety behavior ($t_{(181)} = -0.400$; $p = 0.042$), with men (M = 3.69; SD = 0.71) also present a higher mean than women (M = 3.64; SD = 0.91); and 3) physical safety behavior participation ($t_{(181)} = -0.071$; $p = 0.045$), where the men (M = 3.49; SD = 0.84) present a mean slightly higher than the women (M = 3.47; SD = 1.52).

4 DISCUSSION AND CONCLUSION

This study aimed to analyze the role of personality and risk perception variables as predictors of safety behaviors. The results showed that these variables only demonstrated a predictive effect on the sample of participants with non-risk professions and on the participation dimension of the physical and psychosocial safety behaviors. These non-expressive results may be related to the fact that the study participants presented relatively low means in all variables. That is, when they present low scores on the personality trait sensation seeking, they have little propensity to be involved in situations of risk, so it is presumed that their perception of an accident happening and their concern with the eventuality of the occurrence of something nasty will be more diminutive. Thus, by not engaging in dangerous or risky situations and not having a high-risk perception, individuals do not feel the need to develop safety behaviors.

On the other hand, we can consider that individuals with risk professions, have a better sense of the dangers to which they are subject, and as such the adoption of safety behaviors is already part of their professional profile. Perhaps because of this, these professionals have higher means compared to individuals with non-risk professions. In relation to gender differences, the masculine gender, similar to what has been illustrated in the literature (e.g., Burri, 2017; Drane, Modecki, & Barber, 2017), presented a higher score in the sensation seeking trait. In seeking more risky situations, males tend to engage more in physical security behaviors, as well as in the participation of an environment that supports safety.

The performance of work in a safe manner can be determined by the behaviors associated with achieving and promoting the climate and safety standards. Investment in health and safety is an extremely important factor for organizations, since it allows an increase in the physical and psychosocial well-being of their human resources (Mearns, Hope, Ford, & Tetrick, 2010). Research on safe work environments, particularly in relation to personality, social cognition and safety behaviors, is still underdeveloped. In this sense, the promotion of safety behaviors is an important preventive factor.

ACKNOWLEDGEMENTS

This paper is financed by National Funds provided by FCT—Foundation for Science and Technology through project UID/SOC/04020/2013.

REFERENCES

Adler-Tapia, R. 2014. Early mental health intervention for first responders/protective service workers including fire-fighters and emergency medical services professionals. In M. Luber (Ed.), *Implementing EMDR Early Mental Health Interventions for Man-Made and natural disasters: Models, scripted protocols, and summary sheets* (pp. 343–369). New York, NY: Springer Publishing Co.

Bentler, P. & Wu, E. 2002. *EQS for Windows user's guide.* Encino, CA: Multivariate Software, Inc.

Bratko, D., & Butkovic, A. 2003. Stability of genetic and environmental effects from adolescence to young adulthood: results of Croatian longitudinal twin study of personality. *Twin Res Hum Genet.* 10(1): 151–7.

Bronkhorst, B. 2015. Behaving safely under pressure: The effects of job demands, resources, and safety climate on employee physical and psychosocial safety behavior. *Journal of Safety Research* 55: 63–72.

Burri, A. 2017. Sexual Sensation Seeking, Sexual Compulsivity, and Gender Identity and Its Relationship with Sexual Functioning in a Population Sample of Men and Women. *J Sex Med* 14(1): 69–77.

Buss, D. 2015. *Evolutionary psychology: The new science of the mind* (5th ed.). Boston: Allyn & Bacon.

Clay, L., Trebarne, G., Hay-Smith, E., & Milosavljevic, S. 2014. Are agricultural quad bike loss-of-control events driven by unrealistic optimism? *Safety Science* 66: 54–60.

Curran, P., West, S., & Finch, J. 1996. The robustness of test statistics to nonnormality and specification error in confirmatory factor analysis. *Psychological Methods* 1: 16–29.

Drane, C., Modecki, K., & Barber, B. 2017. Disentangling development of sensation seeking, risky peer affiliation, and binge drinking in adolescent sport. *Addictive Behaviors* 66: 60–65.

Glanz, K., Rimer, B., & Lewis, F. (Eds.). 2002. *Health behavior and health education* (3rd ed.). San Francisco: Jossey-Bass.

Griffin, M.A. & Neal, A. 2000. Perceptions of safety at work: A framework for linking safety climate to safety performance, knowledge, and motivation. *Journal of Occupational Health Psychology* 5 (3): 347–358.

Hoyle, R., Stephenson, M., Palmgreen, P., Puzzles, L., & Donohew, L.R. 2002. Reliability and validity of a brief measure of sensation seeking. *Personality and Individual Differences* 32: 401–414.

Kunadharaju, K., Smith, T., DeJoy, D. 2011. Line-of-duty deaths among U.S. firefighters: An analysis of fatality investigations. *Accident Analysis and Prevention* 43(3): 1171–1180.

LeRoy, Z. 2005. *Personality trait and cognitive ability correlates of unsafe behaviours*. [Master Thesis]. The University of British Columbia.

Mearns, K., Hope, L., Ford, M., & Tetrick, L. 2010. Investment in workforce health: exploring the implications for workforce safety climate and commitment. *Accident and Analysis Prevention* 42(5): 1445–54.

Moen, B.-E. 2007. Determinants of safety priorities in transport – The effect of personality, worry, optimism, attitudes and willingness to pay. *Safety Science* 45: 848–863.

Neal, A., & Griffin, M. 2006. A study of the lagged relationships among safety climate, safety motivation, safety behavior, and accidents at the individual and group levels. *Journal of Applied Psychology* 91(4): 946–953.

Neal, A., & Griffin, M. 2006. A study of the lagged relationships among safety climate, safety motivation, safety behavior, and accidents at the individual and group levels. *Journal of Applied Psychology* 91(4): 946–953.

Podsakoff, P., MacKenzie, S., Lee, J., & Podsakoff, N. 2003. Common method biases in behavioral research: A critical review of the literature and recommended remedies. *Journal of Applied Psychology* 88: 897–903.

Rahemi, Z., Ajorpaz, N., Esfahani, M., & Aghajani, M. 2017. Sensation-seeking and factors related to dangerous driving behaviors among Iranian drivers. *Personality and Individual Differences* 116: 314–318.

Sabey, B., & Taylor, H. 1980. *The known risks we run: The highway* (TRRL Supplementary Report 567). Crowthorn, Berkshire, UK: Transport and Road Research Laboratory.

Salminen, S., Klen, T., & Ojanen, K. 1999. Risk taking and accident frequency among Finnish forestry workers. *Safety Science* 33(3): 143–153.

Sousa, C., Gonçalves, G., Sousa, A., & Pinto, E. 2017. Are the Portuguese sensation seeking? Psychometric properties of the Brief Sensation Seeking Scale in a Portuguese sample. *Journal of Personality Assessment.* (submitted).

Taylor, W.D. & Snyder, L.A. 2017. The influence of risk perception on safety: A laboratory study. *Safety Science* 95: 116–124.

United States Fire Administration 2009. *Firefighter Fatalities in the United States in 2008*. United States Fire Administration, Federal Emergency Management, Agency, Emmitsburg, MD.

Whissell, R., & Bigelow, B. 2003. The Speeding Attitude Scale and the Role of Sensation Seeking in Profiling Young Drivers at Risk. *Risk Analysis* 23(4): 811–820.

Wishart, D., Somoray, K., Rowland, B. 2017. Reducing Reversing Vehicle Incidents in Australian Fleet Settings—A Case Study. In Stanton N., Landry S., Di Bucchianico G., Vallicelli A. (Eds), *Advances in Human Aspects of Transportation. Advances in Intelligent Systems and Computing* (vol. 484). USA, Springer.

Zakletskaia, L., Mundt, M., Balousek, S., Wilson, E., & Fleming, M. 2009. Alcohol-impaired driving behavior and sensation-seeking disposition in a college population receiving routine care at campus health services centers. *Accident Analysis and Prevention* 41(3): 380–386.

Zuckerman, M. 1994. *Behavioral expressions and biosocial bases of sensation seeking*. New York: Cambridge University Press.

Zuckerman, M. 2008. *Sensation Seeking and Risky Behavior*. Washington, DC: American Psychological Association.

Work ability, ageing and musculoskeletal symptoms among Portuguese municipal workers

C.A. Ribeiro
Ergonomics Laboratory, Faculdade de Motricidade Humana, Universidade de Lisboa, Portugal

T.P. Cotrim
Ergonomics Laboratory, Faculdade de Motricidade Humana, Universidade de Lisboa, Portugal
CIAUD, Faculdade de Arquitectura, Universidade de Lisboa, Portugal

V. Reis, M.J. Guerreiro, S.M. Candeias, A.S. Janicas & M. Costa
Department of Health and Safety, Municipality of Sintra, Portugal

ABSTRACT: The demographic ageing has a huge impact in working lives. The legal retirement age is increasing, leading to extending the working life as well as the exposure to occupational risks. This prospective study with municipal workers was conducted aiming at characterizing the associations between work ability, self-reported symptomatology and age. In 2015, the sample consisted of 885 workers and in 2017, of 1167 workers. A self-administered questionnaire was used including the Work Ability Index (WAI) and an adaptation of the Nordic questionnaire. The WAI showed a lower average in 2017. The age had a negative correlation with work ability in both years. The prevalence of low-back symptoms was high for 2015 (45,2%) and 2017 (49,2%) and work ability changes are influenced by musculoskeletal symptoms. The results evidence the need to promote health and to prevent work related disorders aiming at maintaining the work ability over the years.

Keywords: WAI; Municipal Workers; Ageing; Musculoskeletal symptoms; Occupational Health

1 INTRODUCTION

The trend of demographic ageing has been evidenced for several decades in the world population, bringing with it several questions. The workforce in Europe will reach unprecedented age levels: projections indicate that by 2030, workers aged 55–64 will account for 30% or more of the total workforce in many countries. The official retirement age is increasing in many Member States and therefore longer working life as well as occupational risk exposure (EU-OSHA, 2016).

Concomitantly, ageing is associated with a decrease in physical abilities and an increase in musculoskeletal disorders. These changes may reduce an individual's ability to perform their work at the same level of productivity (Oakman, et al., 2016). Several studies showed that the physical demands of work activity are the main determinants of poor ability to work (Lindberg, 2006; Savinainen, at al., 2004).

The reduction of work ability over the years occurs due to the imbalance between work demands and individual resources. The rhythm of biological ageing is frequently not followed by adequate changes in work activity (Ilmarinen, 2012; Bonsdorff et al., 2011). Therefore, it is justified a study that considers the analysis of work ability and self-reported musculoskeletal symptoms, aiming at contributing to the promotion of the permanence of workers in their workplace, healthy, with quality of life and good performance.

In order to characterize the evolution of work ability among municipal workers from 2015 to 2019, a longitudinal study was organized in the Portuguese municipality of Sintra. The present paper aims at characterizing the differences on work ability of the municipal workers related to self-reported musculoskeletal symptoms, comparing the results from 2015 and 2017.

2 METHODOLOGY

2.1 Methods

The analysis was done using the Portuguese version of the Work Ability Index (Silva, et al., 2011) and an adaptation of the Nordic questionnaire.

The Work Ability Index is a self-administered instrument developed in Finland, which evaluates the worker's perception on how well they can perform work in function of work demands, health and physical and mental resources. Seven dimensions compose the WAI, with a final score ranging from 7 to 49 (better score). The results are classified as poor, moderate, good or excellent (Ilmarinen, 2006; Silva, et al., 2011).

The Nordic Musculoskeletal Questionnaire (NMQ) was developed with the purpose of standardizing the measurement of musculoskeletal symptoms and, therefore, facilitating the comparison of the results between studies (Pinheiro, et al., 2002). A general form comprising all anatomical areas was included in the present study. The instrument consists of binary choices regarding the occurrence of symptoms in the last 12 months.

2.2 Population and sample

The population of this study included 1,667 municipal workers. In 2015 the response rate was of 54.7%, with 885 participants. In 2017, the follow up comprised 1167 participants, corresponding to a response rate of 70%. The inclusion criteria was to have a valid WAI, so 20 questionnaires were excluded from the analysis in the first phase and 14 questionnaires in the follow up.

2.3 Procedures

According to the objectives of the study and the literature review, the evaluation instruments used were integrated in a single questionnaire.

The questionnaire was self-administered to municipal workers during the year 2015 and the follow up was done in 2017. Workers were asked about their interest in participating in this study and those who agreed to volunteer signed the Informed Consent Form.

The confidence level assumed for the statistical analysis was 95%.

3 RESULTS

3.1 Socio-demographic characteristics

The sample showed an average age of 46.9 years (sd = 8.3) in 2015, and 48.4 years (sd = 8,7)

Table 1. Socio-demographic characteristics.

		2015		2017	
		N	%	N	%
Gender	Woman	548	65.6	689	61.8
	Man	287	34.4	425	38.2
Civil status	Single	144	16.9	216	19.0
	Married	540	63.3	709	62.5
	Divorced/ Separated/ widowed	169	19.8	209	18.4
Professional category	Operational Assistant	287	33.7	370	32.7
	Technical Assistant	323	37.9	420	37.1
	White Collars	229	26.9	330	29.2
	Municipal police	13	1.5	11	1.0
Qualifications	Elementary to Junior high school	242	28.2	314	27.9
	High school	324	37.8	411	36.5
	Graduated/ Post Graduated	291	34.0	402	35.7

in 2017. The mean age was higher in the follow up and the differences on the mean values of age between the two moments were statistically significant, according to a T-Student test (P ≤ 0,001).

In both years there were a higher percentage of female workers, married, with a High School level of education, and with the professional category of Technical Assistants (Table 1).

3.2 Self-reported musculoskeletal symptoms

The highest frequency of self-referred musculoskeletal symptoms in the last 12 months was reported on Low-back (45.2%), Neck (37.9%), Upper-back (33%) and Shoulders (32%) in 2015. In 2017, Low-back (49.2%), Neck (40.6%), Shoulders (37.8%) and Upper-back (36.2%) were the regions with a higher prevalence (Table 2).

Only for the shoulders region were obtained significant differences in the prevalence of complaints between the two moments of data collection, according to the T-Student test (p = 0.007). The frequency of shoulders' self-reported symptoms was higher in 2017 (Table 2).

3.3 Work ability index

The WAI showed an average of 40.7 (±5.1) points, in 2015, and 40.2 (±5.1), in 2017, corresponding to a "good" work ability classification in both years. The differences in the WAI scores between 2015 and 2017 were statistically significant, according

Table 2. Self-reported musculoskeletal symptoms.

Symptoms in the last 12 months		2015 N	2015 %	2017 N	2017 %
Neck	No	527	62.1	686	59.4
	Yes	321	37.9	469	40.6
Upper back	No	568	67.0	737	63.8
	Yes	280	33.0	418	36.2
Low back	No	465	54.8	587	50.8
	Yes	383	45.2	568	49.2
Shoulders	No	577	68.0	718	62.2
	Yes	271	32.0	437	37.8
Elbows	No	755	89.0	997	86.3
	Yes	93	11.0	158	13.7
Wrists and hands	No	671	79.1	872	75.5
	Yes	177	20.9	283	24.5
Hips	No	741	87.4	1004	86.9
	Yes	107	12.6	151	13.1
Knees	No	654	77.1	879	76.1
	Yes	194	22.9	276	23.9
Ankles and Feet	No	704	83.0	957	82.9
	Yes	144	17.0	198	17.1

Table 3. WAI scores.

Year	N	Mean	Mín.	Máx.	S.D.	P value
2015	885	40,7	14	49	5,1	0,016
2017	1167	40,2	7	49	5,1	

Table 4. Distribution of WAI by categories.

	2015 N	2015 N%	2017 N	2017 N%
Poor	14	1,6%	19	1,6%
Moderate	140	16,2%	240	20,8%
Good	417	48,2%	554	48,0%
Excellent	294	34,0%	340	29,5%

to the T-student test, indicating a decrease in the ability to work during the last two years (Table 3).

When analyzing the distribution of the WAI results by category and by year, the mains differences appear for the moderate and excellent levels, with a decrease of those with excellent work ability from 2015 to 2017 and an increase among those with moderate work ability (Table 4).

Age correlated negatively with WAI, in both moments of the research (2015: r = –0,16; p ≤ 0,000; 2017: r = –0.18; p ≤ 0.000), meaning that when age increases, the ability to work decreases, according to the R-Pearson test.

The Work Ability Index is also influenced by the presence of musculoskeletal symptoms, with those reporting complaints, showing a lower WAI in both years, for all body regions. The differences between those with and without musculoskeletal symptoms were statistically significant for both years and all body regions, according to the T-student test (p ≤ 0.001 for all the analysis) (Table 5).

4 DISCUSSION

The average age of our samples was higher in the follow up, what corresponds to a stable population with low level of workers' turnover and corresponds to the ageing process of the working population. These values were higher than the mean age found in other national studies (Cotrim et al., 2017; Silva et al., 2011). These results can be explained by the changes in Portuguese legislation concerning the extension of the retirement age in the public administration up to the age of 66, thus seeking to concretize measures to address the risks of demographic ageing (DL 187/2007).

The average results of the WAI had a slight decline during the research period and the differences in the WAI mean values between 2015 and 2017 were statistically significant, indicating a decrease in the ability to work during the last two years. Despite the fact that the WAI average remained "Good" in the two moments, classification also found in other Portuguese studies (Cotrim et al., 2017; Silva, et al., 2011), there was an increase of the percentage of the unsatisfactory category from 2015 to 2017.

With respect to the ageing process, a negative correlation was found between age and the WAI what is in line with several national and international studies (Cotrim et al., 2017; Pensola et al., 2016; Bonsdorff et al., 2011).

The ability to work is strongly affected by a set of different interactions between biological ageing, health, lifestyle, skills, values, attitudes, motivation, organizational context, social and work activity requirements (Ilmarinen, 2012). The cumulative exposure to physical and psychosocial risk factors in the workplaces lead frequently to musculoskeletal complaints. But, at the same time, the ageing process is, also, associated with decreased physical resources and increased musculoskeletal disorders, and this may reduce individual's ability to perform the work tasks at the same level of productivity (Nabe-Nielsen et al., 2014; Oakman et al., 2016).

In our study, the presence of musculoskeletal symptoms in the last 12 months determined a lower mean in the WAI scores for both years. Other studies indicate that pain affecting back region is an important source of reduced work capacity. And, likewise,

Table 5. Differences on WAI depending on self-reported musculoskeletal symptoms in the last 12 months.

		2015					2017				
		N	Mean	Min.	Máx.	S.D.	N	Mean	Min.	Máx.	S.D.
Neck*	No	523	41.4	22	49	4.8	680	41.4	24	49	4.7
	Yes	312	39.6	14	49	5.5	465	38.4	7	49	5.3
Upper-back*	No	561	41.4	14	49	4.8	728	41.1	23	49	4.8
	Yes	274	39.3	20	49	5.4	417	38.5	7	49	5.3
Low-back*	No	461	41.9	26	49	4.5	582	41.6	26	49	4.5
	Yes	374	39.3	14	49	5.5	563	38.7	7	49	5.3
Shoulders*	No	567	41.4	23	49	4.8	712	41.2	23	49	4.7
	Yes	268	39.4	14	49	5.5	433	38.6	7	49	5.4
Elbows*	No	743	41.0	14	49	5.0	988	40.5	7	49	5.0
	Yes	92	38.6	20	48	5.8	157	37.9	24	49	5.3
Wirsts and hands*	No	660	41.2	14	49	5.1	863	41.0	7	49	4.8
	Yes	175	39.2	20	49	5.2	282	37.7	20	49	5.4
Hips*	No	730	41.2	23	49	4.8	994	40.6	23	49	4.8
	Yes	105	37.6	14	49	6.0	151	37.1	7	49	6.2
Knees*	No	645	41.1	22	49	4.9	871	40.8	25	49	4.7
	Yes	190	39.6	14	49	5.6	274	38.0	7	49	5.8
Ankles and Feet*	No	695	41.2	14	49	5.0	949	40.7	7	49	4.9
	Yes	140	38.6	20	49	5.3	196	37.6	20	48	5.4

*$p \leq 0.001$ for both years.

workers with chronic, non-specific musculoskeletal symptoms often have an unsatisfactory work ability index (Bethge et al., 2013; Vries et al., 2013).

5 CONCLUSION

Our study found that musculoskeletal symptoms influence the ability to work. These findings are in line with the need to keep workers healthy along their working lives in order to avoid costs for the organizations and early work retirement, and to keep productivity. There is a need to define policies for health promotion and disease prevention, aiming at maintaining work ability over the years.

REFERENCES

Bethge, M., Gutenbrunner, C., & Neuderth, S. (2013). Work ability index predicts application for disability pension after work-related medical rehabilitation for chronic back pain. Archives of Physical Medicine and Rehabilitation, 94(11), 2262–2268. https://doi.org/10.1016/j.apmr.2013.05.003.

Bonsdorff, M.B.M.E., Kokko, K., Seitsamo, J., von Bonsdorff, M.B.M.E., Nygård, C.H., Ilmarinen, J., … Rantanen, T. (2011). Work strain in midlife and 28-year work ability trajectories. Scandinavian Journal of Work, Environment and Health, 37(6), 455–463. https://doi.org/10.5271/sjweh.3177.

Cotrim, T., Carvalhais, J., Neto, C., Teles, J., Noriega, P., & Rebelo, F. (2017). Determinants of sleepiness at work among railway control workers. Applied Ergonomics, 58, 293–300. https://doi.org/10.1016/j.apergo.2016.07.006.

Decreto-Lei Nº187/2007. Portugal.

EU-OSHA. (2016). Healthy Workplaces for All Ages Promoting a sustainable working life. Retrieved from http://www.healthy-workplaces.eu/.

Gould, R., Ilmarinen, J., Järvisalo, J., & Koskinen, S. (2000). Dimensions of Work Ability. https://doi.org/10.1108/02656710210415668.

Ilmarinen, J. (2006). The Work Ability Index (WAI). Occupational Medicine, 57(2), 160–160. https://doi.org/10.1093/occmed/kqm008.

Ilmarinen, J. (2012). Promover o envelhecimento ativo no local de trabalho. Agencia Europeia de Saúde E Segurança No Trabalho. Bruxelas: OSHA, 1–9.

Kumashiro, M., Yamamoto, K., & Shirane, K. (2004). WAI and Job Stress, Five Years of Follow-up Research. Aging and Work, 1–6.

Lindberg, P. (2006). Promoting excellent work ability and preventing poor work ability: the same determinants? Results from the Swedish HAKuL study. Occupational and Environmental Medicine, 63(2), 113–120. https://doi.org/10.1136/oem.2005.022129.

Nabe-Nielsen, K., Thielen, K., Nygaard, E., Thorsen, S.V., & Diderichsen, F. (2014). Demand-specific work ability, poor health and working conditions in middle-aged full-time employees. Applied Ergonomics, 45(4), 1174–1180. https://doi.org/10.1016/j.apergo.2014.02.007.

Oakman, J., Neupane, S., & Nygård, C.H. (2016). Does age matter in predicting musculoskeletal disorder risk? An analysis of workplace predictors over 4 years. International Archives of Occupational and Environmental Health, 89(7), 1127–1136. https://doi.org/10.1007/s00420-016-1149-z.

Pensola, T., Haukka, E., Kaila-Kangas, L., Neupane, S., & Leino-Arjas, P. (2016). Good work ability despite multisite musculoskeletal pain? A study among occupationally active Finns. *Scandinavian Journal of Public Health*, *44*(3), 300–310. https://doi.org/10.1177/1403494815617087.

Pinheiro, F.A., Tróccoli, B.T., & Carvalho, C.V. de. (2002). Validação do Questionário Nórdico de Sintomas Osteomusculares como medida de morbidade. *Revista de Saúde Pública*, *36*(3), 307–312. https://doi.org/10.1590/S0034-89102002000300008.

Savinainen, M., Nygård, C.-H., & Arola, H. (2004). Physical capacity and work ability among middle-aged women in physically demanding work—a 10-year follow-up study. *Advances in Physiotherapy*, *6*(3), 110–121. https://doi.org/10.1080/14038190310017309.

Silva, C., Pereira, A., Martins, A., Amaral, V., Vasconcelos, G., Rodrigues, V., et al. (2011). Associations between Work Ability Index and demographic characteristics in Portuguese workers (pp. 80–88). Tampere University Press: Proceedings of the 4th Symposium on Work Ability.

Silva, C., Rodrigues, V., Pereira, A., Cotrim, T., Silvério, J., Rodrigues, P., Sousa, C. (2011). Índice de Capacidade para o Trabalho-Portugal e Países Africanos de Língua Oficial Portuguesa. 2nd Ed., Portugal.

Vries, H.J., Reneman, M.F., Groothoff, J.W., Geertzen, J.H.B., & Brouwer, S. (2013). Self-reported work ability and work performance in workers with chronic nonspecific musculoskeletal pain. *Journal of Occupational Rehabilitation*, *23*(1), 1–10. https://doi.org/10.1007/s10926-012-9373-1.

Psychological health and wellbeing in intralogistics workplaces—an empirical analysis

M. Hartwig & E. Mamrot
Federal Institute for Occupational Safety and Health (BAuA), Unit Human Factors, Ergonomics, Dortmund, Germany

ABSTRACT: Flexible and time efficient commissioning, sorting and packing of materials provide the access to individualized products goods worldwide. At the same time, these demands and the ongoing digitalization of work can result in working conditions that are not ideal, especially from a psychological work science point of view. To get an overview of psychologic health relevant aspects in modern intralogistics jobs, this paper analyses the results from the working time survey from the German Institute for Occupational Safety and Health regarding intralogistics employees. In order to identify which psychological demands in terms of working conditions lead to high job satisfaction as well as better health status, descriptive analyses and bivariate correlations were conducted, indicating significant differences between intralogistics and other workplaces. Consistent correlations between working conditions and indicators for mental health also highlight the importance for psychological research and interventions to improve working conditions and in turn mental health and wellbeing of employees in the business.

Keywords: psychological demand; intralogistics; psychological stress and strain; job satisfaction; health; work design

1 INTRODUCTION

Considering the progressive social change, e-commerce and increasing economic globalisation, the consumer behaviour in society is changing. The increasing desire for individualised products not only put pressure on production and logistics, but also provokes the requirement of flexible production and logistics chains. Therefore, especially the logistics sector is facing major challenges. This transformation process is also accompanied by a new level of complexity (ten Hompel & Kerner, 2015).

In addition, the digitalisation of the work place also increases, not only in logistics sector but also in other industries like media industry, retail and finance sector or handicraft businesses (BMAS, 2016). These transformation processes ultimately result in the emergence of complete new products, services and work processes, which in turn create complete new value chains and value-added networks. Especially logistics companies lose their autonomy due to the increasing networking of production and logistics themselves (BMBF, 2016).

These developments not influence society as a whole, but also can have concrete and tangible consequences for workplaces and employees. The increased demands in terms of flexibility can for example result in decreased decision latitude for the worker. Extremely optimised working processes in terms of product efficiency can lead to monotonous work tasks. Finally, progressing digitalisation and automated information flow can decrease social interaction.

In contrast, new technologies and new forms of work organisation offer also great potential for new and innovative digital solutions that focus the needs of employees in companies to increase facets like decision latitude, learning at work and qualification at the job (BMBF, 2016).

All these factors are relevant parameters when it comes to healthy work from a psychological perspective. Therefore, the current developments and changing demands in the intralogistics sector can have an impact on psychological health and wellbeing of workers in this business.

This article aims at giving an overview of psychologic relevant aspects in modern intralogistics jobs. The self-assessments of workers in the intralogistics regarding different key aspects of working conditions are compared to those of employees in other industries. In addition, correlations between the working conditions and indicators for wellbeing as well as physical and mental health are analysed.

2 METHOD

2.1 Derivation of sample

To identify relevant psychological stresses in intralogistics, the data set of BAuA (Federal Institute for Occupational Safety and Health) working time survey from year 2015 was used. The telephone survey was conducted between May and October 2015 with N = 20.030 individuals aged 15 years or older.

These individuals were a random sample of German employees, stratified regarding key demographic variables like age and gender, qualifying as a representative sample for the German working population. In a first step, all employees within the sample working in the field of intralogistics were identified.

To do this, the international standard classification of occupations from year 2008 (ISCO-08) was used. This classification is internationally comparable and an accepted standard even for international labor statistics (see ILO, 2012). The ISCO-08 tries to classify the occupational activities of the working population by grouping tasks and obligations that are mandatory and have to be done.

The following two task unit groups were selected to represent employees working in the operational intralogistics:

Stock Clerks (n = 185)
Freight Handlers (n = 130)

These two unit groups carry out similar tasks and obligations usually including receiving, storing and issuing goods or other production materials. Therefore, these work groups embody those As they are very similar to each other with fluid transition in operational practice, the groups were analyzed together. The groups added up to a subsample of 315 employees. The subsample consists of 214 men (67.9%) and 101 woman (32.1%). The average age is approximately 45 years (M = 45.62) and ranged from 16 to 68 years. Almost all of them describe their occupation as worker (52.6%) or employee (46.2%). Only a small part of respondents describe their occupation as self-employed person, freelancer or do not know (1.2%).

2.2 Selected variables

After identification of employees working in the intralogistics, this subsample was compared to the remaining sample of 19.715 employees not working in intralogistics. They were compared regarding (a) key aspects of psychologically relevant working conditions and (b) satisfaction and health aspects. This paper reports about a section of these variables that are either of high relevance to the intralogistics sector or show high correlations between working conditions and health aspects.

The variables classified as psychologically relevant working conditions are divided in three superordinate categories: *demands, social support and leadership* and *general conditions*. Psychological health indicators include job satisfaction, general health status and emotional exhaustion.

The response options for variables classified in the categories *demands* and *social support and leadership* based on a four-point Likert scale. This scale offers a range of survey questions from 1 (never) to 4 (often). The questions concerning the topic *general conditions* were formed on the basis of a five-point Likert scale (1 = not at all, 5 = extremely).

For the rating of *job satisfaction* a four-point Likert scale from 1 (very satisfied) to 4 (dissatisfied) was used. General health was recorded with a five-point Likert scale from 1 (very good) to 5 (very poor). Survey questions concerning emotional exhausting were recorded with a five-point Likert scale from 1 (not at all) to 5 (extremely) as well. All variables fulfill the requirements for interval scales.

3 RESULTS

To gain insights in the existing psychological relevant working conditions in the intralogistics, three research questions were investigated:

1. In which working conditions do workplaces in the intralogistics differ from workplaces in other work sectors?
2. Do employees in the intralogistics differ from other employees regarding the psychological health indicators?
3. Which working conditions correlate with psychological health indicators of employees in the intralogistics?

In order to find out which independent variables relate to job satisfaction and mental as well as physical health of the employees, descriptive analyses and bivariate correlations were conducted.

The calculation of mean values and its comparison aimed the description of relevance of independent variables for intralogistics in comparison with other work sectors.

Figure 1 shows an exemplary selection of working demands that differ between the intralogistics and the remaining total sample. "Frequent repetition of the same work process" is reported considerably higher in the intralogistics group than in the total sample (m = 3.49 vs. 3.05, t = 8.85, $p < 0.01$). The self-report about "unchanging, similar tasks" shows a comparable pattern (m = 1.85 vs. 1.52, t = 6.29, $p < 0.01$). In contrast, the "confrontation

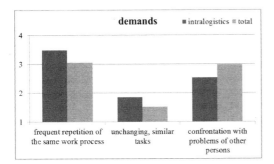

Figure 1. Mean self-assessments of working task demands in the intralogistics and in total.

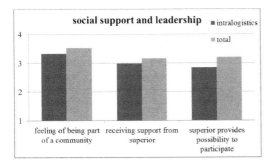

Figure 2. Mean self-assessments of social support and leadership aspects in the intralogistics and in total.

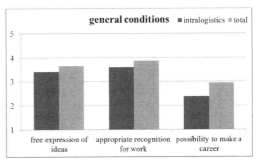

Figure 3. Mean self-assessments of general working condition aspects in the intralogistics and in total.

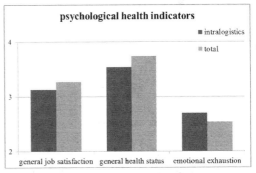

Figure 4. Mean self-assessments of psychological health indicators in the intralogistics and in total.

with problems of other persons" as an example of social working demands is reported significantly lower in the intralogistics than in the total sample (m = 2.53 vs. 2.99, t = 8.29, p < 0.01).

The results for working conditions regarding social support and leadership are shown in Figure 2. A variety of aspects are reported significantly lower in the intralogistics than in the total sample. This is true for "feeling of being part of a community" (m = 3.33 vs. 3.52, t = 3.36, p = 0.01), receiving support from superior" (m = 2.99 vs. 3.16, t = 2.85, p < 0.05) as well as "superior provides possibility to participate" (m = 2.85 vs. 3.20, t = 5.61, p < 0.01).

Figure 3 shows different aspects of general working conditions. The mean self-assessment of allowance of "free expression of ideas" at the workplace is significantly lower in the intralogistics group than in the total sample (m = 3.42 vs. 3.65, t = 3.51, p = 0.01). A similar pattern can be found for "appropriate recognition for work" (m = 3.60 vs. 3.86, t = 3.13, p < 0.05) and "possibility to make a career" (m = 2.38 vs. 2.94, t = 7.33, p < 0.01).

Regarding the second research question, indicators for psychological health of employees in the intralogistics were compared with those of general working population. Several aspects are reported significantly worse by intralogistics-employees compared to the total working population. These aspects include general job satisfaction (m = 3.13 vs. 3.27, t = 2.48, p < 0.05), general health status (m = 3.54 vs. 3.74, t = 3.70, p < 0.01) and emotional exhaustion (m = 2.70 vs. 2.54, t = 3.17, p < 0.05). Figure 4 gives an overview of the assessments.

To answer the third research question, correlations between psychological stress and strain were determined by calculating correlation coefficients according to Pearson. All correlations between the working conditions and the indicators for psychological health are shown in Table 1.

The results show mostly significant, low to medium correlations over the different facets. General job satisfaction as an indicator for emotional wellbeing is significantly linked with all discussed working condition aspects. Highest correlations can be found to the self-reported free expression of ideas within the working environment and the subjective recognition for work from the working environment (r = ,359, p < 0.01; r = ,366, p > 0.01, respectively). Least but still significant negative correlations exist with unchanging, similar tasks

Table 1. Correlations between working condition aspects and indicators for psychological health.

	General job satisfaction	General health status	Emotional exhaustion
Repetition of work process	-,203**	-,150**	,113
Unchanging, similar tasks	-,172**	-,141**	,080
Confrontation with problems	-,172**	-,166**	,171**
Part of a community	,240**	,171**	-,147**
Support from superior	,184**	,136*	-,186**
Possibility to participate	,283**	,095	-,103
Free expression of ideas	,359**	,115**	-,187**
Recognition for work	,366**	,307**	-,189**
Possibility to make a career	,263**	,219**	-,105

* = $p < 0.05$; ** = $p < 0.01$.

and confrontation with other problems of other people ($r = -,172$, $p < 0.05$ for both variables). General health status also shows significant correlations to nearly all working condition variables. Interestingly, the self-reported possibility to participate is the only aspect that shows no significant connection to general health status. Recognition for work shows in contrast the highest correlation ($r = ,366$, $p < 0.01$), just as it was seen regarding general job satisfaction. The same is also true concerning emotional exhaustion as a key indicator for burnout symptomatic. Recognition for work shows the numerical highest connection with $r = -,189$, $p < 0.01$. In contrast, there are several working condition that show no significant correlation with emotional exhaustion, like unchanging working tasks or possibilities regarding participation and career.

In sum, correlations between working conditions and general job satisfaction tend to be somewhat stronger than those with general health status or emotional exhaustion. Across all psychological health indicators, the self-assessed recognition for own work has the highest correlations of all working facets.

4 DISCUSSION AND CONCLUSION

The article deals with the question of the psychological working conditions in intralogistics and their correlations with indicators of mental health, based on a representative survey of 20,000 employees in total. It can be seen that different facets of the work in intralogistics are on average less favorable than in the working world in general. At the same time, it should be noted that the differences in the mean values between the two groups are significant, but rarely massive differences.

In the area of work requirements, an increased monotony in the task design can be noted. This applies to frequent, short-term repetition of identical tasks as well as to unchanging tasks over a longer period of time. Monotonous working tasks are discussed as a risk factor for wellbeing for several decades by now (for example Cox, 1985; Davis, Shackleton, & Parasuraman, 1983). There is a more differentiated situation concerning the social aspects of the work. Overall, the work in intralogistics is accompanied by a comparatively low degree of social interaction. However, on the one hand, the comparatively small degree of social interaction in intralogistics is associated with less positive aspects, such as the shared feeling of community. On the other hand, potential stress factors such as the forced employment with problems of others are also limited to a lesser extent.

In addition to these important aspects of the working task, the work in intralogistics are characterized by comparatively unfavorable various general conditions, such as a lesser assessment of appropriate recognition for work or less possibility to make a career.

When considering the correlations between working conditions and the health indicators, one or more significant correlations with the different health indicators in the expected direction are found for all working conditions discussed in this study. Looking at the strength of the correlations, it is interesting that not the aspects of the working task themselves, but the general framework conditions of the work which have the highest interrelations. This is especially true for the self-assessed possibility of expressing one's own ideas or appreciation for one's own work.

From the perspective of work design, this is an important indication for possible interventions. While the work task itself is often very difficult to change, for example the typical constraints in terms of time and monotony, social frameworks may be at least just as promising foci of intervention, which might in turn affect the mental health of persons in intralogistics.

Overall, the work in intralogistics represents a field characterized by a combination of unfavorable working conditions. The fact that these conditions show a wide range of consistent correlations with both work satisfaction and physical and mental health suggest placing the intralogistics more intensively in the focus of work science and the improvement of working conditions. This is especially true for the psychological perspective. While there are studies dealing with the physical risks at work (for example Kelterborn et al, 2013) and general impacts of digitalization (for example Williams Jimenez, 2016), there is virtually no research to our knowledge about analyzing and optimizing logistic work from an industrial psychology point of view. This is not only inadequate because of the mentioned sub-par working conditions, but also because of the huge number of employees in the branch. In Germany alone, the number of person is estimated on our randomized sample at about 800.000–900.000 persons.

ACKNOWLEDGEMENT

This research and development project "Preventive principles and methods of aging-e and market-oriented work system design in intralogistics (Previlog)" is funded by the German Federal Ministry of Education and Research (BMBF) within the program "Innovations for Tomorrow's Production, Services, and Work" and managed by the Project Management Agency Karlsruhe (PTKA).

We also thank two anonymous reviewer for their comments to improve the paper.

REFERENCES

BMAS (2016). Weißbuch. Arbeiten 4.0. URL: http://www.bmas.de/SharedDocs/Downloads/DE/PDF-Publikationen/a883-weissbuch.pdf?__blob=publicationFile&v=9, Stand: 15.09.2017.

BMBF (2016). Zukunft der Arbeit. Innovationen für die Arbeit von morgen. https://www.bmbf.de/pub/Zukunft_der_Arbeit.pdf, Stand: 15.09.2017.

Cox, T. (1985). Repetitive work: Occupational stress and health. In C. L. Cooper & M. J. Smith (Eds.), Job stress and blue collar work (pp. 85–111). Chichester, England: Wiley.

Davis, D. R., Shackleton, V. J., & Parasuraman, R. (1983). Monotony and boredom. In R. Hockey (Ed.), Stress and fatigue in human performance (pp. 1–32). Chichester, England: Wiley.

ILO (2012): https://www.cbs.nl/NR/rdonlyres/B30EE525-22DB-4C1B-B8D5-6D12934AF00A/0/isco08.pdf, 15.11.2017.

http://www.ilo.org/public/english/bureau/stat/isco/docs/publication08.pdf, 15.11.2017.

Kelterborn, M; Koch, M & Günthner, W.A. (2013) Physische Belastung in der Produktionslogistik.

Ten Hompel, M. & Kerner, S. (2015). Logistik 4.0. Die Vision vom Internet der autonomen Dinge. Heidelberg: Springer.

Williams Jimenez, I. (2016). Digitalisation and its impact on psychosocial risks regulation. Fifth International Conference on Precarious Work and Vulnerable Workers. London, United Kingdom.

ZWF Zeitschrift für wirtschaftlichen Fabrikbetrieb 2013 108, 11, 846–849.

Detailed evaluation of human body cooling techniques through thermo-physiological modelling

António M. Raimundo & Divo A. Quintela
ADAI-LAETA, Department of Mechanical Engineering, University of Coimbra, Coimbra, Portugal

A. Virgílio & M. Oliveira
Department of Mechanical Engineering, Coimbra Institute of Engineering, Polytechnic Institute of Coimbra, Coimbra, Portugal

ABSTRACT: This study aims to test and validate a software—the *HuTheReg* (Human Thermal Regulation) program—that simulates the human body thermophysiological response during intense activities. Emphasis will be given to the identification and analysis of the effectiveness of cooling techniques able to mitigate the risk of heat stress by hyperthermia. Different activity levels, exposure periods, types of clothing and alternative techniques for recovery of body temperatures are considered. The validation of the software is performed by comparing its results with experimental data from the literature. To assess the compliance level of the *HuTheReg* program statistical analysis was performed. Within the range of situations analyzed in the present paper, no limitations were found to the applicability of the software, despite differences that were acknowledge. The results obtained in this study demonstrate the usefulness and applicability of the software to predict the thermophysiological behavior of firefighters.

1 INTRODUCTION

There are several activities that expose the human body to very high levels of heat (industry, military, athletes, firefighters, etc.). This exposure may lead to increases in deep body temperature to values higher than 39°C which may have serious consequences or be even fatal (Carter et al., 1999; Raimundo and Figueiredo, 2009). Temperatures above 39°C represent acute hyperthermic stress. When attained, the person physical and mental performance substantially decreases while the risk of developing heat-related illnesses increases (Mündel et al. 2006, Lopez et al. 2008, Raimundo and Figueiredo, 2009).

The most common hyperthermic disorders reported in the literature are muscle cramping, heat exhaustion (collapse during or after exercise), introversion (violent sweating, loss of judgment, amnesia, delusions, etc.) and heat stroke (fainting, cessation of sweating, central nervous system alteration, etc.) (Raimundo & Figueiredo 2009). Thus, experience recommends the use of cooling strategies to mitigate the rise of body temperature (Carter et al. 1999, Selkirk et al. 2004, Giesbrecht et al. 2007, Chou et al. 2008, Lopez et al. 2008, Barr et al. 2009, Hostler et al. 2010, Colburn et al. 2011).

To establish the guidelines for safety criteria, it is important to have knowledge of the physiological response of the human body before the effects of such extreme environmental conditions prevail. Among the various methods available for this purpose, the use of simulation programs of the thermophysiological behavior of the human body appears to be the most appropriate either when the performance of such activities or the mitigation of the risk of heat-related disorders are foreseen.

The present work addresses this matter with two main goals in mind: (i) evaluation, testing and validation of the ability of a software, the *HuTheReg* program, to predict in detail the human body thermophysiological behavior during passive and active body cooling processes; and (ii) the assessment of the effectiveness of some human body cooling techniques to mitigate the risk of heat stress.

2 METHODS

In this study use was made of the *HuTheReg* software (Raimundo et al. 2012) to predict the human thermophysiological response. This program is composed by several modules, namely for the calculation of the heat and water transport through clothing, the heat and mass exchange between the external surface of clothing (or skin) and the environment, the start and evolution of skin injuries (pain and burn), the detection of specific incidents within the individual and the human body thermophysiological response. Due to its interdependency, all modules run iteratively in each time step until a specific convergence criterion is reached.

The module for simulation of human's thermophysiological response is based on the Stolwijk (1971) thermoregulation model of 25 nodes (6 segments of 4 layers each and the central blood compartment), improved with knowledge of the literature (Fiala et al. 1999; Raimundo and Figueiredo, 2009). This enhanced 111-node model assumes the human body composed by 22 segments (face, scalp, neck, chest, abdomen, upper back, lower back, pelvis, left shoulder, right shoulder, left arm, right arm, left forearm, right forearm, left hand, right hand, left thigh, right thigh, left leg, right leg, left foot and right foot), each one formed by 5 layers (core, muscle, fat, skin and clothing) and the central blood compartment (the 111th node).

The model was implemented for an average man with 74.43 kg and 1.72 m (1.869 m^2 of skin). For an individual with other anthropometric data, the appropriated coefficients are proportionally changed as a function of its body weight and skin area. Each run can simulate up to 20 consecutive scenarios (phases), each one representing different conditions in terms of: posture, orientation, activity, intake of food/drinks, clothing, thermal environment characteristics and impinging of thermal radiation from 8 directions (cardinal, top and bottom).

The loss of heat by respiration is supposed to occur across the elements of the pulmonary tract. The repartition coefficients proposed by Fiala et al. (1999) were considered.

The determination of the outer surface temperature of the clothing (or of the skin of nude segments) involves the model of simulation of the human thermophysiological response and simultaneously the heat and mass balances of the garment both considering the flow and the storage of heat and water in the clothing layer (Raimundo et al. 2012). The global algorithm is applied to each specific human body segment, but always considering its influence and interdependence with the global thermal state of the body. Thus, individual values of clothing insulation (I_{cl}) and of vapor permeability efficiency (i_{vp}) must be specified for each of the 22 human body segments. Each section is either completely clothed or nude. To consider the reduction of insulation due to body movements, the I_{cl} value at each human body segment is adjusted according to the person activity using the relations proposed by Oliveira et al. (2011).

The convective heat and mass transfer phenomena are predicted using the empirical relations proposed by Quintela et al. (2004) for natural convection and by Havenith et al. (2002) for forced convection. The exchange of heat by radiation between the external surfaces of the segments of the human body and between them and the surroundings, is determined by a set of expressions established by Raimundo et al. (2004).

With the *HuTheReg* software a significant number of data can be obtained, both for the human body as a whole and for each of the segments, namely: (i) core, muscle, fat, skin and clothing temperatures; (ii) metabolism, heat stored and flux-rates of heat, of sweat, of water and of work; (iii) thermal comfort evaluation and indexes; (iv) detection of probable appearing of heat-related disorders within the person; (v) skin pain, burn areas and corresponding degree; and (vi) a wide range of other thermo-physiological parameters.

The evaluation, test and validation of the software ability to predict in detail the human body thermo-physiological behavior is performed by comparing the program projections with experimental results from scientific literature. However, in order to be able to perform a detailed validation, an extensive set of data is necessary which was not found in any scientific paper. Therefore, a set of values were assumed. To assess the level of compliance graphical analysis and statistical tools were used, namely averages of mean relative differences (δ) and respective standard deviation (σ), mean square deviation (MSD) and Pearson coefficient (r).

3 MATERIAL

The scientific literature seldom presents a complete description of experimental tests, condition that is required to accurately validate mathematical models. In the absence of information, the values suggested by Raimundo & Figueiredo (2009) were adopted, namely gender (man), height ($h^* = 1.72$ m) and weight ($w^* = 74.43$ kg), the clothing specific heat ($c_p^* = 1\,000$ J/(kg.°C)), water vapor permeability ($i_{vp}^* = 1$ for nude body parts, $i_{vp}^* = 0$ for the body parts dressed with impermeable clothing, $i_{vp}^* = 0.15$ for boots and shoes and $i_{vp}^* = 0.35$ for "normal" clothing) and radiative emissivity ($\varepsilon^* = 0.90$ for "normal" clothing and $\varepsilon^* = 0.93$ for human skin). Note that the "*" symbol indicates that this value was assumed since it was not found in scientific papers.

For situations involving human thermoregulation, the clothing is never characterized with the necessary detail. Therefore, for the *HuTheReg* program, which requires the characteristics of the clothing in each body part, it was necessary to distribute the garment characteristics over the

22 body parts and for this purpose the studies of Oliveira et al. (2008) and Raimundo & Figueiredo (2009) were considered.

The generic protocol of the cases selected for this study is composed by the following steps (phases in the simulations): (i) preliminary thermal stabilization at neutral temperature; (ii) muscle warming; (iii) exercise in a hot environment; (iv) transition period; and (v) thermophysiological recuperation. In some cases, phases (iii) to (v) were repeated more than once. The duration of each phase differs from case to case. Except in the case where the body cooling is driven by drinking water at 3.6°C, in all the other situations the ingested water was at temperatures between 19°C and 21°C.

A preliminary stabilization period is needed to ensure that the "virtual individuals" start with the same thermal state as the "real persons" that participated in the research studies. Unfortunately, this initial period is not always described with enough detail in the literature. In such cases the following parameters were considered: $t^* = 60$ minutes, air temperature $T_{air}^* = 29°C$, mean radiant temperature $T_{mrt}^* = 29°C$, relative humidity $RH^* = 60\%$, air velocity $v_{air}^* = 0.2$ m/s and a resting individual with an activity level $M^* = 0.8$ met and dressed in shorts (overall values $I_{cl}^* = 0.1$ clo, $i_{vp} = 0.98$ and $\varepsilon^* = 0.92$).

Due to lack of essential information only 22 cases reported in the literature were identified as suitable for the present purposes. In terms of timing of application of the body cooling methodology, the cases selected can be gathered in 3 different scenarios.

i. *body cooling during both exercise and recovery phases:*
   – UIJ-ER: use of an ice jacket – 1 case (Chou et al. 2008).
ii. *body cooling only during the exercise phases:*
   – ICW: regular intake of very cold water during exercise phases - 1 case: Mündel et al. 2006.
iii. *body cooling only during the recovery phases:*
   – EPCE: Exposure to a passive cold environment – 9 cases: 1. Hardy and Stolwijk 1966; 2. Bittel 1987; 3. Carter et al. 1999, 4. Giesbrecht et al. 2007, 5. Chou et al. 2008, 6. Lopez et al. 2008, 7. Barr et al. 2009, 8. Hostler et al. 2010, 9. Colburn et al. 2011.
   – HFICW: EPCE reinforced with hands and forearms immersion in cold water – 5 cases: 1. Selkirk et al. 2004, 2 and 3. Giesbrecht et al. 2007 ($T_{water} = 20°C$ and 10°C, respectively), 4. Hostler et al. 2010, 5. Colburn et al. 2011.
   – EACE: Exposure to an active cold environment (EPCE reinforced with forced air movement using fans) – 2 cases: 1. Carter et al. 1999, 2. Hostler et al. 2010.
   – UIJ: EPCE reinforced using an ice jacket – 3 cases: 1. Lopez et al. 2008, 2. Hostler et al. 2010, 3. Colburn et al. 2011.
   – UIJ+HFICW—UIJ reinforced with HFICW – 1 case: Barr et al. 2009.

4 RESULTS AND DISCUSSION

Due to the lack of space, it is not feasible to show graphical representations of the 22 cases tested. Therefore, only two figures are shown (1 from the 1st scenario and 1 from the 3rd) together with a very brief description of each respective case. Despite this limitation, it is important to emphasize that the statistical assessment includes all the 22 cases under analysis, which embraces 40 assessments.

4.1 *UIJ-ER: use of an ice jacket during exercise and recovery phases*

In the laboratory study of Chou et al. (2008), the effectiveness of the body cooling technique of wearing an ice jacket both during the exercise and the recovery phases was analyzed. For this purpose, 8 individuals (all males, $h^{(a)} = 1.68$ m, $w^{(a)} = 62.5$ kg, [a] mean values) were dressed with a fire-fighting personal protective ensemble. These individuals were submitted to the following protocol: (i) preliminary thermal stabilization; (ii) muscle warming (t = 10 min, $T_{air} = T_{mrt} = 25°C$, $RH = 50\%$, $v_{air}^* = 0.2$ m/s, $M^* = 1.2$ met, $I_{cl}^* = 2.54$ clo, $i_{vp}^* = 0.28$ and $\varepsilon^* = 0.90$); (iii) transition (t = 10 min, the rest equal except $T_{air} = T_{mrt} = 28°C$); (iv) exercise (t = 30 min, $T_{air} = T_{mrt} = 30°C$, $RH = 50\%$, $v_{air}^* = 0.2$ m/s, $M^* = 3.0$ met, $I_{cl}^* = 2.54$ clo, $i_{vp}^* = 0.28$ and $\varepsilon^* = 0.90$); and (v) recovery (t = 10 min, $T_{air} = T_{mrt} = 30°C$, $RH = 50\%$, $v_{air}^* = 0.2$ m/s, $M^* = 0.8$ met, $I_{cl}^* = 2.18$ clo, $i_{vp}^* = 0.37$ and $\varepsilon^* = 0.89$).

Figure 1 shows the experimental results obtained by Chou et al. (2008) for the rectal (T_{re}) and mean skin (T_{sk}) temperatures as well as the predicted values. A very good agreement was achieved for both temperatures, particularly for the rectal temperature.

4.2 *HFICW: Recovery phase with passive cooling reinforced with hands and forearms immersion in cold water*

The study of Selkirk et al. (2004) aimed to verify the effectiveness of the passive human body cooling technique in reducing thermal stress in a hot thermal environment and where it was necessary to wear personal protective clothing. Fifteen male firefighters participated in the study ($h^{(a)} = 1.81$ m, $w^{(a)} = 86.9$ kg, [a] mean values).

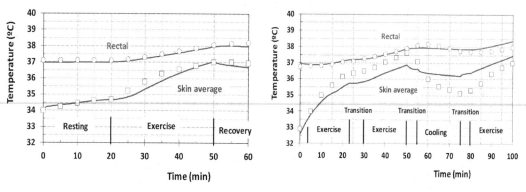

Figure 1. Experimental values (points) and predicted values (lines) for case UIJ-ER (Chou et al. 2008) for body cooling using an ice jacket during the exercise and the recovery phases.

Figure 2. Experimental values (points) and predicted values (lines) for case HFICW-1 (Selkirk et al. 2004) for body cooling by immersion of hands and forearms in cold water during recovery phase.

All the steps of the tests were performed in a room with controlled thermal conditions ($T_{air} = T_{mrt} = 35°C$, RH = 50%, $v_{air}* = 0.1$ m/s). The protocol included the following phases: (i) thermal stabilization in a neutral environment; (ii) muscle warming (t* = 3 min, M* = 1.2 met); (iii) exercise (t = 20 min, M* = 2.8 met); (iv) transition (t = 10 min, M* = 1.5 met); (v) exercise (t = 20 min, M* = 2.8 met); (vi) transition (t = 5 min, M* = 1.2 met); (vii) application of passive recovery reinforced with hands and forearms immersion in water at $T_{water} = 17°C$ (t = 20 min, M* = 0.8 met); (viii) transition (t = 5 min, M* = 1.2 met); (ix) exercise (t = 20 min, M* = 2.8 met).

During the exercise phases the subjects wear a typical forest fire-fighting personal protective ensemble, including individual breathing apparatus ($I_{cl} = 1.55$ clo, $i_{vp}* = 0.28$ and $ε* = 0.84$). During the muscle warming and the recovery phases, they remove the helmet, the flash hood, the gloves, the jacket and the individual breathing apparatus ($I_{cl}* = 1.06$ clo, $i_{vp}* = 0.56$ and $ε* = 0.90$).

Figure 2 shows that the rectal (T_{re}) and average skin (T_{sk}) temperatures experimental results obtained by Selkirk et al. (2004) are in good agreement with the predicted ones, particularly for the rectal temperature.

4.3 Statistical evaluation of the 22 cases

Table 1 shows the comparison between the experimental and the predicted values in terms of the statistical parameters. In the 22 cases tested 40 assessments are considered. The quantification of the global quality of the numerical predictions was performed on basis of the assessment of each of the four statistical parameters. A weighted mean value was obtained in which the Pearson coefficient (r) has a 50% contribution, followed by the mean square deviation (MSD) with 30%, the mean relative difference (δ) with 15% and the respective standard deviation (σ) with 5%. Some predictions were not as good as expected. That shall not be exclusively attributed to the program but most probably to the number of estimations that were necessary to perform due to the insufficient detail and lack of data found in the scientific papers selected for the present analysis.

4.4 Effectiveness of the human body cooling methodologies considered

This paper presents different human body cooling techniques. In order to assess the effectiveness of the various techniques for cooling the human body when exposed to hot thermal environments and/or when performing intense exercise, the same protocol should be considered. That was not the case, so direct comparisons between the results should be avoided. Nevertheless, despite this drawback, a clear outcome of the present analysis is that all the cooling techniques that were considered have advantages.

Techniques that can be used during the exercise phases have the advantage of dissipating the excess of heat as it is being generated. However, the use of ice jackets during exercise periods is not always possible. On the other hand, drinking very cold water, besides being extremely unpleasant, can have unhealthy consequences. The application of body cooling techniques during recovery periods is usually easier to accomplish. All of them have shown to be beneficial in terms of reducing hyperthermic stress, and their efficiency can be even enhanced by simultaneous application of more than one technique.

Table 1. Statistical comparison between experimental and predicted values. Cases identified by its symbol.

Case	Value T [°C] Q [W/m^2]	δ	σ	MSD	r
UIJ-ER	T_{re}	0.108	0.038	0.032	0.996
	T_{sk}	0.149	0.198	0.067	0.986
ICW	T_{re}	−0.362	0.463	0.227	0.991
EPCE-1	T_{re}	−0.005	0.120	0.024	0.905
	T_{sk}	−0.125	0.381	0.079	0.993
	Q_{metab}	1.912	2.012	0.585	0.945
	Q_{evap}	42.438	40.319	11.830	0.950
EPCE-2	T_{re}	0.256	0.070	0.080	0.963
	T_{sk}	−0.497	0.912	0.291	0.957
	T_{body}	−0.718	0.445	0.252	0.921
	Q_{prod}	−6.329	68.438	20.626	0.962
	Q_{stored}	8.198	7.596	3.452	0.968
EPCE-3	T_{re}	0.562	0.094	0.089	0.976
	T_{tymp}	−0.090	0.142	0.026	0.955
	$T_{sk,thighs}$	0.219	0.310	0.059	0.965
EPCE-4	T_{tymp}	−0.919	0.703	0.214	0.854
EPCE-5	T_{re}	0.052	0.064	0.022	0.992
	T_{sk}	0.191	0.227	0.081	0.984
EPCE-6	T_{re}	−0.316	0.275	0.189	0.954
	T_{sk}	−1.444	1.252	0.630	0.916
EPCE-7	T_{core}	0.327	0.394	0.150	0.893
	T_{sk}	−0.667	0.865	0.320	0.880
EPCE-8	T_{core}	0.040	0.384	0.128	0.878
EPCE-9	T_{core}	−0.005	0.188	0.066	0.843
HFICW-1	T_{re}	−0.205	0.248	0.052	0.946
	T_{sk}	−0.500	0.862	0.162	0.731
HFICW-2	T_{tymp}	−0.523	0.516	0.135	0.868
HFICW-3	T_{tymp}	−0.688	0.557	0.163	0.658
HFICW-4	T_{core}	0.224	0.269	0.119	0.769
HFICW-5	T_{core}	0.179	0.147	0.085	0.956
EACE-1	T_{re}	0.299	0.181	0.054	0.991
	T_{tymp}	−0.294	0.204	0.056	0.964
	$T_{sk,thighs}$	−0.242	0.221	0.051	0.959
EACE-2	T_{core}	0.148	0.214	0.088	0.919
UIJ-1	T_{core}	0.049	0.457	0.103	0.968
	T_{sk}	0.503	0.130	0.312	0.922
UIJ-2	T_{core}	0.118	0.218	0.086	0.927
UIJ-3	T_{core}	0.320	0.215	0.142	0.997
UIJ+HFICW	T_{core}	0.216	0.500	0.145	0.837
	T_{sk}	−1.131	0.757	0.405	0.949

5 CONCLUSIONS

In the present paper, the applicability of the *HuTheReg* program was tested with several cases, each involving different periods (phases) with distinct levels of metabolic activity, clothing thermal insulations, environmental conditions and body cooling techniques.

In all the 40 assessments of the 22 cases tested, the quality of the program's forecasts can be classified as very acceptable. This statement is based on the statistical comparison between experimental and predicted values, namely on a weighted mean value calculated with the four statistical parameters considered. A very acceptable outcome was obtained in several cases, namely in the two cases presented in detail in sections 4.1 and 4.2. No limitations were found on the program's ability to predict the thermo physiological behavior of the human body when exposed to a wide range of situations, even in cases involving periods with very different conditions.

The results obtained with this study put in evidence the importance of simulation programs to carry out detailed evaluations of the thermo physiological response of the human body, which allows the anticipation of actions to attenuate thermal stress and thus avoid the advent of heat-related illnesses.

Although with different effectiveness, all body cooling techniques considered proved to be capable to mitigate the level of hyperthermic stress of people performing intense physical activities and/or exposed to hot environments.

REFERENCES

Barr D, Gregson W, Sutton L & Reilly T (2009). A practical cooling strategy for reducing the physiological strain associated with firefighting activity in the heat. *Ergonomics* 52(4), 413–420.

Bittel JHM (1987). Heat debt as an index of cold adaptation. *J Appl Physiol* 62, 1627–1634.

Carter JB, Banister EW & Morrison JB (1999). Effectiveness of rest pauses and cooling in alleviation of heat stress during simulated fire-fighting activity. *Ergonomics* 42, 299–313.

Chou C, Tochihara Y & Kim T (2008). Physiological and subjective responses to cooling devices on firefighting protective clothing. *European J Applied Physiology* 104, 367–374.

Colburn D, Suyama J & Reis S (2011). A comparison of cooling techniques in firefighters after a live burn evolution. *Prehospital Emergency Care* 15, 226–232.

Fiala D, Lomas KJ & Stohrer M (1999). A computer model of human thermoregulation for a wide range of environmental conditions—the passive system. *J Applied Physiology* 87, 1957–1972.

Giesbrecht GG, Jamieson C & Cahill F (2007). Cooling hyperthermic firefighters by immersing forearms and hands in 10°C and 20°C water. *Aviation Space Environmental Medicine* 78, 561–567.

Hardy JD & Stolwijk JAJ (1966). Partitional calorimetric studies of responses man during exposures to thermal transients. *J Applied Physiology* 21, 967–977.

Havenith G, Holmér I & Parsons K (2002). Personal factors in thermal comfort assessment: clothing properties and metabolic heat production. *Energy and Buildings* 581–591.

Hostler D, Reis S, Bednez JC, Kerin S & Suyama J (2010). Comparison of active cooling devices with passive cooling for rehabilitation of firefighters performing exercise in thermal protective clothing: a report from the Fireground Rehab Evaluation (FIRE) trial. *Prehospital Emergency Care* 14, 300–309.

Lopez RM, Cleary MA, Jones LC & Zuri RE (2008). Thermoregulatory Influence of a Cooling Vest on Hyperthermic Athletes. *J Athletic Training* 43(1), 55–61.

Mündel T, King J, Collacott E & Jones DA (2006). Drink temperature influences fluid intake and endurance capacity in men during exercise in a hot, dry environment. *Experim Physiology* 91(5), 925–933.

Oliveira AVM, Gaspar AR & Quintela, DA (2008). Measurements of clothing insulation with a thermal manikin operating under the thermal comfort regulation mode: comparative analysis of the calculation methods. *Eur J Appl Physiol* 104(4): 679–688. doi 10.1007/s00421-008-0824-5.

Oliveira AVM, Gaspar AR, Quintela DA (2011). Dynamic Clothing insulation. Measurements with a thermal manikin operating under the thermal comfort regulation mode. *Appl Ergon* 42(6), 890–899. doi 10.1016/j.apergo.2011.02.005.

Quintela DA, Gaspar AR & Borges C (2004). Analysis of sensible heat exchanges from a thermal manikin. *European J. Appl Physiology* 92, 663–668.

Raimundo AM & Figueiredo AR (2009). Personal protective clothing and safety of firefighters near a high intensity fire front. *Fire Safety Journal* 44: 514–521.

Raimundo AM, Gaspar AR, Quintela DA (2004). Numerical modelling of radiative exchanges between the human body and surrounding surfaces. Climamed 2004–1st Mediterranean congress of climatization, 16–17 of April, Lisbon, Portugal, paper 8/1.

Raimundo AM, Quintela DA, Gaspar AR & Oliveira AVM (2012). Development and validation of a computer program for simulation of the human body thermophysiological response. Portuguese chapter of IEEE-EMBS.

Selkirk G, McLellan TM & Wong J (2004). Active versus passive cooling during work in warm environments while wearing firefighting protective clothing. *J Occup Environ and Hygiene* 1, 521–531.

Stolwijk JAJ (1971). A mathematical model of physiological temperature regulation in man. *NASA Contractor Report CR-1855*, NASA, Washington, DC.

Air quality assessment in copy centers

Marta Cunha & Ana Ferreira
ESTeSC—Coimbra Health School, Environmental Health, Polytechnic Institute of Coimbra, Portugal

Francisco Silva
Technological Center for Ceramics and Glass, Coimbra, Portugal

João Paulo Figueiredo
ESTeSC—Coimbra Health School, Complementary Sciences—Statistics/Epidemiology, Polytechnic Institute of Coimbra, Portugal

ABSTRACT: Exposure to fine and ultrafine particles in copy centres has been studied for some years since they represent a health risk when present in high quantities, especially in indoor air. Studies indicate that ultrafine particles can have more harmful consequences, since they can move from the lungs to other parts of the body. Thus, it was relevant to evaluate the exposure of workers of copy centres to particles and other pollutants, as well as thermal comfort and try to notice which factors influence it. The sample consisted in 2 copy centres, and fine particles, carbon monoxide and carbon dioxide and some parameters of thermal comfort/weather variables were measured. Regarding the variation of pollutant concentrations, these were higher in the afternoon, in the interior and next to the printers and were influenced by the number of times the door was opened.

1 INTRODUCTION

Nowadays there is little doubt about the importance of Indoor Air Quality (IAQ), since in modern society people tend to spend most of the time indoors. Indoor air is influenced by a combination of factors such as the climate, the building's ventilation system, the sources of pollutants inside, the exchange of pollutants with the outside, fumes, building materials and equipment inside the building, the number of occupants and their activities at the location, including work, leisure and cleaning activities (Coelho, 2014; Morawska & Taplin, 2007). Parameters such as air temperature (T_{Air}), relative humidity (H_R) and air velocity (V_{Air}) influence the concentration and distribution of pollutants in indoor air. The strategies and habits adopted by the users of the buildings are fundamental in the management of the IAQ (Ginga et al., 2012). Thus, the IAQ encompasses the concentration of pollutants that affect the occupant's comfort, environmental satisfaction, health and productivity (Brickus & Neto, 2001; Coelho, 2014). Studies claim that the level of air pollution inside buildings is often worse than outside and can reach values two to five times higher, and occasionally up to one hundred times higher than the level of outdoor air pollution (Ferreira & Cardoso, 2013).

An increasing number of studies (Grgic et al., 2016; Khatri et al., 2013; Lee et al., 2006; Lee & Hsu, 2007; Lee et al., 2001; Martin et al., 2015; Singh et al., 2014), refer exposure to high concentrations of some air pollutants in print shops or copy centres and even offices, as a potential health risk for workers in these locations. Among these pollutants airborne particles, particularly coarse, fine and ultrafine (PM_{10}, $PM_{2.5}$ e $PM_{<0.1}$ or UFP), Volatile Organic Compounds (VOC's) and ozone (O_3) gain attention. High levels of carbon dioxide (CO_2), which are sometimes found in places with poor ventilation, often held in a natural way without forced ventilation systems, and some parameters of thermal comfort, like large temperature variations or high temperature combined with low humidity are also relevant (Brickus & Neto, 2001; Ginga et al., 2012; Szigeti et al., 2017). Many studies (Ansaripour et al., 2016; Destaillats et al., 2008; Kagi et al., 2007; Khatri et al., 2013; Morawska & Taplin, 2007; Wang, 2011; Wang et al., 2012; Wensing et al., 2008), refer the printers and photocopiers, as significant emission sources of $PM_{2.5}$ and UFP, relating their operation to high concentrations of these particles.

The aim of this study was to evaluate the QAI in copy centres, to identify pollutants' emitting sources in the interior and evaluate parameters of thermal comfort.

2 MATERIAL AND METHODS

2.1 Sample characterization

To carry out this study, two copy centres from higher education schools in the municipality of Coimbra, located in a semi-urban area were selected, identified as copy centre 1 and copy centre 2. Copy centres have less than 10 m² area and 3 m ceiling height. In copy centre 1 there were 2 workers, operating with 3 laser printers. In copy centre 2 there was only 1 worker and 2 laser printers and 1 ink-jet printer were used. This was a Level III study, from the observational type, prospective cohort, analytical. The sampling was non-probabilistic type and the technique by convenience.

2.2 Methodology and instruments for data collection

Measurements of IAQ physicochemical parameters and outside air quality (OAQ) were carried out during a first period of sampling. Measurements of IAQ and OAQ for air pollutants and thermal environment variables were carried out during 15 min periods, comprised in copy centres opening hours (from 9h00 to 19h00 on weekdays). In each day, 8 aleatory periods of measurement where considered for IAQ and 2 for OAQ, resulting in a total of 42 samples in copy centre 1 (2 different locations) and 17 samples in copy centre 2. The study took place during different weeks and with different weather conditions. UFP sampling was performed later and only in one afternoon, during approximately 90 min in each copy centre. All measurements were made in spring. The hours of the measurements were randomly chosen and the following concentration of the following atmospheric air components were measured: CO, CO_2, $PM_{2.5}$, PM_{10}, total suspended particles (TSP). Simultaneously the meteorological variables T_{Air}, H_R and V_{Air} were also measured. The number of times the exterior access door was opened and the number of prints per day were also recorded. During measurements, in general, the windows of the spaces were closed and the workers were performing their usual tasks.

The parameters were measured with direct measurement portable equipment: *TSI Q-TrakTM Plus—IAQ Monitor*, 8552/8554 model, with direct reading electrochemical cell to evaluate the concentration of CO, CO_2 and H_R; *Lighthouse Handheld 3016 IAQ* model for collecting the quantitative values of $PM_{2.5}$, PM_{10} and TSP, with reading range between 0.3 µm and 10 µm; TSI *VelociCalc*, 9535 model, direct reading to evaluate the V_{Air} and T_{Air}; and the *Condensation Particle Counter (CPC)* meter, *TSI*, 3007 model, direct reading to collect the quantitative values of UFP with diameter between 0,01 µm and 1 µm. The values of the readings performed by these equipment's correspond to the CPC particle saturation value, therefore, the exposure is usually higher than this value. To transfer data from measuring instruments to computer, the specific software of each of the equipment was used.

To verify if there was good ventilation in the space, the CO_2 concentration was used as an indicator of good ventilation rate of the space, as mentioned in Portaria n.º 353-A/2013, of December 4th, and in the *Portuguese Agency of Environment (PAE)* Technical Guide related to IAQ (Agência Portuguesa do Ambiente, 2009), considering a maximum concentration of 1250 ppm.

2.3 Statistical analysis

Statistical analysis of the data was performed using the *software IBM SPSS Statistics*, version 24.0. Through this software it was possible to carry out the statistical treatment of the obtained data, using descriptive statistics, with measures of central tendency as the mean and measures of dispersion as the standard deviation. The statistical tests used for inferential analysis were the *t-Student* one-sample and independent samples, the *Mann-Whitney* Test, the *Kruskal-Wallis* Test and the Multiple Comparisons – *Dunn* Test, the Linear Correlation of *Pearson* and the Ordinal Correlation of *Spearman*. For statistical inference, it was possible to establish a 95% confidence level for a ≤ 5% random error.

3 RESULTS

After data collection and statistical analysis, it was concluded that there were statistically significant differences (p≤0,05) between the mean concentrations of the physicochemical parameters $PM_{2.5}$, CO e CO_2 analysed and their respective protection thresholds (8h). It was found that in most cases the mean concentration was well below the threshold. Only in the case of PM_{10} in copy centre 2, the difference was not so significant.

For thermal comfort parameters, the values were generally kept within the recommended ranges, except the V_{Air} in both copy centres and the T_{Air} in copy centre 2 that was below the range (Table 1).

The concentrations of the control measurements (before the workers started the activity) were compared with the average of weekly evaluations, and it was concluded that particulate matter, CO and CO_2 concentrations were significantly lower before workers started the activity in both copy centres and in both points of measurements. Comparing indoor and outdoor concentrations,

Table 1. Variation of the weekly mean concentrations of the physicochemical parameters in each copy centre.

Parameters	N	Mean	SD	Mean difference	Protection threshold	p-value
Copy centre 1						
$PM_{2.5}$ (µg/m³)	42	9,13	6,19	–15,87	25	<0,0001
PM_{10} (µg/m³)	42	35,39	13,67	–14,61	50	<0,0001
CO (ppm)	42	2,6	0,3	–6,4	9	<0,0001
CO_2 (ppm)	42	675	121	–575	1250	<0,0001
T_{Air} (°C)	42	22,7	1,1	–1,8	22,0–24,5	<0,0001
H_R (%)	42	48,4	7,0	–11,6	40–60	<0,0001
V_{Air} (m/s)	42	0,04	0,02	–0,08	0,10–0,12	<0,0001
Copy centre 2						
$PM_{2.5}$ (µg/m³)	17	11,31	8,12	–13,70	25	<0,0001
PM_{10} (µg/m³)	17	43,65	12,81	–6,35	50	0,058
CO (ppm)	17	2,9	0,6	–6,43	9	<0,0001
CO_2 (ppm)	17	556	101	–694	1250	<0,0001
T_{Air} (°C)	17	21,7	1,1	–0,03	22,0–24,5	<0,0001
H_R (%)	17	46,3	10,5	–13,7	40–60	<0,0001
V_{Air} (m/s)	17	0,06	0,02	–0,07	0,10–0,12	<0,0001

Parametric test: t-Student for one sample; N – Sample; SD – Standard Deviation.

Table 2. Variation of concentrations of physicochemical parameters indoors and outdoors and depending on the time of day in each copy centre.

	Local					Period of the day				
Parameters	N	Local	Mean	SD	p-value	N	Period	Mean	SD	p-value
Copy centre 1										
$PM_{2.5}$ (µg/m³)	42	Indoor	9,13	6,19	0,449	19	Morning	7,57	5,97	0,136
	11	Outdoor	9,61	10,88		23	Afternoon	10,43	6,19	
PM_{10} (µg/m³)	42	Indoor	35,39	13,67	0,188	19	Morning	32,12	14,76	0,056
	11	Outdoor	32,07	22,12		23	Afternoon	38,10	12,38	
TSP (µg/m³)	42	Indoor	63,52	18,61	0,105	19	Morning	55,85	16,88	0,019
	11	Outdoor	50,89	26,88		23	Afternoon	69,86	17,88	
CO (ppm)	42	Indoor	2,6	0,3	0,394	19	Morning	2,6	0,3	0,809
	11	Outdoor	2,5	0,4		23	Afternoon	2,6	0,3	
CO_2 (ppm)	42	Indoor	675	121	<0,0001	19	Morning	619	145	0,055
	11	Outdoor	418	152		23	Afternoon	721	73	
Copy centre 2										
$PM_{2.5}$ (µg/m³)	17	Indoor	11,31	8,16	0,410	7	Morning	9,91	7,90	0,380
	11	Outdoor	9,97	9,89		10	Afternoon	12,29	8,53	
PM_{10} (µg/m³)	17	Indoor	43,65	12,81	0,070	7	Morning	43,79	14,68	0,961
	11	Outdoor	34,74	21,64		10	Afternoon	43,55	12,17	
TSP (µg/m³)	17	Indoor	74,58	21,33	0,025	7	Morning	71,77	25,41	0,845
	11	Outdoor	53,28	24,41		10	Afternoon	76,55	19,18	
CO (ppm)	17	Indoor	2,9	0,6	0,310	7	Morning	2,8	0,3	0,883
	11	Outdoor	2,5	0,5		10	Afternoon	3,0	0,7	
CO_2 (ppm)	17	Indoor	556	101	<0,0001	7	Morning	503	88	0,040
	11	Outdoor	355	61		10	Afternoon	593	97	

Non-parametric test: Mann-Whitney; N – Sample; SD – Standard Deviation.

it was only possible to find statistically significant differences (p≤0,05) in the CO_2 concentrations in both copy centres, where the difference from the outdoor to the indoor was more than 200 ppm in both copy centres. There were also significant differences (p = 0,02) in TSP concentrations at copy centre 2. It was also found that in the morning, concentration of pollutants was lower than in the afternoon, and there were statistically significant differences in the concentrations of CO_2 in both copy centres and TSP at copy centre 1. Only concentrations of PM_{10} at copy centre 2 were slightly higher in the morning (Table 2).

The relation of number of times the door was opened with the physicochemical parameters was determined and it was concluded that in copy centre 1, the more times the door was opened the lower the concentrations of pollutants, since the correlations were mostly low negative, except for CO that had a positive low correlation (r = 0,287). In other words, in this case, it was verified that the more times the door was opened the higher the concentrations. With relation to copy centre 2, it was found that for the parameters $PM_{2.5}$ and CO_2 the correlation was very low negative, which means that, there was a slight decrease of these pollutants with the increase in the number of times the door was opened. For PM_{10}, TSP and CO the correlation was very low positive, which means that there was a slight increase in concentration with the increase in the number of times the door was opened.

Relatively to the presence of UFP in each copy centre, the mean concentrations were higher in copy centre 1, and there were significant differences (p < 0,0001). It was further noted that in copy centre 1, the door was opened 4 times throughout the measurement period of 87 minutes, where the mean UFP concentration was 21126 particles/cm^3. In copy centre 2, the door was opened 16 times during the 89 minutes, and the mean concentration was 19230 particles/cm^3. It was observed that the concentration of UFP was higher in the copy centre 1 than in the waiting area (outside the copy centre 1). There were significant differences between the 2 measurement positions (p < 0,0001). In copy centre 1, it was observed that, when printing equipment was not used, the concentrations were higher than when they were used, and there were statistically significant differences (p < 0,0001). In copy centre 2, UFP concentrations were slightly higher when printing equipment was used, but there were no statistically significant differences (p = 0,549).

4 DISCUSSION

After analysing the results, it was possible to verify that the average concentration of the physicochemical parameters analysed did not exceed the respective protection thresholds (8h). Relatively to the thermal comfort values were generally within the legislated ranges, except for V_{Air}, with mean value 0,04 m/s at copy centre 1 and 0,06 m/s at copy centre 2. The recommended range for V_{Air} is from 0,10 m/s to 0,12 m/s and values above or below the recommended range were found occasionally, possibly related to climate, since the values were similar to those found outdoors. However, Service Medical Enterprises did a risk assessment in this kind of workplace, highlighting the thermal discomfort as one of the existing risks (Falcão, 2013).

Concentrations of the pollutants monitored were generally below the recommended level which may be related to the existence of good ventilation in both copy centres, as verified through CO_2 concentrations. Although the mean daily concentrations of pollutants were below the protection threshold (8h), it is necessary to consider the existence of some concentration peaks found throughout the day, particularly the concentration of particles, where some values above protection thresholds (8h) were found in both copy centres. These peaks can be harmful at long term if they exist regularly, which could not be verified by the present study, carried out in the short term, but may be an indicator that additional long term studies should be performed. The mean values of the thermal comfort parameters found in the present study were similar to those found in a study carried out in Portugal (Ramos et al., 2004), in offices in the Lisbon region with conditions similar to those of the copy centres under study, in which physicochemical parameters were evaluated. Only the temperature values were above the values found in the present study. Concerning to particles, similar mean concentrations to those found in copy centre 1 are referred, but the concentration peaks obtained were about 4 times higher than those found in the present study. Another study developed in several European countries, including Portugal in offices located in Porto (Szigeti et al., 2017) also detected values that fit those found in present study. The fact that the mean particle concentrations $PM_{2.5}$ e PM_{10} were generally higher in copy centre 2, could also be related to the lack of maintenance and cleaning of the air conditioning system of this space, referred by the worker. Copy centre 1 did not have any type of air conditioning or forced ventilation system, having only one window in the workspace that was rarely open and one door outside the work area. The workers of this copy centre reported that there was sometimes a lack of ventilation, particularly in winter or on bad weather, where the window was often not opened for long periods, and the same occurring with the outside access door, since there was another access to copy centre inside the building.

5 CONCLUSIONS

In the present study it was concluded that evaluated pollutants mean values were below legal threshold values, thermal comfort parameters were adequate and there was fair ventilation in the workplaces. However, the detection of peak values above referred thresholds should be object of further research. In what concerns to UFP the detected values could be considered high and further research on workers exposure is also advisable.

It was observed that pollutants concentrations were higher during afternoon and it was not possible to find relation between concentrations of pollutants with neither printing rate nor number of times the access door was opened.

To reduce workers' exposure ventilation with air filtration and improved locations cleaning to reduce particles re-suspension are possible measures. Workers should also leave the workplace during breaks, in order to minimize exposure to high concentrations of these pollutants. During the study, some limitations were found that made it difficult to assess IAQ throughout the study, in copy centre 2, and it was necessary to reduce the number of collections. Also, some parameters of IAQ, such as formaldehyde, VOCs and O_3, were not measured due to equipment malfunctions. As for UFP measurements, it was not possible to perform several measurements, as in the case of the other parameters, since the equipment was not from the same institution. In future studies, measurements of other chemical parameters, considering factors that interfere in the concentrations of pollutants and emitting sources, should be carried out, which were not very well investigated in this study.

REFERENCES

Ansaripour M, Abdolzadeh M, Sargazizadeh S. Computational modeling of particle transport and distribution emitted from a Laserjet printer in a ventilated room with different ventilation configurations. Appl Therm Eng [Internet]. Kerman, Irão: ResearchGate; 2016; 103 (Applied Thermal Engineering): 920–33. Available from: https://www.research gate.net/publication/301691494% 0 AComputational.

Brickus LSR, Neto F. A qualidade do ar de interiores e a química. Rev Bras Toxicol. 2001;14(1): 29–35.

Coelho P. Exposição aos compostos orgânicos voláteis— Trabalhadores em cozinhas escolares [Internet]. Lisboa; 2014. Available from: http://repositorio.ipl.pt/handle/10400.21/3877.

Destaillats H, Maddalena RL, Singer BC, Hodgson AT, McKone TE. Indoor pollutants emitted by office equipment: A review of reported data and information needs. Atmos Environ. 2008; 42(7): 1371–88.

Falcão C. Avaliação de riscos em contexto escolar e industrial [Internet]. Universidade de Lisboa; 2013. Available from: https://www.repository.utl.pt/bitstream/10400.5/6425/1/mestrado.avaliaçãoderiscosemcontextoescolareindustrial.pdf.

Ferreira AMC, Cardoso SM. Estudo exploratório da qualidade do ar em escolas de educação básica, Coimbra, Portugal. Rev Saude Publica [Internet]. 2013; 47(6): 1059–68. Available from: http://www.scielo.br/scielo.php?script=sci_arttext&pid=S0034-89102013000601059&lng=pt&nrm=iso&tlng=en.

Ginja, J., Borrego C, Coutinho M, Nunes C, Morais-Almeida M. Qualidade do ar interior nas habitações Portuguesas. In Aveiro & Lisboa: CINCOS'12 – Congress of Innovation on Sustainable Construction; 2012. p. 1–10. Available from: https://www.ua.pt

Grgic I, Bratec J, Rogac M. Indoor Nanoparticles Measurements in Workplace Environment: The Case of Printing and Photocopy Center. 2016; 327–34.

He C, Morawska L, Taplin L. Particle Emission Characteristics of Office Printers. 2007; 41(17 OP-Environmental Science & Technology. 9/1/2007, Vol. 41 Issue 17, p6039–6045. 7p. 2 Charts, 3 Graphs.): 6039. Available from: http://search.ebscohost.com/login.aspx?direct=true&site=edslive&db=bth&AN=26607300.

Kagi N, Fujii S, Horiba Y, Namiki N, Ohtani Y, Emi H, et al. Indoor air quality for chemical and ultrafine particle contaminants from printers [Internet]. 2007. p. 1949. Available from: http://search.ebscohost.com/login.aspx?direct=true&site=eds-live&db=edselp&AN=S0360132306000965.

Khatri M, Bello D, Gaines P, Martin J, Pal AK, Gore R, et al. Nanoparticles from photocopiers induce oxidative stress and upper respiratory tract inflammation in healthy volunteers. Nanotoxicology [Internet]. 2013; 7(5): 1014–27. Available from: http://www.ncbi.nlm.nih.gov/pubmed/22632457.

Lee C-W, Dai Y-T, Chien C-H, Hsu D-J. Characteristics and health impacts of volatile organic compounds in photocopy centers. Environ Res [Internet]. 2006; 100(2): 139–49. Available from: http://www.ncbi.nlm.nih.gov/pubmed/16045905.

Lee CW, Hsu DJ. Measurements of fine and ultrafine particles formation in photocopy centers in Taiwan. Atmos Environ. 2007; 41(31): 6598–609.

Lee SC, Lam S, Kin Fai H. Characterization of VOCs, ozone, and PM10 emissions from office equipment in an environmental chamber. Build Environ. 2001; 36(7): 837–42.

Martin J, Bello D, Bunker K, Shafer M, Christiani D, Woskie S, et al. Occupational exposure to nanoparticles at commercial photocopy centers. J Hazard Mater [Internet]. Elsevier B.V.; 2015; 298: 351–60. Available from: http://dx.doi.org/10.1016/j.jhazmat.2015.06.021.

Morawska L, He C, Johnson G, Jayaratne R, Salthammer T, Wang H, et al. An investigation into the characteristics and formation mechanisms of particles originating from the operation of laser printers. Environ Sci Technol. 2009; 43(4): 1015–22.

Portaria n.º 323-A/2013 de 4 de dezembro. Relativa aos valores mínimos de caudal de ar novo por espaço, bem como os limiares de proteção e as condições de referência para os poluentes do ar interior dos edifícios de comércio e serviços novos. Diário da República. 2013; 1a série (235): 6644(2)-6644(9).

Ramos CD, Dias CM, Cano MM. Qualidade do Ar Interior em Edifícios de Escritórios e Serviços. Estud Em Destaque. 2004; 1–6.

Singh BP, Kumar A, Singh D, Punia M, Kumar K, Jain VK. An assessment of ozone levels, UV radiation and their occupational health hazard estimation during photocopying operation. J Hazard Mater [Internet]. Elsevier B.V.; 2014; 275:55–62. Available from: http://dx.doi.org/10.1016/j.jhazmat.2014.04.049.

Szigeti T, Dunster C, Cattaneo A, Spinazz?? A, Mandin C, Le Ponner E, et al. Spatial and temporal variation of particulate matter characteristics within office buildings—The OFFICAIR study. Sci Total Environ [Internet]. Elsevier B.V.; 2017;.587–588:59–67. Available from: http://dx.doi.org/10.1016/j.scitotenv.2017.01.013.

Wang H, He C, Morawska L, McGarry P, Johnson G. Ozone-Initiated Particle Formation, Particle Aging, and Precursors in a Laser Printer. Environ Sci Technol [Internet]. Environmental Science & Technology; 2012;46(2 OP-Environmental Science & Technology. 1/17/2012, Vol. 46 Issue 2, p704–712. 9p.):704. Available from: http://search.ebscohost.com/login.aspx?direct=true&site=edslive&db=bth&AN=70786569.

Wang Z-M, Wagner J, Wall S. Characterization of Laser Printer Nanoparticle and VOC Emissions, Formation Mechanisms, and Strategies to Reduce Airborne Exposures. Aerosol Sci Technol. 2011; 45(9): 1060–8.

Wensing M, Schripp T, Uhde E, Salthammer T. Ultrafine particles release from hardcopy devices: Sources, real-room measurements and efficiency of filter accessories. Sci Total Environ [Internet]. Elsevier B.V.; 2008; 407(1): 418–27. Available from: http://dx.doi.org/10.1016/j.scitotenv.2008.08.018.

The contribution of digital technologies to construction safety

Débora Pinto
PROA/LABIOMEP/INEGI, Faculty of Engineering, University of Porto, Porto, Portugal

Fernanda Rodrigues
RISCO, Department of Civil Engineering, University of Aveiro, Aveiro, Portugal

J. Santos Baptista
PROA/LABIOMEP/INEGI, Faculty of Engineering, University of Porto, Porto, Portugal

ABSTRACT: Construction is still a sector with high incidence of accidents. Therefore, it's imperative to implement methodologies to improve safety in this sector. Building Information Modeling (BIM) emerges as a tool that allows the implementation of safety throughout the life cycle of a construction project, minimizing hazards and risks, consequently increasing safety. This paper aims to demonstrate BIM's methodology potential for the implementation of safety measures throughout planning and the application of this information on the worksite management. A case study based on BIM 4D was developed in which safety measures to prevent falls from height were applied. With 4D simulation, it is possible to follow the execution of work giving Safety Technicians the capability to visualize in each phase which measures should be implemented. In conclusion, BIM contributes to safety measures integration from the design phase to the construction phase, making safety management easier throughout all construction project phases.

1 INTRODUCTION

Over the last years a great technological development in different sectors of activity was achieved. However, in the Architecture, Engineering and Construction industry (AEC), new technologies were not fully implemented in the design, planning and construction phases, as well in the integration of safety measures.

According to the European Agency for Safety and Health at Work, worldwide construction workers have three times more probability to be killed and twice to be injured than in other sectors (EU-OSHA 2003). In Europe every year, about 1.300 workers die in construction accidents, 800.000 are injured, and others suffer from occupational diseases (EU-OSHA 2004). Therefore, this scenario confirms the dangerousness of this sector and the lack of development in the component which carries out the estimation of safety factors. This also reveals that the actual methodologies have not been sufficient to reduce the number of accidents.

Recently, there has been growing interest in the use of BIM methodology in the AEC industry, regarding integration of safety measures throughout the building life cycle. With technological advances in BIM, proactive safety solutions have been developed in the planning and management of worksites (Enshassi, Ayyash, and Choudhry 2016).

This study started with the research question: *What can be the contribution of the digital tools of BIM methodology to the planning and management of safety in construction?*

A case study using three software programs was developed: Autodesk Revit, Microsoft Excel and Autodesk Navisworks. A BIM 4D model, was created with the three models overlapped, with the purpose of analyzing the suitability of the 3D model and the safety measures applied to it, to the possible use and management on site.

This work aims to show the potential of BIM methodology to implement safety measures throughout planning and to implement this information in the construction site management.

2 MATERIALS AND METHODS

2.1 *Methodology*

Initially it was essential to understand the technologies that can be applied to the planning phase, namely: BIM 4D, Virtual Reality (VR) and Geographic Information Systems (GIS), being BIM 4D the chosen for this work. It consists on adding construction planning information to a 3D model, establishing collaboration between construction sequences and clear visualization of work sequence. The most common application of BIM 4D for risk management is to establish a comprehensive 4D model, combining all project data on construction objects, construction processes, activities and sequences, building up additional risk analysis based on the model (Zou, Kiviniemi, and Jones 2016). The graphic simulation obtained from the 4D model provides a deep view of the building construction on a daily basis, revealing potential problems and opportunities for possible improvements to the worksite, equipment, space conflicts, and safety problems, among others. This type of analysis is not possible with 2D documents. BIM allows a much more adequate view and follow-up (Eastman 2011).

The case study of this work was based on a Radiotherapy Center. Architecture and stability designs were modeled in 3D software, the Autodesk Revit. Protection measures, with the purpose of avoiding risks throughout the construction phase, were applied to each one of the models, as temporary structures such as guardrails, scaffoldings and platforms, building site fences, definition of paths/traffic directions and hole covers. With the introduction of the 3D models to the Autodesk Navisworks software, the principles of BIM 4D were achieved. In this software, the temporal component was introduced in the models, incorporating the respective scheduling, developed in Microsoft Excel. A simulation of the building construction through the planed time was obtained, in which it is possible to identify the safety measures to be implemented during the construction in due time.

Finally, the results obtained with the simulation were analyzed and subsequent conclusions presented.

2.2 *Case study*

A deep analysis and understanding of the Radiotherapy design was done through AutoCAD (2D) architecture and stability projects. The building consists of two floors: the -1 and 0. The -1 floor is completely buried with a structure composed by reinforced concrete walls around the entire perimeter. Subsequently, and taking into consideration the construction phase, documents were developed containing work instructions and the respective safety measures to be implemented.

Following the 2D architecture and stability projects, the respective 3D models in Revit were developed. Safety measures were applied to these models emphasizing the use of temporary structures. To incorporate the safety measures, it was necessary to use Revit's external libraries, available online (Revit City, Bimstore, BIMobject and BIM&CO), from which elements like scaffolding, platforms and building site fence were imported. Despite a great research, there was a considerable lack of available objects regarding safety components in these libraries. Therefore, some families had to be developed by the author to be incorporated into the models, namely: guardrails with clamping system, guardrails with spike system and hole covers.

The safety elements incorporated in the stability model were two families of guardrails, hole covers, paths/traffic directions and the site fence (Fig. 1). In the architecture model the scaffolding and the working platform in height were used (Fig. 2). This separation between the safety elements was necessary because their use is different throughout the construction phase.

The two completely finished and characterized 3D models were overlapped in Navisworks, generating a composite model (Fig. 3).

The 4D model is created adding the time component to the Navisworks composite model. Time component was created by the activity schedule in Microsoft Excel. This schedule took into account the main construction activities and the temporary tasks, which correspond to the use of safety systems and the expected start and end for each task. Behind all scheduling must be a broad knowledge of the activity to achieve a realistic definition of works sequence. Throughout the planning phase a critical thought must be had regarding the time

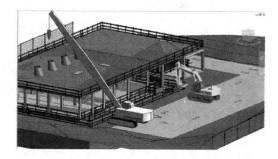

Figure 1. 3D model of structure and safety elements—Revit.

Figure 2. 3D model of architecture and safety elements—Revit.

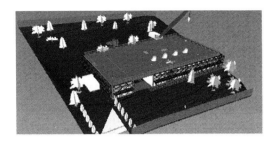

Figure 3. Composite model—Navisworks.

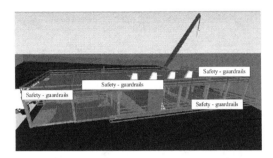

Figure 4. Links to safety guardrails for assembly/disassembly.

Table 1. Maps of quantities—guardrails (extract).

Specialty equipment—guardrails

Type	Length (mm)	Bar spacing (mm)	n° bars	Slab tickness (mm)
Guardrails with clamping	4187	1500	3	200
Guardrails with clamping	30920	2500	12	200
Guardrails with clamping	19880	2500	8	200
Guardrails with clamping	31020	2500	12	200
Guardrails with spike	4260	2000	2	
Guardrails with spike	3940	2000	2	
Guardrails with spike	4986	2000	2	
Guardrails with spike	2980	1500	2	

of placement and removal of temporary structures and safety elements, since they have different permanence periods in the worksite. Within the 4D model, to reinforce safety, were inserted tools like comments and links, adding information of temporary tasks, pointing out the need to consult safety documents about the assembly/disassembly of scaffolding and guardrails (Fig. 4). These documents provide a detailed description of assembly/disassembly of scaffolding and guardrails, hazards, risks and preventive measures.

Associating, in the 4D model, the planned schedule to each task and the sets of information of the corresponding models, creates a simulation to follow all the requirements of construction work in terms of safety and work execution.

3 RESULTS

Concerning safety, it was possible to extract maps of quantities of the created families from the models and work developed, namely guardrails with clamping system, guardrails with spike system and hole covers. Since these were modeled taking into consideration the sharing of parameters information for further quantification. Tables 1 and 2 present some information about the extracted maps.

From the remaining safety elements, like scaffolding and site fence, the extraction of maps of quantities was not possible because these families were not designed taking into consideration the sharing of information at that level. Therefore, it was verified that when working with families that are non-existent in the Revit's library and if external libraries are used, it may happen that the family parameters do not provide the information intended by the user at the level of quantification. Thus, when modeling new families and using families designed by other authors, attention must be paid to the method used to achieve the parameterization, verifying if the parameters are shared.

The 4D model results from the incorporation of 3D models and Excel schedule of work into Navisworks software. When the Navisworks simulation tool is applied to the 4D model, it is possible to keep track of worksite development over time, making possible to visually control safety interventions and management needs at each moment.

The Navisworks simulation of the 4D model produces a color coded graphic where each color is associated with a type of task. Each task type has the following graphic meanings:

- Construct: when the task starts, the elements referring to the task are displayed in green and upon finishing their color turns into the color of the Revit model.
- Demolish: in the beginning of the task the elements to demolish are denoted with the color red and disappear at the end.
- Temporary: for temporary constructions, the elements are coded yellow and disappear at the defined date.

Table 2. Maps of quantities—hole covers.

Specialty equipment—hole cover

Level	Type	Width (mm)	Length (mm)	Tab (mm)	Lower thickness (mm)	Upper thickness (mm)
Floor 0	Cover 2400*1500	1700	2600	100	50	50
	Cover 2400*1500	1700	2600	100	50	50
	Cover 2400*1500	1700	2600	100	50	50
	Cover 2400*1500	1700	2600	100	50	50
	Cover 2400*1500	1700	2820	100	50	50
Ceiling	Cover 300*300	500	500	100	50	50
	Cover 300*300	500	500	100	50	50
	Cover 300*300	500	500	100	50	50
	Cover 1800*800	1000	2000	100	50	50
	Cover 2000*989	1189	2200	100	50	50
	Cover 600*600	800	800	100	50	50
	Cover 1198*600	800	1398	100	50	50

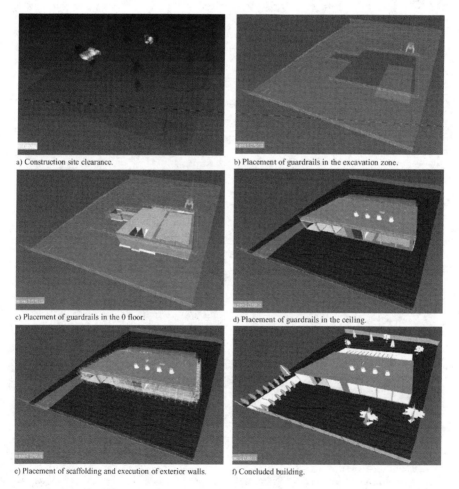

a) Construction site clearance.
b) Placement of guardrails in the excavation zone.
c) Placement of guardrails in the 0 floor.
d) Placement of guardrails in the ceiling.
e) Placement of scaffolding and execution of exterior walls.
f) Concluded building.

Figure 5. 4D simulation.

Figure 5 presents some simulation phases: a) represents the phase of construction site clearance, therefore it is attributed to the soil the color red; b) placement of guardrails in the excavation zone, these elements being temporary structures have the color yellow; c) placement of guardrails in the 0 floor, these elements are temporary consequently are yellow, the completed concrete structure, to that phase, has the final color of the project (color of the Revit model) and the foundations of the 0 floor are represented in green, since they are still being executed; d) placement of guardrails in the ceiling, these elements are represented in yellow; e) placement of scaffolding and execution of exterior walls, the scaffolding is coded yellow and the exterior walls while in execution are green; f) concluded building, all the works are concluded and the final aspect corresponds to the grafic model in Revit.

4 DISCUSSION

With the incorporation of comments and links in the 4D model, and the quantification of some safety elements, it is verified that Safety Technicians have at their disposal a whole set of information related to safety in a single software. It is possible to ascertain, following the simulation and consulting, at same time the safety documents, what measures need to be implemented at a specific time and in advance, throughout the whole construction time period. Specifically it is possible to analyze in detail the need for guardrails and hole covers in the construction phase, by extracting the maps of quantities.

The model is also a good tool to provide professional training to workers, about the whole process of construction, and the safety measures which need to be applied.

It is also verified that the more detailed the planning of work is, the more information is available in the Navisworks simulation, and consequently a greater control over the worksite is possible.

Another aspect that was verified and goes accordingly with BIM methodology was that changes can be made directly to the 3D model and those changes will be automatically updated to the 4D model, avoiding information loss. This means, there is an effective improvement of communication between the different stakeholders throughout the construction process, avoiding delays regarding taking knowledge of changes and its possible implementation on construction site.

All the control created with the use of this technology, makes the safety process more efficient, allowing less space for unforeseen events, and it is expected that this will contribute effectively to reduce accidents and to a more efficient safety implementation.

5 CONCLUSIONS

With this work, several lessons were withdrawn. Effectively, the construction sector has severe hazards and risks for workers, where the occupational accidents figures depict the highest fatalities.

Revit families, regarding safety at work, have flaws since there has been a need to model safety elements (guardrails and hole covers).

In contrast to the widely used 2D technology, the use of 4D models enables faster visualization and perception of the work development and its safety needs, enabling faster and more reliable decision-making.

The use of new methodologies, such as BIM, contributes to the optimization of the construction process, making it possible to apply effective safety measures in space and time, with real-time control and work planning knowledge through 4D simulations. BIM methodology fills the gaps in the traditional methodology (2D), by gathering the information through software, which has the ability to automatically update the changes made to the models, providing fast and automatic communication between all the stakeholders of the construction project.

As construction is a continuous process, it was also verified that there should be no separation between phases regarding safety. Given the current growing complexity of constructions, safety should not be thought and applied at the execution phase. Safety procedures entails costs that must be accounted for when budgeting the construction process. It is understood that the coordination of safety between all phases determines the solutions adopted in the design and execution phase, and that communication between all the stakeholders along all the process, and the use of multidisciplinary teams contribute to a more efficient and controlled construction process.

REFERENCES

Eastman, Chuck et al. 2011. *BIM Handbook—A Guide to Building Information Modeling for Owners, Managers, Designers, Engineers and Contractors*. 2nd Edi. New Jersey: John Wiley & Sons, Inc.

Enshassi, Adnan, Abed Ayyash, and Rafiq M. Choudhry. 2016. «BIM for construction safety improvement in Gaza strip: awareness, applications and barriers» 16 (3). Taylor and Francis Ltd.: 249–65. doi:10.1080/15623599.2016.1167367.

EU-OSHA. 2003. «Factsheet 36 – Accident prevention in the construction sector». https://osha.europa.eu/en/tools-and-publications/publications/factsheets/36/view.

EU-OSHA. 2004. «Factsheet 55 – Achieving better safety and health in construction». https://osha.europa.eu/en/tools-and-publications/publications/factsheets/55/view.

Zou, Yang, Arto Kiviniemi, and Stephen W. Jones. 2016. «A review of risk management through BIM and BIM-related technologies». *Safety Science*, Abril. Elsevier. doi:10.1016/j.ssci.2015.12.027.

Tower crane safety: Organizational preventive measures vs. technical conditions

J.A. Carrillo-Castrillo
Universidad de Sevilla, Sevilla, Spain

J.C. Rubio-Romero & M.C. Pardo-Ferreira
Universidad de Málaga, Málaga, Spain

ABSTRACT: The objective is to analyze the circumstances and causes of the accidents with cranes. Data is based on accident reports and official accident investigations of non slight tower crane accidents notified from 2003 to 2008 in Andalusia. Official accident investigation found that over 80% of the causes were latent. Only six accidents of one hundred and eleven could be directly attributed to the poor technical condition of the crane. Two technical conditions were found to be relevant: electrical and stability conditions. Although regulations tend to focus on technical features, our study shows that they are not the origin of most serious crane accidents. In fact, accidents investigated were mainly due to: poor work organization and planning; non-use of personal protection equipment; and improper work supervision by contractors at the construction site. Safety regulations in certain European countries are in need of improvement. Instead of only focusing on the technical conditions of tower cranes, regulations should also include the safe use and operation of this equipment.

1 INTRODUCTION

1.1 Crane safety

Tower cranes are an indispensable piece of equipment at construction sites but their assembly, maintenance, and operation can be hazardous for workers.

According to the Instituto Nacional de Seguridad e Higiene (National Institute for Health And Safety in Spain), hoisting equipment is the fifth most important source of work accidents, causing an even greater number of fatal accidents than scaffolding and ladders.

The material agent of deviation was a crane in 1.5% in fatal 43 accidents. (Instituto Nacional de Seguridad e Higiene en el Trabajo 2007, 2011).

Technological progress is helping to stem such accidents to a certain degree as global industrialization is leading to the widespread use of prefabricated construction units, which evidently requires more mechanical load handling equipment at building sites, contributing to a corresponding increase in the work accident rate.

Previous research on the safe use and operation of tower cranes is based on data from work accident reports.

Hakkinen (1978, 1993) highlighted that even though a large number of accidents were due to human errors, many could have been prevented by procedures, good pre-planning of lifting operations, and safety management.

Shepherd et al. (2000) quantifies crane serious accidents in the following categories and percentages: crane contact with overhead power lines (36%); falls of people from a height (17%): falls of a suspended load onto a person under the load (10%); crane overturns and falls onto people (7%); falls of lattice boom during dismantling (6%); and person caught between swinging counterweight and crane structure (3%).

Beavers (2006) concluded that crane operators and riggers should be qualified, and crane safety training should be provided to workers before they are allowed to work with or around cranes during lifting operations.

Mohan and Zech (2005), Suruda et al. (1999) and McCann (2003, 2009), among others, studied other types of hoisting equipment and their cost benefits, and also analyzed workplace accidents.

Specific risk factor affecting crane safety such as "overlapping cranes" and "operator proficiency" where deeply analyzed, concluding that it is necessary an intimate understanding of working environment (Shapira and Simcha 2009a). In that purpose, new methods such as AHP-based weighting of risk factors have been applied to provide new evidence of the importance of the human factors and the role of the crane operator and the general superintendent. (Shapira & Simcha 2009b).

1.2 Crane regulations

The safety requirements for this type of machinery are specifically regulated by European Union directives as well as by Spanish legislation.

1. European Union Safety and Health Directives pertaining to the product:
   - Directive 89/392/EEC, amended by Directive 91/368/EEC, on the approximation of the laws of the Member States relating to machinery.
   - Directive 84/528/EEC on the approximation of the laws of the Member States relating to common provisions for lifting and mechanical handling appliances.
2. European Union Safety and Health Directives pertaining to the workplace:
   - Framework Directive 89/391/EEC on the application of measures to promote the improvement of the safety and health of workers at the workplace.
   - Directive 89/655/EEC, amended by Directive 95/63/EEC, concerning the minimum safety and health requirements for the use of work equipment by workers at work.
   - Directive 96/57/EEC on minimum safety and health requirements at temporary or mobile construction sites.
3. Spanish Legislation pertaining to industrial safety:
   - Royal Decree 836/2003, regulating technical requirements of tower cranes in construction work and other applications.

Such regulations tend to focus on the technical conditions of tower cranes rather than on their safe use and operation.

The crane and its description are included by each contractor in the work plan. Even though this document must be approved by an engineer or architect, its exact format and content are not regulated. This work plan is extremely important because it specifies the use of the crane, coordination with other contractors, and the safety procedures to be implemented.

1.3 Crane accident causes

As is well known, the cause of workplace accidents is a very complex issue. According to most authors, it is not only necessary to specify active causes (technical conditions), but also latent causes (those related to organizational and human factors).

Tower crane accidents are due to both active and latent causes (Aneiziris 2008; Swuste 2013) just the same as any other construction accident (Carrillo-Castrillo et al. 2017).

Previous studies have shown the incompleteness of a safety analysis based only in the accident records and the need of finding the root causes of accidents with cranes. (Shapira and Lyachin 2009)

The integration of all these risk factors (technical, organizational and personal) is needed to assess the overall safety index of a crane operation (Shapira & Lyachin 2009).

Although many technical improvements have been proposed to control the risk of crane operation, (Shapira et al. 2008; Zhang & Hammad 2012; AlBahanassi & Hammad 2012; Li et al. 2012) the safety of crane operation in Spain still relies most of the times in the skills, expertise and training of the crane operator and supervisors and in an appropriate organization of the tasks.

2 METHODS

In Spain, accident notifications are electronically collected in Official Workplace Incident Notification Forms. All accidents that result in an absence from work of 1 day or more must be notified.

In terms of severity, accidents can be slight (minor) or non-slight (major). Medical criteria are applied to classify the accident as slight or non-slight, depending on the expected time to recover.

For non-slight accidents the Andalusian Labour Authority decide whether to conduct an official accident investigation after the accident notification has been received. These investigations are carried out according to internal procedures and an official extended investigation report is submitted. Details of notification and investigations are explained elsewhere (Carrillo-Castrillo et al. 2017).

This research study analyzed all non-slight tower crane accidents that occurred during 2003–2008 (105 accidents) in order to identify their main causes and presents the results of the inspections of a sample of cranes in order to monitor the

Table 1. Classification of crane accidents.

Electrical contact
Fall of ground
Falls of boom
Falls of suspended load
Swinging load
Fall of part of crane
Fall of people from aerial lift
Fall of people from construction areas
Fall of people from suspended load
Fall of people from crane cabin
Fall of people from ladder of the crane
Fall of crane/overturn
Caught in between, struck by
Run over

*According to Shepherd (2000).

actual working conditions at construction sites of these pieces of equipment.

The results obtained provided valuable insights into tower crane regulations, the incidence of technical conditions in work accidents involving cranes, and possible measures to reduce the number of this type of accident.

For twenty of these accidents, an official accident investigation report is available, including the cause/s identified by official investigators. Those causes were coded according to the national system established for that purpose (Carrillo-Castrillo & Onieva 2013).

These accidents were then studied with a view to ascertaining whether the technical conditions of the crane had played a role in the accident.

Moreover, the main circumstances of the accident were specified and classified, based on the type of damaging energy involved (see Table 1) adapted from Shepherd (2000).

3 RESULTS

3.1 *Accident circumstances in accident notifications*

The analysis of severe and fatal accidents showed that only six accidents could be directly attributed to the poor technical condition of the crane (see Table 2).

Two of the nine fatal accidents were due to the fall of the crane. Most of the fatal accidents were caused by a swinging load (3 cases) or falling load (3 cases). In regards to the severe accidents,

Table 2. Non-slight accidents reported from 2003 to 2008 in Andalusia involving cranes.

Accident type	No. acc (fatal)	Technical conditions
Electrical Contact	1 (0)	1
Setting	1 (0)	0
Fall of boom	0 (0)	0
Fall of load	16 (3)	0
Swinging load	16 (3)	0
Fall of part of crane	9 (1)	0
Fall of people from lift	0 (0)	0
Fall of people from suspended load	13 (1)	0
Fall of people from crane cabin	0 (0)	0
Fall of people from ladder of the crane	3 (0)	0
Fall of crane/overturn	3 (2)	5
Caught in between, struck by	40 (0)	0
Run over	0 (0)	0
Total	105 (9)	6

*Accident type according to Shepherd (2000).

Table 3. Causes coded in the accidents that were officially investigated from 2003 to 2008 according to Fraile (2011).

Type of cause	Group of cause	No. Acc	No. Causes
Active	Workplace	3	4
	Equipment	5	5
	Materials	1	1
Latent	Organizational	14	17
	Safety management	5	9
	Human factors	8	9
	Other	4	7
Total		40	52

*Accidents can have more than one cause.

the most prevalent type involved the victim being caught in between or being struck by the load or the accessories.

There were also many severe accidents caused either by a swinging or falling load.

Only two technical conditions among the ones included in the technical regulations were found to be relevant: electrical and stability conditions.

3.2 *Accident causes in investigation reports*

Regarding the twenty accidents in our study that were officially investigated, inspectors found that over 80% of the causes were latent (see Table 3).

The most frequent cause of accidents was faulty work method, which is a latent cause because it is one of the set of organizational causes. This cause was identified in eight of the twenty accidents (40%).

4 DISCUSSION

Nevertheless, this analysis of severe and fatal accidents clearly shows that it is not enough to improve the technical conditions of cranes.

The results obtained in our study reflect that most of the crane accidents were caused by unsafe work procedures, lack of coordination, and insecure loading (Shapira & Simcha 2009a, 2009b). Besides, as in most accidents in construction, human factors are one also an important cause (Carrillo-Castrillo et al., 2017).

In order to foment the safe use of cranes, it is thus necessary to regulate how work is carried out at the construction site.

Despite the fact that each contractor has to elaborate a work plan, the fact that the form and content of this document is not regulated may be the latent cause of many accidents.

This study has also identified preventive measures that should be implemented. An effective work plan should guarantee that workers will not be working under a load or at risk of an impact by a swinging load.

Furthermore, crane operators as well as workers handling loads should carry out their work in such a way so as not to be at risk of falling from a height.

These identified measures can be implemented with proper work organization and coordination and should be enforced by regulations. Instead of only focusing on the technical conditions of tower cranes, regulations should also include the safe use and operation of this equipment and proper safety management.

REFERENCES

AlBahnassi, H. and Hammad, A. 2012. Near Real-Time Motion Planning and Simulation of Cranes in Construction: Framework and System Architecture. *Journal of Computers in Civil Engineering* 26(1), 54–63.

Aneziris, O.N., Papazoglou, I.A., Mud, M.L. et al. 2008. Towards risk assessment for crane activities. *Safety Science* 46(6): 872–884.

Beavers, J.E., Moore, J.R., Rinehart, R., & Schriver, W.R. 2006. Crane-Related Fatalities in the Construction Industry. *Journal of Construction Engineering Management* 132(9): 901–910.

Carrillo-Castrillo, J.A., Trillo-Cabello, A.F. & Rubio-Romero, J.C. 2017. Construction accidents: identification of the main associations between causes, mechanisms and stages of the construction process. *International Journal of Occupational Safety and Ergonomics* 23(2): 240–250.

Carrillo-Castrillo, Rubio-Romero, J.C. & Onieva, L. 2013. Causation of Severe and Fatal Accidents in the Manufacturing Sector. *International Journal of Occupational Safety and Ergonomics* 19(3): 423–434.

Fraile A. 2011. *NTP 924: Causas de accidentes: clasificación y codificación*. Spain: Instituto Nacional de Seguridad e Higiene en el Trabajo.

Hakkinen, K. 1978. Crane accidents and their prevention. *Journal of Occupational Accidents* 1: 353–361.

Hakkinen, K. 1993. Crane accidents and their prevention revisited. *Safety Science*, 16(2): 267–277.

Instituto Nacional de Seguridad e Higiene. 2007. *Análisis de la mortalidad por Accidentes de Trabajo 2002–2004*. Spain; Instituto Nacional de Seguridad e Higiene en el Trabajo.

Instituto Nacional de Seguridad e Higiene. 2011. *Análisis de la mortalidad por Accidentes de Trabajo 2005-2007*. Spain: Instituto Nacional de Seguridad e Higiene en el Trabajo.

Li, H., Chan, G. & Skitmore, M. 2102. Multiuser Virtual Safety Training System for Tower Crane Dismantlement. *Journal of Computers in Civil Engineering* 26(5), 638–647.

McCann, M. 2003. Deaths in construction related to personnel lifts. *Journal of Safety Research* 34(5), 507–514.

McCann, M., Gittleman, J. and Watters, M. 2009. Crane-Related Deaths in Construction and Recommendations for Their Prevention. USA: The Center for Construction Research and Training.

Mohan, S. & Zech, M.C. 2005. Characteristics of worker accidents on NYDOT construction projects. *Journal of Safety Research* 36(4): 353–360.

Shapira, A. & Simcha, M. 2009a. Measurement and Risk Scales of Crane-Related Safety Factors on Construction Sites". *Journal of Construction Engineering Management* 135(10): 979–989.

Shapira, A. & Simcha, M. 2009b. "AHP-Based Weighting of Factors Affecting Safety on Construction Sites with Tower Cranes". *Journal of Construction Engineering Management* 135(4): 307–318.

Shapira, A. and Lyachin, B. 2009. Identification and Analysis of Factors Affecting on Construction Sites with Tower Cranes. *Journal of Construction Engineering Management* 135(1): 22–33.

Shapira, A. Rosenfeld, Y. & Mizrahi, I. 2008. Vision System for Tower Cranes. *Journal of Construction Engineering Management* 134(5): 320–332.

Shapira, A., Simcha, M. & Goldenberg, M. 2012. Integrative Model for Quantitative Evaluation of Safety on Construction Sites with Tower Cranes. *Journal of Construction Engineering Management* 138(11): 1281–1293.

Shepherd, G.W., Kahler, R.J. & Cross, J. 2000. Crane fatalities—a taxonomic analysis. *Safety Science* 36(2): 83–93.

Suruda, A., Liu, D., Egger, M., & Lillquist, D. 1999. Fatal injuries in the United States construction industry involving cranes 1984–1994. *Journal of Occupational and Environmental Medicine* 41(12): 1052–1058.

Swuste, P. 2013. A 'normal accident' with a tower crane? An accident analysis conducted by the Dutch Safety Board. *Safety Science* 57: 276–282.

Zhang, C. & Hammad, A. 2012. Multiagent Approach for Real-Time Collision Avoidance and Path Replanning for Cranes. *Journal of Computation in Civil Engineering* 26(6): 782–794.

Occupational and public exposure to radionuclides in smoke from forest fires—a warning

F.P. Carvalho, J.M. Oliveira & M. Malta
Laboratório de Protecção e Segurança Radiológica/Instituto Superior Técnico (LPSR/IST), Universidade de Lisboa, Bobadela LRS, Portugal

ABSTRACT: Vegetation fires release large amounts of smoke particles which carry artificial and natural radionuclides into the atmosphere. Naturally-occurring radionuclides were found to be concentrated from wood to smoke materials by factors of 10^2–10^3 on a mass basis. These radionuclides are present in smoke particles (fly ash) from all sizes, from <0.5 μm to >7 μm. Polonium-210, which is the most volatile among all radionuclides analyzed and the most radiotoxic, becomes concentrated especially in the smallest smoke particles, <0.5 μm, which are inhalable. Prolonged breathing in smoked air, as it occurs often during wild forest fires, may give rise to increased inhalation of radionuclides carried by smoke, and polonium-210 alone may originate a radiation dose to the lungs some 2000 times higher than the natural radioactivity background. Altogether natural radionuclides, especially those from uranium and thorium series including polonium-210, may give rise to radiation doses exceeding the dose limit of 1 mSv/year adopted for protection of members of the public. This risk of radiation exposure from smoke makes highly recommended the protection of respiratory tract when breathing in areas affected by forest fires. Although fire fighters generally wear gear to protect the respiratory tract, members of the public often do not and this is likely to having radiological exposure consequences in the population. Recommendations are made to enhance occupational and public safety.

1 INTRODUCTION

Exposure to smoke from forest fires has confirmed health impacts, such as respiratory and cardiovascular effects and it is responsible for enhanced mortality (Adetona et al. 2016, Liu et al 2016, Kollanus et al. 2017). Health effects, such as inflammatory diseases, asthma, and bronchitis have been related with exposure to particulate concentrations in the air, such as particulate matter with aerodynamic diameter lower than 10 and 2.5 μm (PM10 and PM2.5, respectively) (Vos et al. 2009, Liu et al. 2016). Beyond physical particle effects, chemical composition of smoke and gases released from forest fires into the atmosphere may play a key role. Chemical composition of smoke is complex and a very large number of toxic chemical compounds have been measured, such as polycyclic aromatic hydrocarbons (PAH), formaldehyde, acrolein, dioxin, amongst many others (Liu et al. 2016). Radioactivity has not been investigated as a toxic (radiotoxic) risk factor present in wildfire smoke, as recent reviews of exposure to forest fires clearly highlighted (Liu et al 2016, Kollanus et al. 2017).

First concerns about the presence of radioactivity in the air in association with forest fires occurred in the summer of 2008 with large fires in Belarus and Ukraine. These fires devastated thousands of hectares of forest that had been heavily contaminated by the intense deposition of cesium-137 originated from the nuclear disaster at the Chernobyl power station, on the outskirts of Kiev, Ukraine, in 1986 (Paatero et al. 2009). In the cloud of smoke resulting from these fires and swept into West Europe by winds, cesium-137 was detected, causing again some alarm, but concentrations were lower than those measured in 1986 in the radioactive cloud from Chernobyl (Paatero et al. 2009).

Although the radioactivity released into the atmosphere by the Chernobyl nuclear accident in 1986, and later by the Fukushima nuclear accident in 2011, and deposited onto soils and vegetation in Portugal was much smaller than the radioactivity deposited in countries at North and Central Europe (Masson et al. 2011, Carvalho et al 2012), this topic motivated research which purpose was to investigate concentrations of radionuclides that could be resuspended in the atmosphere by forest fires. The exposure to radiation doses added to the natural radiation background are regulated and shall not exceed 1 mSv/year in order to protect the members of the public from harmful effects of ionizing radiation (EU Directive 59/2013).

Results from our measurements of naturally occurring radionuclides in smoke turned out to be more worrisome in this region than artificial radionuclides originated from nuclear accidents referred above (Carvalho et al. 2014). Very recent research on radiation doses received by firefighters in radioactive contaminated areas still neglects this source of radiation exposure (Vinera et al. 2018).

As Portugal and countries around the Mediterranean basin are often hit by forest fires during summer, an assessment of radioactivity exposure through inhalation of smoke is made herein. Radiation doses from smoke inhalation are compared with dose limits for radiation protection, the results are discussed, and recommendations are made.

2 MATERIALS AND METHODS

Samples of vegetation were collected in the district of Viseu, centre-north of the country, to determine the concentrations of artificial radionuclides and natural radionuclides belonging to the radioactive uranium and thorium series, both accumulated in the plant biomass. Vegetation samples, after washing off dust particles, were dried in the oven, ground, and dissolved in mineral acids (HCl and HNO_3) for radionuclide analysis.

The smoke particles (fly ash) from forest fires were collected using large volume aerosol samplers and glass microfiber filters. These samplers (Andersen and FJ & G) were placed in the proximity to forest fires and in villages in the smoke plume from wild forest fires as described before (Carvalho et al. 2014). Samples of smoke particles collected on filters, generally each one weighing several hundreds of mg, were dissolved in mineral acids for radionuclide analysis. Smoke aerosols were sampled also with a six stage cascade impactor (Andersen) for analysis of radionuclides in six particle size fractions, from <0.5 to >7.6 μm, to assess radioactivity in inhalable aerosol particles.

The radionuclides contained in dissolved samples were then separated by radiochemical procedures to obtain various fractions with elements such as uranium, thorium, radio, radioactive lead and polonium, duly purified. These radioelements were then electrodeposited on stainless steel or silver discs, and the activities and concentrations of the radionuclides determined by alpha spectrometry using procedures validated and published elsewhere (Oliveira & Carvalho 2006). The analytical quality control was performed through the use of isotopic tracers added to each sample at the beginning of the analytical procedure and through analysis of certified reference materials and participation in international inter-laboratory comparison exercises with good results (Carvalho & Oliveira 2007, Oliveira & Carvalho 2006).

Environmental parameters, such as the soil-to-plant transfer factor (TF = (Bq/kg plant)/(Bq/kg of soil)), and computation of doses from the inhalation route through activity-to-dose conversion factors are made according definitions by the International Atomic Energy Agency (IAEA 2010, 2014).

3 RESULTS AND DISCUSSION

Activity concentrations of main naturally-occurring alpha emitting radionuclides in vegetation, forest tree materials, and smoke from fires are given in Table 1. The concentrations of these naturally occurring radionuclides in plant materials were generally low, and most of the times were below 10 Bq/kg (dry weight) as reported before already (Carvalho et al. 2011). Radionuclide concentrations in plant materials are much lower than concentrations of the same radioelements in surface soils, where for example uranium (^{238}U) and radium (^{226}Ra) display activity concentrations of 20–150 Bq/kg (dry weight) depending on the geology of the area and composition of soils (Eisenbud and Gesell 1997, Carvalho et al. 2011). The much lower radionuclide concentrations in plants compared to soils is a result of reduced root absorption of these radioelements, which generally have no known biochemical or physiological function in plants (IAEA 2010). The soil-to-plant radionuclide transfer factor (TF) vary depending on soil type, soil chemistry and plant species but, generally, were at around 10^{-1} to 10^{-2} for natural radionuclides (see Table 1) and are of 1.6×10^{-4} for plutonium and 6.3×10^{-2} for cesium (IAEA 2010).

Analysis of smoke particles released from forest fires into the atmosphere revealed high concentrations for all those radionuclides in smoke as compared to radionuclide concentrations in the vegetation and surface soils. This increase of concentration of radionuclides in smoke particles occurs due to mass reduction of plant materials by combustion (loss of water, destruction of the organic compounds, etc.). Thus, while in pine wood for example the concentration of uranium was 0.1 Bq/kg (dw), its concentration in the smoke from pine wood forest in fire was about 300 Bq/kg (dw), i.e. about 3×10^3 higher on a mass basis, and reaching a concentration of about 1 mBq/m^3 in the air with a particle load of about 1 mg/m^3. Similar increases in radionuclide concentrations (on mass or volume basis) were recorded for all natural radionuclides analyzed. In Table 1, the plant-to-smoke enrichment factors are given for

Table 1. Activity concentrations (±1 SD) of alpha emitting radionuclides in samples from environmental materials including smoke from forest fires, and transfer parameters.

	^{238}U	^{234}U	^{230}Th	^{226}Ra	^{210}Po
A	0.94 ± 0.03	0.94 ± 0.03	0.90 ± 0.05	5.3 ± 0.4	3.45 ± 0.06
B	0.29 ± 0.01	0.30 ± 0.01	0.089 ± 0.08	26 ± 2	2.6 ± 0.06
C	0.114 ± 0.004	0.115 ± 0.004	0.23 ± 0.002	0.96 ± 0.13	1.68 ± 0.05
D	1.99 ± 0.07	1.91 ± 0.07	2.1 ± 0.1	4.9 ± 0.3	3.10 ± 0.08
E	0.105 ± 0.006	0.112 ± 0.006	0.067 ± 0.006	1.14 ± 0.06	0.97 ± 0.03
F	2.4 ± 0.1	2.5 ± 0.1	2.2 ± 0.6	31 ± 6	62 ± 2
G	213 ± 10	220 ± 11	216 ± 17	2404 ± 718	635 ± 34
H	262 ± 8	262 ± 8	230 ± 10	944 ± 150	2390 ± 118
I	133 ± 11	132 ± 4	116 ± 5	475 ± 75	1200 ± 60
J	195	200	170	60	510
L	4.6×10^{-2}	4.6×10^{-2}	4.2×10^{-2}	1.3×10^{-1}	1.2×10^{-1}

Samples and parameters:
A—Concentrations in herbaceous vegetation (Bq/kg dry weight);
B—Concentrations in Eucalyptus bark (from Caldas de Felgueira, Viseu);
C—Concentrations in Eucalyptus wood (from C. Felgueira, Viseu);
D—Concentrations in Pine tree needles (from C. Felgueira, Viseu);
E—Concentrations in Pine tree wood (from C. Felgueira, Viseu);
F—Concentrations in surface air with no forest fires (μBq/m^3), and aerosol particle load = 71 μg/m^3;
G—Concentrations in surface air at a village downwind forest fire (μBq/m^3) and aerosol particle load (smoke) = 0.469 mg/m^3;
H—Volumic activity concentrations (μBq/m^3) in surface air near a forest fire and aerosol particle load (smoke) = 1.99 mg/m^3;
I—H sample but results given in massic activity concentrations (Bq/kg) in surface air near a forest fire and aerosol particle load = 1.99 mg/m^3;
J—Plant-to-smoke radionuclide enrichment factor near the forest fire;
L—Soil-to-plant radionuclide Transfer Factor (from IAEA 2010).

radionuclides based on the averaged results for herbaceous and tree materials shown in the table.

In surface soils of Portugal, artificial radionuclides such as radioactive cesium and plutonium are present due to global radioactive fallout. The concentrations in soils are at about 1 Bq/kg for ^{137}Cs and also at around 1 Bq/kg for $^{239+240}$Pu (measured in Azores). Using known soil-to-plant transfer factors for grasslands in temperate climates of 6.3×10^{-2} for ^{137}Cs and 1.6×10^{-4} for $^{239+240}$Pu, the expected concentrations in vegetation would be 0.063 Bq/kg for ^{137}Cs and 1.6×10^{-4} Bq/kg for plutonium. Combustion of this vegetation would release these radionuclides into the atmosphere with activity concentrations on a mass basis of about 3.7 Bq/kg for ^{137}Cs (calculated using the plant-to-ash enrichment factor of radium, a chemical analog to cesium), and 0.0272 Bq/kg of $^{238+240}$Pu (calculated using the plant-to-ash enrichment factor of thorium, a chemical analog to plutonium, Table 1). In the smoke near a forest fire, with a particle load of 1.99 mg/m^3 as in the case shown in Table 1, the expected concentrations in the air would be 7.3×10^{-6} Bq/m^3 (7.3 μBq/m^3) for ^{137}Cs and 5.4×10^{-8} Bq/m^3 (54 nBq/m^3) for $^{239+240}$Pu. Indeed, radionuclide measurements made in the atmosphere near Lisbon, in the absence of fires, gave results for ^{137}Cs < 2 μBq/m^3, and at about 5 nBq/m^3 for $^{239+240}$Pu (Carvalho et al. 2012).

Calculations and experimental determination of concentrations of the most abundant artificial radionuclides in the atmosphere indicate that they can be expected to be present in smoke from forest fires also, but at concentration levels 8 to 10 orders of magnitude lower than the concentrations of naturally occurring radionuclides shown Table 1. Therefore, these artificial radionuclides with current concentrations in soils, are of little or no radiological concern in the present.

Analysis of the results for natural radionuclides (Table 1) showed that, following biomass combustion, for some radionuclides there was more than a simple reduction of plant mass beyond their activity enrichment in ashes (Bq/kg). Radium-226 and ^{210}Pb, and especially polonium (^{210}Po) whose compounds have a low volatilization point (below 300°C), have been much more concentrated in the smoke particles than the refractory radionuclides such as uranium and thorium (boiling points well above 1000°C). For all radionuclides released into the atmosphere by forest fires, fly ash and smoke are a particulate carrying phase in the atmosphere but, for some radionuclides especially ^{210}Po, volatilization seems to occur also and the radionuclide is released by burning from plants in the gas phase. In the atmosphere, the gaseous radionuclide form(s) rapidly condensate onto smoke particles, which further increases the radionuclide concentration in the small aerosol particles.

Analyzes of smoke particles separated by aerodynamic particle size were performed using a cascade impactor in six fractions, ranging from <0.5 μm to >7 μm in diameter. Analysis of the radionuclides in these fractions revealed that the higher concentrations of ^{210}Po were associated with the smaller particles, while the majority of the radioelements such as uranium and thorium were distributed in nearly similar concentrations throughout all fractions (Figure 1).

The evaluation of radioactivity associated with smoke particles and inhaled with the 24 hours breath in an atmosphere with forest fire smoke, led to the conclusion that the radiation dose from inhaled polonium alone, to the lung, can be about 5 μSv per day (Table 2). This radiation dose is 2000 times higher than the radiation dose received by a person breathing outdoors in the absence of smoke, thus exposed to natural radioactive background only (Carvalho et al. 2011).

The radiation dose levels for the lung, calculated for ^{210}Po from inhalation in a dense smoke atmosphere, may be much greater (about 50-fold) than the radiation dose received by a smoker of 20 cigarettes per day, a dose which is already considered excessive and possibly carcinogenic (Carvalho & Oliveira 2006). Today, there is sound knowledge about bioaccumulation and radio-toxicity of ^{210}Po to humans (Harrison et al. 2007, Carvalho and Oliveira 2009). During the summer period, in regions of central Portugal, firefighters from Fire Brigades have been exposed more than 100 days to dense smoke from forest fires. Members of the public helping firefighting have been exposed for similar periods of time, and members of the community in areas downwind fires experienced at least 25 days of smoke exposure in the summer. Radiation doses through inhalation of radionuclides carried by smoke particles are increased with higher smoke density and with longer exposure time. Calculated effective doses are shown in Table 2 as indicative values. The annual radiation dose received by a firefighter (assuming 100 days exposure in one year) may attain 0.7 mSv/year due to ^{210}Po only. Adding the contribution of other alpha emitting radionuclides inhaled, such as U, Ra, Th isotopes, the occupational radiation dose might largely exceed the dose limit of 1 mSv/year (EU Directive 59/2013).

Exposure to ionizing radiation may trigger lung cancer if radiation dose is sufficiently high and repeated. However, so far no studies were made to relate exposure to forest fire smoke with lung cancer. We hypothesize that radiation exposure of lung may be very high in firefighters and members of the public exposed to smoke inhalation. Radiation doses

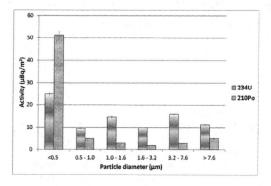

Figure 1. Activity concentration in aerosols near a forest fire, by aerodynamic aerosol particle size classes for a volatile (^{210}Po) and a refractory (^{234}U) naturally occurring radionuclide.

Table 2. Calculated effective radiation doses through inhalation of the key radionuclide ^{210}Po in several circumstances. Activity-to-Dose conversion factor for ^{210}Po: 3.3×10^{-6} Sv/Bq.

	^{210}Po concentration in the air	^{210}Po inhaled per day (22 m³)	Effective dose	
			Sv/day	Sv/year
Member of the public (background exposure)	31 μBq/m³	6.82×10^{-4} Bq/d	2.25×10^{-9} Sv/d	8.21×10^{-7} Sv/a (0.821 μSv/a)
Smoker of one pack cigarettes per day	30 mBq per cigarette	0.6 Bq/d	9.9×10^{-8} Sv/d	3.6×10^{-5} Sv/a (0.036 mSv/a)
Firefighter near the fire without protective mask (exposure 100 day/year)	100 mBq/m³	2.2 Bq/day	7.26×10^{-6} Sv/d	7×10^{-4} Sv/a (0.7 mSv/a)

that might be imparted by prolonged smoke inhalation to the lung tissues, approach radiation dose values per cell turnover causing injury to the lungs, as investigated in other contexts (Puukila et al. 2017).

Often, firefighters use protective masks to avoid smoke inhalation and some members of the community helping to control the flames protect also their respiratory tract with pieces of cloth, thus avoiding to inhaling the entire load of smoke particles and gases released by fires. However, members of the population downwind the fires are often fully exposed to smoke inhalation for long periods of time (Table 1). To abate occupational exposures of firefighters, the use of duly tested respiratory tract protective gear efficient for radioactive aerosols should be mandatory.

4 CONCLUSIONS

The current level of artificial radionuclides from nuclear fallout in vegetation and soil, and their re-suspension in the atmosphere caused by wind or forest fires, does not reached levels of much concern and does not seem to constitute in the present a significant radiological risk in the region. However, the release of natural radionuclides present in the vegetation in low concentrations, followed by re-concentration in smoke particles, raises radioactivity levels in the atmosphere to much higher levels. In particular, polonium-210, an alpha particle emitter known as highly radiotoxic that is volatilized during biomass combustion and condensed on smoke particles, is present in higher concentrations in small and inhalable smoke particles.

The particularity of vegetation and forest fires to concentrate polonium-210, raising its concentration in inhalable smoke particles, has been ignored to date. Although many toxic substances exist in smoke, and the inhalation of the particles is in itself irritating to the pulmonary epithelium, polonium carried by inhaled smoke particles adds a disturbing radio-toxicity component to lungs. Inhalation of smoke particles from forest fires over long periods originates relatively high doses of ionizing radiation to the lung. To avoid exposure to this radiation dose, protection of the respiratory tract should be strongly recommended or even mandatory especially for firefighters and elements of the population in the vicinity of forest fires.

The effectiveness of different types of filters and masks available for respiratory protection should be assessed for this hazard in order to ensure occupational radiation safety to firefighters.

The radiological impact of smoke inhalation by large populations on global public health requires full assessment in order deduce appropriate safety recommendations.

REFERENCES

Adetona, O., Reinhardt, T.E., Domitrovich J., Broyles G., Adetona A., Klieinman M.T., Naeher L.P. 2016. Review of the health effects of wildland fire smoke on wildland firefighters and the public. *Inhalation Toxicology* 28(3), 95–139.

Carvalho, F.P., Oliveira, J.M. 2007. Alpha emitters from uranium mining in the environment. *Journal of Radioanalytical and Nuclear Chemistry* 274, 167–174.

Carvalho, F.P., Oliveira, J.M., Malta, M. 2011. Vegetation fires and release of radioactivity into the air. In: *Environmental Health and Biomedicine, WIT Transactions on Biomedicine and Health* 15, 3–9.

Carvalho, F.P., Oliveira, J.M., Malta, M. 2014. Exposure to radionuclides in smoke from vegetation fires. *Science of the Total Environment* 472, 421–424.

Carvalho, F.P., Oliveira J.M. 2006. Polonium in Cigarette Smoke and Radiation Exposure of Lungs. (Proceedings of the 15th Radiochemical Conference). *Czechoslovak Journal of Physics* 56 (Suppl. D), 697–703.

Carvalho, F.P., Oliveira, J.M. 2009. Bioassay of 210Po in human urine and internal contamination of man. *Journal of Radioanalytical and Nuclear Chemistry* 280, 363–366.

Carvalho, F.P., Reis, M.C., Oliveira, J.M., Malta, M., Silva, L. 2012. Radioactivity from Fukushima nuclear accident detected in Lisbon, Portugal. *Journal of Environmental Radioactivity* 114, 152–156.

Eisenbud, M., Gesell, T. 1997. Environmental Radioactivity from Natural, Industrial, and Military Sources, 4th Edition. *Academic Press*, New York.

EU Directive 59/2013. COUNCIL DIRECTIVE 2013/59/ Euratom of 5 December 2013 laying down basic safety standards for protection against the dangers arising from exposure to ionizing radiation and repealing Directives 89/618/Euratom, 90/641/Euratom, 96/29/ Euratom, 97/43/Euratom and 2003/122/Euratom. *Official Journal* No. L 13 of 17/1/2014, p. 01–72.

Harrison, J., Leggett, R., Lloyd, D., Phipps, A., Scott, B. 2007. Polonium-210 as a poison. *Journal of Radiological Protection* 27, 17–40. doi:10.1088/0952-4746/27/1/001.

IAEA 2010. Handbook of parameter values for the prediction of radionuclide transfer in terrestrial and freshwater environments. Technical reports series no. 472. Vienna, *International Atomic Energy Agency*.

IAEA 2014. Radiation Protection and Safety of Radiation Sources: International Basic Safety Standards", IAEA Safety Standards, General Safety Requirements Part 3. Vienna, *International Atomic Energy Agency*.

Kollanus, V., Prank, M., Gens, A., Soares, J., Vira, J., Kukkonen, J., Sofiev, M., Salonen, R.O., Lanki, T. 2107. Mortality due to vegetation-fire originated PM2.5 exposure in Europe: assessment for the years 2005 and 2008. *Environmental Health Perspectives* 125(1), 30–37.

Liu, Z., Murphy, J.P., Maghirang, R., Devlin, D. 2016. Health and Environmental Impacts of Smoke from Vegetation Fires: A Review. *Journal of Environmental Protection* 7, 1860–1885.

Masson, O., Baeza, A., Bieringer, J., Brudecki, K., Bucci, S., et al. 2011. Tracking of airborne radionuclides from the damaged Fukushima Dai-Ichi nuclear reactors by

European networks. *Environmental Science and Technology* 45 (18), 7670–7677.

Oliveira, J.M., Carvalho, F.P. 2006. A Sequential Extraction Procedure for Determination of Uranium, Thorium, Radium, Lead and Polonium Radionuclides by Alpha Spectrometry in Environmental Samples. (Proceedings of the 15th Radiochemical Conference). *Czechoslovak Journal of Physics* 56 (Suppl. D), 545–555.

Paatero, J., Vesterbacka, K., Makkonen, U., Kyllonen, K., Hellen, H., Hatakka, J., et al. 2009. Resuspension of radionuclides into the atmosphere due to forest fires. *Journal of Radioanalytical and Nuclear Chemistry* 282(2), 473–6.

Puukila, S., Thome, C., Brooks, A.L., Woloschak, G., Boreham, D.R. 2017. The Role of Radiation Induced Injury on Lung Cancer. *Cancers* (Basel). 2017 Jul 12; 9(7).

Vinera, J.B., Jannika, T., Hepworthb, A., Adetonac, O., Naeherd, L., Eddye, T., Domane, E., Blake, J. 2017. Predicted cumulative dose to firefighters and the off-site public from natural and anthropogenic radionuclides in smoke from wildland fires at the Savannah River Site, South Carolina USA. *Journal of Environmental Radioactivity* 182, 1–11 (online on December 2017).

Vos, A.J.M.D, Reisen, F., Cook, A., Devine, B., Weinstein, P. 2009. Respiratory irritants in Australian bushfire smoke: air toxics sampling in a smoke chamber and during prescribed burns. *Archives of Environmental Contamination and Toxicology* 56, 380–388.

Annual school safety activity calendar to promote safety in VET

S. Tappura & J. Kivistö-Rahnasto
Tampere University of Technology, Tampere, Finland

ABSTRACT: Safety management is based on systematic actions towards a safer and healthier work environment. Vocational Education and Training (VET) institutions engage in a variety of activities to manage school safety and Occupational Health and Safety (OHS). In this study, an annual calendar of school safety activities was constructed to promote both school safety and OHS. Six large VET institutions participated in the study. Typical safety activities that could be undertaken by VET institutions to improve school safety are suggested. Yearly school safety activities include safety introduction, safety exercises, internal safety inspections, safety walks, personnel safety trainings and top management reviews. They also include regular update of risk assessments, rescue plans, chemical information and machine inspections. The activity list can be supplemented with the OHS responsibilities for top and middle management, teachers, support services personnel and OHS experts. The results can be utilised to develop school safety at various school levels.

1 INTRODUCTION

1.1 VET in Finland

In 2017, there are 165 Vocational Education and Training (VET) institutions in Finland involving about 250,000 students per year (Ministry of Education and Culture 2017a). Currently, VET institutions are facing major changes due to vocational education reform, economic pressures and a decrease in student numbers among future age groups (HE 39/2017, Pirhonen 2014).

The number of VET providers has decreased since previous years (Pirhonen 2014), and there will be 144 VET institutions from 2018 onward (HE 39/2017). At the same time, VET institutions have become larger and more multifaceted in their respective educational fields (Tappura et al. 2017a, 2017b).

1.2 Promoting safety in VET

VET institutions operate under both education and occupational health and safety (OHS) regulations (A 630/1998, A 738/2002). Regarding personnel and contractors, VET institutions have an obligation to care for their physical and mental health and safety at work (A 738/2002). In addition, students have a right to a safe learning environment (A 630/1998). Moreover, students act as employees when they work under the supervision of a VET provider during their vocational education (A 738/2002, FNAE 2017, FNBE 2011, Tappura 2014).

However, in Finnish VET provider organisations, OHS has often been developed via projects with limited resources and competence and from students' perspective during on-the-job learning (Tappura 2014, 2012). Thus, further research is needed in this area, and a broader consideration of occupational safety is needed in VET.

Previous studies on school safety are often focused on students' safety and school injuries (e.g., Laflamme & Menckel 1997, Langley et al. 1990, Mytton et al. 2008, Sun et al. 2006), whereas the OHS of the school personnel has been considered inadequately in school safety studies (Lopez Arquillos et al. 2017, Tappura et al. 2017a). However, school injuries can also be regarded

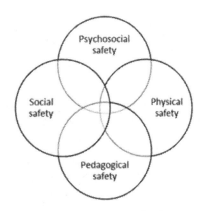

Figure 1. The sections of school safety (modified from FNAE 2017).

as occupational injuries (Laflamme & Menckel 1997). Nevertheless, the number of available studies on occupational safety risks in schools is limited (Lopez Arquillos et al. 2017).

The current school safety models typically focus on student safety. School safety is often defined as a combination of the physical, psychological, social and pedagogical safety of students (see Figure 1, FNAE 2017). VET providers also have a regulatory obligation to provide employees with a healthy and safe workplace so long as it is reasonably practicable (738/2002). However, OHS issues are addressed inadequately in school safety models (Tappura et al. 2017a). Establishing a safety culture with effective safety processes helps VET institutions to meet their duty of care (WorkSafe Victoria 2015).

1.3 Managing OHS in VET provider organisations

OHS issues are under the responsibility of management in VET institutions, while the management of OHS can be integrated into existing management processes and development, rather than achieved through developing separate procedures (EC 2016, Tappura et al. 2017a). Moreover, OHS management frameworks (such as OHSAS 18001:2007) can be utilised in developing the systematic management of OHS.

The motivation behind the development of OHS should arise from ethical and legal objectives (EC 2011). Developing OHS is a good way to demonstrate to employees and students that the employer cares about their well-being. Developing OHS may contribute to decreased occupational and school injuries, ill health, sick leaves, resignations and early retirements, with reduced related costs (Aaltonen et al. 2006, Tappura et al. 2013, 2017a). Moreover, developing OHS may contribute to the realisation of other VET providers' objectives, particularly regarding the quality of education, job satisfaction and operational efficiency (Tappura et al. 2017a).

According to Launis and Koli (2005), OHS risks and problems are studied in the educational sector, but the means to control the risks are rarely identified and developed as a part of organisational development. Hence, organisational measures must be addressed to reduce occupational injuries and ill health among students and personnel by VET institutions.

In a recent study (Tappura et al. 2017a), a conceptual model for managing OHS by VET providers was presented (Figure 2). The model integrates OHS issues into the strategic planning and general management practices of VET organisations. The model includes OHS procedures and activities to control risks and to achieve OHS objectives of a VET provider organisation.

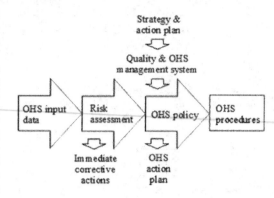

Figure 2. Managing OHS in VET provider organisations (modified from Tappura et al. 2017a).

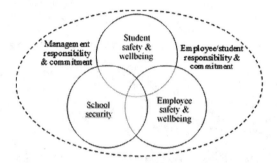

Figure 3. VET providers' definition of school safety (modified from Tappura et al. 2017a).

At the same time, OHS can be considered a part of school safety, emphasising the physical and mental health and safety of both employees and students when working under the supervision of VET providers (see Figure 3). The responsibility and commitment of managers, employees and students to OHS were highlighted as important factors in VET providers' safety.

The purpose of this study is to suggest an annual safety calendar and examples of safety activities to be undertaken during the school year. The prevention focus and proactive approach are emphasised regarding the activities to increase the feeling of safety. The calendar is in line with other education management activities in VET institutions, as well as with national safety campaigns and theme days. The focus is on organisation-level activities; hence, employees and students' individual activities are not explored in this study.

2 MATERIALS AND METHOD

This study is part of a research project whose first part was reported in Tappura et al. (2017a). The study employed qualitative (Denzin & Lincoln 2011) and constructive (Kasanen et al. 1993) approaches due to the contextual and descriptive nature of the study. The constructive research phases (setting of requirements, construction of alternative solutions and validation of the constructions) (Kasanen et al. 1993) were followed to an extent appropriate for the purpose of this study. The constructive approach typically applies goal-directed problem solving through the construction of a practical solution.

Six large Finnish VET providers with 3,300 full-time employees in total participated in the study. The constructive research approach was applied in the design of the annual calendar of safety activities. Here, the problem is the need to plan, inform, visualise and follow up regular safety activities during the school year in VET organisations. Hence, the design herein is based on the VET providers' needs, as well as on both existing knowledge and the heuristic research process.

The construction began with defining requirements and collecting the existing OHS activities from the six participating organisations. Moreover, the national yearly OHS campaign and themes were exploited. Based on this information, a preliminary calendar was constructed. The draft was then presented in a workshop, and it was validated with the OHS professionals of the participating organisations, whose feedback was considered as the model was developed further.

Finally, the research contribution of the model was presented, and its final usefulness was evaluated. The constructed schedule provides a practical solution to be utilised in VET and hence, it is valid for its purpose.

Table 1. The annual school safety activity calendar.

Month	Safety activities
August	Safety introduction to employees
	Safety introduction to students
September	Evacuation exercises
	Traffic safety campaign
	OHS committee meeting
	National traffic safety week
October	Dangerous situation exercise
	Internal OHS inspections
	National well-being week
November	Safety walks
	Internal fire safety inspections
	First-aid extinguishing exercises
	OHS committee meeting
	National fire safety week
December	Risk assessment update
	Employees' well-being meetings
	Occupational health care meeting
January	Management review
	Campus safety group meetings
	OHS committee meeting
	National accident/injury day
February	Personnel well-being inquiry
	Dangerous situation exercise
	National 112-day
March	Inside safeguarding exercises
	Safety trainings
	Safety introduction for on-the-job training
	OHS committee meeting
	Authority notices of young employees
April	Rescue plan update
	Cleaning day
	National safety forum in education
May	Chemical information update
	Occupational health care meeting
June	Periodic machine/equipment inspections update
	Safety calendar update
	OHS committee meeting

3 RESULTS

In Finland, the school year is divided in two semesters. The autumn semester begins in mid-August, and the spring semester begins at the beginning of January. In VET institutions, certain school safety activities are planned and carried out regularly during the school year. Most of the activities are repeated annually, but some (for example, safety training) are repeated less frequently, such as every three years.

The yearly safety activities typically take place at a certain moment during the school year, depending on the other educational management activities in VET institutions. The annual school safety calendar and related school safety activities are presented in Table 1. In the beginning of the school year, safety introduction is arranged for employees and students. In addition, safety exercises, inspections, walks, trainings and campaigns take place. At the same time, OHS committee and occupational health care meetings are held every two or three months. At the end of the school year, the safety calendar update takes place.

The calendar is supplemented with the definition of the safety roles and responsibilities of different parties, such as top management, middle management, teachers, OHS professionals, employees and students. All the employees and students are responsible for complying with the safety instructions and reporting dangerous situations. Moreover, the way all the employees and students participate in other safety activities should be described. Typically, students participate in evacuation exercises, safety walks and risk assessments of their learning environment.

4 DISCUSSION

In VET, school safety denotes a safe learning environment for students and a safe and healthy work environment for employees. However, OHS issues are generally addressed inadequately in current school safety models (Tappura et al. 2017a). Hence, organisational measures and activities to promote the safety of both students and employees in VET are needed. Developing a safety culture requires the systematic management of safety processes and activities (WorkSafe Victoria 2015).

This study presented an example of a calendar for school safety activities during the school year. The calendar should be in line with the other annual education management activities. The calendar can be utilised to plan safety activities, to inform respective parties and to realise the activities.

The calendar includes practicing and other preventive activities to promote school safety, as well as to raise the visibility and feeling of safety. The most apparent safety activities for the students and employees are the evacuation exercises, safety walks and risk assessments of their working environment, in which they participate regularly.

The calendar and related activities list can be utilised as a checklist to ensure that all the necessary safety activities are carried out and that responsible parties are assigned to each activity. They can be used in coordinating and communicating the safety activities to employees and students. They can also be exploited during the introduction of new personnel and in the safety training.

When applied in other VET institutions or at other school levels, the calendar can be completed with organisation-specific safety activities. In the future, the calendar and its applicability to other school levels should be further explored.

5 CONCLUSIONS

This study suggests an annual calendar of school safety activities to promote safety in VET institutions, where school safety denotes the safety of both students and employees. The practical implication of the study is the provision of examples of regular safety activities for the systematic management of school safety. The suggested calendar enables the systematic planning, executing and reviewing of safety activities, and it can be utilised at various school levels to promote the safety of both students and employees.

REFERENCES

A (Act) 630/1998 *Laki ammatillisesta peruskoulutuksesta* (in Finnish).
A (Act) 738/2002 *Työturvallisuuslaki* (in Finnish).
Aaltonen, M., Oinonen, K., Kitinoja, J.-P., Saari, J., Tynkkynen, M. & Virta, H. 2006. Costs of occupational accidents—effects of occupational safety on company business. In M. Hannula, A. Lönnqvist & P. Malmberg (eds.), *European Productivity conf. proc., Helsinki, 30 August-1 September 2006.* Helsinki: Hakapaino.
Denzin, N.K. & Lincoln, Y.S. 2011. Introduction: the discipline and practice of qualitative research. In N.K. Denzin & Y.S. Lincoln (eds.), *The SAGE handbook of qualitative research*: 1–19. Thousand Oaks: SAGE Publications Inc.
EC (European Commission) 2011. *Socio-economic costs of accidents at work and work-related ill health. Key messages and case studies.* Luxembourg: European Union.
EC (European Commission) 2016. *Health and safety at work is everybody's business—Practical guidance for employers.* Luxembourg: Publications Office of the European Union.
FNAE (Finnish National Agency for Education) 2017. *Opetustoimen ja varhaiskasvatuksen turvallisuusopas* (in Finnish).
FNBE (Finnish National Board of Education) 2011. *Performance indicator for initial vocational education and training in Finland 2011.* Helsinki: Finnish National Board of Education.
HE 39/2017 *Hallituksen esitys eduskunnalle laiksi ammatilli-sesta koulutuksesta ja eräiksi siihen liittyviksi laeiksi* (in Finnish).
Kasanen, E., Lukka, K. & Siitonen, A. 1993. The constructive approach in management accounting research. *Journal of Management Accounting Research* 5: 243–264.
Laflamme, L, & Menckel. E. 1997. School injuries in an occupational health perspective: what do we learn from community based epidemiological studies? *Injury Prevention* 3(1): 50–56.
Langley, J.D., Chalmers, D. & Collins, B. 1990. Unintentional injuries to students at school. *Journal of Paediatrics and Child Health* 26(6): 323–328.
Launis, K. & Koli, A. 2005. Opettajien työhyvinvointi muutoksessa (in Finnish). *Työ ja ihminen* 19(3): 350–366.
Lopez Arguillos, A., Martinez Rojas, M., Pardo Ferreira, M.C. & Rubio Romero, J.C. 2017. A review about safety risks in schools. *Proc. intern. symp. on Occupational Safety and Hygiene, Guimaräes, 10–11 April 2017.* Guimareas: SPO-SHO.
Ministry of Education and Culture 2017a. *Ammatillinen koulutus lukuina* (in Finnish).
Ministry of Education and Culture 2017b. *Vocational education and training in Finland.*
Mytton, J., Towner, E., Brussoni, M. & Gray, S. 2008. Unintentional injuries in school-aged children and adolescents: lessons from a systematic review of cohort studies. *Injury Prevention* 15(2): 111–124.
OHSAS 18001:2007. *Occupational Health and Safety Management Systems—Requirements.* London: OHSAS Project Group BSI.
Pirhonen, E.-R. 2014. *Rakenneuudistus—toinen aste* (in Finnish). Helsinki: Ministry of Education and Culture.
Sun, Y.H., Yu, I.T., Wong, T.W., Zhang, Y., Fan, Y.P. & Guo, S.Q. 2006. Unintentional injuries at school in China—patterns and risk factors. *Accident; analysis and prevention* 38(1): 208–214.

Tappura, S. 2012. Occupational safety development in voca-tional education. In A.-B. Antonsson & G.M. Hägg (eds.), *Proc. 44th int. conf. of the Nordic Ergonomics Society, Stockholm, 19–22 August 2012*. Stockholm: KTH Royal Institute of Technology.

Tappura, S. 2014. Vocational education providers' network promoting occupational safety during on-the-job learning. In M. Aaltonen, A. Äyräväinen & H. Vainio (eds.), *Proc. intern. symp. on Culture of Prevention, Helsinki, 25–27 September 2013*: 78–81. Helsinki: Finnish Institute of Occupational Health.

Tappura, S., Pulkkinen, J. & Kivistö-Rahnasto, J. 2017a. A model for managing OHS in Finnish vocational training and education provider organisations. In A. Bernatik, L. Kocurkova & K. Jørgensen (eds.), *Prevention of Accidents at Work: Proc. intern. conf. on Prevention of Accidents at Work (WOS 2017), Prague, 3–6 October 2017*. London: CRC Press/Balkema, Taylor & Francis Group.

Tappura, S., Pulkkinen, J. & Kivistö-Rahnasto, J. 2017b. *Henkilöstön työturvallisuuden ja työkyvyn edistäminen ammatillisissa oppilaitoskissa* (in Finnish). Tampere: Tampere University of Technology.

WorkSafe Victoria 2015. *A handbook for workplaces. OHS in Schools. A practical guide for school leaders*. Edition No. 2. Melbourne: Victorian WorkCover Authority.

Accidents prevention at home with elderly

E.M.G. Lago, B. Barkokébas Jr., F.M. da Cruz, A.R.B. Martins, B.M. Vasconcelos, T. Zlatar, R. Manta & S. Porto
Universidade de Pernambuco, Recife—Pernambuco, Brasil

ABSTRACT: The aging population process is a change in the age structure, causing a larger proportion of elderly people in relation to the total population. The objective of this study was to identify and measure risks and dangers in residences of the over 60 years-old elderly. For the accomplishment of this research, a bibliographical survey was conducted considering scientific articles with statistical data accidents, types and causes. In the second stage, an investigative questionnaire was applied within the elderly population. It surveyed on type of accidents they suffered in their residences, and causes and consequences of those accidents. A checklist was applied in order to identify risks of accidental situations in their home environment. Both results from the field research and literature review indicate falls as the main cause of domestic accidents within the elderly population, which could be avoided with guidance and efforts to make the environment safer.

1 INTRODUCTION

Population aging is a worldwide trend, especially in developed countries.

The Institute for Supplementary Health Studies —IESS (2013) reports that population aging is one of the consequences of the demographic transition. This transition begins with decreasing mortality and fertility rates. Advances in medicine and public health increase the life expectancy of the population. According to data from the Brazilian Institute of Geography and Statistics (IBGE) (2016) from 1940 to 2015, life expectancy in Brazil for both sexes increased from 45.5 years to 75.5 years, an increase of 30 years. In the same period, the infant mortality rate fell from 146.6 deaths per thousand live births to 13.8 deaths per thousand, a reduction of 90.6%.

According to the Global Report of the World Health Organization—WHO (2010) the elderly group, people over 60 years-old, is the fastest growing group in the world. According to WHO (2014), in the coming decades, the world's population of over 60 years-old, will go from current 841 million to 2 billion people by 2050, making chronic diseases and the welfare of the elderly, as new challenges of global public health.

The elderly spend most of their time at home. This environment that may seem as safe as possible can become a very risky environment due to the self-confidence by knowing the environment where they live. Attention is also reduced, as activities made at home are customary.

Although the domestic environment is outside the context of Brazilian safety regulatory standards, it is also important to have an Occupational Safety and Health perspective for this environment.

According to research released by the Albert Einstein Hospital in Brazil (2012), around 30% of the elderly suffer falls at least once a year. For people over 85 years, the risk of this accident may be more than 50%, being 70% risk of falls inside the domestic environment. The WHO Global Report on Falls Prevention in Older Age (2010) reports that falls and subsequent injuries represents 50% of hospitalizations in people over 65 and 40% of injury-related deaths.

In order to identify causes and consequences of falls in elderly people, Ferretti et al. (2013) studied 389 elderly residents in the city of Chapecó, Brazil, from April to May 2012. They concluded that the most frequent place of falls for women and men was the bathroom, with 24.94% and 26.10%, respectively. The majority suffered some type of injury (92.03%). The same authors also verified that the average annual fall in women (1.63) is higher than in men (1.57).

The objective of this study was to identify and measure risks and dangers for the elderly in their residences, through field research and literature review, comparing the results and proposing preventive measures to eliminate or reduce encountered risks.

2 MATERIAL AND METHODS

In this study, a literature review was conducted on domestic accidents with the elderly, their main characteristics and causes. In the database search in Brazil, in total, thirty-six articles were found providing some kind of data about accidents with elderly in their home environment. Among those, eighteen articles were selected as providing statistical data of the accident, type, cause and consequence, and which therefore served as a base for comparison with the field research that was made.

Through a field research, fifty households with elderly from 60 years were visited, residents in the metropolitan area of Recife, Brazil. An investigative questionnaire was applied to the elderly and a checklist with investigative characteristics of risks of accidents in the home environment.

The questionnaire was based on factors associated with accidents with elderly individuals, identified in analyzed scientific papers. The questionnaire model was designed and adapted from the Albuquerque (2014) and Schiaveto (2008) questionnaires, where the age, gender, type, causes, consequences and other factors were investigated.

The data collection for the checklist was performed during the household visits. Further on it was elaborated through accessibility checks of the environments based on Lima et al. (2014), Fernandes et al. (2011), and adapted by the Falls Environmental Scale created by Albuquerque (2014).

Situation of conformity with the Brazilian Regulatory Norms of the Ministry of Labor and Employment that deal about buildings, safety in electrical installations and ergonomics, was verified by using the checklist.

In addition, conformity was verified with technical standards NBR 9050 (Accessibility and buildings, furniture, spaces and urban equipment), NBR 5410 (Low voltage electrical installations) and COSCIP-PE (Pernambuco Fire and Panic Safety Code).

The results obtained in the field research, through the application of the questionnaire and the checklist, were presented in table forms and compared with the results of the database search in Brazil.

Afterwards, a series of recommendations were made for domestic accidents prevention based on the main risks found through this research.

3 RESULTS AND DISCUSSION

In this part are presented and discussed data results on accidents in domestic environments, gathered through the questionnaire and the investigative checklist from elderly residences. Further on, the analysis from the literature review are presented and compared with the results from the field research.

3.1 Data collection from the literature review on domestic accidents

A literature review was conducted, consulting eighteen published articles dealing with falls among elderly. This analysis contains the causes and consequences of falls, which served as a base for comparison with results from the field research.

Through analyzed articles, a total of 2.520 elderly people were interviewed. The majority (66.5%) was female and the age group of the elderly was between 60 and 97 years old. Before the interviews, 31.3% of the elderly had already suffered falls. Most frequent consequences were analyzed, in descending order: the most frequent consequence was the fear of a new fall, followed by fractures, visual problems, injuries, hospitalization and loss of autonomy.

Regarding to causes of falls, they may be related due to extrinsic (inadequate environments) and intrinsic factors (existing diseases).

The extrinsic factors, verified in descending order were: loose carpet; slippery floor; problems with steps; absence of anti-slip material in the bathroom; objects on the floor.

The intrinsic factors, verified in descending order were: difficult to see; arterial hypertension; heart disease; dizziness; diabetes.

It was also verified, that in a household, the highest registered incidence of falls were the bedroom, followed by bathroom and kitchen.

3.2 Results from the questionnaire application

In total, visits were conducted to 50 elderly persons, older than 60 years old, where the questionnaire was applied with the profile of the interviewed and the data collection of accidents suffered by the elderly, their causes and effects.

The Table 1 illustrates the universe of the questionnaire application, number of respondents, gender and age group.

Table 1. Distribution of the visited elderly in relation to the gender and age group.

Age group	Men (%)	Women (%)
60 to 69 years	4 (30.7%)	9 (69.2%)
70 to 79 years	6 (31.6%)	13 (68.4%)
80 or more	8 (44.4%)	10 (55.5%)
Total	18 (36%)	32 (64%)

Table 2. Distribution of the visited elderly in relation to the accident rates.

Accidents/Non accidents	Men (%)	Women (%)
No accidents	15	14
Suffered accidents	3 (14.3%)	18 (85.7%)
Total	18	32

Table 3. Distribution of the elderly according to the type of accident.

Accident	Men (%)	Women (%)	Total (%)
Fall	3 (12.5%)	21 (87.5%)	24 (92.3%)
Eletric shock	0	0	0
Burn	1 (3.85%)	0	1 (3.85%)
Choke	0	0	0
Intoxication	0	0	0
Sliding	0	1 (3.85%)	1 (3.85%)
Total	4	22	26

Table 4. Distribution by the location of the accident.

Accident location	Men (%)	Women (%)	Total (%)
Bedroom	0	8	8 (30.7%)
Living room	2	5	7 (26.9%)
Bathroom	0	6	6 (23.1%)
Kitchen	2	2	4 (15.4%)
Backyard	0	1	1 (3.9%)

Table 5. Distribution of the elderly according to intrinsic factors.

Intrisic factors	Men (%)	Women (%)	Total (%)
Dizziness	0	5 (20.8%)	5 (20.8%)
Fainting	2 (8.3%)	0	2 (8.3%)
Total	2 (8.3%)	5 (20.8%)	7 (29.2%)

Table 6. Distribution of the elderly according to extrinsic factors.

Extrinsic factors	Men (%)	Women (%)	Total (%)
Loose carpets	0	9 (37.5%)	9 (37.5%)
Slippery floor	0	3 (12.5%)	3 (12.5%)
Objects on the floor	0	3 (12.5%)	3 (12.5%)
Irregular floor	1 (4.2%)	1 (4.2%)	2 (8.33%)
Total	1 (4.2%)	16 (66.7%)	17 (70.8%)

Table 7. Accidents consequences.

Consequences	Men (%)	Women (%)	Total (%)
Hospitalization	2 (7.7%)	0	2 (7.7%)
Hospitalization with surgery	0	3 (11.5%)	3 (11.5%)
Fractures	0	7 (26.9%)	7 (26.9%)
Total	2 (7.7%)	10 (38.5)	12 (46.1%)

The Table 2 illustrates the relation of the visited elderly in relation to the accident rates.

From the interviewed elderly, 21 of them suffered accidents, equivalent to 42% of the total number, being the highest occurrence in women.

The Table 3 indicates the type of accident suffered by the interviewed elderly, distributed by gender.

From the total number of accidents it can be noticed that falls represent the most frequent accident, with 92.3% of accidents, being more present in the female gender.

According to Gawryszewski et al. (2004), Filho and Gorzoni (2008), the number of falls is greater among women, because they are the major victims of osteoporosis, and have lower muscle mass and strength when compared with men.

For Pinho et al. (2012), there is a higher incidence of falls among women aged up to 75 years-old. Afterward, the chances of falls are similar in both genders. The cause of this increased incidence is due to the prevalence of chronic diseases, and the fact that women are more present in domestic activities.

The Table 4 illustrates the number of accidents per elderly depending on gender and accident location.

The most frequent places of accidents were bedrooms, followed by the living room and bathroom.

The Table 5 illustrates intrinsic factors, and Table 6 the extrinsic factors which caused falls in the elderly persons.

The intrinsic factors are related to individual and physiological changes happening due to aging. Accident occurrences were recorded due to dizziness and fainting situations.

Extrinsic factors are related to the environment and were responsible for most falls (70.8%). Most registered factors were: presence of loose or folded carpets, followed by floors that are not anti-slippery.

The Table 7 illustrates consequences suffered by the accident.

Regarding to the consequences, it is possible to verify that almost half (46.1%) of the elderly had

Table 8. Results from the checklist for the access to the property, bedroom and living room.

Verification of non-conformity	Acess to property	Living room	Bedroom
Floor	20%	14%	14%
Slope	34%	0%	0%
Exposed eletric wire	0%	6%	0%
Physical arrangement	0%	18%	0%
Corner edges	0	56%	66%
Multi-plug adaptor	0%	52%	0%
Grounding	0%	0%	0%
Bedside lamp	0%	0%	88%
Bed Height (45 to 50 cm)	0%	0%	88%
Chair to sit while dressing	0%	0%	82%

Table 9. Results from the checklist for the access to the bathroom, kitchen and locomotion areas.

Verification of non-conformity	Acess to property	Living room	Bedroom
Grounding	0%	0%	0%
Non-slippery floors	98%	0%	0%
Floor	0%	14%	12%
Non-slippery carpet	16%	0%	0%
Support bars	40%	0%	0%
Residual current protection	0%	84%	0%
Stove	0%	10%	0%
Adequate cabinet height	0%	0%	0%
Corner edges	0%	92%	0%
Free circulation	0%	0%	46%

hospitalization or fractures as a consequence of accidents.

3.3 *Checklist application results*

Table 8 shows the result of the verification checklist observed in the 50 residences, in the areas of access to the property, living room and bedroom.

It was evidenced the presence of slopes and corner edges in bedrooms and living rooms. In over half of the visited residences were found multi-plug adaptors in use. Most of the elderly are unaware that these adapters are inspected by the National Institute of Metrology, Quality and Technology—INMETRO, a Brazilian federal institution.

It is noteworthy that most of the visited residences do not have adequate bed heights, chairs to assist in changing clothes or bedside lamps, important if you need to get up at night, all items important to avoid accidents in the room.

The Table 9 illustrates the result from verified residences on the checklist results for the access to the bathroom, kitchen and locomotion areas.

According to Table 9, most of the homes do not have the residual current protection in circuits and electrical points in wet environments such as outdoor environments and bathrooms. This absence could be explained due to the reason that the majority of the visited residences were old constructions and by knowing that the legal enforceability installation of the residual current protection devices occurred in Brazil only from 1997 through the Brazilian standard NBR-5410.

A positive aspect is that all homes have electrical grounding in bathrooms and kitchens, which reduces the risk of electric shock accidents when using electrical equipment and showers.

Non-slippery floor was verified in only one visited residence. Corner edges were mostly observed on desks in kitchens.

3.4 *Final discussions*

This research was conducted on elderly people, with surveys based on domestic accidents from 50 residents. From a total number of elderly persons, 42% suffered accidents before the interview, and the majority (64%) belonged to the female gender. Similar values were found when comparing to the results with analyzed articles, which found for 66.5% to be females, and that 31.3% of elderly were injured before the interview.

In the questionnaire data collection, it was also observed that the places of greatest occurrence of accidents are in bedrooms, followed by the living rooms and bathrooms. Similar results were found in analyzed articles, where the bedroom was the environment of the house where the greatest number of falls occurred. In the present study, most falls were caused by extrinsic factors, those that depend on the environment where the elderly live, such as loose carpets, slippery floors and objects on the floor. Fractures were the most cited consequence. With regard to analyzed articles, they also agree with this study when indicate loose carpets and slippery floor as extrinsic factors (higher incidence) responsible for the occurrence of falls in the elderly and indicate the fear of falling and fractures as main consequences.

4 CONCLUSIONS

It was concluded that there are some nonconformities that expose the elderly to risk of falling, such as bed with inadequate height, absence of chair and lamp in the room. Almost none of the residences had non-slip flooring in the bathroom. It should be noted the presence of corner edges in bedrooms and living rooms, and presence of carpets in several rooms. Many elderly homes use multi-plug adaptors to connect multiple electronic devices into a single outlet, which can cause short-circuit current and electric shock.

Considering the conducted literature review, the applicable legislation and the field research in residences, it was verified that although the elderly and their family are aware of the risks in their homes, they do not follow the minimum care necessary for making their homes safer.

This research was limited with studying risks and dangers of elderly persons with residences on the Brazilian territory. The intention is to continue the work, conducting comparative studies between the result found in Brazil and results found in other countries. It is necessary to expand and verify the occurrence of same and other risks, main causes of accidents within the elderly, further on, considering habits, culture and particularities of each region.

5 RECCOMENDATIONS FOR ACCIDENTS PREVENTION IN ELDERLY PERSONS

A booklet with recommendations for preventing domestic accidents was developed in order to maintain the house as safe as possible for the elderly population. It is based on standards, already mentioned in the methodology part, on the guidelines of Barros (2011), which developed safe house projects for the elderly, and on main risks encountered through this research.

In order to avoid falls, the residential physical environment for the elderly, should be planned as follows:

All environments should be well illuminated and free from obstacles; use ramps for slopes; the entry doors should be at least 0.80 meters wide; on the stairs, should be used handrails on both sides of the wall; avoid the use of carpets.

In the rooms, the bed should have a height between 0.45 and 0.50 meters, and a chair to assist in changing of clothes. The bathrooms should have non-slippery floors and support bars for the shower, toilet and washbasin. The carpets should have suction cups to attach to the floor and the shower should have a curtain, eliminating the glass door.

To avoid home-accidents with electricity, care must be taken that will be listed below:

Outlets and equipment, such as refrigerator, washing machine, electric shower, and others, must be grounded to prevent current leakage; one must use an adequate electrical outlet; install a residual current circuit breaker system on the electrical switchboard, avoiding shocks in situations involving current leakage; review the house's electrical installation regularly by a qualified professional; avoid the use of multi-plug adaptors.

REFERENCES

ABNT NBR 5410: Instalações elétricas de baixa tensão. Rio de Janeiro: ABNT, 2015.

ABNT NBR 9050: Acessibilidade e edificações, mobiliário, espaços e equipamentos urbanos. Rio de Janeiro: ABNT, 2004.

ALBUQUERQUE, J.P. Prevalência e fatores associados à queda de idosos atendidos por um serviço de atendimento domiciliar privado. Belo Horizonte, 2014. Disponível em: http://www.enf.ufmg.br/pos/defesas/805M.PDF. Acesso em 17.fev.2016.

BARROS, C. Projeto Casa Segura. 2011. Disponível em: <casasegura.arq.br> Acesso em: 25 jan. 2016.

BRASIL, Ministério do Trabalho e Emprego—Normas Regulamentadoras de Segurança e Medicina. NR 8 – Edificações. Disponível em: <http:/trabalho.gov.br> Acesso em: 19 outl. 2017.

BRASIL, Ministério do Trabalho e Emprego—Normas Regulamentadoras de Segurança e Medicina. NR 10 – Segurança em Instalações e Serviços em Eletricidade. Disponível em: <http:/trabalho.gov.br> Acesso em: 19 outl. 2017.

BRASIL, Ministério do Trabalho e Emprego—Normas Regulamentadoras de Segurança e Medicina. NR 17 – Ergonomia. Disponível em: <http:/trabalho.gov.br> Acesso em: 19 outl. 2017.

Código de Segurança contra Incêndio e Pânico do Estado de Pernambuco—COSCIP-PE. Disponível em:<bombeiros.pe.gov.br>. Acesso em 19 out. 2017.

FERNANDES, J.C.F. de A.; CARVALHO, R.J.M. de. Mapeamento da acessibilidade nas instituições de longa permanência para idosos da cidade de Natal-RN. XXXI Encontro Nacional de Engenharia de Produção. Belo Horizonte-MG, 2011. Disponível em: http://www.abepro.org.br/biblioteca/enegep2011_TN_STO_138_277_19075.pdf. Acesso em 14.jun.2016.

FERRETTI, F; LUNARDI, D; BRUSCHI, L. Causas e consequências de quedas de idosos em domicílio. Fisioterapia em movimento. v.26, n.4, Curitiba, set/dez, 2013. Disponível em: http://www.scielo.br/pdf/fm/v26n4/a05v26n4.pdf. Acesso em 20.jul.2016.

FILHO, W.J.; GORZONI, M.L. Geriatria e Gerontologia. O que todos devem saber. Editora Roca, São Paulo, 2008.

GAWRYSZEWSKI, Vilma Pinheiro; JORGE, Maria Helena Prado de Mello; KOIZUMI, Maria Sumie. Mortes e internações por causas externas entre os idosos no Brasil: o desafio de integrar a saúde coletiva e atenção individual. Rev. Assoc. Med. Bras., São Paulo, v. 50, n. 1, p. 97–103, 2004. Disponível em <http://www.scielo.br/scielo.php?script=sci_arttext&pid=S0104-42302004000100044&lng=pt&nrm=iso>. acessos em 25 out. 2017. http://dx.doi.org/10.1590/S0104-42302004000100044.

IBGE, Tábua completa de mortalidade para o Brasil – 2015. Breve análise da evolução da mortalidade no

Brasil. Rio de Janeiro, 2016. Disponível em: <ftp.ibge.gov.br/Tabuas_Completas_de_Mortalidade/Tabuas_Completas_de_Mortalidade_2015>. Acesso em: 25.out. 2017.

IEES-Instituto de Estados de Saúde Suplementar-Envelhecimento populacional e os desafios para o sistema de saúde brasileiro—São Paulo, 2013. Disponível em: <http://www.iess.org.br/html/1apresentacao.pdf> Acesso em 24.mai.2016.

LIMA, M.R.S.; SILVA, R.M.; OLIVEIRA, M.E. de. Avaliação da acessibilidade de uma instituição de longa permanência para idosos no município de Teresina-PI. Revista Interdisciplinar. V.7, n.1, 2014. Disponível em: http://revistainterdisciplinar.uninovafapi.edu.br/index.php/revinter/article/view/147/pdf_92. Acesso em 01.abr.2016.

Pinho, T.A.M.; Silva, A.O.; Tura, L.F.R.; Moreira, M.A.S.P.; Gurgel, S.N.; Smith, A.A.F.; Bezerra, V.P. Avaliação do risco de quedas em idosos atendidos em Unidade Básica de Saúde. Rev.Esc. Enfermagem, v.46, n.2, abr.2012, São Paulo. Disponível em: http://www.revistas.usp.br/reeusp/article/view/40951/0. Acesso em 20.fev.2016.

SCHIAVETO, F.V. Avaliação do risco de quedas em idosos na comunidade. Ribeirão Preto. São Paulo. 2008. Disponível em: http://www.teses.usp.br/teses/disponiveis/22/22132/tde-19122008-153736/pt.br.php Acesso em 22.mar.2016.

Secretaria de Estado de saúde de São Paulo. Relatório Global da OMS sobre prevenção de quedas na velhice. São Paulo, 2010. Disponível em: <http://bvsms.saude.gov.br/bvs/publicacoes/relatorio_prevencao_quedas_velhice.pdf> Acesso em 15.jun.2016.

World Health Organization. 'Ageing well must be global priority', warns UN health agency in new study, 2014. Disponível em: <http://www.un.org/apps/news/story.asp?NewsID=49275#.WfDVfdKnHIV>. Acesso em 25 out. 2017.

Occupational safety and health for workers managing wild animals

A.R.B. Martins, B. Barkokébas Jr., E.M.G. Lago, B.M. Vasconcelos,
F.M. da Cruz, T. Zlatar, T.C.M. de França & L.R. Pedrosa
Universidade de Pernambuco, Recife—Pernambuco, Brasil

ABSTRACT: The trade of wild animals is considered to be the third largest illegal activity in the world. On the other hand, there are some activities dealing wild animals which are permitted by law, being the case of rescue of specimens of fauna due to public works, production of consumer goods with the attributes of wild animals, or even the maintenance of collections, as zoos for recreation and scientific research. In all cases, there is a contact of animals with the worker which is conducting his daily tasks. Aiming at the preservation and physical integrity of workers, a field research was carried out at a Screening Centre for Wild Animals (CETAS), aiming to evaluate occupational risks inherent in the management of these animals. As a result stand out the biological and accident risks. Preventive measures have been proposed to reduce the workers' exposure, thus promoting the safe management of wild animals.

1 INTRODUCTION

The trade of wild animals is considered to be the third largest illegal activity in the world. According to the Law 9605/98 (BRAZIL, 1998), environmental agencies are responsible to restrain such practice and, when possible, to return the specimens to their "natural habitat". According to the study published by the IBAMA's Environment Inspection Coordination, there are many consequences associated with illegal trade of wild fauna, it spread serious diseases, as animals are sold illegally, and without any sanitary control, exposing workers, other animals and the general population to biological hazards, many of which are still unknown (DE-STRO, et al, 2012).

The realization of major works such as hydroelectric plants, potable water-treatment plants, roads, among many other goods for the current world consumption, usually involves the transformation of the natural environment and the loss of natural habitat, being a mitigating measure of such impact rescue and safe disposal of affected wild animals.

On the other hand, there are several activities involving wild animals that are authorized by law, whether for conservation, recreation, scientific research or use of their natural attributes. In the Federal Register of Potentially Polluting Activities, and Users of Natural Resources, there are registered about 1200 enterprises legally authorized to develop techniques with wild animals (IBAMA, 2009).

Although such institutions are registered and hire workers, the notifications and reports on occupational diseases and accidents, involving the management of wild animals are rare. The lack of supervision, research and data, contribute to poor application of concepts and techniques of work safety in the prevention of accidents in these institutions.

Reintroducing wild animals into the wild requires knowledge on how to manage-it, through specific scientific and technical methods to treat the fauna outside its natural habitat, in order for them to return to a free life. In order to do so, it is necessary to have establishments where such techniques are practiced, besides the identification of adequate release areas and specialized professionals to carry out all the tasks involved, such as: veterinarians, biologists, zoo-technicians, caretakers and auxiliaries (The Humane Society Of The United States, 2015).

The objective of this research was to analyze occupational hazards inherent to wild animal management activities, present in a Wild Animals Triage Centre (CETAS), in order to preserve the physical integrity of workers during the development of their daily activities while being in contact with wild animals. Finally, it proposes corrective measures for safe handling of wild animals.

2 METHODOLOGY

The research technique was the exploratory case study whose data and information were obtained

Table 1. Risk category.

Probability	Consequence			
	Insignificant I	Marginal II	Critical III	Catastrophic IV
Probable – A	MR	MR	HR	HR
Reasonably Probable – B	LR	MR	MR	HR
Remote – C	RB	LR	MR	MR
Extremely Remote – D	LR	LR	LR	MR

Acronyms: LR – Low Risk; MR – Moderate Risk; HR – High Risk.

through a literature review, through scientific literature, by consulting the official database, analyzing documentaries, through interviews and a field research. In order to carry out the field research, a CETAS was chosen, located in the city of Petrolina in the State of Pernambuco, Brazil.

During the field research, in April 2014, a survey of was carried out on all wild animal management activities. Also, interviews were conducted with all CETAS's technical and administrative managers. The questions proposed to those professionals had as objective to get information on formal and legal questions related to the Screening Centre, the number of employees and their work regime, data on accidents or incidents occurred in the institution, training and emergency care.

The interviews conducted with all workers aimed to get to know the workers, their level of education, their perception of risks of accidents during the development of their activities, training, use of protection and their opinion on work safety.

The results of the first data collection lead in choosing the Preliminary Risk Analysis (PRA) method, to identify possible risks and dangers, causes and consequences and suggest control measures. For that purpose it was adapted the Cordella (1999) PRA method, while for risk categorization, the American standard MIL-STD-882 (1993) was adapted to take into account the probability of occurrence of certain event and its consequences (Table 1).

As a way of guiding work safety actions, the research results allowed the elaboration of control measures, based on labour legislation, which can be adopted, besides zoos, in other centres, by breeders and other companies that use wild animal management for commercial purposes.

3 STUDY CASE: O CETAS/CEMAFAUNA

The CETAS, in a general form, has as main operations the reception, rehabilitation and destination of wild animals, rescued from nature or due to seizures made by environmental agencies in enforcement actions (PEDOA, ADEODATO, 2014). In particular, the CETAS de Petrolina had one extra focus, due to the installation works for transposition channels of water from the São Francisco river.

The Centre is divided into two technical areas, the Veterinary Clinic and the Management Sector. All animals arriving to the Centre are subject to a veterinary evaluation. After this first screening the specimens are sent to the internal sectors, according to their situation. The veterinary clinic is the gateway where the first evaluations are made, and animals that need special care are kept until they can be removed to the rehabilitation centres or being released.

The purpose of the Management Sector is to rehabilitate healthy and adult animals for their return to the wilderness.

The animals are received the veterinary clinic where is performed, among other tasks, the evaluation of the general condition, taxonomic identification, determination of the area of natural occurrence and, in the case of animals from seizures, the degree of domestication.

The puppies are referred to the clinic (illustrated in Figure 1) where they are submitted to special care. In the clinic remain animals which are debilitated, with health problems, broken limbs or even mutilations. In such cases, medicines are given or small surgeries are performed. The clinic contain a pharmacy where medicines are stored, and a necropsy room for studying and researching in conjunction with other areas of the Centre for Conservation and Management of Fauna of Caatinga (CEMAFAUNA). It integrates installations and a serpentarium, where snakes and their offspring

Figure 1. Veterinary clinic.

are kept adapted boxes. The laundry for washing uniforms in located in the building attached to the clinic, and it serves all employees of the Centre.

The Management Sector, located in an independent building, with housed animals that are in quarantine, awaiting adequate conditions for them to be released or transferred to other sites of the Centre. The quarantine is endowed with fourteen enclosures for small mammals and birds. The quarantine facility has also an area for food preparation, storage and handling.

There are six separated, outdoor enclosures for rehabilitation of mammals, birds and reptiles. One is for medium-sized mammals where one specimen of Puma concolor (jaguar) is kept, and the other for small-sized mammals: *Cebus libidinosus* (bearded capuchin monkeys); *Mazama guoazoubira* (catingueiro deer); *Tayssy pecari* (wild pig); and the reptile *Chelonoides carbonaria* (chelonium). The flight corridor is an enclosed area where birds are kept with the purpose of rehabilitating their natural abilities to fly and to search for food, being the last stage before their release.

The team of workers responsible for the management of wild animals is composed of 12 employees, being two veterinarians, six caretakers and general service aides. The CETAS counts with twelve trainees from courses of Biology and Veterinary Medicine, who also work with CEMAFAUNA researchers and assist in specific activities. To meet the daily work routine, these workers are divided into two teams, each with three handlers and a veterinarian. It is up to handlers to clean cages and enclosures, prepare and deliver food (illustrated in Figure 2) and, when necessary, handle animals. The veterinarians perform the evaluation, screening and clinical follow-up, supervise all activities, train the collaborators, and are legally reponsible to the control institutions. The group of general services works on cleaning the common areas and assist handlers in cleaning of smaller cages and rooms of the veterinary clinic. Trainees carry out research activities and support specific tasks, such as the management of snakes and birds of prey.

During the weekdays, the work starts at 6:00 am, developed by two teams, while during the weekend, on Saturdays and Sundays, remain only one monitor group. The CETAS routine activities are closed around 4:00 pm.

4 RESULTS

4.1 *Questionnaire analysis*

The administrator of CETAS couldn't specify how many animals pass through the Centre in one month, being housed or in transition. He informed that this number varies greatly depending on the working phase of the transposition channel of the São Francisco river, and on regional enforcement operations. Further on, he said that the permanence of the animals depends on the animal's need and seeking solutions whenever possible.

According to veterinarians, the attention to worker's safety is initiated during the selection, through identifying people with greater aptitude for dealing with animals. During the first days of work, through informal conversation, new employees are advised on dangers and the use of Personal Protective Equipment (PPE). In the initial months, a new employee always works together and under the guidance of a more experienced employee. Before starting their activities, workers are required to get anti-rabies and anti-tetanus vaccinations, being treated with verminfuge every six months.

In relation to the accident register, during the past four years of CETAS existence, there were no recordings of accidents leading to sick leave of workers, rather only cases of accident due to attacks and bites of animals. There is no emergency response team, nor risk map in the premises. In the event of any emergency situation, the care this is conducted by veterinarians themselves.

It was also informed that workers are advised to wash their uniforms at the Centre, and that there is a concern in avoiding the contact sick (such as colds and flu) workers with animals, due to the serious damage that such contact may cause to animals.

The facilities are very well sectorized, facilitating the tasks and the organization of the work. The rehabilitation rooms for medium and small mammals are equipped with a double-doored cages, a space with that allows the worker to isolate the animal to enter the enclosure and perform the tasks, a fundamental safety item for dealing with aggressive

Figure 2. Delivering food inside the cages.

animals, as for example the bearded capuchin monkeys and jaguars.

The questionnaire was answered by twelve collaborators and two trainees. All of them had at least medium course as their level of education and work in the Centre for about two years. Only one trainee, a veterinary student, reported that had previously worked in contact with wild animals, the others were trained by CETAS itself. All questioned workers responded to perceive attacks of largest animals as the main danger, no one mentioning exposure to biological agents or any other risk agent. All reported that they have been already victims of animal attacks, from bird bites to bites of small mammals like bearded capuchin monkeys and marmosets. In order to perform the activity safely, they reported that it is important to be careful and to know the behaviour of each animal.

4.2 Preliminary Risk Analysis (PRA)

From the information obtained and the monitoring of the daily routine, eleven different activities were identified in the management of wild animals. To substantiate the analysis of the PRA, each one of these activities was described observing the physical space, characteristics of the animals, and giving a comment on obtained results.

It is also important to note that the used classification of risk, according to Table 1, is based on the results of the field observations and the interviews with the workers, since no studies with scientific data were found available quantifying the degree of exposure to the agent in wild animal management activities.

Activity A: Technical and administrative management

Veterinary activities involving management tasks: selection and supervision of employees, control of materials, organization of teams, contact with external clients; and technical tasks: reception and clinical examinations, execution of veterinary procedures and management of all animals arriving to the centre.

PRA Results: In the veterinary procedures were identified risks of accidents with sharps, needles and scalpels, besides the risk of attack. Biological risk was identified due to exposure to biological fluids and the chemical risk generated by the dust from feathers, coatings and other wastes.

Activity B: Reception and ambulatory management

These are activities carried out by handlers, under the supervision of the veterinarians when the animals are received. They immobilize and manipulate the animals to support the clinical examination, removing them to other spaces, among other necessary tasks.

PRA Results: The handler, when developing his activity, is exposed to biological fluids, feather and coat residue dust, and the possibility of attacks. It can be considered that the degree of exposure of the handler is greater than that of the veterinarian, since they effectively manipulate and immobilize the animals to carry out the clinical procedures.

Activity C: Snake management

The snakes exhibit biological behaviour that does not require daily actions, such as cleaning and feeding, so the direct contact of the specimens with the worker is minimal. PRA Results: Exactly because of the venomous nature of snakes, the risk of accidents from attack is high, as venom from a snake can lead to death.

Activity D: Special care for wild animal offspring's

Activity carried out by handlers and trainees under the supervision of the veterinarian. In addition to the tasks already described in Activity B, it includes the preparation of special foods such as broths and porridges and the direct supply through a bottle, spoon or other type of utensil. Also, they are cleaning the cages, boxes and incubators. PRA Results: The worker, in these activities, is exposed to biological fluids, dusts and feather and coat residue, and the possibility of attacks. As it is a question of puppies, it can be inferred that the risk of attack is smaller; the contact with puppies is direct and more intense, increasing the degree of exposure to the biological and chemical agents.

Activity E: Quarantine

The handlers perform the tasks of: food selection and preparation (cutting and grinding meat, fruit and vegetables) for all the animals of the Centre; cleaning of the enclosures of several birds, from *P. maracana* (maracanã), *Caracara plancus* (carcass hawk), *Amasona aestiva* (parrots); to small mammals such as *Aloutta belzebul* (howler monkeys), *Nasua nasua* (quatis), *C. jacchus* (marmosets), among others.

PRA results: Due to this sector being closed, the worker is exposed to concentration of particulates in the environment, being exposed to biological and chemical risks (dusts composed of excrement, feathers, hairs and others, besides biological fluids).

Activity F: Bird rehabilitation site

These are the tasks developed inside the flight corridor, the screened outdoor space, where only birds and free chelonians are kept, not requiring daily cleaning. PRA Results: Because it is an outdoor space, the degree of exposure to biological agents is lower. However, when the handlers arrive to the enclosure to supply food, the birds are quickly attracted, causing the risk of attack.

Activity G: Rehabilitation site of chelonians

It is an open site with a part of the ground floor where chelonians (*Chelonoides carbonaria*) are kept, in the central area there is a tank for maintenance of these reptiles. A daily activity includes sweeping the entire area and washing-of the places for food placement. PRA Results: The animals are not aggressive, and the risk of attack is insignificant.

Activity H: Small mammalian rehabilitation site M. guoazoubira (deer catingueiro)

The tasks include sweeping the shelter and cleaning the containers for food placement. The enclosure has natural vegetation and does not need daily care. PRA Results: these mammals are docile, so the worker is only exposed to the dust of the land and coat of deer due to the sweeping task. The task is fast and outdoors, which greatly reduces the risk of exposure to chemical agents and accidents.

Activity I: Rehabilitation of small mammals—T. pecari (wild pig)

Twenty-one specimens of this mammal are considered very aggressive. The area is on land, with cement flooring only at the place of food placement. The tasks are performed daily and consist of cleaning the area by sweeping; excrement collection and food placement. PRA Results: Since these animals form groups and are very aggressive, and the worker enters the place without protection barriers, the work is obligatorily carried out by two persons. In this case, there is exposure to dust, due to particulate material and excrement particles, in addition to the danger of attack.

Activity J: Rehabilitation site of small mammals—C. libidinosus (bearded capuchin monkeys)

It includes a large area covered with vegetation simulating a natural environment, keeping a group of five specimens. Such animals exhibit aggressive behaviour, which requires specific procedures for both food placement and cleaning. PRA Results: The enclosure is endowed with a cage area, where the animals are kept to allow the worker to enter the space. Even so, the task requires agility as the monkeys would approach the worker if they perceive his presence. It is an environment with greater noise due to the sound emitted by the monkeys.

Activity L: Medium-sized mammalian rehabilitation site—P. concolar (jaguar)

It is the most isolated enclosure of the Centre, where an adult specimen of a jaguar is kept. They were rescued from the traffic of wild animals, arrived as puppies and grew-up within the CETAS. The space is all screened and the area is similar to the natural environment with shrubs, small pond, hiding places and steps for movement. The food is made by offering live prey. When it is necessary to enter the space, it is planned a day before, by placing the prey in the space of a cage, to attract the animal and to isolate the area of the enclosure. Therefore, the worker performs safely tasks such as tree pruning, weeding and cleaning. PRA Results: The applied working process that for this activity is highly efficient, as it is the risk, which is high and fatal.

The described activities are very similar. However, they differ in the degree of risk depending to the type of animal, its aggressiveness, place where the activity is performed and time of exposure. Therefore, in Tables 2 and 3 are summarized risk agents and classifications of the degree of risk.

4.3 *Proposed control measures*

Aiming at the safety and health of the workers of 1200 national establishments that handle wild animals, this study proposed some control measures which could be adopted, without high costs, but which pretend to promote a safety culture. The main measure is to adopt training routines, in order to give instructions on risks of exposure to the environmental agents identified in the risk analysis, and present in each activity developed

Table 2. Risks agents.

Risk	Agent
Accident	Shear material; attack of animals (pecking, scratching, biting); falls
Biological	Biological fluids (zoonoses)
Chemical	Dust from: earth, feathers and pelage
Physical	Sound of animals
Ergonomic	Transport of materials and food

Table 3. Classification of the degree of risk (C.R.).

Agent	Specimen/Location	C.R
Shear material	Nursery	MR
Attack	Chelonium	LR
	Hawk	MR
	Parrot	MR
	Monkey	HR
	Wild pig	HR
	Jaguar	HR
Falls	In all activities	LR
Biological fluids (zoonoses)	Open space and short exposure time	MR
	Closed space and high exposure time	HR
Dust from: earth, feathers and pelage	Open space (Outdoor)	MR
	Closed space (Indoor)	HR
Sound of animals	In all activities	LR
Cargo transport	In all activities	LR

C.R.: LR – Low Risk; MR – Moderate Risk; HR – High Risk.

Table 4. Appropriated PPE.

Risk agent	PPE
Shear material	Latex glove CA 28324
Attack of animals (pecking, scratching, biting)	Long-sleeved shirt, cowboy gloves, boots
Falls	Levelling of floors and organization of spaces
Biological fluids	Latex gloves, safety glasses, mask PFF 1
Dust from: earth, feathers and pelage	Mask PFF1, safety glasses
Noise	Ear plug protector (when required)
Transport of materials and food	Conveyor Carriers

in the management of wild animals. Further on, perform salvage and rescue trainings. Do not allow the worker to work alone in areas of greater attack risk. Correct and mandatory use of PPE's, appropriated to each activity and agents of risk, as illustrated in Table 4. Rotation of activities among workers is important not only to reduce the risk of exposure but also to promote awareness of the safety culture.

5 CONCLUSIONS

It is concluded from this study that working with wild animals, whether in sorting centers and scientific research, recreation places or in authorized productive activities, it exposes workers to easily controlled risks of accidents, depending on the effectiveness and continuously monitoring of preventive measures. Maior attention should ge given to animal attacks, biological and chemical agents, which are the most significant due to permanent exposure of the worker to these agents. In addition, it is considered fundamental to prepare written procedures to guide the implementation of activities, which contributes to standardize tasks and form a safe work culture.

REFERENCES

BRASIL, Lei n°. 9.605. 1998. "Dispõe sobre as sanções penais e administrativas derivadas de condutas e atividades lesivas ao meio ambiente". Presidência da República Casa Civil. Brasília, 13 fev. 1998.

Cordella, B. 1999. "Segurança no trabalho e prevenção de acidentes: uma abordagem holística". São Paulo: Atlas, 254 p.

Destro, Guilherme Gomes; Pimentel, Tatiana Lucena; Sabaini, Raquel Monti; Borges, Roberto; Barreto Raquel. 2012. "Efforts to Combat Wild Animals Trafficking in Brazil". Biodiversity, Book 1, chapter XX. < http://www.ibama.gov.br/sophia/cnia/periodico/esforcosparao combateaotraficodeanimais.pdf>.

IBAMA, 2009. INSTITUTO BRASILEIRO DO MEIO AMBIENTE E DOS RECURSOS NATURAIS RENOVÁVEIS. "Instrução Normativa n° 31". 03 de dez.

MIL-STD-882. 1993. "Military Standard—System Safety Program Requirements". Department of Defense United States of America.

Pessoa, André; Adeodato, Sérgio. 2014. "Caatinga Selvagem: O legado de um projeto de desenvolvimento para a conservação da fauna". 1ª Edição. São Paulo. Ed. MCLE 226 p.

The Humane Society Of The United States. (2015). The Humane Society Of The United States. Retrieved 2015 йил Novembro Obtido de The Humane Society Of The United States: http://www.humanesociety.org/.

Pulmonary disease due to exposure to nanoparticles

B. Calvo-Cerrada
PrevenControl Prevention Service, Spain

A. López-Guillén
Laboral Advanced Radiology, Assistencial Health Service, Autonomous University of Barcelona, Barcelona, Spain

P. Sanz-Gallen, G. Martí-Armengual & A. López-Guillen
Department of Medicine, Faculty of Medicine and Health Sciences, University of Barcelona, Barcelona, Spain

ABSTRACT: Objective: The objective is to synthesize the available scientific evidence about pulmonary effects in humans when exposed to nanoparticles. Methods: A literature systematic review in PUBMED, Medline, LILACS, Web of Science and Cochrane Library database is performed, using "nanoparticles", "nanotoxicity", "lung disease", "pulmonary disease", "pulmonary effects" and "lung toxicity" as keywords. Papers with human results, full text and published between 2012 June and 2017 July where included. Results: from the initial 208 papers obtained, after the exclusion criteria were applied, 6 papers remained. A clear relation between nanoparticles exposure and human lung disease was found. Discussion: This is the first study that synthesizes the available scientific evidence available about nanoparticle exposure lung disease in humans (between 2012–2017). However, scientific evidence is scarce, and further studies are required.

1 INTRODUCTION

1.1 Background

The fourth industrial revolution has already begun, and it's the result of robotic, nanotechnology and biotechnology, information technology and artificial intelligence convergence. The laws ruling matter at nanometric scale (1 nm = 10^{-9} m) are different to the macroscopic ones, providing us with new physic and chemical properties, separated to the same composition solids or to the individual molecules ones (Mendoza & Rodriguez 2007). The nanouniverse open the doors to great advances in industry or biomedical investigation, but at the same time it raises controversies as to know the possible effects on the health that it could cause.

1.2 Exposure routes

Human exposure to Nanoparticles (NP) can happen mainly through airway (air-suspension NP), dermic (ambient NP, cosmetics) and oral (NP in food, water, etc.) (Gutierrez-Praena et al. 2009). This exposure could also happen through instrumentation or medical practice such as cancer diagnosis and treatment. Once the NP has been absorbed, they are distributed through blood and linfatic circulation, reaching several organs, like liver, kidneys and brain (Donaldson et al. 2005, Obersdorster et al. 2002). NP toxicity depends on their organ persistence and if a biological response to eliminate them is produced, inter alia. Toxicity mechanisms are not well known, although cellular membrane damage, membrane potential disruption, protein oxidation genotoxicity, formation of oxygen reactive molecules and inflammation seem to be included.

1.3 Toxicity effects

As to toxicity effects of NP exposure reported, they could generate dermic toxicity (work exposure to dendrimers), gastrointestinal (burn treatment through bandaging with silver can produce liver toxicity), cardiovascular (Baxter et al. 2010) since inhaled ultrafine particles (UFP) could be a factor risk in coronary pathology (they could even produce AMI in firefighters) and respiratory (Song et al. 2009) where rapidly progressive pulmonary fibrosis causing death of some silica and Nano silicate NP exposed workers of decorative paints in China is reported. However, the existence of recent scientific evidence on pulmonary effects in humans is unknown.

2 OBJECTIVE

The objective of this study is to synthesize the available scientific evidence, during the period

2012–2017, about the pulmonary toxic effects in humans exposed to NP.

3 METHODOLOGY

3.1 Study design

This study is a systematic review of the scientific literature.

3.2 Sources and instruments for collecting information

A systematic search of the literature of the articles published in the PUBMED, Medline, LILACS, Web of Science, Cochrane Library and gray literature (doctoral theses, dissertations, etc.) databases was made. MeSH terms where nanoparticles AND ("lung disease" OR "pulmonary disease" OR "pulmonary effects" OR "lung toxicity" OR "pulmonary toxicity") NOT therapy. As inclusion criteria, we selected those scientific articles (cross-sectional, prospective, retrospective cohorts, control cases, reviews) with results in humans, full text and published between 2012–2017.

Given the obtained results from the search through the aforementioned databases, a first selection of the articles has been made according to the title. Subsequently, a reading of the abstract was made, excluding those whose content did not analyze the objective of interest, all articles of opinion and those whose objective was diagnostic or pulmonary therapy with nanoparticles. Given that our objective is to assess lung damage in humans, all in vitro and in vivo studies in animals have been excluded.

Finally, once selected the items according to the abstract, we proceeded to complete reading and to assess methodological quality. To do this, after reading them, those that met three criteria were selected: to define the concept "pulmonary pathology", to define "exposure to nanoparticles or nanomaterials" and to analyze the possible relationship between exposure to nanoparticles and lung pathology secondary to said exposure.

4 RESULTS

After reviewing the scientific evidence between the nanoparticle exposure and pulmonary effects, of the 208 articles obtained by applying search terms in the selected bibliographic databases, 143 were excluded after reading the title and not be considered consistent with the interest of the study. Of the remaining 65 articles, after reading the abstract, 32 were excluded. Finally, of the 27 articles obtained and after excluding 21 for not meeting the methodological criteria described in the previous section, a total of 6 articles were analyzed (Figure 1).

A clear association between developing lung disease and exposure to nanoparticles in humans has been observed (Table 1). However, in most articles there is an incomplete description of the type of NP, as well as the intensity and duration of exposure. The NPs that appear in the selected studies come from companies engaged in painting, welding, nanotechnology plants, etc. The NPs mainly observed are Carbon Nanotube (CN), nanosilica (Nano-SiO$_2$), nanotitanium (Nano-TiO$_2$), and nanosilver (Nano-Ag).

Numerous articles agree that exposure to nanoparticles cause respiratory alterations of restrictive pattern, which can lead to respiratory failure and even death. In one study it was observed that exposure to nano-TiO$_2$ can reach higher levels of Fractional Nitric Oxide (FENO), an indicator of lung inflammation, especially in workers with a history of asthma or allergic rhinitis. Lung epithelial injury markers as a Clara Cell Protein (CC16) and pulmonary function have been associated with nanomaterials handling.

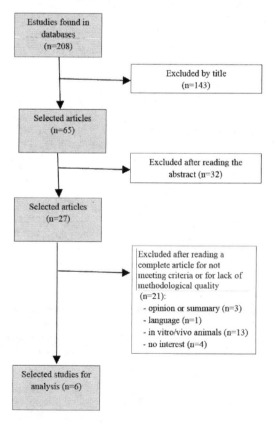

Figure 1. Phases of the study.

Table 1. Characteristics of selected scientific studies: pulmonary effects due to exposure to nanoparticles in humans (2012–2017).

Authors, year	Type of NP	NP form	Exposure mechanism Intensity	Duration	Pulmonary effects Anatomo-pathological	Radiology	Pulmonary function
Ferreira et al. 2012	Poliacrilate ester NP SiO_2, TiO_2, ZnO, nanosilver		8 workers of printing plant (adhesive, paint and decoration). 8–12 hrs/day exposure	5 months	Nonspecific lung inflammation, fibrosis and pleural granulomas. BALF: ↓ macrophages, ↑ leukocytes lymphocytes, neutrophils and eosinophils. NP 30 nmØ in pleural fluid.		Dyspnea, pleural and pericardial effusion. Severe restrictive pattern (CVF 24.8–35.4%, FEV1 24.8–36.5%). 1 moderate (FVC 61.3%; FEV1 63.4%). Respiratory failure and death 2 workers.
Cheng et al. 2012	Carbon nanotube (CNT) Nano-TiO_2 & nano-SiO_2	NT	A 58-year-old worker exposed to polyester powder paint.	3 months	Lung biopsy: opacity and birefringence, macrophages, NP Silica, aluminum silicates, titanium dioxide, talc and rails. Granulation tissue filling the alveolar ducts and alveoli. Chronic inflammation in the surrounding parenchyma. Eosinophilic pneumonia. CNT in 3 patients. Electron microscopy of pulmonary transmission: titanium dioxide and silica.	RX: irregular aire frames, pulmonares infiltrated. HRCT: areas of airspace consolidation and opacity.	Bronchiolitis obliterans organized as pneumonia
Zhang et al. 2014	Carbon black (CB)	Tridimensional nanostructure (30–50 nm)	Average concentration of CB in exposed air: 14.90 mg/m³ (almost 4.26 times higher than the VLA). There are local exhaust ventilation systems. Unlikely use of masks.		↑ interleukin 1β, IL-IL-8, inflammatory protein of macrophages-1β (MIP-1β) and tumor necrosis factor α (TNF-α) in the exposure group (P < 0.05).	No radiological changes are observed	Exposed: significant reduction in FEV1%, FEV1/CVF, PEF%, MMF%. The level of CVF% was not significantly different between exposure group and control group. No significant differences FEV1% and MMF% in smokers.

(Continued)

Table 1. (Continued).

Authors. year	Type of NP	NP form	Exposure mechanism			Pulmonary effects		
			Intensity	Duration	Anatomo-pathological	Radiology	Pulmonary function
Wu WT et al. 2014	CNT, Nano-TiO$_2$, Nano-SiO$_2$, nano Ag	NT	Nanotechnology plant workers. Nano-TiO$_2$. The effect of NP exposure on fractional exhaled nitric oxide (FENO) is examined as a useful diagnostic test of airway inflammation.				The exposed to Nano-TiO$_2$ group had significantly higher levels of FENO. More in workers with asthma or allergic rhinitis
Andujar et al. 2014	NP metálicas: Fe$_2$O$_3$, Fe$_3$O$_4$, MnFe$_2$O$_4$ y CrOOH	NP chain type	Welders exhibition: complex of gases (carbon monoxide, ozone) and dangerous metal fumes. Up to 11% of the total mass, and 80% of the total number of particles emitted in the welding fumes are NP.	27 years median exposure	Macrophages in alveolar lumen and in fibrous regions. Significant overload of Fe, Mn, cromo (Cr) and titanium (Ti) in sections of lung tissue. I increase macrophages in alveolar light and in areas of fibrosis, neutrophils, lymphocytes.	X-ray microfluorescence: increase of sulfur and iron ratio in welders.	Several adverse respiratory outcomes have been described in welders, among whom inflammation and pulmonary remodeling are largely described
Liao et al. 2014	UFP Nanosilver CNT, Fe$_2$O$_3$, nanoiron, Silicon dioxide, Titanium dioxide	NT	Workers handling nanomaterials Average exposure of 2.43 times/week and 2.69 hours/time.	Follow up for 6 months			↓ Maximum expiratory flow medium and forced expiratory flow to 25% between the start and follow-up at 6 months in exposed > than in the control group. Damage markers (CC16 and lung function) were associated with the handling of nanomaterials.

Regarding the anatomopathological data, non-specific lung inflammation, fibrosis and pleural granulomas are shown in the samples obtained in the selected studies. Some studies show an increase of leukocytes, lymphocytes, neutrophils and eosinophils, nanoparticles in pleural fluid, significant overload of iron (Fe), manganese (Mn), chromium (Cr) and titanium (Ti) in sections of lung tissue. Others observe increased levels of interleukin (IL) 1β, IL-IL-8, inflammatory protein of macrophages-1β (MIP-1β) and tumor necrosis factor α (TNF-α) in the exposure group ($P < 0.05$).

In one of the two studies in which radiological data are shown, no changes are observed in exposed to non-exposed to nanoparticles, while in another one irregular and infiltrated air patterns are observed in both pulmonary fields. High Resolution Computed Tomography (HRCT) showed bilateral areas of airspace consolidation and opacity.

5 DISCUSSION

This is the first study that synthesizes the existing scientific evidence regarding pulmonary pathology secondary to exposure to nanoparticles in humans (2012–2017).

The increasing use of NPs for a wide variety of commercial, industrial and biomedical applications has led to some concern about their safety and the possible adverse health for those exposed (Zhang et al. 2017). Most studies on the effects of respiratory exposure to NPs involve lung models and are performed by instillation, aspiration and inhalation of carbon nanotubes (the most studied kind of NPs) (Ferreira et al. 2013). For researchers, the toxicity of nanoparticles and their relation in development of occupational lung diseases it's being studied. There are very few studies that analyze the relationship between occupational exposure to nanoparticles and the emergence of health damage in humans. Currently, several follow-up studies are being carried out on workers exposed to NP in order to detect and/or corroborate the possible toxic effect on lungs.

NPs have characteristics (their size, surface, area, structure, agglomeration and solubility, among others) that make their physical-chemical properties and behavior very different from other types of particles. They can reach different organs of the body, such as the liver, spleen, kidney and brain, through blood circulation after inhalation exposure (Donaldson et al. 2005). Inflammation, fibrosis, genotoxicity and carcinogenicity may also be associated with inhalation (Hubbs et al. 2011).

Respiratory exposure to NP can cause significant adverse respiratory effects, such as multifocal granulomas, peribronchial inflammation, progressive interstitial fibrosis, chronic inflammatory response, collagen deposition, oxidative stress, pleural lesions and genetic mutations, at least in experimental studies in animals (Ferreira et al. 2013). This data overlap with the results in humans, which are reflected in this study.

At this time, it is essential to know more about the toxicological effects of NPs, to address the growing concern about potentially harmful exposures for both the general population and the working population (Ferreira et al. 2013, Sweeney et al. 2016). It is important to note that the NP and UFP were related to cardiopulmonary toxicity, but do not rule out other possible organic effects.

There are recent experimental studies (Sweeney et al. 2016) showing that some CNT can mimic the asbestos (cellular damage, pulmonary inflammation, genotoxicity) biological effects (Genaidy et al. 2009; Jaurand 2015; Kayat et al. 2011). A recent IARC study reviewed the literature on the toxicological effects of the CNT. One type of Multi-Walled Carbon Nanotubes (MWCNT) known as MWCNT-7 was classified as possibly carcinogenic to humans (Group 2B) and Single-Walled Carbon Nanotubes (SWCNT) and MWCNT excluding MWCNT-7 as unclassifiable in terms of their carcinogenicity to humans (Group 3) (Grosse et al., 2014).

This study has a number of limitations that must be taken into account. First, it has been observed that there is little scientific evidence that analyzes the relationship between NP and the pulmonary effects in humans. Evidently, the foundations of ethics allow experimental studies in laboratory animals and in vitro lung tissue, but the scarcity of pulmonary effects in humans documented in the literature causes little external validity. Secondly, it is noted that there may be little consistency in the results given that, in most studies conducted, there is exposure to multiple PNs, which means that it is not possible to have conclusive data on pulmonary effects specific to the type of NP. Likewise, there are no depth studies regarding the type of exposure (intensity and duration of the exposure) as well as the safety of the protective equipment used by the exposed workers.

It is important that future studies take into account factors of exposure to nanoparticles, such as specifying the NP type, time of exposure and preventive measures adopted (both individual and collective). This could be used to detect what type of NP and under what assumptions can lead to the development of certain diseases (not only lung based ones). Having more knowledge about this matter is very important to take appropriate prevention strategies without alarming the population exposed.

6 CONCLUSION

Nanotechnology is a field of science in constant expansion. The study of exposure to NP is a current and important topic considering general population is increasingly more exposed in both work and non-work level. In recent years there has been much progress in the knowledge of exposure to NP and nanomaterials as well as their mechanism of injury, but few studies analyze the effect in humans for ethical and time related reasons.

As established by the SCENIHR (Scientific Committee on Emerging and Newly Identified Health Risks), more risk assessments should be conducted to begin investigating exposure risk in order to assess the impact on workers' health. It would be interesting to detail the NP type, intensity and duration of the exposure, etc.

There are very few publications that study the possible relationship between occupational exposure to nanoparticles and the occurrence of health damage in humans. However, the absence of evidence is not evidence of absence of harm. It is important to take into account the precautionary principle, that is, to consider nanoparticles as dangerous unless there is sufficient information to prove otherwise.

The awareness of the working population with education campaigns and strict prevention, as well as the multidisciplinary approach among the physicians of Occupational Medicine, Radiology, Pulmonologists and Prevention Technicians could minimize the possible occupational diseases derived from risks whose damage is still not known today.

REFERENCES

Andujar P, Simon-Deckers A, Galateau-Sallé F, Fayard B, Beaune G, Clin B, et al. 2014. Role of metal oxide nanoparticles in histopathological changes observed in the lung of welders. Particle and Fibre Toxicology. 11:23.

Baxter CS, Ross CS, Fabian T, Borgerson JL, Shawon J, Gandhi PD, et al. 2010. Ultrafine Particle Exposure During Fire Supression—Is it an important Contributory Factor for coronary Heart Disease in Firefighters. Journal of Occupational and Environmental Medicine. 52: 791–796.

Cheng TH, Ko FC, Chang JL, Wu KA. 2012. Bronchiolitis obliterans organizing pneumonia due to titanium nanoparticles in paint. Annals Thoracic Surgery. 93:666–669.

Donaldson K, Tran L, Jimenez LA, Duffin R, Newby DE, Mills N, et al. 2005. Combustion derived nanoparticles: A review of their toxicology following inhalation exposure. Particle and Fibre Toxicology. 2:10.

Ferreira AJ, Cemlyn-Jones J Cordeiro R. 2013. Nanoparticles, nanotechnology and pulmonary nanotoxicology. Revista Portuguesa Pneumolologia. 19:28–37.

Genaidy A, Tolaymat T, Sequeira R, A-Rehim AD. 2009. Health effects of exposure to carbon nanofibers: systematic review, critical appraisal, meta analysis and research to practice perspectives. Science of the Total Environment. 407: 3686–3701.

Grosse Y, Loomis D, Guyton KZ, Lauby-Secretan B, El Ghissassi F, Bouvard V, et al. 2014. Carcinogenicity of fluoroedenite, silicon carbide fibres and whiskers, and carbon nanotubes. The Lancet Oncology. 15: 1427–1428.

Gutierrez-Praena D, Jos A, Pichardo S, Puerto M, Sánchez-Granados E, Grillo A, Cameán AM. 2009. Nuevos riesgos tóxicos por exposición a nanopartículas. Revista de Toxicología. 26 (2–3):87–92.

Hubbs AF, Mercer RR, Benkovic SA, Harkema J, Sriram K, Schwegler-Berry D, et al. 2011. Nanotoxicology—a pathologist's pespective.Toxicologic Pathology. 39:301–324.

Kayat J, Gajbhiye V, Tekade RK, Jain NK. 2011. Pulmonary toxicity of carbon nanotubes: a systematic report. Nanomedicine. 7:40–49.

Liao HY, Chung YT, Lai CH, Wang SL, Chiang HC, Li LA, et al. 2014. Six-month follow-up study of health markers of nanomaterials among workers handling engineered nanomaterials. Nanotoxicology. Suppl 1:100–110.

Mendoza G, Rodriguez JL. 2007. La nanociencia y la nanotecnología: una revolución en curso. Perfiles Latinoamericanos. 15: 161–186.

Song Y, Li X, Wang L, Rajanasakul Y, Castranova V, Li H, Ma J. 2011. Nanomaterials in human: Identificacion, Characteristics and Potencial Damage. Toxicologic Pathology 2011; 39:841–849.

Sweeney S, Leo BF, Chen S, Abraham-Thomas N, Thorley AJ, Gow A, et al. 2016. Pulmonary surfactant mitigates silver nanoparticle toxicity in human alveolar type-I-like epithelial cells. Colloids and Surfaces B: Biointerfaces. 145:167–75.

Zhang R, Dai Y, Zhang X, Niu Y, Meng T, Li Y, et al. 2014. Reduced pulmonary function and increased pro-inflammatory cytokines in nanoscale carbon black-exposed workers. Particle and Fibre Toxicology. 11:73.

Zhang Y, Li X, Zhang L, Pan W, Zhu H, et al. 2017. Silica dioxide nanoparticles combined with cold exposure induce stronger systemic inflammatory response. Environmental Science and Pollution Research. 24:291–298.

Wu W.-T, Liao H.-Y, Chung Y.-T, Li W.-F, Tsou T.-C, Li L.-A, et al. 2014. Effect of nanoparticles exposure on fractional exhaled nitric oxide (FENO) in workers exposed to nanomaterials. International Journal of Molecular Sciences. 15:878–894.

Assessment of air quality in professional kitchens of the city of Coimbra

A. Ferreira, A. Lança, D. Barreira, F. Moreira & J.P. Figueiredo
IPC, ESTeSC, Coimbra Health School, Saúde Ambiental, Coimbra, Portugal

ABSTRACT: Fumes produced in the confection, represent the most common risk in kitchens, releasing harmful components reflecting on the Indoor Air Quality (IAQ) and the health of the worker. Carbon Monoxide (CO) and Carbon Dioxide (CO_2) are particularly significant, among other gases, considering the legislation, in order to verify if concentrations are within limits. The sample for this study was of 8 restaurants, in the county of Coimbra. This is a study of level II, of descriptive and observational type with cross-sectional study. The observed values are above the limit of protection and standard values, mainly at the time of confection, where even a significant increase of the concentrations was verified. We conclude that the worker is exposed to some concentrations of pollutants harmful to health, such as PM_{10} – Particulate Material with a diameter of 10 mg/m³, $PM_{2,5}$ – Particulate Material with a diameter of 2,5 mg/m³, and high temperatures (°C).

1 INTRODUCTION

Concerns related with the effects of air quality on public health generally takes into account the air pollution outside buildings, even so, in today's society people spend most of their time indoors. In these spaces, both the number of pollutants and concentration are generally much higher than in outside buildings[1]. Inside the buildings are produced a series of pollutants, which arise in the use of cleaning materials, mold, human metabolism, and man's activities such as cooking. Such pollutants compromise the health and efficiency of employees' work[3]. The working conditions in kitchens presents several risks associated with physically demanding work, exposure to high levels of noise, hot or cold environments, falls, cuts, burns, and psychosocial risks, among others, however, some of these are not obvious to the worker, as is the case of exposure to cooking fumes/vapors[7,12]. In kitchens food are prepared, meals are made, just like daily cleaning of equipment and facilities[7]. During these processes there is generation of heat, water vapor, chemical substances, smoke, and smells that affect working conditions[5]. Some examples of hazardous substances contained in the smoke or vapors released may be Volatile Organic Compounds (VOCs), Particulate Matter ($PM_{2,5}$ and PM_{10}) Polycyclic Aromatic Hydrocarbons (PAHs), fats and water vapor[4]. Cooking at High temperatures lead to the emission of large amounts of fumes that can cause eye irritation and the emission of a wide variety of toxic agents, some of which are potentially carcinogenic and mutagenic[5]. Also, the use of fossil fuels releases harmful substances such as heavy metals, PAHs, CO, and CO_2. PAH's were one of the first carcinogens to be identified and arise because carbon or fuel is not converted into CO and CO_2[2]. Atmospheric particulate matter is one of the most serious public health pollutants, according to the World Health Organization, particles affect more people than any other pollutant. They are identified according to their aerodynamic diameter, PM_{10} or $PM_{2,5}$ and the latter are more dangerous, when inhaled, they may reach the peripheral regions of the bronchioles and interfere with the exchange of gases within the lungs[14]. The concern about the existence of these substances in professional kitchens has been increasing, and several studies over the past 30 years have reported an increased risk of cancer among cooks and other workers. Therefore, is necessary to guarantee the extraction of chemical substances, as well as the high thermal loads generated during all the activities. This is made possible by exhausting and venting the space, extracting the indoor air and renewing it with fresh clean air[4]. Concerning IAQ in kitchens, there are not many studies, however, with the evolution of the restoration and with increasing population demands, workers are exposed for a very long time to harmful substances and also to high temperatures. This fact ends up having negative repercussions on the degree of efficiency in the performance, as well as on the health of the worker, and it is important to understand what they are exposed to, so that one can try to find solutions to minimize the exposure.

With this study, we can evaluate the IAQ in professional kitchens, for this, several chemical measurements of components in the atmosphere of the restaurants were carried out, and it was verified if the same values were within the permitted values.

and standard deviation. For the collection of data in the establishments, the consent of the owner of the restaurant was always sought. The purpose of this research study was solely of academic interest, bearing any economic or financial ends or interests.

2 MATERIAL AND METHODS

The study it was held in 8 traditional restaurants in the city of Coimbra. This is a level II study, of descriptive and observational type, with a cross-sectional study. The type of sampling was non-probabilistic, where the convenience sampling technique was applied, with inclusion criteria being traditional restaurants, 2 measurements were taken before and 2 during the confection along with the preparation of a checklist and a questionnaire to the workers of the same kitchens. The study was carried out in 3 phases. The initial phase consisted of filling in a checklist prepared in advance, in order to verify the physical conditions of the kitchens. In the second phase, physical-chemical parameters were measured, being CO, CO_2, PM_{10}, $PM_{2,5}$, relative humidity and air temperature. In a third phase, questionnaires were distributed to workers. The physical-chemical parameters mentioned above were evaluated, to ensure a good evaluation of the studied establishments. During the measurements two different air quality monitoring equipment were used (Lighthouse – Model 3016; Q-Track – Model 8554). The evaluation took place during the normal period of operation, where measurements were taken before the beginning of the confection and during the most intense confection interval, in order to verify the variations of the concentrations of the evaluated pollutants to the kitchens which workers are exposed on a daily basis. Through Administrative Rule no. 353-A/2013 of December 4, we verified the protection threshold for each physical-chemical pollutant, being the limit of CO 9 ppm (10 mg/m³), for CO_2 the limit is of 1250 ppm, the limit of $PM_{2,5}$ it's 25 µg/m³ and for PM_{10} is 50 µg/m³. For the physical parameters such as temperature, according to EASHW (European Agency for Safety and Health at Work) the ideal temperature in kitchens should be between 20 and 22°C, so that above 24°C starts to occur a drop-in productivity and relative humidity is known to range from 50% to 70%. As standard values for comparison, 22°C and Relative Humidity were then used as 70%. After the data collection, a statistical analysis was performed using the IBM SPSS Statistic version 24.0.

The statistical interpretation was based on the significance level of p-value = 0.05 with a 95% confidence interval. Simple descriptive statistics were also used: frequency absolute and relative, mean, variance

3 RESULTS

3.1 Checklist

In the 8 sites under study, it was always made on arrival a brief visit to the facilities to verify the structural conditions of the site, which allowed us to collect some data. In all the places, the question was asked about the type of oil/fat used in food confection where we found that all the restaurants used olive oil, ensuring that the type of oil/fat used was basically the same, not influencing the final values. The structural and operating characteristics were similar in all the kitchens, and all the premises had artificial exhaust systems of the Hotte Parietal type, operating in the area of the stoves and grills and artificial ventilation systems by displacement.

3.2 Questionnaires for workers

19 individuals were questioned, with 4 individuals being male and 15 females. The most present age range was undoubtedly from 46 to 55 years old, and in general we were able to find workers from the interval of 18 to 25 years until of more than 55 years. Regarding the educational level, about 9 individuals have the first cycle of schooling, with the lowest number of individuals with higher education literacy, such as with the 2nd cycle that also verifies 5,3% each. Regarding the health status of the worker, homogeneity was verified, that all workers claim to have no occupational disease, or even chronic diseases. It was noteworthy that 11 individuals had been working in the facilities for less than a year, and 3 individuals working for more than 16 years in the same place. On average, all workers were working approximately 40 hours per week (standard deviation of 5,657). 16 people stated that the hours of greatest inconvenience were from 12 h

Table 1. Symptoms associated with exposure to fumes, vapors, and high temperatures (%).

Headaches	25,0
Dehydration	12,5
Dizziness	6,3
Fatigue	12,5
Fast Heart Rate	25,0
Throat discomfort	6,3
Exhaustion	12,5

to 14 h. Only 16 of the respondents answered the question related to the symptoms that may have felt during the course of their daily tasks, and the most frequently mentioned symptoms were headaches and fast heart rate. It should be noted that 94,7%, 18 of the respondents, consider that exposure to fumes and vapors was harmful to health.

3.3 Evaluation of physical-chemical parameters

The sample studied consisted of 8 traditional restaurants in the city of Coimbra, where were made the measurements of physical and chemical parameters or the style tag List numbers.

Both CO and CO_2 verified concentration which on average do not exceed the values of the protection threshold, so these workers were not exposed to values of concentration harmful to their health, we can confirm this by observing the mean difference of $-5,959$ ppm and $-1561,563$ ppm respectively. However, we can verify that there are average values of concentration of some pollutants, as in the case of $PM_{2.5}$ and PM_{10} that were above the values required by the protection threshold, where we can verify from the difference of the means that the pollutants actually present worrying values for the health of the worker. The pollutant that is present in greater concentration in the evaluated kitchens was the PM_{10}, where the values exceeded on average 375.063 µg/m³, considering that the protection threshold of this pollutant is 50 µg/m³, we can verify that the values actually exceeded this value, this over measure was significant in relation to the limit value of Protection ($p < 0.05$). The difference of averages with respect to the pollutant $PM_{2.5}$ of 9,831 µg/m³ also revealed that the concentration of this pollutant was above the protection threshold, although this increase was not significant ($p > 0.05$). In the case of the temperature, average values of 27,529°C were registered, which are higher than the standard values, the average difference is 5,529°C, which in fact confirmed the increase considering the standard values. The temperature values presented a p-value $p < 0,05$, which shows that the values are statistically significant. The mean relative humidity values were 56%, within the normal parameters, with a mean difference of $-13,988$%, confirming the values were in the normal parameters. The values in this case were statistically significant, which we can affirm due to p-value <0,05. This study was carried out with the purpose of verifying the pollutants and concentrations in which the workers were exposed in their daily work function, for this the concentrations of pollutants before the confection were evaluated, in the beginning of the work, in order to understand if there were already pollutants before the start of working hours and their concentrations, which also allowed us to verify differences in the concentration of pollutants before and during the confection. In the case of CO, it was observed that both before and during confection, there were always statistically significant values (p-value <0,0001) either before or during the confection, being below the protection threshold, what we also observed in the analysis of the mean difference of the pre-confection averages was $-6,419$ ppm and during the confection was $-5,500$ ppm, although there was an increase of concentration of this pollutant during the confection this is not a worrisome value that harms workers' health. In relation to CO_2, the same values are statistically significant at both times (p-value <0,0001), lower than the protection threshold, both before and during the confection, confirming this through the mean difference of the averages, being -1647.1786 ppm and -1475.938 ppm, respectively, we can also confirm that during the confection there was an increase in the mean concentration. Regarding $PM_{2.5}$, when we compare the mean concentration we find that there is an increase in the concentrations during the confection, the concentrations are so different that at the moment before confection the values of the mean find were below the protection threshold value, with an average difference of $-2,062$ µg/m³, not representing a value that damages the health of the worker.

However, in the observation of the values during the confection there are values that already exceed the limits dictated by the mentioned legislation, giving as mean concentration value 46,724 µg/m³ above the stipulated threshold (25 µg/m³). This

Table 2. Results of the physical-chemical parameters.

N = 16	Mean	Stand. Dev.	p-value	Mean Diff.	Protc. T
CO (ppm)	4,0406	1,341	<0,0001	−5,959	9
CO_2 (ppm)	688,438	184,021	<0,0001	−1561,563	1250
$PM_{2.5}$ (µg/m³)	34,831	33,597	0,108	9,831	25
PM_{10} (µg/m³)	425,063	998,622	0,042	375,063	50
Temp. (°C)	27,529	8,598	0,001	5,529	22
RH (%)	56,013	15,522	<0,0001	−13,988	70

is confirmed when the average difference during confection is 21,724 µg/m³, and the concentration of this pollutant in these values is already harmful to health. However, the values at both times were not statistically significant, p-value = 0,053 and p-value = 0,062, respectively. In PM_{10}, the values at the time before confection were considered statistically significant (p-value = 0,002), it was verified that even before the confection there were already average values of the pollutant above the protection threshold, verified in the average difference of 29,475 µg/m³. As with all pollutants, there was an increase in mean concentration at the time of confection, in which case a marked increase actually occurred, which means that the values even exceeded the protection threshold (50 µg/m³). The mean verified concentration was 770,6500 µg/m³, which can be verified through the mean difference of 720,650 µg/m³. At this time values are also considered statistically significant with a p-value of 0,049. During the confection, what also happens to this pollutant is that the standard deviation obtained values are higher than the mean of the measurements, which means that the results of the measurements made were remarkably different. It is concluded that this is the pollutant to which the workers are most exposed, and at concentrations most worrisome to the health of the worker. Regarding temperature, pre-confection values are not statistically significant (p-value = 0,485), while during confection the values are already statistically significant p-value <0,0001. It was verified that the average temperature before confection (23,056°C) was already higher than the standard value, which confirms by the average difference of 1,056°C, in this parameter it is also possible to observe a rise in temperature during confection compared to prior the confection, being that during the confection are registered values of 32,001°C, that are also above the standard value, and the average difference (10,001°C) came to verify the same. The relative humidity parameter was found both before and during confection, within the stipulated values as normal, with the mean values recorded being 52,350% and 59,675, respectively, the mean difference before confection of −17,650% and during of −10,325%. Both values at both moments show statistically significant values (p-value <0,0001 and p-value = 0,033). This study showed that there are statistically significant differences between the mean concentrations of CO, CO_2, $PM_{2,5}$ and temperature. Regarding the two moments of data collection (before and during), with PM_{10} and relative humidity being shown to have no statistically significant differences with respect to the time before and during confection.

From Table 4 we can see that the difference between the mean values, from the moment before to the moment during confection, are statistically significant (p-value ≤0,05) in the pollutants Carbon Monoxide, Carbon Dioxide, $Pm_{2,5}$, and the temperature, while the pollutant values PM_{10} and relative humidity are not considered to be statistically significant (p-value >0,05). It is possible to verify that all values of pollutants evaluated, temperature and relative humidity increased.

Table 3. Results of the physical-chemical parameters before and during confection.

N = 16	Meas.	Mean	Mean Diff.	Stand. Dev.	p-value
CO (ppm)	Before Conf.	3,581	−6,419	1,023	<0,0001
	During Conf.	4,500	−5,500	1,505	<0,0001
CO_2 (ppm)	Before Conf.	602,813	−1647,188	128,983	<0,0001
	During Conf.	774,063	−1475,938	194,202	<0,0001
$PM_{2,5}$ (µg/m³)	Before Conf.	22,938	−2,062	12,914	0,533
	During Conf.	46,724	21,724	43,177	0,062
PM_{10} (µg/m³)	Before Conf.	79,475	29,475	32,514	0,002
	During Conf.	770,650	720,650	1343,552	0,049
Temp. (°C)	Before Conf.	23,056	1,056	5,895	0,485
	During Conf.	32,001	10,001	8,680	<0,0001
HR (%)	Before Conf.	52,350	−17,650	12,606	<0,0001
	During Conf.	59,675	−10,325	17,619	0,033

Table 4. Results of the physical-chemical parameters—difference relative before and during confection.

N = 16	Meas.	Mean. Dif.	p-value
CO	Before Conf. During Conf.	−0,919	0,054
CO_2	Before Conf. During Conf.	−171,250	0,006
$PM_{2,5}$	Before Conf. During Conf.	−23,786	0,049
PM_{10}	Before Conf. During Conf.	−691,175	0,057
Temp.	Before Conf. During Conf.	−8,945	0,002
RH	Before Conf. During Conf.	−7,325	0,187

4 CONCLUSIONS

After analyzing the results obtained, it was possible to verify that on average most of the pollutants such as CO, CO_2 were found below the protection threshold as also the relative humidity were below standard values, the same did not occur with $PM_{2.5}$, PM_{10} or the temperature, where the average values were above the threshold given by the legislation and the standard values. However, it is possible to verify that the concentrations of CO and CO_2 increases from the moment before to the moment during confection, this is understandable due to the fact that during the confection the stove and the grills were already connected, these pollutants are mostly originated by these places. The average temperature found in all establishments were above the standard values. During the cooking it was registered an increase of approximately 9°C, these temperatures are already very high and taking into account the number of hours workers are exposed to these temperatures tend to be detrimental to worker health and productivity. The relative humidity was always within the parameters of 70%, with no danger to the health of workers. In the case of $PM_{2.5}$ (respirable particles) when we check the concentrations we can see that during confection there is a rise, making those values above the protection threshold. In the case of PM_{10} (thoracic particle size), it appears that before the confection there were already values above the limit, but during confection these values increased even more. That being said, we found that PM_{10} and $PM_{2.5}$ were in harmful concentrations to the worker, representing an increased risk for developing asthma, while $PM_{2.5}$ contribute to an increase in mortality and morbidity.

In conclusion by analyzing this study, we can verify that $PM_{2.5}$ and PM_{10} were both above the protective values. Temperature that was also a factor that exceeded the stipulated value as advisable, making these factors harmful to the health of the worker. Through this study it was possible to conclude that there are pollutants inside professional kitchens way above the protection threshold and standard values. Therefore, it is important to ensure a very good ventilation and exhaustion system in these places and there must be a greater concern related to the pollutants to which the workers are exposed, so that there is a greater control and good measures to prevent and/or eliminate the inhalation of pollutants, which in the long run can cause serious respiratory diseases, even causing cancer in extreme cases.

REFERENCES

[1] Agência Portuguesa do Ambiente. (2016). Qualidade do Ar Interior. Retrieved November 1,2016. Disponível em: URL: https://www.apambiente.pt.
[2] Baptista FM. Ventilação de Cozinhas Profissionais (Ambiente Térmico e Qualidade do Ar). Dissertação. Coimbra. Faculdade de Ciência e Tecnologia da universidade de Coimbra; 2011.
[3] Carmo, A. T., & Prado, R. T. A. (1999). Qualidade do Ar Interno, 35.
[4] Carneiro PMCMF. Ambiente Térmico e Qualidade do Ar em Cozinhas Profissionais. Dissertação. Coimbra. Faculdade de Ciência e Tecnologia da universidade de Coimbra; 2012.
[5] Coelho PIC. Exposição Aos Compostos Orgânicos Voláteis—Trabalhadores Em Cozinhas Escolares. Lisboa. Instituto Politécnico de Lisboa Escola Superior de Tecnologia da Saúde de Lisboa. 2014.
[6] Ferreira A. Qualidade do Ar Interior em Escolas e Saúde das Crianças. Tese. Coimbra. Faculdade de Medicina da Universidade de Coimbra.
[7] Ferreira D, Rebelo A, Santos J, Sousa V, Silva MV. Estabelecimentos de Restauração e Bebidas: Estudo sobre a Qualidade do Ar Interior em Cozinhas. International Symposium on Occupational Safety and Hygiene 2012: 189–191.
[8] Gomes M. Ambiente e pulmão. Conferência VIII Congresso de Pneumologia e Tisiologia do Estado do Rio de Janeiro, V Jornada de Pneumologia Pediátrica, I Jornada Luso-Brasileira. Rio de Janeiro. Julho de 2002 Available from: http://www.scielo.br/pdf/jpneu/v28n5/a04v28n5.pdf.
[9] Lin Y. & Lee L. An Integrated Occupational Hygiene Consultation Model for the Catering Industry. Oxford University Press. 23 March 2010. Available from: https://academic.oup.com/annweh/article/54/5/557/163674An-Integrated-Occupational-Hygiene-Consultation.
[10] Portaria n. 353-A/2013. (2013). (in Portuguese: Regulamento de Desempenho Energético dos Edifícios de Comércio e Serviços (RECS) - Requisitos de Ventilação e Qualidade do Ar Interior). Diário Da República.
[11] Santos J. Avaliação experimental dos níveis de qualidade do ar interior em quartos de dormir—Um Caso de Estudo. Dissertação. Lisboa. Universidade Nova de Lisboa. 2008.
[12] Santos M, Almeida A. COFS (Cooking Oil Fumes). Revista Portuguesa de Saúde Ocupacional online 12 de Outubro de 2016; volume 2, 1–2. Disponível em: URL: http://www.rpso.pt/cofs-cooking-oil-fumes/.
[13] Svendsen K, Jensen HN, Sivertsen I, Sjaastad A. Exposure to Cooking Fumes in Restaurant Kitchens in Norway. Oxford Journals. 17 December 2001. Available from: http://annhyg.oxfordjournals.org/content/46/4/395.short.
[14] Vieira S. Caracterização das partículas no ar interior em escolas de Aveiro. Tese. Aveiro. Universidade de Aveiro; 2011.

Analyse of occupational noise and whole-body vibration exposure at bus drivers—case study

A. Filipa Teixeira
Faculty of Engineering, University of Porto, Porto, Portugal

M. Luísa Matos
Faculty of Engineering, University of Porto, Porto, Portugal
LNEG, Laboratório Nacional de Energia e Geologia, São Mamede de Infesta, Portugal

Luciana Pedrosa & Paulo Costa
Faculty of Engineering, University of Porto, Porto, Portugal

ABSTRACT: Exposure of urban bus drivers to occupational noise and Whole-Body Vibrations (WBV) is often the cause of discomfort, inability to focus on tasks, stress, hearing loss, development of musculoskeletal diseases and so on. This paper aims to expose profile of noise and WBV in *standard* bus drivers considering type of pavement and main sources of exposure. This study involved ten drivers and two routes in the city of Porto, Portugal. As for occupational noise, the methodology used was provided in the ISO 9612:2011, in addition to the information available in Law-Decree nº 182/2006. The WBV measurements were taken according to ISO 2631-1(1997) for daily exposure calculation and ISO 2631-5(2004) for the calculation of the SEAT, S_{ed} and R parameters in addition to the information available in Law-Decree nº 46/2006. The calculation of uncertainties was based on the mathematical model defined in the Guia Relacre 21. There were no situations where preventive measures were necessary.

1 INTRODUCTION

Around 83% of workers in the transport sector are men. Workers in this sector are exposed to prolonged sitting, tiring or painful positions, long working hours (average more than 48 hours a week) and non-standard working hours (night and evening work, weekend work and more than 10 hours worked per day) (European Agency for Safety and Health at Work, 2010).

Compared with other occupational groups, drivers of urban and suburban transportation of passengers show high absenteeism and various incapacities for work, either psychological—fatigue, mental overload and tension, sleep disorders—or musculoskeletal disorders—knee, back, neck and shoulders (ACT, 2016).

Noise-induced hearing loss is the most common occupational disease in Europe, accounting for about one third of all work-related diseases. This disease is usually caused by prolonged exposure to loud noise and the damage is permanent (European Agency for Safety and Health at Work, 2005).

The prolonged exposure may lead to non-auditory pathological symptoms, such as: racing pulse, elevated blood pressure, dilated pupils, and increased production of thyroid hormones, and stomach and abdominal cramps. (Portela & Zannin, 2010), as well, discomfort, sleep disturbances, fatigue, depression, impulsive behavior, increased absenteeism and inability to concentrate on tasks (Mondal, Dey, & Datta, 2014; Sanju & Kumar, 2016).

According to Machado (2003) and Radhakrishna et al. (2012) there are several sources of noise in vehicles, such as: engine, exhaust system, transmission and braking, tires, air conditioning, wind and road traffic.

The authors main identified factors are: model and typology of the bus, age, engine and traffic noise, bus routes (type of pavement), human noise and loading and unloading passengers (Mohammadi, 2014; Portela et al., 2013; Silva & Correia, 2013; Nadri et al., 2012). Some authors have also verified the sources of noise in buses, the tires and the type of motor, as well as the location of the latter (Zannin, 2008; Portela & Zannin, 2010).

As this study was developed in Portugal the action limits and exposure limits values to noise exposure are specified in Law-Decree nº 182/2006 of 6 September (Table 1).

Table 1. Action limits and exposure limits values to occupational noise.

	Individual exposure [dB(A)]	Peak sound pressure level [dB(C)]
Exposure limit	$L_{EX,8h} = 87$	$L_{Cpeak} = 140$
Higher action limit	$L_{EX,8h} = 85$	$L_{Cpeak} = 137$
Lower action limit	$L_{EX,8h} = 80$	$L_{Cpeak} = 135$

Table 2. Action limits and exposure limits values to WBV.

	ISO 2631-1	ISO 2631-5	
	$A(8)$ [m.s^{-2}]	S_{ed} [MPa]	R [Probability of an adverse health effect]
Action limit	0.5	0.5	Moderate
Exposure limit	1.15	0.8	High

Across the EU-27, 25% of workers complain of backache and 23% report muscular pains (Podniece & Taylor, 2008).

According to Podniece & Taylor (2008), in a set of 11 studies it was identified that between 18% and 80% of back pain reported by workers were caused by exposure to WBV.

Also several studies with professional bus drivers have established a strong link between the development of low back pain and exposure to WBV (Blood et al., 2015; Alperovitch-Najenson et al., 2010; Bovenzi & Betta, 1994; Magnusson et al., 1996; Okunribido et al., 2007; Tiemessen et al., 2008; Wilder et al., 1994).

Among the main factors that contribute to the exposure to WBV were identified by different authors, the resonance frequency of the body, where different structures in the body have different resonant frequencies. When the frequency content of the WBV matches the resonant frequency of the body or structures within the body, more of the vibration energy content is transferred into the body. Also the suspension system of the vehicle reveals an impact on vibration amount to which the driver is subject, as well as the type of seat used and conditions of the roads (Thamsuwan et al., 2013; Seo & Kim, 2013; Blood et al. 2010; Lewis & Johnson, 2012).

The International Organization for Standardization (ISO) Standard 2631-1, *"Mechanical Vibration and Shock—Evaluation of Human Exposure to Whole-Body Vibration—Part 1: General Requirements"*, defines how to measure and analyze WBV. Exposure limits values and action limits are introduced in Table 2 (Barreira et al., 2015).

2 OBJECTIVES

This study had as objective the analyze the exposure to occupational noise and WBV among urban bus drivers, relating vibration magnitude and noise level variations with the road characteristics. The analyze of the workplaces quality, seat's transmissibility and probability of an adverse health effect was also studied.

3 METHODOLOGY

3.1 *Equipment*

Data collection was performed using a Sonometer 01 dB Blue Solo da 01 dB Metravib to monitored occupational noise and two accelerometers SVANTEK, model SV106, were used for measurements, simultaneously, vibrations magnitude on seat and vehicle's floor. Continuous data were downloaded and analyzed using two softwares. First was dBTrait 5.1 and it was used to analyze occupational noise, second was SVAN PC ++, version 3.1.1 of SVANTEK.

Microsoft Office Excel was used to calculate all the parameters.

3.2 *Experimental procedure*

The methodology used for the treatment of data collected of occupational noise was in accordance with ISO 9612: 2011 – Acoustics: Determination of occupational noise exposure—Engineering method, in addition to the information available in Law-Decree n.º 182/2006.

As for the measurements of exposure to WBV was in accordance with NP ISO 2631-1:1997 and NP ISO 2631-5:2004 standards. Data were collected from the seat pan and the floor of the bus according to three axes (x, y and z). Data obtained on the floor was used to quantify the parameter SEAT (Seat Effective Amplitude Transmissibility).

Both expositions were monitoring during each driver performance for four complete journeys in the typical daily transport service, which is 6.4 hours in this case.

In this study there weren't taken into consideration: (1) if the air conditioning is on or off; (2) if the driver's window is open or closed; (3) number monitoring of passengers inside the bus; and (4) continuous recording the speed at which the vehicle moves.

3.3 *Road type and drivers*

This study involved ten drivers and two routes (five drivers for each route). Routes characterization (asphalt or brick) is shown in Table 3.

The group study is constituted by five drivers of the Transports Company for each route. The driver's characterization is presented in Table 4.

Table 3. Routes characterization.

Route	Asphalt	Brick
1	95.6%	4.4%
2	93.5%	6.5%

Table 4. Drivers characterization.

	Route	
Average	1	2
Gender	1 female 4 males	1 female 4 males
Age (years)	44	40
Weight (kg)	83.0	87.8
Height (cm)	170	172
Bus driving experience (years)	21	15
Daily time exposure (h)	6.40	6.40

4 RESULTS AND DISCUSSION

4.1 Source of exposure

The analysis of the phenomena identified during the courses made it possible to observe that the equivalent sound level to which drivers are exposed is lower in situations where they travel at low speeds or when they are stopped, such as in traffic lights or in traffic situations. On the other hand, it was not possible to verify significant differences in the exposure to occupational noise considering the type of pavement. However, this parameter reveals greater relevance in the exposure to whole body vibrations, where it was found that in brick routes, bumps and uneven asphalt, acceleration levels were higher, especially in the z-axis.

The occurrence of sound phenomena, such as horns, ambulance passage in emergency service and the entry or exit of passengers, also allowed observing peaks in the sound levels to which these professionals are exposed.

4.2 Daily noise exposition

As regards to the assessment of the occupational noise exposure, two parameters were analysed, $L_{EX,8h}$ and L_{Cpeak}. The $L_{EX,8h}$ values ranged from 70.8 dB(A) to 75.6 dB(A) (already considering the uncertainty) and the values of L_{Cpeak} varied between 115.7 dB(C) and 127.8 dB(C).

It isn't necessary to adopt preventive measures as the $L_{EX,8h}$ values were never higher than 80 dB(A), neither the peak values (L_{Cpeak}) were higher than 135 dB(C), which are the lower action limit defined in Law-Decree nº 182/2006 of 6 September.

4.3 Exposure to WBV

Regarding WBV exposure, parameters evaluated were A(8), SEAT and R. In the A(8) parameter it was verified that the values varied between 0.155 m.s^{-2} and 0.226 m.s^{-2} (already considering the uncertainty).

That way the action limit established in Law-Decree nº. 46/2006 of February 24 (0.5 m.s^{-2}) was never exceeded, therefore, it isn't necessary to adopt preventive measures.

For the SEAT parameter, values lower than 100% were observed, with the exception of two drivers in the second route. In these two cases it is verified that the driver's seat is not performing its function correctly, since isn't attenuating vibrations transmissibility, so replacement of the seats is suggested.

Finally, parameter R, which determines the probability of adverse effects on health due to exposure to vibrations, it was verified that, all drivers showed an R lower than 0.8 (low probability), except one driver in the second route, who has an R equal to 0.8 (moderate probability). This result can be justified because this driver is the one with most exposure years and older than most of others drivers.

4.4 Calculation of uncertainties

In order to evaluate the driver's occupational noise exposure, it was proceeded a validation of the parameter $L_{EX,8h}$ and determination of respective uncertainties in accordance with NP EN ISO 9612:2011. For this parameter, the values obtained for the uncertainties varied between ±1.2 dB(A) and ±1.3 dB(A).

For the evaluation of parameter A(8) the associated uncertainties were also calculated, according to Guia Relacre 21. The uncertainties determined have a value of ±0.01 m.s^{-2}.

5 CONCLUSION

This study allowed expanding the knowledge in these topics. Evaluating, occupational noise and WBV in urban bus drivers according to ISO standards, allowing complete assessment of working conditions.

It was evidenced that the exposure to occupational noise and WBV in this sample of urban bus drivers, do not exceed the recommended limit values. Most of the evaluated seats are also doing its role, reducing WBV transmission to drivers.

In the future, this research may be consolidated by overcoming encountered limitations, namely the absence of supplementary data on the field sheet such as: checking if the air conditioning is on or off, verifying if the driver's window is open or closed, monitoring the number of passengers inside the bus and continuous recording the speed at which the vehicle moves.

REFERENCES

ACT. (2016). Autoridade para as Condições de Trabalho. Consult on november 30, 2017. Available at: http://www.act.gov.pt/(pt-PT)/Campanhas/Campanhasrealizadas/SegurancaeSaudenoTrabalhodaConducao-AutomovelProfissional/Paginas/default.aspx.

Alperovitch-Najenson, D., Santo, Y., Masharawi, Y., Katz-leurer, M. & Ushvaev, D.L.K. (2010). Low Back Pain among Professional Bus Drivers : Ergonomic and Occupational-Psychosocial Risk Factors. Israel Medical Association Journal, Vol. 12, 26–31.

Barreira, S., Matos, M. L., & Santos Baptista, J. (2015). Exposure of urban bus drivers to whole-body vibration. Occupational Safety and Hygiene III. Taylor & Francis Group, 321–324.

Blood, R.P., Ploger, J.D., Yost, M.G., Ching, R.P., & Johnson, P.W. (2010). Whole body vibration exposures in metropolitan bus drivers: A comparison of three seats. Journal of Sound and Vibration, 329(1), 109–120.

Blood, R.P., Yost, M.G., Camp, J.E., & Ching, R.P. (2015). Whole-body Vibration Exposure Intervention among Professional Bus and Truck Drivers: A Laboratory Evaluation of Seat-suspension Designs. Journal of Occupational and Environmental Hygiene, 12(6), 351–362.

Bovenzi, M., & Betta, A. (1994). Low-back disorders in agricultural tractor drivers exposed to whole-body vibration and postural stress. Applied Ergonomics, 25(4), 231–241.

Decreto-Lei n°. 46/2006, de 24 de fevereiro. Diário da República n°. 40/2006, Série I-A de 2006-02-24. Lisboa: Ministério do Trabalho e da Solidariedade Social.

Decreto-Lei n°. 182/2006, de 6 de setembro. Diário da República n°. 172/2006, Série I de 2006-09-06. Lisboa: Ministério do Trabalho e da Solidariedade Social.

European Agency for Safety and Health at Work (2010). Promoção da saúde no setor dos transportes rodoviários, Facts 47.

European Agency for Safety and Health at Work (2005). O impacto do ruído no trabalho, Facts 57.

Guia Relacre 21. (2008). Exposição dos trabalhadores às vibrações—Apontamentos sobre estimativa das incertezas de vibração.

ISO 2631-1 (1997). NP ISO 2631-1:1997. (Mechanical vibration and shock—Evaluation of human exposure to whole-body vibration—Part 1: General requirements). International Organization of Standardization.

ISO 2631-5 (2004). NP ISO 2631-5:2004. (Mechanical vibration and shock—Evaluation of human exposure to whole-body vibration—Part 5: Method for evaluation of vibration containing multiple shocks). International Organization of Standardization.

ISO 9612 (2011). NP ISO 9612:2011. (Acoustics: Determination of occupational noise exposure—Engineering method). International Organization of Standardization.

Lewis, C.A., & Johnson, P.W. (2012). Whole-body vibration exposure in metropolitan bus drivers. Occupational Medicine, 62(7), 519–524.

Machado, W.D. (2003). Identificação de fontes de ruído externo de um veículo utilizado a técnica de intensidade sonora. Universidade Federal de Santa Catarina—Programa de Pós-Graduação em Engenharia Mecânica.

Magnusson, M.L., Pope, M.H., Wilder, D.G., & Areskoug, B. (1996). Are occupational drivers at an increased risk for developing musculoskeletal disorders? Spine, 21(6), 710–717.

Mohammadi, G. (2014). Noise exposure inside of the Kerman urban buses : measurements, drivers and passengers attitudes. Iranian Journal of Health, Safety & Environment, Vol.2, No.1, Noise, 2(1), 224–228.

Mondal, N.K., Dey, M. & Datta, J.K. (2014). Vulnerability of bus and truck drivers affected from vehicle engine noise. International Journal of Sustainable Built Environment, 3(2), 199–206.

Nadri, F., Monazzam, M.R., Khanjani, N., Ghotbi, M.R. & Rajabizade, A. (2012). An Investigation on Occupational Noise Exposure in Kerman Metropolitan Bus Drivers. International Journal of Occupational Hygiene, Vol.4, No.1, 1–5.

Okunribido, O., Shimbles, S.J., Magnusson, M., & Pope, M. (2007). City bus driving and low back pain: A study of the exposures to posture demands, manual materials handling and whole-body vibration. Applied Ergonomics, 38(1), 29–38.

Podniece, Z. & Taylor, T.N. (2008). Work-related musculoskeletal disorders: prevention report. A European campaign on musculoskeletal disorders, Vol. 4.

Portela, B.S, Queiroga, M., Constantini, A. & Zannin, P.H.T. (2013). Annoyance evaluation and the effect of noise on the health of bus drivers. Noise and Health, 15(66), 301–306.

Portela, B.S. & Zannin, P.H.T. (2010). Analysis of factors that influence noise levels inside urban buses. Journal of Scientific and Industrial Research, 69(9), 684–687.

Radhakrishna, D.V., Kallurkar Shrikant, P. & Mattani, A.G. (2012). Noise & vibrations mechanics: Review and diagnostics. International Journal of Applied Engineering Research, 7(1), 71–78.

Sanju, H. & Kumar, P. (2016). Self-assessment of noise-induced hearing impairment in traffic police and bus drivers: Questionnaire-based study. Indian Journal of Otology, 22(3), 162–167.

Seo, K.S. & Kim, K.S. (2013). Analysis of friction noise and vibration from the cushion frame of a driver's seat in passenger cars. International Journal of Control and Automation, Vol.6, No.6, 63–72.

Silva, L.F. & Correia, F.N. (2013). Evaluating noise exposure levels inside the buses for urban transport in the city of Itajuba-MG, Brazil. Revista CEFAC, 15(1), 196–206.

Thamsuwan, O., Blood, R.P., Ching, R.P., Boyle, L. & Johnson, P.W. (2013). Whole body vibration exposures in bus drivers: A comparison between a high-floor coach and a low-floor city bus. International Journal of Industrial Ergonomics, 43(1), 9–17.

Tiemessen, I.J.H., Hulshof, C.T. & Frings-Dresen, M.H. (2008). Low back pain in drivers exposed to whole body vibration: analysis of a dose–response pattern. Occupational and Environmental Medicine, 65(10), 667–675.

Wilder, D., Magnusson, M.L., Fenwick, J. & Pope, M. (1994). The effect of posture and seat suspension design on discomfort and back muscle fatigue during simulated truck driving. Applied Ergonomics, 25(2), 66–76.

Zannin, P.H.T. (2008). Occupational noise in urban buses. International Journal of Industrial Ergonomics, 38(2), 232–237.

Evaluation of the infrastructural conditions and practices of the manipulators—case study: Nursing home and day center

I.N. Roxo, C. Santos, J.P. Figueiredo & A. Ferreira
Coimbra Health School, Coimbra, Portugal

ABSTRACT: Sometimes as surfaces where food is prepared they do not meet as adequate conditions for their confection in an innocuous way allowing the existence of several risks for Public Health. In recent years, the number of institutions supporting elderly people (such as life centers and nursing homes) has increased, thus allowing an alternative to the family home. These institutions provide the majority of health care, in addition to everything, it is not as easy as possible, but also to users with conditional mobility. To avoid such as called Foodborne Diseases they can be quite serious in the at-risk group that are the elderly food available at safety rates. In this case, when we talk about the data security section, see the section "Security Devices". This study aims to evaluate how infrastructure conditions of the institutions concerned, such as practices employed by food handlers. In addition, the number of microorganisms in the hands of the manipulators, surfaces, dishes and utensils should be determined to determine whether there are implications for the safety of the risk to the health of users. It is concluded that, according to a selection list, the Nursing Home presents better results (98.7%) than those obtained by the Day Center (96.3%) regarding the infrastructural part. At the microbiological level and when comparing as two companies we can affirm that the Home presents better results regarding the surfaces, dishes and hands of the manipulators (values less than 10 URL), only presenting values higher than the Day Center for utensils.

1 INTRODUCTION

The concept of food security has its origin in the early twentieth century, from the Second World War. The Food and Agriculture Organization of the United Nations (FAO) has stated that food security exists when "all people at all times have physical and economic access to food that is sufficient, safe, nutritious and that meets nutritional needs and Food preferences in order to provide an active and healthy life" (Assão, 2007; Serrazina, 2013).

Having said this, we can say that there are three basic elements that make up the concept of food security: "sufficient quantity, regularity and quality", assuming a unique approach when it comes to guarantee adequate food for the elderly, population group whose growth is a worldwide phenomenon (Rosa et al, 2012).

The concept of food quality is built on the dynamics and the relationship of consumption involving the State, the productive sector and consumers. Quality is a multidimensional trait being a combination of microbiological, nutritional and sensorial attributes (Blanc et al, 2014).

On the other hand, control in all stages of food production aims to ensure quality, promoting consumer health. A food fit for consumption, that is to say with safety, is one that does not cause disease to the consumer. On the other hand a safe food is based on the absence of chemical, physical and microbiological contaminations guarantees food safety. Improper production and hygiene practices contribute to cross-contamination of food, posing health risks not only to the food handler but also to the final consumer (Blanc et al, 2014; Abreu, 2011).

The term food handler is applied to all persons coming into contact with an edible product or part thereof at any stage of production, this is from the field to the consumer's table. The hands thus constitute an important focus of microorganisms, so when poorly sanitized can transmit deteriorating, pathogenic and fecal origin microorganisms. Thus, microorganisms from the gut, mouth, nose, skin, hair, hair and even secretions and wounds are transferred from the manipulators to food (Blanc et al, 2014; Commission, 2003).

In recent years, there has been an excessive increase in occurrences of Foodborne Diseases, such as salmonellosis, hepatitis A, giardiasis, gastroenteritis, worldwide. The elderly are at serious risk of contracting Foodborne Diseases due to the fact that they are not able to recover as the youngest individuals, and may thus go through a hospital stay, or depending on the severity, even being at risk of death (Blanc et al, 2014; Souza, 2006).

In this case, it emphasizes food poisoning, diseases caused by the consumption of food contaminated with bacteria, parasites, viruses or toxic substances. Most cases of food poisoning occur daily in all countries, regardless of their degree of development. Every year, billions of people worldwide become ill due to the ingestion of unsafe food, and food poisoning is still the biggest cause of illness and death worldwide (AMCN, 2007; WHO, 2007).

By analyzing European and North American intervention in the reduction of food-borne diseases, it can be seen that the greatest benefits in reducing foodborne illness arise from the implementation and monitoring of controls from primary production to commercialization to the consumer (Adak, 2005).

As a Public Health issue, Governments sought to address food safety issues by focusing on risk prevention throughout the food chain through the creation and application of Good Practice Manuals, Hazard Analysis Systems and Critical Control Points, Food Safety Management Systems and Traceability Systems[11]. Community Regulation 852/2004 lists the 7 principles of the Hazard Analysis and Critical Control Point (HACCP) system which defines this system as the basis of food safety. Emphasizing the accountability of stakeholders in the food chain through the adoption of codes of good practice (CE, 2004).

On the other hand, cooked foods are often defiled by dirty or improperly sanitized hands, by insects, rodents, or by simple acts such as sneezing or coughing. This situation reinforces the need to demand and promote the professionalization of food handlers and strengthen health education actions and programs[8, 13]. The education, training and motivation of food handlers are indispensable values for a good prevention policy. Proper compliance with good manufacturing practices and the professional training of food handlers limits the number of foodborne toxins (Alimentar, 2014).

However, precautions should also be taken against cross-contamination of food at different stages of preparation, separation and storage, as well as for the use of different equipment and utensils for raw foods and ready-made foods. Never forgetting that storage at room temperature favors the multiplication of pathogenic bacteria, in the vast majority of foodstuffs, and the microorganisms can remain on the surfaces for hours or days after contamination (Serrazina, 2013; Declan et al, 2004).

The present study aims to know if there are significant differences regarding the structural and practical conditions of the manipulators in a home and a day center, located in the district of Coimbra, in order to analyze their impact on food safety/quality levels.

At a more specific level it is intended to assess whether the structural conditions of the premises concerned (home and day-care center) influence the quality of foodstuffs; to evaluate the microbiological levels of the hands of the manipulators and also to evaluate the practices of the manipulators of each institution

2 MATERIAL AND METHODS

The present case study was conducted in two institutions located in the region of Coimbra providing services to senior users, being the same as the observational and cross-sectional cohort. For this study, data collected between December 2016 and February 2017 were used. The food handlers of the institutions concerned were the target population of the study (8 manipulators) and the sample consisted of 5 manipulators (3 manipulators of the Nursing Home and 2 of the Day Center).

The data collected were based on a questionnaire (Supplement 1) and a checklist (Supplement 2) of the structural and operational conditions developed by the research team based on a Eurest questionnaire[16] and an existing checklist within the area[17], taking into account the legislation applied for this activity.

As regards the questionnaire (Supplement 1), the questionnaire is divided into three groups, the first of which was to collect data related to the knowledge of the Handlers on Hygiene and Food Safety (HSA), the second group sought to gather Information about the hygiene habits and behaviors used by the manipulators and the third one allowed the collection of sociodemographic data on the manipulators that integrated the study.

With regard to the checklist (Supplement 2), it was divided into six fundamental parts: the first concerning the characterization of physical facilities and the environment, the second referring to food handlers, the third to equipment and utensils, the fourth Relating to the reception and storage of products, the fifth recipe for preparation, confection, canning and distribution, and the sixth part on quality control.

Regarding the smears and whenever the harvests were made on flat surfaces, whether dishes or trays, we used the use of molds made of paperboard with a total area of 25 cm^2. When the surface to be analyzed was irregular, such as utensils or the hands of the manipulator, a movement consisting of approximately 1 cm^2 was used and repeated the same for 25 times. Therefore, a dry swab was used for the harvest, which was then inserted into a specific solution for reading through the Clean-Trace 3M NGi equipment to measure the levels of contamination through the bioluminescence of

ATP (adenosine triphosphate) expressed in Relative Light Units (RLU) having an error/standard deviation associated with the equipment of ±0.01.

ATP is found in most food residues and in bacteria, mold and yeast cells, its presence indicates that the surface may contain organic material that could allow the growth of bacteria, with subsequent contamination of the products. The intensity of the light generated by the activated swabs is directly proportional to the amount of ATP and therefore to the amount of residues/microorganisms present, the so-called Relative Light Units (Sovnet, 2016).

For each situation to be analyzed, three measurements were made so that the results were as reliable as possible. In order to determine if the values obtained were satisfactory, this is if they were in accordance with what was regulated was established as an appropriate value: 0–10 RLU; Worth to be aware of: 11–30 RLU and inappropriate value: +31 RLU[20]. For the analysis of results it was considered as adequate value: 0–10 RLU and inappropriate value: +11 RLU.

The treatment of statistical data was performed using IBM® SPSS® Statistics software version 24.0. Simple descriptive statistics, such as location (mean and median) and dispersion (variance and standard deviation), as well as statistical tests, were used to perform this procedure. Therefore we used the Mann Whitney U tests and Kruskal-Wallis Non-Parametric Variance Analysis. The interpretation of these statistical tests was performed based on the significance level of p-value ≤ 0.05 for a Confidence Interval (CI) of 95%.

3 RESULTS

Taking into account the checklists applied at each institution, it was noted that in the section "Physical Facilities and Environment" both institutions complied with the parameters "construction and Design", "facilities", "workspace", "changing rooms", "ceilings", "windows", "doors", "water", "Ice" and "waste/food waste". However, the Day Center had some gaps in the "sanitary facilities" parameters (due to the lack of faucets and cisterns with non-manual controls), "walls" (since they had no rounded corners) and Lack of washing procedure, liquid soap and a suitable drying device). Regarding the Nursing Home, the only unfavorable item was also the parameter "sanitary facilities" (no tap and cistern with non-manual control). As such, and for this section the Day Center get a total of 63 at 72 points (87.5%). Therefore, for this section, the institution has a weighting of 68 by 72 points (94%). As for the section "Food Handlers", both institutions met the "clothing/footwear", "absence of adornment and injury", "training" and "health and HST" parameters. However, it should be noted that there was a disagreement regarding the parameters "handwashing" and "hygienic habits" (due to the incorrect use of disposable gloves) with regard to the Day Center. For the Nursing Home no non-conformities were detected. For this section, the Day Center was weighted 27 out of 29 points (93%) and the Nursing Home weighted 29 out of 29 (100%).

In the "Equipment and Utensils" section, it was found that both institutions were in compliance with the parameters "surfaces (contact with food)", "equipment", "utensils and containers", "procedures", thus obtaining a weighting of 20 out of 20 points (100%).

Regarding the "Receiving and Storage" section, it was observed that both institutions again met the parameters "reception", "commissary" (this parameter did not apply to the Day Center), "cold storage", "refrigeration equipment ","thawing" and thus achieving a score of 53 out of 53 (100%).

In the section "Preparation, Confection, Cup and Distribution" it was observed that both institutions fulfilled all specifications, namely "raw food", "frying", "confectionery", "Refrigeration/maintenance/freezing", "transported food", "regeneration or heating", "cleaning and storage area", "fine dishwashing", "thick dishwashing" and "hygiene", thus making it possible to obtain a 69 points (100%).

In the last "Quality Control" section it was found that both establishments again met all the specifications, namely "implementation of procedures and systems", "traceability", "documentation and records", "hygiene plan", "pest control program", "visitors", "food identification", "microbiological records and analysis" and "sample", thus achieving a score of 57 out of 57 (100%).

After the application of the data collection instruments it is now possible to compare the two institutions at the infrastructural level by the total score obtained. In this case, the Nursing Home obtained a total weight of 296 points out of a possible 300 (98.7%). On the other hand, the Day Center obtained a total weight of 289 points in 300 (96.3%).

Looking now at the answers obtained in the questionnaires applied to the manipulators of the institutions, and highlighting some questions, we verified that regarding the questions "It is at the ambient temperature that the microorganisms have their peak of development so these temperatures must be avoided", "The cooking temperature should always be above 65°C in order to eliminate any microorganisms that may exist", "The manipulator can perform its functions even if it is diseased", "Food after thawing can be frozen again from (example bracelets) or excessive make-up" and "Suppose you are using a frozen product, where you must keep the original label of the product", all handles answered correctly (100%). Concerning the question "Waste buckets/containers can be opened during the service to facilitate procedures" only 3 of the respondents responded correctly (60%). Regarding the issue "Freezing of food eliminates all pathogenic microorganisms" only one respondent was correct in the correct answer (20%). We can point out that the highest score obtained was 19.0 values (corresponding to a Nursing Home Handler) and that the lowest score was 17.5 values (for a Day Center handler).

We then tried to understand if there was any relationship between the score obtained in the questionnaire and the sociodemographic data of the manipulators. Taking this into account we found that the highest score was obtained by a respondent in the 36 to 49 age range, as well as the lowest score was obtained by a respondent aged ≤ 35 years. If we correlate the results obtained with the level of schooling, we determined that the majority of the respondents (80%) had the 3rd cycle (7th to the 9th grade) as their level of education and that the best and the worst score obtained was associated with this level.

The manipulator who has been working in this area for a longer time (≥ 20 years) obtained a score of 18.5 values, while the highest score corresponded to a manipulator with 10 to 19 years of activity in the area. Finally, it should be noted that the lowest score corresponded to a manipulator with 4 to 6 formations, and that the highest score refers to a manipulator with more than 7 formations in the food area.

Using statistical analysis and applying the Kruskal-Wallis Nonparametric Variance Analysis test we found that there were no statistically significant differences at a significance level of 5% (p-value > 0.05) between the results obtained Questionnaires and sociodemographic data. In relation to the Nursing Home and in a more detailed way we can affirm that the highest average value recorded at the level of the hands of the manipulators was 4,7 RLU, in the case of surfaces 14 RLU, in the dishes 6,7 RLU and with regard to utensils 22 RLU. On the other hand, the data referring to the Day Center are less positive since the highest average value recorded at the hands level of the handlers was 21,7 RLU, in the case of surfaces 22,3 RLU, In the dishes 11,7 RLU and in respect to the utensils 10,3 RLU.

After the application of the data collection instruments, the estimated analytical mean values for the microorganisms in the analyzed surfaces were compared. Using the statistical test it was verified that the average number of microorganisms (RLU) on the surfaces for the 4 observations (17,150 ± 5, 9691) was higher than the average number of microorganisms (RLU) presented for the other sites 4 observations in utensils (13,900 ± 5, 5480), the 5 observations in the hands of the manipulators (10,880 ± 9, 6110) and the 4 observations in plates (8,500 ± 2, 5140). With the Mann-Whitney U test, it was found that there were no statistically significant differences at a significance level of 5% (p-value > 0.05).

4 DISCUSSION

Based on the results it was possible to conclude that in general the parameters "food handlers", "equipment and utensils", "reception and storage", "preparation, confection, canning and distribution", "control and quality" However, there were some non-compliances, such as the absence of a flush-mounted flushing valve and a non-manual flushing valve in the toilets, the walls had no rounded corners, there was no disinfectant in the flushing system. Handwashing area and it was not equipped with an appropriate drying device, in addition to pointing out the incorrect use of disposable gloves. Preponderant factors to be improved in the installations include the installation of a non-manual control valve in the sanitary facilities, disinfectant and an ad-equate hand-drying device (Declan, 2004).

Taking into account the information described above and also considering the scores obtained by the institutions regarding the checklist, we can affirm that at the infrastructural level the Nursing Home has better results compared to the Day Center, since it complies with 98.7% of the evaluated points, contrasting with the 96.3% of the Day Center.

At the level of the questionnaires applied to the 5 manipulators, it should be pointed out that the highest score obtained was 19, 0 values (corresponding to a Nursing Home Handler) and that the lowest score corresponded to 17, 5 values (regarding a Handler of the Center during the day). There is no significant difference between the results obtained by the manipulators and the sociodemographic data.

Only 53% of the collected samples presented positive results (considered compliant) at the microbiological level, that is, presented values within the recommended limits (0–10 RLU). Considering the amount of microorganisms present in the analyzed situations, it was found that the surfaces were the places that presented the highest microbiological contamination, followed by the utensils, the hands of the manipulators and finally with a lower contamination of the dishes.

The high contamination of the surfaces may be related to the nature of the surface, which must be smooth and impermeable, easy to wash and clean and non-toxic [21]. However, the analyzed surfaces comply with all the previous assumptions, so the results may be related to the hygiene technique used, or, on the other hand, they are indicators of the ineffectiveness of the hygiene process.

The surfaces in contact with the hands are more often contaminated than the surfaces in contact with the food, justifying the necessity of continuous training, giving a greater responsibility to the food handlers with respect to their procedures[22, 23].

At the level of the manipulators hands, the results were slightly below what was expected, since two of the samples had values above the appropriate values, in this case 21 and 21,7 RLU, these results fall into a group of values To keep in mind (11–30 RLUs)[20]. In order to be more rigorous we can consider as limit value the 10 RLUs, so that two samples have non-conforming values (+11 RLU). Thus, hand hygiene plays an important role in controlling the spread of infectious diseases (Laranjeiro et al, 2013).

The manipulators are of great importance in all stages of the meal confection process, as they can facilitate the spread of pathogenic microorganisms in the work environment (Green, 2006). Therefore it should be noted that hands should be frequently cleaned in order to avoid cross-contamination and the use of ornaments should be eliminated as they are a risk factor, inadvertently falling and incorporating food, or favoring bacterial growth in foods with Which come into contact (Laranjeiro, 2013 et al, Borges et al, 2007).

5 CONCLUSION

Support institutions for the elderly should ensure the safety, health and proper maintenance of the meals they provide, but there may be some failures such as the failure to adopt good practices by food handlers for reasons of negligence or lack of knowledge, and (24) Good hygiene practices during food handling are an important means of reducing cross-contamination between surfaces and handlers. That said, it is essential to ensure the hygiene conditions of both the physical facilities and the working environment, and of the manipulators themselves (Laranjeiro et al, 2013; Pérez-Rodríguez, 2008).

At the infrastructural level it can be concluded that the Nursing Home presents better results (98.7%) than those obtained by the Day Center (96.3%). However, the results obtained within this scope are positive, since the corrections to be made are not significant, that is, they do not significantly compromise food safety.

Regarding the answers obtained in the questionnaires applied to the manipulators of the institutions under study, it was concluded that the highest score obtained corresponded to the manipulator of the Home for the Elderly and that the lowest score corresponded to a Handler of the Day Center. These results indicate that the evaluated manipulators are familiar with the concepts and practices of Food Hygiene and Safety.

At the microbiological level, it should be noted that in most of the "places" analyzed, they did not present a very high number of microorganisms. At the individual level, we highlight the number of microorganisms present in household utensils (above 12 RLUs) and in the case of the Day Center the number of microorganisms found on the surfaces and hands of the manipulators that was above the 20 Relative Light Units. With this in mind we concluded that at the microbiological level, the Nursing Home presented, once again, better results.

When comparing the two institutions we can conclude that the Nursing Home presents better results in regards to the surfaces, dishes and hands of the manipulators, only presenting values higher than the Day Center for utensils.

Emphasis is placed on the importance of proper hand hygiene and the training of handlers with a view to making them aware of the repercussions of their role and responsibilities in preventing contamination and to have the knowledge and skills necessary for the performance of their duties. However, we must always bear in mind that in order to guarantee food safety processes and procedures, they must go through an integrated system from their production to the final consumer (Tuominen, 2003).

REFERENCES

Abreu ESd, Medeiros FdS, Santos DA. 2011. Análise Microbiológica De Mãos De Manipuladores De Alimentos Do Município De Santo André. Revista Univap.

Adak GK, Meakins SM, Yip H, Lopman BA, O'Brien SJ. 2003. Disease Risks from Foods, England and Wales, 1996–2000. Emerging Infectious Diseases. 2005;11:8.

Alimentar PH-PdS. 2014. Toxinfecções Alimentares Portal de Segurança Alimentar. [cited 2017 13 de janeiro]. Available from: http://www.segurancalimentar.com/conteudos.php?id=75.

AMCN R. 2007. Particularidades Dos Serviços De Alimentação Em Instituições Geriátricas. Alimentação Humana.

Assão TY. 2007. Práticas e Percepções acerca da Segurança Alimentar e Nutricional entre os Representantes das Instituições Integrantes de um Centro de Referência. Saúde e Sociedade.16:15.

Blanc PA, Azeredo DRP. 2014. A Segurança De Alimentos No Contexto Do Idoso. Revista Brasileira de Tecnologia Agroindustrial.

Borges LFdAe, Silva BL, Filho PPG. 2007. Hand washing: Changes in the skin flora. Am J Infect Control.

Clayton D, Griffith C, Price P, Peters A. 2002. Food handlers' beliefs and self-reported practices. International Journal of Environmental Health Research.12:15.

Clean-Trace™ M. ATP, URL y UFC. 2011.

Commission CA. 2003. Considerations of the draft revised international code of practice—general principles of food hygiene. Food and agriculture organization of the United Unions. Washington: FAO/WHO food Standards. p. 5.

Declan JB, Bláithín M. 2004. Guidelines for Food Safety Control in European Restaurants. Teagasc—The National Food Centre.

DeVita M, Wadhera R, Theis M, Ingham S. 2007. Assessing the potential of Streptococcus pyogenes and Staphylococcus aureus transfer to foods and customers via a survey of hands, hand-contact surfaces and food-contact surfaces at foodservice facilities. Journal of Foodservice. 18:4.

CE. Regulamento (CE) N.º 852/2004 relativo à higiene dos géneros alimentícios. In: Europeu P, editor. Jornal Oficial da União Europeia: Jornal Oficial da União Europeia. p. 54.

Green LR, Selman CA, Radke V, Ripley D, Mack JC, Reimann DW, et al. 2006. Food Worker Hand Washing Practices: An Observation Study. J Food Protect.

Hygiena. 2014. Limites inferior y superior de las URL para un programa de monitoreo de ATP. p. 4.

Laranjeiro C., Santos, C., Figueiredo, J., Ferreira, A. 2013. Avaliação dos Parâmetros Microbiológicos de Superfícies Em Cantinas De Escolas Da Região Centro.

WHO. Several foodborne diseases are increasing in Europe 2003 23 Maio 2007 [cited 2007 23 Maio 2007]. Available from: http://www.euro.who.int/mediacentre/PR/2003/20031212_2.

Pérez-Rodríguez F, Valero A, Carrasco E, García RM, Zurera G. 2008. Understanding and modelling bacterial transfer to foods: a review. Trends in Food Science & Technology. 19(3):131–44.

CE. Regulation (EC) Nº 852/2004. European Parliament and of the Council: European Parliament and of the Council; 2004.

Ribeirete VPB. 2009. Segurança Alimentar na Força Aérea Portuguesa. [80 p.].

Rosa TEdC, Mondini L, Gubert MB, Sato GS, Benício MHDA. 2012. Segurança alimentar em domicílios chefiados por idosos, Brasil. REV BRAS GERIATR GERONTO.10.

Serrazina VF. 2013. Higiene das mãos dos manipuladores de alimentos dos estabelecimentos de restauração e bebidas. Lisboa: Universidade Nova de Lisboa.

Silva AMMC, Mendonça FJM. 2014. Higiene Alimentar em Creches, Infantários, Escolas, e Instituições de Apoio Social. In: Algarve ARd.

Souza LHLd. 2006. A manipulação inadequada dos alimentos: fator de contaminação. Higiene Alimentar.

Sovnet. 2016 Monitoramento de Higiene por Bioluminescência de ATP. [cited 2017 24 Maio 2017]. Availablefrom:http://www.sovnet.com.br/Products/649-monitoramento-de-higiene-por-bioluminescncia-de-atp.aspx.

Tuominen P, Hielm S, Aarnisalo K, Raaska L, Maijala R. 2003. Trapping the food safety performance of a small or medium-sized food company using a risk-based model. Food Control.

Veiros M, Macedo S, Santos M, Proença R, Rocha A, Kent-Smith L. 2007. Proposta de check-list hígio-sanitária para unidades de restauração. Alimentação Humana. 13:11.

Viveiros FCd. 2010. Questionário—Higiene e Segurança Alimentar.

Neuromuscular fatigue assessment in operators performing occupational-transportation/transshipment tasks: A short systematic review

T. Sa-ngiamsak, N. Phatrabuddha & T. Yingratanasuk
Department of Industrial Hygiene and Safety, Faculty of Public Health, Burapha University, Chonburi Province, Thailand

ABSTRACT: Transportation sector is one of key contributors in driving the economy. Muscle fatigue in this sector alone can cause numerous losses, in term of occupational safety and health. This study was conducted to give a short review on the state-of-the-art in neuromuscular fatigue assessment and its related-contributing factors, which played a key role in detecting muscle fatigue onset. The searches were performed over 3 available and widely accepted electronics data bases. There were 710 articles found combined with 18 articles sought from search engines. Only 8 were considered exactly relevant. Time domain (RMS, muscle activity) and frequency domain (MNF, MDF), extracted from surface Electromyography, were practically key assessment parameters. Body parts affected from fatigue during occupational tasks varied among active muscles frequently employed, as well as different types of activities being performed. Moreover, other provoking factors also needed to be accounted such as vibration, stress, age and the medical history.

1 INTRODUCTION

Transportation sector, involving transporting people and goods such as shipping, trucking, airlines, parcel delivery and railroads, is one of key contributors in driving growing economy. With rapid industrialization, the economy and trade can grow exponentially fast. This allows transportation businesses to expand alongside as evident from the revenue obtained from this industry, which in 2015 it could reach to $1.6 trillion (Downie, 2016).

Reviews conducted by Jeffrey & Rosekind (2017) had revealed some remarkable evidences. From 182 major accidents documented since the year 2001 till 2012, the National Transportation Safety Board (NTSB), an independent U.S. government investigative agency, revealed that fatigue accounted for 20% of all cases which included probable causes, contributing factors or findings. In addition, the NTSB itself had addressed since 1972 that safety risks associated with human fatigue must be recommended (Jeffrey H Marcus & Rosekind, 2017).

Muscle fatigue is a phenomenal which is basically caused from repetitive or sustained work, short work cycles, and localized muscle loadings. And as a result, it can be noticeable through whether: decreasing in strength, performance of the task, exercise capacity, person's ability to exert force and power output (Amedeo Troiano et al., 2008; Mellar P. Davis & Walsh, 2010).

Neuromuscular fatigue represents fatigue which can be evaluated through changes in EMG signals properties. This approach works since the central nervous system controls muscle force through varying activity of motor units via action potentials that are transmitted from motor neuron innervation (Sa-ngiamsak, 2016).

An invasive (needles insertion) and non-invasive (surface EMG electrodes) techniques can be applied in this signal detection. The EMG signal evaluation using needles insertion can provide better accuracy and widely utilized as clinical examination tools. However, it is on one hand totally impractical in detecting muscle fatigue while performing activity tasks such as in occupational safety and health or in sport assessment (Roberto Merletti et al., 2008; Roberto Merletti & Parker, 2004).

To deal with muscle fatigue related losses among operators performing transportation/transshipment tasks, the evaluation of muscle fatigue onset via surface EMG seemed to be the method of choice. Thereby, it had led to the quest of this study in obtaining the state-of-the-art of the mentioned assessment and all possible contributing factors to fatigue.

2 METHODS

2.1 Search strategy

This systematic review was conducted in compliance with PRISMA statement (PLoS_Medicine, 2009). The literature searches were carried out during September 2017. The searches were performed over 3 available and widely accepted electronic databases, including: ScienceDirect, Pubmed and Springer_Link. The used key-words were: "Muscle fatigue", "Electromyography", "Transportation", "Transshipment", "Vehicle operators", "Drivers" and were used on all databases with the appropriate Boolean operators (such as And, Or).

In addition, other literature searches were also performed through search engines and reference lists of those relevant articles. Only full papers were considered eligible, meanwhile those with insufficient information formats, such as abstracts published in conference or workshop proceedings were not considered.

2.2 Screening and eligibility criteria

After duplication removal, all remaining articles were then screened based on several criteria. Those that corresponded to the following ones would be dismissed:

- Unrelated to the subject
- Not written in English
- Publication before 01-01-2004;
- Inaccessible for the full-text articles;
- Studies even on neuromuscular fatigue assessment but not exactly relevant to the sectors being focused.

The remaining articles would be considered eligible if they met with the following criteria:

- Studies on neuromuscular fatigue assessment and exactly relevant to the title related to occupation.
- Studies involving with only occupational tasks in transportation/transshipment, not for other purposes such as engineering design, agriculture, medical purpose, signal processing, etc.

3 RESULTS

3.1 Study selection

After screening and eligibility processes, eight articles were considered relevant to the topic being studied. The entire details of the selection criteria processes could be summarized as in Figure 1.

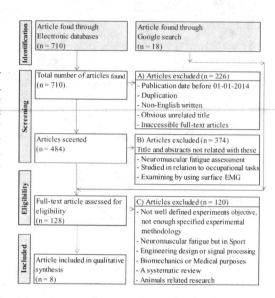

Figure 1. Details of studies selection and inclusion criteria.

3.2 Neuromuscular fatigue assessment in operators performing occupational transportation/transshipment tasks

There were six studies found conducted on land transporting/transshipment-related tasks and other two on air transportation. To describe all in a brief effective way, Table 1 and Table 2 were designed to present all important information, on land occupational transportation/transshipment and air transportation respectively.

4 DISCUSSION

Neuromuscular fatigue assessment related studies were found in various occupational applications, ranging from conventional land transportation (car, bus, truck, trailer truck), transshipment (quay crane) to the air transport (helicopters).

According to the nature of car and bus driving operation, the drivers are essentially required to control the steering wheel, gear shift lever and foot pedals consistently and continually. As a result, neuromuscular fatigues were most likely found developing during driving on *trapezius, deltoid, lumbar paraspinal, splenius capitis, Latissimus dorsi medial, Erector spinae, biceps, forearm flexor* and *frontal* muscle (Balasubramanian & Adalarasu, 2007; Hostens & Ramon, 2005; Jagannath & Balasubramanian, 2014; Katsis C.D. et al., 2004; Leinonen et al., 2005).

Another included land operating task was a container transshipment quay crane. According

Table 1. Found studies of land transporting/transshipment-related tasks (6 studies).

Article title (Location) [vehicle]	Subjects	Investigated muscles (EMG parameters)	Measuring method (Equipment types)	Results considered via fatigue indicators
Back and neck extensor loading and back pain provocation in urban bus drivers with and without low back pain (Finland) [Bus: urban bus in local city-traffic] (Leinonen et al., 2005)	40 voluntary urban bus drivers 25 (22 men and 3 women) with recurrent low back pain and 15 (14 men and 1 women) with healthy back	Lumbar paraspinal and trapezius muscle (RMS/MVC, myoelectric activity)	The maximum voluntary contraction and EMG amplitude of back and trapezius muscles were measured before driving and continuously for the EMG signal while driving with the length of approximately 7 hr. (Bipolar surface EMG, Portable ME 3000p; Mega Electronics Ltd., Kuopio Finland)	Average paraspinal myoelectric activity during driving was approximately 1% of MVC in both groups. Average trapezius myoelectric activities while driving were noticed from 2 to 4% of MVC. Low back and neck-shoulder fatigue increased during driving in both groups especially in those subjects with positive vibration pain provocation. The neck-shoulder pain and fatigue were found more sever in drivers suffering from LBP
Assessment of muscle fatigue in low level monotonous task performance during car driving, Belgium (Belgium) [Car: Fiat scudo 1.9 tD] (Hostens & Ramon, 2005)	22 male subjects participated in the tests. Only 12 were retained for further processing due to EMG signal noise contents	left and right trapezius and deltoid muscles (RMS, MNF)	Every subjects drived 1 hr. on a chosen private route which imposed only left turns providing minimal traffic (ME3000p8 Electronics Ltd., Finland)	Muscle stiffness was observed by more than half of the subjects after 1 hr drive. Only for the active parts, significant decreases of MNF was seen. But also the EMG amplitude decreased significantly
EMG-based analysis of change in muscle activity during simulated driving (India) [Car: Simulation] (Balasubramanian & Adalarasu, 2007)	11 male subjects consisting of 5 professional race car drivers 6 college students	Bilateral of Medial deltoid trapezius type II fibers and splenius capitis (MNF)	Prior to the study, all subjects performed a trial run for 15 min also a given break for 15 min. After that the experimental simulated drive for 15 min was performed with strict preferred posture. EMG signals were extracted at 1st and 15th minute. (EMG acquisition device Bagnoli-8TM, Delsys Inc., USA)	Significant changes in Mean Power Frequency (MNF) between 1st and 15th min were observed in left deltoid, bilatal trapezius and splenius capitis muscle groups. There was no such difference found in other muscle groups.

(Continued)

Table 1. (Continued).

Article title (Location) [vehicle]	Subjects	Investigated muscles (EMG parameters)	Measuring method (Equipment types)	Results considered via fatigue indicators
Assessment of early onset of driver fatigue using multimodal fatigue measures in a static simulator (India) [Car: Simulation] (Jagannath & Balasubramanian, 2014)	20 male participants volunteered for this study. All had license and more than two years of driving experience in light motor vehicle (LMV)	Bilateral of *Trapezius Medial* (TM), *Latissimus Dorsi Medial* (LDM) and *Erector spinae* (ES) (Mean RMS values of sixth level wavelet coefficients)	A 30 min driving practice was allowed for getting accustomed. Then all were required to perform a 60 min driving, while sitting in their comfortable driving posture (16 channel EMG wireless Myomonitor IV, Delsys Inc.)	Significant differences ($r < 0.05$) were found in the TM, LDM and ES muscle groups between the initial of driving period and latter periods of driving. There were no significant differences observed in other muscle groups.
Assessment of Muscle Fatigue during Driving using Surface EMG (Greece) [Truck, Trailer truck, Tractor] (Katsis C.D. et al., 2004)	10 healthy male volunteers including Truck (3), Trailer truck (3), Tractor (4)	Left *Biceps* (LB), Right *Biceps* (RB) left forearm *flexor* (LW) right forearm *flexor* (RW) and *frontal* (L) muscles (MDF, MNF)	All drivers drove twice in a predefined route that was around the city of Torino. The measurements were performed at the starting and the end points of the route where the vehicles were stopped. (A nova Conder V2 Portable PC Polygraph)	The MDF and MNF values decreased by 9.5%–18.9% and 11.3%–18.4% from their initial ones respectively. For the RMS value, it increased from its initial one by 25.1%–47.7%
Multidisciplinary Study of Biological Parameters and Fatigue Evolution in Quay Crane Operators (Italy) [Quay crane: Simulation] (Fadda et al., 2015)	8 quay crane operators working in two different Italian ports	*Trapezius* and *lumbar paravertebral* muscles (*spinal erectors*) (MNF)	All subjects were seated in the cockpit. EMG signals were Collected via the examined session lasting for 4 hr. with variable environmental conditions. mean muscle activation of each muscles was then acquired. (mega ME6000)	Six operators out of eight were found reductions of the mean contractibility in neck-shoulders and lumbar muscles.

Table 2. Found studies of air transportation (2 studies).

Article title (Location) [vehicle]	Subjects	Investigated muscles (EMG parameters)	Measuring method (Equipment types)	Results considered via fatigue indicators
Analysis of muscle fatigue in helicopter pilots (India) [Helicopter: Coast Guard] (Balasubramanian et al., 2011)	20 helicopter pilots only 8 associated with EMG	Bilateral of *Trapezius*, *erector spinae*	EMG signals were collected from pilots while performed Maximal Voluntary Contraction (MVC). This involved lifting of weight kept on the floor during pre—and post-flight (Bagnoli-8TM, Delsys Inc, USA)	Statistical analysis of time and frequency domain parameters indicated significant fatigue in right *trapezius* muscle due to flying. Muscle fatigue correlated with average duration of flight, total service as pilot, pain and total flying hours However, muscle fatigue was found weakly correlated with Body Mass Index (BMI) and age.
Back Muscle EMG of Helicopter Pilots in Flight: Effects of Fatigue, Vibration, and Posture (Brazil) [Helicopter: Sikorsky S-76 or Bell 412 aircrafts] (Oliveira C. & Nadal, 2004)	12 male oil rigs transportation helicopter pilots	Right and left *erector spinae* (ES) (MVC, RMS, MDF)	Pilots were measured Maximal Voluntary Contraction (MVC) while seated upright in a chair for normalization. The major data collection was conducted during the flight for RMS and Median Frequency (MDF) parameters with the averaged flight lasting for 2 hr. (model ME3000P, Mega Electronics, Kuopio, Finland)	The absolute RMS-EMG values were generally very low for all pilots on both left and right sides. When expressed as % MVC, there was a strong evidence of low muscle activity. No significant difference between left and right EMG was detected and no trend was detected on the time series of MDF

to the nature of its operating task, operators need to work in the cockpit high above the ground controlling joysticks and controllers consistently and continually. By this way, they were forced to sit and lean forward for downward operating sights all the time. As a consequence, neck-shoulders and *lumbar* muscles fatigue were most likely found (Fadda et al., 2015).

For other two included studies on air transportations both helicopter, pilots were basically required to control cyclic pitch with right hand, collective pitch with left hand and anti-torque pedals with both legs (Flight-Mechanics, 2017).

In that event, the study results revealed significant fatigue in right trapezius muscle in coast guard helicopter pilots. This obviously suggested that in-flight muscle fatigues were consistent with active muscles.

Two main evaluating options were demonstrated: 1) time domain via amplitude: Root Mean Square (RMS), Maximum Voluntary Contraction (MVC), muscle activity; 2) frequency domain: means of EMG power spectrum including Mean spectral frequency (MNF) & Median spectral frequency (MDF).

The neuromuscular term interpretation of those fatigue developments were possible via differences in EMG signal manifestations, which could be explained as follows: 1) Progressive loss of MVC during the task, resulting from peripheral mechanisms impairment (Mellar P. Davis & Walsh, 2010). 2) The increase in EMG amplitude or muscle activity (represented by RMS) as a result from recruitments of new motor units (MUs). (Edwards R.G. & O.C., 1956; Hostens & Ramon, 2005; Luttmann et al., 1996). 3) The decrease in MNF, MDF due to declination of muscle activations (neural inputs) which are stimulated from Central Nervous System (CNS) or described as the reduction of CNS discharge rate (Abbiss CR & Laursen., 2005; Sa-ngiamsak, 2016).

By the way, in low level monotonous tasks, no significant fatigue in RMS could be found, but still for MNF/MDF. (Hostens & Ramon, 2005).

Muscle fibers being deployed could be responsible for this since: type I (slow-twitch, most fatigue resistant), type IIa (Fast-twitch, fatigue resistant), and type IIb (Fast-twitch, fast fatigable) (Roberto Merletti & Parker, 2004). Along these lines, frequency domains were found using as base lines to ensure the results.

In addition, other exposures to provoking factors such as vibration, stress, operating period, total service, pain history and age were also mentioned affecting the fatigue development.

5 CONCLUSION

Neuromuscular fatigue assessment had proved its capability, accuracy and practicability in detecting muscle fatigue in operators performing occupational-transportation/transshipment tasks. With different exposures to various kinds of repetitive operating activities, muscles frequently deployed as well as muscle fibers type need to be considered. There are other contributing factors such as vibration, stress, age as well as medical history that could potentially enlarge the risks of fatigue development and consequences.

REFERENCES

Abbiss CR, & Laursen., P. 2005. Models to explain fatigue during prolonged endurance cycling. Sports Med, 35(10), 865–898.

Amedeo Troiano, Francesco Naddeo, Erik Sosso, Gianfranco Camarota, Roberto Merletti, & Mesin., L. 2008. Assessment of force and fatigue in isometric contractions of the upper trapezius muscle by surface EMG signal and perceived exertion scale. Gait & Posture, 28, 179–186.

Balasubramanian, V., & Adalarasu, K. 2007. EMG-based analysis of change in muscle activity during simulated driving. Journal of Bodywork and Movement Therapies, 11(2), 151–158.

Balasubramanian, V., Dutt, A., & Rai, S. 2011. Analysis of muscle fatigue in helicopter pilots. Applied Ergonomics, 42(6), 913–918.

Downie, R. 2016. Global Transportation: Exploring Revenue Trends and Fundamentals. Retrieved 24/10/2017, 2017, from http://www.investopedia.com.

Edwards R.G., & O.C., L. 1956. The relation between force and integrated electrical activity in fatigued muscle. J. Physiol., 132, 677–685.

Fadda, P., Meloni, M., Fancello, G., Pau, M., Medda, A., Pinna, C., . . . Leban, B. 2015. Multidisciplinary Study of Biological Parameters and Fatigue Evolution in Quay Crane Operators. Procedia Manufacturing, 3(Supplement C), 3301–3308.

Flight-Mechanics. 2017. Helicopter Axes of Flight—Control Around the Longitudinal and Lateral Axes. Retrieved 28-Jan, 2017, from http://www.flight-mechanic.com.

Hostens, I., & Ramon, H. 2005. Assessment of muscle fatigue in low level monotonous task performance during car driving. Journal of Electromyography and Kinesiology, 15(3), 266–274.

Jagannath, M., & Balasubramanian, V. 2014. Assessment of early onset of driver fatigue using multimodal fatigue measures in a static simulator. Applied Ergonomics, 45(4), 1140–1147.

Jeffrey H Marcus, & Rosekind, M.R. 2017. Fatigue in transportation: NTSB investigations and safety recommendations. Injury Prevention, 23(4), 232–238.

Katsis C.D., Ntouvas N.E., Bafas C.G., & D.I., F. 2004. Assessment of Muscle Fatigue during Driving using Surface EMG. Biomedical Engineering, 2(2004), 4.

Leinonen, V., Kankaanpää, M., Vanharanta, H., Airaksinen, O., & Hänninen, O. 2005. Back and neck extensor loading and back pain provocation in urban bus drivers with and without low back pain. Pathophysiology, 12(4), 249–255.

Luttmann, M. Jäger, J. Sökeland, & W. Laurig. 1996. study on surgeons in urology, Part II: Determination of muscular fatigue. Ergonomics, 39, 298–313.

Mellar P. Davis, & Walsh, D. 2010. Mechanisms of fatigue. The Journal of Supportive Oncology, 8(4), 164–174.

Oliveira C., & Nadal, J. 2004. Back Muscle EMG of Helicopter Pilots in Flight: Effects of Fatigue, Vibration, and Posture. Aviation, Space, and Environmental Medicine, 75, No. 4, Section I April, 6.

PLoS_Medicine. (2009). The PRISMA Statement for Reporting Systematic Reviews and Meta-Analyses of Studies That Evaluate Health Care Interventions: Explanation and Elaboration Guidelines and Guidance.

Roberto Merletti, & Parker, P. 2004. Electromyography Physiology, Engineering, and Noninvasive Application. New York, USA: A John Wiley & Sons, Inc., Publication.

Roberto Merletti, Ales Holobar, & Farina., D. 2008. Analysis of motor units with high-density surface electromyography. J Electromyogr Kinesiol, 18, 879–890.

Sa-ngiamsak, T. 2016. Assessment of Muscle Fatigue in Work Related Musculoskeletal disorders by High-Density Surface Electromyography. (Doctoral degree in Occupational Safety and Health), University of Porto, Porto, Portugal.

# BIM (Building Information Modelling) as a prevention tool in the design and construction phases

M. Tender, J.P. Couto, C. Lopes & T. Cunha
School of Engineering, University of Minho, Guimarães, Portugal

R. Reis
Xispoli Engenharia, Portugal

ABSTRACT: BIM tools have been gaining relevance in the development and management of projects. That is not yet the case when it comes to the Health and Safety Plan (HSP) and the Technical File (TF). The current way of managing prevention often results in a long list of procedures that are hard to understand by those who should be implementing them. This whole scenario creates an environment prone to the downgrading of these problems, and it is urgent to take immediate, easy to understand and efficient management measures. This paper shares the results of a test conducted on the "*BIMSafety*" concept, which is an approach to the use of BIM methodologies in prevention when it comes to the way HSP and TF are presented. To assess the potential of this new approach, the results of a survey conducted on a panel of 42 technicians in the field will be shared. The conclusion reached showed that this methodology streamlines prevention planning, making it easier to understand while integrating prevention and production.

1 INTRODUCTION

1.1 Current picture of prevention in the construction and operations phases

The Health and Safety Plan in the construction phase (HSP) and the Technical File (TF) are legally mandatory documents governing and guiding risk management during the stages of construction and operation of buildings, respectively. Currently, however, apart from a few exceptions, these instruments are merely translated into a long list documents that are not easily understood, which becomes an obstacle to risk analysis and to the implementation of preventive measures (Azhar and Behringer, 2013). The enforcement of these instruments, particularly the TF, falls very short of what was to be expected and what is needed, since they are often perceived as being separate from planning (Sulankivi and Kähkönen, 2010). This makes it hard to decide what preventive measures are to be applied, when and where (Zhang et al., 2015). Also, in many cases, no risk analysis is conducted for that specific case. Instead, a template is used for all projects, regardless of the type of work (Pinto and Reis, 2012). This whole scenario creates an environment prone to the downgrading of these problems. We need to find ways to overcome these gaps, in order to bring risks down to acceptable levels (Reis et al., 2014).

1.2 4th industrial revolution

The Fourth Industrial Revolution is the merging of production methods with the latest developments in information and communications technology. It features "smart" processes, based on virtual templates, streamlining communication between people, machines and equipment (Deloitte, 2017), enabling suppliers to offer a wider range of products adapted to each individual client (Cavalcante e Silva, 2011).

1.3 Building Information Modelling (BIM) and prevention

Architecture, Engineering and Construction are experiencing a technological and organizational revolution, prioritizing information, sustainability and productivity. "*Building Information Modelling*" (BIM), which has taken almost two decades to reach its current state of development (Bargstädt, 2015), enables the digital representation of the physical and functional features of an infrastructure. It is therefore possible to see the Fourth Industrial Revolution embodied in BIM, since it is a way to manage the different expert fields, by facilitating the exchange of information during the construction and operation phases of an enterprise.

According to several authors, BIM presents several advantages, among which: three dimensional viewing, making designs easier to interpret;

decreased potential for contradictions; less time needed to obtain detailed results than with manual design; possibility to create complex views and details; easy comparison between what was planned and what is done; decreased probability for human error in graphic modelling; and use of a computerized database, minimizing manual changes.

Every year, there is an increased interest in integrating prevention issues with BIM (Azhar and Behringer, 2013). Aguilera has analysed papers published in 11 countries regarding the use of BIM methodologies applied to prevention in construction (Aguilera, 2017) and he found that 89% of those papers had been published between 2012 and 2016, with a boost from 2013 onwards.

In September 2016 it was carried out by Fernandes et al. (2017) an experience in the organization of the Marão tunnel using BIM, concluding that the potentialities for optimization and minimization of space risks are significant.

The *"BIMSafety"* concept was forged to help improve the currently inefficient approach to risk prevention. It is the result of a technical and scientific Research & Development partnership between the University of Minho and Xispoli-Engenharia (Reis et al., 2017). The concept links each element or equipment in a worksite to their associated construction and operation risks, throughout the service life of the building. Based on this information, a set of preventive measures are put forth, organizational, collective and personal in nature (Reis et al., 2017). Therefore, *"BIMSafety"* translates into a 3D view of the construction elements, linking them to a set of parametric information that includes the preventive measures associated with that particular construction element. This paper aims to compare the traditional model of prevention to the *"BIMSafety"* concept now presented.

2 METHODOLOGY

2.1 Case study presentation

The case study presented concerns the rehabilitation of a building in the historic centre of Porto.

The software chosen to model the building was *Autodesk Revit*, a BIM modelling software increasingly used to design, construct and manage structures and infrastructures.

The whole building was modelled, by gradually modelling its construction elements: exterior walls, slabs, interior walls, doors, windows and surroundings of the building. First, all of the parameters of units to be worked on in *Revit* were defined in accordance with the *Autocad* units (the software that has been used to deliver the design). The heights of the different storeys were defined, so that the relevant blueprints could be entered. Afterwards, the construction elements were modelled, in stages, following the details on the architecture project design. The last stage was the modelling of the building's surroundings. The modelling of each element creates parametric elements, including all related information, such as their size, their materials and layers, their physical features and their appearance.

Once the modelling of the building was done, the modelling of the prevention planning elements in the HSP and the TF was carried out.

2.2 Integrating the prevention planning into Autodesk Revit

"BIMSafety" foresees the integration of prevention into the model in two stages. The first stage includes the identification of risks and preventive measures, through 3D viewing. The second stage is the input of prevention information into the parameters of each of the construction elements.

This study focuses on what is usually considered to be the most important specific components of the HSP and the TF. For the HSP: site plan; mechanical handling of loads plan; collective protection plan. For the TF: facade works plan; roof works plan; interior works plan.

This paper will test the 3D viewing of these six Plans, along with the input of parametric information into each of the Plans.

2.2.1 Health and safety plan
2.2.1.1 Site plan

The location of the work, in an area with a high density of buildings and pedestrians on the sidewalks adjacent to it, translates into increased difficulties in the assembly and management of the worksite. The solution was to resort to the part of the sidewalk next to the facade for the assembly of the worksite, an option that interferes with the normal circulation of vehicles and pedestrians in the vicinity. The following elements were included in the worksite model (Figure 1): scaffolding; fences; access door to the site; shaded mesh and signage.

Figure 1. Site plan.

Thus, BIM tools allow us to represent the surrounding space, which is very important given the constraints present, and to anticipate risks such as run-overs or fall of materials to lower levels.

2.2.1.2 Mechanical handling of loads plan

The mechanical handling of loads plan includes the handling of loads at heights, not only for the works on the roof, but also for the works on the whole building (Figure 2).

The risk of fall of materials to lower levels and the interference of the crane with obstacles are identified, which proves that the three-dimensional viewing has the potential to plan the movements of the tower crane in detail, anticipating its reaction to obstacles.

2.2.1.3 Collective protection plan

Considering that the work under study is composed of 5 storeys, 3 of them in height, the collective protections assume considerable relevance, given the risk of fall of person to a lower level. The need to adopt preventive measures of a collective nature (e.g. guardrails) is clear. Figure 3 shows a way of planning the assembly of guardrails and obtaining a three-dimensional view of the locations where the guardrails will be applied.

2.2.2 Technical file

2.2.2.1 Facade works plan

Maintenance and repair work on facades generally consist of works on coatings, downspouts, window frames, glazing, etc. For this study, we simulated the maintenance work on the glazing on the facade of the first floor, using a scaffold, which showed the risk of fall of person to a lower level and fall of materials (Figure 4).

2.2.2.2 Roof works plan

The works carried out on roofs in the operations stage are mainly to repair tiles, gutters, chimneys and to service solar panels, etc.. For this study, the simulated works are intended for the servicing of solar panels (Figure 5).

2.2.2.3 Interior works plan

The lack of preventive maintenance in air conditioning systems can cause problems to their users and require works inside their homes, as this study simulates on Figure 6.

2.3 Including preventive measures in the construction elements

The next step in integrating prevention planning is the introduction of written and parametric

Figure 2. Placement of the scaffolds for façade works.

Figure 4. Positioning of guardrails.

Figure 3. Integration of load handling into the BIM model.

Figure 5. Positioning of workers during servicing of solar panels.

Figure 6. Placement of the ladder for servicing of the air conditioning equipment.

Figure 7. Inserting parametric information.

information on preventive measures in each of the modelled elements. The introduction of information relating to prevention was divided in three stages.

Stage I: *Shared Parameters*: Information is introduced by parameters and it is, therefore, necessary to introduce the *Shared Parameters* into the software. This option allows the parameters to be used in several projects. For the characterization of the parameters, properties like the name, the specialty to be used (Architecture, Structures, Mechanics), and the type of parameter are required. It is also required to add a description.

Stage II: Project Parameters: Afterwards, it is necessary to link the parameters to the construction elements. The difficulty here is to decide whether to have sets of parameters by instance or by family. Assigning by family has the advantage of assigning the same information to all of the objects of the family, thus saving time and work. However, it is not feasible, since each structural element may have several risks that require different preventive measures. So, there is the choice of doing the parameter assignment by instance (i.e., for each modelled element), where information differs in all elements, and is independent and unique.

Stage III: Linking the Parameters construction elements: As a result, it is possible to integrate, in written format, the parametric information on risks and measures in their respective element—in this case, a slab of an upper floor –, shown in Figure 7.

3 SURVEY

The survey evaluated the usefulness of the new model regarding the way information is viewed and indexed to each constructive element. Six plans were carried out, presenting two versions for each plan: the traditional format and the new presentation format using BIM methodologies. Two surveys (HSP and TF) were conducted on a panel of 42 civil construction technicians with experience in the sector. Most of the respondents (29%) work as Safety Coordinator or Technician and have an average of 11.5 years of professional experience.

The questions established about *"BIMSafety"* were:

Question 1 (Q1): Does it simulates more effectively the work conditions allowing anticipation of hazards and risks identification?

Question 2 (Q2): Does it assists in identifying the necessary preventive measures for the identified risks?

Question 3 (Q3): Does it improves the quality of information and has the ability to be used in training?

Question 4 (Q4): Does the 3D visualization helps in preventive actions allowing comparisons between the predicted and the real?

Question 5 (Q5): Is it more advantageous than the traditional method?

4 RESULTS

The results of the survey concerning the HSP were as shown (Figure 8).

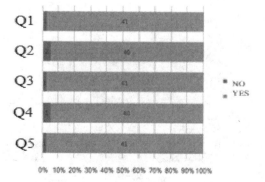

Figure 8. Results concerning HSP.

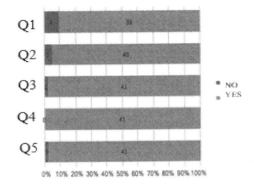

Figure 9. Results concerning TF.

The results of the survey concerning the TF were as shown (Figure 9).

5 DISCUSSION OF RESULTS

Question 1: The results of the survey concerning the HSP show that only one respondent does not agree that the new model is more efficient when compared to the traditional one. This may be justified by the routine work that the current prevention planning model offers. As for the results concerning TF, only 4 of the respondents have denied the efficiency of this model when dealing with this task. This result can be justified with a lack of experience, since the respondents have probably never had the chance to compare both methods in real life.

Question 2: In this question, most of the respondents find that the application of the "*BIMSafety*" concept enables a more efficient perception of the preventive measures, through the new visualization possibilities. Only 2 of the respondents disagree with the usefulness of the new model in the perception of preventive measures, which may have to do with a conservative opinion and a resistance to change.

Question 3: In this question, the respondents generally validate the adequacy of the quality of the information represented, and the model is considered suitable to be used for training. These results reinforce those of several authors who conclude that a 3D model allows the trainer to instruct more easily, in a virtual environment, enabling the simulation of insecure actions, so that the trainees know the hazards that are identified.

Question 4: The answers to the question asking if 3D visualization helps in the prevention of activities are enlightening and show that 3D visualization of the activity plans helps those involved to better plan construction and maintenance tasks.

Question 5: This question ends up summarizing/evaluating all of the work carried out in this research, as well as the whole purpose of the survey. The responses show that almost all of the respondents consider that linking prevention to the new approach allows a more effective simulation of the real working conditions. The minority is identified by one of the respondents who is in the age group between 40 and 50 years, which may also mean that they may have adapted and conformed to the traditional method.

6 CONCLUSIONS

By observing the survey results, it is concluded that:

- The implementation of the new approach to prevention, through the "*BIMSafety*" concept, in relation to the specific plans analysed, is well accepted by the sample of respondents.
- The adoption of the new format in the specific plans studied, allowing 3D viewing and parametric written information for each construction element, is seen as advantageous, compared to the traditional model, optimizing prevention planning, namely at the level of simulation, visualization and understanding of the real working conditions. The new way of visualization is very useful as an instrument to support training actions.
- The study and development of "*BIMSafety*" could revolutionize the elaboration of the Health and Safety Plans and Technical File, leading to a paradigm shift, providing them with a content digitization capability, implying a discussion of the problems in a virtual environment, preventing accidents at work.
- In the future, studies should be conducted to verify the applicability of "*BIMSafety*" to other components of the Health and Safety Plan and the Technical File. In addition, the range of objects available for *Autodesk Revit* should be broadened, with regard to collective and individual protection.

REFERENCES

Aguilera, A. 2017. Review of the state of knowledge of the BIM methodology applied to health. In Arezes et al. (Ed.), *Occupational Safety and Hygiene V:* 447–452. London: Taylor & Francis.

Azhar, S., Behringer, A. 2013. A BIM-based Approach for Communicating and Implementing a Construction

Site Safety Plan. In *49th ASC Annual International Conference Proceedings*. EUA.

Bargstädt, H. 2015. Challenges of BIM for construction site operations. *Procedia Engineering* 117(1): 52–59.

Cavalcante, Z., Silva, M. 2011. The importance of the industrial revolution in technology world. In *VII Encontro Nacional de Produção Científica*. Available in: http://www.cesumar.br/prppge/pesquisa/epcc2011/anais/zedequias_vieira_cavalcante2.pdf.

Deloitte. 2017. *4.0 Industria, COTEC Portugal*. Available in: https://www.industria4-0.cotec.pt/wp-content/uploads/2017/07/industria4_0medidas-pt.pdf.

Fernandes, J., Tender & M., Couto, J. 2017. Using BIM for risk management on a construction site. In Arezes et al. (Ed.), *Occupation Safety and Hygiene V*: 269–272. London: Taylor & Francis.

Pinto, D., Reis, C. 2012. Analysis of Health and Safety plans analysis. In Arezes et al. (Ed.), *Occupational Safety and Hygiene III:* 447–452. London: Taylor & Francis.

Reis, C.. 2014. Preventive Measures in Rehabilitation of Buildings—Facades Shoring. Arezes et al (eds). *Occupational Safety and Hygiene II:* 537–542. Taylor & Francis: London.

Tender, M., Couto, J., Reis, R., Lopes, C., Cunha, T. 2017. A Integração do BIM na Gestão da Prevenção na Construção. *Revista Segurança* 237: 7–8.

Sulankivi, K., Kähkönen, K. 2010. 4D-BIM for construction safety planning. In *W099-Special Track 18th CIB World Building Congress*. Salford, United Kingdom.

Zhang, S. 2015. BIM-based fall hazard identification and prevention in construction safety planning. *Safety Science* 72: 31–45.

Recent trends in safety science and their importance for operations practice

V. Rigatou, K. Fotopoulou, E. Sgourou, P. Katsakiori & Em. Adamides
University of Patras, Rion, Patras, Greece

ABSTRACT: In this paper, we review recent research output in an attempt to identify trends in the evolution of safety science and its relationship with developments in operations management and strategy. The synchronized co-evolution of the two disciplines is very important for operations and safety management alike, since investment in safety is an investment in the productivity of a company's human capital and the operations production. The review revealed that, currently, safety, as a recognized and distinct science, is increasing its scope and becomes heavily dependent on changes in the trends of technology and organizations management. Similarly, by building a safety capability, companies can increase and sustain their operations performance.

Keywords: dynamic capabilities; operations management; safety science

1 INTRODUCTION

While in the occupational and safety context, performance is associated with work accidents and illness prevention, in business operations, performance is associated with efficiency, productivity and profits. Although these two fields—safety and operations—have been developed separately into mature areas of management theory and practice, lately there is evidence of co-operation. Publication trends reveal that occupational safety, while traditionally been treated as a separate function and studied as a standalone field, is now a critical component of production and operations management. Safety is critical for operational excellence and it needs to be considered as a dimension of operational performance in production and operations management research (Pagell et al, 2014). So far, the linkage between safety practices and operations management has not been examined empirically from the safety side, although both are associated with workplace performance and have the potential to be integrated into a joint management system. A typical example in this direction is Safety Management Systems (SMS) that are moving from a prescriptive style to a more 'self-regulatory' and 'performance oriented' model. They become more proactive and participative, as well as better integrated in business activities at a strategic level. Meanwhile, safety of human capital is necessary for operational excellence, so organizations need to develop a capability in safety to be able to motivate and leverage their personnel's capabilities in the pursuit of operational improvements.

Recently, there has been an increased interest of operations management and strategy towards dynamic capabilities, as an extension of the resource-based approach, in markets characterized by fluidity and strong dynamism (Bititci et al, 2011; Eisenhardt & Martin 2000). The logical argument that accompanies this extension is that the resource-based approach cannot interpret in a satisfactory way, why and how specific firms gain operations-based competitive advantage, in situations where the business environment is characterized by rapid and unpredictable change. In these markets, the landscape of competition is constantly changing so dynamic capabilities with which managers "complete, build and reconfigure the company's internal and external capabilities to respond to environmental changes" can be a source of sustainable competitive advantage (Lin et al, 2016). In order to be developed, dynamic capabilities in operations, a core stable base of human capital that actually contributes to the improvement and change of initiatives, when moving between performance objectives is required. Research suggests that occupational safety can provide the basis for the development of such operational capabilities.

In this context, the field of operations management and strategy has already stressed the importance of occupational safety (Pagell et al, 2016). The inevitable question is whether safety science has realized its own importance and contribution to operations in its effort to be established as a science in the context of management and organization studies. In this direction, this paper attempts to capture the state of the art in the evolution of safety science and its relation to operations management and strategy. It becomes clear that an investment in safety is an investment in knowledge/capability-development productivity of a company's human capital. With proper process discipline, companies can build a safety capability and achieve success by jointly managing safety and production outcomes (Pagell et al, 2016). Towards that end, we review recent safety literature to identify themes that connect safety to operations strategy.

2 METHODOLOGY

As it has already been indicated, this paper is based on a review of thirteen recent research papers in an attempt to identify the evolution of safety science and its connection with operations' dynamic capabilities. The selection of these publications was made after a thorough search of the existing literature using as combined keywords the terms "safety", "science" and "dynamic capabilities" in Google Scholar. Initially, the search was conducted without any time restrictions as far as publication dates were concerned. Around one hundred articles came up with apparently related terms such as "cultural safety" and "safety management". The final choice of the thirteen papers was based on their content as being more consistent with the purpose of the study i.e. the historical evolution of safety science and its cooperation with operations strategy studies.

The methodology used to carry out the analysis and the subsequent presentation of the review's results, combined two different approaches: argumentative review and historical review. The argumentative form of review examines selectively the literature to support or refute an argument, a deeply embedded hypothesis or a philosophical problem already established in bibliography. The aim is to develop a body of bibliography that creates a view. The historical form of review focuses on examining the research over a period of time. The aim is to place research in a historical context in order to show familiarity and progress in technology and methods developments and identify possible directions for future research (Sutton, 2016).

Furthermore, in order to depict correctly the evolution of Safety Science, a comparative analysis was accomplished. After reading carefully the thirteen scientific papers and codifying terms, five topics that were mentioned most frequently came up: Definition of Safety and Risk, Accident Analysis & Top Theories, Accident examples, Scientific community and Multidisciplinarity.

3 RESULTS AND ANALYSIS

3.1 *Descriptive chronologic presentation of articles*

A brief summary of each scientific article follows, in order to facilitate their comparative analysis.

The main conclusions of Hopkins (2014) are: (1) the limits of safety science are negotiable, depending on the scientific community; (2) the three theories of accident analysis (normal accident theory, theory of high reliability organizations, resilience engineering) have no practical application but are used to validate the research of the scholars and (3) the conclusions resulting from major accident investigations are not accompanied by a corresponding increase in avoidance recommendations that can be widely applied.

Aven (2014) deals with the notion of safety science, concluding that it can be addressed either theoretically, covering a wide range of educational and research programs, publications, articles, scholars, etc., or practically as a tool for developing and managing methods and theories to avoid risks.

Hollnagel (2014) focuses on the question whether safety alone is a subject of scientific research or whether it is actually a scientific theory. He suggests that safety science should be occupied with the way people work and the functionality of organizations, always in collaboration with other sciences and different research fields.

Almklov et al. (2014) through two case studies in the rail and maritime sectors, investigate whether there is a marginalization of local and specialized knowledge when safety status focuses on accountability and standardization. Their research demonstrates that the old regime of skilled technicians is canceled when safety science results are introduced as a new safety management scheme and it is up to safety scientists to help technicians cope with emerging security knowledge models.

Le Coze et al. (2014) claim that the fragmentation of perspectives makes it difficult for a clear vision of safety science to be produced, but despite this, it should be considered as an applied field of research, obliged to contribute towards improvements and developments of safety.

Haavik (2014) is conducting research at a level where normal accident theory, theory of high reliability organizations and resilience engineering can communicate and be compared with each other without succumbing to ontological conflicts. The author concludes that all three approaches are based on the same socio-technical constructuvist ontology and their difference lies in the way the results of research are adopted and functionalized.

Dekker (2014) reviews the secondary effects of safety bureaucracy, as recorded in the literature in recent years. He notes that the relationship between bureaucracy and regulations on one hand and individual skills, diversity, skills and know-how on the other, might not balance.

Li and Hale (2015) used report analysis to identify the core of the journals concerning safety science in order to map the communication of knowledge among them. The level of interfering is satisfactory despite the multidisciplinary in the field.

Swuste et al. (2016) attempt a bibliographic review at the impact of 1970 and 1979 research into the causes of accidents and disasters in relation to safety management, trying to illuminate historically the evolution of the field of safety. Based on the fact that safety management is also supported by the knowledge of the causes of accidents and basic business management ideas, they also demonstrate the multidisciplinary nature of safety.

Li and Hale (2016) try to visualize and cluster the map of topics covered in six core safety journals. Their analysis of 13.028 articles showed the fast growth of the area of safety science over the last half century and its multidisciplinary nature.

Griffin et al. (2016) take the research a step further, introducing the concept of "dynamic safety capability" to describe an organization's capacity to adapt its systems in changing environments, stressing the importance of safety science in operations management.

Stoop et al. (2017) record their experience of participating for more than 40 years in education, research and development of safety science, proposing three key concepts that emerged during public debates: interdisciplinarity, problem-orientation and systems approach. For the evolution of safety science there is a need for primary data sources, imposition of adaptation and change and creation of a landscape to preserve safety as a science activity.

Finally, Curcuruto et al. (2017) in the framework of safety science, present a multidimensional assessment tool called "Fitness-to-operate" for a better understanding of dynamic safety capability of an organization in high-risk industrial environments.

3.2 Comparative analysis

Table 1 shows the appearance of the aforementioned topics in the papers reviewed. These topics (Definition of Safety and Risk, Accident Analysis & Top Theories, Accident examples, Scientific community and Multidisciplinarity) also reveal a chronological order in the evolution of safety science and are placed in the table accordingly. At first, researchers were occupied to find a suitable definition of safety and risk to base their assumptions about safety as a science (Hopkins, 2014; Aven, 2014; Hollnagel, 2014). Following that initial approach, five researchers (Hopkins, 2014; Haavik, 2014; Swuste et al, 2016; Griffin et al, 2016; Curcuruto et al, 2017) used accident analysis and top theories (normal accident theory, theory of high reliability organizations, resilience engineering) to support their arguments. In several papers, researchers (Hopkins, 2014; Hollnagel, 2014; Almklov et al, 2014; Dekker, 2014; Swuste et al, 2016; Griffin et al, 2016; Stoop et al, 2017) also used accident examples in an attempt to extract knowledge about safety a posteriori, from past events and enhance their theoretical proposition. Finally, this categorization reveals that almost all researchers (Aven, 2014; Hollnagel, 2014; Almklov et al, 2014; Le Coze et al, 2014; Haavik, 2014; Dekker, 2014; Li et al, 2015; Li et al, 2016; Swuste et al, 2016; Griffin et al, 2016; Stoop et al, 2017; Curcuruto et al, 2017) agree on the multidisciplinary nature of safety science, while they point out that it is up to the scientific community to accept safety as a separate science (Aven, 2014; Le Coze et al, 2014; Haavik, 2014; Li et al, 2015; Li et al, 2016; Griffin et al, 2016; Stoop et al, 2017; Curcuruto et al, 2017).

Capturing the flow of developments in safety science through the selected scientific papers, it becomes clear that, initially, there was a great deal of concern about the question of whether safety can be considered as science, and the first three articles attempted to provide a satisfactory answer. Hopkins (2014), Aven (2014) and Hollnagel (2014), using in-depth theories, concluded that safety can be considered as a science that includes several aspects of other sciences and integrates the research and the study of accidents. Then, new general principles of safety management appeared that led to standardized models, which replaced the current specialized knowledge about safety (Almklov et al, 2014). The importance of safety ontology was then emphasized (Haavik, 2014) in the move towards its bureaucratization of procedures (Dekker, 2014). As research evolved through time, a core of scientific journals and publications was formed, while at the same time, attempts from researchers to exchange knowledge in the field were undertaken and bibliographic reviews by analyzing developments in the field of safety science produced more concrete results (Li et al, 2015; Li et al, 2016; Swuste et al, 2016). Ultimately, nowadays, there is a universal recognition of the value and reliability of occupational safety as a science that has a common ground with other sciences, but stands independently in the academic community. Furthermore, safety forms closer relationships with operations management, as terms such as "dynamic safety capability" are introduced (Griffin et al, 2016) and researchers identify the need to measure this capability by creating new assessment tools (Curcuruto et al, 2017).

Overall, the historical evolution of the main themes in safety science that the literature review revealed can be depicted diagrammatically in

Table 1. Categorization of scientific articles.

Author (Year)	Definition of safety and risk	Accident analysis & Top theories	Accident examples	Scientific community	Multi-disciplinarity	Topic
Hopkins, Andrew (2014)	X	X	X			Issues in safety science
Aven, Terje (2014)	X			X	X	What is safety science?
Hollnagel, Erik (2014)	X		X		X	Is safety a subject for science?
Almklov, G. Petter, Rosness, Ragnar, Storkersen, Kristine (2014)			X		X	When safety science meets the practitioners: Does safety science contribute to marginalization of practical knowledge?
Le Coze, J.C, Pettersen, Kenneth, Reiman, Teemu (2014)				X	X	Addressing the foundations of safety science—relevance and benefits
Haavik, K. Torgeir (2014)		X		X	X	On the ontology of safety
Dekker, W.A. Sidney (2014)			X		X	The bureaucratization of safety
Li, Jie and Hale, Andrew (2015)				X	X	Identification of, and knowledge communication among core safety science journals
Swuste, Paul, Gulijk, Van Coen, Zwaard, Walter, Lemkowitz, Saul and Oostendorp, Yvette (2016)		X	X		X	Developments in the safety science domain, in the fields of general and safety management between 1970 and 1979, the year of the near disaster on Three Mile Island, a literature review
Li, Jie and Hale, Andrew (2016)				X	X	Output distributions and topic maps of safety related journals
Griffin, Mark, Cordery, John and Soo, Christine (2016)		X	X	X	X	Dynamic safety capability: How organizations proactively change core safety systems
Stoop, John, Kroes, de Jan and Hale, Andrew (2017)			X	X	X	Safety science, a founding fathers' retrospection
Curcuruto, Matteo, Griffin, Mark and Hodkiewicz, Melinda (2017)		X		X	X	Dynamic safety capability and organizational management systems: an assessment tool to evaluate the "Fitness-To-Operate" in high-risk industrial environments

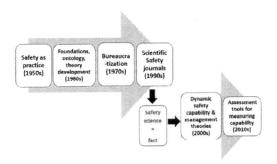

Figure 1. The evolution of safety science.

Figure 1. Recent publications signify a move from an era of introspective existenalistic focus to establish the field as a scientific discipline to a more practical and outwards looking perspective. The term "dynamic safety capability" borrowed from operations strategy, augments the role of safety to the strategic level of operations, making it, in effect, an intrinsic attribute of operations strategy's objectives of cost, flexibility, speed, dependability and quality. The overall goal is to build an organizational capability for safe production, equally for products and services.

Instead of treating safety and productivity as separate entities, there is a need for a single, overarching culture that aligns safety with other priorities, such as productivity, in order to create a competitive advantage (Pagell et al, 2016). In this direction, the operations management (and strategy) field has recognized the importance of safety in operations, as part of the general operations' organizational development decision area (Slack and Lewis, 2007), and operations improvement in particular, investigating the pros and cons of the adoption of novel management systems with different work organization. As an example in this direction is the adoption of lean practices that has been associated in different context with opposing/contradicting results for human safety: positive (Longoni et al, 2013) as well as negative (Mehri, 2006).

4 CONCLUSIONS

Safety science has foundations in organizational science, which is thought to affect attitudes and members' behavior in relation to an organization's current health and safety performance. It is now a recognized science that increases its research scope and importance, since it is heavily dependent on changes in the trends of technology that affect peoples' everyday habits.

In short, safety and risk control, both professional and non-professional, are a multidimensional and highly interesting challenge that should nevertheless be simple to implement and man-based. Further discussions should be held on updating or improving the existing safety models that are robust, in order to allow for the inclusion and exclusion of as many sources of danger as possible for a safer new environment.

REFERENCES

Almklov G.P., Rosness R., Størkersen K. 2014. When safety science meets the practitioners: Does safety science contribute to marginalization of practical knowledge? Safety Science 67: 25–36.

Aven, T. 2014. What is safety science? Safety Science 67: 15–20.

Bititci, S.U., Ackermann, F., Ates, A., Davies, J.D., Gibb, S., MacBryde, J., Mackay, D., Maguire, C., van der Meer, R., Shafti, F. 2011. Managerial processes: an operations management perspective towards dynamic capabilities, Production Planning & Control, 22:2, 157–173.

Curcuruto, M., Griffin & M., Hodkiewicz, M. 2017. Dynamic Safety Capability and Organizational Management Systems: an Assessment Tool to Evaluate the "Fitness-To-Operate" in High-Risk Industrial Environments. Chemical Engineering Transactions 57: AIDIC.

Dekker, S. 2014. The bureaucratization of safety, Safety Science 70: 348–357.

Eisenhardt, K.M. & Martin, J.A. 2000. Dynamic capabilities: what are they? Strategic Management Journal 21: 1105–1121.

Griffin, M., Cordery, J. & Soo, C. 2016. Dynamic safety capability: How organizations proactively change core safety systems. Organizational Psychology Review Vol. 6(3): 248–272.

Haavik, K.T. 2014. On the ontology of safety. Safety Science 67: 37–43.

Hollnagel, E. 2014. Is safety a subject for science? Safety Science 67: 21–24.

Hopkins, A. 2014. Issues in safety science. Safety Science 67: 6–14.

Le Coze, J-C., Pettersen K. & Reiman T. 2014. Addressing the foundations of safety science—relevance and benefits. Safety Science 67: 1–5.

Li, J. & Hale, A. 2015. Identification of, and knowledge communication among core safety science journals. Safety Science 74: 70–78.

Li, J. & Hale, A. 2016. Output distributions and topic maps of safety related journals. Safety Science 82: 236–244.

Lin, C., Tsai, H.L. 2016. Achieving a firm's competitive advantage through dynamic capability, Baltic Journal of Management, Vol. 11 Issue: 3, pp. 260–285.

Longoni, A., Pagell, M., Johnston, D., Veltri, A. 2013. When does lean hurt? – an exploration of lean practices and worker health and safety outcomes. International Journal of Production Research 51: 3300–3320.

Mehri, D. 2006. The darker side of lean: an insider's perspective on the realities of the Toyota production system. Academy of Management Perspectives 20 (2): 21–42.

Pagell, M., Johnston, D., Veltri, A., Klassen, R., Biehl, M. 2014. Is safe production an oxymoron? Production and Operations Management 23: 1161–1175.

Pagell, M., Johnston, D., Veltri, A. 2016. Getting workplace safety right. MIT Sloan Management Review 57: 12–14.

Slack, N. & Lewis, M. 2007. Operations Strategy. London: Financial times/Prentice-Hall.

Stoop J., Kroes de J. & Hale A. 2017. Safety science, a founding fathers' retrospection. Safety Science 94: 103–115.

Sutton, Anthea. Systematic Approaches to a Successful Literature Review. Los Angeles, CA: Sage Publications, 2016.

Swuste P., Gulijk v.C., Zwaard W., Lemkowitz S. & Oostendorp Y. 2016. Developments in the safety science domain, in the fields of general and safety management between 1970 and 1979, the year of the near disaster on Three Mile Island, a literature review. Safety Science 86: 10–26.

Latest efforts aimed at upgrading the IMS-MM

C. Saraiva, P. Domingues, P. Sampaio & P. Arezes
Department of Production and Systems, University of Minho, Braga/Guimarães, Portugal

ABSTRACT: A great deal of research has been published recently by authors that systematically contribute to the scientific domain that deals with the management systems integration phenomenon. This paper aims at reporting the latest efforts to overcome the shortcomings detected in the first version of the Integrated Management Systems-Maturity Model (IMS-MM) by presenting the results collected from a Delphi panel. This Delphi study aimed at rating (according to an importance scale) several potential indicators/metrics that may be suitable to assess the efficiency of integrated management systems by further incorporation into the existing IMS-MM (back office component). The results suggest the existence of several "clusters" each encompassing various indicators/metrics grouped into 11 levels according to its importance aiming at the monitoring of integrated management systems.

1 INTRODUCTION

The Management Systems (MSs) integration phenomenon (or the multiple certification) has been addressed by several authors throughout the last years. A stream of literature within this scientific domain addresses the benefits, motivations and obstacles related to integration and the limitations of managing various sub-systems apart (ex: Mustapha *et al.*, 2017; Moumen & El Aoufir, 2017). Other stream of literature focus on the potential integration levels attained, models, strategies and guidelines to proceed with a proper integration (ex: Gianni *et al.*, 2017; Kania & Spilka, 2016; Rebelo *et al.*, 2017). In addition, the standards usually adopted to implement the MSs are dissected in order to seek for potential improvement opportunities and to propose strategies aiming at its alignment and harmonization. Furthermore, some authors address a critical and relevant topic—the audit function—suggesting how to select the audit team and pointing out which strategies should be implemented in order to optimize the available resources (ex: Rivera *et al.*, 2017). Other authors (in a context not so related with the standardized MSs but more focused on the integration of organizational issues) such as Visser (2017) addressed the issue of integrated value creation, Oke *et al.* highlight the importance of safety guidelines (2017) in an integrated context and Paranitharan *et al.* (2017) reported the validation of the integrated manufacturing business excellence model. Finally, a more recent stream of literature concerns with the maturity of the resulting integrated MS (IMS) and how to assess it (ex: Domingues *et al.*, 2016; Dragomir *et al.*, 2017; Poltronieri *et al.*, 2016). This paper aims at reporting the results from a Delphi panel conducted to assess the importance of various proposed indicators/metrics. Ultimately, it is expected that these indicators/metrics will encompass and improve the existing IMS-MM by minimizing the shortcomings ascribed to it (Domingues *et al.*, 2017c). Although some of the proposed indicators/metrics do not address explicitly the efficiency assessment of the IMS it is believed that, when incorporated in the model, they will be able to provide information regarding this issue notably through the back office component of this hybrid maturity model (Domingues *et al.*, 2016). This paper is structured as follows: section two presents the revision of literature carried out, namely, a revision of the latest studies that addressed various topics within the IMSs domain and a revision of literature of the existing maturity instruments to assess IMSs. The following section (Research Method) describes the procedure adopted to select the experts that encompass the panel, the procedures adopted throughout the Delphi study and the criteria to define the consensus and the consensus zone. The "Results and Discussion" and the "Conclusions" sections present the results collected, discuss some of their potential applications and shortcomings and point out the soundest conclusions.

2 LITERATURE REVIEW

2.1 *IMSs*

Lately, the IMSs scientific domain has been focused explicitly by several authors and relevant contributions were published. At the same time various authors addressed implicitly this scientific domain and assessed the impact of the new requirements brought forward in the new revisions of the ISO 9001 and ISO 14001standards, notably, the High Level Structure (HLS) depicted in the Annex SL (Fonseca & Domingues, 2017). Based on the results reported in this latter paper the authors pointed out that QMS auditors from around the world find the HLS and the Annex SL a suitable instrument to align and harmonize the MSs implemented in companies. Mustapha et al. (2017) addressed the MSs that relate with the sustainability construct and proposed the Sustainable Green Management System (SGMS), i.e., an integrated management framework that, the authors argue, has the potential to '… save ample resources, remove significant redundancies, promote cleaner production and enhance the profitability and efficiency…' of companies. In this line, Ezzat et al. (2017) proposed a framework for the integration of MSs and an evaluation method encompassing simultaneously the stakeholders' requirements, sustainability management and the level of integration of the MSs. Bernardo et al. (2017) reported the results from a comparative study (between Greek and Spanish companies) aiming at the identification of common patterns emergent during the integration process and found a great deal of similarities encompassing strategy, methodology and integration level. The authors found dissimilarities between the audits conducted by Greek and Spanish companies and concluded that Spanish companies integrate the audit function at higher levels than Greek companies. Gianni et al. (2017) addressed concurrently the IMS and the corporate sustainability topics and pointed out that the triple bottom line accountability has the potential to provide the metrics to assess IMS effectiveness. The integrated audits was a topic addressed by Rivera et al. (2017) who identified several shortcomings regarding the development of the audit team. In this paper the authors, in addition to the identification of the shortcomings of the current audit teams formation, propose a methodology (encompassing a set of indicators aiming at the evaluation of the efficiency of the team) to optimize the process of selection of the auditors. Rebelo et al. (2017) dissected the integration of MSs and the management of business risk. Supported on insights both from the existing literature and case studies the authors concluded that the integration of MSs has the potential to minimize the risks that may potentially impact key aspects of the business which highlights the potential of integration and alignment of standardized management and business management. Domingues et al. (2017b) reported the results from a survey conducted among companies managed through an integrated approach and pointed out a set of common characteristics: an effective integration of policies, the existence of an integrated system manager and the alignment of tools, methods and objectives of the individual MSs. The same authors (Domingues et al., 2017a) dissected some dimensions of the Portuguese certified OHSMSs and concluded that the majority of these MSs (up to 95% of the certified OHSMSs) coexists with other implemented MSs, i.e., Portuguese OHSMSs should be analysed from an integrated perspective.

2.2 *Maturity models in the domain of IMSs*

As of today, and to the authors' knowledge, three instruments were proposed to assess IMSs and/or maturity. Domingues et al. (2016) proposed the IMS-MM a (1+5)-level maturity model, supported on three axes and encompassing two components, Dragomir et al. (2017) proposed an instrument that take into account the information collected in process audits and Poltronieri et al. (2016) developed an instrument supported on the sustainability pillars. The shortcomings of the IMS-MM were evidenced when conducting case studies in several companies (Domingues et al., 2017c). The qualitative nature of the information required to populate the IMS-MM (Figure 1) and the fact that the model do not consider some relevant dimensions external to the standardized MSs were identified as the major limitations of the IMS-MM. Concerning the instrument proposed by Dragomir et al. (2017) it should be pointed out that it is designed to assess an IMS that comprises three subsystems maximum and, in the version proposed, relies on information from the audit reports and do not take into account other dimensions or peculiarities than those reflected by the requirements of the standards that are being audited (similarly to the IMS-MM). Regarding the IMS-MM (Domingues et al.,

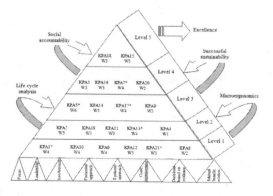

Figure 1. The front office component of the IMS-MM.

2016) it should be emphasized that it is supported on three axes: the Key Process Agents (KPAs), the externalities and the principles of quality management. It is intended, due to the abovementioned shortcomings of the current version, to introduce changes (enabling the incorporation of quantitative information) in the three axes to be assessed. The information collected from the present study will be incorporated in the axis "KPAs" in order to decrease the bias of the information supported, in the current version, in a structured questionnaire. Additionally, some changes are planned to occur in the two other axes that encompass the model and the collection of data is being carried out. However, it is not the scope of the current study to disclose the specifics of such update.

3 RESEARCH METHODS

3.1 Delphi method

The Delphi method is usually adopted to validate some research relevant topics. This method is supported by four main characteristics: anonymity, iteration, controlled feedback and the statistical treatment of the collected information. In the current study the selection of the experts (32 total) that comprise the panel took into account their experience in the IMSs domain (both academic and industrial—at least 5 years) and their contact with different geographical contexts (to assure a diversified sample). Experts from Portugal, Spain, Brazil, Sweden, Romania, Italy, UK, Greece, Denmark, Poland, Pakistan, Trinidad and Tobago and Netherlands encompassed the Delphi panel. The identification of the 29 indicators/metrics (Table 1) to be assessed by each expert was supported in the existing literature and in two case studies previously conducted.

The assessment of the importance of each indicator/metric was carried out through the adoption of a 5-point Likert scale by the means of an online survey. This assessment procedure was carried out throughout 3 iterations (or till the consensus criterion was fulfilled for each indicator/metric). After each iteration statistical treatment of the information took place to check if the consensus criterion was fulfilled or to define the consensus zone and identify the median of the results of each indicator/metric (Section 3.2). The statistical analysis of the collected data was carried out by IBM SPSS v. 24 software.

3.2 Establishment of the consensus zone and the consensus criterion

The establishment of the consensus zone and the consensus criterion were based on those reported in previous studies (Ramos, 2013). Although the existence of several definitions of consensus zone and different reported consensus criteria the option relied on those established in a study with similar goals of the current one. The consensus

Table 1. Indicators/metrics to be assessed by the Delphi panel.

Dimension	ID_{Ind}	Indicator
Costs	ID 20	Total costs ascribed to the IMS.
	ID 21	Cost reduction achieved with the implementation of the IMS.
Audits	ID 24	% of non-conformities detected and ascribed, simultaneously, to the various MSs.
	ID 3	% of audits conducted adopting an integrated approach.
	ID 23	Average time to close corrective actions derived from external and internal audits.
Stakeholders/ Suppliers	ID 1	N° of complaints from the stakeholders.
	ID 2	N° of suppliers holding more than one certification.
	ID 18	% of integrated requirements demanded to suppliers.
	ID 19	N° of suppliers assessed in the dimensions of quality, environment and OHS.
Employees	ID 10	N° of improvement proposals originated from the employees.
	ID 14	N° of meetings conducted to provide employees with information concerning the IMS.
	ID 16	% of employees informed about the relevance of the IMS.
	ID 17	% of employees who attended training courses about the IMS.
	ID 22	N° of training hours addressing the topic "Integration of MSs".
	ID 12	% of training courses addressing the IMS.

(Continued)

Table 1. (Continued).

Dimension	ID$_{Ind}$	Indicator
Standardized Organizational Structure	ID 13	% of IMS procedures improved due to corrective actions.
	ID 8	Existence of integrated procedures.
	ID 5	N° of policy references to all stakeholders.
	ID 4	Existence of an integrated policy.
	ID 9	N° of meetings, addressing IMS issues, with the participation of Top Management.
	ID 11	Existence of organizational functions with responsibilities and duties in the IMS.
Monitoring	ID 15	N° of integrated goals established.
	ID 6	Existence of integrated indicators.
	ID 7	Existence of integrated objectives.
	ID 25	Effectiveness rate of preventive actions.
	ID 26	Effectiveness rate of corrective actions.
	ID 27	Effectiveness rate of training sessions.
Integration Process	ID 28	N° of guidelines, frameworks or standards adopted throughout the integration process.
	ID 29	N° of integrating concepts adopted during the integration process.

criterion was established considering the difference between the 1st and 3rd quartile, *i.e.*, the interquartile range (IQR) ≤1. The indicator/metric that fulfilled the consensus criterion was removed from the assessment survey in the next Delphi iterations. Those indicators/metrics that did not attained consensus according to the abovementioned criterion were assessed in the following iterations. In this case, both the median and the consensus zone of the results collected were reported to the experts in the 2nd and 3rd iterations. The consensus zone was established considering the items in the 5 point scale that encompassed, at least, 50% of the responses from the experts.

4 RESULTS AND DISCUSSION

The Delphi study was carried out throughout 3 iterations (rounds). A total of 32 experts agreed to encompass the Delphi panel and, at the end of the first iteration, 30 responses were collected. In the following iterations (2nd and 3rd) 29 and 28 responses were collected, respectively. The successive abandonment of experts during the Delphi study is a well reported shortcoming of this research method but, it is our belief, do not invalidate the results attained (3.3% abandonment rate). A total of 29 indicators/metrics were proposed to the experts in the first round to be assessed according to a 5-point importance scale (1 – Without any importance; 2 – Little importance; 3 – Important; 4 – Very important; 5 – Extremely important). Figure 2 presents the evolution of the number of indicators/metrics that attained consensus throughout the successive rounds according to the criterion previously established.

Figure 3 presents the box-plot diagram of the results from the 1st iteration.

Based on Figure 2 one should stress that all the indicators/metrics attained consensus among the experts that encompassed the panel concerning their importance to be adopted in IMSs. In order to prioritize the assessed indicators/metrics according to their importance the following guidelines and decision vectors were adopted cumulatively:

1st – Importance Level (Median).
2nd – Iteration (round) at which the consensus criterion was fulfilled and/or IQR score (the lowest the IQR the better the definition of consensus level).
3rd – Evolution of the Median score throughout the successive rounds (for those indicators/metrics without consensus).
4th – Analysis of the opinion of the experts round after round.

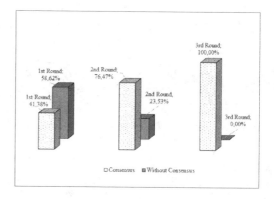

Figure 2. Evolution of the indicators/metrics that attained consensus.

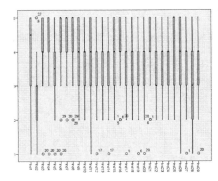

Figure 3. Box-Plot diagram from the results of the 1st round.

Table 2. Indicators/metrics clustered according their importance.

Level	Line-up	ID_{Ind}	Comments
1	1st	ID 4	Indicators/metrics that scored 5 (Median) and fulfilled the consensus criterion in the 1st iteration (round). All of them scored IQR = 1.
	2nd	ID 6	
	3rd	ID 7	
	4th	ID 8	
2	5th	ID 25	Indicator/metric that scored 5 (Median) and fulfilled the consensus criterion in the 2nd iteration (round).
3	6th	ID 1	Indicator/metric that scored 5 (Median) and fulfilled the consensus criterion in the 3rd iteration (round).
4	7th	ID 9	Indicator/metric that scored 4 (Median) and fulfilled the consensus criterion in the 3rd iteration (round) with IQR = 0. The score of the Median kept unchanged throughout all the iterations of the Delphi study.
5	8th	ID 16	Indicator/metric that scored 4 (Median) and fulfilled the consensus criterion in the 2nd iteration (round) with IQR = 0.5. The score of the Median kept unchanged throughout all the iterations of the Delphi study.
6	9th	ID 13	Indicator/metric that scored 4 (Median) and fulfilled the consensus criterion in the 3rd iteration (round) with IQR = 0.75. The score of the Median kept unchanged throughout all the iterations of the Delphi study.

(Continued)

Table 2. (Continued).

Level	Line-up	ID_{Ind}	Comments
7	10th	ID 14	Indicators/metrics that scored 4 (Median) and fulfilled the consensus criterion in the 1st iteration (round) with IQR = 1.
	11th	ID 17	
	12th	ID 18	
	13th	ID 20	
	14th	ID 23	
	15th	ID 24	
	16th	ID 29	
8	17th	ID 3	Indicators/metrics that scored 4 (Median) and fulfilled the consensus criterion in the 2nd iteration (round) with IQR = 1.
	18th	ID 5	
	19th	ID 10	
	20th	ID 11	
	21st	ID 12	
	22nd	ID 15	
	23rd	ID 21	
	24th	ID 26	
	25th	ID 27	
9	26th	ID 19	Indicator/metric that scored 3 (Median) and fulfilled the consensus criterion in the 1st iteration (round).
10	27th	ID 2	Indicators/metrics that scored 3 (Median) and fulfilled the consensus criterion in the 2nd iteration (round). IQR = 1
	28th	ID 22	
11	29th	ID 28	Indicator/metric that scored 3 (Median) and fulfilled the consensus criterion in the 3rd iteration (round).

Considering these guidelines the following 11 clusters of indicators/metrics were identified (Table 2) and grouped according to the importance level ascribed by the experts that comprised the Delphi panel. In addition, Table 2 presents the main justification (according to the abovementioned guidelines and decision vectors) to cluster the indicators in a generic Level.

5 CONCLUSIONS

A Delphi study was conducted among 32 selected experts aiming at the assessment of the importance level that may be ascribed to 29 proposed indicators/ metrics with the potential to be adopted in the monitoring of IMSs. To sum up, it should be highlighted that indicators/metrics such as the "Existence of an integrated policy", "Existence of integrated indicators/metrics", "Existence of integrated objectives" and "Existence of integrated procedures" were rated as the most important. Moreover, the indicator/metric "Number of guidelines, frameworks or standards adopted to integrate the MSs" was ranked at the 11th (and lowest) level of importance. Further work will encompass the incorporation of the proposed indicators/metrics in the IMS-MM as well as a more detailed analysis of the responses considered as outliers in this study.

ACKNOWLEDGEMENTS

Acknowledgements are due to all the experts that comprised the Delphi panel. The authors acknowledge the contribution of Acácio Costa (DPS) throughout the development of the online survey. FCT (Fundação para a Ciência e Tecnologia of Portugal) supported this study under the project ID/CEC/00319/2013. P. Domingues is supported by FCT scholarship Ref. SFRH/BPD/103322/2014.

REFERENCES

Bernardo, M., Gianni, M., Gotzamani, K. & Simon, A. 2017. Is there a common pattern to integrate multiple management systems? A comparative analysis between organizations in Greece and Spain. *Journal of Cleaner Production* 151: 121–133.

Domingues, J.P.T., Sampaio, P & Arezes, P.M. 2016. Integrated management systems assessment: A maturity model proposal. *Journal of Cleaner Production* 124: 164–174.

Domingues, J.P.T., Sampaio, P. & Arezes, P.M. 2017a. Analysis of certified occupational health and safety management systems in Portugal. *International Journal of Occupational and Environmental Safety* 1(1): 11–28.

Domingues, J.P.T., Sampaio, P. & Arezes, P.M. 2017b. Management systems integration: Survey results. *International Journal of Quality and Reliability Management* 34(8): 1252–1294.

Domingues, J.P.T., Sampaio, P., Arezes, P.M., Inácio, I. & Reis, C. 2017c. Management system maturity assessment based on the IMS-MM: Case study in two companies. In: Arezes P. *et al.* (eds) *Occupational Safety and Hygiene V*. SHO 2017. CRC Press—Taylor and Francis group, London, pp. 235–239. ISBN: 978-1-138-05761-6.

Dragomir, M., Popescu, S., Neamțu, C., Dragomir, D. & Bodi, Ș. 2017. Seeing the immaterial: A new instrument for evaluating integrated management systems' maturity. *Sustainability* 9(9): 1643.

Ezzat, A., Bahi, S. & Nasreldeen, T. 2017. Towards better environmental performance: A framework for IMS. *International Journal of Scientific and Engineering Research* 8(2): 105–130.

Fonseca, L. & Domingues, J.P.T. 2017. The results are in—What auditors around the world think of ISO 9001:2015. *Quality Progress* 50(10): 26–33.

Gianni, M., Gotzamani, K. & Tsiotras, G. 2017. Multiple perspectives on integrated management systems and corporate sustainability performance. *Journal of Cleaner Production* 168: 1297–1311.

Kania, A. & Spilka, M. 2016. Analysis of integrated management system of the quality, environment and occupational safety. *Journal of Achievements in Materials and Manufacturing Engineering* 78(2): 78–84.

Moumen, M. & El Aoufir, H. 2017. Quality, safety and environment management systems (QSE): Analysis of empirical studies on integrated management systems (IMS). *Journal of Decision Systems* 26(3): 207–228.

Mustapha, M.A., Manan, Z.A. & Alwi, S.R.W. 2017. Sustainable Green Management System (SGMS)- An integrated approach towards organisational sustainability. *Journal of Cleaner Production* 146: 158–172.

Oke, A., Aigbavboa, C. & Seemola, M. 2017. Importance of safety guidelines on South African construction sites. In: *Arezes P. (eds) Advances in Safety Management and Human Factors. AHFE 2017. Advances in Intelligent Systems and Computing*, Vol. 604, Springer, Cham, pp. 152–160. https://doi.org/10.1007/978-3-319-60525-8_16.

Paranitharan, K.P., Babu, T.R., Pandi, A.P. & Jeyathilagar, D. 2017. An empirical validation of integrated manufacturing business excellence model. *The International Journal of Advanced Manufacturing Technology* 92(5–8): 2569–2591.

Poltronieri, C.F., Gerolamo, M.C., Dias, T.C.M. & Carpinetti, L.C. 2016. An instrument that evaluates the IMS and sustainable performance. In *proceedings of the 2nd ICQEM*, Guimarães, Portugal, pp. 19–31.

Ramos, D.G.G. 2013. Análise custo-benefício em avaliação de risco ocupacional. PhD Thesis, University of Minho, Braga, Portugal, 271 pgs.

Rebelo, M.F., Silva, R. & Santos, G. 2017. The integration of standardized management systems: managing business risk. *International Journal of Quality & Reliability Management* 34(3): 395–405.

Rivera, D.E., Villar, A.S. & Marrero, J.I.S. (2017). Auditors' selection and audit team formation in integrated audits. *Calitatea: Acces la Success* 18(157): 65–71.

Visser, W. 2017. Innovation pathways towards creating integrated value: A conceptual framework (September 30, 2017). Available at SSRN: https://ssrn.com/abstract=3045898.

Exposure to particules in the poultry sector

F. Capitão, A. Ferreira, F. Moreira & J.P. Figueiredo
Department of Environmental Health, IPC—Coimbra Health School, Coimbra, Portugal

ABSTRACT: The health and well-being at the workplace are crucial to get a good performance in order to achieve productivity in a company. Specifically in the poultry sector, the worker is exposed to a series of risk factors in the execution of his tasks. The objective of this study was to evaluate workers' exposure to interior air quality, such as egg laying pavilions and in the sorting center, taking into account particulate matter, as well as to analyze the data collected associating with occupational risk factors in the employees of the poultry sector. The investigation lasted for 8 months (from September 2016 to May 2017). Concerning the collection of the analytical data of the particulate matter in the different jobs, this was carried out from February to April 2017. The study was developed in an activity company of poultry (CAE) located in the district of Santarém. As a sample, three workers were considered in the company, two in egg classification and one in the cleaning and disinfection of egg laying pavilions, being evaluated the quantitative data of the insoluble particles to which they were exposed. Taking into account the results obtained, it was shown that the particulate matter did not in fact present an increased risk to the health and safety of the poultry farmers studied. The levels of exposure to particulate matter were below the reference values, having reduced potential to trigger cardiac and respiratory diseases. As such, the measures that allowed the improvement of the conditions in the work places within the reference values were: general ventilation of the work areas, reduction of the times of exposure and individual protection.

1 INTRODUCTION

The various air pollutants have an effect on workers' health, leading to or aggravating illnesses, particularly in the more debilitated age groups, such as individuals with respiratory and cardiovascular problems, children and the elderly (1). However, there are scientific studies showing that the levels of ambient air pollution in Europe have been harmful to health (studies carried out under CAFE—Clean Air For Europe by the European Commission) (1).

The concerns about the effects of ambient air quality on public health are only focused on air pollution outside buildings. Despite that, people currently spend most of their time indoors, especially in their homes, workplaces, shopping and leisure areas within buildings, among others. However, indoor air quality may turn out to be lower than ambient air quality by several negative factors such as, types of construction material, poor ventilation and air renewal in conjunction with human occupation and misuse of cleaning products (2).

In order to increase performance and productivity in a company, reduce the health problem, it became important that all conditions at the level of well-being and health of the worker are present at the workplace. However, when workplaces are closed and air quality is inadequate, worker's health may be affected and therefore productivity decreases (3).

The business structure in Portugal, specifically in the agricultural sector, is mainly made up of micro and small enterprises where the exploitation itself is carried out by a small number of professionals, being mainly family managing the entire process in this sector, however, professional specialization has been reduced (4).

Specifically in the poultry sector, the worker was exposed to a series of risk factors in the execution of his tasks (4). In order to create the bird more efficiently, through a controlled environment (dubbed as avian ambience), therefore, there is a varied exposition of factors, such as the differences between the interior and the exterior temperatures and the high concentrations of particulate matter (PM, from the expression "Particulate Matter") originated within the premises (5,6,7).

The particulate matter are the most harmful air pollutant for human health. In this investigation was carried out the study of the Total Suspended Particles (TSP). Some of these particles are so tiny (1/30 to 1/5 of the diameter of a human hair), which does not only penetrate our lungs but also, like oxygen, pass into our Circulatory System (breathable particles) (8,9). Inhalable particles only remain between the nasal tract and the pulmonary alveoli (9).

The air quality in animal production environments, specifically in aviaries, has been the topic of many researches in the agricultural sector. In this context, the amount of dust is a determining factor, capable of making the environment adverse for the worker and for animal production (10).

The present study had as an objective to evaluate the workers' exposure to air quality, such as egg laying pavilions and in sorting and classification center, taking into account particulate matter (dust, chemical products), as well as to analyze the collected data associating it with occupational risk factors in the workers of the poultry sector company.

2 MATERIAL AND METHODS

The investigation lasted for 8 months (from September of 2016 to May of 2017). As for the collection of the analytical data of the particulate matter in the different jobs, it was executed in the months of February to April 2017.

The study was developed in an activity company located in the district of Santarém. As a target population, three workers were considered in the company, two in egg classification and one in cleaning and disinfection of egg laying pavilions.

Regarding the definition of the study, it was classified as observational with a transverse cohort, level II and non-probabilistic sampling type, and the technique was of convenience.

The present study was divided into 3 phases: sample selection (decision of the company and its application for authorization); on-site data collection (measurements of concentrations of dust levels to which workers were exposed) and the treatment and analysis of the data collected using the IBM SPSS statistical software version 24.

The data was collected through the SIDEPAK Personal Aerosol Monitor, Model AM510, with serial number 11406031 and with the authenticated certificate. This equipment displays counting data of respirable particulate and in-halable particulate, during the 8 working hours. The collected data was stored in a software called TSI TrackPro Data Analysis.

The appliance was placed so that the harvest was performed as close as possible to the airways, and each worker used the equipment during the 8 hours of the workday.

The measurements were carried out in two workstations, that is, in the sorting and egg classification and in the cleaning of posture pavilions. In the first workstation, the workers perform constant and fixed tasks. The main task was to identify any anomalies found on the conveyor belt of eggs, checking whether the eggshell is partially or totally damaged, as well as the presence of foreign objects. As for the second workstation, the worker performs his tasks in the egg laying pavilions, that is, washing and disinfecting facilities and equipment.

The development of the research was supported by the NP 1796 of 2014, which establishes the Occupational Exposure Limit values (OELs) for all substances that may have an effect on the health of workers. The particles present in these two workstations are called as bioaerosols and fall into the definition of particles with no other classification. These particles are not biologically inert, may not cause fibrosis but may cause adverse effects. As such, the NP recommends for these particles values below 3 mg/m^3 in respirable particles and 10 mg/m^3 for inhalable particles (9).

In terms of evaluation, the statistical analysis of the data was based on the application of the parametric tests (independent t test) and simple descriptive statistics, that is, absolute dispersion measures (standard deviation and variance) and measures of central tendency (mean and median for each work station and for different particulate matter).

It should be noted that this type of research did not present economic or financial interests but rather the benefit to the elaboration of this academic study. Measurements were made in a private company with the consent of the employer and employees, where the objectives of this research were duly clarified, anonymity and confidentiality being respected.

3 RESULTS

Three workers were a part of this study, two females worker's in the sorting center and one male, whose task consisted in cleaning and disinfecting egg laying pavilions. Of the three workers, the male worker repeatedly uses a mask as personal protective equipment (PPE-FFP3). The remaining workers, who work at the sorting center as a fixed task, did not use any type of PPE.

In the following table, the values in the work places were expressed according to the mean concentration of particulate matter, taking into account the number of evaluations captured during the 8 working hours, the mean, the standard deviation and the "p-value". The group cases present were the places where the evaluations were made, that is, the "Sorting Center" and the "Cleaning and Disinfection".

Table 1 shows a statistically significant mean difference in respirable and inhalable particle values between the sites under study (p ≤0,05). We can also affirm that the places identified as "Cleaning and Disinfection" revealed average levels (on both

Table 1. Identification of mean values and "p-value" in workstation and as a function of particle concentration.

Group statistics and Levene's Test for equality of variances

	Workstation	N	Mean (mg/m³)	Standard deviation	p-value
Respirable particles	Sorting Center	123891	0,02294	0,71690	0,006
	Cleaning and Disinfection	77869	0,51842	49,913442	
Inhalable particles	Sorting Center	124658	0,43406	49,060962	0,054
	Cleaning and Disinfection	76704	1,03493	77,186795	

Legend: N – sample number (number of evaluations taken during the 8 working hours), Independent t test.

Table 2. Identification of the statistical values for the test value 3 mg/m³.

One-sample test

		Test value = 3 mg/m³		
Workstation		Mean	Sig. (2-tailed)	Standard deviation
Sorting center	Respirable particles	0,02294	p < 0,0001	−2,977060
Cleaning and disinfection	Respirable particles	0,51842	0,006	−2,481578

T test.

Table 3. Identification of the statistical values for the test value 10 mg/m³.

One-sample test

		Test value = 3 mg/m³		
Workstation		Mean	Sig. (2-tailed)	Standard deviation
Sorting center	Inhalable particles	0,43406	0,006	−9,5665492
Cleaning and disinfection	Inhalable particles	1,03493	p < 0,0001	−8,965074

T test.

types of particles) superior compared to the "Sorting Center" site.

Tables 2 and 3 demonstrate the test for a single sample (t-test). According to the results presented in the tables, we can see that the values were significant, since the mean values are far from the reference values.

Taking into account the results obtained, the average values are within the recommended values. However, these present an expressive variation, since the activity carried out in the different work stations leads to different concentrations of exposure to particulate matter and the type of activity performed at the time (different mean difference for each workplace).

4 DISCUSSION

The poultry worker was exposed to risk factors that could cause health effects, including exposure to dust, subjecting him to particle concentration, composition and size, and the time of exposure to it (4,11).

There has been confirmation that scientific studies have shown that exposure to particulate matter (short and long term) has had undesirable effects on human health (11). Of the poultry farmers evaluated in this investigation, only one used personal protective equipment (PPE-FFP3), namely FFP3 classification mask. Therefore, this mask presented a protection not applicable to the execution of its tasks, since according to the National Institute of Occupational Safety and Health (NIOSH), the correct use of disposable masks was characterized by dust protection, being of classification FFP1 and FFP2, shell type.

Among poultry farmers, the most frequent symptom was dryness of the skin and eyes, followed by bouts of sneezing, difficulty breathing, allergies (rhinitis) and runny nose or stuffy nose (when they do not have flu or constipation). The above signs were caused by excessive levels of particles (12). Epidemiological studies have effectively demonstrated associations between air pollutants and a range of adverse health effects, such as particulate matter (11).

A clinical trial of observation of dust exposure and concentration in avian workers found a decrease in lung function (13). These studies provided scientific evidence on the impact of particulate matter on average life expectancy, as well as on the development of chronic diseases, which corresponded to the worker exposed to these particles without protection (11,14).

After analyzing the data collected, the tests performed between the mean values of the different particles (respirable and inhalable particles) according to the type of site, revealed higher levels in the inhalable particles than in the respirable particles. Previous studies have indicated that particles can penetrate to the level of gas exchange in the lungs (respirable particles) and can trigger diseases through accumulation in the upper respiratory tract, below the vocal cords (inhalable particles) (15). Another investigation compared particle levels between agricultural and traffic areas and concluded that the concentration of inhalable particulates was indeed higher in agricultural areas (16).

In this investigation it was evidenced that there are statistically significant differences, since the significance associated with the test was not higher than 0.05, so there were differences of concentration between the "Sorting Center" and "Cleaning and Disinfection" sites. So the exposure to particulate matter was different in the two places/workstations, since the worker of the place "Cleaning and Disinfection" was in direct contact with the birds and was exposed to larger amounts of suspended particles. At the "Sorting Center" site, the workers were only in contact with the eggs coming from the laying pavilions.

In relation to the values established by the standard for respirable particles at 3 mg/m^3 and at 10 mg/m^3 (inhalable particulates), it was found that the mean values obtained did not exceed the reference values, the highest value being presented in the "Cleaning and Disinfection" in the inhalable particles (9). It should be noted that the mean values identified before in the different workplaces were different, and could mean the increase if the activity performed occurs in a larger period of time.

Several investigations in aviaries indicated that the level of dusting was above that recommended for humans (17). A study done to farmers indicated that the group most exposed to dust was the poultry farmers, 76% of whom reported two or more types of dust during poultry farming. As such, the prevalence of exposure to particulate matter was higher among poultry farmers compared to non-avian farmers (14).

Studies performed with humans and other species have assumed exposures to particles with changes in cardiac function, namely a higher incidence of myocardial infarction and the induction of arrhythmias. Other long-term studies of particulate matter exposures have indicated associations between exposures and reduced survival (reduction in life expectancy by 1 to 2 years). However, it was important to promote measures to improve working conditions, including implementing the concept of "operational hygiene" and conducting periodic medical examinations and admission tests (18,19).

5 CONCLUSION

Taking into account the results obtained, it was shown that the particulate matter did not in fact present an increased risk to the health and safety of the poultry farmers studied. The levels of exposure to particulate matter were below the reference values, having reduced potential to trigger cardiac and respiratory diseases.

In order to eliminate or reduce risk factors, it was necessary to act at the source, in the layout phase and methods and manufacturing processes. It's important to have a balance between technology and the human and social factors in order to avoid accidents/incidents of work and occupational diseases.

In view of the great expansion of poultry farming and the specific risk observed by employees, the daily use of respiratory protection masks (FFP1 or FFP2, dust/powder use), shell type, was recommended for all workers. Training in health and safety at work would be essential to clarify in a simple way the importance of measures necessary to maintain safety and health in the workstation.

The implementation of the concept of "operational hygiene" would be essential in this sector, since it indicated possible measures that allowed the maintenance of the environmental conditions within the limits not harmful to the health of the worker. The guidelines or criteria applied in this concept to achieve its objective was: general ventilation of the work areas (constructive measure); Reduction of exposure times (organizational measure) and individual protection.

It was important that workers perform medical examinations, whether periodical or admission by the medicine of the company's work. Periodic examinations are necessary not only to regularly check the health status of workers but also to allow, indirectly, the confirmation of other control measures that may have been adopted.

REFERENCES

[1] Decreto Lei no 102/2010 de 23 de Setembro. (2010), 1–35.
[2] Agência Portuguesa do Ambiente. (n.d.). APA – Políticas – Ar – Qualidade do Ar Interior. Retrieved November 24, 2016, from http://www.apambiente.pt/index.php?ref=16&subref=82&sub2ref=319.
[3] Afonso E. Importância da Qualidade do Ar Interior na Manutenção de Boas Condições de Trabalho. Grupo 4work. 2011 [cited 2016 Nov 24]. Available from: http://www.4work.pt/cms/index.php?id=98&no_cache=1&tx_ttnews%5Btt_news%5D=122&tx_ttnews%5BbackPid%5D=1&cHash=8678c9adfc.
[4] Santos MBG, Silva CH da, Almeida LF, Monteiro LF, Nascimento JWB do. Avaliação da higiene, saúde e segurança no trabalho em galpões para criação de frangos de corte. XXXI Encontro Nacional de Engenharia de Produção. 2011 [cited 2016 Nov 23]. Available from: http://www.abepro.org.br/biblioteca/enegep2011_tn_sto_138_877_18404.pdf.
[5] Vigoderis RB. Sistemas de aquecimento de aviários e seus efeitos no conforto térmico ambiental, qualidade do ar e perfomance animal, em condições de inverno, na região sul do Brasil. Universidade Federal de Viçosa; 2006. p. 1–114.
[6] Santos MR. Segurança no trabalho avícola e meio de prevenção e controle na saúde ocupacional. Universidade Tecnológica Federal do Paraná; 2012. p. 0–36.
[7] Menegali I. Diagnóstico da qualidade do ar na produção de frangos de corte em instalações semi-climatizadas por pressão negativa e positiva, no inverno, no sul do brasil. Universidade Federal de Viçosa; 2005. p. 0–78.
[8] Agência Portuguesa do Ambiente. (n.d.-a). APA – Políticas – Ar – Qualidade do Ar Ambiente – Partículas em Suspensão. Retrieved November 23, 2016, from http://www.apambiente.pt/index.php?ref=16&subref=82&sub2ref=316&sub3ref=383.
[9] Norma Portuguesa 1796 – Segurança e Saúde no Trabalho. Valores limite de exposição profissional a agentes químicos. Instituto Português da Qualidade. 2007.
[10] Barreto, B., Nääs, I. A., Paz, A., Garcia, R. G., Felix, G. A., & Docente, 2. (n.d.). VII SIMPÓSIO DE CIÊNCIAS DA UNESP – DRACENA VIII ENCONTRO DE ZOOTECNIA – UNESP DRACENA Qualidade do ambiente aéreo na produção de frangos de corte: poeira e amônia.
[11] Dias DSO. Avaliação de risco para a saúde humana associado a partículas inaláveis. Universidade de Aveiro; 2008.
[12] Agência Portuguesa do Ambiente, Laboratório Referência do Ambiente. Qualidade do Ar em Espaços Interiores—Um Guia Técnico. 2009; (Agência Portuguesa do Ambiente).
[13] Fernandes FC. Poeiras em Aviários. Rev Bas Med Trab, Belo Horiz. 2004;2(n°4, out-dez):253–62.
[14] Faria NMX, Facchini LA, Fassa AG, Tomasi E. Trabalho rural, exposição a poeiras e sintomas respiratórios entre agricultores. Rev Saúde Pública. 2006;40(5):827–36.
[15] Santos CS, Viegas C., Almeida A., Clérigo A., Veríssimo C., Carolino E., et al. Avaliação da exposição a fungos e partículas em explorações avícolas e suinícolas:estudo. 2013;(ACT):0–62.
[16] Gomes MJM. Ambiente e pulmão. J Pneumol. 2002;28:261–9.
[17] Naas IDA, Miragliotta MY, Baracho MDS, Moura DJ De. AMbiência aérea em alojamento de frangos de corte: Poeira e Gaases. 2007;27:326–35.
[18] Salgado PE de T. Informações gerais e ecotoxicológicas de material particulado. Série Cad Ref Ambient. 2002;14 (Magnesita S.A.).
[19] Miguel ASSR. Manual de Higiene e Segurança do Trabalhador. 13ª ed. Porto Editora, editor. 2014.

Scale psychometrics of the Portuguese short and middle length variants of the Copenhagen Psychosocial Questionnaire III pilot study

D.A. Coelho
Universidade da Beira Interior, Covilhã, Portugal

M.L. Lourenço
Instituto Politécnico da Guarda, Guarda, Portugal

ABSTRACT: The Copenhagen Psychosocial Questionnaire (CoPsoQ) has evolved into its third version, as a consequence of the development process of the assessment tool. A cross-sectional study was undertaken of administrative collaborators in 5 higher education institutions in Portugal spread throughout the country. Total number of valid responses was 116, with a low response rate (estimated at 8%), but unveiling the psychometric properties of the scales across the two lengths of the questionnaire. For most scales, the results of internal reliability consistency, support the reductions in scale items between the middle and short lengths of the Portuguese CoPsoQ III. The short version is hence appropriate for use in practitioner assessments, while the middle and long versions are deemed appropriate as research instruments, with a choice between both representing a trade-off between granularity of the results and time required to complete the questionnaire. Scores reported in the article can serve as reference points during future evaluation of results using the CoPsoQ III.

1 INTRODUCTION

1.1 Background

Psychosocial factors at work have been studied for a long time (Utzet, Navarro, Llorens & Moncada, 2015). Since the 1960s, several study models have emerged, each of which evaluates different aspects and seeks to explain the impacts that the labor situation may have on the quality of life of workers. In this scenario, in the 2000s, COPSOQ was designed to meet the legal requirements in Denmark for assessing the psychosocial factors of work in occupational risk surveys. Thus, its authors incorporated theoretical models established in the literature with a view to a multidimensional instrument (Kristensen et al., 2005; Kristensen, 2010), such as: Demand-Control-Social Support (DCS, Johnson & Hall, 1988) and Effort-Reward Imbalance (ERI, Siegrist, Peter, Junge, Cremer & Seidel, 1990). The DCS model advocates a relationship between well-being and a triad of labor factors. The ERI model, however, suggests that the imbalance between the effort made and the compensation offered by the work (e.g., status, esteem, salary) can be a risk factor for health and well-being (da Silva, Wendt & de Lima Argimon, 2017).

1.2 Aims and method

The main objectives of the study, part of a large international collaborative research effort, were: to adapt and psychometrically characterize the short and middle length Portuguese translation of the third version of the CoPsoQ, by means of a pilot cross-sectional study. Adapting the original questionnaire and developing the Portuguese version proved to be a necessary process, which was based on the translation and reverse translation method. Moreover, as part of the collaborative effort of defining the scope of the short version of the questionnaire, psychometric analysis of the scales was undertaken, with a view to compare the short and middle length pilot variants of the questionnaire and ascertain the representativeness of the scope of the short variant.

Authorization was sought and once granted, the questionnaire was advertised to administrative (non-teaching and non-researching) public servants of five public higher education institutions in Portugal, and disseminated via electronic version, as a short invitation text, accompanied by a text link. The scales included (core and middle versions with pilot additional questions, for a total of 63 items) were: QD—Quantitative Demands, WP—Work

Pace, ED—Emotional Demands, HE—Hiding Emotions, IN—Influence at Work, PD—Possibilities for Development, CT—Control over Time, MW—Meaning of Work, PR—Predictability, RE—Recognition, CL—Clarity of Work Role, CO—Role Conflicts, IT—Illegitimate Tasks, QL—Quality of Leadership, SC—Social support from Colleagues, SS—Social Support from Supervisors, SW—Community at Work Perception, JI—Job Insecurity, IW—Insecurity at Work, QW—Quality of Work, TE—Horizontal Trust, TM—Trust in Management, JU—Justice in Organization, WF—Work Family Conflict, JS—Job Satisfaction and GH—General Health. Scales were analyzed for internal reliability consistency, as well as correlation, using IBM SPSS v. 23.

2 COPSOQ III IN PORTUGUESE

The question items that make up the middle and short length variants of the third version of the Portuguese CoPsoQ are depicted in Tables 1 through 4. Scale items that are only included in the middle length variant are marked in the aforementioned Tables. The middle variant contains the short one.

Table 1. Items making up scales: QD—Quantitative Demands, WP—Work Pace, ED—Emotional Demands, HE—Hiding Emotions and IN—Influence at Work.

Scale	Item	Answer options
QD - *Exigências quantitativas*	*A sua carga de trabalho acumulase por ser mal distribuída?	S-F-P-R-N/Q
	O seu trabalho fica atrasado?	S-F-P-R-N/Q
	Com que frequência não tem tempo para completar todas as tarefas do seu trabalho?	S-F-P-R-N/Q
WP - *Ritmo de trabalho*	Tem de trabalhar muito rapidamente?	S-F-P-R-N/Q
	Trabalha a um ritmo elevado ao longo de toda a jornada?	MG-G-A-P-MP
ED - *Exigências Emocionais*	*O seu trabalho coloca-o em situações carregadas de emoção e perturbadoras?	S-F-P-R-N/Q
	No seu trabalho tem de lidar com os problemas pessoais de outras pessoas?	S-F-P-R-N/Q
	O seu trabalho exige de si emocionalmente?	MG-G-A-P-MP
*HE - *Exigência de esconder as emoções*	*O seu trabalho exige que esconda os seus sentimentos?	MG-G-A-P-MP
	*É lhe exigido que seja gentil e aberto perante qualquer pessoa— independentemente de como se comportam consigo?	MG-G-A-P-MP
	*É lhe exigido que não manifeste a sua opinião no seu trabalho?	S-F-P-R-N/Q
IN - *Influência no Trabalho*	Tem um elevado grau de influência nas decisões respeitantes ao seu trabalho?	S-F-P-R-N/Q
	*Pode influenciar a quantidade de trabalho que lhe é atribuída a si?	S-F-P-R-N/Q
	*Tem alguma influência sobre aquilo que faz quando trabalha?	S-F-P-R-N/Q
	*Tem alguma influência sobre o MODO como executa o seu trabalho?	S-F-P-R-N/Q

*This item is not part of the short length variant.
S-F-P-R-N/Q: *Sempre, Frequentemente, Por vezes, Raramente, Nunca/Quase nunca*; MG-G-A-P-MP: *Em muito grande medida, em grande medida, de algum modo, em pequena medida, em muito pequena medida*.

Table 2. Items making up scales: PD—Possibilities for Development, CT—Control over Time, MW—Meaning of Work, PR—Predictability, RE—Recognition, CL—Clarity of Work Role, CO—Conflicts and IT—Illegitimate Tasks.

Scale	Item	Answer options
PD - *Possibilidades de desenvolvimento*	Tem possibilidade de aprender coisas novas através do seu trabalho?	MG-G-A-P-MP
	Pode usar as suas habilidades ou perícias no seu trabalho?	MG-G-A-P-MP
	*O seu trabalho dá-lhe a oportunidade de desenvolver as suas habilidades?	MG-G-A-P-MP
*CT - *Controlo sobre o tempo de trabalho*	*Pode decidir quando fazer um intervalo?	S-F-P-R-N/Q
	*Pode ir de férias mais ou menos quando deseja?	S-F-P-R-N/Q
	*Pode deixar o seu trabalho para conversar com um colega?	S-F-P-R-N/Q
	*Se tem um assunto privado para resolver, é possível deixar por meia hora a tarefa que tem em mãos sem ter de pedir autorização especial?	S-F-P-R-N/Q
MW - *Significado do trabalho*	O seu trabalho tem significado?	MG-G-A-P-MP
	Acha que o trabalho que faz é importante?	MG-G-A-P-MP

(*Continued*)

Table 2. (*Continued*).

Scale	Item	Answer options
PR - *Previsibilidade*	*No seu local de trabalho, é informado com antecedência sobre decisões importantes, mudanças ou planos para o futuro?*	MG-G-A-P-MP
	Recebe toda a informação de que necessita para fazer bem o seu trabalho?	MG-G-A-P-MP
RE - *Reconhecimento*	*O seu trabalho é reconhecido e apreciado pela gerência?*	MG-G-A-P-MP
CL - *Clareza do papel laboral*	*O seu trabalho tem objetivos claros?*	MG-G-A-P-MP
	*Sabe exatamente quais são as áreas da sua responsabilidade?	MG-G-A-P-MP
	*Sabe exatamente o que se espera de si no trabalho?	MG-G-A-P-MP
CO - *Conflitos no papel laboral*	*No seu trabalho é objeto de exigências contraditórias?*	MG-G-A-P-MP
	Por vezes tem que fazer coisas que deveriam ter sido feitas de outra maneira?	MG-G-A-P-MP
*IT - *Tarefas ilegítimas*	**Por vezes tem que fazer coisas que parecem desnecessárias?*	MG-G-A-P-MP

*This item is not part of the short length variant.
S-F-P-R-N/Q: *Sempre, Frequentemente, Por vezes, Raramente, Nunca/Quase nunca*; MG-G-A-P-MP: *Em muito grande medida, em grande medida, de algum modo, em pequena medida, em muito pequena medida.*

Table 3. Items making up scales: QL—Quality of Leadership, SC—Social Support from Colleagues, SS—Social Support from Supervisors, SW—Community at Work Perception, JI—Job Insecurity and IW—Insecurity at Work.

Scale	Item	Answer options
QL - *Qualidade de liderança*	*Em relação à sua chefia direta até que ponto considera que…*	–
	*Oferece aos colaboradores laborais boas oportunidades de desenvolvimento?	MG-G-A-P-MP
	É bom no planeamento do trabalho?	MG-G-A-P-MP
	É bom a resolver conflitos?	MG-G-A-P-MP
SC - *Apoio social dos colegas*	*Com que frequência é que recebe ajuda e apoio dos seus colegas, se for necessário?*	S-F-P-R-N/Q
	*Com que frequência é que os seus colegas estão dispostos a ouvir falar dos seus problemas no trabalho, se for necessário?	S-F-P-R-N/Q
SS - *Apoio social dos supervisores*	**Com que frequência é que o seu superior imediato fala consigo sobre os seus problemas no trabalho, se for necessário?*	S-F-P-R-N/Q
	Com que frequência é que o seu superior imediato ajuda e apoio, se for necessário?	S-F-P-R-N/Q
SW - *Perceção duma comunidade no trabalho*	*Existe bom ambiente entre si e os seus colegas?*	S-F-P-R-N/Q
	*No seu local de trabalho sente-se parte de uma comunidade?	S-F-P-R-N/Q
JI - *Insegurança no emprego*	*Sente-se preocupado com a possibilidade de ficar desempregado?*	MG-G-A-P-MP
	Sente-se preocupado com a dificuldade em encontrar outro trabalho se ficar desempregado?	MG-G-A-P-MP
IW - *Insegurança sobre as condições de trabalho*	*Está preocupado com a possibilidade de ser transferido para outro emprego contra a sua vontade?*	MG-G-A-P-MP
	*Está preocupado com a possibilidade do horário ser alterado (turno, dias laborais, hora de entrada e de saída…) contra a sua vontade?	MG-G-A-P-MP
	*Está preocupado com a diminuição do seu salário (redução, introdução de remuneração variável…)?	MG-G-A-P-MP

*This item is not part of the short length variant.
S-F-P-R-N/Q: *Sempre, Frequentemente, Por vezes, Raramente, Nunca/Quase nunca*; MG-G-A-P-MP: *Em muito grande medida, em grande medida, de algum modo, em pequena medida, em muito pequena medida.*

Table 4. Items making up scales: QW—Quality of Work, TE—Horizontal Trust, TM—Trust in Management, JU—Justice in Organization, WF—Work Family Conflict, JS—Job Satisfaction and GH—General Health.

Scale	Item	Answer options
QW - *Qualidade do trabalho*	Está satisfeito com a qualidade do desempenho no seu local de trabalho?	MG-G-A-P-MP
*TE - *Confiança horizontal*	*Os funcionários confiam uns nos outros?	MG-G-A-P-MP
TM - *Confiança vertical*	A gerência confia que os seus funcionários fazem bem o seu trabalho?	MG-G-A-P-MP
	Os empregados podem confiar na informação que lhes é transmitida pela gerência?	MG-G-A-P-MP
	*Os empregados estão autorizados a expressar os seus pontos de vista e sentimentos?	MG-G-A-P-MP
JU - *Justiça organizacional*	Os conflitos são resolvidos de uma forma justa?	MG-G-A-P-MP
	O trabalho é distribuído de forma justa?	MG-G-A-P-MP
WF - *Conflito entre o trabalho e a vida pessoal*	Há alturas em que tem de estar em casa e no trabalho ao mesmo tempo?	S-F-P-R-N/Q
	As próximas quatro perguntas dizem respeito às formas como o seu trabalho afeta a sua vida privada:	–
	Sente que o seu trabalho lhe exige tanta energia que acaba por afetar negativamente a sua vida privada?	MG-G-A-P-MP
	Sente que o seu trabalho lhe exige tanto tempo, que acaba por afetar a sua vida privada negativamente?	MG-G-A-P-MP
	As exigências do meu trabalho interferem com a minha vida privada e familiar?	MG-G-A-P-MP
	Devido a obrigações relacionadas com o trabalho, tenho de mudar os meus planos de atividades privadas e familiares.	MG-G-A-P-MP
JS - *Satisfação com o trabalho/ com o emprego*	Relativamente ao seu trabalho em geral, quão satisfeito está com: – as suas perspetivas de ter trabalho?	–
	– o seu emprego como um todo, tomando tudo em consideração?	MS-S-N-I-MI
	– o seu salário?	
GH - *Saúde auto-avaliada*	Em geral, acha que a sua saúde é:	E-MB-B-R-F

*This item is not part of the short length variant.
S-F-P-R-N/Q and MG-G-A-P-MP: *please see legend for Table 3*; MS-S-N-I-MI: *Muito satisfeito, satisfeito, nem satisfeito nem insatisfeito, insatisfeito, muito insatisfeito*; E-MB-B-R-F: *Excelente, Muito Boa, Boa, Razoável, Fraca*.

3 RESULTS AND ANALYSIS

3.1 Cross-sectional study

Administrative collaborators in 5 higher education institutions in Portugal spread throughout the country participated in the cross-sectional study filling in the middle version. Total number of valid responses was 116, from a universe estimated at 15 hundred, yielding a very low response rate (estimated at 8%).

Responses were received during the Spring of 2017. 58% of respondents were women, 9% aged 25 to 34, 29% were aged between 35 and 44 and 46% were aged between 45 and 54. 16% of respondents were 55 or older.

Scales are scored 0–100 points. The five response options are scored 100, 75, 50, 25, 0. The total score on a scale for a respondent is the average of the scores on the individual items. A person is considered missing if less than half of the questions in a scale have been answered. In most cases high levels are "good" or "healthy". The exceptions are Quantitative Demands (QD), Work Pace (WP), Emotional Demands (ED), demands for Hiding Emotions (HE), Role Conflicts (CO),

Table 5. Scores and Cronbach's alpha obtained for selected scales of the middle length variant.

Scale	M	sd	Alpha
QD	41.7	21.2	0.85
WP	56.6	21.1	0.74
ED	49.8	23.3	0.81
HE	47.7	21.2	0.58
IN	58.8	19.1	0.79
PD	68.2	25.3	0.89
CT	61.0	17.9	0.68
MW	83.2	19.4	0.92
PR	56.2	25.7	0.84
RE	57.8	27.9	n/a

Job Insecurity (JI) and Work-Family conflict (WF). Mean and standard deviation of the scores across the sample are shown in Tables 5 and 6, considering the middle length variant of the scales. Mean and standard deviation of the scores across the sample are shown in Table 7, considering the short length variant of the scales.

Table 6. Scores for additional scale set from middle length variant showing Cronbach's alpha.

Scale	M	sd	Alpha
CL	74.6	18.6	0.83
CO	45.7	24.0	0.80
IT	47.4	26.0	n/a
QL	56.2	25.3	0.87
SC	65.4	20.7	0.81
SS	55.5	26.9	0.88
SW	74.7	22.1	0.81
JI	49.8	30.4	0.82
IW	41.9	25.3	0.68
TE	59.3	21.2	n/a
TM	66.0	20.4	0.83
JU	52.4	24.4	0.81
WF	40.3	24.0	0.90
JS	57.9	21.1	0.74
GH	53.0	22.4	n/a

n/a: not applicable (single item scale).

Table 7. Scores, Cronbach's alpha obtained for scales of the short length variant, as well as correlation factor with applicable middle length scale.

Scale	M	sd	Alpha	Correlation
QD	41.4	23.2	0.85	0.96
WP	56.6	21.1	0.74	1
ED	51.0	24.3	0.73	0.96
IN	53.9	25.4	n/a	0.82
PD	69.7	25.0	0.78	0.98
MW	82.1	20.5	n/a	0.97
PR	56.2	25.7	0.84	1
RE	57.8	27.9	n/a	1
CL	69.8	23.0	n/a	0.85
CO	45.7	24.0	0.80	1
QL	57.2	26.5	0.87	0.96
SC	67.0	23.7	n/a	0.92
SS	58.0	28.5	n/a	0.94
SW	78.4	20.8	n/a	0.90
JI	49.8	30.4	0.82	1
IW	33.6	32.0	n/a	0.75
TM	67.0	19.8	0.78	0.95
JU	52.4	24.4	0.81	1
WF	40.3	24.0	0.90	1
JS	67.4	24.7	n/a	0.86
GH	53.0	22.4	n/a	1

n/a: not applicable (single item scale).

3.2 *Psychometric characteristics of the scales*

In addition to scores obtained, the characterization of the scales from a statistical perspective benefits from additional measures. In the current report, only a part of those measures are presented. In particular, Cronbach's alpha, a measure of internal consistency is computed for all the scales with more than one item, and shown in Tables 5 and 6, in regard to the middle length variant of the CoPsoQ III scales, as well as in Table 7, with regards to the short length variant. In addition, the Pearson correlation coefficient between the scales co-existing in the short and middle length variants are included in Table 7.

4 DISCUSSION

Mean scores on these scales and standard deviations given in the article can serve as reference points during future evaluation of received results. Internal reliability of the scales was very satisfactory, across the board, with Cronbach's alpha above 0.67 in every case in the short length variant. Moreover, discriminant validity could be seen in some of the scales with less items in the short variant, where lower correlations are shown (JS—job satisfaction, IW—insecurity over working conditions and IN—influence at work). For practitioner surveys where there is interest in exploring these scales to a greater extent, it would be advisable to use the middle length variant of the CoPsoQ III.

Moreover, ED (Emotional Demands), PD (Possibilities for Development) and TM (vertical trust) are the scales where Cronbach's alpha is smaller (and can be calculated) in the short length scale variants compared to the middle scale variant. This indicates a successful reduction in number of items, going from the middle to the short length variant. For the other cases, where Cronbach's alpha is unchanged, there is an indication of the potential to reduce the number of scale items further. However, reducing to one scale item only, on a short length variant of the scale, impedes the internal consistency reliability analysis, due to the loss of ability to compute Cronbach's alpha.

Compared to earlier studies, conducted in the private sector in Portugal (Coelho et al., 2015, 2017), the current results show a more positive outlook. The differences overall may be attributable to both the difference in sector (the current paper reports on public sector workers and the previous studies focused on private utility workers), but also on a less bleak economic outlook in the country.

5 CONCLUSION

The study reported in this paper sought the adaption and validation of the short and middle length Portuguese translation of the third version of the CoPsoQ, by means of a pilot cross-sectional study. The process has been reported and the results of the adaptation have been included in the Portuguese language within this paper (Tables 1 to 4). Variance of the results can be estimated from standard deviation, which had a minimum value for CT (control over working time) and a maximum for IW (insecurity over working conditions). Scales where the mean value have the strongest departure from the middle of the rating scale (50) are MW (meaning of work), CL (role clarity) and SW (sense of community at work). These results suggest a positive outlook in these dimensions for public servants working in administrative roles in Portuguese higher education institutions.

The short variant of the CoPsoQ III is deemed appropriate for use in practitioner assessments, while the middle and long variants are deemed appropriate as research instruments, with a choice between both representing a trade-off between granularity of the results and time required to complete the questionnaire. Scores reported in the paper can serve as reference points during future evaluation of results using the CoPsoQ III. Future studies include the expansion of responses, and sectorial analysis, considering industry as well as demographic variables, including across sexes.

GLOSSARY (SCALE ABBREVIATIONS)

QD – Quantitative Demands; WP – Work Pace
ED – Emotional Demands
HE – Demands for Hiding Emotions
IN – Influence at Work
PD – Possibilities for Development
CT – Control over Working Time
MW – Meaning of Work; PR – Predictability
RE – Recognition; CL – Role Clarity
CO – Role Conflicts; IT – Illegitimate Tasks
QL – Quality of Leadership
SC – Social Support from Colleagues
SS – Social Support from Supervisors
SW – Sense of Community at Work
JI – Insecurity over Employment
IW – Insecurity over Working Conditions
TE – Horizontal Trust; TM – Vertical Trust
JU – Organizational Justice
WF – Work Life Conflict
JS – Satisfaction with Work – Job Satisfaction
GH – Self Rated Health

REFERENCES

Coelho, D.A., Tavares, C.S., Lourenço, M.L. and Lima, T.M., 2015. Working conditions under multiple exposures: A cross-sectional study of private sector administrative workers. *Work*, *51*(4), pp. 781–789.

Coelho, D.A., Tavares, C.S., Lima, T.M. and Lourenço, M.L., 2017. Psychosocial and Ergonomic Survey of Office and Field Jobs in a Utility Company. *International Journal of Occupational Safety and Ergonomics*, pp. 1–25.

da Silva, M.A., Wendt, G.W. & de Lima Argimon, I.I., 2017. Propriedades psicométricas das medidas do Questionário Psicossocial de Copenhague I (COPSOQ I), versão curta. *REGE-Revista de Gestão*.

Johnson, J.V. and Hall, E.M., 1988. Job strain, work place social support, and cardiovascular disease: a cross-sectional study of a random sample of the Swedish working population. *American Journal of Public Health*, *78*(10), pp. 1336–1342.

Kristensen, T.S., Hannerz, H., Høgh, A. and Borg, V., 2005. The Copenhagen Psychosocial Questionnaire—a tool for the assessment and improvement of the psychosocial work environment. *Scandinavian Journal of Work, Environment & Health*, pp. 438–449.

Kristensen, T.S., 2010. A questionnaire is more than a questionnaire. *Scandinavian Journal of Public Health*, *38*(3_suppl), pp. 149–155.

Siegrist, J., Peter, R., Junge, A., Cremer, P. and Seidel, D., 1990. Low status control, high effort at work and ischemic heart disease: prospective evidence from blue-collar men. *Social Science & Medicine*, *31*(10), pp. 1127–1134.

Utzet, M., Navarro, A., Llorens, C. & Moncada, S., 2015. Intensification and isolation: psychosocial work environment changes in Spain 2005–10. *Occupational Medicine*, *65*(5), pp. 405–412.

Evaluation of occupational safety and health conditions in a Brazilian public hospital

C.F. Machado
Environmental Engineer, M.S. in Safety Engineering, Federal University of Rio de Janeiro, Brazil

J.S. Nóbrega
Safety Engineer, MSc
Department of Industrial Engineering, Veiga de Almeida University, Rio de Janeiro, Brazil
DVSST, Federal University of Rio de Janeiro, Rio de Janeiro, Brazil

ABSTRACT: This study aims to evaluate the occupational health and safety conditions of a public hospital located in the city of Rio de Janeiro based on the provisions of the Brazilian legislation. Motivated by the recurrence of fire outbreaks at different sectors of the hospital's facilities, this work was developed through technical inspections, analysis of documentation and interviews with the hospital's senior leadership and staff. Deviations related to fire protection, electrical installations, ergonomics and occupational medical control and risks prevention programs were found. Based on these results, an action plan has been recommended, using the Gravity, Urgency and Tendency (GUT) method, to prioritize actions.

1 INTRODUCTION

Accidents in hospitals are well known to health professionals. According to Hospital Safety Manual. developed by National Health Surveillance Agency (Anvisa) last issued in 2002, a large number of accidents in health care establishments are caused by negligence or lack of training, misuse or error in the operation of equipment, and inadequate working conditions.

The employees who work in this sector are exposed to many occupational hazards, including: chemicals, noise, high/low temperatures, bacteria, viruses, and ergonomic risks. This type of environment is subject to the occurrence of occupational illnesses and other undesirable outcomes; and, therefore, it is critical that hospital managers implement an effective system that promotes safe work environments. Machado (2016).

This scientific article seeks to present and analyze the occupational health and safety conditions of a Brazilian public hospital, where fire outbreaks have frequently occurred.

1.1 Objectives

The objective of this study is to evaluate the occupational health and safety conditions of a Brazilian public hospital, based on the requirements of the Regulatory Norms of the Brazilian Ministry of Work and Social Security (NR in Portuguese) and other applicable legislation.

The study also aims to propose recommendations for the findings identified, to encourage accident preventive practices, minimize risks and create a safer work environment for all hospital employees.

To achieve the main goal of this project, the following targets were established:

- Identification of activities performed at the health facility;
- Application of relevant labor standards and legislation;
- Identification of occupational hazards and risks;
- Verification of compliance with legal requirements;
- Critical analysis of the Health Institute documentation provided for this work;
- Proposal of recommendations and suggestions to comply with regulations and prevent potential accidents, including fire;
- Elaboration of strategic action plan, using the GUT method, for deviations and opportunities for improvement observed.

1.2 Methodology

The study was conducted by means of inspections at the hospital's facilities; informal interviews with the senior leadership and staff; and verification of the hospital's documentation, highlighting the Occupational Health Control Program, Occupational Risk Prevention Program, Fire Emergency Plan and Waste Management Program.

For the ergonomic analysis, a brief evaluation was performed through the Rapid Upper Limb Assessment (RULA) method. To evaluate the application of the Brazilian legal requirements, checklists were elaborated and applied during the inspections and interviews. Based on the results, a strategic plan was drawn up to prioritize the actions according to the GUT method.

1.3 Delimitation of the study

The public hospital in Rio de Janeiro was chosen as study case due to its size, complexity of services and variety of hazards and risks to which the workers are exposed. The main motivation was the recurrence of fire outbreaks at different sectors of the hospital's facilities, which led the hospital's senior management to accept and collaborate with this study. The requirements of the Regulatory Norms of the Brazilian Ministry of Work and Social Security assessed were 1, 4, 5, 6, 7, 8, 9, 10, 12, 15, 17, 23, 24, 26 and 32. (MTPS, 1978 and beyond).

Despite the receptivity of the health institution, which enabled the author to have access to the hospital facilities; use photographic devices; conduct positive interviews with employees; and to have access to relevant information, there were inherent difficulties to performing this study. Some of the difficulties encountered included obtaining documents, records of training, maps and plants, which limited the research. Additionally, access to surgical rooms, the Intensive Care Unit, and some classified sectors were not allowed.

To protect the name of the Health Institute under study, the location of the establishment, photos of its building's façade, internal programs or information that identify the hospital's brand logo are not presented.

2 PRESENTATION OF THE HEALTH CARE ESTABLISHMENT

2.1 General aspects

The Health Care Establishment under study is a highly complex public hospital unit. It is renowned for its specialty in clinical and surgical cardiology and has been in operation in the city of Rio de Janeiro for more than 40 years. The hospital currently has 170 beds, out of which 51 are from the intensive care unit. With approximately 4,000 hospitalizations, more than 1.2 thousand surgeries and about 30 thousand highly complex procedures performed annually.

The patients come exclusively from the Brazilian Unified Health System (SUS in Portuguese), which means the hospital serves the local, state and national population with prevention programs, outpatient and hospital care. The Hospital stands out in hemodynamic procedures and cardiac surgeries, being recognized as a reference center for training, research, and formulation of Brazilian health policies.

2.2 Building

The Health Institute is located in a building that was built in the early 1940s and that has undergone several renovations over the years, but the hospital only began serving patients in this building in the 1970s. The area where the building is located has become a saturated urban environment, surrounded with buildings, roads, traffic signs, and electric networks.

The hospital is composed of a main building and an accessory building, with a total area of 15.653.29 m^2. The main building, which has 15 floors and 12.268.49 m^2 of built area, houses the administrative sector, and the ambulatory and hospital services. The accessory building, also known as the annex, is made up of 4 floors with 3.384.80 m^2, and it is used for ergometry and information technology.

2.3 Technical staff

The Hospital's workforce is constituted of public service employees, who are affiliated to the Brazilian Ministry of Health or hired by a Brazilian public institute; and employees of outsourced companies.

The Health Institute selects its professionals through a federal public tender for the positions of doctor, nurse, physiotherapist, nutritionist, speech therapist, among others; and administrative and technical level employees, totalizing 1,578 (one thousand five hundred and seventy-eight) public service employees.

The staff members contracted by the Brazilian public institute are operators, receptionists, messengers, occupational safety technicians, among others; totalizing 175 (one hundred and seventy-five) employees.

The hospital hires contractors for specialized and complementary services, which is outsourced to companies that provide transportation, catering, maintenance and property security services. Currently, the staff responsible for carrying out these activities totaled 389 (three hundred and eighty-nine) employees.

In addition, there are 41 (forty-one) trainees, 32 (thirty-two) residents, and 52 (fifty-two) agents from other public health institutions. The total staff of the hospital counts with 2,267 (two thousand two hundred and sixty-seven) professionals.

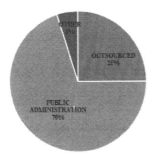

Figure 1. Hospital's workforce per hiring category. Machado (2016). Adapted.

The Figure 1 shows the distribution of the hospital employees by hiring category. Figure 1 shows that approximately 70% of the employees are hired by the hospital, 25% are outsourced, and the remaining 5% of the workers have affiliation with another health institute.

3 RESULTS AND DISCUSSION

In the study conducted by Machado (2016), 76 (seventy-six) observations of nonconformity or opportunities for improvement were identified and are summarized below.

The author noticed lack of motivation in the hospital staff to strengthen the occupational Health, Safety and Environment (HSE) sector within the others, even with the occurrence of the fire outbreaks. Currently, the hospital does not have an implemented HSE Policy, and it neither has a strategic plan for the coming years, nor an objective to promote safety as a top priority.

The robust size of the building, the location of the critical sectors in the upper floors and the lack of escape routes by ramps or elevators, threatens the safety of patients in case of an emergency. The author's recommendation is to assess another building for future installations of the hospital and to immediately comply with the requirements of the Rio de Janeiro Fire and Panic Safety Code (COSCIP in Portuguese); the NR 23, which provides fire protection legal guidance; and the Technical Instructions of the Brazilian Association of Technical Standards (ABNT in Portuguese).

Based on the observations *in loco* and the non-conformities found, it is strongly recommended to draw up a strategic work plan for the whole team responsible for the HSE department, adopting the monitoring of leading and lagging indicators. The target of this recommendation is to implement the principle of continuous improvement developed in the PDCA.

It is important to highlight the necessity of investigating incidents, finding root causes and proposing corrective, preventive and immediate actions; and to disseminate lessons learned, which is something that the current HSE team does not carry out.

Two of the most important HSE management programs foreseen in the Brazilian legislation, therefore mandatory, have not been implemented until this evaluation. Those are the Occupational Health Medical Control Program (PCMSO in Portuguese) and the Risk Prevention Program (PPRA in Portuguese). These programs should be elaborated and implemented in the hospital, particularly the medical examinations requested to determine the health status of all employees.

Regarding the care and safety of energized equipment, presented in NR 10, which provides safety guidance in electrical installations and services; the main recommendations are: the immediate elaboration of an Electrical Facilities Guidebook (set of information regarding plants, schemes and other information necessary for any authorized person to safely maintain the system); the application of flame retardant substance in the foam lining of the generators' rooms; the repair of exposed wires, and to promote the NR 10 training.

Machado (2016) also proposes the further investigation of the unsanitary conditions, by means of measurements and analyzes of the concentrations of chemicals. Furthermore, the author recommends verifying the adequacy of the exhaust and ventilation of the autoclave room and the promotion of noise level measurements and analysis in the Material and Sterilization Center.

To conduct an ergonomic analysis of the hospital, a brief evaluation was carried out of the clothes transportation activity through the RULA method. For Chiasson et al. 2012 apud Pereirinha et al. 2016, this technique gives a global score that considers the postural load, the period this posture is maintained, as well as the force used and whether there are repetitive movements. As result, it is recommended the implementation of control measures immediately to prevent potential ergonomic risks to the worker and further investigation.

Related to the HSE management system, it was verified that the hospital doesn't have an adequate documentation control structure, with noticeable failures in the process of documents registration, and problems in the control of the activities carried out in the workstations. It is therefore necessary to review all the current system and propose solutions more efficient and effective for the general control of the hospital.

Regarding the health and safety at work particularly in health services, it is important to monitor the radioactive areas and the environmental measurements of radiation and keep the chemicals descriptions at the places of use of these products, since such activities are not currently being carried out and are considered critical. Moreover, it is recommended the elaboration of a Rules Manual and good practices in the workplace, including the prohibition of wearing adornments and open shoes in risk areas and utilizing the hospital's external area to smoke.

It was verified that the Waste Management System is well implemented and the employees are well trained in the procedures. It should be highlighted the awareness campaigns carried out on selective collection. However, an opportunity for improvement can be implemented by addressing the reduction of waste generation as a subject in the future campaigns, due the many issues related to health waste discarding. (NR 32 and RDC 306/04).

Regarding the safety conditions in the machines and equipment, a study was carried out in the autoclaves room through inspections *in loco* and interviews with the technical staff. The results were positive for the system operation and maintenance, and observations were highlighted for the exhaust system, noise levels and lack of signaling.

The hospital's Personal Protective Equipment (PPE) control system is not well managed, and it needs to implement a policy and procedures standards to streamline its internal usage requirements, a system of records of PPE delivery to the employees, a control of the PPE certificates, among other related processes. Moreover, the HSE team should be involved in the PPE recommendation process and have greater autonomy to directly address outsourced employees who are not meeting the requirements of NR 6, which provides safety guidance on PPE.

4 ACTION PLAN

Machado (2016) proposes a strategic action plan for the correction or implementation of the 76 global findings observed in this work, using the GUT methodology. This tool application results in the classification and prioritization of recommended actions according to the calculation of gravity, urgency and tendency, and it is therefore applied to strategic planning.

Severity is understood as "possible damage or loss that may arise from a situation", for example, the impact of not complying with a particular recommendation; the urgency is related to the "pressure of the time that exists to solve a certain situation", considering legal items, eminence of occurrence of accidents or damages to the health of the workers; while the tendency refers to the "pattern or tendency of the situation's evolution",

Table 1. Gravity, urgency and tendency classification.

Gravity	Urgency	Tendency
1 = no serious	1 = do not hurry	1 = it will not get worse
2 = not serious	2 = it can wait a bit	2 = it will get worse in a long term
3 = serious	3 = as early as possible	3 = it will get worse in a medium term
4 = very serious	4 = with some urgency	4 = it will get worse in a short term
5 = extremely serious	5 = immediate action	5 = it will get worse quickly

*Machado (2016). Adapted.

linked to the aggravation of the problem raised over time. Therefore, the GUT calculation $(G + U + T)$ assists in indicating the priorities to a greater or lesser degree of a given situation (DGRH, Unicamp, 2009).

In this study, Table 1 is used to classify the findings evidenced throughout this work.

It is important after the classification of the findings, to follow a unique orientation for the interpretation and prioritization of the results. To perform such feat, four groups were created based on the scores derived from the calculation of the GUT method.

For results with higher scores, meaning high severity, high urgency and tendency to worsen the situation, immediate action is necessary; for the situations with lower score, meaning low severity, low urgency and tendency of worsening situation non-existent or in the medium long term, the action's deadline is 90 days; while for the intermediary groups, deadlines were established according to urgency, ranging from 30 to 60 days (Figure 2).

After applying the GUT method, 55 actions were classified as of low criticality, with a 90 days' deadline, corresponding to 72% of the plan; twelve have intermediate criticality, five with 30 days' deadline and seven with 60 days' deadline, corresponding to 7% and 9%, respectively; and nine critical actions were classified, for immediate action, representing 11% (Figure 3).

Regarding the Regulatory Norms, it is verified through Figure 4 that the great majority of the global recommendations raised are related to fire protection (NR 23) and health and safety at work in health services (NR 32), totaling 39% of the total shares. The building safety conditions (NR 8), the safety in electrical installations (NR10) and sanitary and comfort conditions (NR 24) add up to 11%, grouping 22 actions, and the other standards are divided more similarly.

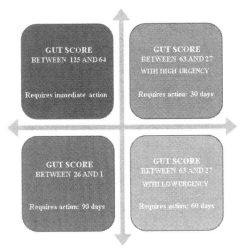

Figure 2. GUT matrix. Machado, 2016. Adapted.

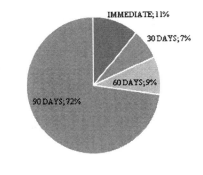

Figure 3. Prioritization of actions. Machado (2016). Adapted.

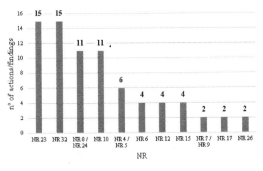

Figure 4. Number of actions/findings per NR. Machado (2016). Adapted.

In relation to the immediate actions, there are a total of nine proposed actions; it should be highlighted that five of them are related to fire protection (NR 23), totalizing 63% of the global critical actions of the strategic plan. Thus, it can be verified that the hospital urgently needs to improve its fire system and to devise an effective management system to control the deadlines and other recommendations proposed in this work.

5 CONCLUSION

This study has presented an evaluation of the occupational health and safety conditions in a public hospital located in the city of Rio de Janeiro based on the provisions of the Brazilian legislation. Motivated by the recurrence of fire outbreaks at different sectors of the hospital's facilities, this work was developed through technical inspections, analysis of documentation and interviews with the hospital's senior leadership and staff.

The author found significant deviations in fire protection, electrical installations, ergonomics, and occupational medical control and risks prevention programs. Based on these results, an action plan has been recommended, using the Gravity, Urgency and Tendency (GUT) method, to prioritize actions.

As result of the strategic plan elaborated by Machado (2016), nine actions were classified as critical and should be addressed immediately. Of these, five are related to fire protection (NR 23), which further aggravates the safety situation in the Health Institution once it was verified that the hospital and its staff are not properly prepared to prevent a fire.

In this way, the hospital should improve its fire system as a matter of urgency and develop an effective management system to control the deadlines of the other critical recommendations proposed in this work.

Thus, it is expected that this work will be applied, and the proposed action plan used as a tool to improve the health and safety conditions of the public hospital, ensuring a comfortable, healthy, and safe environment for all employees.

REFERENCES

Brazil – Brasilia, 2002. National Health Surveillance Agency – ANVISA 2002. *Hospital Safety Manual*. Available at: <http://www.anvisa.gov.br/servicosaude/manuais/manual_seg_hosp.htm >.

Brazil – Brasilia, 2004. National Health Surveillance Agency – ANVISA. Diário Oficial da União. Resolução RDC n° 306 – *Regulamento técnico para o gerenciamento de resíduos de serviços e de saúde*.

Brazil – Rio de Janeiro, 1976. Military Fire Department of the State of Rio de Janeiro – CBMRJ. *Fire and Panic Safety Code for the State of Rio de Janeiro*.

Brazil – São Paulo, 2009. Human Resources General Directorate – DGRH. UNICAMP – *Metodologia GUT – Gravidade, Urgência e Tendência*. Available at: <http://www.dgrh.unicamp.br/documentos/oficios-circulares/anexos/ofcirc092009-anexo.pdf>.

MACHADO, C. F. 2016 *Avaliação das Condições de Segurança e Saúde Ocupacional em um Hospital Público*. Dissertation (Pos Graduation in Safey Engeneering) – Universidade Federal do Rio de Janeiro, Rio de Janeiro.

MCATAMNEY, Lynn; CORLETT, Nigel – *RULA – A Rapid Upper Limb Assessment Tool*. Available at: <http://www.rula.co.uk/brief.html>.

MIDDLESWORTH, Mark. Ergonomic Plus – *A Step-by-Step Guide to the RULA Assessment Tool*. Available at: <http://ergo-plus.com/rula-assessment-tool-guide/>.

MTPS, NR 6 – Personal Protective Equipment, 1978 Brasilia, Brazil: Ministry of Work and Social Security.

MTPS, NR 8 – Buildings, 1978. Brasilia, Brazil: Ministry of Work and Social Security.

MTPS, NR 10 – Safety in Electrical Installations and Services in Electricity, 1978. Brasilia, Brazil: Ministry of Work and Social Security.

MTPS, NR 17 – Ergonomics, 1978. Brasilia, Brazil: Ministry of Work and Social Security.

MTPS, NR 24 – Sanitary and Comfort Conditions in Workplaces, 1978. Brasilia, Brazil: Ministry of Work and Social Security.

MTPS, NR 32 – Safety and Health at Work in Health Services, 2002. Brasilia, Brazil: Ministry of Work and Social Security.

Pereirinha, D.; Cravo, A.; Teixeira, P. Lança, A.; Almeida, J. 2016. *Metodologias de avaliação de risco ergonómico: Da seleção à aplicação*. Coimbra, Portugal. 1º Congresso Luso-Brasileiro de Segurança e Saúde Ocupacional e Ambiental. Coimbra Health School (ESTeSC):161–166.

Levels of non-ionizing radiation in vertical residences in João Pessoa (PB)

A.A.R. Silva, L.B. Silva, R.M. Silva & R.B.B. Dias
Department of Industrial Engineering, Federal University of Paraíba, João Pessoa, Brazil

ABSTRACT: Previous studies have – suggested an association between exposure to electromagnetic fields, at frequencies between 50 and 60 Hz, which includes the frequency of the local power grid in many countries, and childhood leukemia. This study presents an analysis of the residents' exposure to non-ionizing radiation in vertical residences located in the city of João Pessoa (PB). Magnetic field measurements were made over 24 hours in apartments located on the first floor of buildings with transformer substations at the street level. Measurements were made in two environments, chosen according to the layout of the residences, in the frequency range of 1–100Hz. Through the statistical analysis of the data, it was verified that the residences have a high index of exposure to extremely low frequency magnetic fields ($>> 0.4$ μT, in the 60 Hz range), which is potentially harmful to health of its inhabitants according to recent epidemiological studies.

1 INTRODUCTION

Electromagnetic radiation refers to energy propagation as a result of an electromagnetic field varying in time. Radiation in the form of radio waves, cellphone emissions and invisible light constitute electromagnetic radiation. Electromagnetic fields (EMF), have a 60 Hz frequency in USA, Canada and South America and 50 Hz frequency in Europe; they are described as extremely low frequency (ELF EMF) in terms of electromagnetic frequency (Henshaw 2002).

Static and extremely low electromagnetic energy naturally occurs or it is associated with the generation and transmission of electrical energy and use of some devices. Electric and magnetic fields (static and from low frequency) are created by natural and artificial sources. Artificial sources are powered by electrical energy and they are the main emitters of ELF EMF, which are defined as frequencies up to 300 Hz. Yet the natural fields are static or vary very slowly (Brodić & Amelio 2015).

Depending on the effects that they cause in human organism, low frequency electromagnetic radiation can be categorized in: ionizing radiation (group that also covers high radiation, such as gamma rays and X rays) and non-ionizing radiation (low or extremely low frequency). Ionizing radiations are capable of—directly or indirectly—damaging the tissues, as they have sufficient energy to extract electrons from atoms and molecules. On the other hand, non-ionizing radiations (NIR) do not have sufficient energy to ionize, but they are capable of promoting the breakdown of molecules and chemical bonds (Calvente et al. 2010).

Overall, the sources of electromagnetic fields in residences are divided into two categories: area sources, which produce fields that extend themselves apart from the immediate area that surrounds the source (for example, supply and distribution lines); and local sources, in which the field is essentially confined to a restricted space and a short distance from the source (for example, electro-electronic devices, electrical panels) (Câmara 2014).

Possible effects resulting from exposition to emitter sources of non-ionizing radiation in residences and its potential hazards for health have raised several studies. A few investigations acknowledge the occurrence of some symptoms in individuals who have had contact with generating sources, such as the electromagnetic hypersensitivity (Brodić 2015).

As the hypothesis raised by epidemiological studies that the childhood leukemia may be related to magnetic fields exposure, there is a consensus that the reduction of scientific uncertainties in studies that measure the level of chronic exposure to ELF EMF is a relevant objective, raising new studies, different from the ones held until now (Hareuveny et al. 2011).

Some of epidemiological associations between ELF-EMF and its carcinogenic potential may be explained by the application of a selection bias. The international epidemiological study TransExpo, that focus on magnetic field intensity in apartments located above the internal transformer station, aims to minimize this tendency to a selection bias. Magnetic fields in these apartments are significantly higher than in apartments located in different floors of the same building (Kandel et al. 2013).

An internal transformer station (IT) consists of a room, typically with an area between 15 and 25 m^2 and height varying from 2.5 to 3.0 meters, in which are installed one or more distribution transformers (Hareuveny et al. 2011). In many countries, ITs are commonly installed in vertical buildings, in the same level as the urban streets and constitute an important emission source of magnetic fields to the adjacencies residences.

This article aims to evaluate the exposure to non-ionizing radiation in vertical residences, as the result of extremely low frequency magnetic fields levels emitted by adjacencies internal transformer stations, other external sources, as well as electro-electronic presented inside the studied homes.

2 NON-IONIZING RADIATION

Transformer stations located inside buildings may offer a baseline for the development of epidemiological studies that avoid selection bias, minimizing the influence of combined factors, and that include people exposed to relatively intense extremely low frequency magnetic fields (Ilonen et al. 2008).

Researches who performed investigations, having the TransExpo international study as methodology, have verified exposure level from individuals living in vertical buildings surrounding internal transformer stations over the levels described as harmful in different epidemiological studies about the subject (>> 0.4 µT, in a range from 50 to 60 Hz). It is the case of Ilonen et al. (2008), Hareuveny et al. (2011) and Huss et al. (2013).

3 METHODOLOGY

Based on investigations developed by some of the authors previously mentioned, ELF EMF levels were assessed in the apartments located on the first floor of vertical residences, in virtue of the presence of internal transformer substations, as well as external transformers in the level of urban streets, in the adjacencies of these apartments, and electro-electronic devices on the inside of the rooms studied.

Data collection was performed in residences located on the first floor of buildings in the city of João Pessoa, one in Bancários neighborhood, another one in Cabo Branco neighborhood and a third one located in Tambaú neighborhood. All three residences are characterized by the presence of internal transformer substations in the level of urban streets. The total measurement time in each one of them was 24 hours, starting in the morning and continuing throughout afternoon and night.

According to the apartments' inside layout, measurements of magnetic field were performed in two rooms: bedroom and living room in Cabo Branco and Bancários neighborhood and bedroom and kitchen in Tambaú neighborhood. This choice arose from the fact that the apartment in Tambaú neighborhood was the only one that had a large kitchen, whereas the other two had embedded kitchens and living rooms. In order to simplify, all measurements performed in the living room from the two apartments and the kitchen of the apartment in Tambaú neighborhood will be referred just as measurements made in the kitchens, as those ones constitute the major interest for this study, since there are many electro-electronic devices frequently operating throughout the day.

The Aaronia Spectran NF-5035 magnetic field analyzer was used, properly calibrated, whose values were collected in the unit of microtesla (µT), in the frequency of 60 Hz. The device was placed 0.5 m from floor in the rooms where it was fixed, where possible close to the entrance door of each room, in order to capture the low frequency magnetic field spread all over the space throughout the day.

Descriptive and statistical data analyses were performed through R Project 3.1.1 software. Mann-Whitney and Kruskal-Wallis nonparametric hypothesis tests were applied, in order to compare the intensity levels of magnetic field from the atmospheres between the residences, considering the significance level of 0.05.

The behavior from the collected data was evaluated considering its histograms, as well as the smoothing and cumulative probability curves generated from the values obtained in each one of the residences' rooms located in Cabo Branco, Tambaú and Bancários neighborhoods.

4 RESULTS AND DISCUSSION

The mean and standard deviation of the magnetic field intensity measured in each room, in the three apartments, evaluated in the local electric grid (60 Hz) are shown in Table 1.

The magnetic field intensity in all of the rooms in the residences was higher than 0.4 µT, in the frequency of 60 Hz. The highest values were found in the two rooms evaluated in the same residence in Tambáu neighborhood, which presented a mean value of 1.4186 µT and 1.5545 µT, in the bedroom and kitchen, respectively. The probable explanation for the ascertainment of such high rates is the influence of ELF EMF originating not only from

Table 1. Mean and standard deviation of the magnetic field intensity per room in the frequency of 60 Hz.

Location	Rooms	Mean (μT)	Standard deviation (μT)
Bancários	Bedroom	1.214	0.374
	Kitchen	1.191	0.348
Cabo Branco	Bedroom	1.252	0.398
	Kitchen	1.228	0.396
Tambaú	Bedroom	1.418	0.456
	Kitchen	1.554	0.052

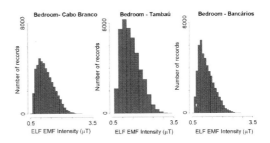

Figure 1. Histogram from the bedrooms of the residences located in Tambaú, Cabo Branco and Bancários neighborhoods.

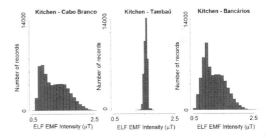

Figure 2. Histogram from the kitchens situated in Tambaú, Cabo Branco e Bancários neighborhoods.

ITS, but also from heat sources located in their inside.

The highest value observed pertained to the bedroom of the same apartment, which reached 3.3656 μT at 18:30, a value 8 times higher than the value 0.4 μT used by many authors of this topic as the exposition limit reference value in order to guarantee that there are not negative effects to the individuals' health. This bedroom is exposed to ELF EMF emitted by high-voltage cables of the power grid, as well as an aerial-type external transformer, which is 10 meters distance from the room.

Similarly, Ilonen et al. (2008) determined that the intensity values of magnetic field obtained in the apartments located immediately above the transformer stations overcame the marks of 0.2 μT and 0.4 μT in 97% and 63%, respectively, from the apartments above the transformer stations. Hareuveny et al. (2011) reported results that comply with the ones found by Ilonen et al. (2008) – which has found 0.40 μT as a mean value of magnetic field—in apartments located above ITS.

Data collected in the experiments present lognormal distribution and its number of records to the rooms evaluated is illustrated by the histograms in Figures 1–2.

It is observed that the values of magnetic field intensity concentrate, mostly, in the approximate range from 0.75 μT to 1.75 μT for the bedrooms studied in the three apartments.

Regarding to the kitchens, in the apartments located in Cabo Branco and Bancários, the values of magnetic field concentrate between 0.7 μT and 1.7 μT, whereas the residence in Tambaú neighbourhood presented a lower variation level, predominantly in a range from 1.45 μT to 1.65 μT. It was verified that, in this way, the kitchen in Tambaú has been exposed to higher rates of ELF EM than similar rooms from the other apartments, which may represent a sign of the presence of an emitter source in this room—associated with the several household appliances operating in its interior—that is not common to the other apartments that were studied. Greater values than the maximum ELF EMF value verified in the kitchen, equivalent to 1.8267 μT, may be observed in the bedroom from the apartment in Tambaú, but occasionally, not repeating for a significant period of time.

The Figures 3–4 show the comparison of the behavior of intensity values of magnetic field obtained in the residences, based on their central tendency curves, for the bedrooms and kitchen, respectively. As the experiments were all performed over 24 hours, starting in the morning e continuing throughout afternoon and night, the collect order is associated with the transition of these turns.

The increase in ELF EMF levels from the bedrooms in the course of the collection, point to an occurrence of higher values during the night, to the detriment of lower levels during the day.

Among the kitchens studied, the kitchen from Tambaú neighborhood was the only one that presented intensity levels of magnetic field approximately constant and, simultaneously, much higher than the ones from other apartments, as it was also verified in Figure 2.

The cumulative probability curve allows determining the interval of intensity levels of magnetic field that each room from the apartments is exposed throughout the experiment. Cumulative

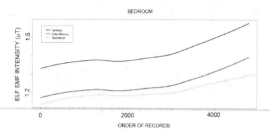

Figure 3. Comparison of exposition to ELF EMF between the bedrooms from the different apartments.

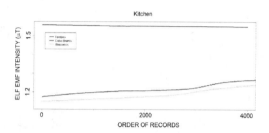

Figure 4. Comparison of exposition to ELF EMF between the kitchens from the different apartments.

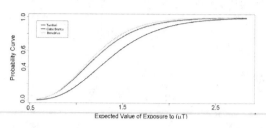

Figure 5. Cumulative probability from the bedrooms of the apartments located in Tambaú, Cabo Branco and Bancários neighborhood.

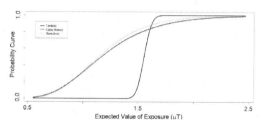

Figure 6. Cumulative probability from the kitchens of the apartments located in Tambaú, Cabo Branco and Bancários neighborhoods.

probability curves are shown in Figures 5–6, for the bedrooms and kitchens from the three apartments, respectively.

Figure 5 confirms that the values of magnetic field in the bedrooms of each one of the vertical buildings evaluated have a higher probability of being situated in a range from 0.75 µT (cumulative probability around 1%) to 1.75 µT (cumulative probability around 90%), as it was previously discussed from the histogram of Figure 1.

Figure 6 reinforces not only that the ELF EMF intensity from the kitchens of the apartments in Cabo Branco and Bancários assumed values between 0.7 µT (cumulative probability around 1%) and 1.7 µT (cumulative probability around 90%), as well as the kitchen in Tambaú, which had values variation much lower than the other neighborhoods, in the range from 1.45 µT to 1.65 µT, with cumulative probabilities around 10% and 90%, respectively.

The Kruskal-Wallis nonparametric hypothesis test was used to test if data groups of magnetic field intensity from the rooms present the same distribution. As an alternative to analyze the variance, this test is similar to Mann-Whitney test and it verifies the null hypothesis (Vieira 2010).

After that, the Mann-Whitney nonparametric hypothesis test was performed to compare, by pairs, collected data groups in the rooms from the residences evaluated, in order to test if the magnetic field's intensity has equal distribution between bedrooms and kitchens from the different locations studied.

For Kruskal-Wallis nonparametric hypothesis test, with data from the bedrooms, the values $\chi^2 = 5512.6$ and p-value $= 2.2*10^{-16} < 0.05$ were obtained, so that the hypothesis was rejected. In pair comparisons between bedrooms by Mann-Whitney test, it was obtained $W = 1248600000$ and p-value $= 2.2*10^{-16} < 0.05$ for Cabo Branco and Bancários, and $W = 896100000$ and p-value $= 2.2*10^{-16} < 0.05$ for Cabo Branco and Tambaú, and finally, $W = 14989000$, p-value $= 2.2*10^{-16} < 0.05$ for Tambaú and Bancários. This way, it is verified that the distribution of magnetic field's intensity values measured in the bedrooms of the different residences do not present similar distribution.

Analogously, in Kruskal-Wallis test with data from the kitchens, $\chi^2 = 33069$ was obtained with $p = 2.2*10^{-16} < 0.05$, so that the hypothesis was rejected. In pair comparisons between the kitchens by Mann-Whitney test, it was obtained $W = 2032700000$ and p-value $= 2.2*10^{-16} < 0.05$ for Cabo Branco and Bancários, $W = 1765100000$ and p-value $= 2.2*10^{-16} < 0.05$ for Cabo Branco e Tambaú, and finally, $W = 2261500000$ and p-value $= 2.2*10^{-16} < 0.05$ for Tambaú and Bancários. This way, it is verified that the distribution of magnetic field's intensity values measured in the kitchens of the different residences, as well as in the bedrooms, do not present similar distribution.

These results show that analyzing ELF EMF levels considering only the level of the residences related to ITS and using an evaluation method of exposition similar to the one used in the TransExpo epidemiological study, in other words, just classifying the apartments located immediately above a transformer station as highly exposed to ELF EMF, is not sufficient to obtain evidences that this phenomenon occurs similarly in different locations.

As suggested by Hareuveny et al. (2011), it is proposed that new investigations emphasize more on ITs, requiring an extensive technical description of them—considering the amount of ITs around, their potency and the distance to the rooms—before performing any evaluation study for exposition.

It is possible to include an evaluation of cumulative distribution pattern of the magnetic's field intensity values emitted around the rooms studied, according to ITs's position and the low voltage cables that power the distribution panels, in which ELF EMF peaks occur, accordingly to the investigations of Kandel et al., 2013.

Likewise, it is suggested that other variables, surroundings to the presence of external transformers and high-voltage cables from urban transmission grids (number of transformers, power and distance to the apartments' rooms), may be considered.

5 CONCLUSION

The measurements of extremely low frequency magnetic field's intensity registered throughout the experiment in the apartments evaluated revealed values considered high, as most of them exceeded 0.4 µT, in the 60 Hz grid frequency of the city of João Pessoa. These results draw the attention for possible effects issues due to the exposition to these non-ionizing radiation levels for the inhabitants of the residences located surrounding internal transformer stations. Such concern is justified by the existence of investigations that study the connection between exposition to ELF EMF and the manifestation of diseases (Henshaw 2002; Mofrecola et al. 2003; Sage 2012; Câmara 2014), including the childhood leukemia (Albom et al. 2000; Calvente et al. 2010).

The evaluation of ELF EMF intensity in vertical residences located on the first floor of building with ITS has been the aim of studies in many regions around the world (Ilonen et al. 2008, Hareuveny et al. 2011, Huss et al. 2013, Kandel et al., 2013). However, considering the associated variables to internal and also external transformers and high-voltage cables presented surroundings the buildings proves to be necessary so that rooms whose individuals are highly exposed may be selected, in different areas. This would facilitate the comparisons between rooms located in different regions, but that presented similar pattern, bearing in mind the differences among the countries with the regard to ELF EMF exposure.

REFERENCES

Ahlbom, A., Day, N., Feychting, M., Roman, E., Skinner, J., Dockerty, J., Linet, M., McBride, M., Michaelis, J., Olsen, J. H. 2000. A pooled analysis of magnetic fields and childhood leukaemia. *British Journal of Cancer*, London 83(5): 692–698.

Brodić, D. & Amelio, A. 2015. Classification of the Extremely Low Frequency Magnetic Field Radiation Measurement from the Laptop Computers. *Measurement Science Review* 15(4): 202–209.

Brodić, D. 2015. Measurement of the extremely low frequency magnetic field in the laptop neighborhood. *Revista Facultad de Ingeniería* (76): 39–45.

Calvente, I., Fernandez, M. F., Villalba, J., Olea, N., Nuñez, M. I. 2010. Exposure to electromagnetic fields (non-ionizing radiation) and its relationship with childhood leukemia: A systematic review. *Science of the Total Environment* 408: 3062–3069.

Câmara, P. R. S. 2014. Effect of exposure to non-ionizing radiation (electromagnetic fields) on the human system: A literature review. *Journal of Interdisciplinary Histopathology*, Wilmington 4(2): 187–190.

Hareuveny, R., Kandel, S., Yitzhak, N., Kheifets, L., Mezei, G. 2011. Exposure to 50 Hz magnetic fields in apartment buildings with indoor transformer stations in Israel. *Journal of Exposure Science and Environmental Epidemiology* 21(4): 365–371.

Henshaw, D. L. 2002. Does our electricity distribution system pose a serious risk to public health? *Medical Hypotheses* 59(1): 39–51.

Huss, A., Goris, K., Vermeulen, K., Kromhout, H. 2013. Does apartment's distance to an in-built transformer room predict magnetic field exposure levels? *Journal of Exposure Science and Environmental Epidemiology* 23(5): 554–558.

Ilonen, K., Markkanen, A., Mezei, G., Juutilainen, J. 2008. Indoor Transformer Stations as Predictors of Residential ELF Magnetic Field Exposure. *Bioelectromagnetics* 29(3): 213–218.

Kandel, S., Hareuveny, R., Yitzhak, N., Ruppin, R. 2013. Magnetic Field Measurements near Stand-Alone Transformer Stations. *Radiation Protection Dosimetry* 157(4): 619–622.

Mofrecola, G., Moffa, G., Procaccini, E. M. 2003. Nonionizing electromagnetic radiations, emitted by a cellular phone, modify cutaneous blood flow. *Dermatology* 207: 10–14.

Sage, C. 2012. The similar effects of low-dose ionizing radiation and non-ionizing radiation from background environmental levels of exposure. *Environmentalist* 32: 144–156.

Vieira, S. 2010. Bioestatística: tópicos avançados. In Elsevier (3ª ed). Brasil: Rio de Janeiro.

Human error prevention by training design: Suppressing the Dunning-Kruger effect in its infancy

E. Stojiljković, A. Bozilov, T. Golubovic & S. Glisovic
Faculty of Occupational Safety in Nis, University of Nis, Nis, Serbia

M. Cvetkovic
Department of Occupational Safety and Health, Faculty of Engineering, University of Porto, Porto, Portugal

ABSTRACT: This paper aims to discuss possibilities for early detection of potential human errors rooted in limited metacognitive abilities of operators and managers alike, as well as to explore conclusions reached by Dunning-Kruger in terms of safety and emergency response management, considering the fact that human errors are often rooted in people's erroneous perceptions of their own skills and capabilities. It also reflects on how this issue may affect the workplace and emergencies, particularly in regards to risk and safety. The paper briefly summarizes a portion of findings from a pilot survey performed under a wider framework of an investigation aimed to reveal and define Performance Shape Factors (PSFs) applying Success Likelihood Index Method (SLIM), in order to identify common sources of cognitive-bias-based human errors in energy supply sector.

1 INTRODUCTION

Industrial managers and operators are the key factors of every industrial system. Consequently, human error is the most common cause of vast majority of occupational accidents, and an important factor in every occupational risk analysis and assessment. The modern approach to human reliability assessment suggests that human error reflects deeper issues within the system, and that it results from system relationships between people, tools, tasks, and work environment (Woods et al. 2010). The performance of an individual depends on a variety of, so called, Performance Shaping Factors (PSFs). Since PSFs profoundly affect human behaviour and decision making, they can affect both the increase and the decrease of human error probability. The typical PSFs used in risk assessment and human error analysis in complex systems are: operator experience, training, stress intensity, information flow, task complexity level, team work factor, etc. (Stojiljkovic 2013). PSFs have a considerable impact on operator performance, and they can be both external and internal. Internal PSFs, among other, also include performance capacity, associated with individual factors, such as psychological capacity (mental state, education level, level of training), as well as psychological preparedness, which involves internal motivation (interest, social integration) and external motivation (opportunities, work conditions, organizational climate).

Individuals respond to available information based on their experience, training, knowledge and motivations. Operators actively select and interpret information or data, judge it and decide on appropriate actions. A cognitive bias is a deviation in judgment, since humans tend to create their appropriate subjective reality based on their perception of their relations with others. There are numerous cognitive biases and effects that affect judgments and decision making in working environments. To deal with an occupational risk it is necessary to recognize that it is subjective and that individual perception of it is biased in many different ways.

Dunning-Kruger effect is one of common biases, along with Confirmation bias, Availability bias, and many others. Almost two decades ago, American scientists Dunning and Kruger have launched hypothesis, and subsequently confirmed it by three consecutive experiments, that a lot of people with limited abilities in certain fields seem to be overly confident in their own capabilities, although the evidences usually disapproved their sincere belief (Kruger & Dunning 1999, Dunning et al. 2003). The researchers revealed that the gap between competences and self confidence really exists, and that the least skilled people are often significantly more confident in their own capabilities than true experts are.

The authors attributed this bias to a metacognitive incapacity of the unskillful to recognize their own incompetence and accurately evaluate their own abilities. The phenomenon, noticed in slightly different forms centuries ago by some (e.g. Confucius, Plato), and expressed in the works of several renowned authors (i.e. Bertrand Russell, Charles Darwin), upon scientific confirmation by the two became globally recognized as Dunning-Kruger effect.

2 METACOGNITIVE ABILITIES AND SAFETY

In terms of safety and emergency response management, it seems that what one don't know, and what one is not aware of, might (and often does) put lives in risk and/or jeopardize relief efforts. Apart from safety design features of the equipment applied and procedures aimed to safeguard operators, the genuine safety problem might be partly related to the fact that those least capable to make knowledgeable decisions are also most likely to deem that they can. It has been confirmed that, when put to the test, those that score low are typically exactly those who rate their own abilities rather highly. Conversely, competent operators and experts are more likely to undervalue their own skills and to assume that the most of their peers are skilled nearly as they are. The study of Burson et al. (2006) reconfirmed that in many tasks, both skilled and unskilled did not accurately assess their own potential performance, so the authors claim that it is cognitive bias that determines subjective assessments.

It is obvious that an appropriate strategy to deal with misjudgments of a kind would be necessary in order to suppress this potential source of human errors. It is of utmost importance that supervisors, crew leaders and operators occupying responsible positions should be adequately confident in their abilities to respond unexpected safety challenges. In many areas, misjudgments might be costly or time consuming, but not lethal. However, in safety management and emergency operations, overestimation of skills and abilities might lead to disastrous consequences.

The effect is spontaneous and relies on so called positive confirmation bias. Thus, decisions made by less skilled individuals can provoke serious consequences, merely because they do not appropriately comprehend that their knowledge is inadequate and experience insufficient. While in many areas of human activities the lack of awareness of personal shortcomings might lead to social and sometimes legal conflicts, when it comes to safety issues, the lack of knowledge on own skill level might pose a major risk to decision maker and anyone else involved in a particular operation.

3 METHOD

Hereby we briefly summarize a portion of findings from a pilot survey performed in 2013 under a wider framework of an investigation aimed to reveal and define Performance Shape Factors (PSF) by applying Success Likelihood Index Method (SLIM) in a branch of the Powers Supply Utility Service in South Serbia (e.g. Elektrokosmet) (Stojiljkovic et al. 2015). There were evidences of various near-miss incidents where human errors were the primary cause of the episodes. Many of them could be traced back to Dunning-Kruger effect, such as some of those described in Elektrokosmet incident reports from 2009–2011. Therefore, a brief, pilot, after training test was performed to reveal if there were miscalibrations between test results and perceptions on achieved scores among trainees. This condensed report is based on an ad hoc Occupational Safety Questionnaire, a survey tool that revealed some underlying, non-technical issues that beyond doubt could influence safety. The results were subjected to descriptive statistics and tested by inferential statistics tools. Every attempt was made to exclude influence of organizational climate on paper-based diagnostic applied. The importance and applicability of the SLIM for quantification of human errors in complex systems should be emphasized, as it was used to confirm that PSFs are the key element for adequate human error reduction (Stojiljkovic et al. 2015).

4 RESULTS

It has been revealed that participants at the bottom of the scale had significantly inflated estimation of their own qualities. Thus, the findings in large extent confirm and somewhat replicate Dunning-Kruger findings on metacognitive bias on both sides of the achievement scale. Some of the bottom quartile participants estimated their skills above 52% of their colleagues, although they actually were better than just 16% of them. When put in safety management perspective, this large misjudgment sounds rather alarming. Simplified, it would be like misjudgment in attempt to jump over 3 m wide gap by one capable to long jump less than one meter. In sudden decision making, that kind of wrong reasoning or lack of perception might easily lead to catastrophic events.

On other hand, top quartile achievers (as expected due to Dunning-Kruger effect) slightly underestimated their performance, putting themselves just over third quartile, although their test results indicated that they all achieved over 82 percentile. The fact that good skilled individuals (who are often being assigned chief positions) are getting modest doesn't help safety operations either, especially in cases of emergency. Decision makers full of doubt could easily hinder prompt response in situations when timely action is of utmost importance.

There were two recognizable groups of the safety training attendees affected by Dunning-Kruger syndrome. The first one consisted of those that lack self-awareness about significant gaps in their knowledge or experience. Another comprises top achievers that lack self-confidence that might prevent them to make educated choice in an emergency situation.

The latest experiences with employees from Elektrokosmet lead to a conclusion that it is worth confronting the bottom quartile participants with test results, and subsequently give them opportunity to attend extra training based on wrong results from the previous test. When repeatedly subjected to similar test, the group tended to improve not only their scores, but also their perception on achieved results. It seems that their metacognitive abilities have slightly risen along with acquired knowledge, as opposed to findings from some archival research, e.g. Ehrlinger (2008). On other hand, it is impression of the authors that upper quartile group required special treatment/training of psychological nature, which design is still subject of debate and investigation, and this issue largely requires expertise which is out of the realm of safety engineering, at least in its narrow, technical form. Composing a multidisciplinary team, consisting of experts of both technical and psychological background, is strongly advised when designing comprehensive training programs for personnel exposed to harsh/hectic environments and/or increased risks. A system safety performance shouldn't be decided just on the basis of a probabilistic risk analysis, but rather through a process that takes into consideration the complex socio-technical interactions (Kontogiannis et al. 2017). Almost two decades ago the same author noticed that corporate training, staff selection and team organization entail collaboration of disciplines such as cognitive science, organizational psychology, ergonomics and computer science (Kontogiannis 1999).

5 POSSIBLE SOLUTIONS

There are several training approaches that could facilitate avoiding over-confidence trap in safety management decision making, such as:

- making sure that trainees are capable to perform proficiently the jobs and procedures as described in their working sheets and safety instructions;
- making sure that trainees receive open and objective performance rating upon every test and assignment;
- performing after-action reviews and assessments of recent events and near-miss incidents;
- analyzing assignments given, and actions performed under untypical circumstances;
- collecting anonymous descriptions and opinions on critical events to shape training content and course design;
- organizing critical thinking rehearsals based on case studies;
- insisting on self-assessment of perceived achievements among participants upon every significant course assignment;
- performing regular appraisals of self-confidence through specially designed tests and comparing those results with actual achievements in particular assignments.

As stated in prominent paper of Kontogiannis (1999), "feedback after a training segment allows people to develop error detection and recovery skills, while after-action reviews can provide opportunities for self-assessment and mental simulation of alternatives". It is quite obvious that quality performance of one such ambitious training system should be constantly monitored and improved. A model of a recurring human error management system, designed in accordance with the model of the Deming PDCA cycle (Stojiljkovic 2017), and fully applicable for development of a cognitive bias suppressing training module, is shown in the following figure.

Trainees should be taught to assume that they do not comprehend the problem when confronted with an atypical situation, to apply critical thinking

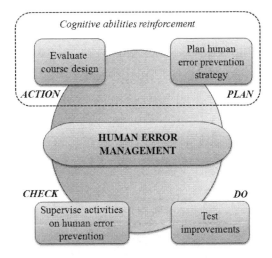

Figure 1. Recurrent human error management system.

(which requires suitable training per se) and to test their assumptions and conclusions whenever circumstances permit.

There is, unfortunately, certain number of individuals working on challenging positions in terms of safety, who are very much suffering from the syndrome of excessive confidence. Such persons not only keep deceiving themselves that they are suitably skilled to conduct entrusted tasks, they also continue to delude their superiors who engaged them thus putting their peers at significant risk of injuries or fatalities. What majority of working professionals often fail to notice is that what they see while watching others often depends on the clarity of their own sight.

A safety audit is only way to ensure that both workers and community are safe, and that health and safety requirements are fulfilled. Comprehensive risk assessments are not only a legal requirement, but they are also essential for maintaining workplace safety. However, it should be kept in mind that while one might believe that assessments are done correctly, there is a chance that some of decision makers are affected by the Dunning-Kruger effect, which is of particular concern when it comes to unexpected events or untypical conditions. The most decision making at work is non-rational, and many decisions are made under pressure. Individuals tend to make non-rational decisions when constrained by time, or when influenced by particular arrangements. Under pressure, human decision making is based on implicit rather than rational knowledge. In such circumstances, inadequate awareness of skill level might lead to devastating outcomes.

Occupants of particular working places and members of certain professional bodies are bound by their rules for continuing professional education and development, which should provide a degree of assurance that required and desirable skill level was achieved. However, the trainings should be designed in such manner that cognitive biases such as Dunning-Kruger effect are timely recognized, suppressed and possibly overcame.

6 CONCLUSION

In the essence of safety management is to deal with risk and reduce it to the reasonably practical level. An important part of safety management activities is organizing comprehensive and specialized trainings aimed to improve skills and knowledge of employees, and to reveal numerous shortcomings in comprehension of contemporary safety issues.

Confronting the bottom quartile achievers with test results, and giving them opportunity to attend an extra training based on wrong results from the previous test might prove beneficial. Secondary test results indicate that metacognitive abilities slightly rise along with acquired knowledge. However, impression is that upper quartile achievers require somewhat different approach, i.e. specially designed psychological training and/or treatment to boost their self-confidence. Designing comprehensive training programs for personnel exposed to increased risks obviously require multidisciplinary teams that comprise experts of both technical and psychological background.

An appropriate training design should consider decision making in the bounds of workplace complexity and particular organizational structure. The questionnaires should be organized in such a manner to avoid asking for straightforward responses to predictable statements, particularly if the inquiry is being made in a strict regulatory environment. An analytics able to assess implicit knowledge would be a good predictor of risk decision making and judgment. Also, a diagnostic capable to assess the psychology of the workplace would be a good indicator of factors that shape safety related decision making. In a modern approach to risk management and human reliability assessment, it is necessary to merge external data with the data on PSFs and psychological error mechanisms. It is also of crucial importance to comprehend circumstances and mechanisms that limit human perception.

Obviously, ignoramus is never doubtful, and an old proverb emphasizes that little knowledge might be risky. With advancement of technologies, and ever increasing complexity of technical systems and relations within them, this is worth today more than ever before.

ACKNOWLEDGEMENT

This paper was written as part of the projects No. III 43014 and No. III 43011, financed by the Ministry of Education, Science and Technology of the Republic of Serbia.

REFERENCES

Burson, K. A. et al. 2006. Skilled or unskilled, but still unaware of it: How perceptions of difficulty drive miscalibration in relative comparisons, *Journal of Personality and Social Psychology*, 90 (1): 60–77.

Dunning, D. et al. 2003. Why People Fail to Recognize Their Own Incompetence, *Current Directions in Psychological Science*, 12 (3): 83–87.

Ehrlinger, J. 2008. Why the unskilled are unaware: Further explorations of (absent) self-insight among the incompetent, *Organizational Behavior and Human Decision Processes*, 105 (1): 98–121.

Kontogiannis T. 1999. Training Effective Human Performance in the Management of Stressful Emergencies, *Cognition, Technology & Work*, 1 (1): 7–24.

Kontogiannis, T., Malakis, S., McDonald, N. 2017. Integrating operational and risk information with system risk models in air traffic control, *Cognition, Technology & Work*, 19 (2–3): 345–361.

Kruger, J., Dunning, D. 1999. Unskilled and unaware of it: How difficulties in recognizing one's own incompetence lead to inflated self-assessments, *Journal of Personality and Social Psychology*, 77 (6): 1121–1134.

Stojiljkovic, E. 2013. The application of an event tree for human error analysis in the Electric power company in Serbia, *Facta Universitatis, Series: Working and Living Environmental Protection*, 10 (2): 135–142.

Stojiljkovic, E. 2017. Knowledge management for the purpose of human error reduction. Proceedings of M&S 2017: 12th International Conference Management and Safety, Neum and Mostar, Bosnia and Herzegovina, 9–10. June 2017. Zagreb, Croatia: The European Society of Safety Engineers, pp. 1–8 (invited paper).

Stojiljkovic, E., Glisovic, S., Grozdanovic, M. 2015. The role of human error analysis in occupational and environmental risk assessment: a Serbian experience, *Human and Ecological Risk Assessment: An International Journal*, 21 (4): 1081–1093.

Woods, D., Dekker, S., Cook, R., Johannesen, L., Sarter, N. 2010. *Behind human error*. Farnham, UK: Ashgate.

Workload and work capacity of the public agent at a department of a federal institution of higher education

L.L. de F. Cavalcanti & L.B. Martins
U.F.P.E., Recife-PE, Brazil

ABSTRACT: The objective of this study was to identify the health of the federal public agent, at a department of physical education and sports of a federal institution of higher education. The research was exploratory and descriptive type, with participatory approach. The adapted questionnaires used were WCI - Work Capacity Index and NASA-TLX. It was noted that the work capacity had low and moderate, besides the workload was on high level. The results showed problems in the work environment that affect the public agent's health, and this is generating labor disabilities in the short/medium term. Ergonomic actions need to be developed, with the central administration Institution.

1 INTRODUCTION

Labor relations are imposed by the rules and regulations in force in the country, and these are aimed at balancing the interests of employers and employees (Guimarães 2009). The health, safety, and well-being of workers are the main aspects of this relationship. To ensure compliance with these legal provisions, the oversight exercise of some institutions has a fundamental assignment in the work system. For example, there are auditors of the Ministry of Labor and Employment, who do on-site verification visits at companies that have workers contracted by the Consolidation of Labor Laws (CLT) rules, and may even interfere with the functioning of the labor activities, if risks to the worker are identified (Brazil 2002).

The Ergonomic actions are carried out to meet the labor demands of public initiative in order to meet a social demand (Vidal 2003). However, most Brazilian companies have in their ergonomic actions a perspective that seeks primarily to comply with current legislation only to avoid suffering sanctions, rather than the real need of the worker.

In order to understand the current work situation, it was made a systematic literature review, as well as knowledge of the current legislation about the ergonomic conditions of work, such as Regulatory Norm nº17 of 1978 and its updates, the Brazilian labor laws as CLT and the legal rules of civil public agents of the direct administration, municipalities and federal public foundations. Also, the Ordinances and Normative Guidelines of the Ministry of Planning, Budget and Management were consulted, because they establish some parameters to address Ergonomics in the work environment of public service.

On the public institutions, productivity and efficiency are fundamental aspects and are increasingly demanded by its managers, following the trend of private initiative, since the public service must meet the demands of society with quality and agility. However, the implementation of ergonomic measures in public institutions is deficient due to lack of specific legislation. In order to regulate labor relations in the public service, the so-called Unified Judicial Regime (RJU) was established through Law nº 8.122 of December 11, 1990, but the issues of promotion and maintenance of workers' health remained pending regulations, preventing the effective implementation of these actions.

Although it has been observed a movement of the federal public administration pointing to changes in the attention of the health-work relation of the server, as affirmed Andrade (2009), in the public organizations, a great divergence between the existent problems and the presented solutions, which offer in general specific assistance activities (Ferreira et al. 2009). The Federal University of Pernambuco (UFPE), as a public institution, is also part of this reality. Still, Ricart (2011) points out that the ergonomic actions in the scope of the public institutions are still incipient and also did not reach the institutional maturation necessary to be considered effective.

By concept, a university is an environment of building knowledge and skills that contribute to the sustainability of society, and from its basic vocational training objectives, it acts in the generation of new knowledge and dissemination of this knowledge. Thus, it is essential to study the ergonomic conditions of this productive environment.

The Nucleus of Physical Education and Sports of UFPE (NEFD-UFPE) is responsible for the development of physical activities for the academic community and the surrounding community, and is also committed to promoting actions to improve the health of the worker, which can reduce rates of illness, work-related accidents and disability pensions.

In this context, this research aims to contribute to a better knowledge of the organizational functioning of the public administrative service of an department of physical education and sports of a IFES - Federal Institution of Higher Education regarding the ergonomic organizational aspects, from the perspective of its technical/administrative employees/public agents. The aim of this study was to use WCI and NASA-TLX tools that identify the workload and work capacity index, to assess the health of the federal public agent who works at a IFES.

2 ERGONOMICS IN THE CONTEXT OF THE PUBLIC SERVICE OF A FEDERAL INSTITUTION OF HIGHER EDUCATION GETTING STARTED

In some organs or entities of the direct or indirect Public Administration, although there is no specific legislation that determines the adoption of ergonomic measures, there are some actions directed at the public agent health. For Ricart (2011), the absence of a national health policy for the server elaborated by the Brazilian State, and also the lack of an information system, made it difficult to stipulate the demands, and led the elaboration of server health policies with diverse criteria and organizations, which resulted in divergent attendances, diagnoses and procedures.

Legal labor law is a global concern and its principles and fundamental rights are defined by the International Labour Organization (ILO). The ILO is an agency of the United Nations (UN), and aims to guide actions to promote the dignity of work.

In agreement with the ILO, the Brazilian legal system dealing with workers' health is very extensive, since it covers the Federal Constitution of 1998 (CF), the Consolidation of Labor Laws of 1943 (CLT), the Organic Health Law of 1990 (LOS) that creates the Unified Health System (SUS), and goes to the Regulatory Norms - NR of the Ministry of Labor and Employment, until ministerial decrees and ordinances, but only those regulations and norms can not ensure an adequate work environment Effective interventions are needed to ensure compliance with these ordinances, as well as frequent studies and surveys that may contribute to the advancement of these laws.

Work environments that do not offer an ergonomic suitability to workers cause occupational diseases, even if they represent the lowest percentage of occupational accidents, are factors that keep workers away from their activities, causing a decrease in productivity and directly affecting the worker's quality of life. In this context, some strategic measures should be considered to benefit workers, who will suffer less with occupational diseases, as well as employers, who will have their full manpower.

In all these measures, there is a multiplicity of actions and decisions that must be taken by the manager in the job planning and deployment of tasks, in order to provide a healthy working environment for their employees, adapting the workplace to satisfactory standards of comfort and safety.

3 METHOD

The initial search strategy was to make a documentary analysis of the legislation of the Health Care Policy of the Server. After completion of this phase, it were applied systematic procedures for quantitative and qualitative identification of the real working conditions of federal public agents.

The tools used were: the adapted questionnaire of the WCI – Work Capacity Index; and the adapted NASA-TLX (National Aeronautics and Space Adiministration/Task Load Index) questionnaire. All these tools have been applied to the interviewee in the same day and sequentially.

The WCI questions/inquiries are geared to the physical and mental demands of the work developed, and the health status of the worker. With the application of WCI, it is possible to identify, at an early stage, if there is workers and work environments that need support measures (Tuomi et al. 2010).

The results of the WCI questionnaire responses can reach a score ranging from seven to 49 points, and this number indicates the worker's perception of their own ability to work. According to the classification of results, measures are suggested such as restoring, improving, supporting or maintaining work capacity, respectively according to the following classifications: low, moderate, good or optimal for the capacity of the work.

The choice of the WCI was relevant to this study because it was observed that more than ¾ of the respondents has more than 45 years old, and according to Tuomi (2010), the factors that lead to the decrease of the work capacity begin to accumulate in the "middle age" and are observed in workers aged around 45 years. The ability to work is the basis of well-being, and this will not remain satisfactory if there is no search for the balance of the factors that influence it.

As for measuring the workload, there was the option to adopt the NASA-TLX questionnaire, because it is a tool that allows subjectively, through the personal perception of the interviewee, to obtain a general workload score based on the measurement of some components related to work.

The NASA-TLX method uses six factors to define a multidimensional value of overall workload. In order to clarify the interviewees on the dimensions considered, the definition of each component was explained in the questionnaire.

The questionnaire was structured in three blocks, the first of them focused on surveying the sociodemographic characteristics of the interviewees; the second block of questions made a comparison between the influencing factors in the workload, in which the participant chooses one of each pair of factors that is most significant in the development of his work, in a total of fifteen comparisons between pairs of factors. In the third block of the questionnaire, the interviewee chooses a point on a continuous scale of 15 cm, which is relative to the level of influence or contribution of each of the factors to the workload. NASA-TLX was adapted and validated by Diniz & Guimarães (2004), and this adaptation had the objective to facilitate the filling of the questionnaire and to increase the possibilities of analysis.

All technical-administrative workers of NEFD-UFPE, in a total of fourteen, who in the last twelve months have worked at the NEFD-UFPE and who perform or have performed their activities at three different sectors of the same supplementary department, were invited to voluntarily participate in this research during a time interval of work. At this department, the minimum level of education required for the positions currently occupied is elementary education, at the positions of Cleaner, Office porter, Administrative helper, in sequence the requirement is of High school, with the positions of Administrative assistant and Nursing technician, and goes to the upper level, in charge of Doctor; with four, nine and one worker respectively in each level, so this study covers various levels of schooling. The objectives of the research, as well as instructions for filling, have been clarified by a brief presentation of the survey, conducted in expository form and also during the beginning of the questionnaire. The data were collected in a period of 15 days in two shifts, morning and afternoon, during the first half of 2016. To validate the questionnaires, it is required that they be completely answered; therefore, as there was no absence of answers in any of the fourteen questionnaires, all of them were validated and considered for this research.

4 WCI AND NASA-TLX QUESTIONNAIRES RESULTS

To compose the analysis of the results of the NASA-TLX questionnaire, we counted how many times each workload component was mentioned in the choice between the pairs, and this characterized the weight of each of the factors. The number of times of each component mentioned was multiplied by the intensity marked on a continuous scale from zero to fifteen centimeters.

The results of the six components were added, which presented the workload in the worker's perception, as shown in Figure 1. The workload can range from zero to fifteen.

As can be seen, a quasi-uniform trend analysis is observed in which twelve of the fourteen respondents support a workload that is well over half possible, with some approaching near the maximum of the workload. Considering that the closer to the 15 score, the higher the workload, so we have established that loads above 7.5 have a high workload.

Only two of the interviewees had presented load values below the possible half (7.5). These workers are in opposite situations: while one has the

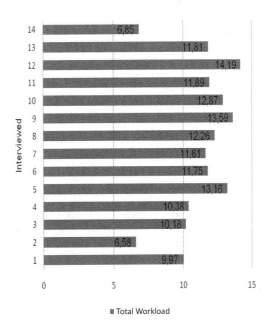

Figure 1. Workload. Note: Standard deviation of 2.27. Source: Author.

shortest period of service in the department (one year and eight months) and the highest level of education required for the job, and has the temporal demand as the main factor influencing his workload, the other one has the longest service in the sector (36 years) with the lowest level of education required for the job, and the factors that most influence him are physical demand and effort, representing more than half of his work load.

After analyzing the questionnaires in which the capacity index could return in four conditions: low, moderate, good and optimal, we have seen that, according to results, Tuomi et al. (2010) recommended that for low capacity index, this capacity must be restored, already moderate capacity indices indicate the need to improve worker capacity, good index needs to be supported and finally the optimal index needs to maintain the capacity for work.

Other assessments associated with WCI can help to identify the factors that influence the work capacity index. Only two of the interviewees presented an excellent work capacity index, four obtained a good index, therefore less than half of the interviewees are with index that we can consider acceptable; more than half (eight out of a total of fourteen) had low or moderate indexes, which needs support to evolve this work capacity index, without forgetting the others, because if there is no follow-up of these workers, the index that was considered acceptable may worsen and becomes an alert index.

The work capacity indexes of the fourteen participating workers, with results that could range from low, moderate, good until optimal, are shown in Figure 2.

The high demand, along with the reduction of staff with poorly defined work processes, promote an overload which was identified with the application of NASA-TLX questionnaire. Regarding the time demand factor, the workers feel pressured to perform their tasks and appointments in adverse conditions, which is characterized by excess demand and few streamlined procedures.

Measures to support those with low capacity for work should be adopted in order to avoid temporary or permanent disabilities, according to Tuomi et al. (2010). Many of these workers who have their ability to work compromised, presenting low ability, may thunder up unable to work in short period of time.

With the lack of planning, evaluation of working conditions and systematization of tasks and productive processes, workers suffer to achieve a good development of activities, and increasingly distance themselves from the quality of life and work, transforming the well-being at work in rare or almost intangible moments.

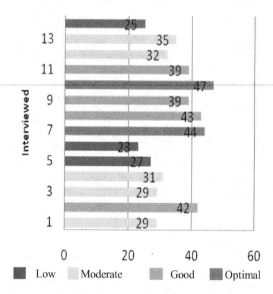

Figure 2. Work Capacity Index. Note: Standard deviation of 7,71. Source: Author.

In view of these results, it is essential that the institution, through its central administration, make use of the information obtained in this research in order to undertake the commitment to improve measures and promote ergonomic actions, as well as follow the effective application of health care standards of the public agent, to raise the quality of the service provided and to prosper the humanization of the work of federal civil agents.

5 FINAL CONSIDERATIONS

This study aimed to identify the workload and work capacity in which the federal public agent is inserted, in order to contribute to the policy of Attention to Health and Safety of the worker, their working conditions and circumstances, thust raise the question of what measures can be effective to improve the well-being of public agents at their work environment and extend their social life.

In applying the proposed method, based on the participatory approach of macroergonomics, which focuses on the human-organizational interface, taking into account the work process and the organizational form and is structured in four elements that form the technological, personal, external environment subsystems and organizational architecture. The interactions and harmony of these elements contribute to the development and quality of work activities, workers' safety and health and their productive results, according to Hendrick & Kleiner (2006). By attributing to the actors an effective participation in the process, it is allowed that the knowledge is constructed and evolves from the interrelation of the academic/scientific knowledge with the experience of the work and the practice.

Based on the evaluation of the Workload and Work Capacity, an excessive number of workers with a high workload was identified, in which the main factors contributing to this load are Frustration and Temporary Demand. Yet as a result, the WCI questionnaire showed that more than half of the participating workers are at low or moderate work capacity. These two indicators warn that there are problems in the work environment that are directly affecting the health of the server, which can affect short and medium-term work disabilities, such as dismissals, accidents and even early retirements, in addition to leaving the working environment pernicious.

In view of these facts, it points out that ergonomic actions need to be developed, but to ensure their development, implementation and follow-up of these actions, it is necessary the commitment of the Central Administration of the Institution.

REFERENCES

Andrade, E. T. 2009. *O Processo de Implantação da Política de Atenção à Saúde do Trabalhador em Instituições Públicas Federais: o desafio da integralidade* (Master's dissertation). Escola Nacional de Saúde Pública Sérgio Arouca. Rio de Janeiro.

Brazil. 6 December 2002. *Lei nº 10.593 - Dispõe sobre a reestruturação da Carreira Auditoria do Tesouro Nacional, que passa a denominar-se Carreira Auditoria da Receita Federal - ARF, e sobre a organização da Carreira Auditoria-Fiscal da Previdência Social e da Carreira Auditoria-Fiscal do Trabalho, e dá outras providências.*

Diniz, R. L. & Guimarães, L. B. M. 2004. Avaliação da carga de trabalho mental. In L. B. M. Guimarães (ed.), *Ergonomia cognitiva*. Porto Alegre: FEENG.

Ferreira, M. C. et al 2009. Gestão de Qualidade de Vida no Trabalho (QVT) no Serviço Público Federal: o descompasso entre problemas e práticas gerenciais. *Psicologia: Teoria e Pesquisa* 25 (3): 319–327.

Guimarães, M. C. 2009. Transformações do trabalho e violência psicológica no serviço público brasileiro. *Rev. bras. Saúde ocupacional* 34 (120): 163–171.

Hendrick, H; Kleiner B. 2006. *Macroergonomia: uma introdução aos projetos de sistemas de trabalho*. M.C. Vidal & J.R. Mafra (trad.). Rio de Janeiro: Virtual Científica.

Ricart, S. L. S. I. 2011. *Avaliação e Controle de Ações Ergonômicas no Serviço Público Federal: O Caso da Fiocruz – RJ* (Master's dissertation). Federal University of Rio de Janeiro.

Tuomi, K. et al. 2010. *Índice de capacidade para o trabalho*. F. M. Fischer (trad.). São Carlos: EdUFSCar.

Vidal, M. C. 2003. *Guia para Análise Ergonômica do Trabalho (AET) na empresa: uma metodologia realista, ordenada e sistematizada*. Rio de Janeiro: Editora Virtual Científica.

Exposure to psychosocial risks in footwear industry workers: An exploratory analysis

D. Sousa
IPMAIA, Polytechnic Institute of Maia, Avioso S. Pedro, Maia, Portugal

M.M. Sá
CIDESD—ISMAI, University Institute of Maia, Avioso S. Pedro, Maia, Portugal

R. Azevedo
CIDESD—ISMAI, University Institute of Maia, Avioso S. Pedro, Maia, Portugal
ALGORITMI—University of Minho, Braga, Portugal

O. Machado, J. Tavares & R. Monteiro
CATST—Technical Support Centre for Work Safety, ISMAI, University Institute of Maia, Avioso S. Pedro, Maia, Portugal

ABSTRACT: The footwear industry has showed continuous development, imposing as an essential industry in Portuguese social and economic growth. Three shoe factories in northern Portugal were subjected to psychosocial risk assessment, using COPSOQ II. Results show that diverse psychosocial factors are detrimental to workers' health and well-being, namely: work pace, cognitive demands, emotional demands, influence and job insecurity. This negative exposure requires urgent psychosocial and organizational interventions in order to reduce the malicious effects of the work context on employees' health. The results provide meaningful and important information about the footwear industry specific psychosocial risk factors and their impact on health and well-being. Preventive measures should be taken in order to improve health at work.

1 INTRODUCTION

Nowadays there has been severe changes in the working environment, in work processes with an increase in working time, ageing in active population, changes in organizational structures, increased intensity and workload, and workers have to deal with growing imperatives of demands and performance (EU-OSHA, 2010).

In the field of Occupational Safety and Health, the psychosocial risks have imposed as a critical field of evaluation and study, due to its repercussions at organizational and individual levels (Moncada, Pejtersen, Navarro, Llorens, Burr, Hasle, & Bjorner, 2010; Salanova, Martínez, Cifre & Llorens, 2009). Accordingly to ILO (1984, p. 3) psychosocial factors are defined as the "interactions between and among work environment, job content, organizational conditions and workers' capacities, needs, culture, personal extra-job considerations that may, through perceptions and experience, influence health, work performance and job satisfaction". Over the past decade increasing attention has been devoted to the study of psychosocial work characteristics (e.g. work pace), work-related states (e.g. commitment) and its implications on workers' health (Clausen, Christensen & Borg, 2010). The European Agency for Safety and Health at Work (2014) estimates that the cost on medical treatment and production losses due to psychosocial risks reaches 240 billion Euros per year.

The Portuguese footwear industry, has demonstrated a continuous growth, exporting more than 81.6 billion pairs of shoes with a total value of more than 1.9 million euros (APICCAPS, 2017). In 2017 there are 2294 companies with a growth production of 2222 million, with a total of 41418 employees.

The campaign for the improvement of work conditions in the footwear industry launched by the Portuguese Authority for Working Conditions (2013) defined the strategic aim of promoting the continuous improvement of working conditions in this industry. In this following, ACT (2013) also underlined the noxious effects that psychosocial risks impose at psychological, physical and social levels. These risks result from the relation between work facets (e.g. conception, organization, management), social interactions and workers' individual characteristics (e.g. skills, needs).

The campaign brochures emphasized occupational stress, harassment (moral and sexual) and violence at work, calling for the need to address this sector actively (ACT, 2013).

A poor psychosocial work environment can lead to illness due to long-term exposure to poor working conditions. An extensive number of papers have been published, showing an association between psychosocial factors and health outcomes, but few have focus on the footwear industry. The studies available refer high work rhythm, lack of recognition by management, difficulty to express feelings, job insecurity, low support from supervisors and co-workers, low job satisfaction and stress (Silva, Silva & Gontijo, 2017; Afonso, 2013).

The focus on the psychosocial risks and the lack of studies in footwear sector, points towards the need for evaluation of the psychosocial risks present in the work environment. The first step to prevention is the evaluation and identification of work demands/stressors that affect workers (Salanova, Cifre, Martínez, & Llorens, 2007; Leka, Cox, & Zwetsloot, 2008). As stressed by Silva (2006, p. 5) "The identification of these risks is not an end itself neither a legal imposition. It is an effective signaling strategy, necessarily preliminary to adequate and effective prevention. The law requires a logical sequence: first evaluate and then prevent". Therefore, it is our aim to identify the prevalence of the diverse aspects of the psychosocial environment imposing as risks to the health and well-being of footwear industry workers.

2 MATERIALS AND METHODS

The psychosocial work characteristics were measured with the short version of the COPSOQ II (Kristensen, & Borg, 2005) validated for Portuguese version by Silva (2006). The evaluation comprised eight factors: Demands at work, Work organisation and job contents, Interpersonal relations and leadership, Work-individual interface, Values in the workplace, Health and well-being and Offensive behaviour. In addition, 23 dimensions of the psychosocial work environment were evaluated: quantitative demands, emotional demands, influence, possibilities for development, degree of freedom at work, meaning of work, commitment to the workplace, predictability, quality of leadership, social support, feedback, social community at work, job insecurity, and job satisfaction. For most items, five response categories were used either with intensity (from "to a very large extent" to "to a very small extent") or frequency (from "always" to "never/hardly ever").

The sample included 240 workers, 20.3% males and 79.7% females, age range from 20 to 57 years old, mean age of 41.1 years. The majority worked on production (89.4%) and 10.6% were administrative staff.

For the data collection, questionnaires were administered among three companies in the footwear industry, located in the northern Portugal. A total of 240 workers delivered the questionnaire completely answered.

The relevance and objectives of the study were explained to the participants and requested informed consent. Questionnaires and informed consent were delivered in separate envelopes, to be closed after its completion, thus ensuring the confidentiality and anonymity of respondents.

The data was processed with IBM SPSS Statistics 24 software. For initial analysis a descriptive analysis to characterize the sample and the prevalence of psychosocial risks was performed. Also, independent t-test was subsequently conducted to examine whether there was a difference in our population scores and general industry samples.

3 RESULTS

The prevalence and presence of working context factors that pose as a risk to health or that are intermediate or in a favourable situation to workers' health were identified. The factors that represented major impairment to their well-being were: work pace, cognitive demands, emotional demands, influence and job insecurity. In fact, the health impairment association is highly manifested: job insecurity represents a risk to health for 81% workers, work pace for 51%, cognitive demands for 56%, emotional demands for 36% and influence for 41% (Table 1).

In turn, the analysis of the effect of the indicators of psychosocial risks reveals low job satisfaction (49%), 44% of workers perceive repercussions on their general health, 30% sleep difficulties, 21% depressive symptoms, 32% symptoms of burnout, and 28% occupational stress (Table 1).

In addition, we must underline the positive characteristics that should be monitored, in order to sustain its resourceful character, buffering the adverse effects of work demands. In this way, the factors in a favourable situation are possibilities for development (54%), role clarity (92%), quality of leadership (45%), social support from supervisors (51%), trust regarding management (63%) low offensive behaviour (93%) and low work-family conflict (53%).

At last, referring to workers' personal resources, results reveal the presence of positive protector factors as commitment (53%), meaningful work (87%) and self-efficacy (72%).

Comparison results with general industry sector are presented in the following table.

Table 1. Percentage of exposure indicators.

	Risk for health %	Inter-mitate %	Favourable for health %
Demands at work			
Quantitative demands	12	42	46
Work pace	51	34	15
Cognitive demands	56	19	24
Emotional demands	36	35	29
Work organisation job contents			
Influence	41	31	28
Pos. Development	16	30	54
Meaning of work	11	11	87
Commitment	25	22	53
Interpersonal relations and leadership			
Predictability	34	30	37
Recognition	28	31	41
Role clarity	4	5	92
Quality of leadership	23	32	45
Social support	22	27	51
Work-individual interface			
Job insecurity	81	7	12
Job satisfaction	49	39	12
Work-family conflict	25	20	53
Values in the workplace			
Trust management	12	25	63
Justice	25	38	37
Social inclusiveness	7	25	68
Health and well-being			
General health	44	38	19
Burnout	32	37	30
Stress	28	36	36
Sleeping troubles	30	32	38
Depressive symptoms	21	37	41
Self-efficacy	5	23	72
Offensive Behaviour			
Offensive Behaviour	2	5	93

4 DISCUSSION

Diminished health among workers has been associated with psychosocial working variables such as long working hours and high job demands (Niedhammer, Chastang, Sultan, Taïeb, Vermeylen, & Parent-Thirion, 2012; Niedhammer, Sultan, Taïeb, Chastang, Vermeylen, Parent-Thirion, 2012; Gil-Monte, 2012).

Slany, Scüttte, Chastang, Parent-Thirion, Vermeylen and Niedhammer (2014) with a sample of 32708 European workers, observed that 30% workers perceived high insecurity at work. In addition, the results of Survey on Working Conditions in Continental Portugal, reported that 17% of workers felt job insecurity (ACT, 2016). Also, Silva, Silva and Gontijo (2017) observed that 54.6%, of women and 59% men, perceived job insecurity in footwear industry at Brazil. These results underline a higher level of perceived job insecurity in our sample (mean 4.15) revealing significant differences with the Portuguese general industry mean of 3.57 (Silva, 2006) ($p = 0.000$) (see Table 2).

In turn, high work pace is present in 51% of the sample, which is congruent with data from ACT (2016), having observed that 68.4% of workers perceived their work as elevated. Emotional and cognitive demands although presenting higher levels of risk for health didn't evidence significant differences from the general industry mean (see Table 2). In contrast, 28% of footwear industry workers stated having influence over work tasks, being this data distant from results reported for the general population, where 64% of employees declared having the ability to choose and alter tasks (ACT, 2016).

Table 2. T-test comparison with general industry sector.

	General Industry Mean ± Std. Dev.	Foot Industry Mean ± Std. Dev	T-Test	p-value
Quantitative demands	2.51 ± 0.91	2.52 ± 0.87	0.09	0.993
Work pace	3.34 ± 0.95	3.58 ± 1.21	3.01	0.003
Cognitive demands	3.67 ± 0.74	3.60 ± 0.80	−1.34	0.183
Emotional demands	3.12 ± 1.18	3.12 ± 1.31	−0.01	0.992
Influence	2.86 ± 1.00	3.19 ± 1.32	3.76	0.000
Pos. devlopment	3.77 ± 0.83	3.67 ± 1.02	−1.49	0.137
Meaning of work	4.07 ± 0.77	4.14 ± 0.63	1.78	0.08
Commitment	3.57 ± 0.92	3.46 ± 1.19	−1.42	0.16
Predictability	3.41 ± 0.96	3.08 ± 1.11	−4.53	0.000
Recognition	3.79 ± 0.87	3.26 ± 1.26	−6.37	0.000
Role clarity	4.30 ± 0.71	4.64 ± 0.81	6.49	0.000
Quality leader	3.57 ± 0.88	3.40 ± 1.15	−2.19	0.030
Social support	3.26 ± 0.96	3.48 ± 1.26	2.72	0.007
Job insecurity	3.57 ± 1.35	4.15 ± 1.10	7.98	0.000
Job satisfaction	3.58 ± 0.75	3.48 ± 0.95	−1.52	0.130
W-F conflict	2.44 ± 1.02	2.63 ± 1.24	2.29	0.023
Trust manager	3.61 ± 0.64	3.85 ± 0.99	3.76	0.000
Justice	3.41 ± 0.84	3.21 ± 1.10	−2.76	0.006
Social inclusiv.	4.02 ± 0.82	4.05 ± 1.03	0.46	0.648
Self-eficcacy	3.88 ± 0.71	4.02 ± 0.88	2.45	0.015
General health	3.44 ± 0.97	3.23 ± 0.89	−3.54	0.000
Stress	2.67 ± 0.91	2.90 ± 1.07	3.30	0.001
Burnout	2.68 ± 0.99	3.03 ± 1.11	4.76	0.000
Sleeping troub.	2.47 ± 1.09	2.88 ± 1.24	4.97	0.000
Depressive symp.	2.27 ± 0.93	2.73 ± 1.22	5.70	0.000
Offensive Beh.	1.19 ± 0.45	1,20 ± 0.37	0.426	0.670

When comparing with data from general industry it's observed a significant difference between means (p. = 0.000). Considering job satisfaction, there is also a discrepancy between our data (12%) and the 89.9% of general workers that are satisfied with their work (ACT, 2016). Also, although our results show employees more dissatisfied (49%) than workers from the general industry (Silva, 2006), the difference is not statistically significant.

Finally, concerning the health problems related to work, our findings are congruent with ACT (2016) data, namely: 46.9% of workers referred that in the last 12 months had suffered at least one health problem due to work, 10% depression or anxiety, sleep problems (9%). However, stress, burnout, sleeping problems and depressive symptoms present significantly higher means comparatively to general industry (p = 0.000) (Silva, 2006).

5 CONCLUSION

The present study indicates the presence of psychosocial risks at footwear industry workers and emphasizes its harmful effects on workers' health.

Our results identified positive characteristics that must be monitored, in order to sustain its resourceful character: possibilities for development, role clarity, quality of leadership, social support from supervisors, trust regarding management, low offensive behaviour and low work-family. The following psychosocial risks were detected: work pace, cognitive demands, emotional demands, influence and job insecurity. Comparatively to general industry, footwear industry employees present significant higher means of stress, burnout, sleeping problems and depressive symptoms.

Preventive measures should be implemented at organizational and individual levels. This study provides useful information for development of prevention strategies in order to improve better psychosocial working conditions.

Further studies may be needed to better understand the dynamic relation and effect between working facets, worker characteristics and well-being.

REFERENCES

ACT. 2016. *Inquérito às Condições de Trabalho em Portugal Continental: Trabalhadores/as*. Lisboa: ACT.

ACT. 2013. *Campanha para a melhoria contínua das condições de trabalho na indústria do calçado*. Lisboa: ACT.

Afonso, L. 2013. *Comparativo da prevalência de sintomas músculo-esqueléticos em trabalhadores de duas empresas da indústria do calçado: setor da costura*. Dissertação apresentada para obtenção do grau de Mestre Engenharia de Segurança e Higiene Ocupacionais. Porto: Faculdade de Engenharia da Universidade do Porto.

APICCAPS. 2017. *Facts & Numbers 2017*. Maia: APICCAPS.

Clausen, T., Christensen, K.B., & Borg, V. 2010. Positive work-related states and long-term sickness absence: a study of register-based outcomes. *Scandinavian Journal of Public Health*, 38(3 Suppl), 51–58.

EU-OSHA—European Agency for Safety and Health at Work. 2010. *European Survey of Enterprises on New and Emerging Risks: Managing safety and health at work*. Luxembourg: Publications Office of the European Union.

EU-OSHA—European Agency for Safety and Health at Work. 2014. *What are stress and psychosocial risks? Healthy workplaces manage stress*. Luxembourg: European Agency for Safety and Health at Work.

Gil-Monte, P.R. 2012. Riesgos psicosociales en el trabajo y salud ocupacional. *Revista Peruana de Medicina Experimental y Salud Publica*, 29(2), 237–241.

ILO 1984. *Psychosocial factors at work: Recognition and control*. Geneva: International Labour Office.

Kristensen, T., Hannerz, H., Hogh, A. & Borg, V. 2005. The Copenhagen psychosocial Questionnaire—a tool for the assessment and improvement of the psychosocial work environment. *Scandinavian Journal of Work, Environment, and Health*, 31(6), 438–449.

Leka, S., Cox, T. & Zwetsloot, G. 2008. The European Framework for Psychosocial Risk Management (PRIMA-EF). In S. Leka & T. Cox, (eds.). *The European Framework for Psychosocial Risk Management: PRIMA-EF* (1st ed.) (pp. 1–16). Nottingham: WHO Publications.

Moncada S., Pejtersen J., Navarro, A., Llorens, C., Burr, H., Hasle, P., Bjorner, J. 2010. Psychosocial work environment and its association with socioeconomic status. A comparison of Spain and Denmark. *Scandinavian Journal of Public Health*, 38, 137–148.

Niedhammer, I., Chastang, J., Sultan-Taïeb, H., Vermeylen, G., Parent-Thirion, A. 2012. Psychosocial work factor and sickness absence in 31 countries in Europe. *European Journal of Public Health*, Vol, 23, 4, 622–628.

Niedhammer, I., Sultan-Taïeb, H., Chastang, J., Vermeylen, G., Parent-Thirion, A. 2012. Exposure to psychosocial work factors in 31 european countries. *International Journal of Occupational and Environmental Health*, vol. 62, 196–202.

Salanova, M., Cifre, E., Martínez, I.M. & Llorens, S. 2007. *Caso a caso en la prevención de riesgos psicosociales: Metodología WONT para una organización saludable*. Bilbao: Lettera Publicaciones.

Salanova, M., Martínez, I.M., Cifre, E., & Llorens, S. 2009. La salud ocupacional desde la perspectiva psicosocial: aspectos teóricos y conceptuales [Occupational health from the psychosocial perspective: Theoretical and conceptual aspects]. In M. Salanova (Dir.), *Psicología de la Salud Ocupacional* [Occupational Health Psychology] (pp. 27–62). Madrid: Editorial Síntesis.

Silva, C. 2006. *Copenhagen Psychosocial Questionnaire—COPSOQ: Versão portuguesa*. Aveiro: Análise Exacta.

Silva, J., Silva, L., & Gontijo, L.A. 2017. Relationship between psychosocial factors and musculoskeletal disorders in footwear industry workers. *Production*, 27, 1–13.

Slany, C., Scüttte, S., Chastang, J., Parent-Thirion, A., Vermeylen, G., & Niedhammer, I. 2014. Psychosocial work factors and long sickness absence in Europe. *International Journal of Occupational and Environmental Health*, vol. 20, 16–25.

Inclusive, interdisciplinary and digital-based OSH resources: A booklet for students

F. Rodrigues & F. Antunes
RISCO, Department of Civil Engineering, University of Aveiro, Aveiro, Portugal

A.M. Pisco Almeida, J. Beja & L. Pedro
DigiMedia, Department of Communication and Art, University of Aveiro, Aveiro, Portugal

M. Clemente, R. Neves & R. Vieira
Department of Education and Psychology, University of Aveiro, Aveiro, Portugal

ABSTRACT: This paper addresses Occupational Safety and Health (OSH) under the scope of the Erasmus+ project "Mind Safety—Safety Matters!" which intends to increase students' awareness of OSH in their daily routine and professional life. Following an integrated approach of Health Education and also a Design for all perspective, an interactive booklet for young students (14–18 years old) is under development, covering a variety of topics divided into general OSH topics and OSH hazards, accessible for blind and low vision students. Developing this innovative educational resource addressing a large diversity of usages has shown to be a challenge not only in what concerns the design of the resources but also considering the nature of the learning strategies and contents included in the booklet. "Mind Safety—Safety Matters!" represents thus an opportunity to explore OSH topics in the education field, developing inclusive (for blind and/or low vision students), multimedia and interdisciplinary contents.

1 INTRODUCTION

Analysing the statistics of accidents at work by age collected by the Portuguese Authority for Working Conditions in 2016, it is possible to verify that almost 27% of the injured workers have less than 34 years old. Such percentage, although it may seem low, needs to take into account the proportion of the total number of workers and the incidence rate, which is above the average of the remain age class (EU-OSHA, 2006).

Research also indicates a set of challenges, constraints and lack of consistency in the delivery of OSH in schools:

"While there is evidence that school-based educational programs have the potential to increase knowledge about safety (Lerman et al., 1998; Linker et al., 2005), it cannot be assumed that this knowledge will translate to safe behavior and ultimately, injury reduction" (Pisaniello, Stewart & Jahan, 2013, p. 53). These gaps include lack of information and training, problems concerning the availability/quality of educational resources and poor results on effectiveness concerning how can they be prevented and the long term effects (Pisaniello, Stewart & Jahan, 2013).

An occupational hazard is a threat that can occur in a given workplace. According to Liu & Tsai (2012) and Bellamy (2014), there are 3 main reasons why occupational hazards keep happening: occasional occurrence of unpredictable and random accidents, unsafe conditions and unsafe behaviour, mostly related to the worker's education, age and experience, since the majority of the accidents are with single fatalities.

OSH concerning young workers is a subject that needs not only the attention of employers and policy makers but also educators, since it is only after they start working that they learn the notions of safety. Moreover, people with disabilities face a wide range of physical and virtual barriers, and technology can be a powerful tool in supporting education and inclusion, improving their quality of life (UNESCO, 2013).

In this paper we will focus on one of the key dimensions of "Mind Safety—Safety Matters!" project, which is the development of "OSH! What a bright idea", an interactive booklet for students that addresses the issue of Occupational Safety and Health (OSH) among youth. The main purpose of the interactive booklet is to increase students' awareness of OSH in their daily personal and professional lives and, in the long term, to raise levels of safety at work and to decrease the number of occupational accidents and diseases among young workers.

2 DESIGN AND DEVELOPMENT

Following an integrated approach of Health Education (EU-OSHA, 2004), which includes risk education, health promoting schools and safe learning environment dimensions, an interactive booklet concerning the field of OSH is being created. The booklet is aimed at students from 14 to 18 years old.

The development of the booklet follows the methodology of Developmental Research, directed towards the creation of innovative instructional environments and innovative design and development process (Richey, Klein & Nelson, 2004; Richey & Klein, 2005). In the specific case of "Mind Safety—Safety Matters!" the research problem is focused on the development a digital-based educational resource that also fulfils the usability needs of blind and low vision students while ensuring its nuclear educational approaches (see 2.3).

The next few sections describe the booklet creation process concerning the target audience preferences, the booklet's themes, structure, contents and activities.

2.1 Youth preferences and the design of the interactive booklet

To better understand the main digital visual environment and media uses and preferences of our target population, an online questionnaire was applied to a group of 12 students in order to develop further knowledge about their preferences of school activities, access to social networks, YouTube channels that they follow and even their daily life routine.

The results revealed preferences to music contents and humour videos typically performed by teenager youtubers, such as Wuant or Pewdiepie. The results also showed preferences to school activities like school trips and video presentations, and daily access to a range of social networks (Figs. 1–2) using predominantly their smartphones. These results confirmed what several studies reveal about the media preferences of this particular population segment (Regulatory Entity for Social Communication, 2016).

Based on these results, the team recorded 8 original videos performed by a teenager with funny short stories of his daily life and occupational situations. The videos' main purpose is to act as triggers for each OSH theme, trying to create a language that sounds familiar to the target audience. In total, 16 videos were produced, which includes 8 regular videos and 8 audio described video versions for blind or with low vision students.

The visual language and interaction intends to answer to their preferences too, using images and topics with small blocks of content (Figs. 3–4),

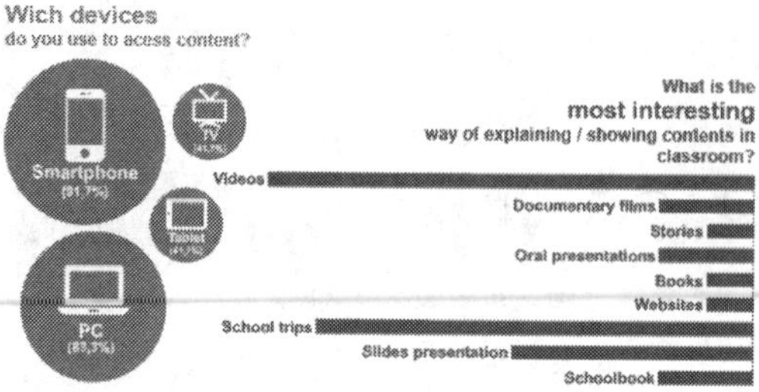

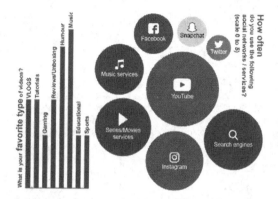

Figure 1. Results of favourite type of videos and most used social networks/services by survey respondents.

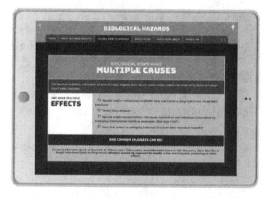

Figure 2. Results of most used devices and most interesting ways to explain contents in classroom context by survey respondents.

Figure 3. Screen example of Biological Hazards ("causes, effects, injuries" page).

enhancing the preference of access to short and in time information.

2.2 The interactive booklet themes, menu's structure and contents

The Occupational Safety and Health Administration (OSHA) considers the existence of six types of

Figure 4. Screen example of Biological Hazards ("video" page).

hazards in a workplace: Safety, Physical, Biological, Ergonomic, Chemical and Work Organization related (European Commission, 1996). Considering these OSHA references, "Mind Safety—Safety Matters!" selected the above-mentioned types of hazards adopting them as OSH themes. It was also considered relevant to adopt the designation Work Processes and include Safety issues in this theme's scope. Nevertheless, three other themes were assessed during the development of the booklet: General OSH topics, that comprises the general principles of risk prevention, safety and health concepts, protective equipment, safety signs, rights and responsibilities, emergency plans, history, OSH and Healthy Schools; and New Hazards and Risks that explores emerging hazard and risk situations, such as nanotechnology. In brief, the booklet covers the following 8 themes:

- General OSH topics;
- Chemical Hazards;
- Biological Hazards;
- Physical Hazards;
- Work processes Hazards;
- Ergonomic Hazards;
- Work organization Hazards;
- New Hazards and Risks.

Each theme has a common navigation structure and has been divided into six menus:

- Video: an original video production aimed to introduce each theme and inspired by survey respondents' preferences;
- What you need to know: key information about each theme;
- Causes, effects, injuries: the main characteristics of the selected hazards;
- Safety rules/wear: essential safety and health recommendations concerning each hazard context, including personal and collective protective equipment;
- Know more about: curiosities and/or advanced knowledge about theme;
- Hands on: set of varied activities about each theme.

2.3 *The booklet's "OSH! What a bright idea!": Educational strategies and activities*

All the educational contents are integrated into the school curriculum of the consortium countries of the "Mind Safety—Safety Matters!" project (Portugal, Spain, Romania, Netherlands, Czech Republic). Bearing in mind that the interdisciplinary approach became a requisite for developing OSH educational and/or training solution, to foster a safety and health culture among youth (Limborg, 2001; Rosen et al., 2011), the booklet also gives preference to this educational approach.

In addition to the interdisciplinary approach, the interactive booklet "OSH! What a bright idea" has drawn on a set of other educational approaches that influenced its structure, contents and the activities:

- Technology-based activities and active learning (OECD, 2016);
- Multimodal learning (Kress & van Leeuwen, 2006);
- Critical and creative thinking (Gregory, 1991; Morgensen, 1997; Simovska, 2005);
- Design for all (UN General Assembly, 1972).

The activities were designed to be carried out using the interactive booklet tools or in the classroom/outdoor contexts, filling a wide variety of educational strategies: team and project work; quizzes: role play and problem solving; moral reasoning using dilemmas; map drawing; digital checklist; field trips and outdoor context; digital storytelling; storyboard design; and, finally, interactive data and pictures.

2.4 *The booklet development and accessibility guidelines*

The interactive booklet is being designed and developed in a web-based approach, typically to be used with tablets in classroom context. The booklet development takes into consideration the adequate navigation of regular users and of blind and low vision users, following the application of Web Content Accessibility Guidelines (W3C, 2008) with implemented features like: text zoom in and zoom out, grayscale option, high contrast, colour invert, and highlight links. However, to ensure proper navigation several challenges have been overcome, at both the strategies and activities conceptualization and implementation, trying to keep the same contents in a Design for all perspective without separated versions of contents.

3 CONCLUSIONS

Developing educational resources able to address the diversity of usages and users while meeting the requirements of an innovative educational resource has a major influence not only in the design process of the resource but also in the nature of the learning strategies and activities. The booklet represents thus an opportunity to present an approach to OSH education that is mainly concerned with the design and development of inclusive solutions for blind and/or low vision students and with the development of multimedia and interdisciplinary contents throughout the different themes. These pillars allow an holistic view of the themes, unveiling different perspectives of OSH problems, challenging the students to present creative solutions with the contribution of different school subjects, enriching the discussion moments and contributing thereby to promote critical thinking skills, values and citizenship education. It is thus expected that the students actively engage in the activities and in the discussion, understanding the relevance of OSH themes in their daily and (future) occupational lives.

The project's impact on the participant organizations can be analysed so far in two dimensions: on the one hand, the international consortium dynamics and the process, and on the other hand the challenges and opportunities that mainstreaming OSH into education means today. The dynamics of the partners has further contributed to boost the discussion about the genuine need and benefits of joining different perspectives and national realities to improved OSH in the educational sector. Hence, this very particular process and outputs' objectives also impact on the innovation need, challenging the partners to present digital-based solutions and educational resources able to emphasize the relevance of OSH for young people in contemporary society.

ACKNOWLEDGMENTS

With the support of the Erasmus+ programme of the European Union, grant n.º 2015-1-PT01-KA201-013082.

REFERENCES

ACT, 2016. Autoridade para as Condições do Trabalho. Available at: http://www.act.gov.pt/(pt-PT)/CentroInformacao/Estatistica/Paginas/default.aspx.

Bellamy, L. Exploring the relationship between major hazard, fatal and non-fatal accidents through outcomes and causes. Safety Science 71, 93–103.

EC—European Commission, Guidance on Risk Assessment at Work, Luxembourg, 1996. Available at: http://osha.europa.eu/en/topics/riskassessment/guidance.pdf.

EU_OSHA, European Agency for Safety and Health at Work. (2004). *Mainstreaming occupational safety and health into education: good practice in school and vocational education*. Luxembourg: Office for Official Publications of the European Communities.

EU-OSHA (2006). E-fact 8 - A statistical portrait of the health and safety at work of young workers. Avaliable at: https://osha.europa.eu/sites/default/files/publications/documents/en/publications/e-facts/efact08/E-fact_08_-_A_statistical_portrait_of_the_health_and_safety_at_work_of_young_workers.pdf.

Gregory, R. (1991). Critical Thinking for Environmental Health Risk Education. In *Health Education Quarterly*, *18*(3), 273–284.

Jensen (Eds.) (2005). *The health promoting school: International Advances in Theory, Evaluation and Practice* (pp. 173–192). Copenhagen: Danish University of Education Press.

Kress, G. & Leeuwen, T. (2006). Reading images. The grammar of visual design. London: Routledge.

Limborg HJ. (2001). The professional working environment consultant—A new actor in the health and safety arena. In Human Factors and Ergonomics in Manufacturing, 11, 159–172.

Liu, H., Tsai, Y. A fuzzy risk assessment approach for occupational hazards in the construction industry. Safety Science 50, 1067–1078.

Morgensen, F. (1997). Critical thinking: a central element in developing action competence in health and environmental education. In *Health Education Research, Theory & Practice*, *12*(4), 429–436.

OECD (2016). Insights from the Talis-PISA link data: Teaching strategies for instructional quality. OECD.

Pisaniello, D., Stewart, S. & Jahan, N. (2013). The role of high schools in introductory occupational safety education—Teacher perspectives on effectiveness. In *Safety Science*, 55, 53–61.

Poniewozik, J. (2000, November). TV makes a too close call. Times, 20, 70–71. DOI: 10.15476/dona.v83n197.54590.

Regulatory Entity for Social Communication. (2016). *As novas dinâmicas do consumo audiovisual em Portugal 2016.* Retrieved from http://www.erc.pt/documentos/Estudos/ConsumoAVemPT/ERC2016_AsNovasDinamicasConsumoAudioVisuais_web/assets/downloads/ERC2016_AsNovasDinamicasConsumoAudioVisuais.pdf.

Rosen, M., Caravanos, J., Milek, D. & Udasin, I. (2011). An Innovative Approach to Interdisciplinary Occupational Safety and Health Education. In AMERICAN JOURNAL OF INDUSTRIAL MEDICINE, 54, 515–520.

Simovska, V. (2005). Participation and learning about health. In S. Clift and B.

UN General Assembly (1972). *Stockholm Declaration*.

UNESCO (2013). *Global Report Opening New Avenues for Empowerment ICTs to Access Information and Knowledge for Persons with Disabilities*. Paris: UNESCO.

When the working hours become a risk factor: The debate about 3 × 8h and 2 × 12h shifts

L. Cunha & D. Silva
Centro de Psicologia da Universidade do Porto, Porto, Portugal

I. Ferreira & C. Pereira
Faculdade de Psicologia e de Ciências da Educação da Universidade do Porto, Porto, Portugal

M. Santos
Centro de Psicologia da Universidade do Porto, Porto, Portugal

ABSTRACT: The interest in 8-hour and 12-hour shifts is not entirely recent; however, it earns an emergent status on the research hypotheses due to a conjuncture that leads more workers into such schedules, as well as due to the existence of evidence about their negative consequences on the workers' health. Based on the analysis of the change from an 8-hour to a 12-hour schedule in a Portuguese company, this study aims at discussing the factors and conditions considered as aggravated risks in 12-hour days. This study brings to light a resizing of the difficulties felt by the workers, both in their health and at work. Moreover, this study points out that the way each worker (immediately and in the long term) thinks through the sustainability of a 12-hour shift brings to debate the specificities of the work content, the demands of life outside work and the stage of the professional trajectory.

1 INTRODUCTION

1.1 The renewed interest in the organisation of work in 12-hour shifts

Today's estimates indicate that the "normal" working hours – 8 hours, Monday to Friday – is a condition encompassing an ever-decreasing number of workers, given the multiplication of schedules designed in such a way that contradicts the rhythms of life: shift schedules, night work, work that forces getting up before 5 a.m., long working hours, "split shifts" or at weekends.

Certain practices these days define the working hours, both in their extension and in their distribution throughout the 24h, so to favour either the organisations' continuous operation or the continuity of the work in the same teams. This choice is possible when the work is organised in rotating shifts, for instance in the systems 3 (teams) × 8h (hours) and 2 × 12h. Though it is not an entirely new option, the systems 2 × 12h enable an extension of the daily working hours (50% additional time), making it possible to reduce the working weeks to 3/4 days.

1.2 A commitment that does not "fit" everyone

In Europe, recent data (Eurofound 2016) indicate that shift workers report, more often than the typical "9 to 5" workers, that their work has an impact on health (1.9 more often) and that they work at high speed most of the time (1.8 more often). Despite this and other evidence, the tendency to extend the work shift is "encouraged" by European and national regulations, gradually less normative (Barthe 2009, 2015), in line with the time flexibility that defines the labour market over the last decades.

Given a legal framework that "legitimates" the increase of extra 4 hours in the working day, the 2 × 12h shifts are back in the choices about the time organisation of work, which can be fostered by the company's desire or by the workers' will. From the company's point of view, the 12-hour shifts are expected to facilitate the management of the demands posed by round-the-clock work, hence they support the continuity of the tasks. In turn, the workers face the 12-hour shifts as a better commitment to attend the demands of life outside work, due to the higher number of days off. However, this "apparent enthusiasm" (Barthe 2009) about the 12h shall not exclude from the agendas of the organisational decision makers an analysis from "the work point of view", which takes into consideration the work demands or the risks that are defined according to a variable schedule geometry. It is this approach in particular that will shed light on the commitments that each worker assumes by working 12h a day (the management of possibilities and boundaries) or the costs (in terms of health) the working hours impose.

1.3 The effects on health

The change of 3 × 8h by 2 × 12h systems became a systematic research-object by the end of the 1980s (Rosa et al. 1989). The purpose was threefold: analyse the system's levels of performance, safety and reliability; the workers' evolution in health and well-being (in the short, medium and long terms); or considering the management of the family and social domains (Tucker et al. 1996, Axelsson et al. 1998, Johnson & Sharit 2001, Loudoun 2008, Wagstaff & Lie 2011). Regarding health in particular, the research underlines the resizing (or the "amplification") of the difficulties the workers feel and the negative consequences that increase when the work lasts 12 hours a day (Barthe 2015, Vincent 2017).

Among researchers and social partners, Loudoun (2008) points out that it is precisely this awareness that causes the concerns related extended 12-hour schedules to emerge, in particular regarding the accumulation of fatigue and the risk of accidents or injuries in the short term and the negative consequences on health in the long term (Weibel et al. 2014, Cabon 2017). The decision to take on a 2 × 12h regime shall not be made exclusively based on the idea of a "unique solution", taken as the one that is valid in any given continuous operation context and for all the workers (Vincent 2017). Consequently, it is up to the researchers to state the reasons and the commitments regarding the length of the working hours, so that all the choices are made in "total awareness, and all the existing possibilities to ease the negative consequences may be thoroughly analysed" (Quéinnec et al. 1985 p. 110, free translation).

2 METHOD

2.1 Objective and the organisational context

The research took place in one of the production centres of a large agro-industry where the 3 × 8h shifts were recently replaced by 2 × 12h shifts. In this production centre, the change in the working hours resulted from an initiative presented by the workers themselves, stating that the 2 × 12h would give them more leeway for the work-life balance.

The company's decision to implement the 12-hour shift, after a negotiation phase, was considered as "experimental". For that reason, and in order to support this decision, a group of work psychologists was asked to study the transition process between schedules and the workers point of view regarding the 12h work experience.

It is still an on-going research that aims at analysing what is noticed in terms of impacts of the schedule on the work activity and on the productivity indicators, as well as on health, conferring visibility to a set of conditions which, according to the workers, may be risk factors of 8h and 12h work shifts.

2.2 Material, procedure and participants

In order to meet the company's request, a research and analysis plan was designed based on Work Psychology scientific framework. Regarding the working hours issue, the contributions from this scientific field are set within the interdisciplinary action of Activity-centered Ergonomics (Lacomblez et al. 2007).

For the analysis of the change in the working hours at this company, a context-based approach was favoured. Such approach was possible due to ergonomic analysis of work (EAW) (Lacomblez & Teiger 2007). The purpose was to get to know the work content, the constraints at stake and the risks aggravated with the new schedule, bearing in mind the commitment between the demands of the work activity and the health preservation (Barthe et al. 2004, Bessa & Lacomblez 2013, Knauth 2007). Consequently, in the research plan, two main data sources were used: qualitative analysis (work activity analysis based on observations and interviews) and quantitative analysis (questionnaire).

To begin with, the work analysis was sustained in a set of on-the-spot observations (in a total of 108 hours) and in the record of statements from workers from all the production lines. Once this body of knowledge about the activity was gathered, the next phase encompassed the interviews to a sample of workers ($n = 20$) from all the production lines and work teams.

The third research phase relied on the data collected during the first two phases and it included the creation of a questionnaire to be filled in online. The questionnaire was sent to all the workers from the production lines ($n = 92$) and the response rate was 66% (average age: 37 years old; average seniority: 12.5 years).

2.3 Data collection and analysis

As a primary source of results, privilege will be conferred to the results of the questionnaire and the reasons thereto are twofold: they represent the perception of a higher number of workers; and the data was collected when the experience of working in 12-hour shifts was more consolidated (9 months after the working schedule was changed).

Three sections of the questionnaire were chosen for the presentation of the findings: (1) the physical risk factors and the risks related to the work organisation (40 items); (2) the impacts of the schedule felt on health and on life outside work (24 items); (3) the evaluation of the will to keep the 2 × 12h system now and until retirement age (2 items).

3 RESULTS AND DISCUSSION

As a starting point, it was raised the question whether the 2 × 12h schedule should be kept beyond the "experimental" period. Most workers (80.3%) give a positive answer, and the majority (54.3%) justifies such favourable opinion with the free time provided by this schedule (in particular given the concentration of days off) to attend the demands of life outside work. 19.6% indicate they would rather keep the 2 × 12h schedule at this point when compared to the previous regime (3 × 8h), specifically due to the reduction in the number of consecutive working days (three at most, now). Table 1 shows the tendency for a global approval of the 2 × 12h schedule is common to all age brackets included in the analysis (20–29, 30–39, 40–49 and 50–59 year olds).

However, the approval rate of a given work schedule cannot be taken into consideration *per se* (Bourdouxhe et al. 1999). In other words, the question about the working hours cannot be disassociated from a deeper analysis of the workload, the physical risk factors, the risks related to the work organisation, the evolution in the workers' health condition or the problems caused by the previous schedule (3 × 8h). Consequently, the higher percentage of workers who wish to keep the 2 × 12h schedule cannot be interpreted as an exclusive satisfaction criterion with the schedule and simultaneously as an effort to proceed with the 2 × 12h arrangement in the future. Actually, the evaluation that was carried out also considered other relevant elements: i) the problems caused by the previous schedule, now resized; ii) the risks felt in 12 working hours per day; iii) or the health problems (that are already felt and those which may be developed in the future). The workers express the desire to continue to work in 12-hour shifts, but by doing so they give a signal that the current schedule's administration is not risk-free, either physically or considering the work organisation. Table 2 identi-

Table 1. Workers who wish to keep the 2 × 12h schedule, per age bracket.

		20–29	30–39	40–49	50–59
After 9 months working 2 × 12h, are you in favour of keeping this schedule?	Yes	4	32	9	4
	No	0	11	0	1

Table 2. Factors and conditions considered risks that may be aggravated in the 2 × 12h schedule by comparison to the 3 × 8h schedule.

Physical	%
Noise	73.8%
Stand	67.2%
Go up and down very often	54.1%
Intense physical efforts	52.5%
Work organisation and teams management	
Insufficient number of team members	73.8%
Use of temporary workers	72.1%
Provide training to temporary workers	60.7%
Work at an intense pace	50.8%
Work organisation × Physical	
Insufficient number of team members × Work at an intense pace × Intense physical efforts	57.4%

fies the factors and conditions the workers refer as "highly risky"; such factors and conditions were already felt in 8-hour schedules, but the perception is that they may be worse in 12-hour schedules.

Among the physical factors, more than half of the workers mention certain risks that may be aggravated in 12h: the "noise", the need to "stand for a long time", "go up and down very often" (for instance, stairs) and the "intense physical efforts".

A common feature of these risk factors is precisely their amplification reported by the workers in 2 × 12h, i.e., the noise level is not higher or lower in 12h than in 8h, however, the extension of one's exposure (extra 4 hours a day) resizes the workers' constraint.

Considering work organisation and team management, the following factors are highlighted: "insufficient team members", and two factors related to one another, the "use of temporary workers" (72.1%) and "provide training to temporary workers" (60.7%).

A cross-analysis between physical risk factors and risk factors associated with work organisation reveals that most workers (57.4%) consider the "number of insufficient team members" and, simultaneously, "work at an intense pace" and make "intense physical efforts" as risks in 2 × 12h schedule. Therefore, apparently, certain issues related to the work organisation and to the management and composition of the teams influence the perception of aggravation about the physical constraints felt by the workers over a 12-hour shift (for example, the absence of a team member requires an intensification of the surveillance/control of two machines for a longer period).

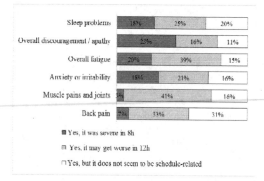

Figure 1. Main perceived health problems according to the schedule.

Addressing health, considering the 2 × 12h schedule should be kept does not mean the workers are not feeling the health problems caused by the organisation of the work in shifts, whether it is 3 × 8h or 2 × 12h, or that they are not fearing their health will deteriorate in the future. Figure 1 identifies the main health issues pointed out by the workers according to their severity, in the 3 × 8h schedule, in the 2 × 12h schedule, or apparently not schedule-related.

Considering the average number of working hours is the same both in 3 × 8h and in 2 × 12h (42h per week), a direct comparison between the two schedules underlines the perspective on the aggravation of the complaints of "muscles pains and joints" in the 2 × 12h schedule (41% of the workers say it is a current problem that may get worse).

This finding is in line with the main factors the workers consider risky: "stand for too long", "go up and down (the stairs) very often" or "make intense physical efforts" for longer in a working day. On the other hand, the issue of "overall discouragement/apathy" is considered more severe in the previous work schedule than in the current one, which may be related to the number of consecutive days along the week (5 working days in 3 × 8h vs. 3 working days in 2 × 12h).

It is also relevant to address the indicators regarding the "back pain" issue. Though 33% of the workers acknowledge they feel the problem (and that it may get worse in 12h), 31% considers it may not be schedule-related. This finding raises awareness for the need to take into account in the analysis the effects of the workload on health, the frequency of the technical processes, the organisation of the production process, the formal and informal breaks or even the strategies for job rotation, in addition to the characteristics of a given work schedule.

In line with recent researches (Vincent 2017), the workers acknowledge the negative effects of a 12h time organisation on their health. Still, when weighting this option with the previous 3 × 8h schedule, they choose to support today a commitment to accept the 12h so to achieve greater flexibility to meet the family demands (for example, to spend more time with the children), to expand the decision-making latitude to reach work-life balance or to save in the commuting.

However, when the reflection about the 2 × 12h schedule considers a broader time span, the workers take on less unanimous positions. When questioned whether they imagine themselves working in a 2 × 12h shift until retirement age, 49.2% of the workers say yes, and 50.8% answer no. Table 3 reflects the distribution of responses to this item, per age bracket.

Overall, the perception about the continuation of the current work in 2 × 12h shifts until retirement age gets more approval from the older age brackets (40–49 years old and 50–59 years old). Regarding this topic, age is an important variable, precisely as Molinié (2005) pointed out, because the closer one is from retirement age, the shorter the perspective over the professional path, which might enhance an affirmative ability to keep in the same job until retirement (Barros-Duarte et al. 2015).

Table 3. Distribution of the workers who imagine themselves working 2 × 12h until retirement age per age bracket.

		20–29	30–39	40–49	50–59
Do you imagine yourself working 2 × 12h until retirement age?	Yes	2	18	6	4
	No	2	25	3	1

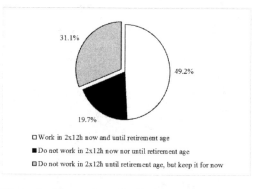

Figure 2. The workers distribution according to the desire to keep the 12-hour schedule now and until retirement age.

A cross-analysis to the responses to the items *after 9 months working in 2 × 12h, shall the schedule continue* and *do you imagine yourself working in 2 × 12h until retirement age* shows that among the workers who do not imagine themselves working in this schedule until retirement age (50.8%), 19.7% already disagree with the 12h schedule and 31.1% agree with the continuity of the 2 × 12h schedule in the present (cf. Figure 2). The time frame is once again a relevant factor inasmuch as, after 9 months working in 2 × 12h, in the weighing of the current adverse effects on health stemmed from work and its schedule, the workers anticipate their unavailability to deal with such problems in a broader time span (until retirement age).

The design of a working schedule that suits everybody requires the acknowledgement that more than a "time envelope" (Barthe et al. 2004) of the activity, the schedule may be a risk factor even when its impact on health is not significant yet.

4 CONCLUSIONS

The work organisation in 12h periods, though presenting advantages for companies and workers, resizes the risks, as indicated in this study's findings. Thus, it is important to study the risks resulting from the schedules bearing in mind their associated effects with the task demands and the particular conditions where the work is developed.

The arbitration between health preservation and the guarantee of a greater flexibility for the work-life balance, because its conciliation is not unanimous, is incommensurable. Choosing a 12-hour shift may be particularly useful in certain moments of the workers' lives and when the work effects on health keep a discreet expression, infrapathological. The analysis under the scope of a longer time span reveals the work performance up until retirement age is less consensual.

For that reason, it is fundamental to think thoroughly about the professional paths, about the impact of the exposure to different schedules in the duration of these workers' professional trajectories. It is argued that a possible adjustment shall also be implemented in advance—in the public space of the norms that regulate the temporal dimension of work, whose surveillance only supervises the work length, omitting the risks related to the intensification of the work performed in extended work schedules.

ACKNOWLEDGEMENTS

This work was funded by the Portuguese Science Foundation (CPUP UID/PSI/00050/2013; FEDER/COMPETE2020 POCI-01-0145 FEDER-007294).

REFERENCES

Axelsson, J., Kecklund, G., Akerstedt, T. & Lowden, A. 1998. Effects of alternating 8 and 12-hour shifts on sleep, sleepiness, physical effort and performance. *Scandinavian Journal Work Environment Health* 24(3): 62–68.

Barros-Duarte, C., Carnide, F., Cunha, L., Santos, M. & Silva, C. 2015. Will I be able to do my work at 60? An analysis of working conditions that hinder active ageing. *Work* 51: 579–590.

Barthe, B. 2009. Les 2 × 12h: une solution au conflit de temporalités du travail posté? *Temporalités* 10. Retrieved from http://temporalites.revues.org/1137.

Barthe, B. 2015. La déstabilisation des horaires de travail. In A. Thébaud-Mony, P. Davezies, L. Vogel & S. Volkoff (eds), *Les risques du travail:* 223–232. Paris: Découverte.

Barthe, B., Quéinnec, Y. & Verdier, F. 2004. L'analyse de l'activité de travail en postes de nuit: bilan de 25 ans de recherches et perspectives. *Le Travail Humain* 67(1): 41–61.

Bessa, N. & Lacomblez, M. 2013. Endiguer le travail du temps de travail sur la santé: un débat de normes et de valeurs dans l'aménagement des horaires postés. In F. Hubault (ed), *Y a-t-il un age pour travailler?:* 69–86. Toulouse: Octares.

Bourdouxhe, M., Toulouse, G. & Quéinnec, Y. 1999. Les défis et mirages de la recherche intervention sur les temps de travail. *Perspectives interdisciplinaires sur le travail et la santé* 1(1): 1–20.

Cabon, P. (2017). Effets de l'age sur la tolérance aux horaires decalés et la sécurité. Communication présenté au Séminaire du Créapt, *Des heures et des années: les horaires de travail au fil du parcours professionnel*; Paris, Mai 2017. France.

Eurofound 2016. *Sixth European Working Conditions Survey*. Luxembourg: Publications Office of the EU.

Johnson, M. & Sharit, J. 2001. Impact of a change from an 8-h to a 12-h shift schedule on workers and occupational injury rates. *International Journal of Industrial Ergonomics* 27: 303–319.

Knauth, P. 2007. Extended work periods. *Industrial Health* 45: 125–136.

Lacomblez, M. & Teiger, C. 2007. Ergonomia, formações e transformações. In P. Falzon (ed), *Ergonomia*: 587–601. São Paulo: Blucher.

Lacomblez, M., Bellemare, M., Chatigny, C., Delgoulet, C., Re, A., Trudel, L. & Vasconcelos, R. 2007. Ergonomic Analysis of Work Activity and Training: Basic Paradigm, Evolutions and Challenges. In R. Pikaar, E. Koningsveld & P. Settels (eds), *Meeting Diversity in Ergonomics:* 129–142. Oxford: Elsevier.

Loudoun, R. 2008. Balancing shiftwork and life outside work: do 12-h shifts make a difference? *Applied Ergonomics* 39: 572–579.

Molinié, A. 2005. Se sentir capable de rester dans son emploi jusqu'à la retraite? *Perspectives interdisciplinaires sur le travail et la santé* 7: 1–33.

Quéinnec, Y., Teiger, C. & Terssac, G. 1985. *Repères pour négocier le travail posté*. Toulouse: Octares Editions.

Rosa, R., Colligan, M. & Lewis, P. 1989. Extended workdays: effects of 8-hour and 12-hour rotating shift schedules on performance, subjective alertness, sleep patterns and psychosocial variables. *Work & Stress* 3(1): 21–32.

Tucker, P., Barton, J. & Folkard, S. 1996. Comparison of 8 and 12 hour shifts: impacts on health, wellbeing, and alertness during the shift. *Occupational and Environmental Medicine* 53: 767–772.

Vincent, F. 2017. Penser sa santé en travaillant en 12 heurs: les soignants de l'hôpital entre acceptation et refus. *Perspectives interdisciplinaires sur le travail et la santé* 19(1):1–20.

Wagstaff, A. & Lie, J. 2011. Shift and night work and long working hours—a systematic review of safety implications. *Scandinavian Journal of Work Environment Health* 37(3): 173–185.

Weibel, L., Herbrecht, D., Imboden, D., Junker-Mois, L. & Bannerot, B. 2014. Organisation du travail en 2 × 12h: les risques pour la santé et la sécurité des travailleurs. *Références en Santé au Travail* 137: 143–149.

Occupational risks in hospital mortuary

A. Ferreira, A. Lança, C. Mendes, M. Sousa & S. Paixão
IPC, ESTeSC, Coimbra Health School, Saúde Ambiental, Coimbra, Portugal

ABSTRACT: Certain categories of health work are perceived, as having a greater risk inherent in health. For this reason, the present study focused on hospital mortuary services, a work with multiple risk factors. The present study is under development, and its main objective is to evaluate the occupational hazards to which mortuary workers are exposed. The risks associated with the work environment in these workplaces are perceptible, and according to the risks, there are consequences that may possibly lead to the occurrence of occupational accidents and, in turn, occupational diseases. It is also noticeable, that the task in which the workers is most likely to acquire risks, is, in the sanitization of the corpse, in which specific procedures and techniques are implicit. According to the research, the risks that are most meaningful, and which require greater intervention, are biological, chemical, ergonomic and mechanical risks.

1 INTRODUCTION

The hospital environment is complex and presents a great number of risks to its workers. This is due, to the fact that, the hospital has directed its design and allocation of resources in meeting the needs of the patients, without adequate consideration of the working conditions of their professionals (Brás et al. 2006).

Because of this complexity, professional risk assessment and control is one of the most comprehensive areas of intervention of the Technician of Safety and Health at Work. Generally, in a hospital there is exposure to nonspecific risk factors (falling from a staircase or on a slippery floor, exposure to ozone and UV rays from a photocopier) that may arise in any other work environment and more specific risk factors (punctures, cuts or perforations with sharp-perforating material, exposure to hepatotoxic anesthetic gases, ionizing and nonionizing radiation, exposure to biological agents and mechanical risk factors related to ergonomic issues, such as continuous manual movement of dependent patients) (Lança et al. 2006).

The risk of a health professional contracting work-related diseases is about 1.5 times higher than the risk of all other workers. The need to prevent occupational hazards associated with the handling of hazardous equipment, exposure to infectious agents, physical factors, among others, is the determining reason for all efforts being made to ensure safer environments in health facilities and to ensure that work can be carried out under healthier and safer conditions (Graciela et al. 2016).

Several authors demonstrate that, the risk factors compromise the health and safety of workers. These are subdivided into four groups, according to their origin, in physical, chemical, biological and psychosocial risk factors (Batawi 2001, Lança et al. 2006 & Mauro et al. 2004).

For other authors, risk factors are subdivided into five groups, adding risk factors related to work or activity (Graciela et al. 2016).

Certain categories of health work are perceived, as having a greater risk inherent in health and with specific characteristics. For this reason, the present study focused on hospital mortuary services, a work with multiple risk factors.

2 MATERIALS AND METHODS

The present study is under development, and its main objective is to evaluate the occupational hazards to which mortuary workers are exposed. Being the specific objectives, to identify hazards and to evaluate professional risks, to carry out semi-quantitative ergonomics studies, analytical studies, namely, lighting, thermal comfort and air quality, as well to elaborate professional risk control plans.

Also, a checklist was made to support the assisted observation study. The study was classified as a "transverse" type.

3 HOSPITAL MORTUARY

The work has an important role in the life of the man and fulfills some objectives, such as: respect the life and health of the worker, prioritizing the problem of safety and wholesomeness of places of work activity. Concerning the health of the worker, it is understood as a set of interdisciplinary theoretical practices: technical, social, human and interinstitutional.

Therefore, the study of the work environment should comprise the following aspects: the different types (characteristics); the factors that condition it; the changes of these factors and their causes; the technique for exploring such changes; the measures that must be adopted to avoid the aggression of the environment on the individual (Mauro et al. 2004).

The hospital mortuaries are considered places of mystery, sadness, mourning and repulsion. The general environment of a mortuary is depressing and gloomy, being therefore the most neglected and ignored place by almost every hospital (Sirohiwal et al. 2011).

However, it is still an important part of each hospital, insofar as it deals with the preservation of the corpse, so that forensic technicians, as well as pathologists and other professionals can investigate the cause of death and do scientific research. A carefully planned mortuary complex is of great benefit to all those, who come in contact with it, such as police officers, doctors, students, health professionals and parents of deceased persons (Gajuryal, 2016).

Mortuary workers play an important and challenging role, providing an efficient, safe and sanitary service, guaranteeing the care, respect and sensitive treat with families. By exercising very judicious and careful work, they are particularly vulnerable to professionally acquired infectious diseases (Ruslin, 2004).

These workers face hazards at work, such as biological, chemical, ergonomic and mechanical risks, among others (Sepkowitz et al. 2005).

Health, safety and infection prevention are vital in hospital mortuaries, these places must ensure that the risks are eliminated, or if they are not possible, be reduced, providing and maintaining a safe working environment, ensuring that all workers are protected (Department of health, 2006).

4 OCCUPATIONAL RISKS IN HOSPITAL MORTUARY

4.1 *Phases of the evaluation of exposure to risks factors*

Hazard identification, risk analysis and risk assessment are the three stages of the overall risk assessment process and are presented in Figure 1 according to ISO 31000: 2003. This assessment process allows assessing the risk, for the health and safety of workers at work, resulting from the circumstances of occurrence of a hazard in the workplace. The result of this assessment, allows to

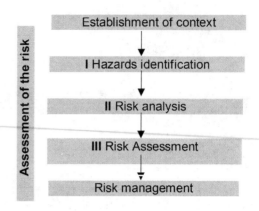

Figure 1. Generic risk management flowchart, according to ISO 31000: 2013.

make decisions, about the necessary measures, to be provided for the purpose of protecting workers for those same risks. Such measures can be of different orders, from measures that include prevention of occupational hazards, training of workers and organization of the necessary means to implement preventive measures (Guimarães et al. 2016).

It is intended that this assessment be a systematic examination, which includes all aspects of the work, considering all hazards and factors contributing to this hazard and which may result in injury or damage, also, it is important to check whether these same factors can be eliminated and whether preventive measures are effective or not in risk control (Jorge, 2013).

4.2 *Risk factors in hospital context*

In terms of exposure and work environment, the hospital is an inexhaustible source of health risks for workers, patients, companions and visitors (Lança et al. 2006).

In the hospital context, the identification of risk factors is excessively important, since it is an essentially descriptive step about the elements and processes of work and relative to the understanding of the professional activity performed. Its description is performed from the perspective of potential diversity (negative effects on the health and safety of exposed workers). It is a procedure that requires rigor in the analysis, encompassing beyond the observation, description and interpretation of the work to identify potential risk factors. In this sense, professional risk factors in the hospital context may be of nature: mechanical, physical, chemical, biological, ergonomic, psychosocial, electrical, order and cleaning, and fire and explosion.

They result, therefore, from the interaction between the individual, living conditions at work and living conditions outside work, being likely to influence the health, safety and well-being of the worker, with possible repercussions on labor productivity and worker satisfaction (Graciela et al. 2016).

4.3 Generic description of hospital mortuaries and equipment

Generally, hospital mortuaries, according to the norms, disposes a rest room for family members, a chapel, corpse storage/storage room, corpse preparation room, storage room for materials and equipment, sanitary facilities separated by sex for chaperones and relatives, and sanitary and social facilities for workers. In relation to work equipment, in a generic way, a hospital mortuary must have mortuary refrigerators and freezers, body trays, mortuary trolley and stretchers for the deposit of corpses.

4.4 Main tasks of hospital mortuary workers

There is permanent occupational exposure for these workers, in relation to the tasks performed. It should be noted that, according to the research, they can be performed by mortuary workers, by funeral workers, or both (in collaboration).

The tasks of these workers are very specific, however for the present study it is important to mention the main tasks, namely: Transfer the corpse to the electric lift truck (The corpse is removed from the mortuary refrigerator and transferred to the electric lift truck, in order to clean the corpse); Remove the sheet and the body bag where the corpse is placed (The corpse comes from the original service and the process of the mummy, where it is later placed in a body bag and with a sheet covering the corpse); Raise the lift car, to a height appropriate to the, height of the worker to begin the process of sanitizing the corpse (already with the corpse placed); Sanitize the corpse (Process of sanitization and preservation of the corpse); Transfer the corpse to the coffin; Handling biological waste from group III and sharp objects and sanitize the space.

4.5 Working environment and associated professional risks

To identify occupational risks, it is essential to analyze the work environment. Like any other workplace, hospital mortuary workers are exposed to hazards and risks associated with the work environment. In this sense, it is important to identify occupational risks according to their nature, as well as the possible consequences for workers. Occupational risk factors associated with the work environment of the hospital mortuaries, may be of nature: Physics, mechanics, electrical and psychosocial.

4.5.1 Physical

1. Thermal environment
Exposure to low temperatures, due to the characteristics of the space, namely the presence of the mortuary refrigerators and freezers, where the possible consequences are: skin changes, eye irritation, chills, fatigue, dehydration and decreased productivity.

2. Lighting
Work spaces usually with insufficient lighting levels, providing workers with: Visual fatigue, inappropriate posture and difficulty concentrating.

3. Noise
Exposure to noise due to the operation of the mortuary refrigerators and freezers, being the possible consequences: Deconcentration, irritability, decreased auditory acuity.

4.5.2 Mechanical

1. Shock or impact
There may be a shock or impact, for example, in the transfer of corpses, manipulation of the electrical lift truck, lack of space, in the mortuary refrigerators and freezers, which could lead to their workers, fractures or crushing.

2. Fall of objects on workers
For example, there may be a fall of objects on workers when there is manipulation of the body trays, when it is necessary to remove them from the upper mortuary refrigerators and freezers, causing fractures or bruises.

4.5.3 Electrical

The consequences can be: electrocution, burns or electric shocks, when cleaning procedures take place, near to the electric wires and to the mortuary refrigerators and freezers.

4.5.4 Psychosocial

1. Work overload
Preparation of corpses without work breaks, providing workers with stress, fatigue, lack of concentration, irritability, emotional disorders and repercussions in their personal lives.

2. Dealing with death
Assistance to families, sensitizing families, awareness of religious beliefs, which can lead to stress, depression, suffering, anguish. Identification of hazards and risks associated with the tasks performed.

4.5.5 *Transfer the corpse to the electric lift truck and to the coffin*
1. Hazards: Mobilization and transfer of corpses and manual handling of loads.
2. Risks: Effort and inadequate postures, tension in some areas of the body, tiredness, discomfort, musculoskeletal disorders, and shock and blows against objects and equipment.
3. Nature of risk factor: Ergonomic and mechanical.

4.5.6 *Remove the sheet and the body bag where the corpse is placed*
1. Hazards: Mobilization and transfer of corpses and manual handling of loads.
2. Risks: Effort and inadequate postures, tension in some areas of the body, tiredness, discomfort, musculoskeletal disorders, and shock and blows against objects and equipment.
3. Nature of risk factor: Ergonomic and mechanical.

4.5.7 *Raise the lift car, to a height appropriate to the, height of the worker to begin the process of sanitizing the corpse*
1. Hazards: Mobilization/Sustenance and transfer of corpses and manual handling of loads.
2. Risks: Effort and inadequate postures, tension in some areas of the body, tiredness, discomfort, musculoskeletal disorders, and shock and blows against objects and equipment.
3. Nature of risk factor: Ergonomic and mechanical.

4.5.8 *Sanitize the corpse*
1. Description: Combing hair, nose and mouth tamponade with cotton, dressing the corpse, cleansing facial and wrists.
2. Hazards: Manipulation of the corpse, standing posture during the tasks performed, exposure to biological agents, manipulation of sharp objects (used in the tamponade process), handling of chemicals products used to clean the corpse and continued exposure to chemical agents.
3. Risks: Contamination/contact with biological agents (virus, bacteria, fungi and parasites), possibility of infection, musculoskeletal disorders, tension in some areas of the body, contamination/contact with chemical agents and inhalation of chemicals.
4. Nature of risk factor: Ergonomic, mechanical, biological and chemical.

4.5.9 *Handling biological waste from group III and sharp objects*
1. Hazards: Manipulation of biological waste and sharp objects.
2. Risks: Contact/contamination with biological agents, possibility of infection, cross-contamination.
3. Nature of risk factor: Biological.

4.5.10 *Sanitize the space*
1. Hazards: Continued exposure to chemical agents.
2. Risks: Inhalation of chemicals, release of harmful aerosols to the worker and to the environment compromising indoor air quality.
3. Nature of risk factor: Chemical.

4.6 *Prevention strategies*

In safety and health/occupational health, the prevention encompasses a set of measures aimed at: reduce the impact of disease-determining factors or other health problems; prevent its occurrence; contain its progression or limit its consequences. Workers who perform mortuaries duties, should be formed about care and risks present and should know how to avoid or minimize such risks (Department of health, 2016).

The culture of prevention must be to: Combat the risk at source by replacing what is dangerous with what is safe from danger (or less dangerous); Personal protective equipment (PPE); Adapt the work to the individual and, technical progress; To promote training, information and awareness to its workers on hazards and prevention measures and to provide them with appropriate instructions on protective measures and their use.

Regarding the work environment and the main tasks performed by hospital mortuary workers, the essential prevention strategies are: Make request for improvement of the lighting system; Reorganize the workplace, in order to comply with the requirements mentioned; Perform tasks where there is no need to use noisy equipment on an interval basis, reducing the time of continuous exposure to noise; Modify the organization of work to reduce the duration and/or intensity of workers' exposure to risks; Suppress manual movements as much as possible; Privileging mechanical aids and ergonomic solutions; Program and intervention plans and psychological support;

Training workers on ergonomics and manual handling of loads; Training workers on good practice in the context of safety and health and prevention of infection; Compliance with safety standards and use of Personal Protective Equipment; Assistance of another worker for collaboration (transfer of corpses).

In short, the implementation of the Basic precautions of infection control, hand hygiene, proper use of PPE and sanitation plans are essential to minimize risk.

5 CONCLUSION

From the bibliographical research carried out, we conclude that, there is no precisely a universal definition of the tasks, and procedures in these places of work, nor standardized concepts. Because there are no data collection instruments for this area, and studies according to other authors are scarce in relation to this subject, this study has become a major limitation. However, from the bibliographic research performed and comparing with the following authors, Sirohiwal et al. 2011, Sharad 2016, George 2015, Cardoso & Navarro 2012, it is verified that, because of this occupational exposure, workers' health is compromised. In this sense, the risk can be estimated to investigate the necessity and possibility of their elimination, reduction and control.

According to the previous analysis, the risks associated with the work environment in these workplaces are perceptible, and according to the risks, there are consequences that may possibly lead to the occurrence of occupational accidents and, in turn, occupational diseases. It is also noticeable, that the task in which the workers is most likely to acquire risks, is, in the sanitization of the corpse, in which specific procedures and techniques are implicit. It is also possible to perceive that, the risks that are most meaningful, and which require greater intervention, are biological, chemical, ergonomic and mechanical risks.

According to all these factors, it should be noted that, the assessment of risk levels for procedures with increased risk of exposure, agents transmissible by blood and organic fluids should be carried out and recorded, and a plan with improvement actions for the control of the risks identified.

The implementation of the Basic Precautions of Infection Control is a great ally for the protection of workers. Hand hygiene is, one of the most important measures to reduce the transmission of infectious agents, particularly after the handling of corpses. The use of PPE provides adequate protection to health professionals, according to the risk associated with the procedure to be performed (George, 2014).

After the use of PPE, these should be discarded as group III waste (biological waste).

The work environment should be free of unnecessary objects and equipment to facilitate the process of hygiene (previously described in specific plans) and the occurrence of accidents related to mechanical risk factors (Cardoso et al. 2012).

REFERENCES

Batawi, M. 2001. *Occupational Health, A manual for primary health care workers, World Health Organizations,* p. 168.
Brás, D. & Gonçalves, C. 2006. *Avaliação de Riscos no Laboratório de Urgência,* p. 132.
Cardoso, T., Costa, F. & Navarro, M. 2012. *Biossegurança e desastres: Conceitos, prevenção, saúde pública e manejo de cadáveres, Physis Revista de Saúde Coletiva* 22(4): 1523–1542.
Department of Health. 2006. *Care and Respect in Death, good practice guidance for NHS Mortuary Staff,* p. 44.
Gajuryal, S. 2016. *Mortuary Service in Hospital,* p. 57.
George, F. 2015. *Organização e funcionamento do Serviço de Saúde Ocupacional/Saúde e Segurança do Trabalho dos Centros Hospitalares/Hospitais. Orientação da Direção Geral da Saúde* 2017(008/2014): 1–15.
Graciela, S., Mário, C. & Sandra, M. 2010. *Gestão dos Riscos Profissionais em Estabelecimentos de Saúde O, Ministério da Saúde, Administração Regional da Saúde de Lisboa e Vale do Tejo, IP,* p. 52.
Guimarães, H., Baptista, J. & Nunes, O. 2016. *Avaliação Do Risco De Exposição a Agentes Biológicos: Reprodutibilidade Dos Métodos Em Matadouros,* p. 89.
Jorge, A. & Oliveira, C. 2013. *Princípios Da Gestão De Risco Da NP Iso 31000,* p. 105.
Lança, A., Cardoso, M. & Morais, A. 2006. *Avaliação do Risco de Manuseamento de Resíduos Hospitalares,* pp. 10–156.
Mauro, M., Muzi, C., Guimarães, R. & Mauro, C. 2004. *Riscos Ocupacionais em Saúde, Revista Enfermagem,* 12, pp. 338–345.
Mendes, J. & Lima, P. 2014. *Avaliação De Riscos Em Restauração,* p. 64.
Nordin, R. 2004. *Occupational risk in healthcare,* p. 66.
Sepkowitz, K.A. & Eisenberg, L. 2005. *Occupational deaths among healthcare workers. Emerging Infectious Diseases* 11(7): 1003–1008.
Sirohiwal, B., Paliwal., Sharma, L. & Chawla, H. 2011. *Design and Layout of Mortuary Complex for a Medical College and Peripheral Hospitals. Forensic Research* 2(6): 4.

Occupational exposure in fitness clubs to indoor particles

K. Slezakova & M.C. Pereira
LEPABE, Departamento de Engenharia Química, Faculdade de Engenharia, Universidade do Porto, Porto, Portugal

C. Peixoto, C. Delerue-Matos & S. Morais
REQUIMTE–LAQV, Instituto Superior de Engenharia do Porto, Instituto Politécnico do Porto, Porto, Portugal

ABSTRACT: The aim of this study was to evaluate occupational exposure to indoor particulates in fitness clubs. $PM_{2.5}$ were continuously monitored during the opened periods of ten consecutive days at four fitness clubs (C1–C4) situated in Porto Metropolitan Area, Portugal. Various indoor spaces (main body building areas, studios and rooms for group classes) were assessed. Across the clubs the thermal parameters were within the indicated limits. The obtained $PM_{2.5}$ ranged between 5–777 µg m^{-3} with average concentrations (13–43 µg m^{-3}) exceeding the Portuguese limit of 25 µg m^{-3} at C1–C2, thus highlighting the possible risks for the respective staff. Clubs with mechanical ventilations (C3–C4) demonstrated considerably cleaner air, despite the higher number of clients per day. Long period of the monitoring activities (in the main areas) contributed to the majority of the occupational exposure of the staff; however, dose rates due to class teaching accounted between 30% (males at C3) and 47% (females at C1). These results demonstrate that intense physical activity, even though of short duration, may considerably increase the daily inhalation dose. In addition, female instructors have shown 5–20% higher inhalation doses, thus highlighting the necessity of the gender-consideration when assessing personal exposure.

1 INTRODUCTION

Exposure to airborne Particulate Matter (PM) is one of the most relevant environmental risks that nowadays society faces, whether it is indoors or outdoors. PM refers to an air pollutant that represents a mixture of solid and/or liquid particles suspended in the atmosphere (WHO 2010). These particles can be formed from other gaseous pollutants (i.e. secondary) or directly emitted from primary sources, both of natural or man-made origin (such combustions of solid and or liquid fuels). Indoors, combustion and human activities are a significant source of PM (Morawska et al. 2013, 2017). Specifically, fine particulate mater (i.e. $PM_{2.5}$) designates particles with aerodynamic diameter below 2.5 µm. The small size of this fraction allows for penetration to deep parts of human respiratory tract, even up to lungs (Wang et al. 2011). Exposure to fine particles has been associated with various adverse short-term health effects (such as eye, nose, throat and lung irritation, coughing, etc.; WHO 2005). Chronic exposure to fine PM has been linked with increased hospital admissions and mortality from lung cancer and heart disease (Wang et al. 2011; WHO 2006).

Despite the known benefits of exercising, there is a worldwide trend towards less total daily physical activity. Such decline may lead to deterioration of health predictors, such as obesity, and the associated increase of cardiovascular and all-cause mortality (Hallal et al. 2012; Lee et al. 2012). Thus, interventional strategies promote an increased physical activity. For adults (including seniors), World Health Organization (WHO) recommends 150 min (minimum) of moderate to vigorous physical activity per week (WHO, 2016). In order to improve the physical fitness (Storer et al. 2014), people often join gyms and fitness clubs, but the positive effects of exercising can be countered if these facilities have insufficient quality of indoor air. In addition, for the respective staff that spends longer periods in these places there might be some occupational exposure risks. Whereas some information in regards to PM has been emerging, primarily focusing on educative gymnasiums and or sport halls (Alves et al. 2013, 2014; Braniš et al., 2009, 2011; Braniš & Šafránek, 2011; Buonanno et al. 2012; Castro et al. 2015; Filipe et al. 2013; Kic, 2016; Saraga et al. 2014; Ward et al. 2013; Weinbruch et al. 2012; Žitnik et al. 2016), lesser attention was given to profit fitness clubs (Almeida et al. 2016; Ramos et al. 2014; Onchang & Panyakapo, 2016), with no-data regarding the occupational exposure of the instructors.

Considering the lack of existent information, this work aimed to assess occupational exposure to indoor particulates in fitness clubs.

2 MATERIAL AND METHODS

2.1 Sample collection

Four fitness clubs (C1–C4) located in Oporto Metropolitan area, Portugal, were selected. All clubs were situated in urban zones influenced mainly by vehicular emissions. Whereas C1 and C2 were smaller gyms (118–265 clients/day) frequented mostly by local fitness enthusiasts, C3 and C4 were internationally recognized large health clubs (410 up to 1000 clients/day), which installations were more sophisticated and included swimming pools, health and spa cares.

In each club, monitoring was conducted for a duration of 10 days (continuously during opened hours, typically between 7 a.m. and 11 p.m.) and in various spaces that included: (i) Main Areas (MA), *i.e.* a combined space for body-building (machines and free weights) and aerobic activities (cardiovascular equipment such as treadmills, ellipticals, stationary bikes, and rowing machines); and (ii) in studios and rooms for group classes (SR). TSI DustTrak DRX photometers (model 8533; TSI Inc., MN, USA; flow rate of 3.0 L min^{-1}; reading accuracy ± 0.1% of 1 μg m^{-3}) was used to sample PM$_{2.5}$. Physical parameters, namely Temperature (T) and Relative Humidity (RH) were monitored by a multi-gas sensor probe (model TG 502; Gray-Wolf Sensing Solutions, Shelton, USA). Logging intervals were 1 min (*i.e.* a mean value over 1 min interval were registered). The equipments were calibrated (at manufacturer facilities) prior to the sampling. In addition, on-site, the photometer was daily zero-checked (with use of external module) to minimize the occurrence of sudden artefact jumps (Rivas et al. 2017).

Finally, the staff of the respective clubs provided each day details concerning the relevant activities and/or untypical occurrences (cleaning and maintenance measures, occupancies, uncommon incidences and *etc.*).

2.2 Inhalation dose calculation

Inhalation doses (1) were calculated (Cavaleiro Rufo et al. 2016; Slezakova et al. 2015):

$$D = \left(\frac{IR}{BW}\right) \times C \times t \quad (1)$$

where D = age-specific dose (μg kg^{-1}); C = average concentration of PM$_{2.5}$ in the respective spaces (μg L^{-3}); t = time spent in the microenvironment (min); IR = inhalation rate (L min^{-1}); and BW = body weight (kg). Gender- and age-specific parameters for three age groups (21 to <31 yrs. old; 31 to <41; and 41 to <51) were adapted from US Environmental Protection Agency exposure handbook (USEPA 2011) with the following values: IR of 12–13 (female) and 12–14 (male) L min^{-1} considering light physical activity (walking and etc.), 23–25 and 29–32 L min^{-1} for females and males with moderate physical activity, 46–47 and 54–57 L min^{-1} for intense physical exercising (cardio). BW of 85–91 kg were for male population (USEPA 2011) and 61 kg for women (Ramos et al. 2015). The exposure time was 8 h and consisted of 420 min (monitoring activities in the main areas) and 50 min of instructing/teaching activities (*i.e.* giving group classes).

3 RESULTS

3.1 PM levels

PM$_{2.5}$ concentrations obtained at the four clubs are presented in Figure 1.

Overall, across the four clubs the respective levels demonstrated large variations of the obtained data. At C1 PM$_{2.5}$ ranged between 5 and 285 μg m^{-3} (with means of 34 and 49 μg m^{-3} at MA and

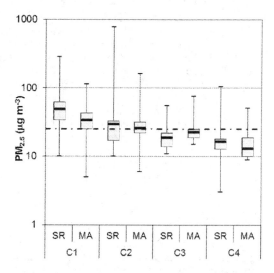

Figure 1. PM$_{2.5}$ levels (■ mean, □ 25–75%, and ⊥ range) at four fitness clubs (C1–C4). PM$_{2.5}$ distributions were significantly different ($p < 0.05$) across the four clubs. Horizontal dash line identifies limit value of 25 μg m^{-3}, as set by the national legislation (Decreto-Lei 118/2013). Y axis is scaled to logarithmic mode to adequately display the concentration ranges; SR = studios and rooms for classes; MA = main bodybuilding areas.

SR, respectively), while at C2 this range was even higher (6–777 µg m^{-3}; means of 26 µg m^{-3} at MA and 29 µg m^{-3} at SR). These ranges were significantly higher ($p < 0.05$) than in clubs C3–C4, where the respective concentrations were much lower: 11–76 µg m^{-3} (MA: 19 µg m^{-3}; SR: 26 µg m^{-3}) at C3 and 3–104 µg m^{-3} (MA: 13 µg m^{-3}; SR: 17 µg m^{-3}) at C4. Although clubs C3 and C4 were much larger facilities (daily frequented by 2–5 times more clients), they were equipped with mechanical ventilation systems that resulted in overall cleaner indoor air (approximately up to 4 times lower PM levels).

Typically, higher variations in PM levels were observed in studios, rooms and spaces for group classes rather than in the main areas (Figure 1). This is justified by the "movement" and the high number of exercising people (and corresponding emissions) in the main areas. $PM_{2.5}$ means were statically different between the two spaces of each club ($p < 0.05$), with the respective concentration means typically higher in SR than in MA (15–50%). It is assumed that this was due to the confinement of studios and intensive human activities conducted in these spaces (Ramos et al. 2014; Sack & Shendell, 2014a,b; Žitnik et al. 2016), in combination with overcrowding of these rooms; the larger volume (2–11 times more in comparison with SR) of the main areas allowed for dilution and spreading of respective PM emissions.

Finally, to better understand the obtained ranges, $PM_{2.5}$ was compared with the existent standards. WHO recommendation for PM levels in indoor air is based on the values set for outdoors (as there is no conclusive evidence concerning the hazardousness of indoor-emitted PM in a comparison of the ambient one; WHO 2010). The value of WHO $PM_{2.5}$ recommendation (25 µg m^{-3}) corresponds to the one designated in the Portuguese legislation on indoor air quality of public buildings (Decreto-Lei 118/2013). However, the given time-duration is set to 8 h (unlike WHO that considers 24 h). As shown in Figure 1, average concentrations in C1 and C2 exceeded the respective value in both types of spaces (SR and MA). Thus, in these clubs there might be possible risks to the respective staff.

Temperature and relative humidity are two of several parameters that affect thermal comfort of the respective occupants. The recommendation for room temperature inside sport facilities is set (summer season) at 18–25°C, whereas the relative humidity of the air should be maintained between 55 and 75% (SEJD, 2008). The results in Table 1 show that in general, temperature was within the recommended guidelines. Higher exceedance (maxima value) was observed in MA of C2, which occurred during the late-afternoon period. Considering the gym position and the orientation of the rooms, MA was exposed to the direct sunlight entering through the windows that might warm up the entire room by few degrees. In addition, the number of people can also increase the air temperature. The occupancies of the centers typically peaked at around midday (approx. at 12–13 h) and at early evening hours (approx. at 20–21 h), whereas at early mornings and early afternoons (approx. at 8–9 and 15–16 h) the centers were the calmest. Another possible mechanism for warming-up the gym atmosphere would be mixing of the 'hot' water vapor, characterized by a large specific heat (Žitnik et al. 2016).

Regarding the relative humidity, the indoor conditions of C1 and C4 were within the recommended ranges, whereas in MA of C2 and C3, average RH were somewhat lower. Though it is necessary to highlight that RH was within the recommendations of the American Society of Heating, Refrigerating, and Air-Conditioning Engineers (ASHRAE) that suggests for indoor spaces 30–60% (for temperature range of 23–28°C; Alves et al. 2013).

Table 1. Comfort parameters at four fitness clubs (C1–C4).

T(°C)	C1		C2		C3		C4	
	MA	SR	MA	SR	MA	SR	MA	SR
Mean	21.1	21.1	22.3	21.2	23.5	24.2	20.8	20.4
Min	19.3	17.9	17.1	18.3	22.0	21.0	18.4	18.3
Max	22.5	22.4	32.7	23.7	25.6	26.2	24.4	23.6
25th	20.8	20.8	20.2	20.8	23.2	23.0	20.0	19.1
75th	21.6	21.4	24.0	21.6	23.7	25.2	21.2	21.6

RH(%)	C1		C2		C3		C4	
	MA	SR	MA	SR	MA	SR	MA	SR
Mean	61.9	58.7	44.9	55.8	48.6	55.8	67.5	67.2
Min	59.6	44.6	23.4	41.1	40.4	45.6	50.2	53.0
Max	68.2	81.5	69.7	70.4	64.3	72.0	80.4	81.0
25th	60.9	53.1	36.5	53.5	45.7	52.3	64.5	59.8
75th	62.5	64.3	52.5	58.3	50.0	58.9	70.2	74.7

3.2 *Inhalation doses*

Figure 2 illustrates $PM_{2.5}$ inhalation doses for female and male staff of four fitness clubs. C1 exposure doses were the highest ones (1.2–1.7 times higher than in C2) for all considered age groups and both genders, most likely due to the highest $PM_{2.5}$ levels. Considering the long period (7 h) the monitoring activities represented the majority of the occupational exposure; however despite the short period (50 min) dose rates due to exposure in studios/classrooms accounted between 30% (males at C3) and 47% (females at C1). Class

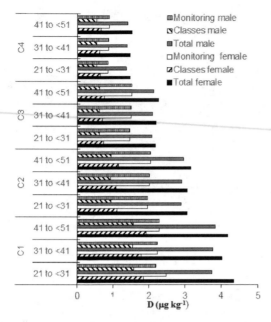

Figure 2. Estimated doses of $PM_{2.5}$ for female and male staff at four fitness clubs (C1–C4) during various activities.

instructors often co-exercise with the group and clearly, the increased inhalation rates might lead to elevated particle intakes. It is though necessary to highlight that values estimated in this work represent a "baseline" once the respective exposures can be much higher. Fitness instructors commonly give more than a class per day; in addition, they also perform they own training. Lastly, women professional exhibited 5–20% higher dose rates than males, mostly during the exercise period exposures. These results were in agreement with previous work (Ramos et al. 2015) that reported larger limitation in expiratory flow of females and increased efforts to breathe when intensively exercising.

4 CONCLUSIONS

Exposure to airborne particulate matter is one of the most relevant environmental risks that nowadays society faces, whether it is indoors or outdoors. Since people spent most of their time in indoor environments, better understanding of the indoor particles is needed in order to mitigate the respective risks. In the four studied fitness clubs $PM_{2.5}$ concentrations varied greatly (5–777 µg m^{-3}) with significantly lower levels observed in clubs with mechanical (i.e. controlled) ventilations. $PM_{2.5}$ Portuguese limit of 25 µg m^{-3} for indoor air was exceeded at clubs with natural ventilation, thus demonstrating possible risks for the respective staff. Therefore, in order to reduce the exposure to air pollutants (and to reach the thermal comfort) optimizing air systems (in terms of ventilation rate, air drying, and air filtering) in these places would be crucial.

Monitoring activities of the respective staff contributed around 50–70% of the occupational exposure whereas class teaching accounted between 30% (males at C3) and 47% (females at C1). Intense physical activity, even though of short duration, considerably increased the daily inhalation dose, with higher (5–20%) doses for female instructors indicating the need for the gender-specific personal exposure studies.

ACKNOWLEDGMENTS

This work was supported by European Union (FEDER funds through COMPETE) and National Funds (Fundação para a Ciência e Tecnologia) through projects UID/QUI/50006/2013 and UID/EQU/00511/2013-LEPABE, by the FCT/MEC with national funds and co-funded by FEDER in the scope of the P2020 Partnership Agreement. Additional financial support was provided by *Fundação para Ciência e Tecnologia* through fellowship SFRH/BPD/105100/2014.

REFERENCES

Almeida, S.M., Ramos, C.A. & Almeida-Silva, M. 2016. Exposure and inhaled dose of susceptible population to chemical elements in atmospheric particles. *Journal of Radioanalytical and Nuclear Chemistry* 309(1): 309–315.

Alves, C., Calvo, A.I., Marques, L., Castro, A., Nunes, T., Coz, E. & Fraile, R. 2014. Particulate matter in the indoor and outdoor air of a gymnasium and a fronton. *Environmental Science and Pollution Research* 21(21): 12390–12402.

Alves, C.A., Calvo, A.I., Castro, A., Fraile, R., Evtyugina, M. & Bate-Epey, E.F. 2013. Indoor air quality in two university sports facilities. *Aerosol and Air Quality Research* 13(6): 1723–1730.

Braniš, M. & Šafránek, J. 2011. Characterization of coarse particulate matter in school gyms. *Environmental Research* 111(4): 485–491.

Braniš, M., Šafránek, J. & Hytychová, A. 2009. Exposure of children to airborne particulate matter of different size fractions during indoor physical education at school. *Building and Environment* 44(6): 1246–1252.

Braniš, M., Šafránek, J. & Hytychová, A. 2011. Indoor and outdoor sources of size-resolved mass concentration of particulate matter in a school gym-implications for exposure of exercising children. *Environmental Science and Pollution Research* 18(4): 598–609.

Buonanno, G., Fuoco, F.C., Marini, S., & Stabile, L. 2012. Particle resuspension in school gyms during physical activities. *Aerosol and Air Quality Research* 12(5): 803–813.

Castro, A., Calvo, A.I., Alves, C., Alonso-Blanco, E., Coz, E., Marques, L., Nunes, T., Fernández-Guisuraga, J.M. & Fraile, R. 2015. Indoor aerosol size distributions in a gymnasium. *Science of the Total Environment* 524–525: 178–186.

Cavaleiro Rufo, J., Madureira, J., Paciência, I., Slezakova, K., Pereira, M.C., Aguiar, L., Teixeira, J.P., Moreira, A. & Oliveira Fernandes, E. 2016. Children exposure to indoor ultrafine particles in urban and rural school environments. *Environmental Science and Pollution Research* 23(14): 13877–13885.

Decreto-Lei 118/2013. O sistema de certificação energética dos edifícios, o regulamento de desempenho energético dos edifícios de habitação e o regulamento de desempenho Energético dos edifícios de comércio e serviços (in Portuguese). *Diário da República* 1.ª série—N.º 235: 6644(1)–6644(10).

Filipe, T.S., Vasconcelos Pinto, M., Almeida, J., Alcobia Gomes, C., Figueiredo, J.P. & Ferreira, A. 2013. Indoor air quality in sports halls. *Occupational Safety and Hygiene—Proceedings of the International Symposium on Occupational Safety and Hygiene* SHO 2013: 175–179.

Hallal, P.C., Andersen. L.B., Bull, F.C., Guthold, R., Haskell, W., Ekelund, U. & Lancet Physical Activity Series Working Group 2012. Global physical activity levels: surveillance progress, pitfalls, and prospects. *Lancet* 380(9838): 247–257.

Kic, P. 2016. Dust pollution in the sport facilities. *Agronomy Research* 14(1): 75–81.

Lee, I.M., Shiroma, E.J., Lobelo, F., Puska, P., Blair, S.N., Katzmarzyk, P.T. & Lancet Physical Activity Series Working Group 2012. Effect of physical inactivity on major non-communicable diseases worldwide: an analysis of burden of disease and life expectancy. *Lancet* 380(9838): 219–229.

Morawska, L., Afshari, A., Bae, G.N., Buonanno, G., Chao, C.Y.H., Hänninen, O., Hofmann, W., Isaxon, C., Jayaratne, E.R., Pasanen, P., Salthammer, T., Waring, M., & Wierzbicka, A. 2013. Indoor aerosols: from personal exposure to risk assessment. *Indoor Air* 23: 462–487.

Morawska, L., Ayoko, G.A., Bae, G.N., Buonanno, G., Chao, C.Y.H., Clifford, S., Fu, S.C., Hänninen, O., He, C., Isaxon, C., Mazaheri, M., Salthammer, T., Waring, M.S. & Wierzbicka, A. 2017. Airborne particles in indoor environment of homes, schools, offices and aged care facilities: The main routes of exposure. *Environment International* 108: 75–83.

Onchang, R. & Panyakapo, M. 2016. The physical environments and microbiological contamination in three different fitness centres and the participants' expectations: Measurement and analysis. *Indoor and Built Environment* 25(1): 213–228.

Presidency of the Council of Ministers, Secretariat of State for Youth and Sport (SEJD), 2008. *Ginásios: Diploma relativo à construção, instalação e funcionamento*. Lisbon, Portugal: Secretariat of State for Youth and Sport, available at http://www.cd.ubi.pt/artigos/Gin%C3%A1sios.pdf.

Ramos, C.A., Reis, J.F., Almeida, T., Alves, F., Wolterbeek, H.T. & Almeida, S.M. 2015. Estimating the inhaled dose of pollutants during indoor physical activity. *Science of the Total Environment* 527–528: 111–118.

Ramos, C.A., Wolterbeek, H.T. & Almeida, S.M. 2014. Exposure to indoor air pollutants during physical activity in fitness centers. *Building and Environment* 82: 349–360.

Rivas, I., Mazaheri, M., Viana, M., Moreno, T., Clifford, S., He, C., Bischof, O.F., Martins, V., Reche, C., Alastuey, A., Alvarez-Pedrerol, M., Sunyer, J., Morawska, L. & Querol, X. 2017. Identification of technical problems affecting performance of DustTrak DRX aerosol monitors. *Science of the Total Environment* 584–585: 849–855.

Sacks, D.E. & Shendell, D.G. 2014a. Case study: particle concentrations at a local gym dependent on mechanical ventilation in a retrofitted industrial building in Central NJ. *Proceedings, Indoor Air 2014, July 7–12, Hong Kong, China*.

Sacks, D.E. & Shendell, D.G. 2014b. Case study: ventilation and thermal comfort parameter assessment of a local gym dependent on mechanical ventilation in a retrofitted industrial building in Central NJ. *Proceedings, Indoor Air 2014, July 7–12, Hong Kong, China*.

Saraga, D.E., Volanis, L., Maggos, T., Vasilakos, C., Bairachtari, K. & Helmis, C.G. 2014. Workplace personal exposure to respirable PM fraction: A study in sixteen indoor environments. *Atmospheric Pollution Research* 5(3): 431–437.

Slezakova, K., Texeira, C., Morais, S. & Pereira, M.C. 2015. Childrens indoor exposures to (Ultra)fine particles in an Urban Area: Comparison between school and home environments. *Journal of Toxicology and Environmental Health—Part A: Current Issues* 78 (13–14): 886–896.

Storer, T.W., Dolezal, B.A., Berenc, M.N., Timmins, J.E. & Cooper, C.B. 2014. Effect of supervised, periodized exercise training vs. self-directed training on lean body mass and other fitness variables in health club members. *Journal of Strength and Conditioning Research*, 28(7): 1995–2006.

Wang, Y., Hopke, P.K., Chalupa, D.C. & Utell, M.J. 2011. Long-term study of urban ultrafine particles and other pollutants. *Atmospheric Environment* 45: 7672–7680.

Ward, T.J., Palmer, C.P., Hooper, K., Bergauff, M. & Noonan, C.W. 2013. The impact of a community-wide woodstove changeout intervention on air quality within two schools. *Atmospheric Pollution Research* 4(2): 238–244.

Weinbruch, S., Dirsch, T., Kandler, K., Ebert, M., Heimburger, G. & Hohenwarter, F. 2012. Reducing dust exposure in indoor climbing gyms. *Journal of Environmental Monitoring* 14(8): 2114–2120.

World Health Organization 2005. *WHO Air quality guidelines for particulate matter, ozone, nitrogen dioxide and sulfur dioxide. Global update 2005: Summary of risk assessment*. Geneva, Switzerland: World Health Organization.

World Health Organization 2010. *WHO guidelines for indoor air quality: selected pollutants*. Copenhagen, Denmark: WHO Regional Office for Europe.

World Health Organization 2016. *Physical activity strategy for the WHO European Region 2016–2025*. Copenhagen, Denmark: WHO Regional Office for Europe.

Žitnik, M., Bučar, K., Hiti, B., Barba, Ž., Rupnik, Z., Založnik, A., Žitnik, E., Rodriguez, L., Mihevc, I. & Žibert, J. 2016. Exercise-induced effects on a gym atmosphere. *Indoor Air* 26(3), 468–477.

Occupational risks in hospital sterilization centers

A. Ferreira, A. Lança, C. Mendes, M. Sousa & S. Paixão
IPC, ESTeSC, Coimbra Health School, Saúde Ambiental, Coimbra, Portugal

ABSTRACT: A Sterilization Center is a vital unit in a hospital context. Its function is to provide contamination-free materials for use in a variety of hospital procedures. The main objective of the present work is to evaluate the occupational risks to which the workers of these centers are exposed to. This study is classified as a "transverse" type. We conclude that there is a clear undervaluation of the workers affected in this sector, as well as the activities carried out. This might be due to the nature of this closed sector, in which people do not interact so much with the external environment, or even since people do not know the relevance of the work in the Sterilization Centers. In addition, it was verified that there are many complaints and health problems reported by workers in this sector, depending on the activities performed there.

1 INTRODUCTION

The term quality of life was first used by US President Lyndon Johnson in 1964 when he stated that "goals can't be measured through the banks' balance sheet. They can only be measured by the quality of life they provide to people". Interest in concepts such as "standard of living" and "quality of life" was initially shared by social scientists, philosophers and politicians. The increasing technological development of medicine and related sciences brought with it a negative consequence to its progressive dehumanization. Thus, concern with the concept of "quality of life" refers to a movement within the human and biological sciences in order to value parameters broader than controlling symptoms, reducing mortality or increasing life expectancy (Fleck et al. 1999).

The work plays an important role in the life of the man and fulfills some objectives, such as: respect the life and health of the worker, prioritizing the problem of safety and wholesomeness of places of work activity; leave him free time for rest and leisure, highlighting the question of the duration of this journey and its coordination for the improvement of the living conditions outside the place of the occupational activity; and should allow the worker his own personal fulfillment, while providing services to the community, considering the problem of the type of activity and the organization of work (Mauro et al. 2004).

It is known that work brings feelings of pleasure and satisfaction as well as exposes the worker to risks. Workers who work in the health area are exposed to a variety of risks when performing their profession (Ribeiro et al. 2012).

Occupational diseases result from exposure to physical, ergonomic, chemical and biological agents present at the workplace, and more recently psychosocial risk. (Tipple et al. 2007) It is observed that several adverse situations began to be part of the routine of human life, interfering negatively in their quality of life. This situation is evidenced by the extremely serious morbidity and mortality of workers (Silva, 2007).

In this aspect, the articulation between the work process and health is a subject of constant scientific investigation, and in the course of the historical evolution of the societies has been object of observation and reflection of the men, with respect to the ways of learning to deal with this relationship (Silva, 2007).

It's of fundamental importance to remember that health professionals do not always work in direct areas of care; can perform their tasks in support and care services. Work on these services can be just as unhealthy as those for direct care, exposing workers to a considerable number of risks. Among these areas is the Material and Sterilization Center (Silva, 2007).

The Material and Sterilization Center is a vital and fundamental unit of the hospital context, whose function is to provide materials free of contamination to be used in a wide variety of hospital procedures. Those who don't work in the sector often don't know the complexity of their activities (Talhaferro et al. 2006).

This service is responsible for receiving, cleaning, decontaminating, preparing, sterilizing, storing

Figure 1. CME unidirectional flow chart.

and distributing the materials used in the various units of a health facility, which characterizes it as a closed and critical sector in which contaminated and infected materials are handled (Ferreira et al. 2004).

It is recommended that articles that enter the Sterilization Centers have a continuous flow, without retrocession and without crossing the cleaned material with the contaminated one (Daniel, 2011).

The planning of the physical area is intrinsically linked to the activities of the work process (Daniel, 2011): Receiving and classifying dirty items from hospital areas; Wash and dry the materials; Receive clothes from the laundry room; Prepare and pack materials and clothing; Sterilize or disinfect materials and clothing, by methods indicated according to the compatibility of articles, indications of use and cost-benefit assessments; Store processed materials; Distribute processed materials and clothing; Look after the protection and safety of operators.

This service consists of many machines and equipment necessary to perform the tasks, such as: Washing machines/thermal disinfection; Washing machines/thermal and chemical disinfection; Ultrasonic machines; Manual washing; Sterilization machines for wet heat, dry heat, ethylene oxide, steam formaldehyde under pressure 73°, plasma gas of hydrogen peroxide and steam at low temperature with formaldehyde 2%.

2 MATERIAL AND METHODOLOGIES

The main objective of this work is to evaluate the occupational risks to which the workers of the Hospital Sterilization Centers are exposed. The specific objectives are to identify hazards and evaluate occupational hazards, to apply semi-quantitative studies of ergonomics and analytical studies, namely, lighting, thermal environment and air quality, and to carry out professional risk control plans.

Verification checklists have been performed to support the assisted observation study, designed based on existing legislation. At the same time, a study is being carried out through worker consultation processes (distributing questionnaires).

The study was classified as a "transverse" type.

3 WHAT ARE THE RISK FACTORS THAT INFLUENCE THE SAFETY AND HEALTH OF WORKERS?

The workers concerned highlight the risks of fire, contact with chemical and biological substances, exposure to noise, physical effort and cutting and perforating injuries, as well as the risk of falling materials, discomfort due to adopted posture and work overload (Aquino et al. 2014).

3.1 Thermal environment

The heat is widely used in this service, in the operations of cleaning, disinfection and sterilization of articles and hospital areas. Equipment such as autoclaves, disinfectants or ultrasonic machines are considered sources of heat, as they contribute to the increase of the ambient temperature, providing physical discomfort to the workers. Excessive heat can cause undesirable effects on the human body such as transient fatigue, some sweat gland diseases, edema or edema of the extremities, increased susceptibility to other diseases, decreased ability to work, and others (Possari, 2003).

3.2 Noise

The machines and equipment used in the Sterilization Centers produce noises that can reach excessive levels, which in the short, medium and long term can cause considerable damage to health (affects the nervous system, digestive system and circulatory system). The higher the noise level, the shorter the occupational exposure time (Possari, 2003).

3.3 Lighting

Good lighting in the workplace provides high productivity, better final product quality, reduced number of accidents, reduced waste of materials, reduction of eye fatigue and overall, more order and cleanliness of areas and less animosity of workers (Possari, 2003).

3.4 Chemical

Chemicals are widely used in this service for various purposes, such as cleaning agents, disinfection and sterilization. In sterilization systems, machines and equipment require chemicals that present hazards to employees, for example: ethylene oxide is a toxic, mutagenic, carcinogenic and potentially explosive product; formaldehyde under steam at 73°C is a toxic and carcinogenic product; low temperature steam with 2% formaldehyde is a product that does not totally eliminate toxicity and is potentially carcinogenic (Possari, 2003).

3.5 Biological

When we consider the biological risk, these workers are exposed to organic secretions, when washing and handling contaminated articles; and can be sources of transmission of microorganisms to patients by preparing an article to be sterilized and handling an already sterilized article. Thus, the adoption of Individual Protection Equipment, although for individual use, in some situations lends itself to collective protection. In health facilities, multiple use articles that aren't decontaminated between care, may become vehicles for infectious agents, as well as the places where these items are reprocessed and the people who handle them (Daniel, 2011).

3.6 Mechanical

The mechanical risks are related to the insecure conditions existing in the workplace, capable of causing injuries to the physical integrity of the worker. They are due to improper movements, slips, falls, improper handling of equipment and other conditions of insecurity existing in the workplace.

The worker must be instructed about the operating modes and risks associated with the equipment. The necessary personal protective equipment must be demonstrated to the worker as well as the damages that the inadequate uses of the equipment can cause. (Possari, 2003) (eg.: burns caused by contact with machines, failure to use suitable gloves).

3.7 Ergonomics

The prevalence of spinal pain among the workers in the Material and Sterilization Centers is not surprising, since there are many activities performed there that involve excessive manipulation of weight and the adoption of inappropriate and uncomfortable postures. Factors such as work rhythm, performance of activities that cause overload of certain muscle groups, the use of uncomfortable furniture and equipment are responsible for the high number of musculoskeletal disorders in workers (Silva, 2007).

4 WHAT IS THE IMPORTANCE OF PERSONAL PROTECTIVE EQUIPMENT IN THE PROCEDURES USED IN STERILIZATION CENTERS?

The manipulation of devices contaminated by biological material requires the adoption of safety measures by workers. Standard precautions should be taken regardless of the level of soiling/contamination of the article. Therefore, it's indispensable to use Personal Protective Equipment, which should be used to ensure the safety of the worker exposed to the risk of drilling or cutting, preventing occupational accidents or diseases. It should be noted that even using all recommended Individual Protection Equipment, accidents can occur, and measures should be taken to minimize the risk of infection and/or early detection of possible diseases (Ferreira et al. 2004).

The use of personal protective equipment in the soiled area minimizes the risk of direct contact of the skin and mucous membranes with any contaminated material and with the chemicals required for the process. It should be emphasized that the initial cleaning process performed in hospitals is predominantly manual, which increases occupational risk; and even when machines for minimizing the risk of accidents with biological material are used for the cleaning of articles, the recommendation for the use of Personal Protection Equipment remains (Tipple et al. 2007) and yet exposes workers to other risks, such as elevated temperatures or noise.

Often, workers don't consider biological risk when carrying out their activities as important as they are, not being able to identify the consequences of the lack of use of Personal Protective Equipment, so adherence to the use of these equipment's is directly related to the perception of risk to which the worker is exposed (Florêncio et al. 2003).

5 CONCLUSIONS

According to some authors, in relation to the dynamics of work in the Material and Sterilization Centers, it's evident the presence of routine and repetitive tasks, especially due to the sequential way of processing the materials and the necessary productivity, which resembles it to an industry.

Although the technological evolution demands from the workers who work in this area, the knowledge of new raw materials, new packaging, new sterilization mechanisms, among others that make possible the safe and effective use of all the existing technology and advancement (Silva, 2007).

Even so, it's evident the undervaluation of the workers in this sector, as well as the activities performed there, perhaps because it is a closed sector in which people don't interact so much with the external environment, or even because people don't know the relevance of the work of the Sterilization Centers for an efficient functioning of the other units of the hospital. In addition, there are many complaints and health problems reported by workers in this sector, depending on the activities performed there (Silva, 2007).

Satisfied workers tend to carry out their activities with more attention, hospitality and cordiality, which contributes to the humanization of relationships. The quality of the work can't be neglected, since the safety of the user is also derived from the correctly processed materials of this service (Espindola et al. 2014).

It's believed that studying the work from the worker's point of view can reduce evaluation misunderstandings. The rationality of researchers and technicians is not enough, it's necessary to involve those who live the work situation so that there is a collective construction (Espindola et al. 2014).

The bibliographic research found and used for this study is directed to specific risks and generally directed towards biological risks.

REFERENCES

Aquino, J., Barros, L., Brito, S., Ferreira, E., Medeiros, S. & Santos, E. 2014. *Centro de material e esterilização: acidentes de trabalho e riscos ocupacionais*, 19(3), pp. 148–154.

Daniel, K. 2011. *Riscos ocupacionais durante a higienização de materiais em uma Central de Material e Esterilização*, pp. 1–34.

Espindola, M. & Fontana, R. 2014. *Occupational risks and self-care mechanisms used by the sterilization and materials processing department workers*. doi: 10.1590/S1983-14472012000100016.

Ferreira, A., Souza, A., Almeida, A., Sousa, S. & Siqueira, K. 2004. *Acidente com material biológico entre trabalhadores da área de expurgo em centros de material e esterilização*, pp. 271–278.

Fleck, M., Leal, O., Louzada, S., Xavier, M., Chachamovich, E., Vieira, G., Santos, L. & Pinzon, V. 1999. *Desenvolvimento da versão em português do instrumento de avaliação de qualidade de vida da OMS (WHOQOL-100) Development of the Portuguese version of the OMS evaluation instrument of quality of life*, pp. 19–28.

Florêncio, V., Rodrigues, C., Pereira, M. & Souza, A. 2003. *Adesão às precauções padrão entre os profissionais da equipe de resgate pré-hospitalar do corpo de bombeiros de goiás*, pp. 43–48.

Mauro, M., Muzi, C., Guimarães, R. & Mauro, C. 2004. *Riscos ocupacionais em saúde*.

Possari, J.F. 2003. *Centro de Material e Esterilização—Planejamento e Gestão*.

Ribeiro, R. & Vianna, L. 2012. *Uso dos equipamentos de proteção individual entre trabalhadores das centrais de material e esterilização*. pp. 199–203. doi: 10.4025/cienccuidsaude.v10i5.17076.

Silva, A. 2007. *Morbidade referida em trabalhadores de enfermagem de um centro de material e esterilização*, 6(1), pp. 95–102.

Talhaferro, B., Barboza, D.B. & Domingos, N.A.M. 2006. *Qualidade de vida da Equipe de Enfermagem da Central de Materiais e Esterilização*, pp. 495–506.

Tipple, A., Aguiliari, H., Souza, A., Pereira, M., Mendonça, A. & Silveira, C. 2007. *Equipamentos de proteção em centros de material e esterilização: disponibilidade, uso e fatores*, 6(4), pp. 441–448.

Chemical risks associated with the application of plant protection products

Tatiana Neves, Ana Ferreira & Ana Lança
ESTeSC—Coimbra Health School, Environmental Health, Polytechnic Institute of Coimbra, Coimbra, Portugal

Ana Baltazar
ESTeSC—Coimbra Health School, Dietetics and Nutrition, Polytechnic Institute of Coimbra, Coimbra, Portugal

João Paulo Figueiredo
ESTeSC—Coimbra Health School, Complementary Sciences, Statistics and Epidemiology, Polytechnic Institute of Coimbra, Coimbra, Portugal

ABSTRACT: Exposure to particular types and concentrations of plant protection products causes adverse effects on public health, workers and the environment, and it's therefore essential to assess the impact of these substances on their use by authorized or unauthorized entities. The sample was constituted by 6 companies of Coimbra, 3 AE and 3 UE, and by 16 workers in these study places. We conclude that there are significant differences between workers with and without training in the area with regard to knowledge and handling of PPP. It should also be noted that younger workers demonstrate greater knowledge rather than older ones. Significant changes in the concentrations of air components during the 3 study periods were observed in relation to the sampling places, which leads to the conclusion that the methods of preparation and application of the syrup are carried out more safely in the AE that in UE.

1 INTRODUCTION

Pesticides are a category of chemicals that are in growing use and development and which can have a very large impact on society. The continued growth of the world's population generates the need to increase food supplies. As a result, the use of pesticides in the world has increased rapidly, particularly in more developed countries where there's a strong concern for pest extermination, but relatively little attention is paid to the health of the user, the consumer and the environment (Pinto, 2015). PPP are thus an indisputable tool in agricultural production but their application requires exposure to serious risks to workers, public health and the environment (Lança, A. & Ferreira, A., 2016).

Applying a plant protection product aims to solve a particular plant health problem. The achievement of this goal depends on several factors, which can't in any way be forgotten. Incorrect application, in addition to wasting product, can cause additional problems in the crop, contaminate the applicator and the environment (ANEPC, 2007). It's in this context that studies have been developed to evaluate the effects of the use of these products on agricultural production, both for the target worker population and for the final recipient/consumer of the product. These studies consider the elimination of their use to be as much as possible, with the risks associated being more or less severe depending on the way they are used (Paixão, S., Ferreira, A., Lança, A., 2016). In a study on the consequences of pesticide exposure on human health, it was concluded that the intensive use of organophosphorus pesticides and the lack of use of personal protective equipment by farmers preparing and applying these products anticipate dermal exposure, which could be Avoided through the adoption of relatively simple and inexpensive measures such as the use of gloves and long-sleeved shirts (Delgado, I. & Paumgartten, F., 2004). To ensure worker protection, appropriate Personal Protective Equipment (PPE) and work clothing must be worn (OIT, 2014). However, we shouldn't neglect collective protection, where simple measures such as the use of exhaust systems for fumes and gases can be central to guarantee the workers' safety when preparing the syrup (OIT, 2015).

With the increasing demand and use of PPP in the agricultural sector, and taking into account the dangers inherent in their use as well as the producers who handle them, there was a need to draw up Law 26/2013 of April 11. The aim of this program is to regulate the distribution, sale and application of PPP use and of adjuvants of PPP, defining the procedures for monitoring the use of plant protection products (Lei n.° 26/2013). On the other hand, Decree-Law No. 101/2009 of May 11 aims to regulate the non-professional use of plant protection products in the domestic environment, establishing conditions for their authorization, sale and application.

The objective of this investigation was thus to establish a possible relationship between AC and UC, and the different impact that each can have on the environment and the product applicator.

2 MATERIAL AND METHODS

2.1 Sample characterization

This analysis was carried out in 6 companies located in the municipality of Coimbra, where 3 of them were authorized, that is, AC for the terrestrial application of PPP in urban areas, leisure zones and communication routes according to the General Directorate of Food and Veterinary, and another 3 that were considered UC. This study was presented as level II of the observational type and with the type of sectional cohort. The type of sampling was non-probabilistic, where the convenience sampling technique was applied, with inclusion criteria being the AC and UC.

2.2 Methodology and instruments for data collection

This study was based on the collection of data in three distinct phases, namely before the preparation of the syrup, during the preparation of the syrup, and during the application of the product, having made two samples in each phase, together with the elaboration of a questionnaire delivered to workers at each place and to the completion of a checklist on the general safety and hygiene conditions of the company and its safety in the application of PPP.

According to the first evaluation moment, the parameters previously mentioned were evaluated according to an analytical data collection, respecting all the necessary procedures for a good evaluation of the study site. Each measurement was performed using 4 air quality monitoring devices (Lighthouse—Model 3016 - $PM_{2.5}$ e PM_{10} of the work atmosphere in general; Q-Track—Model 8554 - CO, CO_2, relative humidity and temperature of the air; Sidepak™ Personal Aerosol Monitor—Two Models AM510 - $PM_{2.5}$ and PM_{10} of the area that comes directly into contact with the respiratory tract of the worker). The place where the data was collected was based on the ideal conditions for the application of product that is in the morning and with mild climatic conditions (without the presence of strong currents of air or humidity in the air). Measurements of the chemical parameters varied with respect to the duration of the measurement, since three different equipment's were used (Lighthouse—Model 3016; Q-Track—Model 8554; Sidepak™ Personal Aerosol Monitor—Model AM510). To evaluate CO, CO_2, relative humidity and temperature of the air there was an average time of 1 minute per sample, while to evaluate $PM_{2.5}$ and PM_{10} of the overall working atmosphere there was a time of about 15 minutes. To evaluate $PM_{2.5}$ and PM_{10} of the area directly in contact with the respiratory tract of the worker, the duration of the measurement varied according to the time necessary for each worker to prepare the syrup and to apply the product. It should be noted that all measurements were carried out in the same chronological order, prior to the preparation of the syrup, during the preparation of the syrup, and during the application of the product, respectively.

2.3 Statistical analysis

Statistical analysis of the data collected was performed using the IBM SPSS Statistic software version 24.0 for Windows. The tests used were Student's t-test for two independent samples, the Mann Whitney test and the Kruskal-Wallis test. For statistical inference it was possible to establish a 95% confidence level for a random error of less than or equal to 5%.

3 RESULTS

Questionnaires: Regarding the sample presented, the 6 companies were located in the county of Coimbra where applications of PPP were realized. Due to the contribution of the workers, it was verified that of the 16 individuals who completed the questionnaire, 75% were male (12 workers) and 25% were female (4 workers). Regarding the years of service in this area, the majority of respondents (37.5%) were in the active between the last 10 and 14 years. Of the workers under study, about 75% had training to handle this type of products, and in the case of literacy, 87.5% of respondents had 9th grade (9 workers). Only 1 worker reported having some symptoms after contact with PPP (6.25%), namely "allergies" or "headaches". According to the list of occupational hazards of the International Labor Organization, the chemical hazard associated with contact with the skin, inhalation or ingestion of herbicides (93.8%) was the highest among workers, but risks such as exposure to sunlight, overload And on effort, inadequate work postures or individual stress were also mentioned in the questionnaires. When analysing the questionnaires, it was possible to verify that all the workers used the PPE in the course of the work activity, regardless of having training or not. It was found that all workers with training in the area always read the product packaging label (58.3%) or almost always (16.7%), which was not observed in the workers without training, where about 75% of these exercises declare not to carry out this reading.

Analyses the general knowledge of workers on the management and application of PPP through their training in this subject, comparing their answers to the questionnaires filled by a total of 16 workers addressed in the 6 companies under study. Here it was verified that all workers stated that they did not apply PPP when there was strong winds, as they ensure maintenance and cleaning of the application material and PPE. Regarding the washing of the clothes before removing it, it was verified that all the workers without formation carried out this procedure, which isn't verified in the workers with formation, where only 58,3% affirms to carry out this washing. However, the remaining trained workers (41.7%) stated that they did not perform this procedure because they wear disposable clothing. This was also the case in the responses concerning the packaging of PPE in a proper place after washing, since 58.3% of the trained workers use disposable PPE. We could observe that in the two situations described above, workers always performed the mentioned procedures, and those who didn't do are explained by the fact that they use disposable PPE.

It was verified that the majority of the workers didn't apply the products in the hours of greater heat (87.5%), as well as they affirmed to comply with the rules of the label referring to the periods of re-entry in the culture treated and respective intervals of security (81.3%). The question where there was some discrepancy in responses by the workers refers to the storage of PPE in the same place of the PPP, where some workers with and without training (50% and 25%, respectively) admitted that they stored these two objects in the same place local.

Chemical and physical parameters of air quality: After data collection and subsequent statistical analysis, the estimated mean values of the concentrations obtained from the chemical and physical parameters of the air at each place are compared with the time before the preparation of the juice in the preparation of the syrup and at the time of application of the product (Table 1). Regarding the air components at each point of evaluation, there were considerable differences in CO and CO_2 concentrations, As well as the maximum values of PM2.5 and PM10 found in the proximity of the respiratory tract of workers. It should be noted that, in the mean values of PM2.5 and PM10 to which workers were exposed in these three phases, it was also possible to observe statistically significant differences (p-value < 0,05).

The average values of the concentrations obtained from the chemical and physical parameters of the air of each location were then compared with the three moments previously mentioned, but taking into account the typology of each company (Table 2).

When comparing the physical and chemical parameters of air between AC and UC for PPP application, values such as CO_2, $PM_{2.5}$ and PM_{10} of the atmosphere in general presented significant differences before and during the preparation of

Table 1. Results of chemical and physical parameters of air at each evaluation moment.

		n	CO_2 (mg/m³)	CO (mg/m³)	Relative humidity (%)	$PM_{2.5}$ (ug/m³)	PM_{10} (ug/m³)	Air temperature (°C)	$PM_{2.5}$ Individual (ug/m³)			PM_{10} Individual (ug/m³)		
									Max	Min	Avg	Max	Min	Avg
Before preparation	M	14	330.43	3.49	63.00	18.70	78.17	21.95	0.02	0.00	0.01	0.04	0,00	0,01
	SD		39.08	0.32	8.43	14.03	45.03	2.82	0.03	0.01	0.01	0.04	0,00	0,02
During preparation	M	14	409.14	3.69	66.84	20.01	83.06	28.21	0.85	0.00	0.06	1.20	0,01	0,07
	SD		102.64	0.29	8.67	13.82	35.55	16.82	1.04	0.01	0.07	1.97	0,02	0,10
Product application	M	14	1264.50	4.02	63.38	61.16	302.72	28.14	1.31	0.00	0.04	1.63	0,00	0,04
	SD		1246.60	0.48	10.21	51.27	285.52	13.28	1.16	0.00	0.05	2.59	0,00	0,06
p-value			0,002	0,007	0.263	0.150	0.174	0.217	0.001	0.911	0.046	0.002	0.911	0.043

Test: Kruskal-Wallis; M = Mean; SD = Standard Deviation.

Table 2. Results of the chemical and physical parameters of the air in each moment of evaluation against the typology of the company.

Parameters	Type of company	n	Before preparation M	SD	p-value	During preparation M	SD	p-value	Product application M	SD	p-value
CO_2 (mg/m^3)	AC	6	358.50	37.83	0.028	473.17	124.92	0.024	1391.00	1369.79	0.519
	UC	8	309.38	25.14		361.13	47.63		1169.63	1233.60	
CO (mg/m^3)	AC	6	3.35	0.44	0.149	3.62	0.26	0.602	3.85	0.54	0.215
	UC	8	3.60	0.14		3.75	0.31		4.15	0.43	
Relative Humidity (%)	AC	6	57.38	10.49	0.091	61.20	8.39	0.071	55.38	8.39	0.010
	UC	8	67.21	2.46		71.06	6.89		69.38	6.89	
$PM_{2.5}$ (ug/m^3)	AC	6	8.22	1.01	0.032	10.44	2.32	0.034	69.16	51.90	1.000
	UC	8	26.56	14.18		27.19	14.77		55.16	57.87	
PM_{10} (ug/m^3)	AC	6	31.11	6.47	0.032	48.59	24.15	0.034	413.90	328.87	0.157
	UC	8	113.47	12.34		108.92	7.68		219.33	263.27	
Air Temperature (°C)	AC	6	23.52	3.60	0.121	24.03	3.59	0.121	24.72	2.60	0.302
	UC	8	20.78	1.37		31.35	22.13		30.71	17.47	
$PM_{2.5}$ Ind. Maximum (ug/m^3)	AC	6	0.01	0.00	1.000	1.36	1.56	0.289	2.49	0.50	0.034
	UC	8	0.03	0.03		0.47	0.32		0.42	0.34	
$PM_{2.5}$ Ind. Minimum (ug/m^3)	AC	6	0.00	0.00	0.180	0.00	0.00	0.186	0.00	0.00	0.180
	UC	8	0.01	0.01		0.01	0.01		0.00	0.00	
$PM_{2.5}$ Ind. Average (ug/m^3)	AC	6	0.00	0.00	0.711	0.09	0.10	0.212	0.07	0.07	0.372
	UC	8	0.02	0.02		0.03	0.03		0.02	0.01	
PM_{10} Ind. Maximum (ug/m^3)	AC	6	0.03	0.04	0.289	2.08	3.08	1.000	3.19	3.69	0.289
	UC	8	0.05	0.05		0.54	0.032		0.46	0.24	
PM_{10} Ind. Minimum (ug/m^3)	AC	6	0.00	0.00	0.180	0.00	0.00	0.186	0.00	0.00	0.186
	UC	8	0.01	0.01		0.02	0.02		0.00	0.00	
PM_{10} Ind. Average (ug/m^3)	AC	6	0.00	0.00	0.467	0.12	0.14	0.372	0.08	0.08	0.372
	UC	8	0.02	0.02		0.03	0.03		0.02	0.01	

Subtitle: AC—Authorized Company; UC—Unauthorized Company; M = Mean; SD = Standard Deviation; Test: Mann-Whitney.

the syrup. There were also significant differences in the application phase of the product, namely in the parameters of relative humidity and the maximum value of PM2.5 found in the proximity of the respiratory tract. During this study 4 different PPP were used, namely 2 herbicides, 1 insecticide and 1 fungicide. When applying the COSHH method, a simplified methodology for assessing the risk of exposure to chemicals, the risk level of each of the products under study was assessed, ranging from level 1 (light risk) to level 4 (Maximum risk).

A checklist was used in each of the 6 study companies in order to analyse the general conditions of hygiene and safety of them, as well as their safety in the application of PPP. It was found that, regarding the conditions of the infrastructures of the companies and their methods of handling and applying PPP, there were differences between AC and UC. The major difference was focused on the safety part in the application of PPP, where the AC had better structural conditions and a more developed knowledge on the handling and application of these products in relation to UCs.

4 DISCUSSION

All workers reported using PPE, not applying PPP in the hottest hours, as well as complying with the label rules regarding re-entry periods in the treated culture and respective safety intervals (81.3%). However, it's notorious that trained workers give more importance to reading the label than those who aren't trained, since 75% of untrained workers don't read the packaging labels of the PPPs they use. Thus, as has already been observed in other studies[10], It can be seen that the application of PPP is still carried out by workers who don't have the required and compulsory education by law, and it has also been found that the concern of reading the labels is sometimes neglected (Moura, S., Vasconcelos, M., Ferreira, A., et al., 2016). As such, workers should be aware of the hazards of the products, that is, how to interpret the packaging label in its entirety so as to prepare the syrup in a safe and correct way. In addition they should know all the personal protective equipment to be used in relation to the product to be applied, as well as the correct way to handle it (Baltazar, A., Ferreira, A., Barreira, D., et al., 2017). The question where there was some discrepancy of responses by the workers refers to the storage of PPE in the same place of the PPP, where some workers with and without training (50% and 25%, respectively) admit that they store these two objects in the same place. Training is therefore an important aspect to be taken into account in work where PPP is manipulated since applicators with appropriate training know how to act and the risks they are exposed to when manipulating them and may present better working practices that diminish the risks associated with the various tasks (Costa, C. & Teixeira, J.P., 2012). As for the level of knowledge of the workers, according to their training and age, it was possible to observe that with the increase of the age, the level of knowledge decreases. It is thus well-known that the low education of many workers still remains one of the obstacles for a change of mentality that potentiates agriculture (Peixoto, S., 2015). As such, the entry of new, more skilled and innovative farmers can greatly benefit the agricultural sector by implementing innovative ideas that can help the development and growth of the activity, thus allowing agriculture to develop in a more sustainable and safe way for workers and for the environment (Peixoto, S., 2015).

It was also verified that the maximum values of $PM_{2.5}$ and PM_{10} found in the proximity of the respiratory tract of the workers had very significant differences, as well as the average of these same components throughout the 3 study moments. The exponential increase of these particles in the preparation phase and the application of the same are justified by the fact that the workers, in these study phases, have direct and close contact of the product, since it's here that the product is very concentrated and diluted in water to be applied later on the ground. Since it's up to the worker to make this dilution and to apply the same in the future, it's natural that the values of these particles rise exponentially in these phases compared to the phase prior to the preparation of the syrup. This preparation should be carried out outdoors or in well-ventilated areas, where dust must be handled with care to avoid the release of dust, in addition to the worker being kept facing away from the wind to avoid exposure to dusts and splashes (Guerra, V.L., 2012). The values of $PM_{2.5}$ and PM_{10} of the atmosphere in general presented higher values in the UC in the phase of preparation of the syrup and before it. The handling of pesticides, including dilution and preparation of syrups, are operations which are particularly susceptible to producing unwanted exposures of persons and the environment (Diretiva 2009/128/CE). These values can thus be justified by the fact that UC's have their own infrastructures to prepare the syrups, that is, they have ventilation and retention basin, some of the key aspects for the safe handling of these products, both for the environment and for the worker. When analysing the CO and CO_2 compounds present in ambient air in the application of risk level 1 products in the study sites, there were higher values of CO_2 in the urban area compared to the agricultural holding, values observed due to the traffic in the area Urban area compared to the farm area, since one of the most polluting gaseous pollutants emitted by diesel engines is carbon oxides (CO and CO_2) (Portal do Meio Ambiente, 2009).

5 CONCLUSIONS

When comparing AC and UC, it should be noted that these are governed by different laws and, as such, this may have repercussions on the safety of PPP handling. The AC are governed by Law no. 26/2013 of April 11, while UC by Decree-Law No. 101/2009 of May 11. In UC their field of application passes through domestic environment, such as indoor plants, gardens and family gardens. Here it isn't mandatory for workers to have training, just as it isn't necessary to install a warehouse with all the necessary conditions to store PPP. On the other hand, in AC its field of application may be agricultural/forestry, urban or leisure, and workers are required to have specific training in the area, just as it's necessary to set up a warehouse with all store PPP. It is concluded that the concentrations of air components are higher in the UC than in the AC during the three evaluation moments, which leads to the conclusion that the methods of preparation and application of the syrup are performed more safely in the AC than in the UC. As far as the knowledge of officials is concerned, it's found that those who are trained are more enlightened and have all the necessary information regarding the safe handling of the PPP, as well as the procedures to be adopted during the application of the product, after application and in case of emergency.

REFERENCES

Assembleia da República. Lei n.º 101/2009 de 11 de Maio. Diário da República. I série. Páginas 2806–2809. Portugal. 2013.

Assembleia da República. Lei n.º 26/2013 de 11 de Abril. Diário da República. I série. Páginas 2100–2125. Portugal. 2013.

Associação Nacional e Europeia de Proteção das Culturas (ANEPC). 2007. Manual Técnico de Produtos Fitofarmacêuticos. (p. 36).

Autoridade para as Condições de Trabalho. 2014. Utilização de Pesticidas Agrícolas.

Autoridade para as Condições de Trabalho. Abril de 2015. Segurança e Saúde no Trabalho no Setor Agro-Florestal.

Costa C, Teixeira JP. Efeitos genotóxiccos dos pesticidas. Revista Ciências Agrárias [Internet]. 2012;35:19–31. Disponível em: http://www.scielo.gpeari.mctes.pt/scielo.php?pid=S0871018X2012000200002&script=sci_arttext.

Delgado, I. Paumgartten, F. (Janeiro de 2004). Intoxicações e uso de pesticidas por agricultores do Município de Paty do Alferes, Rio de Janeiro, Brasil. (p. 185).

Guerra, Vera Lúcia Oliveira. O uso do Material Informativo e Promocional dos Fitofármacos. 2012.

Lança, A., Ferreira, A. Management of chemical risks in plant protection applicators. 3rd World Congress of Health Research. Outubro de 2016. (p. 51).

Moura, S., Vasconcelos, M., Ferreira, A., Moreira, F. (2016). Uso de produtos fitofarmacêuticos na agricultura.

Paixão, S., Ferreira, A., Lança, A. & Leitão, A. Gestão de Riscos em Aplicadores de Produtos Fitofarmacêuticos. Revista portuguesa de Saúde Ocupacional (RPSO). Setembro de 2016.

Parlamento Europeu e do Conselho. Diretiva 2009/128/CE de 21 de Outubro. Jornal Oficial da União Europeia. Página L309/71-L309/86.

Peixoto, S. Mestrado, Dissertação. Universidade de Trás-Os-Montes E Alto Douro Sistema de Informação Para Explorações Agrícolas—Proposta de Um Modelo Estrutural Universidade Trás-Os-Montes E Alto Douro Sistema de Informação Para Explorações Agrícolas—Proposta de Um Modelo Estrutural. (2015).

Pinto, G.M. (Julho de 2015). Os Pesticidas, Seus Riscos e Movimento no Meio Ambiente. (p. 1).

Portal do Meio Ambiente. Emissão do escapamento de veículos. REBIA—Rede Brasileira de Informação Ambiental. Outubro de 2009. Disponível em: http://portal.rebia.org.br/cidadania/2095-emissao-dos-escapamentos-dos-veiculos.

Levels of urinary biomarkers of exposure and potential genotoxic risks in firefighters

M. Oliveira, C. Delerue-Matos & S. Morais
REQUIMTE-LAQV, Instituto Superior de Engenharia, Instituto Politécnico do Porto, Porto, Portugal

K. Slezakova & M.C. Pereira
LEPABE, Departamento de Engenharia Química, Faculdade de Engenharia, Universidade do Porto, Porto, Portugal

A. Fernandes
Escola Superior de Saúde, Instituto Politécnico de Bragança, Bragança, Portugal

S. Costa & J.P. Teixeira
Instituto Nacional de Saúde Pública, Departamento de Saúde Ambiental, Porto, Portugal

ABSTRACT: This study characterizes the levels of six urinary monohydroxyl-polycyclic aromatic hydrocarbons in Portuguese firefighters directly involved in firefighting activities. Median concentrations of urinary total monohydroxyl-PAHs were predominantly higher in exposed subjects comparatively with non-exposed firefighters (3.12 *versus* 1.59 µmol/mol creatinine, respectively). Urinary 1-hydroxynaphthalene and 1-hydroxyacenaphthene were the most predominant metabolites (87–90% of total monohydroxyl-PAHs), being followed by 2-hydroxylfluorene (2.4–6.3%), 1-hydroxyphenanthrene (4.1–5.4%), and 1-hydroxypyrene (2.0–2.6%). Firefighters who were directly involved in fire combat activities presented increased percentages of DNA damage (9.85 *versus* 10.9%; $p \leq 0.01$) and oxidative stress (0 *versus* 2.14% TDNA; $p \leq 0.01$) comparatively with non-exposed subjects, thus revealing the impact of fires on the health of exposed firefighters.

1 INTRODUCTION

Firefighting is one of the most dangerous professions, being this occupation classified as possible carcinogen to humans (IARC, 2010; NIOSH, 2007). During fire combat, firefighters inhale unknown concentrations of complex mixtures of gaseous and particulate pollutants that are released during fires, namely particulate matter, carbon monoxide, nitrogen dioxide, and volatile organic compounds including Polycyclic Aromatic Hydrocarbons (PAHs) (IARC, 2010b). Among these pollutants, PAHs constitute one of the most health relevant groups, since some of the individual congeners are classified as possible/probable carcinogens to humans (IARC, 2002, 2010a). PAHs are mostly formed during incomplete combustion of organic materials; its exposure has been associated with reproductive, developmental, hemato-, cardio-, neuro-, and immuno-toxicities (ATSDR, 1995). Some authors have described a higher incidence of respiratory health effects, specifically exacerbations of asthma and chronic obstructive pulmonary disease and some genotoxic effects in firefighters directly involved in fire combat activities (Reid et al., 2016; Semmens et al., 2016; Youssouf et al., 2014). It is also known that mixtures of PAHs are capable of induction of reactive oxygen species that will promote cell damage mostly through lipid peroxidation or enzymatic disruption, thus causing oxidative stress. Exposure to PAHs has been also associated with inflammation induction and worsening of cardio-respiratory diseases (Alshaarawy et al., 2016; Kamal et al., 2015; Kim et al., 2013).

Firefighters' exposure to PAHs may occur via inhalation, ingestion, and dermal contact (by skin contamination or via clothing and undergarments). Therefore, the assessment of total internal dose of PAHs is only possible through biomonitoring. Biomarkers of exposure to PAHs, namely monohydroxyl-PAHs (OH-PAHs) reflect the total exposure to PAHs while biomarkers of effect reveal the body's response to that exposure, sometimes being indicative of some early subclinical changes (Links et al., 1995). Since fire combat activities are usually unpredictable, firefighter's biomonitoring is very difficult. So far, limited number of studies on firefighter's total exposure to PAHs is available (Caux et al., 2002; Edelman et al., 2003; Laitinen et al., 2010; Oliveira et al., 2016, 2017; Robinson et al., 2008) and the comparison between the studies is very difficult since the urinary levels of OH-PAHs are usually not normalized with creatinine concentrations of the involved subjects.

This study aims to assess the levels of six OH-PAHs (1-hydroxynaphthalene and 1-hydroxyacenaphthene (1OHNaph, 1OHAce), 2-hydroxylfluorene(2OHFlu),1-hydroxyphenanthrene (1OHPhen), 1-hydroxypyrene (1OHPy), and 3-hydroxybenzo(a)pyrene (3OHBaP)) in the urine of non-exposed and exposed firefighters to fire combat activities in the North of Portugal. The quantification of early genotoxic effects was determined through DNA damage and oxidative stress in whole blood of Portuguese firefighters.

2 MATERIAL AND METHODS

2.1 Firefighters subjects

The professional selected firefighters were directly involved in fire combat activities during the summer of 2014. A similar group of non-exposed subjects was also considered. All firefighters filled a questionnaire to collect information on physical characteristics (age, gender, weight, and height), work history (number of years as firefighter) and smoking habits (WHO, 2002). Table 1 summarizes the general characteristics of the firefighters considered in this study. Each fireman described in detail all the personal protective equipment used during fire combat. Only firefighters who reported the absence of chronic diseases and no history of chronic exposure to known carcinogenic substances were considered for this work. The study was approved by the Ethic Committee of University of Porto and National Institute of Health Doutor Ricardo Jorge; all subjects provided informed consent forms.

2.2 Sample collection

Each firefighter provided an urine sample at the end of the work shift using a sterilized container. At the same day, venous blood samples from each firefighter were collected in Ethylenediamine Tetra-Acetic Acid (EDTA) medium. All blood samples were suspended in cryopreservation medium of DMSO:RPMI (1:4, v/v). Urine and blood samples were frozen at −20°C and −80°C, respectively, until further analysis.

2.3 PAH metabolites chromatographic analysis

PAH metabolites present in the urine of firefighters were extracted according to Chetianukornkul et al. (2006) and Oliveira et al. (2016, 2017). Urine samples were quantified by liquid chromatography with fluorescence detection (Oliveira et al., 2016, 2017). The urinary concentrations were normalized with the levels of creatinine (μmol/mol) (Kanagasabapathy & Kumari, 2000).

2.4 Comet assay analysis

The alkaline and formamidopyrimidine DNA glycosylate (FPG) comet assay were performed as previously described by Abreu et al. (2017). The % of DNA in the comet tail (%TDNA) corresponds to the estimated DNA damage. Net FPG-sensitive sites were determined by subtracting the values of %TDNA for the slide incubated with buffer from the score for the slide incubated with enzyme.

2.5 Statistical analysis

SPSS (IBM SPSS Statistics 20) and STATISTICA (v. 7, StatSoft Inc., USA) were used to perform statistical analysis. Data does not follow a normal distribution, reason why medians were compared using nonparametric tests. Statistical significance was defined as $p \leq 0.05$.

3 RESULTS AND DISCUSSION

Levels of monohydroxyl-PAHs detected in the urine of Portuguese non-exposed and exposed firefighters are presented in Table 2. Although there are no international and national reference guidelines for any of the urinary PAH biomarkers, Jongeneelen (2001, 2014) proposed the value of 1.4 μmol/mol creatinine of 1-hydroxypyrene in urine

Table 1. Characterization of firefighters' study population.

Data	Firefighters	
	Non-exposed	Exposed
Number of firefighters	15	18
Age (mean, range; years)	33 (23–40)	32 (21–38)
Weight (mean, range; kg)	82 (62–98)	76 (59–98)
Years as firefighter (mean; min-max; years)	10 (6–20)	12 (2–20)
Smoking habits	No	No
Exposure to fires (mean; min-max; hours)	0	3.6 (2.0–5.0)

Table 2. Urinary monohydroxyl-PAHs levels[a] (µmol/mol creatinine; median; min-max) quantified in Portuguese firefighters.

Compound[a]	Non-exposed	Exposed
1OHNaph+1OHAce	1.38* (0.58–2.28)	2.75* (0.60–121)
2OHFlu	0.11 (1.5×10^{-3}–0.13)	0.06 (8.2×10^{-4}–0.19)
1OHPhen	0.09 (0.02–0.17)	0.06 (0.03–0.28)
1OHPy	0.04 (0.02–0.10)	0.02 (1.7×10^{-3}–0.19)
ΣOH-PAHs	1.59* (0.76–2.57)	3.12* (0.86–121)

[a]Whenever a concentration was below the detection limit, the value of LOD/√2 was considered (Hornung and Reed, 1990).
*Statistically significant differences between non-exposed and exposed firefighters ($p \leq 0.05$).

of exposed workers as the minimum concentration at which genotoxic effects were not found.

Urinary concentrations of total monohydroxyl-PAHs (ΣOH-PAHs) varied from 0.76 to 2.57 µmol/mol creatinine and between 0.86 to 121 µmol/mol creatinine in non-exposed and exposed firefighters, respectively ($p = 0.05$).

Regarding the urinary concentrations of individual OH-PAHs, significant differences were found only for 1OHNaph + 1OHAce (1.38 versus 2.75 µmol/mol creatinine; $p = 0.05$), however higher maxima of individual and total PAH metabolites were always observed in exposed firefighters (Table 2). At a fire scenario, Portuguese firefighters use personal protective equipment for firefighting activities (structural helmet with eye protection, mask with filter for particulate matter, flash hood, gloves, boots and fire-resistant clothes), however heat stress and difficulty in communications may lead to misuse of the masks, with some firefighters being more exposed than others to the pollutants released during fires.

The compounds that contributed the most for ΣOHPAHs were, by descending order: 1OHNaph +1OHAce (87% and 90% of ΣOHPAHs for non-exposed and exposed firefighters, respectively), 2OHFlu (6.3% versus 2.4%), 1OHPhen (5.4% versus 4.1%), and 1OHPy (2.6% versus 2.0%).

The urinary concentrations of 1OHNaph + 1OHAce, 2OHFlu, and 1OHPy were slightly higher (146–601%, 34–487%, and 25–190%, respectively) than the levels reported in some firefighters who were directly involved in fire combat activities (Oliveira et al., 2016). Urinary 1OHPy is the metabolite that is the mostly used to estimate occupational exposure to PAHs (Caux et al., 2012; Edelman et al., 2003; Jongeneelen, 2014; Laitinen et al., 2010; Oliveira et al., 2016, 2017; Robinson et al., 2008). The urinary concentrations of 1OHPy ranged between 0.02–0.10 µmol/mol creatinine in non-exposed subjects and from 1.7×10^{-3}–0.19 µmol/mol creatinine in individuals who were actively involved in fire combat (Table 2). The urinary concentrations of 1OHPy in exposed firefighters were lower than the levels reported by Caux et al. (2002) in 43 Canadian firefighters (0.08–3.63 µmol/mol creatinine; participation in knockdown and overhaul activities). The American Conference of Governmental Industrial Hygienists established the urinary concentration of 1OHPy (0.5 µmol/mol creatinine) as indicative of occupational exposure to PAHs (ACGIH, 2010). The urinary concentrations of 1OHPy for all firefighters (non-exposed and exposed) were always well below these proposed limits (Table 2).

Figure 1 and Figure 2 present the percentages of genotoxic biomarkers in the form of basal damage and oxidative stress in the non-exposed and exposed firefighters. Basal DNA damage was significantly higher in peripheral blood cells of exposed firefighters compared to non-exposed subjects (9.85 versus 10.9%; $p \leq 0.01$). Regarding the oxidative damage, it was also found that firefighters directly involved in fire combat activities presented higher percentage of NET-FPG than non-exposed individuals (0 versus 2.14% TDNA; $p \leq 0.01$). Thus, this work contributes to a better knowledge on firefighters' occupational exposure to PAHs after an active participation in firefighting activities and proved the existence of genotoxic effects (basal damage and oxidative stress) in firemen.

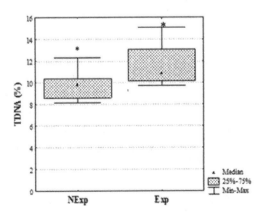

Figure 1. Percentage of DNA damage (TDNA,%) among non-exposed (NExp) and exposed (Exp) firefigters. *Statistically significant differences between the two groups ($p \leq 0.05$).

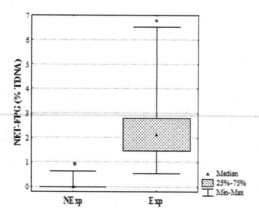

Figure 2. Percentage of oxidative stress (NET-FPG,%) among non-exposed (NExp) and exposed (Exp) firefigters. *Statistically significant differences between the two groups ($p \leq 0.05$).

4 CONCLUSIONS

The findings achieved in this study suggest that firefighters who were directly involved in fire combat activities presented predominantly higher concentrations of total urinary PAH metabolites. In addition, exposed subjects exhibited increased levels of DNA damage and higher oxidative stress comparatively with non-exposed ones, thus suggesting potential adverse health risk at genetic and molecular levels. Still, more studies with firefighters that actively participate in fire combat activities are needed to validate the findings of this study.

ACKNOWLEDGMENTS

Authors thank to *Escola Superior de Saúde-Instituto Politécnico de Bragança* and to all firemen involved in this work, for their collaboration. This work received financial support from European (FEDER funds through COMPETE) and National (*Fundação para a Ciência e Tecnologia* project UID/QUI/50006/2013) funds and in the scope of the P2020 Partnership Agreement Project POCI-01-0145-FEDER-006939 (Laboratory for Process Engineering, Environment, Biotechnology and Energy—LEPABE) funded by FEDER funds through COMPETE2020 - *Programa Operacional Competitividade e Internacionalização* (POCI). K. Slezakova is grateful for her fellowship (SFRH/BPD/105100/2014).

REFERENCES

Abreu, A., Costa, C., Pinho e Silva, S., Morais, S., Pereira, M.C., Fernandes, A. et al. 2017. Wood smoke exposure of Portuguese wildland firefighters: DNA and oxidative damage evolution. Journal of Toxicology and Environmental Health, A, 00: 1–9.

ACGIH, 2010. *Documentation for a Recommended BEI of Polycyclic Aromatic Hydrocarbons*. American Conference of Governmental Industrial Hygienists. Cincinatti, Ohio, USA.

Alshaarawy, O., Elbaz, H.A., Andrew, M.E. 2016. The association of urinary polycyclic aromatic hydrocarbon biomarkers and cardiovascular disease in the US population. Envronemnt International 89–90: 174–178.

ATSDR, 1995. Toxicological Profile for Polycyclic Aromatic Hydrocarbons, Atlanta, Agency for Toxic Substances and Disease Registry. Retrieved from /http://www.atsdr.cdc.gov/toxprofiles/tp69.html (last accessed in October 2017).

Caux, C., O'Brien, C. & Viau, C. 2002. Determination of firefighter exposure to polycyclic aromatic hydrocarbons and benzene during fire fighting using measurement of biological indicators. *Applied Occupational and Environmental Hygiene.* 17: 379–386.

Chetiyanukornkul, T., Toriba, A., Kameda, T., Tang, N., Hayakawa, K. (2006) Simultaneous determination of urinary hydroxylated metabolites of naphthalene, fluorene, phenanthrene and pyrene as multiple biomarkers of exposure to polycyclic aromatic hydrocarbons. Anal. Bioanal. Chem. 386, 712–718.

Edelman, P., Osterloh, J., Pirkle, J., Caudill, S.P., Grainger, J. & Jones, R. et al. 2003. Biomonitoring of chemical exposure among New York City firefighters responding to the World Trade Center fire and collapse. *Environ. Health Perspect.* 111: 1906–1911.

Hornung, R.W. & Reed, L.D. 1990. Estimation of average concentration in the presence of nondetectable values. *Appl. Occup. Environ. Hyg.* 5: 46–51.

IARC, 2002. Monographs on the Evaluation of the Carcinogenic Risks to Humans. *Naphthalene.* World Health Organization, International Agency for Research on Cancer. 82, Lyon, France.

IARC, 2010a. Monographs on the Evaluation of the Carcinogenic Risks to Humans. *Some non-heterocyclic polycyclic aromatic hydrocarbons and some related exposures.* International Agency for Research on Cancer. 92, 1–853.

IARC, 2010b. Monographs on the Evaluation of Carcinogenic Risks to Humans. *Painting, firefighting and shiftwork.* International Agency for Research on Cancer. 98, Lyon, France.

Jongeneelen, F.J. 2001. Benchmark guideline for urinary 1-hydroxypyrene as biomarker of occupational exposure to polycyclic aromatic hydrocarbons. *Ann. Occup. Hyg.* 45: 3–13.

Jongeneelen, F.J. (2014) A guidance value of 1-hydroxypyrene in urine in view of acceptable occupational exposure to polycyclic aromatic hydrocarbons. Toxicol. Lett. 231, 239–248.

Kamal, A., Cincinelli, A., Martellini, T., Malik, R.N. (2015) A review of PAH exposure from the combustion of biomass fuel and their less surveyed effect on the blood parameters. Environ. Sci. Pollut. Res. 22, 4076–4098.

Kanagasabapathy, A.S. & Kumari, S. 2000. *Guidelines on standard operating procedures for clinical chemistry.* World Health Organization, Regional Office for South-East Asia, New Delhi: 25–28.

Kim, K.-H., Jahan, S.A., Kabir, E. & Brown, R.J.C. 2013. A review of airborne polycyclic aromatic hydrocarbons (PAHs) and their human health effects. *Environ. Int.* 60: 71–80.

Laitinen, J., Mäkelä, M., Mikkola, J. & Huttu, I. 2010. Fire fighting trainers' exposure to carcinogenic agents in smoke diving simulators. *Toxicol. Lett.* 19: 61–65.

Li, Z., Romanoff, L., Bartell, S., Pittman, E.N., Trinidad, D.A., McClean, M. et al. (2012) Excretion profiles and half-lives of ten urinary polycyclic aromatic hydrocarbon metabolites after dietary exposure, Chem. Res. Toxicol. 25, 1452–1461.

Links, J.M., Kensler, T.W., & Groopman, J.D. (1995). Biomarkers and mechanistic approaches in environmental epidemiology. Annual review of public health, 16(1), 83–103.

Marie, C., Bouchard, M., Heredia-Ortiz, R., Viau, C., Maitre, A. (2010) A toxicokinetic study to elucidate 3-hydroxybenzo(a)pyrene atypical urinary excretion profile following intravenous injection of benzo(a)pyrene in rats. J. Appl. Toxicol. 30, 402–410.

NIOSH, 2007. *NIOSH Pocket Guide to Chemical Hazards*. U.S. Department of Health and Human Services, Public Health Service, Centers for Disease Control and Prevention. National Institute for Occupational Safety and Health, Cincinnati, Ohio.

Oliveira, M., Slezakova, K., Alves, M.J., Fernandes, A., Teixeira, J.P. & Delerue-Matos, C., et al. 2016. Firefighters' exposure biomonitoring: impact of firefighting activities on levels of urinary monohydroxyl metabolites. *Int. J. Hyg. Environ. Health.* 219: 857–866.

Oliveira, M., Slezakova, K., Alves, M.J., Fernandes, A., Teixeira, J.P. & Delerue-Matos, C., et al. 2017. Polycyclic aromatic hydrocarbons at fire stations: firefighters' exposure monitoring and biomonitoring, and assessment of the contribution to total internal dose. *J. Hazard. Mater. 323: 184–194.*

Reid, C.E., Brauer, M., Johnston, F.H., Jerrett, M., Balmes, J.R., Elliott, C.T. 2016. Critical review of health impacts of wildfire smoke exposure. Environ. Health Perspect. 124: 1334–1343.

Robinson, M.S., Anthony, T.R., Littau, S.R., Herckes, P., Nelson, X. & Poplin, G.S., et al. 2008. Occupational PAH exposures during prescribed pile burns. *Ann. Occup. Hyg.* 52(6): 497–508.

Semmens, E.O., Dmitrovich, J., Conway, K., Noonan, C.W. 2016. A Cross-sectional survey of occupational history as a wildland firefighter and health. American Journal of Industrial Medicine 59: 330–335.

WHO, 2000. *Air Quality Guidelines*, second ed. WHO Regional Publications, European Series No. 91, Copenhagen.

Yamano, Y., Hara, K., Ichiba, M., Hanaoka, T., Pan, G., Nakadate, T. (2014) Urinary 1-hydroxypyrene as a comprehensive carcinogenic biomarker of exposure to polycyclic aromatic hydrocarbons: a cross sectional study of coke oven workers in china. Int. Arch. Occup. Environ. Health. 87, 705–713.

Youssouf, H., Liousse, C., Roblou, L., Assamoi, E.M., Salonen, R.O., Maesano, C., et al. 2014. Quantifying wildfires exposure for investigating health-related effects. Atmospheric Environment. 97: 239–251.

Whole body vibration and acoustic exposure in construction and demolition waste management

M.L. de la Hoz-Torres
Department of Applied Physics, University of Granada, Granada, Spain

M.D. Martínez-Aires & M. Martín-Morales
Department of Building Construction, University of Granada, Granada, Spain

D.P. Ruiz Padillo
Department of Applied Physics, University of Granada, Granada, Spain

ABSTRACT: Waste management is a problem in Europe and has been identified in the lines of action of Horizon 2020. The processes involved in the waste management of the construction and demolitions sectors cause workers to be exposed to physical risks such as whole body vibrations and noise. This case study presents the analysis of a construction waste management plant. In order to analyse the physical risks to which workers are exposed in waste recovery processes, exposure to Whole-Body Vibrations (WBV) and noise has been identified, measured and evaluated in operations carried out by workers. The measurements have been made according to the Physical Agents (Vibration) Directive 2002/44/EC and the Physical Agents (Noise) Directive 2003/10/EC. This study shows that the workers are not exposed to WBV exposure level above the Exposure Action Value (EAV). However, on the sorting belt the noise level exposure is above the EAV.

1 INTRODUCTION

The future of the Construction Industry is under the EU framework programme Horizon 2020. The European Commission aims to enhance global competitiveness, promote sustainable economic growth and create new jobs (European Commission, 2017).

The Waste Framework Directive, or Directive 2008/98/EC of the European Parliament and of the Council of 19 November 2008 (Directive 2008/98/EC, 2008), includes one target to be achieved by 2020, which is to increase the weight of materials prepared for re-use, recycling and other material recovery by a minimum of 70%.

In order to achieve this, the construction industry in Europe must undergo some transformation. It will be necessary to optimize and improve waste management, and as a result, new technology and resources can be implemented. However, as a result of these changes workers will be exposed to new risks.

The construction sector is among the most hazardous industries, with a high number of workers affected by occupational diseases which are caused by vibrations and noise (MEYSS, 2017). The Construction and Demolition Waste (CDW) has been linked to dust, noise and vibration generation (Aguilar, 2016) and the exposure of these risk factors is found in a wide range of activities carried out in the process (use of vehicles, earth-moving machinery, etc.)

The European Directive 2003/10/CE (Directive 2003/10/EC, 2003) specifies that particular attention should be given to the effects on workers' health and safety resulting from interactions with noise and vibrations.

Workers can be exposed to these physical risks simultaneously and this may result in different health problems other than those caused by a single isolated exposure (OMS, 1981).

Temporary threshold shifts in hearing (Zhu et al, 1997) as well as permanent ear damage, may be caused by the combined exposure to noise and vibrations (Wasserman & Badger, 1973). It has affected workers psychologically, and the peripheral nervous and osteomioarticular systems of those who have been exposed (Mugica et al, 2004). As consequence, the exposure produces a higher number of stress-related diseases and work-related musculoskeletal disorders (MSDs) (OIT, 2001).

It is more common to find studies investigating the isolated effect of these agents than the combined effect. However, the great majority of the studies concluded that the there is a synergic effect associated with the combined exposure (Duarte et al, 2009).

Against this backdrop and taking into account the effects on the health and quality of life of workers, as well as the impact on productivity and economic growth (OMS, 2007), there is a clear need for control and preventive strategies (INSHT, 2014).

The objective of the present case study was to collate information regarding the whole body vibration (WBV) and noise risks in the different stages of the waste management process in order to determine the exposure levels, and subsequently compare them to the exposure limits laid out in the legislation.

2 MATERIAL AND METHOD

2.1 Legislation

One of the objectives of the Treaty on European Union and of the Treaty Establishing the European Community (2002/C 325/01) is the improvement, in particular, of the working environment to protect the health and safety of workers. To this end, twenty specific regulations have been developed, including those related to noise and vibration exposure.

The Directive 2003/10/EC (Directive 2003/10/EC, 2003) aims to ensure the health and safety of the worker and to create a minimum protection for all community workers, and in particular the risk to hearing, from exposure to noise.

As regards vibration exposure, the Directive 2002/44/EC lays down minimum requirements for the protection of workers from risks to their health and safety arising from exposure to mechanical vibrations (Directive 2002/44/EC, 2002), especially MSDs. The directive distinguishes between hand-arm-system vibrations (HAV) and WBV.

Both Directives define exposure limit values respectively on the basis of an eight hour reference period.

2.2 Equipment used

Both physical agents were measured in the CDW recycling plant. The equipment chosen to measure noise was a sound level meter (Model 2260 Investigator, Brüel & Kjaer). This machine allows simultaneous sound pressure level measurement and storage of 1/3 and 1/1 octave band filters.

WBV exposures during normal operations were measured with the Human Vibration Meter and Analyser SVAN 106 equipment (with the seat accelerometer SV 38V manufactured by SVANTEK). This equipment meets the requirements of ISO 8041-1:2017, ISO 2631-1,2&5 and ISO 5349-1&2 standard.

2.3 Description of the plant and sample collection

The chosen CDW recycling plant was located in Granada (Spain). The waste management process required the use of different machines capable of transmitting WBV and noise levels above the exposure limit levels.

There were six workers in charge of carrying out the recycling process at the plant, five of whom claim to have more than one year's experience in the company, and only one has less than three months' experience.

The CDW process begins with collecting and transporting waste to the recycling plant. Trucks deliver CDW to the plant, the material is then visually inspected and depending on its nature (masonry, concrete, etc.), it is emptied into a dumping area and then redirected to different areas. KOMATSU WA380 is used to move the material.

The process continues with the loading of the trommel using the Caterpillar 321 LCR. The machine is positioned between the storage area and the trommel, and rotates in order to load the material on to the trommel. If at some point the material cannot be reached, the KOMATSU WA380 is responsible for bringing it closer.

The material is separated in the trommel according to diameter, and any waste greater than 10 mm is sent to the sorting belt. On this belt, four workers stand on a platform in order to sort material manually (mainly paper, plastic, wood, etc.). Ferrous metals can be extracted from CDW with an electromagnet.

Finally, the material goes to a conveyor belt that feeds an impact crusher, to produce a recycled aggregate of a different size.

All equipment capable of transmitting noise and WBV to workers has been considered in the study (see Table 1).

2.4 Calculation of vibration exposure

After the observation of working practices and the information regarding the probable magnitude of

Table 1. Machines selected for measurement.

Type	Manufacturer	Model	Year	Wheels or tracks
Backhoe loader	Komatsu	WA380	2012	Wheels
Trommel	Trommelscreen	615	1995	–
Mill	Supertrack	1110	2005	–
Screening	Powercreen Selt time	400	2007	–
Excavator	Caterpillar	321 LCR	2006	Tracks

the vibration corresponding to the equipment, the levels of mechanical vibration to which workers are exposed were measured.

The assessment of the exposure level to WBV is based on the calculation of daily exposure, which can be measured using two different methods (ISO2631-1:2017):

The Daily exposure A(8) expressed as equivalent continuous acceleration over an eight-hour period, calculated as the highest (rms) value (equation 1).

$$A_i(8) = k_i \cdot a_{wi} \sqrt{\frac{T_{exp}}{T_0}} \qquad (1)$$

where a_{wi} is the effective value of acceleration weighted in frequency according to the orthogonal axes x, y, z.; T_{exp} is the exposure time; T_0 is the reference time of 8 hours; k_i is the factor of multiplication (ISO 2631-1:2017). The maximum of A_i calculated in the three directions is the value A(8).

The Vibration Dose Value (VDV) of the frequency weighted accelerations, calculated as the maximum of the value in the three directions (see equation 2).

$$VDV = \left\{ \int_0^T (a_W(t))^4 dt \right\}^{\frac{1}{4}} \qquad (2)$$

where $aw(t)$ is the frequency-weighted instantaneous acceleration; T is the total measurement period in seconds. Daily VDV is calculated as the highest value among the VDV calculated in the three directions.

For both methods the exposure limit values are defined in the Directive 2002/44/E (Directive 2002/44/CE, 2002) (see Table 2).

2.5 *Calculation of noise exposure*

Before the assessment and measurements were carried out, the length of exposure and ambient factors were considered in order to define the methods and the apparatus used.

The Daily noise exposure level (L_{Aeqd}) and the Peak sound pressure (P_{peak}) are the parameters

Table 2. WBV exposure limit values an action value (Directive 2002/44/EC, 2002).

Value	A(8) [m/s²]	VDV [m/s^{1.75}]
Exposure Action Value (EAV)	0.5	9.1
Exposure Limit Values (ELV)	1.15	21

Table 3. Noise ELV and EAV (Directive 2003/10/EC, 2003).

Value	L_{Aeqd} [dBA]	P_{peak} [dBC]
Upper EAV	85	137
ELV	87	140

used as a risk predictor. The L_{Aeqd} is calculated as the time-weighted average of the noise exposure levels for a nominal eight-hour working day (ISO 1999:2013) (see equation 3).

$$L_{Aeq,T} = 10\log\left[\frac{1}{N}\sum_{N}^{n=N} 10^{L_{Aeq,Tm,n}/10}\right] dB(A) \qquad (3)$$

The Peak is calculated as the maximum value of the 'C'-frequency weighted instantaneous noise pressure (see equation 4).

$$L_{Aeq,d} = L_{Aeq,T} + 10\log\left[\frac{T}{8}\right] dB(A) \qquad (4)$$

The EVL and the EAV with respect to the L_{Aeqd} and P_{peak} are fixed in the Directive 2003/10/EC (see Table 3).

3 RESULT

3.1 *WBV exposure*

Three samples have been measured. In two of them the workers were seated (Sample 1 and Sample 2), so the accelerometer disc was attached to the seat.

In the third sample the workers were standing, so the disc was placed on the surface of the floor, where they most often placed their feet.

The working conditions were the standard ones, so the measurements of the WBV were representative of the worker's daily exposure. Once measurement data had been obtained it was possible to calculate the workers' exposure levels. These levels have been compared to exposure limits laid out in the legislation (EAV; ELV).

3.1.1 *Sample 1: Trommel feeding with KATERPILLAR 321LCR*

The task was to feed the trommel using the CATERPILLAR 321LCR. The worker usually uses the machine (see Table 4).

The results obtained show that the acceleration values are higher on the X and Y axes than on the Z axis. This is due to the fact that the machine did not move; the process only requires the machine to

Table 4. Level obtained from WBV samples.

Exposure level	A(8) [m/s^2]	VDV [m/s$^{1.75}$]
Sample 1	0.201	5.012
Sample 2	0.059	3.733
Sample 3	0.031	0.688

Table 5. Level obtained from Noise samples.

Sample	L$_{Aeqd}$ (dBA)
1	81.0
2	78.9
3	82.8
4	77.2
5	76.7
6	79.2
7	73.5
8	89.2
9	88.7
10	77.6

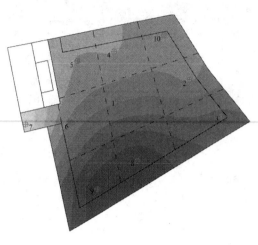

Figure 1. Noise zone of the recycling plant.

rotate to load the trommel. Therefore, this explains why the values on the Z axis are lower.

3.1.2 *Sample 2: KOMATSU WA3820 moving the material during the process*

The KOMATSU WA3820 was constantly moving during the process: loading trucks and moving material to different areas (see Table 4). The driver was the worker who usually undertook this task.

The Z axis has the highest acceleration, which is due to the fact that during the process the Backhoe loader is moving on a very irregular surface. The X and Y axes do not reach an acceleration value of the z axis, but both are similar.

3.1.3 *Sample 3: Sorting belt zone*

Four workers were performing this operation, two on each side of the belt. They had to select from the sorting belt the elements which were to be redirected to other containers for a different treatment. The Z axis has the highest acceleration (see Table 4).

3.2 *Noise exposure*

Ten different measurements have been made to determine the noise exposure level. The Evaluator Type 7820 program has been used to process the data obtained in the measurement (see Table 5).

For acoustic level readings, a grid (zoning of the plant) was established to ensure an accurate and well distributed reading (see Figure 1). To obtain consistent readings, the sound level meter was placed 1.5 m above ground level at every point.

4 DISCUSSION

Workers are exposed to a WBV level that does not exceed the EAV level and ELV level laid down in the European Directive.

However, the results of two samples obtained in noise exposure measurements exceed the ELV (sample 8 and 9). This is the area where the sorting belt is located and four workers remain there. In addition to being exposed to a noise level upper EAV (87 dBA), they are also exposed to WBV.

Although the WBV level is lower than the EAV level, the combined exposure to WBV and noise may affect the workers' health with different consequences than an isolated exposure to a single risk factor (OMS, 1981; Mugica Cantelar et al. 2004). The potential risk of this situation should therefore be taken into account and these factors should be considered in health surveillance.

The area with the lowest L$_{Aeqd}$ is the area furthest from where the machinery operates most frequently. These are material storage areas and are further away from the area where the trommel and the sorting belt are located. The constant presence of workers is not required in this area during the process.

The generation of noise and vibrations is a problem in the waste recovery process which has been highlighted in numerous research studies (Del Río et al, 2010; Aguilar, 2016), and obtaining values close to the limit levels for these physical risks was therefore expected.

5 CONCLUSIONS

Workers in the construction and waste management industry are those who are most exposed to vibrations and noise in their workplaces. This situation constitutes a risk for occupational diseases.

Any action that contributes to lowering these exposure levels will reduce the risk of disease, such as MSDs or hearing related diseases.

None of the workers are exposed to a WBV exposure level above the EAV level. However, on the sorting belt (where there are 4 workers) the noise level exposure is above the EAV. It is necessary to limit the noise level in that area.

Based on the results obtained from WBV measurements, the exposure level does not reach the EAV. However, the combination of noise and WBV may affect the workers' health and it should be considered in health surveillance.

REFERENCES

AENOR, 2004. UNE-EN 14253:2004+A1 Vibraciones mecánicas. Medidas y cálculos de la exposición laboral a las vibraciones de cuerpo completo con referencia a la salud. Guía práctica. AENOR.
AENOR, 2008. UNE-ISO 2631-1:2008 Mechanical vibration and shock. Evaluation of human exposure to whole-body vibration. AENOR.
Aguilar, A. "Reciclado de materiales de construcción". Boletín CF+S. Especial sobre Residuos. Ed. Instituto Juan Herrera. Madrid España. 2016.
Del Río, M., Izquierdo, P., Salto, I., & Santa Cruz, J., 2010. La regulación jurídica de los residuos de construcción demolición (RCD) en España. El caso de la Comunidad de Madrid. Informes de la Construcción, 62(517), 81–86.
Directive 2002/44/EC of the European Parliast and of the Council of 25 June 2002 on the minimum health and safety requirements regarding the exposure of workers to the risks arising from physical agents (vibration)
Directive 2003/10/EC of the European Parliament and of the Council of 6 February 2003 on the minimum health and safety requirements regarding the exposure of workers to the risks arising from physical agents (noise).
Directive 2008/98/EC of the European Parliament and of the Council of 19 November 2008, on waste and repealing certain Directives.
Duarte, M.L.M., Dornela, J.G., & Izumin, R. (2009). Combined Effects of Noise and Whole-body Vibration (WBV) on Human Hearing: Bibliographic Review and Proposed Study Methodology. Cobem.
European Commission. (10 de 10 de 2017). Horizon 2020. Obtenido de The EU Framework Programme for Research and Innovation: http://ec.europa.eu/programmes/horizon2020/en/what-horizon-2020.

European Union consolidated versions of the treaty on European Union and of the treaty establishing the European Community 2002 (2002/C 325/01).
Eurostat. (21 de 04 de 2015). Statistics explained. Obtenido de Waste generation by economic activities and households, EU-28, 2012 (%): http://ec.europa.eu/eurostat/statistics-explained/index.php/File:Waste_generation_by_economic_activities_and_households,_EU-28,_2012_(%25)_YB15.png
INSHT, 2014. Aspectos ergonómicos de las vibraciones, Madrid: (INSHT).
ISO 1999:2013. Acoustics—Estimation of noise-induced hearing loss.
ISO 2631-1:2017. Mechanical vibration and shock-Evaluation of human exposure to whole-body vibration-Part 1: General requirements. The Organization.
ISO 2631-2:2003. Mechanical vibration and shock—Evaluation of human exposure to whole-body vibration—Part 2: Vibration in buildings (1 Hz to 80 Hz).
ISO 2631-5:2004. Mechanical vibration and shock—Evaluation of human exposure to whole-body vibration —Part 5: Method for evaluation of vibration containing multiple shocks.
ISO 5349-1:2001. Mechanical vibration—Measurement and evaluation of human exposure to hand-transmitted vibration—Part 1: General requirements.
ISO 5349-2:2001. Mechanical vibration—Measurement and evaluation of human exposure to hand-transmitted vibration—Part 2: Practical guidance for measurement at the workplace.
ISO 8041-1:2017 Human response to vibration—Measuring instrumentation—Part 1: General purpose vibration meters.
MEYSS. (14/06/2017). Estadísticas de accidentes de trabajo. Madrid: MEYSS.
Mugica Cantelar, J., Román Hernández, J., & Cádiz García, A. (2004). Noise and vibration effects in heavy equipment drivers. Rev Cubana Salud Trabajo.
OIT. (2001). Enciclopedia de salud y seguridad en el Trabajo. Volumen 50: Ginebra: OIT.
OMS. (1981). Efectos sobre la salud de las exposiciones combinadas en el medio de trabajo. Ginebra: Serie de informes técnicos 662.
OMS, 2007. Salud de los trabajadores: plan de acción mundial – 60ª Asamblea mundial de la salud.
Wasserman, D.E., & Badger, D.W., 1973. Vibration and its relation to occupational health and safety. Bulletin of the New York Academy of Medicine, 49(10), 887.
Zhu, S.K., Sakakibara, H., & Yamada, S.Y. (1997). Combined effects of hand-arm vibration and noise on temporary threshold shifts of hearing in healthy subjects. International archives of occupational and environmental health, 69(6), 433–436.

Combined effects of exposure to physical risk factors: Impact on blood oxygen saturation (SpO$_2$) in construction workers

F.M. Cruz
Escola Politécnica da Universidade de Pernambuco, Brazil

P. Arezes
Centro ALGORITMI, Escola de Engenharia da Universidade do Minho, Portugal

B. Barkokébas Jr. & E.B. Martins
Escola Politécnica da Universidade de Pernambuco, Brazil

ABSTRACT: The construction industry workers are often exposed to more than one physical risk factor. However, common methods of risk assessment do not consider their combined effects. This paper aims to study the possible synergy between risk factors among construction workers, and its effect on their blood oxygen saturation (SpO$_2$). For that purpose, this study assessed the level of SpO$_2$ in workers conducting 15 full-time operations including usage of excavators, tractors, bucket trucks and cranes. Obtained results show that over the time, each risk factor had distinct and variable influence on oxygen saturation. This analysis is part of a broader project which includes other effects. Therefore, it is expected that at the end of the project would be possible to define the degree of synergy between studied physical risk factors studied and its impact on worker's health indicators.

1 INTRODUCTION

Workers in the construction industry are exposed to various occupational hazards such as noise, vibration, dust, heat and others (Barkokébas, Jr. et al., 2009). Machine operators have greater exposure to physical risk factors (Cruz et al., 2013).

These factors influence on human bodies (Evgen'Ev et al., 2014), and may, for example, cause changes in hearing thresholds and changes in the circulatory system (Manninen, 1984), or loss of muscle tone and fatigue (Li et al., 2012), or an increase feeling of irritability (Ljungberg et al. 2004).

Traditional risk assessment methodologies examine factors in isolation. However, in working environments of the construction industry, it is common to find simultaneously present more than one risk factor (Van Der Molen et al., 2016).

The combined effects of risk factors on workers' health are still poorly studied. Medical research shows that the human body processes environmental stimuli through the nervous system (Celani & Vergassola., 2012). This system is also responsible for sending responses for organic stabilization, called homeostasis (Schulkin, 2015).

The level of oxygen in the blood is called the oxygen saturation level (abbreviated as O$_2$ sat or SpO$_2$). The SO$_2$ is the percentage of oxygen which the blood is carrying, compared with its maximum transport capacity. Ideally, oxygen should be carried by more than 89% of red blood cells. Oxygen is the "gas" making the body functioning. If the level of this gas is low, the body works badly. Having low levels of blood oxygen can overburden the heart and the brain.

For maintaining cells healthy, people need saturation levels of at least 89%. According to Potthast, S et al., (2009) short time of lower levels will not cause damage. However, with constant low levels of oxygen the cells are battered and suffer damage. Oxygen saturation levels in blood below 89% are indicative of respiratory and pulmonary problems (Liu, J et al., 2010). The effects on the organism of low blood oxygenation are sleepiness, deficiency in the functioning of the nervous system and deficiency in cellular metabolism (Xu et al., 2012). This directly affects the performance of cognitive and muscle performance during the work.

Some authors studied combined effects of environmental stimuli on organic responses, especially when it had greater impacts on health of the individual (Kaas, 2010).

However, the experiments had no equivalent assessment of the 8 hours working day, participants were not workers already adapted to the environmental stimuli of workplaces, and assessments were conducted in controlled environments (chambers) and not with real working activities.

This article aims to explain the influence of combined exposure to physical risk factors and its combined effect on the saturation of O_2 in the blood-SpO_2 (%), among operators of construction machines, during a workday of 8 hours.

2 METHODOLOGY

2.1 Sample

Fifteen healthy, between 25 and 62 years old, machine operators (elevators, excavators, trucks and tractors) were monitored in the construction industry.

2.2 Structure of data collection

At the beginning of the working day the employees were interviewed and introduced to the research objectives. Initial data (weight, age, experience time and physical activity habit) were collected through a form. Subsequently, it was explained the operation of the necessary equipment for the collection of the noise of the whole body vibration and the thermal stress. Then the equipment was installed on workers and their workplaces. Data on levels of blood oxygen saturation (SO_2) were collected at the beginning of the working day.

The SO_2 data collection was carried out in 6 moments during the working day: beginning of the working day (T_0), 1st break (T_1), lunch (T_2), return (T_3), 2nd break (T_4) and end (T_{end}), featuring 6 SpO_2 collecting moments, as shown in Figure 1. The time interval between each period was 2 hours.

This configuration allowed the analysis of exposure time on the responses of the organism.

2.3 Occupational noise evaluation method

Noise was measured using the methodology of the National Institute for Occupational Safety and Heath—NIOSH (1998). An audio dosimeters were installed near the operator's hearing zone (model: The Edge5-Quest/3M, properly calibrated). The occupational exposure was registered during the minimum period of 6 hours and 30 minutes (75% of the working day), as required by the standard. The equipment was in compliance with international guidelines.

2.4 Whole body vibration assessment method

Whole Body Vibration was measured using the ISO 2631 recommendations (ISO 2631, 1997). An accelerometer was installed on the operator's seat (model: HAVpro-Quest/3M, properly calibrated). Data samples were collected in total for 10 minutes throughout the working day. Data was integrated through software provided by the equipment manufacturer.

2.5 Thermal stress assessment method

Thermal stress was measured using the methodologies provided by ISO 7243/1989 recommendations (ISO 7243, 1989). A set of thermometers (model: Quest Temp 34, properly calibrated), was used near the workplace.

2.6 Oxygen saturation measurement-SpO_2 (%)

For heart rate measurements, a pulse oximetry (model: CMS50DL-CONTEC) was installed on the index finger of the operator.

2.7 Statistical treatment

The statistical analysis tools of the MS EXECEL software were used for data treatment. Multiple linear regressions were performed. The input data (noise, vibration and thermal stress) were crossed with data on blood oxygen saturation.

3 RESULTS AND DISCUSSIONS

The data was organized into tables according to the period of the day in which they were collected. Table 1 shows the data collected in the interval T_1. First collected information were organized, then it was performed a statistical analysis of multiple linear regression for each range of data (T_0, T_1, T_2,

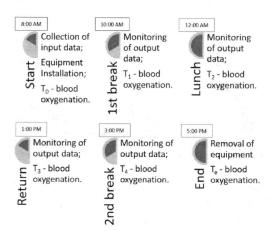

Figure 1. Measuring intervals T_0–T_{end}.

Table 1. Input data for the moment T_1.

Sample	Noise (dB)	WBV (m/s^2)	Heat (°C)	SpO$_2$ (%)
1	69.60	0.338	33.40	98
2	79.80	0.351	30.80	98
3	58.30	0.517	26.70	98
4	81.10	0.258	29.70	97
5	60.90	0.300	31.10	99
6	76.60	0.355	28.80	98
7	77.70	0.454	31.60	97
8	70.40	0.556	28.40	97
9	82.60	0.514	29.20	98
10	84.60	0.543	36.80	99
11	76.40	0.701	29.60	94
12	82.10	0.260	25.30	98
13	81.50	0.314	26.80	98
14	79.20	0.379	27.50	99
15	75.20	0.320	27.20	99

Table 2. Statistical analysis of SpO$_2$ for each risk factor.

	R^2 Adjusted	P-values Noise	WBV	Heat	"Significance-F"
T_0	0.910	0.001	0.625	<0.001	<0.001
T_1	0.914	0.002	0.224	<0.001	<0.001
T_2	0.910	0.013	0.928	0.013	<0.001
T_3	0.914	<0.001	0.872	<0.001	<0.001
T_4	0.913	<0.001	0.804	0.001	<0.001
T_{end}	0.911	0.003	0.368	0.015	<0.001

*Significant at a p-value <0.05.

Table 3. Influence of risk factors on blood oxygen saturation throughout the working day.

Coefficient of factors on SpO$_2$

	Noise	WBV	Heat
T_0	0.501	−18.607	2.464
T_1	0.484	18.197	2.011
T_2	0.608	1.567	1.714
T_3	0.715	−1.775	1.481
T_4	0.768	2.624	1.412
T_{end}	0.684	9.284	1.563

T_3, T_4 and T_{end}), and finally the input data (noise, vibration and thermal stress) were crossed with the given output data (blood oxygen saturation-SpO$_2$).

3.1 *Analysis of blood oxygenation-SpO$_2$ (%)*

Table 2 presents results of linear regression between the input data (noise, WBV and heat) and blood oxygen saturation throughout the working day.

The R-squared (R^2) set indicates the percentage of the output data (SpO$_2$) that can be explained by multiple linear regression of the input data for the 15 samples. In this way, the "adjusted R-squared" ranged between 0.910 (T_0 and T_2) and 0.914 (T_1 and T_3). This means that from 91.0% to 91.4% of the data collected can be explained by multiple linear regression, showing that the SpO$_2$ and physical risk factors have linear relationship above 75% during the working day.

The linear regressions have p-values associated with each data input. These values correspond to the level of statistical significance of each input data (noise, WBV and heat) on the SpO$_2$. Table 2 presents the p-values throughout the working day.

The statistical reference for defining that one variable is significant in relation with another is to be less than 0.05. That is, the input data is significant at the 5% level in relation to the output data. In this context, it is possible to verify that the noise and the heat between the T_0 and T_{end} has statistical significant on the SpO$_2$, while the occupational vibration did not prove significant.

However, the p-values represent only the significance of the input data (isolated) on the regression result. Not being a good parameter for the explanation of the model with a whole. Therefore, another relevant data of regression is the "Significance F" generated by the ANOVA variability analysis test. This is the standard deviation that measures dispersion around the regression line, also considering a level of 0.05. That is, if F is less than 0.05 the linear relationship is significant at the 5% level. In this context, Table 2 presents the values of F of signification of linear regression for each period of the working day.

As the values of F are below 0.05 it is possible to affirm that noise, vibration and heat generated by operations with construction machines have multiple significant linear relationship with SO$_2$ level operators 5%.

Linear regression produces a coefficient (weight) for each physical risk factor generated by operations with machines. This number corresponds to a degree of influence that noise, heat and vibration affect on the SpO$_2$ of operators. Table 3 shows these coefficients (weights) for each period of the working day.

The data organized in this way allow to understand how the influence of the physical risk factors generated by machines in construction operations vary throughout the day, and also to check the degree of influence (weight) of noise, vibration and heat on the SpO$_2$ of operators. Figure 2 shows the graph generated from the data gathered from Table 3, illustrating the behaviour of the WBV on the SpO$_2$ of operators.

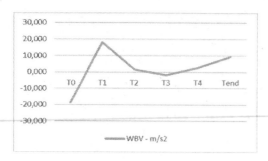

Figure 2. Influence of WBV on blood oxygen saturation during the working day.

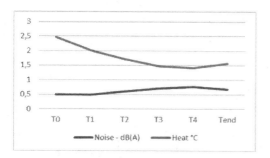

Figure 3. Influence of risk factors on blood oxygen saturation during the working day.

The influence of whole body vibration increases in the first 2 hours of work (T_0–T_1), then its influence reduces until the interval T_3, where it increase till the end of the working day. Therefore, the vibration at the beginning of the day is negative, after 2 hours it exerting high influence on SpO_2. However, it is assimilated by the body, reducing its influence during the day.

The graph on Figure 3 contains coefficients of noise and heat. Values of vibration coefficients have greater variations that those of noise and heat, therefore they were removed from the chart in order to improve the chart display.

The influence of noise presents slow growth trend along the journey. This demonstrates that the noise exert influence on cumulative SpO_2 as far as the working day progresses.

Thermal stress was shown to be more influential in the beginning of the working day, reducing its influence throughout the day. In this context, as well as vibrations, thermal stress is assimilated by the body, reducing its influence on SpO_2 during the day.

4 CONCLUSIONS

Operations using construction machines generate various physical risk factors (noise, vibration and thermal stress). When these risk factors are combined, they significantly influence the level of 5% on oxygen saturation of the operators. The physical risk factors generated by operations using construction machines have different degrees of influence. That is, each factor has a different weight on the operators SpO_2 level, and this influence varies during each period of the working day. This is because coefficients (weights) generated by the linear regression of data are different in each period of the day and different for each risk factor. Therefore, the research shows evidence that for explaining SpO_2 level variations, it is not enough to separately analyse noise, WBV and heat.

In this study, encountered blood oxygen levels were greater than 89%, representing a healthy range according to the existing literature. Although it is possible to affirm that there is influence of risk factors on the SpO_2, the influence is not damaging health of machine operators.

No studies were found on health problems related to high levels of blood oxygenation.

This preliminary research shows there is need to further studies on synergistic effects of occupational hazards on the health of workers.

REFERENCES

Barkokébas, B. J., Lordsllem Junior, A. C., & Vasconcelos, B. (2009). Sistema de Gestão em Segurança e Saúde do Trabalho. Recife-PE: EDUPE.

Celani, A., & Vergassola, M. (2012). Nonlinearity, fluctuations, and response in sensory systems. Physical Review Letters.

Cruz, F. M., Lago, E. G., & Barkokébas Jr, B. (2013). Evaluation of Noise Generated by Propagation Equipment Beat Stakes Construction Sites. Em P. M. Arezes, J. S. Baptista, M. P. Barroso, P. Carneiro, P. Cordeiro, N. Costa, . . . G. Perestrelo, Occupational Safety and Hygiene. Guimarães-Portugal: CRC Press.

Evgen'Ev, M. B., Garbuz, D. G., & Zatsepina, O. G. (2014). Heat shock proteins and whole body adaptation to extreme environments. Moscow: Springer Netherlands.

International Organization for Standardization. (1989). Hot environments—Estimation of the heat stress on working man, based on the WBGT-index (wet bulb globe temperature). Switzerland: ISO.

International Organization for Standardization. (1997). ISO 2631-1 – Mechanical Vibration and Shock—Evaluation of Human Exposure to Whole Body Vibration—Part 1: General Requirements. Switzerland: ISO.

Kaas, J. H. (2010). Evolution of nervous systems. Nashville: Elsevier Inc.

Li, Z., Zhang, M., Chen, G., Luo, S., Liu, F., & Li, J. (2012). Wavelet analysis of lumbar muscle oxygenation signals during whole-body vibration: Implications for the development of localized muscle fatigue. European Journal of Applied Physiology, pp. 3109-3117.

Liu, J., Chen, R., Zhong, N.S. (2010). Application of capnography and SpO$_2$ measurement in the evaluation of respiratory failure in patients with chronic obstructive pulmonary disease. Journal of Southern Medical University, 30 (7), pp. 1565–1568.

Ljungberg, J., Neely, G., & Lundstrom, R. (2004). Cognitive performance and subjective experience during combined exposures to whole-body vibration and noise. Int Arch Occup Environ Health, pp. 217–221.

Manninen, O. (1984). Hearing threshold and heart rate in men after repeated exposure to dynamic muscle work, sinusoidal vs stochastic whole body vibration and stable broadband noise. Int Arch Occup Environ Health, pp. 19–32.

National Institute for Occupational Safety and Health. (1998). Occupational Noise Exposure—Criteria for a Recommended Standard. Cincinnati: NIOSH.

Potthast, S., Schulte, A., Kos, S., Aschwanden, M., Bilecen, D. (2009). Blood oxygenation level-dependent MRI of the skeletal muscle during ischemia in patients with peripheral arterial occlusive disease RoFo Fortschritte auf dem Gebiet der Rontgenstrahlen und der Bildgebenden Verfahren, 181 (12), pp. 1157–1161.

Schulkin, J. (2015). Allostasis, homeostasis, and the costs of physiological adaptation. Washington, DC: Cambridge University Press.

Van Der Molen, H. F., De Vries, S., Jill Stocks, S., Warning, J., & Frings-Dresen, M. (2016). Incidence rates of occupational diseases in the Dutch construction sector, 2010–2014. Occupational and Environmental Medicine, pp. 350–352.

Xu, F., Liu, P., Pascual, J, M., Xiao, G., & Lu, H. (2012). Effect of hypoxia and hyperoxia on cerebral blood flow, blood oxygenation, and oxidative metabolism. Journal of Cerebral Blood Flow & Metabolism, pp. 1909–1918.

Evaluation of professional exposure to nanoparticles in automotive repair activities

Gonçalo Oliveira, Hélder Simões & Ana Ferreira
ESTeSC—Coimbra Health School, Environmental Health, Polytechnic Institute of Coimbra, Coimbra, Portugal

Francisco Silva
Technological Center of Ceramics and Glass, Coimbra, Portugal

João Paulo Figueiredo
ESTeSC—Coimbra Health School, Complementary Sciences, Statistics and Epidemiology, Polytechnic Institute of Coimbra, Coimbra, Portugal

ABSTRACT: Occupational exposure to nanoparticles is a recent and simultaneously increasing risk, which classifies it as an emerging risk. At work, inhalation is the most common route of exposure. Inhaled nanoparticles can be deposited in the airways and lungs depending on their shape and size. The objective of this project was to evaluate the professional exposure of this emerging risk to the level, productivity and sustainability of the automotive repair industry responsible for the production of nanoparticles, or whose production process implies the release of ultrafine particles. Exposure to nanoparticles varied depending on the workplace. The jobs that presented the greatest risk in terms of exposure to large concentrations of nanoparticles were welding, followed by the process of storage/office work and car repair.

1 INTRODUCTION

Nanomaterials have unique and more pronounced characteristics compared to the same material with no nanometric characteristics. Therefore, the physico-chemical properties of the nanomaterials may be different from the properties of the substance at macroscopic scale or larger scale particles. Nanotechnology is rapidly expanding. Nanomaterials offer unique technical possibilities, but can pose risks to the environment and raise health and safety concerns (EC, 2012; EU-OSHA, 2009) HSST.

According to the European Agency for Safety and Health at Work an "emerging OSH risk" is any simultaneously new and increasing risk (AESST, 2008). Because of their small size (size ranging from 1 to 1000 μm) and other unique characteristics, nanoparticles (NP) have the ability to injure humans and wildlife through the interaction of various mechanisms. In this regard, it has been previously reported in lung toxicity studies that exposure of the lungs to ultrafine particles or nanoparticles (defined here as a 100 μm sized particle in one dimension) produce greater adverse inflammatory response, when compared to larger particles of similar composition (Warheit, D. B.; Sayes, C. M.; Reed, K. L. et al., 2008). The situation regarding pollutants such as particulate matter, for which a lower threshold for health effects has not yet been established, remains a particular concern. Vehicles, industry, power stations, agriculture and homes contribute to Europe's air pollution. Transport remains a major contributor to poor air quality levels in cities and to related health impacts (AEA, 2015).

The main objective of the study was to contribute to the knowledge of the nanoparticle exposure in a car repair unit and to determine if the exposure to these particles varied according to the work station and working conditions, so as to be able to characterize the work places where there were emission of the particles, evaluate and analyze the risk of exposure.

2 MATERIAL AND METHODS

2.1 *Sample characterization*

This study was developed in the immediate vicinity of company specialized in nutrition and plant protection, specifically in the sector of the automotive repair shop. The target population was the employees exposed to the car industry, namely those who were involved in any work-related activity in the repair shop. The study was an observational prospective analytical cohort.

2.2 Methodology and instruments for data collection

For the determination of the professional exposure to nanoparticles in the working atmosphere, the collection, measurement, analysis and evaluation were performed using a traditional sampling method in Hygiene and Safety at Work (HSW), a portable particle counter (Model TSI CPC 3007), which read the values in real time. The analyses were always carried out with the care of keeping the device at the airway level. In order to verify the existence of minimum hygiene and safety at work conditions, a workspace analysis was performed, identifying the main aspects to be considered and identified as potential sources of danger in the car repair shop, based on a checklist. The information collection was carried out by direct observation and was applied the Administrative Rule No. 987/1993 of 6 October that establishes the minimum safety and health requirements in the workplace.

All data collected was used solely for academic purposes, and was not disclosed for other purposes other than that previously mentioned, without any financial or economic interests. It should also be noted that the analyzes carried out were made with the consent of the head of the HSW Department of the company.

2.3 Statistical analysis

Statistical analysis of the data was performed using the *software IBM SPSS Statistics*, 21.0 version. Through this software it was possible to carry out the statistical treatment of the obtained data, using descriptive statistics, with measures of central tendency as the mean and measures of dispersion as the standard deviation.

The statistical tests used for inferential analysis were the the *Mann-Whitney* Test, the *Kruskal-Wallis* Test and the Multiple Comparisons – *Dunn*. For statistical inference, it was possible to establish a 95% confidence level for a ≤ 5% random error.

3 RESULTS

For the determination of the professional exposure to nanoparticles in the working atmosphere, an initial evaluation of the total concentrations of nanoparticles inside the repair shop was carried out, comparing them with the existing concentrations in the external environment. After the statistical analysis of the collected data the results presented in Table 1 were obtained:

There was a significant change between the mean values for the sites where the measurement was performed (p-value < 0.05). The interior space, on average, had higher values in the order of 26486.20 part/cm^3 compared to the values found in the external environment.

Since there was a large difference between the mean levels of total NP concentration versus all indoor workplaces when compared to the outdoor environment, it was important to assess whether there was generally a significant difference between indoor space values when there was activity and when there was (control phase—without industrial activity). After the statistical analysis of the collected data the following results were obtained.

There was a significant difference between the mean values for the sites where the measurement was performed (p-value <0.05). At the time of activity, the mean had higher values with a difference of 18464.01 part/cm^3 compared to the values found in the control phase (Table 2).

Next, we tried to match each place of internal activity (Workshop Center, Office/Warehouse, Welding Zone), with the external environment being the following tables (Tables 3 and 4) representative of the results obtained for the various zones.

Table 1. Comparison of the average concentration of nanoparticles in the interior and exterior of the workshop.

	Sites	N° Samples (N)	Mean	Std. Deviation	p-value
Nanoparticle Concentration (part/cm^3)	Interior	2667	35738.04	16848.104	<0.0001
	Exterior	199	9251.84	1397.384	

Table 2. Comparison of the average concentration of nanoparticles in the interior when in activity and in control phase.

	Measurement Phases	N	Mean	Std. Deviation	p-value
Nanoparticle Concentration (part/cm^3)	Control Phase	1654	26090.75	13108.318	<0.0001
	Activity	1212	44554.76	17350.797	–

Table 3. Comparison of the average concentration of nanoparticles in the workshop center and the external environment.

	Workstation	N	Mean	Std. Deviation	p-value
Nanoparticle Concentration (part/cm^3)	Workshop Center	622	30114.42	8437.133	<0.0001
	External Environment	199	9251.84	1397.384	

Table 4. Comparison of the average concentration of nanoparticles in the office/warehouse and in external environment.

	Workstation	N	Mean	Std. Deviation	p-value
Nanoparticle Concentration (part/cm^3)	Office/warehouse	1449	34046.87	18649.035	<0.0001
	External Environment	199	9251.84	1397.384	

Table 5. Comparison of the average concentration of nanoparticles in the welding zone and in the external environment.

	Workstation	N	Mean	Std. Deviation	p-value
Nanoparticle Concentration (part/cm^3)	Welding Zone	596	45718.56	14539.607	<0.0001
	External Environment	199	9251.84	1397.384	

Table 6. Mean of the total concentration of nanoparticles between labor activities.

Nanoparticle Concentration (part/cm^3)	N	Mean	Std. Deviation
Automobile Repair	622	30114.42	8437.133
Office/Storage	1449	34046.87	18649.035
Welding	596	45718.56	14539.607
Total	2667	35738.04	16848.104

When compared, we found a significant difference (p-value < 0.05) between the mean values of the concentration levels of nanoparticles in the workshop center and in the external environment in the order of 20862.58 part/cm^3.

For the results obtained in the above table, we found again a great disparity between the concentrations in the office/warehouse zone and the outdoor environment (p-value <0.05), being the first one higher with a difference of 24795.03 part/cm^3, from the second one.

As for the average values with respect to the welding zone compared to the outside environment, we found a significant difference (p-value < 0.05) between the average values of the concentration levels of nanoparticles in the order of 36466.72 part/cm^3.

Since a significant difference was obtained after comparing the mean values of the nanoparticle concentration at the moment of activity and at the control phase, a comparison of the values resulting from the different activities developed in the interior space was relevant, in order to understand in which activity we would find the highest concentration levels of nanoparticle emission. Let's look at the results in the following table:

From the above table it can be concluded that the highest value of the total average concentrations belongs to the welding activity with about 45718.56 part/cm^3 with a significant difference of 9980,52 part/cm^3 with respect to the total (p-value < 0.05).

Since different mean values of total nanoparticle concentration were found in the different activities, it was necessary to carry out a multiple comparison to understand if there was a significant difference between them and if it existed, in which activity this was more accentuated. As can be seen, at the level of the car repair activity, this activity had significantly lower mean values than the identified in the welding activity (mean difference = 15604.14; p-value = < 0.0001) as well as the activity identified as office/storage (mean difference = 3932.45, p-value = < 0.0001). We also observed that the storage section showed values significantly lower than the welding section (mean difference = 11671.69; p-value = < 0.0001).

Table 7. Comparison of the mean concentration of nanoparticles between the measurement phases.

Work activity	Measurement phases	N	Mean	Std. Deviation	p-value
Automobile Repair	Control Phase	119	31432.15	3916.567	<0.0001
	Activity	503	29802.67	9162.156	
Office/Storage	Control Phase	841	18690.99	1588.170	<0.0001
	Activity	608	55287.49	6895.089	
Welding	Control Phase	495	44148.34	5017.545	<0.0001
	Activity	101	53414.18	32579.800	

In a last step it was important to compare the average concentration of nanoparticles in the control phase, with the average concentrations of nanoparticles in the different activities. Let's look briefly at the results in the following table (Table 7).

Considering the measurements obtained, it was verified that in all activities there was a significant difference between the average concentration of nanoparticles obtained in the control phase compared to the activity phase (p-value < 0.05). The largest mean difference belonged to the storage activity with a value of 36596.50 part/cm^3. In the automotive repair activity the average difference of the nanoparticle concentration was 1629.48 part/cm^3 and the highest mean concentration corresponded to the control phase. As for the welding activity, the value obtained as the mean difference of the nanoparticle concentration between the measurement phases was in the order of 9265.84 part/cm^3.

4 DISCUSSION

Measurements were taken in four different zones—office/warehouse, workshop center, welding area and outdoor environment—taking samples in a control phase, where there was no work activity, and at an operational stage. Three work activities were identified—welding, automobile repair and office/storage work. The results obtained in the initial phase of this study, where the average levels of NP concentration per cm^3 in the interior and exterior of the workshop were compared, suggest that these values were higher in the interior compared to the values found in the external environment. These indicators were expected since in the external environment there were no significant sources of emission of nanoparticles such as automobile traffic, other industrial processes or even influence by natural sources (CCDRN, s.d.).

This significant difference between the average concentrations of nanoparticles in the interior and exterior of the workshop served as control value for the pairing of the various locations inside the workshop, where measurements were taken, with the outside environment. It should also be noted that at this stage the average concentrations of nanoparticles there is no distinction between the values collected during the working and stopping periods.

After analysis of the previously mentioned pairing, it was possible to notice that, once again, there was a significant difference of the mean concentrations of nanoparticles, and of the three zones analyzed, the one that presented a greater disparity in relation to the concentration found in the external environment, was the welding zone, followed by the storage area and the workshop center, in descending order of values.

As expected, the mean concentrations when the sectors were in activity were higher than those found in the control phase. The next step was to understand how the average concentrations of nanoparticles varied among the work activities identified in the repair workshop. Once again, the highest levels of average nanoparticle concentration were obtained for the welding zone, which is due to the fact that the welding machine does not have a dust or smoke extraction system, positioned below the airways of the worker, which is exposed during the execution of the task (Moniz, A., & Albuquerque, P.; 2013). However, and unexpectedly, since some studies point to the products of vehicle exhaustion as the dominant source of fine and ultra-fine particles (Gertler, 2005), the values obtained in the car repair activity, which, in this case, can be defined as an activity with emission of particles resulting from mechanical and combustion processes—defined in Table 3 - were lower than those obtained in the storage activity. This result can be justified because the jobs and their tasks are neither isolated nor independent of the other areas of the workshop. This, complemented by the lack of ventilation, storage and sanitization of the storage/office area, meant that the concentration of the fumes in one place, as well as the concentration of toxic gases released in the welding process, increased by accumulation. Particles may have been deposited on walls, floors and other surfaces, as well as transported by workers themselves (Gomes, 1992).

Finally, it was important to compare the mean concentration of nanoparticles between the measurement phases for each work activity, which meant to see if all the values obtained during the control phase were actually lower than those obtained during the working phase. Once again all values corresponded to the expected one, except for the car repair activity that presented higher values during the control phase. This circumstance can be justified, again due to the lack of isolation of the work spaces and the lack of ventilation, obtained only through the entrance of the workshop, since without it the particles resulting from the exhaust and other processes can accumulate on existing surfaces in that area. In many cases, the local aspiration of a workshop is the most effective system to protect workers in adjacent areas. From all this it follows that the concentration of fumes will have to be controlled by means of ventilation. In fact, adequate ventilation is the key to effective control of the gas and smoke content in this environment (Gomes, 1992). It is also worth mentioning the temperature fluctuations presented in the repair shop, either very hot in the summer or very cold in the winter, which may constitute a factor for the accumulation of the emitted nanoparticles since the thermal environment in the work places must be adequate to the human organism, taking into account the production process, the working methods used and the physical burden imposed on workers (Russo, 2012). This study allowed us to conclude that the combination (workplaces); (sectors); (type of Nanomaterials) creates a three-dimensional array (Schulte, P., Geraci, C., Hodson, L., et al., 2010) with workplaces with different exposures according to task functions and activity (Schulte, P., Geraci, C., Kuempel, E.., et al., 2010).

5 CONCLUSIONS

The results obtained may be relevant indicators for the work to be carried out in the scope of the surveillance of exposed workers in the automotive repair industry, pointing to new strategies to promote safety and health in these professional groups. However, it is important to remember that it is a study carried out with a small sample and the size of the sample does not allow generalizations to be made for other exposed populations.

It is also important to note that the measurements were carried out with equipment that required special authorization and monitoring by a competent technician, which led to a limitation in the use of the equipment and it was not possible to carry out more than one measurement with a duration of approximately one hour. The short duration of this measurement and the fact that the sample is restricted to only one repair shop in a particular fiscal year and of a particular company was another obstacle that limited the generalization of these results. As suggestions for future research, the following can be pointed out: (a) the use of workers from different industries, ceramics, other technological areas, cosmetics and food, social and the arts, with the aim of finding differences between them; (B) include workers at the research and industrial scale of colleges, research organizations and industry; (C) to add other types of evidence and, in the case of cases with pathology, to carry out case-control studies.

REFERENCES

AEA, O Ambiente na Europa: Estado e perspetivas 2015 - Relatório síntese. 2015.

Agência Europeia para a Segurança e Saúde no Trabalho (AESST). (2008). Expert forecast on emerging chemical risks related to occupational safety and health. Bélgica.

Comissão de Coordenação e Desenvolvimento Regional do Norte. (s.d.). Principais Fontes de Emissão de Poluente.

European Agency for Safety and Health at Work (EU-OSHA), Workplace Exposure to Nanoparticles, European Risk Observatory, literature review, 2009. Available at: http://osha.europa.eu/en/publications/literature_reviews/workplace_exposure_to_nanoparticles.

European Commission (EC), Commission Staff Working Paper: Types and Uses of Nanomaterials, Including Safety Aspects. Accompanying the Communication from the Commission to the European Parliament, the Council and the European Economic and Social Committee on the Second Regulatory Review on Nanomaterials, SWD(2012) 288 final, Brussels, 3 October 2012. Available at: http://eurlex.europa.eu/LexUriServ/LexUriServ.do?uri = SWD:2012:0288:FIN:EN:PDF 3.

Gertler, W., (2005). Diesel vs. gasoline emissions: Does PM from diesel or gasoline vehicles dominate in the US? Atmospheric Environment, 39, 2349–2355.

Gomes, J. (1992). Higiene e segurança em soldadura. Oeiras: ISQ.

Moniz, A., & Albuquerque, P. (2013). Exposição profissional a nanopartículas na indústria farmacêutica—estudo exploratório.

Russo, I.C. (2012). Influência das Condições de Trabalho na Exposição aos Fumos de Soldadura e Efeitos na Saúde, 123.

Schulte, P., Geraci, C., Hodson, L., Zumwalde, V., Castranova, V., Kuempell, E., ... Murashov, V. (2010). Nanotechnologies and Nanomaterials in the Occupational Setting. Ital J. Occup Environ.Hygi 2010,1(2), 63–68.

Schulte, P., Geraci, C., Kuempel, E., Zumwald, R., Hoover, M., Castranova, V., ... Savolainen, K. (2008). Sharpening the focus on occupational safety and health in nanotechnology. Scand J Work Environ Health, 2008;34(6), 471–478.

Warheit, D.B., Sayes, C.M., Reed, K.L., and Swain, K.A. "Pharmacology & Therapeutics Health effects related to nanoparticle exposures : Environmental, health and safety considerations for assessing hazards and risks," vol. 120, pp. 35–42, 2008.

Occupational exposure to bioaerosols in the waste sorting industry

V. Santos, J.P. Figueiredo & M. Vasconcelos Pinto
ESTeSC—Coimbra Health School, Instituto Politécnico de Coimbra, Coimbra, Portugal

J. Santos
Escola Superior de Tecnologia e Gestão, Instituto Politécnico de Viana do Castelo, Viana do Castelo, Portugal

ABSTRACT: Biological agents may be present in the occupational environment and become a risk to the workers' health. The main source of exposure are small particles carried by air, the bioaerosols. The study of occupational exposure is still under development, and this investigation contributes to the knowledge of viable bacterial and fungal contamination in the waste sorting industry. For such purpose, air samples were analyzed in 5 Portuguese industries. The mean concentration of bacteria in the air was positively and significantly correlated with the type of sorting carried out, indicating that the glass sorting industry contamination with bacteria was higher, namely *Enterobacter cloacae*, *Micrococcus* spp. and *Staphylococcus* (negative coagulase). In the case of fungi, although the average concentration was not significantly correlated with the waste sorting performed, it was in the plastic sorting industry that the higher contamination was found, namely with species of *Penicillium* spp.

1 INTRODUCTION

There are several environmental factors that affect the individual in his work environment, among which, biological agents can be identified (L. Conceição Freitas, 2011). Resulting from their presence in workplaces, there may be risk situations for workers (A. Guedes, 2006).

At the workplace, bioaerosols are the main source of exposure of workers to biological agents. According to Goyer et al. bioaerosols are defined as airborne particles containing living organisms, such as metabolite microorganisms, toxins or microorganism fragments (Goyer, Lavoie, Lazure, & Marchand, 2001). These are defined as airborne particles, including spores and fungal hyphae, bacteria, endotoxins, β (1 $\rightarrow$ 3)-glucans, mycotoxins, high molecular weight allergens and particles (particulate matter) that generally are composed of biological agents (Oppliger, 2014).

The waste sorting industry is a value-added activity with a strong growth trend marked with an increase in the number of workers devoted to this sector. Therefore, the evolution of waste management technologies has resulted in an increase in the risks to workers responsible for the collection, sorting, treatment and disposal of waste. The workers involved in the sorting activities are identified as having the highest risk of exposure to biological agents, as the debris are perfect for microorganism's proliferation. The health consequences of exposure to biological agents depend on the substance involved, the level of exposure and individual susceptibility of the exposed worker (European Agency for Safety and Health at Work, 2007). The health symptoms observed in workers involved in waste management range from infections caused by parasites, viruses or bacteria, allergies caused by exposure to organic dust, which are recognized as a problem related to exposure to bioaerosols, the main responsible for the mobility of microorganisms by air. Finally, poisoning or toxic effects as well as cancer or damage to the fetus may also be the consequence of some biological risks (European Agency for Safety and Health at Work, 2003).

Portuguese legislation does not cover limit values for exposure to biological agents, braced by the recognized lack of dose/effect ratio of exposure levels. As a result, it is relevant the study of environmental contamination of the paper, packaging and glass sorting industries.

Therefore, it was the main goal of this study to know the risk of occupational exposure to biological agents (viable bacteria and fungi) in five waste sorting industries, of which four are sorting units for paper and plastic packaging and one for selective sorting of packaging glass.

2 MATERIAL AND METHODS

2.1 Sample characterization

In this study, 5 companies were involved in the treatment of waste representing the fractions of the selective collection (plastic/packaging, paper/carton and glass), consisting of 170 workers, were reproduced.

2.2 Methodology and instruments for data collection

Sample collection and microbiological analysis were performed according to the methodology described by Vanconcelos Pinto and colleagues (Vasconcelos Pinto et al., 2015) in three distinct areas: Critical Area (CA—packaging, paper and glass sorting), Noncritical Area (NCA—administrative services) and Control Point (CP—outside). Sampling sites were selected taking into account the possibility of comparison and consequently the confrontation between the most critical exposure areas (where waste is handled) and the least critical (where workers are not directly exposed to the process of sorting waste), in parallel with the control point (outside).

Statistical analysis of results was based on multiple comparisons using the Corrected Bonferroni Test (DUNN).

2.3 Statistical analysis

The results with reference to environmental microbiological contamination were analyzed using statistical tools. The results for the air samples were entered into an IBM SPSS Statistics version 24 database for Windows. The interpretation of the statistical tests was performed based on a significance level $p = 0.05$ (confidence interval of 95%). Due to the nature of the data, the non-parametric Kruskal-Wallis test was applied followed by a Non-Parametric Multiple Comparison Test—Bonferroni correlation (DUNN).

3 RESULTS

The mean count of bacteria and fungi in the CA (1.2×10^3 ufc/m^3 e 1.6×10^4 ufc/m^3, respectively) was always higher than that obtained in the CP (2.3×10^2 ufc/m^3 e 1.0×10^3 ufc/m^3, respectively) and in NCA (3.9×10^2 ufc/m^3 e 5.2×10^2 ufc/m^3, respectively). The multiple comparisons between the three sampling points for the total bacterial count, using the Bonferroni test, showed significant differences. For total fungal counts, significant differences were found, except between NCA and CP (Tables 1 and 2). It was found that in the CA, on average, the estimated values of total bacteria in the air showed significantly higher estimates comparatively in CP and the first one with NCA ($p < 0.05$).

Regarding the average concentration variation of *Bacillus* spp. and *Bacillus cereus* among the evaluation points, there were statistically significant differences ($p < 0.05$). CA presented higher values compared to NCA and CP ($Z = 6.324$; *p-value* < 0.0001) and ($Z = 5.925$; *p-value* < 0.0001), respectively. There were significant statistically differences ($p < 0.05$) among average counts of bacteria of the species *Micrococcus* spp. and *Staphylococcus* (negative coagulase). CA presented higher values compared to NCA and CP ($Z = 4.671$; *p-value* < 0.0001) and ($Z = 4.863$; *p-value* < 0.0001), respectively. However, the remaining species showed to be homogeneous, in relation to their concentration, among the different collection points ($p > 0.05$).

It was found higher values on airborne total fungi in CA showed comparatively in CP and NCA ($p < 0.05$).

There were statistically significant differences between *Aspergillus niger* specie among the sampling points ($p < 0.05$). CA presented higher values compared to NCA and CA ($Z = 6.023$; *p-value* < 0.0001).

Among *Aspergillus* genus, the designations actually represent sections (or complexes) of closely related species (also referred to as cryptic species) that cannot be clearly distinguished morphologically.

The average variation of *Cladosporium* spp. and *Penicillium* spp. showed significant differences between the analyzed sampling points ($p < 0.05$). CA presented higher values compared to NCA and CA ($Z = -3.759$; *p-value* < 0.0001) and ($Z = 10.411$; *p-value* < 0.0001), respectively. Though, the remaining species were found to be homogeneous in relation to their concentration, among the different sampling points ($p > 0.05$).

Total Bacteria concentration was statistically significant between the sampling points ($p < 0.05$). Glass Sorting presented higher values than Paper and Plastic Sorting ($Z = 3.946$, *p-value* < 0.0001). The average variation of *Bacillus cereus* bacteria between the analyzed points showed statistically significant differences ($p < 0.05$). The Glass Sorting presented higher values compared to the Plastic Sorting ($Z = 2.630$; *p-value* < 0.0001). The mean concentration of bacteria in the air was positively correlated and statistically significant with the type of selective sorting carried out. The concentration of airborne bacteria was higher in the glass sorting industry (Table 3).

In the case of fungi, the average concentration found in the air was not positively correlated and statistically significant in relation to the type of selective sorting performed. It was the plastic sorting industry that presented the higher concentration of airborne fungi (Table 4). At the three

Table 1. Bacteria count considering the three points analyzed.

Biological agent	Collecting point			Mean Differences	p-value
	Critical area (n) Mean ± Standard deviation	Noncritical area (n) Mean ± Standard deviation	Control point (n) Mean ± Standard deviation		
Total Bacteria	(79) 1209.11 ± 1512.02	–	(57) 228.25 ± 265.53	980.87	<0.0001
Total Bacteria	–	(62) 393.87 ± 385.36	(57) 228.25 ± 265.53	165.63	0.007
Total Bacteria	(79) 1209.11 ± 1512.02	(62) 393.87 ± 385.36	–	815.24	<0.0001
Acinetobacter lwoffi	(4) 630.0 ± 877.19	(5) 72.0 ± 90.39	(1) 400.0	–	0.163
Arthrobacter spp.	(3) 253.33 ± 337.24	(1) 10.00	(2) 25.00 ± 21.21	–	0.258
Bacillus spp.	(68) 253.97 ± 293.66	(52) 66.54 ± 131.61	(40) 55.00 ± 63.65	–	<0.0001
Bacillus cereus	(50) 131.8 ± 143.45	(21) 30.95 ± 29.65	(32) 26.56 ± 34.51	–	<0.0001
Bacillus subtilis	(9) 126.67 ± 160.93	(2) 310.00 ± 395.98	(6) 28.33 ± 31.25	–	0.087
Bacillus licheniformis	(5) 136.0 ± 248.35	(3) 20.00 ± 10.00	(1) 30.00	–	0.523
Bacillus megaterium	(7) 140.00 ± 168.52	(1) 10.00	(1) 10.00	–	0.112
Bacillus sphaericus	(3) 33.33 ± 23.09	(2) 15.00 ± 7.07	(2) 10.00 ± 0.00	–	0.122
Brevibacterium spp.	(3) 73.33 ± 50.33	(2) 30.00 ± 0.00	(1) 10.00	–	0.294
Cellulomonas spp./ *Microbacterium* spp.	(4) 70.00 ± 47.61	(7) 28.57 ± 26.1	(7) 35.71 ± 44.29	–	0.173
Corynebacterium spp.	(8) 176.25 ± 176.14	(8) 36.25 ± 39.98	(4) 70.00 ± 101.00	–	0.158
Enterobacter cloacae	(5) 360.00 ± 682.64	(1) 20.00	(0)	–	0.235
Gardnerella vaginalis	(2) 20.00 ± 0.00	(2) 10.00 ± 0.00	(4) 17.50 ± 9.57	–	0.279
Klebsiella oxytoca	(5) 92.00 ± 65.73	(1) 130.00	(0)	–	0.373
Micrococcus spp.	(31) 318.71 ± 697.86	(17) 68.82 ± 69.36	(13) 16.92 ± 11.09	–	<0.0001
Pantoea agglomerans	(6) 118.33 ± 61.45	(2) 25.00 ± 7.07	(1) 10.00	–	0.058
Pseudomonas aeruginosa	(4) 25.00 ± 10.00	(1) 510.00	(0)	–	0.114
Serratia marcescens	(3) 1386.67 ± 2246.09	(1) 70.00	(0)	–	0.655
Staphylococcus negative coagulase	(73) 421.37 ± 848.88	(59) 257.97 ± 268.83	(42) 135.00 ± 168.71	–	<0.0001
Staphylococcus positive coagulase	(1) 100.00	(4) 70.00 ± 53.54	(1) 10.00	–	0.493
Stenotrophomonas maltophilia	(4) 82.5 ± 69.46	(1) 10.00	(0)	–	0.157

Note: n = Samples Number. Testes: Wilcoxon-Mann-Whitney; Kruskal-Wallis.

sampling points it was observed that the mean total counts of viable fungi and bacteria in the CA were always higher than the mean counts obtained at the CP and NCA (Table 5). The CA values (paper, packaging and glass) had total bacterial counts between 1.2×10^2 ufc/m^3 and 8.3×10^3 ufc/m^3, with total fungal counts varying between 2.0×10^1 ufc/m^3 and 2.8×10^4 ufc/m^3. While the values obtained in NCA showed total bacterial counts between 3.0×10^1 ufc/m^3 and 1.9×10^3 ufc/m^3, the total fungal counts varied between 5.0×10^1 ufc/m^3 and 5.0×10^3 ufc/m^3. The values obtained in the CP showed total bacterial counts between 1.0×10^1 ufc/m^3 and 1.2×10^3 ufc/m^3, varying the total fungal counts between 5.0×10^1 ufc/m^3 and 2.3×10^4 ufc/m^3. The environmental microflora for viable microorganisms was mostly composed by fungi.

4 DISCUSSION

In the waste management sector, environmental fungi colonize and grow in organic matter beyond human control. They play an important role in the decomposition of it and are heavily involved in waste treatment (Oppliger & Duquenne, 2016). The emission of fungal aerosols during waste recycling operations has been reported in several studies. The average environmental concentration of airborne fungi in Korean recycling centers, for example, was measured in 1.8×10^4 ufc/m^3 (Oppliger & Duquenne, 2016). Another study carried out at two municipal solid waste treatment plants in Finland revealed environmental concentrations of 470 to 2.9×10^5 ufc/m^3 for cultivable airbone fungi (Oppliger & Duquenne, 2016). These values are in line with those obtained in our study where

Table 2. Fungi count considering the three points analyzed.

Biological agent	Collecting point			Mean Differences	p-value
	Critical area	Noncritical area	Control point		
	(n) Mean ± Standard deviation	(n) Mean ± Standard deviation	(n) Mean ± Standard deviation		
Total Fungi	(80) 16414.25 ± 9951.90	–	(57) 228.25 ± 265.53	980.87	<0.0001
Total Fungi	–	(62) 519.36 ± 803.64	(63) 1016.98 ± 2977.06	–497.63	0.206
Total Fungi	(80) 16414.25 ± 9951.90	(62) 519.36 ± 803.64	–	15894.90	<0.0001
Filamentous fungi					
Aspergillus flavus	(0)	(6) 20.00 ± 10.95	(2) 365.00 ± 417.19	–	0.039
Aspergillus niger	(65) 38.31 ± 41.82	(21) 16.67 ± 22.44	(16) 15.00 ± 6.32	–	<0.0001
Cladosporium spp.	(25) 326.00 ± 429.52	(56) 144.64 ± 139.51	(60) 470.33 ± 735.18	–	0.001
Penicillium spp.	(79) 16419.24 ± 10033.82	(62) 289.84 ± 687.20	(60) 107.67 ± 151.92	–	<0.0001
Streptomyces spp.	(1) 80.00	(3) 13.33 ± 5.77	(8) 35.00 ± 30.24	–	0.212
Trichophyton spp.	(3) 33.33 ± 11.55	(10) 16.00 ± 8.43	(13) 26.15 ± 18.05	–	0.120
Trichophyton equinum	(1) 40.00	(2) 10.00 ± 0.00	(9) 17.78 ± 8.33	–	0.115
Yeasts					
Candida spp.	(4) 120.00 ± 121.11	(11) 89.09 ± 176.49	(10) 29.00 ± 28.07	–	0.089
Candida famata	(2) 50.00 ± 42.43	(4) 132.50 ± 201.06	(1) 60.00	–	0.979
Cryptococcus humicola	(0)	(1) 40.00	(3) 13.33 ± 5.77	–	0.157
Cryptococcus laurentii	(2) 30.00 ± 14.14	(8) 20.00 ± 11.95	(14) 31.43 ± 28.52	–	0.562
Rhodotorula spp.	(5) 508.00 ± 1080.06	(18) 206.11 ± 650.37	(21) 234.76 ± 598.75	–	0.264
Saccharomyces spp.	(0)	(1) 60.00	(3) 20.00 ± 0.00	–	0.083

Note: n = Samples Number. Tests: Wilcoxon-Mann; Kruskal-Wallis.

Table 3. Assessment of biological risk (bacteria) in the critical area.

Biological agent	Critical area			p-value
	Paper sorting	Plastic sorting	Glass sorting	
	(n) Mean ± Standard deviation	(n) Mean ± Standard deviation	(n) Mean ± Standard deviation	
Total bacteria	(30) 896.00 ± 755.63	(42) 774.76 ± 436.31	(7) 5157.14 ± 2380.01	<0.0001
Bacillus spp.	(25) 242.00 ± 272.46	(36) 215.56 ± 278.12	(7) 494.29 ± 371.43	0.066
Bacillus cereus	(19) 140.00 ± 135.01	(26) 86.15 ± 90.29	(5) 336.00 ± 230.39	0.022
Bacillus subtilis	(4) 235.00 ± 200.91	(5) 40.00 ± 20.00	(0)	0.133
Bacillus licheniformis	(3) 206.67 ± 323.32	(2) 30.00 ± 14.14	(0)	1.000
Bacillus megaterium	(3) 46.67 ± 46.18	(4) 210.00 ± 200.33	(0)	0.271
Corynebacterium spp.	(5) 130.00 ± 170.59	(3) 253.33 ± 190.09	(0)	0.177
Enterobacter aerogenes	(1) 150.00	(3) 113.33 ± 113.72	(0)	0.655
Enterobacter cloacae	(0)	(4) 55.00 ± 34.16	(1) 1580.0	0.157
Klebsiella pneumoniae spp.	(0)	(4) 55.00 ± 57.46	(1) 300.00	0.147
Micrococcus spp.	(17) 267.06 ± 373.09	(11) 114.55 ± 113.17	(3) 1360.00 ± 2061.36	0.188
Pantoea agglomerans	(2) 95.00 ± 63.64	(4) 130.00 ± 66.33	(0)	0.475
Pseudomonas aeruginosa	(0)	(3) 26.67 ± 11.54	(1) 20.00	0.564
Staphylococcus negative coagulase	(29) 311.72 ± 324.32	(39) 345.64 ± 267.38	(5) 1648.00 ± 3087.32	0.382

Note: n = Samples Number. Tests: Wilcoxon-Mann; Kruskal-Wallis.

Table 4. Assessment of biological risk (fungi) in the critical area.

	Critical area			
	Paper sorting	Plastic sorting	Glass sorting	
Biological agent	(n) Mean ± Standard deviation	(n) Mean ± Standard deviation	(n) Mean ± Standard deviation	p-value
Total Fungi	(29) 14385.52 ± 11121.25	(40) 18370.50 ± 9011.95	(11) 14649.09 ± 9417.98	0.509
Filamentous fungi				
Aspergillus niger	(23) 35.22 ± 22.94	(33) 29.70 ± 26.04	(9) 77.78 ± 88.00	0.098
Cladosporium spp.	(13) 351.54 ± 299.88	(9) 140.00 ± 133.42	(3) 773.33 ± 1069.08	0.139
Penicillium spp.	(29) 14155.86 ± 11284.59	(39) 18777.95 ± 20.00	(11) 14023.64 ± 10021.64	0.165
Yeasts				
Rhodotorula spp.	(3) 26.67 ± 11.55	(1) 20.00	(1) 2440.00	0.264

Note: n = Samples Number. Tests: Wilcoxon-Mann; Kruskal-Wallis.

Table 5. Total counts of viable bacteria and fungi per collecting point.

	Total bacteria count (ufc/m^3)			Total fungi count (ufc/m^3)		
Collecting point	Minimum	Maximum	(n) Mean ± Standard deviation	Minimum	Maximum	(n) Mean ± Standard deviation
Critical Area	120.0	8300.0	(79) 1209.11 ± 1512.02	20.0	28020.0	(80) 16414.25 ± 9951.90
Noncritical area	30.0	1860.0	(62) 393.87 ± 385.36	50.0	4950.0	(62) 519.36 ± 803.64
Control Point	10.0	1230.0	(57) 228.25 ± 265.53	50.0	23330.0	(63) 1016.98 ± 2977.06

Note: n = Samples Number. Tests: Wilcoxon-Mann; Kruskal-Wallis.

–1.6×10^4 ufc/m^3. Other environmental levels were reported in another study conducted during glass bottle recycling in Canada where individual exposure to airborne fungi in waste collection ranged between 4.8×10^3 and 1.0×10^5 ufc/m^3 (Oppliger & Duquenne, 2016). In a recent study conducted in Germany, individual airborne fungi exposures were measured at 1.8×10^6 ufc/m^3 during paper sorting and recycling, and up to 3.2×10^6 ufc/m^3 throughout sorting and recycling of plastic waste (Oppliger & Duquenne, 2016). These values fall within the environmental concentrations in the CA—plastic sorting (1.8×10^4 ufc/m^3) higher than the concentrations in CA—paper sorting (1.4×10^4 ufc/m^3).

Penicilium has been described as being the genus of fungi present in aerosols generated during waste treatment, however there is no information available on the occupational exposure to this species (Madsen, Alwan, Ørberg, Uhrbrand, & Jørgensen, 2016).

According to (Schlosser, Déportes, Facon, & Fromont, 2015) levels of personal exposure in the sorting areas were 13 000 ufc/m^3 for fungi and 1800 ufc/m^3 for bacteria in four Plastic Materials Recovery Facilities during six campaigns of sampling.

Staphylococcus, *Bacillus* e *Micrococcus* were previously found as predominant airborne genus in a waste packaging sorting station (Madsen, Alwan, Ørberg, Uhrbrand, & Jørgensen, 2016), as in the waste sorting units studied.

This study presented limitations, namely the fungal load underestimation bias due to rely of results only on culture based-methods and the use of active methods as the only approach to assess fungal contamination.

The results of the different studies are difficult to compare. In fact, there are a wide variety methods used in the collection and analyzes of viable bioaerosols. The results are often influenced by the methodologies used. Collection of airborne viable bacteria and fungi is based on impaction or filtration. Thus, the development of international standards would be a very important step in order to allow a reliable comparison between different occupational exposures (Goyer et al., 2001). The first step should be the concise knowledge of the different forms of exposure to biological agents.

5 CONCLUSIONS

The results obtained in this research point's the existence of a significant environmental contamination of viable bacteria and fungi in the working atmosphere of waste treatment industries. Occupational exposure to viable microorganisms was considerably different in the studied facilities, depending on the type of waste treated.

In terms of environmental contamination, the glass sorting industry proved to be the most critical, being the environmental microflora of the CA is mainly composed of bacteria. The environmental microflora of the other sectors dedicated to waste sorting was mostly composed of fungi, predominantly characterized by the coexistence of *Penicillium* spp. Regarding the bacteriological characterization of the environmental samples, there was a predominance of *Staphylococcus* spp. (negative coagulase) and *Bacillus* spp.

The results of this research reinforce the need for specific training plans related to occupational exposure to biological agents adopted in parallel with engineering and organizational measures, as well as collective protection measures reinforced by the implementation of individual protection measures.

REFERENCES

Conceição Freitas, L. (2011). Manual de Segurança e Saúde do Trabalho. (Robalo Manuel, Ed.) (2nd ed.). Lisbon.

European Agency for Safety and Health at Work. (2003). FACTS Biological Agents.

EU-OSHA (European Agency for Safety and Health at Work). (2007). Expert forecasts on emerging biological risks related to occupational safety and health. https://doi.org/ISSN 1830–5946.

Guedes, A.B. (2008). Agentes biológicos no Trabalho: Perigos ocultos! Autoridade para as Condições do Trabalho. 1–3.

Goyer, N., Lavoie, J., Lazure, L., & Marchand, G. (2001). Bioaerosols in the workplace: Evaluations, Control and Prevention Guide. Quebec.

Madsen, A.M., Alwan, T., Ørberg, A., Uhrbrand, K., & Jørgensen, M.B. (2016). Waste Workers' Exposure to Airborne Fungal and Bacterial Species in the Truck Cab and during Waste Collection. Annals of Occupational Hygiene, 60(6), 651–668. https://doi.org/10.1093/annhyg/mew021.

Oppliger, A., & Duquenne, P. (2016). Chapter 8 - Highly Contaminated Workplaces. Environmental Mycology in Public Health. https://doi.org/http://dx.doi.org/10.1016/B978-0-12-411471-5.00008-9.

Schlosser, O., Déportes, I.Z., Facon, B., & Fromont, E. (2015). Extension of the sorting instructions for household plastic packaging and changes in exposure to bioaerosols at materials recovery facilities. Waste Management, 46, 47–55. https://doi.org/10.1016/j.wasman.2015.05.022.

Vasconcelos Pinto, M., Veiga, J.M., Fernandes, P., Ramos, C., Gonçalves, S., Vaz Velho, M., & Santos Guerreiro, J. (2015). Airborne Microorganisms Associated with Packaging Glass Sorting Facilities. Journal of Toxicology and Environmental Health, Part A, 78(11), 685–696. https://doi.org/10.1080/15287394.2015.1021942.

Nanomaterials *versus* safety of workers—a short review

Daniela da Conceição Peixoto Teixeira
School of Management and Technology, Polytechnic of Porto, Portugal

Paulo A.A. Oliveira
CIICESI—School of Management and Technology, Polytechnic of Porto, Portugal

ABSTRACT: Nanomaterials are particles invisible to human eye. These materials are present in everyday life. Currently, there are several working places, where workers are exposed to this kind of agents, this causes a health and safety concern. The current work aims to disclose the importance of the work safety on the protection of workers that are exposed to nanomaterials.

Keywords: Nanomaterials, nanotechnology, work environment, worker health, safety at work

Presentation Preference: Poster

1 INTRODUCTION

Nanomaterials (NM) are particles invisible to human eye. The dimensions of these materials are in the order of 10^{-9} m or nanometre, which corresponds to a millionth of a millimetre, or one billionth of a meter. These orders of magnitude are comparable to the size of atoms and molecules (Filho, 2014). The dimension of nanomaterials confers physical, chemical and biological properties different from materials with the same chemical composition with dimensions above the nanometre scale, such as fine powders (Louro & Silva, 2011).

Nanomaterials are present in everyday life (Filho, 2014). These materials can have a natural origin, they can, for example, be found in volcanoes, or anthropogenic origin, which is related with human activity, for example particles originated from automotive combustion and tobacco smoke. Manufactured nanomaterials are the most sought after and are required for specific applications, such as concrete (Louro, Borges, & Silva, 2013). Being an emerging technology, the risk associated with the manufacture and use of these agents is poorly understood (UGT, 2016). However, there are currently many workers who are exposed to nanomaterials, mainly during the production stage, which brings safety and health associated risks. The harmful effects depend on the type of nanomaterial, but not all of them are toxic. In order to reduce worker's exposure to these agents is essential for the employer to be aware and to comply with current legislation. For the protection of workers is essential to apply preventive measures, such as, the application of technical measures to control risks at source, organizational measures and, as last resource, the use of individual protection equipment (OSHA, 2013). Employers should also keep their employees up-to-date on the chemical hazards present in their workplace by making known the chemical products handled and the existing preventive measures for the protection of workers' safety and health (NIOSH/CDC, 2016).

2 NANOMATERIALS

According to the European Commission, a nanomaterial is "*A natural, incidental or manufactured material containing particles, in an unbound state or as an aggregate or as an agglomerate and where, for 50% or more of the particles in the number size distribution, one or more external dimensions is in the size range 1 nm – 100 nm*" (Europeu, 2011).

The International Organization for Standardization (ISO) defines "*material with any external dimension in the nanoscale or having internal structure or surface structure in the nanoscale*" (Calderón-Jiménez et al., 2017). For ISO two categories for NM (Castillo, 2013; Calderón-Jiménez et al., 2017):

- Nano-objects: "discrete piece of material with one, two or three external dimensions in the nanoscale". This is classification into tree categories: "*Nanoparticles: nano-object with all external dimensions in the nanoscale (where the lengths of the longest and the shortest axes of the nano-object do not differ significantly*";

"*Nanofiber: nano-object with two similar external dimensions in the nanoscale and the third dimension significantly larger*" and "*Nanoplate: nano-object with one external dimension in the nanoscale and the other two external dimensions significantly larger*";
- Nanostructured material: with an internal or superficial structure at nanoscale.

NM can have natural origins, for example, they can be produced by volcanoes and forest fires, or can be anthropogenic, originated by human activity, such as automotive combustion or tobacco smoke (Louro, Borges, & Silva, 2013). Those who raise more interest are the manufactured nanomaterials, which are already used in a wide range of products. Some of them have already been used for decades, such as synthetic amorphous silica with application in concrete, tires and food products. Other type of nanomaterials were discovered more recently, for example: nanotitanium dioxide, used as a Ultra Violet (UV) ray blocking agent in paints or sun blockers; nanosilver, used as an antibacterial agent for textile and medical applications; carbon nanotubes, a light material with high mechanical resistance, and interesting heat dissipation and electrical conductivity properties, finding its use in electrotechnology, energy storage, spacecraft and vehicle structures, and in the manufacture of sport equipment (Louro, Borges, & Silva, 2013).

As shown in Figure 1, there is a great progress in the presence of manufactured nanomaterials (MNM) on consumer products, stating the growing significance of the introduction of this technology in product manufacture (Wiesner, Lowry, Alvarez, Dionysiou, & Biswas, 2006; Vance et al., 2015). FNM are becoming the center of attention, because they have the potential to improve the quality and performance of many types of products (Castillo, 2013).

Table 1. Application of products containing nanomaterials by sector (Adapted from Castillo, 2013; Louro et al., 2013).

Sectors of activity	Task/Activity type
Automotive industry	Paint and coatings; Reinforced automotive parts, fuel additives, batteries, durable and recyclable tires; Optical sensors.
Biomedicine; pharmaceutical	hospital textiles coating, masks, surgical gowns, catheters; dressings for wounds; molecular image; additive in polymerizable dental materials, additive in bone cement; Implant coating for joint replacement.
Chemistry and materials	Pigments; Ceramic powders; corrosion inhibitors; Antibacterial surfaces; Thermal insulation; Paints.
Cosmetic and personal care	Sunscreens; Facial moisturizers; Dental paste, lipstick, acne treatments and hair care products.
Electronics and communications	Molecular electronics; Hardware; Memories and high-density information storage; Multifunctional catalysts; Microchips; Sensors; Nanorobots, nanoscale automatics operations.
Energy	Photovoltaic cells; Batteries; Insulating materials.
Environment	Climatic modulation; Fertilizers; Water filters; Catalysts for air quality.
Food	Plastic packaging to block UV rays and provide antibacterial protection; Bottles;
Sports	Textiles; Boat liner; Fishing rods; Tennis rackets; Golf clubs; Baseball Bats; Ski Equipment.

Nanoscale materials improve industrial development, making these materials applicable in a wide range of industrial and consumer products. According to the Table 1, NMs are being used in several sectors of activity, thus causing a rise in the human exposure.

Production, usage and disposal of nanomaterials cause their appearance on air, water, soil and organisms. (Wiesner et al., 2006).

3 NANOMATERIAL EXPOSURE *VERSUS* WORKER'S SAFETY

Humans are exposed to NM during different phases of their life cycle, that is, during synthesis, production, incorporation in products, and usage (Figure 2). Disposal of NM also originates

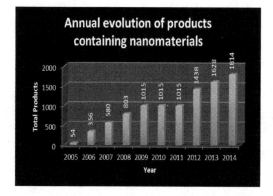

Figure 1. Evolution of marketed products containing nanomaterials (Adapted from Vance et al., 2015).

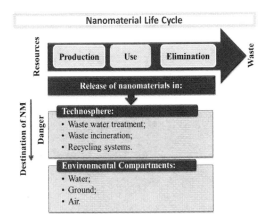

Figure 2. Nanomaterial Life Cycle (Adapted from Louro, Borges, et al., 2013).

human exposure, due to its accumulation in the environment.

The phase where there is a higher probability of exposure is during production, however the human being is also exposed by the inhalation of NM released in the atmosphere and by the ingestion/use of products that contain NM in their composition. (Louro, Tavares, Leite, & Silva, 2013; Louro, Borges, et al., 2013; Wiesner et al., 2006).

Particles on the nanoscale are very active and have various forms such as, spheres, fibres, tubes, rings and sheets. They can enter the human body by inhalation, through the skin, by injection or ingestion.

The main route of exposure is inhalation. Once inhaled, a substantial portion of nanoparticles are able to deposit evenly in the respiratory tract, and may persist for a long period of time. While present in the human body, these particles may be translocated to different places via different routes. They can enter the bloodstream and distribute to different organs, like the liver, kidneys and heart.

The second main route of exposure is ingestion. Nanoparticles enter the digestive tract through the mouth. Here, the particles can move through the digestive tract and intestine, and enter the bloodstream.

The absorption through the skin is another via of exposure. For that reason, skin contact must be avoided during the manipulation of these materials.

Transdermal exposure, may occur during the use of biomedical applications for diagnosis and therapy that require intravenous, subcutaneous, intramuscular administration (Oberdörster et al., 2005; Castillo, 2013).

Since the possibility of increasing the exposure to nanomaterials is during the production phase, concern for the health and safety of workers is a priority. Workers are the first to be exposed to the products of technology, which demands that a set of good labor practices is instilled in the workplace (Louro & Silva, 2011).

In order to protect the safety and health of workers, legislative regulations have been created. European legislation referring to worker protection can be applied to nanomaterials, but does not expressly refer to them.

The current applicable legislation is listed below (OSHA, 2013):

- Directive 2007/30/CE, which alters the directive 89/391/CEE, of the Council, of June 12nd of 1989, relative to the application of measures destined to promote the improvement of safety and health of workers in the working place;
- Directive 98/24/CE, of the Council, of April 7th of 1998, relative to the protection of the safety and health of workers from risks from the exposure to chemical agents in the workplace;
- Directive 2004/37/CE of the European Parliment and the Council, of April 29th of 2004, relative to relative to the protection of the safety and health of workers from risks related to exposure to carcinogens or mutagens at work;
- Legislation about chemical substances REACH.

Employers are the ones responsible for risk assessment and management of nanomaterials in the workplace. If the use and production of nanomaterials cannot be cut out, or substituted by less hazardous processes or materials, the worker's exposure must be minimized. This can be achieved by adopting preventive measures hierarchically structured, establishing the following priority order:

1. Technical measures of risk control on the source;
2. Organizational measures;
3. Use of personal protective equipment.

In order to help in the prevention of safety and health of workers, the employer should analyze and register their exposure to nanoparticles. Nevertheless, identifying nanomaterials and their emission sources is not an easy task, which requires the search for the available tools for its execution.

The evaluation of worker's exposure to nanomaterials, can be organized by different stages (Gibson et al., 2010):

- Identification of potentials sources of nanoparticle emission. The meaning of "emission source" differs from "exposure source", in the sense that there is not necessarily a correlation between working with a source of emission and been exposed;

- Qualitative evaluation of potential exposure. This includes register the number of workers that handle nanomaterials, material quantities, handling procedures and if nanoparticles are dispersed or as a powder during handling;
- Measuring the possible exposure parameters.

Workers in activity sectors that handle nanomaterials must be completely involved and properly trained to perform the job in safety. Employers should ensure that each worker receives appropriate information and training, and that medical examinations are carried out periodically to check the worker's health (Castillo, 2013).

4 CONCLUSION

Nanomaterials are increasingly more present in everyday life. Currently, there are various workstations where workers are effectively exposed to these agents. This situation causes concerns related to the health and safety of workers. For that reason, is of utmost importance to study the type of nanomaterials existing in each workplace and assess their harmfulness. The current legislation (Framework Directive 89/391/CEE, the Directive 98/24/CE and Directive 2004/37/CE) and the guidelines and tools, are available to help manage the risks of nanomaterials in the workplace.

Despite many uncertainties, there are concerns related with the risks of nanomaterials to worker's health and safety. As such, employers should, alongside workers, determine preventive measures based on prudent risk management. Identification of nanomaterials, their emission sources and exposure levels, may not be an easy task. For that reason, is essential to create regulations and specific legislation for NMs, to achieve greater efficiency in risk assessment.

REFERENCES

Calderón-Jiménez, B., Johnson, M. E., Montoro Bustos, A. R., Murphy, K. E., Winchester, M. R., & Vega Baudrit, J. R. (2017). Silver Nanoparticles: Technological Advances, Societal Impacts, and Metrological Challenges. *Frontiers in Chemistry*, 5(February), 1–26. http://doi.org/10.3389/fchem.2017.00006.

Castillo, A. M. P. Del. (2013). *Nanomaterials and workplace health & safety What are the issues for workers?*

Europeu, P. (2011). O que são nanomateriais? Comissão Europeia abre novos caminhos com uma definição comum, 1–2.

Filho, J. de M. (2014). Ética no ambiente de Trabalho. *Boletim Comissão Interna de Prevenção de Acidentes*, 8.

Gibson, R. M., Adisesh, A., Bergamaschi, E., Berges, M., Bloch, D., Hankin, S., ... Riediker, M. (2010). Strategies for Assessing Occupational Health Effects of Engineered Nanomaterials. *NanoImpactNet Reports—Http://www.nanoimpactnet.eu/index.php?page=reports*, (September 2010), 1–27.

Louro, H., & Silva, M. J. (2011). Nanomateriais manufaturados—um risco para a saúde dos trabalhadores?

Louro, H., Borges, T., & Silva, M. J. (2013). Nanomateriais manufaturados: novos desafios para a saúde pública. *Revista Portuguesa de Saude Publica*, 31(2), 145–157. http://doi.org/10.1016/j.rpsp.2012.12.004.

Louro, H., Tavares, A., Leite, E., & Silva, M. J. (2013). Nanotecnologias e saúde pública. *TecnoHospital: Revista de Gestão E Engenharia Da Saúde*, 59–Nanot, 12–15. Retrieved from http://repositorio.insa.pt/handle/10400.18/1973.

NIOSH/CDC. (2016). Building a Safety Program to Protect the Nanotechnology Workforce: A Guide for Small to Medium-Sized Enterprises.

Oberdörster, G., Maynard, A. A., Donaldson, K., Castranova, V., Fitzpatrick, J., Ausman, K. K., Heyder, J. (2005). Principles for characterizing the potential human health effects from exposure to nanomaterials: elements of a screening strategy. *Particle and Fibre Toxicology*, 2(1), 8. http://doi.org/10.1186/1743-8977-2-8.

OSHA. (2013). A gestão dos nanomateriais no local de trabalho. *Segurança E Saúde No Trabalho*, 1–4.

Vance, M. E., Kuiken, T., Vejerano, E. P., McGinnis, S. P., Hochella, M. F., & Hull, D. R. (2015). Nanotechnology in the real world: Redeveloping the nanomaterial consumer products inventory. *Beilstein Journal of Nanotechnology*, 6(1), 1769–1780. http://doi.org/10.3762/bjnano.6.181.

Wiesner, M. R., Lowry, G. V., Alvarez, P., Dionysiou, D., & Biswas, P. (2006). Assessing the Risks of Manufactured Nanomaterials. *Environmental Science & Technology*, 40(14), 4336–4345. http://doi.org/10.1021/es062726m.

Use of the 5S methodology for improving work environment

A. Górny
Poznan University of Technology, Poznan, Poland

ABSTRACT: The use of the 5S methodology may be viewed as a way to boost a company's competitiveness. Competitive advantages result from improvement measures. To ensure that the changes one makes are indeed effective, it is essential to identify the nature of the irregularities faced and the available opportunities for improvement. Improvement measures must reflect the specific operational character of the organization. The article refers to a manufacturing company in which significant work environment issues were discovered. Such issues affected the functioning of its employees ultimately compromising the company's efficiency and competitiveness. The proposed improvements reflect the findings of an assessment of the current situation. Accordingly to their nature, the improvements were ascribed to appropriate action categories, i.e. preparatory, implementing and improving, which can be linked to individual stages of 5S deployment.

1 INTRODUCTION

A substantial factor for the competitiveness of enterprises is their effectiveness in controlling excessive expenditures. In the majority of cases, overspending results from mismanagement. It requires changes to either eliminate the cause of the problem or minimize its impact (Górny 2016a, b, Kawecka-Endler, Mrugalska 2011). An essential part of such actions is the improvement of the work environment in response to the presence of factors that either pose potential health hazards or are already found to have had deleterious effects on human health (Dahlke et al. 2016, Helund et al. 2016, Lasota, Hankiewicz 2016).

For the changes to be effective, the underlying issue needs to be identified. Such identification helps select remedies adequate to the specific nature of the problems faced. In doing so it is important to take into account man. (Górny 2011, Lasota, Hankiewicz, 2016, Mrugalska et al. 2016). The nonconformities discovered may be viewed as impediments to the safe and healthy performance of work. Further aims of the effort are to find solutions that best enhance process efficiency and effectiveness and define improvement goals to ensure the proper functioning of workers in their environment.

The key purpose behind measures intended to improve the work environment is to eliminate the root causes of issues (Lasota, Hankiewicz, 2016, Górny, 2017b). This should be viewed as a way to enable the employer to satisfy its fundamental obligation relating to working conditions, which is to eliminate hazards affecting employees (Dahlke et al. 2016, Górny 2015).

Desirable improvements can be achieved with the use of the Kaizen philosophy focused on eliminating waste, standardization and ensuring orderliness in the workplace (Jaca et al. 2014, Budzynowska et al. 2014, Folejowska 2012, Żuchowski, Łagowski 2004). This requires good management and employee self-discipline. In actual practice, desirable effects are accomplished by employing practices that support improvements in work culture and increase process efficiency. The literature describes such practices as 5S (Imai 2012, Kakkar et al. 2015, Mazur, Gołaś 2015). They can be seen as a tool for ensuring order and rationally managing the resources available at workstations (Zimmerman, 1991). Kaizen guidelines suggest that the key principle applying to occupational safety is to make improvements by unconditionally insisting on having workstations kept in good order at all times (*Tool...*, 2007, Jiménez et al. 2015).

2 DESCRIPTION OF THE 5S METHOD

5S is a set of actions which rely on the creative thinking of workers and their commitment to quality and productivity improvements (Jaca et al. 2014, Mĺkvaa et al. 2016, Kiran 2017, Żuchowski, Łagowski 2004). The resulting changes are aimed at setting in order and precisely utilizing an organization's workspace. When adopting 5S, one must bear in mind that the process should be continual, engage all employees, and affect the functioning of the entire organization (Jiménez et al. 2015). 5S elements may be divided into two categories of measures (Agrahari 2015, Górny 2016a, Imai 2012, Żuchowski, Łagowski 2004):

– Sort, systematize and shine, which make it possible to adopt rules,

- Standardize and employ self-discipline to sustain and improve the effects of employing the first three pillars.

The meanings of these element along with the benefits they provide are summarized in Table 1. The improvement measures adopted in response to a negative assessment of the current situation should be ascribed to one of the three phases of preparation, implementation and improvement.

The design of the 5S methodology makes it a useful tool for managing safety and creating employee-friendly working conditions (Górny 2016a). Safety can be seen as working conditions that protect workers from occupational accidents and diseases (Górny, 2017b, Genaidy et al. 2007, Mĺkvaa et al. 2016). Safety audits should note the availability of equipment that protects employees' health and lives.

It is increasingly common to add one more element of 5S practices to assessing and creating effective working conditions and optimizing loads.

Once added, the element, which is safety, creates a 5S+1 system (Górny 2016a, Tool..., 2007). The 5S principles are also increasingly evolving towards a system that brings together the protection of occupational health and safety and the creation of a highly efficient and clean workplace. The fundamental idea in 5S with respect to such systems is to keep the workspace neat and tidy. Such an approach is largely aligned with the principles of ergonomics and good management at and around workstations (Imai 2012, Misztal et al. 2014).

Table 1. The role of individual elements of 5S.

5S elements	Scope of implementation/Sample benefits
1S seiri sort/separate	– Sort all items into necessary and unnecessary, – Clear any unwanted and unused items from shop floor. This will create ample room for work performance.
2S seiton systematize/ set in order	– Arrange useful items in an orderly and structured manner. This ensures having "the right item at the right place at the right time". It will help use tools fast and avoid errors
3S seiso sweep/shine	– Clean and refresh the workstation and the area around it. This helps eliminate hazards associated with dust, dirt and foreign objects from the workstation limiting their adverse impacts on workers and their disrupting effects on workflows
4S seiketsu standardize	– Document and standardize all practices and solutions you apply. This strengthens the positive perception by employees of the way work is performed and boosts productivity.
5S shitsuke self-discipline/ sustain	– Adhere to rules, ordinances, guidelines, instructions and procedures, – Improve applicable rules, ordinances, guidelines, instructions and procedures. This helps sustain the principles of working together in work environment, etc.

Table 2. Issues considered in assessing behaviors and work organization in a manufacturing company.

5S elements	Requirement area covered by safety audit
1S	– Placement of tools and equipment, – Clear assignment of tools and equipment to specific storage locations, – Accessibility of any tools and equipment required for work performance, – Ability to find tools and equipment easily, – Items left behind from previous task.
2S	– Labels on locations for tool, equipment and material storage, – Preparation and use of classifications employed to sort tools and other items, – Ensuring ways to easily find tools and equipment, – Signage on safety-critical solutions (e.g. emergency escape routes).
3S	– Cleanliness and orderliness at specific workstation, – Cleanliness and orderliness on shop floor, – Impact of cleanliness on technical condition of equipment, – Recognition that every worker is responsible for keeping his/her workstation clean, – Use of cleanliness and orderliness as shop floor control factor.
4S	– Access to properly prepared work instructions and manuals, – Understanding and effective use of sorting, systematizing and cleaning by employees, – Checks and supervision over employees' conformity with requirements pertaining to sorting, systematizing and cleaning
5S	– Extent of 5S use in the organization, – Workers volunteering ideas for improving working conditions, – Honing of 5S principles in the organization, – Worker engagement in the implementation and improvement of 5S principles, – Worker approval of the adopted rules of behavior conforming to 5S guidelines.

3 PRACTICAL ASPECTS OF THE 5S USED

3.1 *The 5S use in a manufacturing company*

The enterprise assessed against 5S criteria carries out production involving the mechanical working of structural materials. It employs 7 workers on the production floor, who operate machine tools of various sizes and types. Each worker is fully capable of working at any of the workstations. Production is supported by Production Management, whose responsibilities include the development of machine tooling technologies and planning execution.

The primary aim behind the 5S deployment in the organization was to reduce hazards to the health and lives of its employees and create opportunities for the improvement of working conditions and load optimization, thus helping the organization to boost work efficiency. The level of conformity with the solutions required under 5S was evaluated against a checklist covering the issues presented in Table 2.

The checklist targeted occupational health and safety, conformity with ergonomic requirements, workplace cleanliness and tidiness and manage-

Table 3. Proposed improvements based on assessment of current situation.

Area of 5S	The preparation phase	The implementation phase	The improvement phase
1S	– Identification of necessary tools and equipment, – Identification of tools and equipment posing hazards and resulting in strenuousness, – Identification of tools and equipment that compromise work efficiency.	– Sorting of tools and equipment at and around workstations, – Removal of tools and equipment posing hazards, resulting in strenuousness and adversely affecting work ergonomics.	– Development of criteria for the future identification of tools and equipment that are unnecessary, pose threats or hamper work performance, – Definition of rules for clearing tools and equipment from workstations.
2S	– Place workstation items as prescribed, – Properly describe places used to store workstation items, – Properly secure items that may pose threats and cause strain.	– Keep areas and items posing potential threats in order, – Eliminate causes of untidiness.	– Take care to put items away to prescribed places, – Checks whether items have been put away properly, – Improve ways of securing items that pose potential threats and strain
3S	– Draw up cleaning schedule for areas critical for occupational safety, – Identify areas in which untidiness may pose significant threats, – Identify causes of the kinds of untidiness that pose threats and causes strain.	– Keep areas and equipment contributing to threats clean and tidy, – Eliminate causes of untidiness.	– Check locations of items and cleanliness for occupational safety, – Perform cleaning regularly upon completion of work, – Tidy items regularly after completion of work.
4S	– Identify conditions that are necessary for the effective selection, systematic arrangement and cleaning duly.	– View the adopted rules for selection, systematic arrangement and cleaning as part of the standard way in which the company is operated, – Ensure proper supervision to ensure that selection, systematic arrangement and cleaning become routine behaviors.	– Constantly update and improve the adopted standards.
5S	– Find ways to maintain and improve the standards in place, – Define intervals for sorting, systematic arrangement and cleaning tasks, – Assess effectiveness of sorting, systematic arrangement and cleaning tasks.	– Assess and monitor outcomes of sorting, systematic arrangement and cleaning, – Present outcomes of adopting sorting, systematic arrangement and cleaning procedures to concerned parties.	– Continuously improve sorting, systematic arrangement and cleaning efforts, – Duly assess the feasibility of applying further improvement measures.

ment engagement in improvement measures. The checklist helped assess working conditions and identify areas for improvement. The proposed improvements will make it possible to implement and hone the selected measures.

3.2 Working conditions improvement guidelines identified by the 5S methodology

Compliance with safety requirements was assessed on the basis of the responses received. Furthermore, areas in which improvement measures were necessary were identified. The proposed measures can be linked mainly to the first three 5S principles. They partially reflect the two remaining ones, thereby ensuring that principles one through three are implemented continually. It is also advisable to employ solutions for improving worker safety in the work environment. Such solutions should envision.

- Changing in the placement of workstation tools and equipment with a view to ensuring their safe use without employees having to search for such items or resort to makeshift measures,
- Setting tools and equipment in order to allow workers to move safely around the workspace,
- Clearing any unnecessary tools and equipment from the workstation to make more room available and increase worker comfort during the performance of production tasks,
- Clearing items left over from previous assignment to remove clutter, which constitutes a common cause of accidents,
- Employing ways to identify and place tools and equipment in put-away areas that increase worker comfort and convenience at the workstation,
- Placing exits and emergency escape routes where they can be used safely and efficiently,
- Securing access to working documents and manuals to ensure proper performance of work, eliminate improper operation of work equipment and reduce work performance without proper process preparation and in breach of agreed job descriptions,
- Adopting 5S principles continuous and unconditional to ensure they are perceived as standard operating procedures that apply to all tasks,
- Changing employee remuneration and benefit plans to provide incentives for finding and proposing improvements.

The proposed system should be viewed comprehensively without assigning its elements rigidly to any selected 5S area. These actions are based on the guidelines of the legislation (e.g. 89/391/EEC, 89/656/EEC, 2009/104/EC) and literature sources (Górny 2015, 2017a, 2017b, Roriz et al. 2017, Mazur, Gołaś 2010, Mwanza, Mbohwa 2017). Sample measures assigned to the phases of preparation, implementation and improvement are shown in Table 3.

4 CONCLUSIONS

Shaping work environments should be viewed as vital for securing the ability of the concerned company to perform its production tasks. Care for the environment, working conditions and employee health are critical areas of improvement in any organization. Without a doubt, an equally crucial aim is to reduce the organization's operating expenses by creating an environment conducive to minimizing the waste that results from failures to keep the work environment in its optimal condition (Dahlke et al. 2016, Jaca et al. 2014, Tool... 2006, Genaidy et al. 2007).

The use of 5S has helped identify and address a number of issues and assign them to individual phases of the effort.

The above demonstrates that adherence to 5S principles helps ensure that workers benefit from conformity with ergonomic requirements at the workplace if the company is capable of keeping it clean, efficient and safe. These effects can be achieved by eliminating irregularities (hazards and strenuousness) and mitigating risks. The benefits thus achieved result from improvement measures which affect not only work efficiency but also worker satisfaction with work. Such measures include the use of protective equipment issued to employees.

To inspire workers to embrace 5S, they must be made aware of the need to behave in ways prescribed by 5S and enabled to achieve specific benefits. By creating a healthy and worker-friendly work environment and keeping it clean in the company in question, work became easier while the time devoted to non-productive activities was shortened. Equally important is the reduction of worker loads and increase in the available work space, both of which ultimately drive work efficiency.

REFERENCES

Agrahari, R.S., Dangle, P.A., Chandratre, K.V. 2015. Implementation of 5S Methodology in the Small Scale Industry A Case Study. *Int. J. of Scientific & Technology Research* 4(4):180–187.

Budzynowska, M., et al. 2012. *Metoda 5S. Zastosowanie, wdrażanie i narzędzia wspomagające*. Warszawa: Verlag Dashofer.

Council Directive 89/391/EEC of 12 June 1989 on the introduction of measures to encourage improvements in the safety and health of workers at work, OJ L 183, 29.6.1989, p. 1–8, as amended.

Council Directive of 30 November 1989 on the minimum health and safety requirements for the use by workers of personal protective equipment at the workplace (third individual directive within the meaning of Article 16 (1) of Directive 89/391/EEC) (89/656/EEC), OJ L 393, 30.12.1989, p. 18–28, as amended.

Dahlke, G., Drzewiecka, M., Butlewski M. 2016. The impact of work on body composition changes in workers. In: P.M. Arezes, et al. (eds.), *Occupational Safety and Hygiene, SHO 2016*: 61–63. Guimarães: Portuguese Society of Occupational Safety and Hygiene (SPOSHO).

Directive 2009/104/EC of the European Parliament and of the Council of 16 September 2009 concerning the minimum safety and health requirements for the use of work equipment by workers at work (second individual Directive within the meaning of Article 16(1) of Directive 89/391/EEC), OJ L 260, 3.10.2009, p. 5–19.

Folejowska, A. 2012. *Kaizen—dążenie do doskonałości. Filozofia działania, której istotę stanowi doskonalenie.* Warszawa: Verlag Dashofer.

Genaidy, A., Salem, S., Karwowski, W., Paez, O., Tuncel, S. 2007. The work compatibility improvement framework: an integrated perspective of the human-at-work system. *Ergonomics* 50(1): 3–25.

Górny, A. 2011. The Elements of Work Environment in the Improvement Process of Quality management System Structure. In: W. Karwowski, G. Salvendy (eds.), *Advances in Human Factors, Ergonomics and safety in Manufacturing and Service Industries*: 599–606. Boca Raton: CRC Press, Taylor and Francis Group.

Górny, A. 2015. Man as internal customer for working environment improvements. *Procedia Manufacturing* 3: 4700–4707.

Górny, A. 2016a. Shaping a work safety by use the 5S methodology. In: P.M. Arezes, et al. (eds.), *Occupational Safety and Hygiene IV*: 111–116. London: Taylor and Francis Group.

Górny, A. 2016b. Work environment in quality assurance. In: P.M. Arezes, et al. (eds.), *Occupational Safety and Hygiene, SHO 2016*: 105–107. Guimarães: Portuguese Society of Occupational Safety and Hygiene (SPOSHO).

Górny, A. 2017a. Choice and assessment of improvement measures critical for process operation (in reference to the requirements of ISO 9001:2015). In: G. Oancea, M.V. Drăgoi (eds.), *MATEC Web of Conferences*, 94: 04006.

Górny, A. 2017b. The use of working environment factors as criteria in assessing the capacity to carry out processes. In: G. Oancea, M.V. Drăgoi (eds.), *MATEC Web of Conferences*, 94: 04011.

Helund, A., Gummesson, K., Rydell, A., Anderson, I-M. 2016. Safety motivation at work: Evaluation of changes for six intervention. *Safety Science* 82: 155–163.

Imai, M. 2012. *Gamba Kaizen. Zdroworozsądkowe podejście do strategii ciągłego rozwoju.* Warszawa: MT Business.

Jaca, C., Viles, E., Paipa-Galeano, L., Santos, J., Mateo, R. 2014. Learning 5S principles from Japanese best practitioners: case studies of five manufacturing companies. *Int. J. of Production Research* 52(1): 4574–4586.

Jiménez, M., Romero, L., Dominguez, M., del Mar Espinosa, M. 2015. 5S methodology implementation in the laboratories of an industrial. *Safety Science* 78: 163–172.

Kakkar, V., Dalal, V.S., Choraria, V., Pareta, A.S., Bhatia, A. 2015. Implementation Of 5S Quality Tool In Manufacturing Company A Case Study. *Int. J. of Scientific & Technology Research* 4(2):208–213.

Kawecka-Endler, A., Mrugalska, B. 2011. Contemporary Aspects in Design of Work. In: W. Karwowski, G. Salvendy (eds.), *Advances in human factors, ergonomics and safety in manufacturing and service industries*: 401–411. Boca Raton: Taylor and Francis Group.

Kiran, D.R. 2017. *Total Quality Management. Key Concepts and Case Studies.* Butterworth-Heinemann: Elsevier.

Lasota, A.M., Hankiewicz, K. 2016. Assessment of risk to work-related musculoskeletal disorders of upper limbs at welding stations. In: P.M. Arezes, et al. (eds.), *Occupational Safety and Hygiene, SHO 2016*: 105–107. Guimarães: Portuguese Society of Occupational Safety and Hygiene (SPOSHO).

Lasota, A.M., Hankiewicz, K. 2016. Evaluation of ergonomic risk in the production line of frozen food products, *Proc. of the 2016 International Conference on Economics and Management Innovations*, 57: 272–278.

Mazur, A., Gołaś, H. 2010. *Zasady, metody i techniki wykorzystywane w zarządzaniu jakością.* Poznań: Wydawnictwo Politechniki Poznańskiej.

Misztal, A. Butlewski, M., Belu, N., Ionescu, L.M. 2014. Creating involvement of production workers by reliable technical maintenance. In: *2014 International Conference on Production Research-Regional Conference Africa, Europe and the Middle East and 3rd International Conference on Quality and Innovation in Engineering and Management* (ICPR-AEM 2014): 322–327.

Mrugalska, B., Wyrwicka, M., Zasada, B. 2016. Human-Automation Manufacturing Industry System: Current Trends and Practice. In: F. Rebelo, M. Soares (eds.), *Advances in Intelligent Systems and Computing*, 485: 137–145.

Mwanza, B.G., Mbohwa Ch. 2017. Safety in Maintenance: An Improvement Framework. *Procedia Manufacturing* 8: 657–664.

Mlkvaa, M., Prajováa V., Yakimovichb B., Korshunovb A., Tyurinc, I. 2016. Standardization—one of the tools of continuous improvement. *Procedia Engineering* 149: 329–332.

Roriz, C., Nunes, E., Sousa S. 2017. Application of Lean Production Principles and Tools for Quality Improvement of Production Processes in a Carton Company. *Procedia Manufacturing* 11: 1069–1076.

Tool for productivity, quality throughput safety. *Management Service* 50(3): 16–18.

Zimmerman, W.J. 1991. Kaizen: The Search for Quality. *The J. of Continuing Higher Education* 39(3): 7–10.

Żuchowski, J., Łagowski, E. 2004. *Narzędzia i metody doskonalenia jakości.* Radom: Politechnika Radomska.

Investigation and analysis of work accidents at the Urban Hygiene Department in a Portuguese Municipality

Paola G. da Fonseca
Laboratório de Ergonomia, Faculdade de Motricidade Humana, Universidade de Lisboa, Estrada da Costa, Cruz Quebrada, Portugal

Rui B. Melo & Filipa Carvalho
Laboratório de Ergonomia, Faculdade de Motricidade Humana, Universidade de Lisboa, Estrada da Costa, Cruz Quebrada, Portugal
CIAUD (Centro de Investigação em Arquitetura, Urbanismo e Design), Faculdade de Arquitetura, Universidade de Lisboa, Rua Sá Nogueira, Lisboa, Portugal

ABSTRACT: This work intended to investigate and analyze the work-related accidents that occurred in the Municipality of Oeiras. The investigation focused on the accidents that took place within the Division of Waste Collection from the Department of Urban Hygiene. It covered a three-year period (2014 to 2016) using a formal and structured process—RIAAT (Recording, Investigation and Analysis of Accidents at Work). Therefore, the workers' characterization, the accidents' characterization and accidents' investigation were, as well, included. The application of the RIAAT process was made up of the reclassification and recoding of 109 accidents and the in-depth investigation of 19 accidents, by means of the application of semi-structured interviews, to the victims. The results of the study reported the type of accident that occurs most frequently and the factors (human errors, contributory individual factors, workplace factors, and the organizational and management factors) that effectively contributed to the occurrence of the accidents.

1 INTRODUCTION

The occurrence of accidents at work is perhaps the oldest topic in safety science, and still is a central focus of today's agenda due to the ever-increasing social pressure towards accident prevention (Jacinto et al. 2011). As known, accidents at work cause many damages to the worker, to the organization and to the State itself, so knowing their causes proves to be extremely important.

According to the Gabinete de Estratégia e Planeamento (GEP), in 2014, 203,548 work accidents occurred in Portugal and the numbers have been increasing since 2012. In 2014, around 22.1% of the occurred accidents produced more than 30 lost days/person/year (GEP 2016).

In Portugal, accident analysis and investigation are not only a legal obligation (article 98° in Lei n° 102/2009 de 10 de setembro), but also aim to improve procedures and work practices as well as the risk assessment and control processes (Nunes et al. 2015). Nevertheless, Jacinto (2005) states that not many organizations investigate the occurred accidents and incidents, particularly when severity is low. Additionally, she reinforces that when workplace accidents are analyzed, the investigation is usually superficial and made in an *ad hoc* basis.

Throughout the years, accident causation models have evolved and no longer rely on a linear sequence of events. In other words, they reveal a gradual shift from searching for a single immediate cause to the recognition of multiple causes (Katsakior et al. (2009). This new trend is accepted by several authors and highlights that both managerial and organizational failures and their interactions with the working activities have become an important issue for the understanding of accidents (Jacinto et al. 2009).

This study was made at the Municipality of Oeiras (CMO), in Portugal, and intended to analyze a set of workplace accidents. It aimed to go beyond the legal requirements in force in the country, integrating the investigation, applying formal and structured methodologies capable of systematizing research and improving communication and transparency of the process. In fact, this is a relevant step towards the optimization of existing control measures.

2 METHODOLOGY

2.1 *RIAAT—Recording, Investigation and Analysis of Accidents at Work*

To analyze the workplace accidents we have relied on the RIAAT method, which was developed by the project CAPTAR (Learn to prevent), in 2010, to be applied in the organizations, regardless of its size or its activity sector (Jacinto et al. 2011). Though, its main targets are SME.

The decision to use this particular method was supported by the following facts:

- It was tested by the Portuguese Authority for the Working Conditions (ACT);
- It has the ability to find causes at different levels of organization, which makes the analysis much more realistic and robust;
- It is considered a practical and structured tool, applied to learn from accidents and develop new prevention strategies (Nunes et al. 2015).

According to the authors of the method, it should be applied by occupational safety and health professionals, which make the necessary recommendations to improve safety afterwards (Jacinto et al. 2010).

The RIAAT method is composed of four parts.

The first part is harmonized with the European Statistics of Accidents at Work variables (ESAW) (Eurostat 2012) and is named Recording. In this part, the accident analysis integrates the variables resulting from the active failures that caused it, which allows defining an accident Type.

The second part integrates the identification of the latent failures, which are the causes of human errors, work place factors and management factors that contributed to the accidents. In-depth level of this analysis could range from medium to high, depending on the severity of the accident.

The third part integrates the action plan, which defines where changes and improvement decisions are to be made, when and how they will be implemented and who will be responsible for them.

Finally, the fourth and last part is named Organizational learning and is the most important part to strengthening the loop of continuous improvement. A complete description of the RIAAT method can be found in articles written by Jacinto et al. (2011; 2010).

2.2 Procedures

It was defined that Part I and Part II of the RIAAT would be applied with a variable in-depth level investigation, according to the learning potential of each accident analyzed.

After a global analysis of the work-related accidents that occurred in the Municipality of Oeiras, the most critical area was selected to be analyzed with the RIAAT method.

The accidents chosen to integrate the reclassification and the in-depth investigation were those registered at the Division of Urban Hygiene (DHU), between 2014 and 2016.

In terms of methodology, the RIAAT method was applied in two moments:

First moment—included the reclassification of the work accidents that happened at DHU, particularly at the RRU (Urban Garbage Collection) service. 109 of the 117 accidents that happened during the period of the study were reclassified. Eight accidents were not analyzed because they did not match the criteria defined to the reclassification of RIAAT, as they were considered incidents without any loss to the worker nor to the organization.

The interviews of the victims were made in an informal way. To facilitate the analysis, the conversation was recorded, with the victim's permission, and after that, there was a detailed analysis of the recordings' content.

Second moment—comprised the application of part II of RIAAT which included the in-depth investigation of 19 work accidents, from the 109 accidents reclassified in part I.

2.3 Data analysis

Data processing was performed with the Statistical Package for the Social Sciences (SPSS©) (version 23) and Excel 2016.

For the interpretation of data collected from the application of the RIAAT (Part I and II), the Pareto principle was used as a reference allowing the identification of the main causes/consequences (depending on the analyzed variable) responsible/associated for/to approximately 80% of the accidents.

To conclude, considering the nature of the variables (ordinal and nominal) and the reduced dimension of the sample (N = 19) the non-parametric Kruskal-Wallis test was used, when relevant, in order to verify if there were significant differences of:

- The number of days lost between the work shifts in which the accidents occurred.
- The type of Failure between the age group of the victims.

In all cases, a significance level of 0.05 was adopted as a criterion to reject the null hypothesis.

For the reclassification of work accidents, the following variables were considered in Part I of the RIAAT form: Age of the victim (A); Deviation (D); Contact-mode of injury (C); Part of Body injured (PB); Days Lost (DL). Regarding the variable Age, it was analyzed taking into account the following six age groups, following the methodology adopted by GEP (2016): 18 to 24 years; 25 to 34 years; 35 to 44 years; 45 to 54 years; 55 to 64 years and over 65 years.

It is important to highlight that the Gender variable was not considered, taking in account that the RRU workers are predominantly male. The same procedure was adopted for other variables that had no variation within the sample, such as: nationality;

occupation, employment status and working environment.

Whenever appropriate, comparisons were made between the obtained results and data published by GEP (2016) concerning the economic activities defined as the letter E of the CAE/REV3, which include *"collection, treatment and distribution of water, sanitation, waste control and remediation"*.

3 RESULTS AND DISCUSSION

The DHU presented a labor force of 396 workers, of which 69% were male. Most workers' age was between 35 and 54 years old and their seniority at work was above 10 years.

3.1 *Work accidents data & RIAAT results (Part I)*

At the CMO, 636 accidents were recorded between 2014 and 2016. From these accidents, 46.6% (297) documented at the DHU, approximately 40% (117) happened at the RRU (Urban Garbage Collection) service. Considering that the RRU was the area where the most severe accidents happened, it was decided that it should be the chosen area for the application of the RIAAT method.

Figure 1 shows the distribution of the victims' age, in the RRU service. The mean age of these workers was 46 years (range: 27–64 years). Compared to the data published by GEP (2016), it is possible to notice that the age groups with the highest rate of accidents were the same, even though the order is inverted.

The results for the variables Deviation (D), Contact-mode of injury (C), Part of Body injured (PB) and Days Lost (DL) are summarized in Table 1.

To facilitate the presentation of the results, family groups were considered within each variable analyzed, following the methodology adopted in others studies (Faria 2016; Batista 2013).

The results presented are based on the analysis made according to the Pareto Principle.

Table 1. Distribution of accidents considering the reclassified variables (Part I—RIAAT).

Variable	RRU (N = 109)	GEP (N = 2678)
	Family Group Code (%)	
Deviation (D)	F-D40 (47.7%)	F-D70 (35.4%)
	F-D70 (23.9%)	F-D40 (16.1%)
	F-D50 (20.2%)	F-D50 (14.3%)
		F-D60 (10.6%)
Contact-mode of injury (C)	F-C40 (30.3%)	F-C 70 (35.4%)
	F-C70 (27.5%)	F-C 30 (21.6%)
	F-C30 (25.7%)	F-C 40 (20%)
Part of Body injured (PBI)	F-PBI 50 (29.4%)	F-PBI 60 (60%)
	F-PBI 60 (29.4%)	F-PBI 50 (29.6%)
	F-PBI 10 (18.3%)	F-PBI 30 (21.3%)
	F-PBI 30 (16.5%)	F-PBI 10 (14.2%)
Days Lost (DL)	Without DL (33%)	Without DL (29%)
	14–20 days (17.4%)	30 days (24.2%)
	7–13 days (16.5%)	7–13 days (17.5%)
	1–3 months (13.8%)	14–20 days (10.2%)

LEGEND
F-D40 – *"Loss of control (total or partial) of machine, means of transport or handling equipment, hand-held tool, object, animal – not specified"*;
F-D50 – *"Slipping – Stumbling and falling – Fall of persons – not specified"*;
F-D60 – *"Body movement without any physical stress (generally leading to an external injury) – not specified"*;
F-D70 – *"Body movement under or with physical stress (generally leading to an internal injury) – not specified"*;
F-C30 – *"Horizontal or vertical impact with or against a stationary object (the victim is in motion) – not specified"*
F-C40 – *"Struck by object in motion, collision with – not specified"*.
F-C70 – *"Physical or mental stress – not specified"*
F-PBI10 – *"Head, not further specified"*
F-PBI30 – *"Back, including spine and vertebra in the back"*;
F-PBI50 – *"Upper Extremities, not further specified"*;

The comparison between the results of this study and GEP's data (2016) shows a similar pattern for "Contact-mode of injury", "Part of Body injured" and "Days Lost" variables.

3.2 *RIAAT results (Part II)*

RIAAT's part II comprised a detailed investigation of 19 accidents at work. This part of the investigation intended to identify the active and the latent failures. The following items were considered: Human failures (HF); Individual Contributing Factors (ICF); Workplace Factors (WF) and Organizational and Management Factors (OMF).

During the interviews of the victims of the 19 accidents, that are part of the detailed investigation, it was noticed that more than one item (ICF, WF and OMF) contributed for the accidents to happen and influenced the total of items that were considered for the analysis.

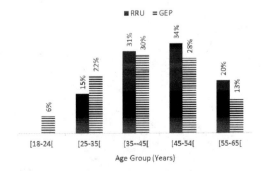

Figure 1. RRU's Workplace Accidents' distribution and GEP's data according to age group.

In order to correlate the Type of Failure with the age of the victim, it was decided to analyze the human failure by age group (Figure 2):

- For the age range 45–54 years old, human failures related to Violation were the most representative (26.3%). For that same group, the second most common failure was "None", with 15.8%. And last "Slips or Lapse" with 10.5%.
- For the group between 35 and 44 years old, Violation was also the most common (10.5%), followed by "Misunderstanding" and "None" with the same proportion (5.3%). It is important to say that for that age group "Slips or Lapse" were not documented, which can be explained because that age group workers are considered to be mature and have a high cognitive ability level.
- Additionally, for the age group from 55 to 64 years old, the most common type of failure was "Slips or Lapse" (10.5%); which can be ascribed to memory loss and some slips that may affect the attention level of these workers, and contribute to the accidents occurrence.

Considering the non-parametric Kruskal-Wallis ($\chi^2 = 0,653$ e p-value = 0.756) test results, it is possible to conclude that there were no significant differences for Type of Failure between the age groups of the victims.

As for the Individual Contributing Factors, the Psychological or Mental Stress (ICF-18) resulting from the time pressure was considered the most critical. Regarding the Workplace Factors the highest rate is related to Technical Factors, particularly poor quality of man-machine interface and/or bad equipment and structure.

After analyzing the results from the application of the first part of RIAAT we can affirm that at the RRU sector there isn't a well-defined accident type. However, the typical accident can be described as the one that involved one man (100%), between 45 and 54 years old (34%), that was hit by a moving object (30%), possibly due to the loss of control of—materials, objects, products, machine parts, etc. (47%), that caused a trauma (47%), at the upper limbs (29%) and lower limbs (29%) forcing the worker to be absent from the workplace for one month.

When comparing the results related to the Part I of RIAAT from this study to those of GEP (2016), it is noticeable that the variables "Contact-mode of injury", "Part of Body injured" and "Days Lost" showed a very similar behavior.

Summing up the main conclusions drawn from the detailed investigation (Part II of RIAAT) it can be said that:

- The most common human failure was violation (42%) and it was present in all the age groups. Even though there were other types of failure within the age groups of the victims, there were no significant statistical differences of the type of failure between the age groups.
- Referring to the other contributing factors (ICF, WF and OMF), Psychological/Mental Stress (ICF-18), Task and Work (F-WF 30) and the

Table 2. Distribution of accidents according to contributing factors—In-depth investigation (RIAAT—Part II).

Variable	RRU (N = 19) Family Group Code (%)
Human failures $N_{HF} = 19$	Violation (42.1%)
	None Failures (26.3%)
	Slip or Lapse (21.1%)
Individual Contributing Factors (ICF) $N_{ICF} = 20$	ICF-18 (30%)
	ICF-99 (20%)
	ICF-13 (15%)
	ICF-16 (15%)
Workplace Factors (WF) $N_{WF} = 26$	F-WF 30 (42.3%)
	F-WF 20 (23.1%)
	F-WF 50 (11.5%)
	F-WF 60 (11.5%)
Organizational and Management Factors (OMF) $N_{OMF} = 27$	F-OMF 30 (37%)
	F-OMF 20 (22.2%)
	F-OMF 40 (18.5%)
	F-OMF 99 (14, 8%)

LEGEND
ICF-18 – "Psychological stress, under pressure";
ICF-99 – "Other factors ICF not included this classification";
ICF-13 – "Distraction – changing pay attention";
ICF-16 – "Intrinsic human variability. It's usually always related to simple "run" errors, in "automatic" mode"";
F-WF 30 – "Task and Job";
F-WF 20 – "Safety Equipment Tools";
F-WF 50 – "Information and Communication";
F-WF 60 – "Ambient, Environment";
F-OMF 30 – "Technical Factor";
F-OMF 20 – "Procedure and Rules";
F-OMF 40 – "Training";
F-OMF 99 – None or others factors.

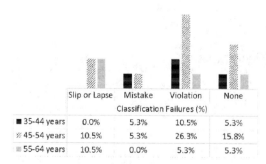

Figure 2. Distribution of work accidents by Type of Failure, according to age group (N = 19).

technical factors (F-OMF 30) were identified as the most critical.

4 CONCLUSIONS

Using the RIAAT method to analyze the accidents that happened at the RRU allowed us, not only to characterize the immediate causes, but also to find the latent causes (Human failures, Organizational and management failures) that directly contributed to the occurrence of these accidents.

The lack of a well-structured methodology used at CMO to analyze the accidents, as well as the discrepancy found in different documents, made the analysis and the reclassification difficult and time-consuming.

Following the application of RIAAT, it was possible to present a list of suggestions for improvement. For the Work Accidents associated to latent failures such as damages on vehicles/containers, short staff, or lack of appropriate training for some specific skills, the following measures were suggested:

- To improve the vehicles' maintenance management through a check-list where the workers (drivers and garbage collectors) could report, in a more effective and systematic way, the problems found. In this way, it would be possible to check, in the beginning of each shift, the adequacy of the vehicle to be used.
- To implement periodic and effective maintenance of the garbage containers that end up being degraded, not only by their usage but also by delinquents.
- To implement an incidents' journal, in which the incidents that are happening in the department could be reported by the workers (before ending their shifts).
- To promote actions aiming at raising the workers' awareness periodically: To reinforce the importance of complying with all safety procedures, to inform and alert the workers about the hazards and the prevention/protection measures to be adopted, to develop skills to behave safely in deviating situations (when necessary).
- Finally, it is known that the reduced number workers at the RRU causes an overload of work. A hydraulic system could help the most physical demanding activities, reducing the workers' physical effort and reducing the time to accomplish the task.

The RIAAT method, revealed itself as an appropriate tool to analyze and investigate work accidents. Its application would be desirable in a consistent way, whether in the analysis of all the other accidents that occurred in the CMO, or in other municipalities. Thus, it will be possible to make comparisons within the same activity sector and between different activities sectors. Actually, this could be appointed as one of the limitations of this work. Despite all efforts, we have not found any study applying the RIAAT in this activity sector.

REFERENCES

Batista, G.A.T., 2013. *Estudo aprofundado da sinistralidade numa empresa da indústria gráfica em Portugal.* Universidade Nova de Lisboa. Available at: http://search.ebscohost.com/login.aspx?direct=true&site=eds-live&db=edsrca z&AN=rcaap.openAccess.10362.10906.

Eurostat, 2012. *European Statistics on Accidents at Work (ESAW) Summary methodology,* Luxembourg: European Union.

Faria, D., 2016. *Estudo RIAAT: investigação e análise de acidentes de trabalho numa indústria gráfica.* Universidade Nova de Lisboa. Available at: http://search.ebscohost.com/login.aspx?direct = true&site = eds-live&db=edsrca&AN=rcaap.openAccess.10362.19104.

GEP, 2016. *Acidentes de trabalho 2014: coleções estatísticas,* Lisboa: Gabinete de Estratégia e Planeamento. Available at: http://jus.uol.com.br/revista/texto/1211%5Cnhttp://jus.uol.com.br/.

Jacinto, C. et al., 2011. The Recording, Investigation and Analysis of Accidents at Work (RIAAT) process. *Policy and Practice in Health and Safety,* 9(1), pp. 57–77. Available at: http://www.tandfonline.com/doi/full/10.1080/14774003.2011.11667756.

Jacinto, C., 2005. Metodologias para análise de acidentes de trabalho. In C.G. Soares, A.P. Teixeira, & P. Antão, eds. *Análise e Gestão de Riscos, Segurança e Fiabilidade.* Lisboa: Edições Salamandra, pp. 183–202.

Jacinto, C., Canoa, M. & Soares, C.G., 2009. Workplace and organisational factors in accident analysis within the Food Industry. *Safety Science,* 47(5), pp. 626–635. Available at: http://dx.doi.org/10.1016/j.ssci.2008.08.002.

Jacinto, C., Soares, C.G. & Fialho, T., 2010. *RIAAT: Registo, Investigação e Análise de Acidentes de Trabalho – Manual do Utilizador* Revisão1.1., Portugal: Projecto "CAPTAR—Aprender para prevenir" – Fundação para a Ciência e Tecnologia (FCT), PTDC/SDE/71193/2006. Available at: http://www.mar.ist.utl.pt/captar/images/Manualdoutilizador_RIAAT_revisão 1.1_Maio 2010.pdf.

Katsakiori, P., Sakellaropoulos, G. & Manatakis, E., 2009. Towards an evaluation of accident investigation methods in terms of their alignment with accident causation models. *Safety Science,* 47(7).

Nunes, C. et al., 2015. Case Study Comparison of different methods for work accidents investigation in hospitals : A Portuguese case study., 51, pp. 601–609.

An investigation of forearm EMG normalization procedures in a field study

S. Aia & M. Reinvee
Institute of Technology, Estonian University of Life Sciences, Tartu, Estonia

ABSTRACT: This electromyographic field study investigates the need for several different ways to normalize the myoelectric activity of forearm muscles. The normalization procedure must allow for reliable comparisons between subjects, muscles and study conditions. Therefore, the contractions used for normalization have essential role in the interpretation of the results. Meanwhile, normalization in different postures increases temporal demands. In the study, five different normalization procedures were compared; three procedures involved squeezing the hand tool's handle, one squeezing a dynamometer and the last one consisted of resisting self-induced wrist flexion or extension. The results of the study group ($n = 10$) showed no need for normalization in handle specific postures. Based on the results, resisting investigator- or self-induced wrist flexion or extension is the preferred way to carry out forearm EMG normalizations in the field.

1 INTRODUCTION

Electromyography (EMG), the method for recording and assessing the muscle's myoelectric activity, is one of the most useful methods in ergonomics. EMG allows to (Marras, 1990): determine the duration of muscular activity; evaluate the muscular effort in terms of relative or in some cases absolute values; and assess the muscular fatigue. Out of these potential uses, the indirect assessment of effort seems most desired in the eyes of a practitioner. Although, the EMG force/torque relationship is a complex problem (Hof, 1984), the EMG does correlate with the Borg RPE and Borg CR-10 scales (Grant et al. 1994). Therefore, EMG could be a complementary method to the observational ergonomic assessment methods like Strain index (Moore & Garg, 1995), HAL (Latko et al. 1997) or OCRA (Occhipinti, 1998). EMG measurements might be especially desired when conducting analyzes in small and medium sized enterprises where each worker might be responsible for a single task. Although EMG is perceived in the scientific community as one of the essential methods, it is far from being an everyday tool for the practitioner. So far one of the limiting factors has been the cost of the apparatus (Dempsey et al. 2004). According to Reinvee et al. (2015), the lack of financial resources should not be considered as a restriction, as the contemporary low-cost hardware functions on a sufficient level. One of the components in the relevant low-cost hardware is a universal microcontroller, such as the Arduino (Smart Projects Srl, Strambino, Italy), which allows to build electronic devices for interactive sensing and controlling of physical objects. Although the contemporary market for low-cost hardware for ergonomic assessment comes in various stages of readiness, it still requires the user to appreciate the *zeitgeist* of the do-it-yourself movement. This and the lack of appropriate software for data analysis are still restrictive factors for the usage of low-cost EMG devices (Reinvee & Pääsuke, 2016). Therefore, the temporal demands on data analysis might not survive the cost and benefit analysis when choosing between EMG and subjective scaling methods. On the one hand, the loss of restrictions also calls for attention and critical analysis of the EMG investigations in practice. Decades old warnings stress that conducting the EMG investigation is easy, but the quality of the results depends on the competency of the investigator (Cavanagh, 1974; De Luca, 1997). On the other hand, a wider base of EMG users might also create a community to improve the skills and increase the appreciation and implementation of EMG-related scientific research.

Obviously in order to communicate efficiently every community needs a common language in the sense of terminology and procedures. Although there are standards for instrumentation (Bischoff et al. 1999), reporting the data (Merletti & Torino, 1999) and there are efforts to standardize the electrode locations (Hermens et al. 2000), the utmost issue seems to be the normalization procedure (Mirka, 1991; Mathiassen et al. 1995; Burden & Bartlett 1999; Al-Qaisi & Aghazadeh, 2015; Ngo & Wells, 2016). The normalization is a procedure which allows comparing the subjects, muscles and study conditions (Marras, 1990) as the absolute value of the signal tends to vary both between and within the subjects (Cram & Kasman, 2010). The normalization (equation 1) compares the signal of interest with the values of myoelectric activity during rest and a reference contraction and scales the value in the range of 0 to 100%. It is common to use subjects' maximum voluntary effort (MVE) for the reference value.

$$nEMG = \left(\frac{EMG_i - EMG_{min}}{EMG_{ref} - EMG_{min}} \right) \times 100 \qquad (1)$$

where EMG_i = muscle's myoelectric activity of an action, mV; EMG_{min} = muscle's myoelectric activity during rest, mV; EMG_{ref} = muscle's myoelectric activity during reference contraction, mV.

It is evident from equation 1 that the value of normalized myoelectric activity (nEMG), as well as the decision on the potential ergonomic risk, depends on the myoelectric activity value of the reference contraction. The myoelectric activity values of both submaximal (Roman-Liu & Bartuzi, 2013) and maximal effort (Ngo & Wells, 2016) depend on the posture, although, there are claims (Gall et al. 2014) that deviations from neutral posture have considerable effect only in the near extreme postures. The effect of posture can be easily controlled in laboratory studies where the upper limb may be fixed in a certain posture (Staudenmann et al. 2005) or the involuntary movement is reduced by supporting the forearm (Duque et al. 1995); meanwhile, in field studies such options are limited, as it is common to interfere with the process as less as possible. In order to avoid unnecessary interference in the studies where the object of interest is forearm EMG, the reference contraction might be the participants' resistance to investigator- or self-induced wrist flexion or extension. An alternative is to squeeze the dynamometer or in the hand tool studies to squeeze the tool handle.

This study compares five alternatives of EMG normalization: squeezing in three tool handle specific postures; squeezing the dynamometer and resistance to self-induced wrist flexion and extension. The goal of the study was to find the best normalization procedure and analyze the need for multiple EMG normalizations, related to the tool handle specific postures.

2 METHODS

2.1 Participants

Ten able bodied men with a mean age of 31.5 years (SE 2.2), mean height of 182.7 cm (SE 2.2), mean weight of 81.9 kg (SE 3.5) and mean hand grip force of 413 N (SE 21) participated in the study. The participants were free from any neuromuscular conditions and they provided a signed informed consent after being briefed with the experimental procedure.

2.2 Procedure

Myoelectric activities of MVE in three forearm muscles, m. flexor carpi ulnaris (FCU); m. flexor carpi radialis (FCR) and m. extensor digitorum (ED), were the dependent variables of the study. On these muscles, bipolar dual Ag/AgCl disposable electrodes with inter-electrode distance of 2.0 cm (Noraxon Inc, Scottsdale, USA) were attached (Figure 1). The electrodes and cables were secured with elastic tubular net to avoid noise from cable movements. The number of muscles investigated in the study was limited by the configuration of the EMG acquisition apparatus. This led to the exclusion of of m. palmaris longus and m. flexor digitorum superficialis, which are often included in the hand tool studies.

The independent variable was the EMG normalization condition; the independent variable had five levels: A, B, C, D and R, hereafter referred to as conditions. The first three conditions represent the wrist positions, specific to the rotatable tool handle. In the case of condition D the EMG activity of MVC was obtained while pressing the hand dynamometer and in the case of condition R the myoelectric activity was obtained while resisting wrist flexion or extension, which was induced by the participants other hand. See Figure 2 for graphic representation of the conditions.

The MVEs were obtained in accordance to the Caldwell's regimen (Caldwell et al. 1974). The participants were instructed to increase their effort without any jerking movements during a period of one second and then hold their maximum effort during four seconds. From this four second period the average two seconds in the middle was used as a sample. A two minute break followed each exertion to diminish the effect of fatigue. In order to ease the compliance with the instructions an in-house developed visual feedback system was provided for the participants. The feedback system consisted of three preprogrammed LEDs, which allowed the participants to follow the pace of rest, increase and exertion periods.

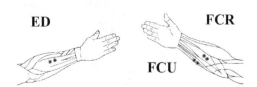

Figure 1. Electrode locations on ED, FCR and FCU muscles.

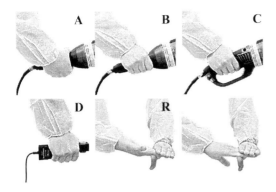

Figure 2. Conditions in which MVE was obtained: A – squeezing the handle, wrist significantly flexed; B – squeezing the handle, wrist just noticeable extension; C – squeezing the handle in near-neutral position; D – squeezing the dynamometer; R – resisting the self-induced wrist flexion and extension.

The process was repeated twice and the average value of the exertions was used.

2.3 Equipment

A 10-bit BITalino telemetric microcontroller, with EMG sensors v.151015 (PLUX wireless biosignals S.A., Portugal) was used to collect the EMG signals of the muscles. The EMG sensors were hardware filtered 10–400 Hz, the gain, input impedance and CMRR of the sensors were 1000x, 100 GΩ and 110 dB respectively. The OpenSignals (r)evolution software (PLUX wireless biosignals S.A., Portugal) was used to acquire the data from the microcontroller with the sampling rate of 1000 Hz. The force data was collected with an electronic I-type dynamometer (Vernier Software & Technology, Beaverton, USA), which was connected to a 10-bit Arduino Uno microcontroller (Smart Projects Srl, Strambino, Italy).

2.4 Data processing and analysis

A custom Matlab (MathWorks, Inc., Natic, USA) script was written for the EMG data analysis. For data interpretation the linear envelope was calculated with a Butterworth second order low pass filter, the cut-off frequency was set to 2 Hz. For statistical analysis the R software (R Development Core Team) was used in order to conduct the ANOVA, one sample t- and Tukey HSD post hoc tests.

3 RESULTS

The myoelectric activity was the highest in the case of conditions D or R. Therefore, these two conditions were used as the reference for maximum electrical activation value in normalizations. Figure 3 refers to the nEMGs normalized to condition D and Figure 4 refers to the nEMGs normalized to condition R.

When the MVEs of conditions A, B and C were normalized to the MVE of condition D the handle specific posture did not have an effect on the corresponding nEMG value in the case of the FCU $[F(2, 27) = 0.758, p = 0.478]$ nor in the case of the FCR $[F(2, 27) = 0.285, p = 0.754]$. In the case of the ED the handle specific posture did have a statistically significant effect on the corresponding nEMG value $[F(2, 27) = 3.398, p = 0.048]$. However, from the post hoc pairwise comparison it became evident that condition A differs from condition B ($p = 0.038$) but there was no statistically significant difference between conditions A and C ($p = 0.354$) nor between B and C ($p = 0.462$).

When the MVC-s of conditions A, B and C where normalized to the MVE of condition R, the handle specific posture did not have a statistically significant effect in the case of the FCU

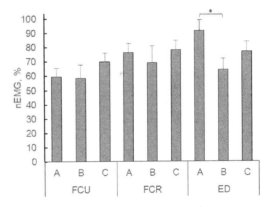

Figure 3. nEMGs (mean + SE) normalized to condition D. See Figure 2 for description of the conditions.

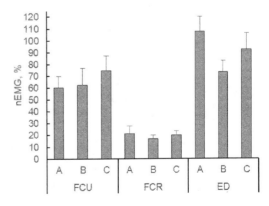

Figure 4. nEMGs (mean + SE) normalized to condition R. See Figure 2 for description of the conditions.

$[F(2, 27) = 0.402, p = 0.673]$ nor in the case of FCR $[F(2, 27) = 0.225, p = 0.800]$ or ED $[F(2, 27) = 1.948, p = 0.162]$.

It can be derived from Figures 3 and 4 that the MVEs of the conditions D and R differ at least in the case of the FCR. Therefore single sample t-tests were run on the absolute differences (in mV) between the myoelectric activities of conditions D and R (D minus R). As expected, in the case of the FCR, the absolute difference between the two conditions (M = –0.323, SD = 0.169 mV) was statistically significant $[t(9) = -6.060, p \ll 0.001]$. However, this was not true in the case of the FCU (M = –0.025, SD = 0.192 mV) nor in the case of the ED (M = 0.036, SD = 0.158 mV), as the differences were not statistically significant $[t(9) = -0.419, p = 0.685$ and $t(9) = 0.728, p = 0.485$ respectively].

4 DISCUSSION

First, on average the myoelectric activities of MVEs in the three tool handle specific postures were lower than their equivalents in conditions D and R. This can also be seen in Figures 3 and 4, as in most cases the nEMGs are below 100%. This also means that using a tool handle specific posture for reference MVE could potentially lead to an overestimation of ergonomic risk. Secondly, the only statistically significant difference between the tool handle specific postures was found in the ED muscle when normalized to condition D but not to condition R. However, the absolute difference of ED muscle's myoelectric activity was not statistically significant between the D and R conditions. The reason here is mainly the variations between participants. As it was a field study, the wrist postures were not restricted and there were minor differences in wrist postures as participants squeezed the tool handle in a posture which was in accordance to their previous experience. However, the finding that myoelectric activity of ED is greater in the case of wrist flexion than in the case neutral or extended wrist is in line with previous work of Mogk & Keir (2003). Thirdly, comparing normalizations in conditions D and R showed only one statistically significant difference. The recent work of Ngo & Wells (2016), showed also statistically significant differences in the case of FCU and ED muscles. The main difference between the two studies is the way the wrist extension and flexions were induced. In the study of Ngo & Wells (2016) wrist flexion and extension was induced by the investigator but in the current study it was self-induced. As psychological and motivational factors affect the muscular effort (Kumar, 1996), it is possible that in the case of self-induced flexion and extension the participants might not induce a true maximal effort on the muscles. In order to induce maximum effort the investigator needs to be stronger than the participant or the investigator needs a mechanical advantage like in the study of Ngo & Wells (2016), where the investigator used an experimental apparatus to induce the wrist flexion and extension. There is currently no study that compares the investigator and self-induced efforts. However, it is certain that in the field the manual laborers are probably stronger than the practitioner. If investigator induced effort needs to be favored then research is needed for the design of apparatus which can provide a mechanical advantage for the investigator.

5 CONCLUSIONS

This study compared five EMG normalization procedures for three forearm muscles. Based on the results the three tool handle related postures are not favored, as their myoelectric activity was not significantly affected by the posture. In addition, each posture requires extra time spent on the effort and rest periods between exertions. From the other two conditions, it is preferred to use the myoelectric activity of maximum voluntary effort from resisting the self-induced wrist flexion or extension as it outputs the highest myoelectric activity.

Further research is needed to determine whether investigator and self-induced efforts produce comparable myoelectric activity.

REFERENCES

Al-Qaisi, S. & Aghazadeh, F. 2015. Electromyo-graphy analysis: Comparison of maximum voluntary contraction methods for anterior deltoid and trapezius muscles. Procedia Manufacturing 3: 4578–4583.

Bischoff, C., Fuglsang-Fredriksen, A., Vendelbo, L. & Sumner, A. 1999. Standards of instrumentation of EMG.G. Deuchl & A.Eisen (eds), *Recommendations for the practice of clinical neurophysiology: guidelines of the International Federation of Clinical Neurophysiology*: 199–211. Amsterdam: Elsevier.

Burden, A. & Bartlett, R. 1999. Normalisation of EMG amplitude: An evaluation and comparison of old and new methods. *Medical Engineering and Physics* 21(4): 247–257.

Caldwell, L.S., Chaffin, D.B., Dukes-Dobos, F.N., Kroemer, K.H.E., Laubach, L.L., Snook, S.H., & Wasserman, D.E. 1974. A proposed standard procedure for static muscle strength testing. American Industrial Hygiene Association Journal 35(4): 201–206.

Cavanagh, P.R. 1974. Electromyography: its use and misuse in physical education. *Journal of Health, Physical Education, Recreation* 45(5): 61–64.

Cram, J.R. & Kasman, G.S. 2010. Comparison of quantified surface electromyography values across muscles and individuals. E. Criswell (ed), Cram's introduction to surface electromyography: 49–51. Sudbury: Jones & Bartlett Publishers.

De Luca, C.J., 1997. The use of surface electromyography in biomechanics. *Journal of Applied Biomechanics* 13(2): 135–163.

Dempsey, P.G., McGorry, R.W. & Maynard, W.S. 2004. Industrial ergonomics tool use by certified professional ergonomists. *Proceedings of the Human Factors and Ergonomics Society Annual Meeting* 48(12): 1373–1377.

Duque, J., Masset, D. and Malchaire, J., 1995. Evaluation of handgrip force from EMG measurements. *Applied ergonomics* 26(1): 61–66.

Gall, J.L., O'Donnell, C.L., Dembinski, K.A. & Kapellusch, J.M. 2014. The Effects of Posture on Force Estimations Using Surface Electromyography. *Proceedings of the Human Factors and Ergonomics Society Annual Meeting* 58(1): 1085–1088.

Grant, K.A., Habes, D.J. & Putz-Anderson, V. 1994. Psychophysical and EMG correlates of force exertion in manual work. *International Journal of Industrial Ergonomics* 13(1): 31–39.

Hermens, H.J., Freriks, B., Disselhorst-Klug, C. & Rau, G. 2000. Development of recommendations for SEMG sensors and sensor placement procedures. *Journal of Electromyography and Kinesiology* 10(5): 361–374.

Hof, A.L. 1984. EMG and muscle force: An introduction. *Human Movement Science* 3(1): 119–153.

Kumar, S. 1996. Electromyography in Ergonomics. S. Kumar & A. Mital (eds), *Electromyography in Ergonomics*: 1–50. London: Taylor & Francis.

Latko, W.A., Armstrong, T.J., Foulke, J.A., Herrin, G.D., Rabourn, R.A. & Ulin, S.S. 1997. Development and evaluation of an observational method for assessing repetition in hand tasks. *American Industrial Hygiene Association Journal* 58(4): 278–285.

Marras, W.S. 1990. Industrial electromyography (EMG). *International Journal of Industrial Ergonomics* 6(1): 89–93.

Mathiassen, S.E., Winkel, J. & Hägg, G.M., 1995. Normalization of surface EMG amplitude from the upper trapezius muscle in ergonomic studies—A review. *Journal of electromyography and kinesiology* 5(4): 197–226.

Merletti, R. & Torino, P. 1999. Standards for reporting EMG data. *Journal of Electromyography and Kinesiology* 9(1): 3–4.

Mirka, G.A. 1991. The quantification of EMG normalization error. *Ergonomics* 34(3): 343–52.

Mogk, J.P.M. & Keir, P.J. 2003. The effects of posture on forearm muscle loading during gripping. *Ergonomics* 46(9): 956–975.

Moore, J.S. & Garg, A. 1995. The strain index: a proposed method to analyze jobs for risk of distal upper extremity disorders. *American Industrial Hygiene Association* 56(5): 443–458.

Ngo, B.P.T. & Wells, R.P. 2016. Evaluating Protocols For Normalizing Forearm Electromyograms During Power Grip. *Journal of Electromyography and Kinesiology* 26: 66–72.

Occhipinti, E. 1998. OCRA: a concise index for the assessment of exposure to repetitive movements of the upper limbs. *Ergonomics* 41(9): 1290–1311.

Reinvee, M. & Pääsuke, M., 2016. Overview of Contemporary Low-cost sEMG Hardware for Applications in Human Factors and Ergonomics. *Proceedings of the Human Factors and Ergonomics Society Annual Meeting* 60(1): 408–412.

Reinvee, M., Vaas, P., Ereline, J. & Pääsuke, M. 2015. Applicability of Affordable sEMG in Ergonomics Practice. *Procedia Manufacturing* 3: 4260–4265.

Roman-Liu, D. & Bartuzi, P. 2013. The influence of wrist posture on the time and frequency EMG signal measures of forearm muscles. *Gait and Posture* 37(3): 340–344.

Staudenmann, D., Kingma, I., Stegeman, D.F. and van Dieën, J.H., 2005. Towards optimal multi-channel EMG electrode configurations in muscle force estimation: a high density EMG study. *Journal of Electromyography and Kinesiology* 15(1): 1–11.

The impact of pesticides on the cholinesterase-activity in serum samples

Alessia Teixeira, Gertie Oostingh, Ana Valado, Nádia Osório, Armando Caseiro & António Gabriel
ESTeSC—Coimbra Health School, Laboratory Biomedical Sciences, Polytechnic Institute of Coimbra, Coimbra, Portugal

Ana Ferreira
ESTeSC—Coimbra Health School, Environmental Health, Polytechnic Institute of Coimbra, Coimbra, Portugal

João Paulo Figueiredo
ESTeSC—Coimbra Health School, Complementary Sciences—Statistics/Epidemiology, Polytechnic Institute of Coimbra, Coimbra, Portugal

ABSTRACT: Pesticides can be lethal to humans and one of the known pathways affected by pesticides is the acetylcholine pathway. Pesticides are highly lipo-soluble and are thus easily absorbed through the mucosa, respiratory system, skin and gastrointestinal tract. Since the Alto Douro Vinhateiro is a vineyard where the application of pesticides is performed with high intensity, it is of interest to study the effects of pesticides on the acetylcholine pathway in the winegrowers of this region. The study of the effects of pesticides on the activity of AChE contributed to the alert of the risks these products may cause and to the defense of security regulations in the agricultural activity, minimizing the dangers to which workers are exposed. Recognizing precociously this sort of incident, as well as a higher vigilance in health related to the usage of pesticides, are extremely relevant attitudes to diminish the number of cases and the severity of the intoxications.

1 INTRODUCTION

Pesticides are widely used in occupational settings and residential areas to prevent and control pests and pest-induced diseases. Based on the target, pesticides are grouped into herbicides, insecticides, fungicides, bactericides and rodenticides. Based on their chemical properties, pesticides can also be categorized into organochlorines, organophosphates (OP), carbamates (CM), dithiocarbamates, pyrethroids, as well as phenoxyl, triazine, amide and coumadin compounds (Ye, M., Beach, J., Martin, J.W., et al, 2013). OP, phosphoric acid derivatives and CM, derived from carbamic acid or N-methylcarbamic acid, are important insecticides that are used in developing countries (Sun, I.O., Yoon, H.J., Lee, K.Y., 2015). These compounds can be lethal to humans. One of the known pathways affected by pesticides is the acetylcholine (ACh) pathway.

ACh is a neurotransmitter of the central nervous system and peripheral nervous system and is involved in the regulation of immune and other physiological functions. ACh regulates motor function, sensory perception, cognitive processing, arousal and sleep/wake cycles. In the periphery, ACh controls the heart rate, gastrointestinal tract motility and smooth muscle activity (Jones, C.K., Byun, N., Bubser, M., 2012). It is synthesized via the acetylation of choline by the cytosolic enzyme choline acetyltransferase and then pumped into synaptic vesicles by the vesicular acetylcholine transporter (Da Silva, S., Pimentel, V., Fiorenza, A., et al. 2011).

Cholinesterases are the enzymes activating the neurotransmitter ACh. Cholinesterases are present in cholinergic and non-cholinergic tissues as well as blood and other body fluids. Cholinesterase activity is due to the action of two enzymes, an acetylcholinesterase (AChE) and a butyrylcholinesterase (BChE) (Krsti, D.Z., Lazarevi, T.D., Bond, A.M., et al., 2013; Chatonnet, A., Lockridgetl, O. 1989). BChE can hydrolyze acetylcholine and differs from AChE with respect to the tissue distribution and sensitivity to substrates and inhibitors (Sonmez, F., 2017). AChE levels are highest in the brain, followed by skeletal muscle tissue. Comparatively low levels of AChE are expressed by other tissues. BChE levels are highest in the liver, adipose-visceral, esophagus, colon, Fallopian tube, uterus, cervix and lung (Lockridge O, Norgren RB, Johnson RC, et al., 2016; Sonmez, F., 2017). AChE is primarily responsible for the modulation of acetylcholine in cholinergic synapses. BChE acts as a co-adjuvant of AChE and compensates the action of AChE in the regulation of acetylcholin in the control of rhythmic movements and tonicity.

Cholinergic receptors specific for acetylcholine are divided into muscarinic cholinergic receptors and nicotinic cholinergic receptors. They participate in a variety of vital functions such as the frequency and strength of cardiac contraction and cognitive functions. Muscarinic receptors are in all target cells of the parasympathetic nervous system, as well as in target cells in sympathetic postganglionic cholinergic neurons. In contrast, nicotinic receptors are found in synapses between pre—and post-ganglion neurons. These receptors are ligand-gated ion channels that mediate diverse physiological responses, including pain processing and cognitive functions. The muscarinic cholinergic receptors are members of the family-A G-protein-coupled receptors and provide slower and more sustained synaptic responses through second messenger systems (Jones, C.K., Byun, N., Bubser, M., 2012).

Respiratory inhalation and dermal absorption are considered the primary routes of exposure to pesticides in occupational settings (Ye, M., Beach, J., Martin, J.W., et al, 2013). Respiratory exposure usually occurs when applying highly volatile pesticide products, especially for those working with no respiratory protective equipment. Pesticides inhibit AChE activity, causing ACh to accumulate at the cholinergic synapses. This accumulation of ACh depolarizes the postsynaptic cell and renders it refractory to subsequent ACh release, causing, among other effects, neuromuscular paralysis (Krsti, D.Z., Lazarevi, T.D., Bond, A.M., et al., 2013). Sustaining moderate ACh levels is crucial for maintaining homeostasis (Ventura, A.L.M., 2010). ACh has a lower rate of blood turnover, about ninety days after the last contact with organophosphates, than the BChE that is an enzyme produced in the hepatic tissue and exported continuously into the bloodstream. Acute intoxications may occur in a mild, moderate or severe manner, depending on the amount of pesticide, the route of contamination, the time of absorption, the toxicity of the product and the time elapsed between exposure and medical care. Symptoms occur acutely within minutes to hours (Gilboa-Geffe, A., Hartmann, G., Soreq, H., 2010). These include vomiting, eye irritation, headache, nausea and allergic reactions. The more chronic symptoms include an imbalance in gait, tearing of the eyes, chronic dermatitis, neurologic problems, or even cancer. The first manifestations are usually muscarinic, such as sweating, nausea, vomiting, abdominal cramps, diarrhea, hypotension, urinary incontinence, chest pain and cough (Peter J, Sudarsan T, Moran J, et al. 2014). Thereafter, nicotinic manifestations occur, such as cramp, muscle weakness, muscle paralysis, hypertension, tachycardia and pallor. Some manifestations arising from the central nervous system can be anxiety, tremors, convulsions and coma (Del Prado-Lu, J.L., 2007). Most of these manifestations can be linked to deregulations of the ACh pathway.

Currently, it is mandatory by law in Portugal that pesticide applicators have received training for their purchase and use of pesticides, according to the law No. 26/2013 of April 11 in the "Diário da República" (Schmitz, M.K., 2003). However, the application of pesticide by wine-growers is often not performed according to the regulations. Therefore, a high exposure to these compounds occurs and this could have major effects on the health status of the farmers.

Therefore, the main objective of the current investigation is to determine the impact of pesticides on the cholinesterase activity.

2 MATERIAL AND METHODS

2.1 Sample characterization

The target population was from the parish of Valdigem, in the municipality of Lamego, district of Viseu, Portugal. It is the Alto Douro Vinhateiro where viticulture predominates and the application of pesticides is made with high intensity. The study cohort was composed of twenty-seven individuals, twenty-five male and two female, with ages above eighteen years old. A questionnaire was also conducted to relate the behaviors of the study population to the laboratory results. The applicators who responded to the enrollment questionnaire are the focus of this study. According to the law, only accredited persons authorized to apply pesticides and who apply during the mentioned periods participated in the study. It excludes all those that do not meet these criteria.

2.2 Methodology and instruments for data collection

A venous blood sample was taken by venipuncture at two different points of time (T0, T1). The first blood sample collection was realized prior to the application of pesticides in the month of January and a second collection was realized in the period of application of pesticides in the month of May. They were collected in test tubes without preparation. All the samples were centrifuged 1800 xg at 4°C for 10 min, following by separation of the serum from the cells.

The activity of cholinesterase was measured using the Ellman's method (Ellman G, Courtney K, Andres V, et al., 1961). Thiols react with this compound, cleaving the disulfide bond to give 2-nitro-5-thiobenzoate (TNB−), which ionizes to the TNB2− dianion in water at neutral and alkaline pH. This TNB2− ion has a yellow color. This reaction is rapid and stoichiometric, with the addition of one mole of thiol releasing one mole of TNB.

The TNB2– is quantified in a spectrophotometer by measuring the absorbance of visible light at 412 nm.

2.3 Statistical Analysis

All statistical analyses were performed using the statistical package IBM SPSS *Statistics* 22.0. To carry out the study, two moments of data collection were necessary. Thus, the Student's test is used for paired samples and the Wilcoxon test. At the level of statistical inference, a 95% confidence level was considered for a random error of less than or equal to 5%. Values are presented as mean ± standard deviation and the values of AChE and BChE was expressed as U/L and kU/L, respectively.

3 RESULTS

In both moments of the collection of the blood (First in January before pesticide application and the second in May during pesticide application), done to twenty-seven pesticide handlers, twenty-five male and two female, with ages above eighteen years old, there were determined activities of AChE and BChE.

AChE activity

An average value of the AChE activity of 74.69 ± 16.58 U/L was observed at the first moment, while in the second moment the average values obtained were 67.62 ± 22.70 U/L.

Comparing the two moments, we found a significant reduction in the values of the cholinesterase activity at the time of application of the pesticides, relative to the first moment coinciding with the non-application period (p = 0.039) (Table 1).

Based on the survey data showing that among the study population, twenty of which are usual workers in viticulture and the remaining seven perform this activity occasionally, a variation between the first and second moments was observed. This variation resulted in a significant decrease in the AChE activity in the group of winegrowers who are habitual workers (p = 0.052).

As for the rest, who are occasional workers in this agricultural activity, there was a non-significant decrease in the AChE activity (p = 0.230). (Table 2).

Table 1. The moments of pre-exposition (T0) and the ongoing exposition (T1) on the activity of AChE AChE and BChE.

	AChE (U/L)	BChE (kU/L)
T0	74.685 ± 16.575	3.818 ± 0.589
T1	67.621 ± 22.700	3.743 ± 0.623
p-value	0.039	0.605

We evaluated the pesticides time of use too. Based on survey data showing that among the population under study, seventeen of which are usual pesticide applicators and the remaining ten perform this work occasionally, we observed a variation between the first and the second moment.

This variation resulted in a significant decrease in the AChE activity in the group of wine growers who are habitual applicators of these products (p = 0.019).

As for workers who use pesticides occasionally, there was a non-significant diminish in the AChE activity (p = 0.764) (Table 3).

Similarly, based on survey data showing that, among the study population, sixteen of these showed that they had already shown changes in well-being, while the remaining eleven did not reveal these well-being.

There was a significant decrease in the AChE activity in the winegrower group that showed changes in well-being (p = 0.030), whereas in those who reported never having felt these changes, there was a non-significant decrease in the AChE activity activity (p = 0.657) (Table 4).

BChE activity

At the first moment a mean value of BChE activity of 3.82 ± 0.59 kU/L was observed and in the second moment we obtained the mean values of

Table 2. The moments of pre-exposition (T0) and the ongoing exposition (T1) and the AChE and BChE activity in ordinary and occasional workers.

	AChE (U/L)	BChE (kU/L)
Ordinary workers T0	78.369 ± 16.448	4.042 ± 0.386
Ordinary workers T1	70.432 ± 21.450	3.777 ± 0.664
p-value	0.052	0.391
Occasional workers T0	64.162 ± 12.624	3.179 ± 0.632
Occasional workers T1	59.588 ± 25.967	3.650 ± 0.522
p-value	0.237	0.063

Table 3. The moments of pre-exposition (T0) and the ongoing exposition (T1) and the AChE and BChE activity in ordinary and occasional pesticides applicators.

	AChE (U/L)	BChE (kU/L)
Ordinary applicators T0	78.018 ± 18.762	3.904 ± 0.556
Ordinary applicators T1	67.080 ± 25.514	3.737 ± 0.643
p-value	0.019	0.687
Occasional applicators T0	69.020 ± 10.545	3.672 ± 0.645
Occasional applicators T1	68.540 ± 18.162	3.755 ± 0.623
p-value	0.445	0.508

Table 4. The moments of pre-exposition (T0) and the ongoing exposition (T1) and the AChE and BChE activity in workers with changes in well-being and without changes.

	AChE (U/L)	BChE (kU/L)
Workers with changes in well-being T0	78.915 ± 15.567	3.813 ± 0.599
Workers with changes in well-being T1	67.966 ± 25.394	3.696 ± 0.644
p-value	0.030	0.679
Workers without changes in well-being T0	68.533 ± 16.745	3.825 ± 0.604
Workers without changes in well-being T1	67.118 ± 19.287	3.813 ± 0.616
p-value	0.657	0.424

3.74 ± 0.62 kU/L. Comparing the two moments, we found a non-significant decrease in the values of BChE activity at the moment of application of the pesticides (p = 0.605). (Table 1).

Based on the survey data showing that among the study population, twenty of which are usual workers in viticulture and the remaining seven perform this activity occasionally, a variation between the first and second moments was observed. This variation resulted in a non-significant decrease of BChE in the group of winegrowers who are habitual workers (p = 0.391). Concerning the other winegrowers, there was a little trend increase in the BChE activity (p = 0.063) (Table 2).

We evaluated the pesticides time of use too. Based on survey data showing that among the population under study, seventeen of which are usual pesticide applicators and the remaining ten perform this work occasionally, we observed a variation between the first and the second moment. This variation resulted in a non-significant decrease in the BChE activity in the group of winegrowers who are habitual applicators of these products (p = 0.677). As for workers who use pesticides occasionally, there was a increase, not significant in the BChE activity (p = 0.508) (Table 3).

Similarly, based on survey data showing that, among the study population, sixteen of these showed that they had already shown well-being changes, while the remaining eleven did not reveal these well-being changes. There was a non-significant decrease in the BChE activity in the group of winegrowers who showed changes in well-being (p = 0.679), while in those who reported never having felt these changes, there was a non-significant decrease in the BChE activity activity (p = 0.424) (Table 4).

4 DISCUSSION

In this study, the determination of the AChE and BChE enzymatic activity, the obtained results in the first moment of blood collection were compared to the results obtained during pesticide application. Thus, a decrease of the AChE activity was observed, when compared to the moments of pre-exposition and the ongoing exposition of the viticultures, in a significant proportion. This diminution confirms the inhibitory effects of the pesticides on AChE activity in individuals who revealed that had felt wellbeing alterations in relation to those who referred never having felt these alterations (Sereniki A, Vital M., 2008).

The time of exposure to pesticides influences the AChE activity. Individuals who perform this work usually have greater changes from workers who occasionally do this work. Another thing observed during this investigation is that the individuals who indicated that they felt changes in their well-being were those who had the lowest AChE activity values, showing a greater decrease between the two moments of evaluation. This may be due to the exposure time, the amount of pesticides used and their handling. This effect was not manifested in the BChE activity. This is due to the fact that some pesticides such as organophosphates have a higher affinity for the AChE than for the BChE, in addition BChE is an enzyme produced in hepatic tissue and exported constantly into the blood, it is active for only a few days whereas AChE has a half-life equal to red blood cells (about three months) (Falconiere, C., 2006). Thus, the decreased of the BChE activity is an indicator of acute exposure (Michael, E., Peter, E., Franz, W., et al., 2009).

BChE is a co-adjuvant of the AChE and compensates its action in the regulation of acetylcholine in the rhythmic movements control and tonicity. Thus, the BChE values are not significantly diminished and in some cases even increased due to this compensation (Fulton, M. & Key, P. 2001).

Our data show that the use of these pesticides has the effect of inhibiting cholinesterase activity, especially of AChE activity, which it is important do have safety precautions when handling these products. These safety norms are recommended by Law No. 26/2013 of April 11, 2013, in the "Diário da República" (Schmitz, M.K., 2003).

5 CONCLUSIONS

The number of cases of intoxication enfolding cholinesterases inhibitor pesticides may be directly associated to the frequent usage of these products in wine production. According to the legislation, it is mandatory in Portugal for the users of pesticides to have proper training in buying and applying the products. Despite this law, security measures are not always applied. The inquired claimed that it is not practical to use the personal protection equipment (PPE), especially in Spring and Summer, due to the high temperatures. One aspect that contributes for the occurrence of intoxication is the inappropriate usage of the products and the non-usage of the PPE. It is necessary to raise awareness among viticultures, in order to minimize the risks.

Nausea, chest compression sensation, and pale skin are indicators of intoxication by organophosphorates diagnosis. These indicators, observed in the majority of the intoxicated patients, and its frequency depend mostly of the exposure intensity.

Moreover, the study of the effects of pesticides on the activity of AChE contributed to the alert of the risks these products may cause and to the defense of security regulations in the agricultural activity, minimizing the dangers to which workers are exposed.

Recognizing precociously this sort of incident, as well as a higher vigilance in health related to the usage of pesticides, are extremely relevant attitudes to diminish the number of cases and the severity of the intoxications.

REFERENCES

Chatonnet A, Lockridgetl O. Comparison of butyrylcholinesterase and acetylcholinesterase. Biochem J 1989;260:625-634.

Da Silva S, Pimentel V, Fiorenza A, et al. Activity of cholinesterases and adenosine deaminase in blood and serum of rats experimentally infected with Trypanosoma cruzi. Ann Trop Med Parasitol 2011;105:385-391.

Del Prado-Lu JL. Pesticide exposure, risk factors and health problems among cutflower farmers: a cross sectional study. J Occup Med Toxicol 2007;2:9.

Ellman G, Courtney K, Andres V, et al. A new and rapid colorimetric determination.

Falconiere C. Padronização dos Valores Séricos da Enzima Butirilcolinesterase em Crianças de 7 a 11 anos. 2006.

Fulton M, Key P. Acetylcholinesterase inhibition in estuarine fish and invertebrates as an indicator of organophosphorus insecticide exposure and effects. Environmental Toxicology and Chemistry 2001;20:37-45.

Gilboa-Geffe A, Hartmann G, Soreq H. Stressing hematopoiesis and immunity: an acetylcholinesterase window into nervous and immune system interactions. Front Mol Neurosci 2010;5:1-10.

Jones CK, Byun N, Bubser M. Muscarinic and nicotinic acetylcholine receptor agonists and allosteric modulators for the treatment of schizophrenia. Neuropsychopharmacol 2012; 37:16-42.

Krsti DZ, Lazarevi TD, Bond AM, et al. Acetylcholinesterase Inhibitors : Pharmacology and Toxicology. Current Neuropharmacology 2013;11:315-335.

Lockridge O, Norgren RB, Johnson RC, et al. Naturally Occurring Genetic Variants of Human Acetylcholinesterase and Butyrylcholinesterase and Their Potential Impact on the Risk of Toxicity from Cholinesterase Inhibitors. Chem Res Toxicol 2016;29:1381-1392.

Michael E, Peter E, Franz W, et al. Predicting Outcome using Butyrlcholinesterase Activity in Organophosphorus Pesticide Self-Poising. 2009;101:476-474.

Peter J, Sudarsan T, Moran J, et al. Clinical features of organophosphate poisoning: A review of different classification systems and approaches. Indian Journal of Critical Care Medicine 2014;18:805.

Schmitz M K. Intoxicação por agrotóxicos inibidores da colinesterase. 2003.

Sereniki A, Vital M. Alzheimer's Disease: Pathophysiological and Pharmacological Features. 2008.

Sonmez F. Journal of Enzyme Inhibition and Medicinal Chemistry Design, synthesis and docking study of novel coumarin ligands as potential selective acetylcholinesterase inhibitors 2017;1: 285-297.

Sonmez F. Journal of Enzyme Inhibition and Medicinal Chemistry Design, synthesis and docking study of novel coumarin ligands as potential selective acetylcholinesterase inhibitors. J Enzyme Inhib Med Chem 2017.

Sun IO, Yoon HJ, Lee KY. Prognostic Factors in Cholinesterase Inhibitor Poisoning. Med Sci Monit 2015;21:2900-4.

Ventura ALM. Sistema colinérgico: Revisitando receptores, regulação e a relação com a doença de Alzheimer, esquizofrenia, epilepsia e tabagismo. Rev Psiquiatr Clin 2010; 37:74-80.

Ye M, Beach J, Martin JW, et al. Occupational Pesticide Exposures and Respiratory Health. Int J Environ Res Public Health 2013;10:6442-6471.

Comparative assessment of work-related musculoskeletal disorders in an industrial kitchen

D.M.B. Costa
FEUP—Faculty of Engineering, University of Porto, Porto, Portugal

R.V. Ferreira
UFRJ—Federal University of Rio de Janeiro, Rio de Janeiro, Brazil

E.B.F. Galante
IME—Military Institute of Engineering, Rio de Janeiro, Brazil

J.S.W. Nóbrega
UFRJ—Federal University of Rio de Janeiro, Rio de Janeiro, Brazil

L.A. Alves
CEFET—Federal Center of Technological Education Celso Suckow da Fonseca, Rio de Janeiro, Brazil

C.V. Morgado
UFRJ—Federal University of Rio de Janeiro, Rio de Janeiro, Brazil

ABSTRACT: This study investigates the ergonomics aspects of work in an industrial kitchen. This work environment is dynamic yet repetitive and often leads workers to adopt inappropriate postures. To assess these work conditions, an ergonomics survey was applied to identify work-related musculoskeletal disorder symptoms, their intensity and affected areas. An evaluation of the most critical activities was performed using the Rapid Upper Limb Assessment (RULA), the Rapid Entire Body Assessment (REBA) and the Ovako Working Posture Assessment (OWAS) methods, whose results were compared. The results of the three methods failed to coincide with each other with the RULA method indicating a greater severity in all tasks evaluated and the OWAS method determined lower severity.

1 INTRODUCTION

Industrial kitchens are highly complex, dynamic, and stressful workplaces. The activities carried out in this environment involve, for instance, handling pressurized and hot equipment as well as cutting tools. In addition, most of activities performed require the execution of repetitive movements, weight lifting, and long periods of time spent standing.

Despite such conditions and the relevance of the performed activities, these aspects are often neglected in kitchens. This study assesses the ergonomic conditions in an industrial kitchen located in Brazil with an ergonomics analysis of the most critical tasks following different methods.

Out of the three domains of specialization of ergonomics (IEA 2017), this study focused on physical ergonomics related to the activities of industrial kitchens to assess exposure to risk factors for work-related musculoskeletal disorders (MSDs). Therefore, two methods were applied: (i) data collection in the workplace through a questionnaire and (ii) application of different observational methods.

This article is divided into five sections. Following this introduction, the background and case study are presented in Section 2. Section 3 discusses the ergonomics survey and the methods applied. Results and discussion are presented in Section 4. Finally, Section 5 brings the final remarks and recommendations.

2 BACKGROUND AND CASE STUDY

2.1 Background

MSDs are the most frequently reported health problems among workers (NRC 2001, Woolf & Pfleger 2003). In Brazil, statistical data shows that about 2.4% of working population is diagnosed with MSDs, of which 58.16% report some limitation of daily activities (IBGE 2013, Maia et al. 2015).

The use of observational methods is still the most commonly used by occupational safety and health practitioners to assess work conditions that may contribute to the development of MSDs, since these are simple and inexpensive to apply (Takala et al. 2010, David 2005). With the advent of a number of new observational methods in literature several efforts have been made to compare their results (Chiasson et al. 2012, David 2005).

2.2 *Case study*

The case study examined is an industrial kitchen which is located inside an industrial complex in Brazil. The 16 members (11 men and 5 women) of the kitchen staff produce about 170 meals daily to 100 employees at five different meal times.

Activities in this work environment can be summarized as: (i) reception of materials, (ii) cold or dry storage, (iii) meal pre-preparation, (iv) meal preparation, (v) meal distribution, (vi) utensils cleaning and sanitation, and (v) waste disposal. Workers have different work shifts and can be divided into the following positions: nutritionist (1 person), cleaners (4 people), kitchen assistants (3 people), bus persons (3 people), supervisor (1 person), stockroom attendant (1 person) and cooks (3 people).

The kitchen is annexed to the dining hall and is divided into two rooms. The first room is air conditioned and is used for pre-preparation of foods (cutting, washing, and sanitizing). The second room has natural ventilation and an exhaust fan for stoves and ovens. This unit is reserved for food preparation and actual cooking.

There are three storage areas for materials used in the processes. Food storage is done in a dry storage area, for products such as bread and canned goods, and in an air-conditioned area, with refrigerators and freezers. Another area is reserved for storage of chemical products such detergent, soap, sanitizer and other cleaning products.

3 MATERIALS AND METHODS

Exposure to MSDs risk factors can be assessed through three categories: subjective judgment, systematic observation and direct measurement (Burdorf & van der Beek 1999). In this study, a subjective judgment was formed through a questionnaire formulated to characterize and identify existing MSDs symptoms of the workers.

Following this, physical ergonomics was assessed by evaluating workers postures while performing different tasks. The action level was used as a parameter for comparison of the results obtained in each method.

3.1 *Ergonomics survey*

An ergonomics survey based on the literature was applied to all employees of the kitchen (Couto 2007, Kathy Cheng et al. 2013). Workers were initially asked to identify if they had any MSDs symptom. In cases where a symptom was reported, they were asked to: (i) identify the symptoms and the body zones affected, (ii) indicate if such symptoms are related to work activities, (iii) indicate when symptoms were first felt, (iv) classify the symptom (tiredness, pain, tingling or numbness, limitation of movements or others), (v) rate the intensity of the symptom (mild, moderate or severe), (vi) identify if pain increases during the working day, (vii) identify if pain decreases after rest, and (viii) identify critical activities in the workday and suggest improvements.

The questionnaire was applied to all workers, except the nutritionist and supervisor. These workers were not considered once they do not perform activities in the kitchen itself, being exposed to different ergonomic risks.

3.2 *Ergonomics assessment*

The activities undertaken in the kitchen were split into four main critical tasks: cutting, dishwashing, food preparation and cleaning—regarded by the workers as the most time consuming and repetitive in this workplace. A brief description of each method and their action levels are presented below.

3.2.1 *RULA method*

The Rapid Upper Limb Assessment (RULA) method allows posture analysis of the upper limbs, neck and trunk, along with the muscular contraction and external force received by the body (McAtamney & Corlett 1993). This method makes use of postural diagrams and three scoring charts for posture evaluation.

The RULA method divides the human body into group A, composed of upper arm, lower arm, wrist and hand, and group B, composed of neck, trunk, and legs. Two tables are used to evaluate posture scores for each group, also considering muscle use and force scores. Finally, these results are grouped in a final table, obtaining a grand score that determines the action level associated with the evaluated tasks (Table 1).

Table 1. RULA grand scores and requirements for action. Source: Adapted from McAtamney et al. (1993).

Score	Action level	Recommendation
1–2	1	Acceptable posture if it is not maintained or repeated for long periods
3–4	2	Further investigation is needed, and changes may be required
5–6	3	Investigation and changes are required soon
7	4	Investigation and changes are required immediately

Table 2. REBA grand scores and requirements for action. Source: Adapted from Hignett et al. (2000).

Score	Action level	Recommendation
1	0	None necessary
2–3	1	May be necessary
4–7	2	Necessary
8–10	3	Necessary soon
11–15	4	Necessary now

3.2.2 REBA method

The Rapid Entire Body Assessment (REBA) method (Hignett & McAtamney 2000) proposes that posture should be evaluated and recorded based on body segment diagrams. This method is based on the body part diagrams from RULA and is considered more suitable to evaluate whole body working postures (Kong et al. 2017).

The first step of the method consists of registering the postures for group A (neck, leg and torso) and for group B (arms, wrists and hands) considering the proposed diagrams. For each of the obtained values, load/force scores are added. Subsequently, the values are cross-referenced, and an activity score is assigned to identify the final score (Table 2).

3.2.3 OWAS method

The Ovako Working Posture Analysis System (OWAS) is a method originally developed in Finland to analyze posture in a steel mill (Karhu et al. 1977). The main function of this method is to record and analyze whole body postures (David 2005).

In this method, individual scores for the position of the back, upper limbs and lower limbs are obtained, generating a three or four-digit code for each position. After each of the postures had been rated, a reclassification is performed considering four classes (Table 3), also referred as operative

Table 3. OWAS operative classes and recommendations. Source: Adapted from Karhu et al. (1977).

Action level	Recommendation
1	Normal postures which do not need any special attention, except in some special cases
2	Postures must be considered during the next regular check of working methods
3	Postures need consideration in the near future
4	Postures need immediate consideration

classes. This classification can be understood as an action level for obtained results.

4 RESULTS AND DISCUSSION

4.1 Ergonomics survey

The sampled work population is predominantly male (71%) and the age group varies between 18 and 48 years old. All workers that participated in the survey reported to have MDSs symptoms and account it to their work activities. All of them report that symptoms caused discomfort for, at least, six months.

Spine, hip and back were the body areas where MSDs symptoms were reported and also the areas when symptoms were first felt by workers. For 64.3%, the most common discomfort reported was fatigue, while for 21.4% it was pain and the remaining 14.3% reported tingling or numbness. A total of 71% of the workers rated the discomfort they felt as moderate and 29% as mild.

Finally, 64% of the interviewees reported an increase of discomfort/pain during working hours, while 36% mentioned that they feel more discomfort after the end of the work day. All workers reported a decrease of discomfort levels after resting.

When questioned about possible improvements in the work environment, all workers stated that a greater turnover between the activities and the existence of more rest periods would bring an improvement of the symptoms of discomfort. These results indicate the need to further assess ergonomic conditions.

4.2 Ergonomics assessment

Each task was analyzed using the RULA, REBA and OWAS methods. The results are presented in Table 4.

The RULA method showed greater severity in the postures, resulting in scores never lower than 3

Table 4. Description of considered tasks and analysis of results.

Task	Description	Position	Task description	Results RULA	REBA	OWAS
1	Cutting	Kitchen Assistant	The activity is performed in the standing position and requires repetitive movements, especially with the arms and wrists. In addition, the workers remain with the neck curved downwards.	Immediate changes (4)	Intervention and further analysis required (2)	No corrective action is required (1)
2	Dishwashing	Cleaners	The height of the sink for washing utensils is low and the workers have adopted some strategies. The task requires repetitive upper limb movements and the use of force for long periods of time, especially when washing heavier or dirtier objects.	Investigate and make changes (3)	Intervention may be required (1)	No corrective action is required (1)
3	Food preparation	Cook	During the preparation of food, the cook performs constant repetitive movements with the upper limbs and remains with the neck slightly inclined and the arms raised.	Immediate changes (4)	Intervention and further analysis required (2)	No corrective action is required (1)
4	Floor cleaning	Kitchen Assistant	The cleaning of the floor requires that the employee is bent over for longs periods of time. This posture is adopted because the length of the broom handle is not adequate.	Immediate changes (4)	Intervention and further analysis required (2)	Corrections are needed in the near future (2)

in any task. This is due to the fact that the method only assesses upper limbs movements.

The results from REBA and OWAS methods were more consistent with each other and resulted in less critical scenarios. This resemblance can be attributed to the consideration of lower limbs in both methods, which does not occur in the RULA method.

Tasks 1 and 3 were the ones in which the methods varied the most. For tasks 1 and 3, REBA produced higher action levels in comparison to OWAS. This can be explained by the inclusion of movement frequency in the REBA method, which is not evaluated in OWAS.

For tasks 1, 2 and 3, both RULA and REBA results are consistent, indicating the need for changes in the tasks, but with different action levels. Conversely, OWAS results for these tasks do not indicate a need to take any corrective measures.

Task 4 was the one resulting in recommended changes by all methods, which may be due to a high spine angle during the task. However, RULA presented a more critical evaluation when compared to the others.

Even though the use of observational methods provides insights about the exposure to risk factors for work related MSDs the discrepancies observed in the use of different methods demonstrate that whenever greater accuracy is necessary, advanced observational techniques and direct methods should be considered.

5 FINAL REMARKS

A qualitative assessment of kitchen workers was performed through an ergonomic survey. The results demonstrated high prevalence of work-related MSDs symptoms among workers. This translates into the need to perform further investigation, changes in work conditions, and the adoption of corrective measures.

Four tasks in the workplace were assessed considering three different observational methods (RULA, REBA and OWAS). The obtained results have demonstrated that the choice of observational method affects the results of the ergonomic assessment.

Therefore, it is recommended that the evaluation of tasks in work environments may be performed considering different methods to account for such differences. This approach may allow a better identification of critical tasks.

The choice of methods to be applied may also consider the nature of activities under investigation. Particularly to the activities evaluated, the REBA method seems to provide more accurate results, once it considers the whole body and the frequency of movements. However, the comparison among methodologies have demonstrated that all work postures expose the work to some level of risk.

The findings also demonstrate that the prevalence of MSDs in the kitchen studied is particularly critical to the upper limbs and related to repetitive movements. Therefore, based on the results the following improvements are recommended:

– Improvement of work instruments, particularly for task 1 and task 4.
– Adapt work conditions, such as the case of task 2.
– Comply with minimum rest times during the shifts and introduce other smaller breaks during the working day, particularly for task 3.
– Introduce mechanized/automated processes that reduce working loads.
– Reorganize the productive process to allow a greater diversification of tasks.
– Promote awareness on work postures in the workplace.

REFERENCES

Burdorf, A. & Van Der Beek, A. (1999). Exposure assessment strategies for work-related risk factors for musculoskeletal disorders. *Scandinavian Journal of Work, Environment & Health*(4), 25–30.

Chiasson, M.-È., Imbeau, D., Aubry, K. & Delisle, A. (2012). Comparing the results of eight methods used to evaluate risk factors associated with musculoskeletal disorders. *International Journal of Industrial Ergonomics*, 42(5), 478–488. doi: https://doi.org/10.1016/j.ergon.2012.07.003.

Couto, H. d. A. 2007. *Ergonomia aplicada ao trabalho: conteúdo básico: guia prático*. Brazil: Ergo.

David, G.C. (2005). Ergonomic methods for assessing exposure to risk factors for work-related musculoskeletal disorders. *Occupational Medicine*, 55(3), 190–199. doi: https://doi.org/10.1093/occmed/kqi082.

Hignett, S. & McAtamney, L. (2000). Rapid Entire Body Assessment (REBA). *Applied Ergonomics*, 31(2), 201–205. doi: https://doi.org/10.1016/S0003-6870(99)00039-3.

IBGE (Instituto Brasileiro de Geografia e Estatística). 2013. *Pesquisa Nacional de Saúde* Brazil.

IEA. *Definition and Domains of Ergonomics*. IEA. 2017. Available at http://www.iea.cc/whats/index.html.

Karhu, O., Kansi, P. & Kuorinka, I. (1977). Correcting working postures in industry: A practical method for analysis. *Appllied Ergonomics*, 8(4), 199–201.

Kathy Cheng, H.-Y., Cheng, C.-Y. & Ju, Y.-Y. (2013). Work-related musculoskeletal disorders and ergonomic risk factors in early intervention educators. *Applied Ergonomics*, 44(1), 134–141. doi: https://doi.org/10.1016/j.apergo.2012.06.004.

Kong, Y.-K., Lee, S.-y., Lee, K.-S. & Kim, D.-M. (2017). Comparisons of ergonomic evaluation tools (ALLA, RULA, REBA and OWAS) for farm work. *International Journal of Occupational Safety and Ergonomics*, 1–6. doi: https://doi.org/10.1080/10803548.2017.1306960.

Maia, A., Saito, C., Oliveira, J., Bussacos, M., Maeno, M., Lorenzi, R. & Santos, S. 2015. *Accidents at work in Brazil in 2013 – comparison between selected data within two data sources: IBGE National Household Health Survey and Statistical Yearbook of the Social Security by Ministry of Social Welfare*. Brazil.

McAtamney, L. & Corlett, N. (1993). RULA: a survey method for the investigation of world-related upper limb disorders. *Applied Ergonomics*, 24(2), 91–99.

NRC (National Research Council). 2001. *Musculoskeletal Disorders and the Workplace: Low Back and Upper Extremities*. Washington, DC: The National Academies Press.

Takala, E.-P., Pehkonen, I., Forsman, M., Hansson, G.-Å., Mathiassen, S.E., Neumann, W.P., Sjøgaard, G., Veiersted, K.B., Westgaard, R.H. & Winkel, J. (2010). Systematic evaluation of observational methods assessing biomechanical exposures at work. *Scandinavian Journal of Work, Environment & Health*(1), 3–24. doi: https://doi.org/10.5271/sjweh.2876.

Woolf, A.D. & Pfleger, B. (2003). Burden of major musculoskeletal conditions. *Bulletin of the World Health Organization*, 81, 646–656.

Work-related musculoskeletal disorders in the transportation of dangerous goods

M.V.T. Rabello
UFRJ—Federal University of Rio de Janeiro, Rio de Janeiro, Brazil

D.M.B. Costa
FEUP—Faculty of Engineering, University of Porto, Porto, Portugal

C.V. Morgado
UFRJ—Federal University of Rio de Janeiro, Rio de Janeiro, Brazil

ABSTRACT: The objective of this study was to investigate the prevalence of work-related musculoskeletal disorders in the population of drivers involved in the transportation of dangerous goods. The case study investigated a company performing such activities since 2010, for which risks have been evaluated in previous studies. The investigation consisted of a survey among drivers to verify the prevalence of work-related musculoskeletal disorders and perceive risks in the work routine, followed by the application of the Rapid Upper Limb Assessment (RULA) method to the activities of fuel loading and unloading. Half of the drivers reported musculoskeletal disorder symptoms, particularly in feet and ankles, hands and wrists, lower back and neck regions. RULA results demonstrated a need for immediate changes in activities and that the bottom loading system should be prioritized.

1 INTRODUCTION

Transportation of liquid fuels derived from oil, ethanol and biodiesel can be performed by all transportation modes except by air. In Brazil, due to the inexistence of other alternatives, the majority of fuel transportation is performed by road in tank trucks (ILOS, 2013).

According to the Occupational Safety and Health Administration (OHSA), the activities of loading and unloading tank trucks containing flammable liquid fuels is one of the most dangerous activities in manufacturing or storage (OSHA 2017). Therefore, it is necessary for studies to consider this aspect besides the transportation phase alone.

In this work, the psychological and physiological conditions of tank truck drivers transporting fuels were assessed. An ergonomic evaluation based on the Rapid Upper Limb Assessment (RULA) method was performed considering working conditions, besides material lifting, transportation and unloading.

This study is further divided into five sections. Following this introduction, the company that is the focus of this case study is presented in Section 2. Section 3 discusses the methodology and is followed by results and discussion in Section 4. Finally, Section 5 presents the final remarks and recommendations for improvement of the observed conditions.

2 BACKGROUND AND CASE STUDY

2.1 Background

Work-related musculoskeletal disorders (MSDs) are a major cost burden to individuals, businesses and society and are the most frequently reported health problem among workers (NRC 2001, Woolf & Pfleger 2003). However, intervention strategies for their prevention face challenges, such as lack of awareness of ergonomic issues, organizational attitudes, and political, social and contextual issues (Rothmore et al. 2017).

Although the literature presents plentiful information on risks associated with the transportation of dangerous goods, the ergonomics of tasks performed by drivers is not often evaluated (ECTA 2009, HSE 2014). Therefore, we conducted a qualitative assessment of existing symptoms of MSDs, followed by a quantitative assessment of fuel loading and unloading activities.

2.2 Case study

The target of the case study is a company engaged in road transportation of dangerous goods in Brazil. The transported goods are mostly petroleum-based fuels (except ethanol), delivered to service stations, seaports and vessels.

Operations began in 2010. Currently the company employs 23 people, 15 of whom are tank truck drivers, and possesses 24 tank trucks (TTs). A risk assessment of this company integrating environmental and occupational factors was presented in a previous study (Costa et al. 2017), which demonstrated ergonomics and accidents as the major risk factors in the company's operational activities.

Fuel is delivered in the modalities of FOB (free on board) to service stations and of CIF (cost, insurance and freight) to seaports and ships. Operational activities are only performed by the TT drivers and the workflow can be summarized as follows: (i) the TT enters the facility and heads toward the loading platform, (ii) the TT enters the loading platform, (iii) the TT is inspected, (iv) the hatch is opened for inspection, (v) the fuel is pumped in, and (vi) the fuel is transported to the customer/receiver where it is unloaded.

2.3 Description of activities

We evaluated the physical ergonomic aspects of fuel loading and unloading activity and its options. TT loading and unloading operations can be performed considering two methods: top or bottom loading. In top loading, the driver attaches the fuel hose to the inlet located in the upper part of the TT while in in bottom loading the inlet valve is located in the bottom of the TT.

Bottom loading is intrinsically safer than top loading since it eliminates the need for work at heights. In addition, this method tends to be faster since more than one compartment can be filled simultaneously, reducing the exposure of workers to fuel vapors.

Although the use of bottom loading presents more advantages and gains in productivity, top loading is still the most used method. This occurs due to the cost associated with the installation of specific bottom loading equipment in each TT. In the studied company, 25% of the fleet has a bottom loading system.

3 MATERIALS AND METHODS

To assess the working conditions of drivers, a questionnaire was applied to characterize these workers and to understand their perception of risks in their daily activities, as described in Section 3.1. Following this, physical ergonomics in this work environment were qualitatively assessed based on the evaluation of postures while performing tasks and the physical environment of the activity, described in Section 3.2.

3.1 Survey

As a preliminary step to evaluate ergonomic conditions, a survey was performed among drivers following the recommendations from the literature (Kathy Cheng et al. 2013). The questionnaire consisted of three sets of questions: general characterization of the drivers, existing MSDs and risk perception during activities performed in the work context.

The first part of the questionnaire contained questions related to demography. Questions identified age, sex, height, weight, time in position, education and exercise habits.

The second part of the questionnaire aimed to identify the occurrence of MSDs in different parts of the body (elbow, feet & ankles, hands & wrists, knee, lower back, neck, shoulder, thighs, and upper back), their severity and frequency. Workers also identified how long they had symptoms, if any, how they felt after a working day and if they believed these symptoms affected their work performance.

Finally, risk perception related to activities performed in the work environment was assessed. Workers were asked which stage of their activity they consider most critical and why, and which risks they identify in their work environment.

3.2 The RULA method

RULA (Rapid Upper Limb Assessment) is a survey method developed in 1993. Its use is recommended to investigate exposure to risk factors in workplaces where work-related upper limb disorders are reported (McAtamney & Corlett 1993). Its use is widely recommended for its quick and easy application and moderate to high reliability in different case studies (Dockrell et al. 2012).

The RULA method divides the human body into two groups: (A) upper arm, lower arm, and wrist, and (B) neck, trunk, and legs. A score is calculated using three tables. The first two tables give the posture scores for each, which are correlated according to the frequency of the operations and the load placed on the limbs. These results are grouped in a final table, obtaining a final score that determines the action level associated with the evaluated tasks (Table 1).

Table 1. Total scores and requirements for action. Source: adapted from McAtamney et al. (1993).

Score	Recommendation
1–2	Acceptable posture if it is not maintained or repeated for long periods
3–4	Further investigation is needed and changes may be required
5–6	Investigation and changes are required soon
7	Investigation and changes are required immediately

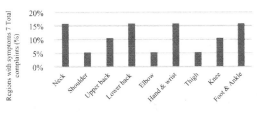

Figure 1. Reported body regions with MSD symptoms.

4 RESULTS AND DISCUSSION

4.1 Survey

The survey was applied to 13 out of the 15 drivers currently working for the company. The first part of questionnaire showed that all drivers were male and had average age of 44 years and 13 years working in this position. Regarding the educational level, 69% declared they finished secondary school.

Only 3 drivers reported exercising regularly outside of the work environment. MSDs were reported by 50% of the workers, totaling 19 complaints distributed in different body regions. The most affected body regions were: (i) feet & ankles, (ii) hands & wrists, (iii) lower back, and (iv) neck. All of them represented 15% of total complaints (Figure 1).

Regarding severity, 32% of MSDs were classified as mild pain and 47% as moderate. Only two cases were classified as unbearable pain, related to upper and lower back. Despite the relatively low numbers of complaints, 85% of the workers declared feeling tired or very tired after a working day.

In terms of identification of risks, 85% declared concern with risks in the work environment, particularly the risk of traffic accidents (54%) and ergonomic risk (31%). Most workers regarded the loading and unloading as the most critical activity performed (69%), followed by driving (31%).

4.2 RULA method

According to the ergonomic evaluation of TT loading activity through the RULA method, the following upper limb movements were recorded: (i) raising arms: greater than 90 degrees; (ii) lowering arms: from 0 to 60 degrees; (iii) wrist angle: between −15 and +15 degrees; (iv) wrist twisting: extreme; (v) neck angle: greater than 20 degrees; (vi) trunk angle: greater than 60 degrees; and (vii) legs: legs and feet are not properly supported and balanced. The load recorded for Group A and Group B was between 2 and 10 kg intermittently.

For TT unloading activity, the following movements were recorded: (i) raising arms: from 45 to 90 degrees; (ii) lowering arms: from 60 to 100 degrees; (iii) wrist angle: between −15 and +15 degrees; (iv) wrist twisting: extreme; (v) neck angle: greater than 20 degrees; (vi) trunk angle: greater than 60 degrees; and (vii) legs: legs and feet are not properly supported and balanced. As for the loading, between 2 and 10 kg intermittently was recorded for groups A and B.

As a result, the final score was 7, and an action level was obtained for both activities. These results demonstrate that these activities require further investigation and immediate changes.

Particularly for TT loading, the bottom system should be prioritized to reduce efforts exerted by the drivers and avoid working at heights. Therefore, a major recommendation is to install the bottom loading system in all vehicles.

At an organizational level, workers should also be encouraged to report MSD symptoms and to take short pauses during their work day. In addition, it is recommended to set up and carry out proper training programs for drivers to make them aware of the proper posture and better positioning to perform these activities.

In terms of general practices, fuel loading and unloading should be avoided without assistance of a site operator and waiting times should be reduced whenever possible, to minimize ergonomic and accident risks. A special focus on observing procedures can be given to avoid accidents in this work context.

5 FINAL REMARKS

This study assessed qualitatively and quantitatively the activities of drivers in the transportation of dangerous goods. Considering that the transportation of such goods presents a major risk to the population in general and particularly to workers exposed to these products, this activity requires continuous monitoring and evaluation.

The survey demonstrated that several of the workers consider loading and unloading to be critical points of their activities. These activities were evaluated through the RULA method, which demonstrated that this activity should be revised to establish immediate changes. Organizational practices recommended are training and executing loading or unloading with assistance. In addition, the bottom loading system should be prioritized and installed in the entire fleet.

This article reports a preliminary evaluation of work conditions in the transportation of dangerous goods. Therefore, future studies can extend this analysis by applying other ergonomic methods to assess the exposure to ergonomic risks and allow comparison among results. In addition, the relationship between MSDs and psycho-sociological aspects in such activities is still poorly evaluated and is of great relevance.

REFERENCES

Costa, D.M.B., Rabello, M.V.T., Galante, E.B.F. & Morgado, C.V. 2017. "Risk evaluation in the transportation of dangerous goods." In *Occupational Safety and Hygiene V*, 13–16. CRC Press/Balkema.

Dockrell, S., O'Grady, E., Bennett, K., Mullarkey, C., Mc Connell, R., Ruddy, R., Twomey, S. & Flannery, C. (2012). An investigation of the reliability of Rapid Upper Limb Assessment (RULA) as a method of assessment of children's computing posture. *Applied Ergonomics*, 43(3), 632–636. doi: https://doi.org/10.1016/j.apergo.2011.09.009.

ECTA (European Chemical Transport Association). *How to reduce time spent by drivers on site and improve their treatment: Recommendations for loading and unloading sites*. ECTA. 2009. Available at https://www.ecta.com/resources/Documents/Best%20Practices%20Guidelines/how_to_reduce_time_spent_by_drivers_on_site_and_improve_their_treatment.pdf.

HSE (Health and Safety Executive). *Unloading petrol from road tankers: Dangerous Substances and Explosive Atmospheres Regulations 2002*. HSE/GOV.UK. 2014. Available at http://www.hse.gov.uk/pUbns/priced/l133.pdf.

Kathy Cheng, H.-Y., Cheng, C.-Y. & Ju, Y.-Y. (2013). Work-related musculoskeletal disorders and ergonomic risk factors in early intervention educators. *Applied Ergonomics*, 44(1), 134–141. doi: https://doi.org/10.1016/j.apergo.2012.06.004.

McAtamney, L. & Corlett, N. (1993). RULA: a survey method for the investigation of world-related upper limb disorders. *Applied Ergonomics*, 24(2), 91–99.

NRC (National Research Council). 2001. *Musculoskeletal Disorders and the Workplace: Low Back and Upper Extremities*. Washington, DC: The National Academies Press.

OSHA (Occupational Safety and Health Administration). *Industry Hazards: Loading and Unloading* 2017. Available at https://www.osha.gov/SLTC/trucking_industry/loading_unloading.html.

Rothmore, P., Aylward, P., Oakman, J., Tappin, D., Gray, J. & Karnon, J. (2017). The stage of change approach for implementing ergonomics advice—Translating research into practice. *Applied Ergonomics*, 59, Part A, 225–233. doi: https://doi.org/10.1016/j.apergo.2016.08.033.

Woolf, A.D. & Pfleger, B. (2003). Burden of major musculoskeletal conditions. *Bulletin of the World Health Organization*, 81, 646–656.

Scaffold use risk assessment model: SURAM

M. Rebelo, P. Laranjeira & F. Silveira
CIICESI, ESTG, Politécnico do Porto, Portugal

K. Czarnocki, E. Blazik-Borowa & E. Czarnocka
Lublin University of Technology, Lublin, Poland

J. Szer
Lodz University of Technology, Lodz, Poland

B. Hola
Wroclaw University of Technology, Wroclaw, Poland

K. Czarnocka
Warsaw University of Technology, Warsaw, Poland

ABSTRACT: Problems concerning occupational health and safety are commonly found in the construction industry, including falling of materials or people from height, stepping on objects and injuries by hand tools handling. Important factor in the occupational safety in construction industry is the use of scaffold. All scaffolds used in construction, renovation, repair (including painting and decorating), and demolition shall be erected, dismantled and maintained in accordance with safety procedure. Therefore, it is crucial to deal with scaffolds safety and risk assessment in construction industry; thus, way on doing assessment and liability of assessment seems to be essential for professionals. However, it is found that those professionals prone to heavily rely on their own experiences and knowledge on decision-making on risk assessment, lack a systematic approach and methodology to assess the reliability of the decisions. Methods: A Scaffold Use Risk Assessment Model (SURAM) has been developed for assessing risk levels as various construction process stages with various work trades. The SURAM is the result of research project realized at the above 60 construction sites, both in Poland and Portugal where 450 observations have been completed including both harmful physical and chemical factors, stress level, worker habits, as well as a hundreds ex-post reconstruction of construction accidents scenarios. Genetic modeling tool has been use for develop the SURAM. Results: Common types of trades, accidents, and accident causes have been explored, in addition to suitable risk assessment methods and criteria. We have found the initial worker stress level is more direct predictor for developing of the unsafe chain leading to the accident than the workload, and concentration of harmful factors at the workplace. Conclusions: The developed SURAM seems to be benefit for predicting high-risk construction activities and thus preventing accidents to occur, based on a set of historical accident data.

1 INTRODUCTION

The construction industry is a booming sector of Polish economy, however, according to Eurostat, the industry is classified among the sectors of the economy presenting high occupational risks and an unsatisfactory level of occupational safety. Although some safety programs have been developed in the country the observed accident reduction rate seems to be rather weak. Employees in the construction sector are exposed to biological, chemical factors, as well as the effects of noise, vibrations, insufficient illumination and temperature. Also the peak workload and especially frequent changes of workload level have been observed in many investigations. More than 45% of workers in the construction sector say that their work has a negative impact on their health (Dabrowski 2015). The construction industry is subjected to high occupational risk and high rates of occupational accidents, occupational diseases and absenteeism at work. In Europe, according to Eurostat data for 2011 (for 28 European Union [EU] countries), the fatal accident rate was 6.39 (per 100,000 persons in employment) in the construction industry (Eurostat 2016). The majority of serious accidents have taken place at the scaffolds or at the construction sites with scaffolds. Taking into consideration the frequency of accidents and high occupational risk in the construction industry with scaffold use, it is important to take the necessary steps to reduce this exposure. In these conditions the research project Scaffold Use Risk Assessment Model for Construction Process Safety "ORKWIZ" have been developed in Poland from early 2016. The project focused on the introduction of a system of new/ additional procedures and tools for monitoring safety on construction sites (Forteza 2017, Törner 2009, Gao 2016). This system built as model could impose strict rules regulating the conduct of contractors in a comprehensive manner to ensure an elimination of hazards from the construction site or an effective reduction of associated risks.

The construction of SURAM is the core part of the ORKWIZ project. The research also shows that many accidents can be avoided by developing a proper concept of safety assurance at the preparation stage (Behm 2005, Namian 2016, Pingani 2016).

2 METHODS

2.1 Population

The study was conducted in Poland. Five different regions of the country have been selected for the project research. The regions have been selected by virtue of economic development level, unemployment rate, technical culture of employees, construction processes intensity, and infrastructure level, among other factors. Accordingly, the study was conducted in the different construction sites, representing typical (more frequent) scaffold size, scaffold system types and technical equipment. Such a diversity of regions, sites and employee praxis wonts and customs is required to achieve universal safety climate for the proposed safety model. At least 120 construction sites with scaffold use will be examined during the research project period. Subsequently, a random sampling procedure was conducted to select individual workers at each construction site; 234 individual workers of those sites potentially exposed to occupational hazards were selected in the first year of the project. For the purpose of the SURAM 800 individuals should be interviewed. An original questionnaire for risk perception and safety climate assessment at the construction site has been developed. At the beginning of our investigation, we have verified several existing questionnaires including NOSACQ-50, Quality of Worklife Module (NIOSH), Contractor OH&S Evaluation, as well as some polish ones and we prepared original tool that better fit to the construction site occupational environment and construction workers perception. Before using the original questionnaire among the selected population, we ran a pilot study among 60 workers. The trial and first run exploratory factor analysis confirmed that the original 45-item questionnaire,

could be used as a risk perception and safety climate scale in polish construction industry. A 5-point Likert-type scale (1 = strongly disagree, 5 = strongly agree) was used to collect the workers' responses. Yes/no responses, lists of options, check-the-box responses, quantity choice etc., were used to self-report incident involvement and demographic data. Since the questionnaire used questions and answers based on scales, a not-applicable option for situations in which the respondents did not know what to respond or did not have an opinion on the issue have also been added. In order to ensure greater objectivity both questionnaire and model and mostly to increase flexibility of SURAM the control group has been recruited at the Portuguese construction sites, considering similarities and differences between countries. Characteristics of the study and control groups have been presented in Table 1. The difference in the drug (including alcoholic beverages) use between groups seems to be

Table 1. Group characteristics.

Variable	Polish study group		Portuguese control group		p
	N	%	N	%	
Monitored Subjects	243	100	38	100	>0.05
Gender					
Male	239	98.3	36	94.7	>0.05
Female	4	1.7	2	5.3	>0.05
Position in Company					
Construction Workers	177	72.8	28	73.7	>0.05
Other Workers	43	17.7	6	15.8	>0.05
Administrative Workers	9	3.7	2	5.25	>0.05
Managers and Supervisors	14	5.8	2	5.25	>0.05
Age (years)					
<25	38	15.6	6	15.8	>0.05
25–34	48	19.8	8	21.05	>0.05
35–44	59	24.3	9	23.7	>0.05
45–54	42	17.3	8	21.05	>0.05
≥55	56	23.0	7	18.4	>0.05
Work experience (years)					
<1	28	11.5	4	10.5	>0.05
2–5	87	35.8	14	36.8	>0.05
6–10	65	26.8	10	26.3	>0.05
11–20	33	13.6	5	13.2	>0.05
≥21	30	12.3	5	13.2	>0.05
Drug Use	191	78.6	20	52.6	<0,05
Smoking	202	83.1	23	60.5	>0.05
Accident involvement or witness	107	44.0	11	28.9	>0.05

the result of different level for acceptance of some alcoholic beverages use in the shift period between Polish and Portuguese workers. In the detail questions regarding type of drug and frequency of use we find no significance except popularity of spirits in Poland and wine or beer in Portugal.

2.2 *Questionnaire validity and reliability*

In the implemented version of the questionnaire, the questions focused on six factors: S_{SOC} (life coherence and social associations): 11 questions; S_{LOC} (sense of control): 10 questions; S_{LKZ} (health state): 5 questions; S_{LWO} (value hierarchy): 11 questions; S_{IZZ} (occupational praxis and psychical attitude): 10 questions; S_{PR} (risk perception): 8 questions. There were also 7 predictors not involved to any of six scales. Sampling adequacy have been measured using the Kaiser-Meyer-Olkin test. Bartlett's test of sphericity was involved for evaluating correlations among safety climate items (Allen 2008, Hair 1998, Saga 2013) Construct validity was tested with exploratory factor analysis, and discriminant validity has been checked by comparing the safety climate scores among groups varying in age, work experience, accident involvement, position in the company, educational and the type of the organization. To evaluate the internal consistency of the questionnaire, Cronbach's α have been used, Speraman-Brown coefficient and Ω. Cronbach's α is used when questions are rated on 5-point Likert scale, used in this investigation; it represents mean correlations among items. Spearman Brown coefficient represents the reliability coefficients that can be attained from possible combinations of split-half questions. The minimal proposed value of these coefficients is 0.70. The data obtained using our questionnaire was analysed Statistica 12.5 StatSoft Inc. The comparison of the difference in accident risk perception and safety climate scores among different demographic groups (age, work experience, occupational experience, position in the company, education, accident involvement, type of construction site) was done with the multiple analyses of variances (MANOVA).

To define questionnaire utility for the final SURAM model, the principal component analysis was performed retaining all the factors with Eigenvalue greater than one. Once the factors were extracted, the Varimax rotation was performed. The analyses showed that the Kaiser-Meyer-Olkin measure of sampling adequacy was 0.81 indicating that these data were appropriate for factor analysis (Dada 2016, Kim 2016, Buica 2017). Bartlett's test of sphericity was significant ($\chi^2 = 1270.6$, $p < 0.01$), which indicated that there were correlations among safety climate items and the correlation matrix was not a unit matrix.

3 RESULTS

The reliability of the measurement method depends on its internal consistency. As already indicated, the consistency was assessed with Cronbach's α, Speraman-Brown coefficient, and Ω. According to Cronbach's α, internal consistency was 0.79 for the entire population. Spearman Brown coefficient was 0.78 and W = 0.70. Most coefficients were higher than 0.70 and adequate for psychometric requirements for a measurement. Thus, the method for measuring occupational hazards, risk perception and the contractor safety climate was appropriate (Lin 2007, Mohamamadfam 2017, Mitropoulos 2009). Table 2 shows each coefficient of the accident risk and safety climate scales. Figure 1 presents the results of the SURAM structural analysis. To make it clearer, it shows only the values of the structural equation, but not the measuring models. Except the questionnaire scales the SURAM have been developed including worker psycho-physiological parameters monitoring before the shift as well as the part of the shift after a break corresponding to the workload during the shift (WL). The wide range of demographic factors (DF) collected both at the construction site as well as from local statistical offices have also been used for model construction. Environmental parameters at the construction site have been monitored on 2 up to 3 levels of the scaffold (depending of its size) during at least five days working week including sound level, illumination, microclimate (EF). Then the diversity from the standard levels has been evaluated as the measure for the matrix construction. To evaluate worker visual concentration on the critical areas or elements of working zone the mobile eye-tracking

Table 2. Questionnaire scales inter-consistency coefficients.

Scale	No. of items	Cronbach's α	Spearman-Brown coefficient	Ω
S_{SOC}	11	0.763	0.771	0.703
S_{LOC}	10	0.694	0.712	0.614
S_{LKZ}	5	0.631	0.680	0.622
S_{LWO}	11	0.895	0.899	0.811
S_{IZZ}	10	0.744	0.752	0.728
S_{PR}	8	0.707	0.719	0.631

Legend:
S_{SOC} – life coherence and social associations,
S_{LOC} – sense of control,
S_{LKZ} – health state,
S_{LWO} – value hierarchy,
S_{IZZ} – occupational praxis and psychical attitude,
S_{PR} – risk perception.

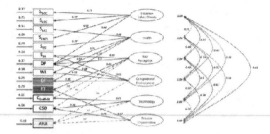

Figure 1. Structural model of SURAM.
Legend:
S_{SOC} – life coherence and social associations,
S_{LOC} – sense of control,
S_{LKZ} – health state,
S_{LWO} – value hierarchy,
S_{IZZ} – occupational praxis and psychical attitude,
S_{PR} – risk perception,
DF – demographic factors,
WL – work load,
EF – environmental factors,
ET – eye tracking,
$C_{scaffold}$ – scaffold construction,
CSO – construction site organization,
AHA – historical accidents analysis.

Table 3. Model fit values.

Statistics	Recommended values	Achieved level
χ^2/df	<3.0	2.87
GFI	>0.90	0.91
AGFI	>0.90	0.92
NFI	>0.90	0.93
CFI	>0.90	0.93
IFI	>0.90	0.91
RFI	>0.90	0.92

Legend:
GFI – goodness-of-fit index,
AGFI – adjusted goodness-of-fit index,
NFI – normed fit index,
NNFI – non-normed fit index,
CFI – comparative fit index,
IFI – incremental fit index,
RFI – relative fit index.

equipment have been used (ET). Stability and quality of scaffold set-up and maintenance have been evaluated ($C_{scaffold}$) as well as construction site organization level (CSO). The complementary element of SURAM especially for model teaching period was Historical Accident Analysis module (AHA). In the Aha module the accidents from past 10 years of the construction industry have been decomposed to the elementary factors. As the model presented at the Figure 1 is a beta one prepared after first year of the projects some of the relations could not be calculated precisely *nd* values. Therefore, even in those partial data it shows potential for use in improving safety at the construction sites with scaffold use.

As to several authors, the goodness-of-fit (GF) model had to be considered first (Hair *et al.* 1998, Ho DCK 1999, Pingani, 2016, Amiri 2017). Within a GF model, it is required to consider three indicators: the measure of absolute fit, the measure of increased fit and the measure of decreased fit. Table 3 presents the results for the proposed model together with the recommended values for satisfactory fit (Ho DCK 1999, Buica 2017, Liao 2016).

Due to the absolute correspondence of the models, the indicators that can be applied in an incompetent strategic analysis are GFI (goodness-of-fit index) and the index of corresponding values. In GFI, the higher the value, the higher the correspondence. In this case, the obtained value was 0.92. This indicator is acceptable since it is over 0.90 (WHO 2010, Molina 2007, Rubio-Romero 2015).

In our model, this indicator has the value of 0.9 which, according to the above-mentioned academics, is an indicator of good correspondence. Table 2 shows inter-correlations among the six scales that were entered into the final model. Because of the comparatively small sample size, each correlation coefficient was significant at 0.05. As a step in model construction this research focused on investigation whether there is any significant difference in risk perception and the safety climate in working teams as well as construction enterprises among the demographic subgroups (Carillo-Castrillo 2016, Choudhry 2014). There significant differences in demographic subgroups in questionnaire scales have been observed. For example, for practicum, there were significant differences on all scales (S_{SOC}, S_{LOC}, S_{LKZ}, S_{LWO}, S_{IZZ}, S_{PR}), but for the education level there were significant differences on (S_{SOC}, S_{LOC}, S_{IZZ}, S_{PR}) scales and not for S_{LKZ} and S_{LWO} scales. Gender did not influence opinions on questions on analyzed factors as more than 97% of employees (94.7% and 98.3% in study and control groups respectively) were male. However, presenting all the results in this manuscript would require too much space. Therefore, they will be discussed in detail in another paper at the end of the project where we could observe larger subgroups. At this stage of the research and project development we have found the initial worker stress level (monitored by the bio-physical parameters at the beginning of shift and after the break + ET) (Fruchter 2011) is more direct predictor for developing of the unsafe chain leading to the accident than the work load (WL), and concentration of harmful factors (EF) at the workplace (Lopez Arquillos 2008).

4 CONCLUSIONS

A study of risk perception, occupational hazards and safety climate at the construction sites with scaffolds in Poland, like the one in this paper, had never been conducted before. We attempted to monitor risk perception, understand the value and beliefs about the safety among Polish workers or precisely workers' teams at polish construction sites as the growing number of migrant workers (mostly Ukrainian) have been noticed during the first stage of research project. The study presented evidence that the perception of the accident risk and safety climate in polish construction sites can be reliably measured with a 45-item questionnaire, involving six factors (S_{SOC} (life coherence and social associations), S_{LOC} (sense of control), S_{LKZ} (health state), S_{LWO} (value hierarchy), S_{IZZ} (occupational praxis and psychical attitude), S_{PR} (risk perception)). The previous research results posited, construction workers put more emphasis on safety training, organizational environment, safety awareness and competency, and management support. Although, the previous thesis are still actual, our recent study shows, that initial stress level could be crucial for developing risky or potentially prone to accident situation. To establish a general model SURAM, our subjects came from several economically, historically and technologically diversified polish regions and the control group came from Portugal. Thus, the developed 45-item questionnaire can be used as a safety measurement tool for the whole construction sector with the scaffold use. This tool was based on the results from different parts of the world and then modified to fit polish construction sites. Further research will focus on a structural equation model, which will result from the structural analysis presented in this work. An additional factors), DF (demographic factors), WL (work load), EF (environmental factors), ET (eye tracking), $C_{scaffold}$ (scaffold construction), CSO (construction site organization), AHA (historical accidents analysis) will have to be included. It will determine the workers' attitude towards the risk level at their workplace, hazardous situations and real occupational accidents that took place there. Subsequently, the six factors from this study will be used to develop a hypothetical frame of the SURAM model. Additionally, as already indicated, each demographic subgroup had strong influence on some of the six factors. This will be analyzed in detail and discussed in next project stages. This study considered workers from six regions (including Portugal), so the level of technical culture and type of organization was one of the variables (CSO). Consequently, the influence of this variable on all six scales will be studied in future. The prognostic validity of the SURAM model developed in this work will be assessed in this way on next stages. Moreover, the results will have practical value for occupational health prevention in construction sector. The developed SURAM, even at this initial stage are found to be useful for predicting high-risk construction activities and thus preventing accidents to occur, based on a set of historical accident data.

REFERENCES

Allen, K., Reed-Rhoads, T., Terry, R.A., Murphy, T.J., Stone, A.D. Coefficient alpha: An engineer's interpretation of test reliability (2008) Journal of Engineering Education, 97 (1), pp. 87–94.

Amiri, M., Ardeshir, A., Fazel Zarandi, M.H. Fuzzy probabilistic expert system for occupational hazard assessment in construction (2017) Safety Science, 93, pp. 16–28.

Behm M. Linking construction fatalities to the design for construction safety concept. Safety Sci. 2005;43(8):589 611.

Buica, G., Antonov, A.E., Beiu, C., Pasculescu, D., Remus, D. Occupational health and safety management in construction sector—The cost of work accidents (2017) Quality—Access to Success, 18, pp. 35–40.

Choudhry, R.M. Behavior-based safety on construction sites: A case study (2014) Accident Analysis and Prevention, 70, pp. 14–23.

Dabrowski A., An investigation and analysis of safety issues in Polish small construction plants. International Journal of Occupational Safety and Ergonomics. 2015;21(4):498–511. doi:10.1080/10803548.2015.1085206.

Forteza, F.J., Carretero-Gómez, J.M., Sesé, A. Occupational risks, accidents on sites and economic performance of construction firms (2017) Safety Science, 94, pp. 61–76.

Fruchter, R., Cavallin, H. Attention and engagement of remote team members in collaborative multimedia environments (2011) Congress on Computing in Civil Engineering, Proceedings, pp. 875–882.

Gao, R., Chan, A.P.C., Utama, W.P., Zahoor, H. Multilevel safety climate and safety performance in the construction industry: Development and validation of a top-down mechanism (2016) International Journal of Environmental Research and Public Health, 13 (11), art. no. 1100,

Hair JF, Anderson RE, Tatham RL, Black WC. Multivariate data analysis with reading. 5th Ed. (1998) Englewood Cliffs, NJ, USA: Prentice Hall.

Ho DCK, Duffy VG, Shih HM. An empirical analysis of effective TQM implementation in the Hong Kong electronics manufacturing industry. (1999) Human Factors & Ergonomics in Manufacturing. 9(1) pp. 1–25.

Lin, S.-H., Wang, Z.-M., Tang, W.-J., Liang, L.-H., Wang, M.-Z., Lan, Y.-J. Development of safety climate measurement at workplace: Validity and reliability assessment (2007) Journal of Sichuan University (Medical Science Edition), 38 (4), pp. 720–724.

Mitropoulos PT, Cupido G. The role of production and team work practices in construction safety: a cognitive model and an empirical case study.(2009) J Safety Res., 40(4):265 275.

Mohammadfam, I., Ghasemi, F., Kalatpour, O., Moghimbeigi, A. Constructing a Bayesian network model for improving safety behavior of employees at workplaces (2017) Applied Ergonomics, 58, pp. 35–47.

Molina ML, Lloréns-Montes J, Ruiz-Moreno A. Relationship between quality management practices and knowledge transfer. Journal of Operations Management. 2007;25(3):682 701.

Namian, M., Albert, A., Zuluaga, C.M., Behm, M. Role of safety training: Impact on hazard recognition and safety risk perception (2016) Journal of Construction Engineering and Management, 142 (12)

Pingani, L., Evans-acko, S., Luciano, M., Del Vecchio, V., Ferrari, S., Sampogna, G., Croci, I., Del Fatto, T., Rigatelli, M., Fiorillo, A. AF, Psychometric validation of the Italian version of the Reported and Intended Behaviour Scale (RIBS), (2016), Epidemiology and Psychiatric Sciences vol. 25(5), pp. 485–492.

Rubio-Romero, J.C., Carrillo-Castrillo, J.A., Gibb, A. Prevention of falls to a lower level: evaluation of an occupational health and safety intervention via subsidies for the replacement of scaffolding (2015) International Journal of Injury Control and Safety Promotion, 22 (1), pp. 16–23.

Saga, R., Fujita, T., Kitami, K., Matsumoto, K. Improvement of factor model with text information based on factor model construction process (2013) Frontiers in Artificial Intelligence and Applications, 254, pp. 222–230.

Törner M, Pousette A. Safety in construction a comprehensive description of the characteristic of high safety standards in construction work, from the combined perspective of supervisors and experienced workers. (2009) J Safety Res., 40(6):399–409.

The response of phagocytes to indoor air toxicity

L. Vilén, J. Atosuo, E. Suominen & E.-M. Lilius
Department of Biochemistry, University of Turku, Turku, Finland

ABSTRACT: This Perspective presents a viewpoint on potential methods assessing toxicity of indoor air. Until recently, the major techniques to document mouldy environment have been microbial isolation using conventional culture techniques for fungi and bacteria as well as in some instances polymerase chain reaction to detect microbial genetic components. It has become increasingly evident that bacterial and fungal toxins, their metabolic products and volatile organic substances emitted from corrupted constructions are the major health risks. Here, we illustrate how phagocytes, especially neutrophils can be used either in vitro as probe cells, directly exposed to the toxic agent studied, or they can act as in vivo indicators of the whole biological system exposed to the agent. There are two convenient methods assessing the responses: to measure chemiluminescence emission from activated phagocytes and to measure quantitatively by flow cytometry the expression of complement and immunoglobulin receptors on the phagocyte surface.

1 INTRODUCTION

Indoor air problem is a tremendous health hazard, especially when a living building or working place is infested with toxic moulds (Campbell et al. 2004). Although microbial communities on surfaces do not directly correlate with the health trouble of the occupants, in many countries conventional techniques to document moldy environment due to dampness has been traditionally based on culture techniques using conventional isolation media and quantitation of colony forming units per e.g. cubic meter (Verdier et al. 2014). However, well-designed and well-conducted international so-called HITEA (Health Effects of Indoor Pollutants: Integrating microbial, toxicological and epidemiological approaches) studies proved that quantities of moulds *per se* were not as good health correlates as the level of microbial markers (Huttunen et al. 2016, Jacobs et al. 2014, Borràs-Santos et al. 2013). Indeed, the toxicity of moulds and other toxic compounds emitted from damaged surfaces had a detrimental health effect as unambiguously shown in teaches in mould-infected schools (Salin et al. 2017). Therefore, it is imperative to adopt new thinking to search and exploit novel methods that will be robust enough, inexpensive and reliable to be implemented into routine. Indoor air toxicity detection instead of colony detection is the future of environmental research.

Mycotoxins are usually analyzed with mass-spectrometry-based techniques that are coupled with pre-separation by a gas or liquid chromatography. These methods allow for the highly sensitive and accurate determination of tested samples. They are, however, expensive and time-consuming, taking from days up to weeks to obtain the results, and still providing only the detection of a single compound or a group of structurally related compounds at any given time. (Eltzov et al. 2015). Moreover, the conventional methods do not take into account the possible synergistic effects of the mycotoxins and other microbial structure components as well as toxic compounds emitted from damaged surfaces which may enhance the toxicity. New technologies are becoming available that may enable the better assessing of the total toxicity. We have developed a test system which assesses rapidly and cost-efficiently the total indoor air toxicity using the *E. coli*-lux method (Suominen et al. 2016a). While correlating well with the building-related symptoms of users this assay can be criticized for using prokaryotic cells as probes. Below, we present ideas and directions of research based on mammalian neutrophils that could be exploited as probe cells for toxicity studies.

2 PRINCIPLES OF PHAGOCYTE ACTIVATION

The process of phagocytosis is recently reviewed by Gordon (2016). The generation and measurement of phagocyte chemiluminiscence (CL) are thoroughly described by Lilius & Marnila (1992). Briefly, the activation of phagocytes produce electronically excited states, which on relaxation to the ground state emit photons, referred to as phagocyte CL. The commonly used activators include various opsonized or unopsonized particles and immune complexes. In the opsonization process particles attach to complement compounds and immunoglobulins. Opsonized antigens are recognized partly by the complement receptor CR3, partly by the complement receptor CR1, and partly by FcγRII and FcγRIII receptors, which bind to the Fc portion of the IgG molecules attached to antigens.

Phagocytic cells are generally isolated from blood treated with anticoagulants. Leukocytes obtained after erythrocyte sedimentation are normally sufficient without further separation. Phagocytic cell activities can also be measured in the *ex vivo* state simply by diluting the whole blood enough to get rid of the inhibitory amounts of plasma and red cells. The *ex vivo* cells are, however, not necessarily in the same functional state as the cells after isolation steps where activation processes may take place. Luminol amplifies the CL emission by the factor of 10^3–10^4 and it has been shown to be oxidized in the myeloperoxidase reaction. When using luminol in the millimolar range, one needs less than a thousand phagocytic cells (as in the case in whole blood tests) to get reliable signals. The number of isolated cells used in routine tests varies, generally around 10^5. Hank's balanced salt solution (HBSS) is the most frequently used buffer. Luminometers with strict temperature controls, multiple sample capabilities (up to 96 in microtiter plate readers) and computerized data processing are the instruments of choice. (Lilius & Marnila 1992).

3 MEASUREMENT OF *IN VITRO* TOXICITY

Routinely, *in vitro* toxicity testing was made as luminol-amplified CL assay by adding 25µl of opsonised zymosan suspension (20 mg/ml) in HBSS buffer supplemented with gelatin (1 mg/ml) (gHBSS), 20µl of luminol (10mM in borate buffer, pH 9.0), and 100µl of leukocyte suspension (varying number of neutrophils depending on the isolation method) to the wells of a white 96-well microtiter plate. Finally, toxic samples in various concentrations were added. The final reaction volume was 200 µl.

The CL signals of the microtiter plate wells were continuously recorded 0.5 sec/well for 200 min in Hidex Sense multimode reader (Hidex Ltd, Turku, Finland) at 37°C. Three parallel wells were prepared from every reaction mixture. The background signal was measured from a well containing only the buffer and this reading was subtracted from the readings of the experimental wells. Figure 1 illustrates the principles of this technique. A strong luminol-amplified CL signal peaking at about 30 min was detected when opsonized zymosan was added to the reaction mixture containing 5×10^4 neutrophils. When, in addition to opsonized zymosan, also toxic samples were added the CL signal was dose-dependently reduced. EC_{50} value was determined from the dose curve where the CL signal was reduced 50%. T-2 toxin from

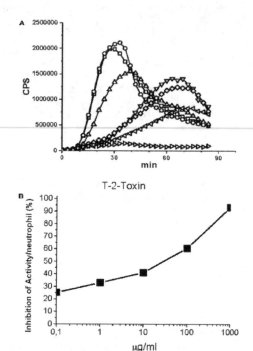

Figure 1. Effect of T-2 toxin on neutrophil chemiluminescence was induced by opsonized zymosan. The reaction mixture contained 25 µl of opsonised zymosan suspension (20 mg/ml) in HBSS buffer supplemented with gelatin (1 mg/ml), 20 µl of luminol (10 mM in borate buffer, pH 9.0), and 100 µl of neutrophil suspension (50000 cells). Finally, toxic samples in various concentrations were added. The final reaction volume was 200 µl. A. The kinetic curves of the CL emission of neutrophils. B. Inhibition of the peak CL emission of neutrophils by the T-2 toxin T-2 toxin: 0 µg/ml (□), 0.01 µg/ml (○), 0.1 µg/ml (△), 1 µg/ml (▽), 10 µg/ml (◇), 100 µg/ml (◁), 1000 µg/ml (▷).

Fusarium species appeared to have an EC_{50} value of 30 µg/ml calculated from peak CL values (Fig. 1b). EC_{50} is the concentration of a toxin that kills 50% of the bacterial cells. CPS is counts per second registered by the luminometer. The other tested mycotoxins deoxynivalenol, moniliformin, antimycin A, and chloramphenicol had EC_{50} values ranging from 20 µg/ml to a few hundreds of µg/ml. The same mycotoxins gave similar or roughly 10-fold lower EC_{50} values in our *Escherichia coli*-lux (*E. coli*-lux) toxicity test (Suominen et al. 2016a).

It should be noted that cytochalasin D was very weakly toxic in *E. coli*-lux toxicity test while chloramphenicol, being very toxic in *E. coli*-lux toxicity test, was weakly toxic in the neutrophil toxicity test. A dust sample (Suominen et al. 2016b) from a moisture-damaged object had EC50 values of 72 µg/ml and 15 µg/ml in neutrophil toxicity test and *E. coli*-lux toxicity test, respectively.

By careful choosing the reaction conditions one is able to differentiate between the binding and ingestion phases of phagocytosis in CL measurements. In these experiments the ingestion of zymosan by neutrophils was measured flow-cytometrically as described by Nuutila & Lilius (2005). Cytochalasin D inhibited the ingestion of opsonized (0.5% serum) and nonopsonized zymosan dose-dependently. The CL response induced by 25 µg zymosan consisted of two peaks with peak times of about 10 min and 90 min, and 10 min and 65 min in the absence and presence of serum (0.5%), respectively. The second peak in the CL response is clearly the ingestion peak since cytochalasin D gradually inhibited it. The first peak is then the adhesion peak, which increases with increasing the concentration of cytochalasin D. The ingestion is dependent on the cytoskeleton and thus on ATP produced by mitochondria. We anticipate that well-known fungal and bacterial toxins affecting mitochondria such as antimycin A (Han et al. 2008) reduce the ingestion phase but not the binding phase of phagocytes being expressed as a decrease of the second peak of the CL response.

4 QUANTITATION OF FUNGAL SPORE ANTIBODIES

Fungal and bacterial spores are one of the main predisposal elements in contaminated indoor air (Kildeso et al. 2003, Haverinen-Shaughnessy et al. 2008, Schmechel et al. 2006, Andersen et al. 2003). The virulence of many pathogens relates in part to their ability to evade phagocytosis by the virtue of certain surface antigens. The function of serum antibodies is to react with these antigens and make the microorganisms more susceptible to ingestion by neutrophils. When serum is incubated with spores, and if specific antibodies are present, the spores are rapidly and efficiently opsonized and phagocytosed by neutrophils. Thus, when this reaction is measured as a CL emission the exposure of an individual to spores is manifested as an elevated CL peak compared with the non-exposed controls. This CL emission correlates with the level of antibodies. The system is functional with both isolated cells and with the whole blood dilutions in both heterologous and autologous systems (Lilius & Nuutila 2006).

The spore assay was operated as described in the section "The measurement of phagocyte CL" but instead of opsonized zymosan, test serum (1%) and intact spores (~ 10^{10}) were added. *Streptomyces albus* spores were cultured and harvested from the mannitol salt agar and *Aspergillus versicolor* spores from malt agar by scraping and by diluting them into the distilled water. This suspension was then sonicated. Spores were isolated by filtration trough the cotton wool and re-suspended in 20% glycerol (Kieser et al. 2000). When *S. albus* or/and *A. versicolor* spores and test serums were introduced to neutrophils the CL emission was significantly higher in individuals exposed to these microbes in damaged buildings (data not shown).

The autologous system utilizing neutrophils from the fingertip whole blood enables the development of in situ quick test systems to be operated with a portable luminometer. By comparison, spore specific IgG and IgM antibody titers were assayed using Enzyme-Linked-Immunosorbent–Assay (ELISA) with intact spores as antigens (Jaakkola et al. 2002, Schmechel et al. 2006) Spores were fixed to the solid face of the microtiter wells, incubated with the test serum, and then with the anti-human immunoglobulin tagged with enzyme horseradish peroxidase (HRP). The enzyme activity from the solid phase was related to the bound antibody. Serum samples of the exposed individuals from the microbe damaged buildings expressed elevated serum spore specific IgG and IgM antibody titers but IgM was in better correlation with the increased CL response from the same test serums. It is noteworthy that according to literature (Jaakkola et al. 2002) and our study the fungal and bacterial spore antigens may share common structures since there is substantial cross-reactivity in antibody responses regardless of the origin of the spore. If this is true, a single test antigen instead of panels containing multiple microbe antigens can be used.

5 ESTIMATION OF TOXICITY *IN VIVO*

We are aiming to collect human blood samples of the users of moisture-damaged buildings for establishing a biomonitoring assay for the toxic effects on exposed users. This assay is based on the measurement of the capability of phagocytes to emit photons when stimulated with opsonized zymosan. This has been shown to reflect remarkably well the pathophysiological state of the host. In many cases even the magnitude of the stress, the presence of pathogen in the body, or the activity of the disease can be estimated. (Lilius & Marnila 1992) Therefore, we believe that neutrophils can be utilized as a toxicological probe acting as an indicator of the whole biological system exposed to the agent. Here we show that the concept is indeed valid with test animals. How it will work with humans remains to be elucidated later.

Previous studies have shown that moniliformin is acutely toxic to rats with an LD50 cut-off value of 25 mg/kg bw (Jonsson et al. 2014). Here are the results of a subacute oral toxicity study. Rats were daily exposed to moniliformin of low doses from

0 to 15 mg/kg bw for 28 days. Two satellite groups were kept alive for an additional 14 days without treatment to detect possible delayed effects and to follow up recovery. The neutrophil CL measurements revealed the toxic effect of moniliformin on the rat innate immunity. The phagocytic activity of the rat neutrophils was dramatically reduced in all dose groups and did not recover. Even the lowest dose (3 mg/kg bw) caused a substantial decrease in the neutrophil activity reducing the final activity by 70% of the initial activity. Moreover, the decrease in neutrophil activity continued in the satellite groups subsequent to the cessation of moniliformin exposure, with a mean activity of 28 ± 18% in the final samples, compared with the negative control group. Also the neutrophil activity of the final samples of the negative control group was reduced by 29% from the initial samples. The reduction in the control group was most likely derived from stress caused by housing in metabolic cages, handling, blood collection and daily i.g. administration of the vehicle by gavage. In spite of the reduced phagocytic activity of neutrophils, the number of neutrophils and total leucocyte counts in the blood samples of the dosed rats remained normal, suggesting that moniliformin has a functional effect on myeloid cells, rather than a lymphoid one. Additionally, the lymphoid organ weights (thymus, spleen) were unaffected, excluding dystrophic and dysplastic effects.

6 ALGORITHMS OF EXPRESSION OF DIFFERENT SURFACE RECEPTORS ON PHAGOCYTES

Do the infectious and other inflammatory diseases or exposure to indoor air pollutants induce alterations in the expression of the opsonin receptors of phagocytes? Receptor expression measurements are described in Lilius & Nuutila (2012). Briefly, erythrocytes were lysed and leukocytes were separated by centrifugation. Leukocytes (3×10^5) were incubated in 50 µL of gHBSS with receptor specific monoclonal antibodies (0.4 µg) for 30 min at +4°C. A relative measure of receptor expression was obtained by determining the Mean Fluorescence Intensity (MFI) of 5000 leukocytes. In the case of neutrophil FcγRI, the percentage of fluorescence positive cells (%) was also determined. Measurement of leukocyte receptor expression was performed using fluorescently (FITC or PE) labelled receptor-specific monoclonal antibodies.

We have performed a few studies where we have measured the receptor expression in various patient groups (Leino et al. 1997, Isolauri et al. 1997, Hohenthal et al. 2006, Salminen et al. 2001). The summary of the results of these studies is

Table 1. Receptor expression changes in various diseases compared to health controls.

Receptor	Bacterial infection	Viral infection	Kidney cancer	Atopic dermatitis
Neutrophils				
CR1/CD35	+++	(−)	n.c.	+
CR3/CD11b	+++	+	++	(+)
FcγRI/CD64	+++	+++	(+)	n.c.
FcγRII/CD32	+	(−)	n.c.	(+)
FcγRIII/CD16	(−)	n.c.	n.c.	(−)
Monocytes				
CR1/CD35	+++	++	+	(+)
CR3/CD11b	+++	++	+++	(+)
FcγRI/CD64	+++	+++	+	(−)
FcγRII/CD32	(+)	n.c.	++	(+)

The +/− without parentheses indicates a significant increase/decrease in the expression of receptor in question compared to healthy control. The +/− in parentheses indicates an insignificant increase/decrease in the expression of receptor in question compared to healthy control. + or − = 0–50% increase or decrease compared to healthy control, ++ = 50–100% increase compared to healthy control and +++ = more than 100% increase compared to healthy control.
*n.c. = no change.

presented in Table 1. In monocytes, all the receptors were upregulated in bacterial and viral infections. In neutrophils, CR1, CR3, FcγRI, and FcγRII were upregulated, while FcγRIII was downregulated in bacterial infections. CR1 and FcγRII were downregulated, while CR3 and FcγRI were upregulated in viral infections. These results led us to conclude that the receptor expression could be used as a basis for the differential diagnosis of bacterial and viral infections. (Lilius & Nuutila 2012). Whether the exposure to molds causes a specific pattern in opsonin receptor expression will be studied in a near future if resources are available.

7 CONCLUSIONS

This Perspective highlights potential of neutrophils as a toxicological tool for studies of indoor air toxicity. Neutrophils are mammalian cells, therefore any toxicity imposed on them can be easily to extrapolated on the whole human body. They can be used either as probe cells, directly exposed to samples collected from damaged buildings or they can act as indicators of toxic effects on humans exposed to the toxic compounds of damaged buildings. The future shows whether the receptor expression measurements from leukocytes bring additional evidence for the indoor air toxicity assessments. Part of this study is presented in the Indoor Air 2016 Conference in Ghent, Belgium, July 3–8, 2016. (Lilius et al. 2016).

ACKNOWLEDGEMENT

The Finnish Work Environment Fund funded the project.

REFERENCES

Andersen, B., Nielsen, K.F., Thrane, U., Szaro, T., Taylor, J.W. & Jarvis, B.B. 2003. Molecular and phenotypic descriptions of Stachybotrys chlorohalonata sp. nov. and two chemotypes of *Stachybotrys chartarum* found in water-damaged buildings. *Mycologia* 6: 1227–1238.

Borràs-Santos, A., Jacobs, J.H., Täubel, M., Haverinen-Shaughnessy, U., Krop, E.J., Huttunen, K., Hirvonen, M.R., Pekkanen, J., Heederik, D.J., Zock, J.P. & Hyvärinen, A. 2013. Dampness and mould in schools and respiratory symptoms in children: the HITEA study. *Occupational and Environmental Medicine* 70: 681–687.

Campbell, A.W., Thrasher, J.D., Gray, M.R. & Vojdani, A. 2004. Mold and mycotoxins: effects on the neurological and immune systems in humans. *Advances in Applied Microbiology* 55: 375–406.

Eltzov, E., Cohen, A. & Marks, R.S. 2015. Bioluminescent liquid light guide pad biosensor for indoor air toxicity monitoring. *Analytical Chemistry* 87: 3655–3661.

Gordon, S. 2016. Phagocytosis: An Immunobiologic Process. *Immunity* 3: 463–475.

Han, Y.H., Kim, S.H., Kim, S.Z. & Park W.H. 2008. Antimycin A as a mitochondrial electron transport inhibitor prevents the growth of human lung cancer A549 cells. *Oncology Reports* 20(3): 689–693.

Haverinen-Shaughnessy, U., Hyvarinen, A., Putus, T. & Nevalainen, A. 2008. Monitoring success of remediation: seven case studies of moisture and mold damaged buildings. *The Science of the Total Environment* 399(1–3):19–27.

Huttunen, K., Tirkkonen, J., Täubel, M., Krop, E., Mikkonen, S., Pekkanen, J., Heederik, D., Zock, J.P., Hyvärinen, A. & Hirvonen, M.R. 2016. Inflammatory potential in relation to the microbial content of settled dust samples collected from moisture-damaged and reference schools: results of HITEA study. *Indoor Air* 26: 380–390.

Hohenthal, U., Nuutila, J., Lilius, E.M., Laitinen, I., Nikoskelainen, J. & Kotilainen, P. 2006. Measurement of complement receptor 1 on neutrophils in bacterial and viral pneumonia. *BMC Infectious Diseases* 6: 11.

Isolauri, E., Pelto, L., Nuutila, J., Majamaa, H., Lilius, E.M. & Salminen, S. 1997. Altered expression of IgG and complement receptors indicates a significant role of phagocytes in atopic dermatitis. *Journal of Allergy and Clinical Immunology* 99: 707–713.

Jacobs, J., Borràs-Santos, A., Krop, E., Täubel, M., Leppänen, H., Haverinen-Shaughnessy, U., Pekkanen, J., Hyvärinen, A., Dockes, G., Zock, J.P. & Heederik, D. 2014. Dampness, bacterial and fungal components in dust in primary schools and respiratory health in schoolchildren across Europe. *Occupational and Environmental Medicine* 71: 704–712.

Jaakkola, M.S., Laitinen, S., Piipari, R., Uitti, J., Nordman, H., Haapala, A.M. & Jaakkola, J.J. 2002. Immunoglobulin G antibodies against indoor dampness-related microbes and adult-onset asthma: a population-based incident case-control study. *Clinical and Experimental Immunology* 129: 107–112.

Jonsson, M., Atosuo, J., Jestoi, M., Nathanail, A.V., Kokkonen, U.M., Anttila, M., Koivisto, P., Lilius, E.M. & Peltonen, K. 2014. Repeated dose 28-day oral toxicity study of moniliformin in rats. *Toxicology letters* 233: 38–44.

Kieser, T., Bibb, M.J., Buttner, M.J., Chater, K. & Hopwood, D.A. 2000. *Practical Streptomyces Genetics.* Norwich: John Innes Foundation.

Kildeso, J., Wurtz, H., Nielsen, K.F., Kruse, P., Wilkins, K., Thrane, U., Gravesen, S., Nielsen, P.A. & Schneider, T. 2003. Determination of fungal spore release from wet building materials. *Indoor air* 13: 148–155.

Leino, L., Sorvajärvi, K., Katajisto, J., Laine, M., Lilius, E.M., Pelliniemi, T.T., Rajamäki, A., Silvoniemi, P. & Nikoskelainen, J. 1997. Febrile infection changes the expression of IgG Fc receptors and complement receptors in human neutrophils in vivo. *Clinical and experimental immunology* 107: 37–43.

Lilius, E.M., Suominen, E. & Atosuo, J. 2016. Indoor air toxicity assessments using neutrophils. *Indoor Air proceedings*, paper ID 831.

Lilius, E.M. & Marnila, P. 1992. Photon emission of phagocytes in relation to stress and disease. *Experientia* 48: 1082–1091.

Lilius, E.M. & Nuutila, J. 2012. Bacterial infections, DNA virus infections, and RNA virus infections manifest differently in neutrophil receptor expression. *The Scientific World Journal* 2012:527347.

Lilius, E.M. & Nuutila J. 2006. Particle-induced myeloperoxidase release in serially diluted whole blood quantifies the number and the phagocytic activity of blood neutrophils and opsonization capacity of plasma. *Luminescence* 21: 148–158.

Salin, J.T., Salkinoja-Salonen, M., Salin, P.J., Nelo, K., Holma, T., Ohtonen, P. & Syrjälä, H. 2017. Building-related symptoms are linked to the in vitro toxicity of indoor dust and airborne microbial propagules in schools: A cross-sectional study. *Environmental Research* 154: 234–239.

Salminen, E., Kankuri, M., Nuutila, J., Lilius, E.M. & Pelliniemi, T.T. 2001. Modulation of IgG and complement receptor expression of phagocytes in kidney cancer patients during treatment with interferon-alpha. *Anticancer Research* 21: 2049–2055.

Schmechel, D., Simpson, J.P., Beezhold, D. & Lewis, D.M. 2006. The development of species-specific immunodiagnostics for *Stachybotrys chartarum*: the role of cross-reactivity. *Journal of Immunological Methods* 309: 150–159.

Suominen, E., Atosuo, J. & Lilius, E.M. 2016a. Indoor air toxicity assessment using bioluminescent *E.coli*-lux. *Indoor Air proceedings,* paper ID 620.

Suominen, E., Atosuo, J. & Lilius E.M. 2016b. Indoor air toxicity assessment from dust samples. *Indoor Air proceedings*, paper ID 625.

Verdier, T., Coutand, M., Bertron, A. & Roques, C. 2014. A review of indoor microbial growth across building materials and sampling and analysis methods. *Building and Environment* 80: 136–149.

Modelling the effect of human presence in a single room with computational fluid dynamics simulation—a short review

P.H. Moreira, J.C. Guedes & J. Santos Baptista
PROA/LABIOMEP/INEGI, Faculty of Engineering, University of Porto, Porto, Portugal

ABSTRACT: The purpose of this paper is to contribute with a short review to the development of a numerical model that reflects the needs of ventilating a space due to the number of its workers and their working period. The 3D model aim is to consider a better wellness satisfaction and productivity for workers, guaranteeing thermal comfort, optimizing the energetic efficiency of a room for better indoor air quality by addressing the minimum ventilation rates proposed. Through detailed analysis and different researching criteria, 24 experimental articles were included. All articles have a validated model that is significant to study the interaction between human presence in a single room with CFD methodology, which can be hopeful for humans to create a more comfortable, health living environment and to help parametrizing a 3D CFD model in the future.

1 INTRODUCTION

By 2020, concerns of indoor environmental quality, energy efficiency, ventilation, building materials and their composition will be one of the main targets of action. High energy costs are accelerating pressures to decrease ventilation rates in buildings. Improving the quality of construction and the resident's quality of life have emerged as one of the most important reasons for 2020 vision main goals.

Moreover, a poor Indoor Air Quality (IAQ) is currently affecting human's health by placing the population at risk for serious health problems such as asthma and respiratory disease, cancer, cardiovascular disease and others. Therefore, it becomes crucial to study the needs of ventilation of an indoor space of work. The ventilation of a space is understood as an exchange of internal air by external air, with the main functions of keeping the environment free of impurities and undesirable odors, besides supplying O_2 and reducing CO_2 concentration, which is not harmful to health at low conditions but can cause headaches, dizziness, confusion, loss of consciousness and even asphyxiation in workplaces with of high concentrations of CO_2, which is a gas naturally present in the breathable air as it is produced by humans.

Computational Fluid Dynamics (CFD) has been adopted as a powerful and useful tool for predicting air movement in ventilated spaces, including heat sources and temperature variations. Moreover, it is used routinely in civil engineering when a large demanding ventilation system must be projected.

The construction of buildings has been subjected to a perpetual development over time. Nowadays, the building sector requires more than 40% from the total energy consumption in Europe (Eurostat 2016), as represented by Figure 1. An automatic question is: How can we reduce this percentage?

Conventional Indoor Air Quality (IAQ) ventilation systems are responsible for a significant loss of energy. The IAQ is an existing parameter critical to workers in office buildings and it is affected by different sources of pollution that are related not only to the occupants and their activities, but also construction materials, equipment, furniture, heating, ventilation and air conditioning. Some buildings with low IAQ have the profile case of SBS (Sick Building Syndrome) with the increase of the risk of diseases at the work environment. Thermal discomfort has also been known to lead

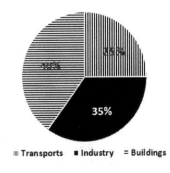

Figure 1. Final energy consumption by sector in Europe.

to sick building syndrome symptoms (Myhren & Holmberg 2008).

The combination of high temperature and high relative humidity serves to reduce thermal comfort and indoor air quality (Fang et al. 2004). On the other hand, a good IAQ helps to provide workers better conditions of hygiene and health which will improve their daily comfort, satisfaction and productivity.

The problem that needs to be solved is the air replacement needs (air quality guarantee) and air conditioning due to the occupation and residence time of workers allowing a rationalization of energy resources. It is crucial to measure the need to ventilate the space, ensuring comfort parameters and legal requirements for occupational health and safety are met. CFD technique can simulate the needs of ventilation that can help the administrator optimize the efficiency of an existing ventilated infrastructure and predict the energetic effectiveness of some equipment, as well as helping to understand what adjustments would be need to be made by changing different boundary conditions. Fluid flows are governed by partial differential equations which represent conservation laws for the mass, energy and momentum. Therefore, the bases of CFD methodology are the Navier-Stokes equations which uses algorithms to predict how liquids and gases behave and how they work with the products that people design.

2 METHODS

2.1 Search strategy

The first aim was to locate all relevant studies. The method to find qualitative studies relied mostly on electronic searches of databases, as well as «Google Scholar». These databases were selected by their relevance in the subject of engineering. The 7 data bases selected were: *Academic Search Complete, American Society of Civil Engineers (ASCE), Scopus, Springer, Web of Science, Inspec* and *Science Direct.*

Literature search strategies were developed considering different types of keywords that can be divided into 5 groups, as follows: Group A: "CFD Simulation" or "Computational Fluid Dynamics Simulation; group B: "Indoor Air Quality" or "IAQ"; group C: "Thermal comfort" or Climatization; group D: Thermoregulation and group E: Efficiency.

In addition to that, 5 combinations were performed between those groups, fixing the most important keyword that corresponds to group A. The combinations were the following: A & B & C; A & B & D; A & B & E; A & C & D and A & C & E.

The previous combinations consisted on searching only abstract, title and keywords in the field camp. In the databases where that search field was not available, the camp "full text" was considered (ASCE, Springer and Academic Search Complete).

In this systematic review, the final studies were selected per 8 criteria, 4 of exclusion and 4 of inclusion, which were chosen to provide a good quality, decision-making and goal-oriented survey/research.

Articles obtained by this systematic search were exported to *Mendeley*'s reference manager and then organized by types of combinations in different directories.

2.2 Screening criteria

In the initial research, all articles resulting of the 5 combinations mentioned above related with the topic were screened for title, abstract and keywords.

After that, exclusion criteria were applied. These criteria were simple as limiting the systematic research by a period of 10 years, ranging from January 2007 until January 2017. Moreover, all studies were filtered by English. Another exclusion criterion was the limited type of documents that were included in the analysis: articles, reviews, articles in press and journals. Lastly, there was the exclusion criteria of our database documents not having the "full text" available or simply not having access to the entire document.

2.3 Eligibility criteria

Considering the initial research, the studies which possessed mathematical models of ventilation systems, indoor air quality and thermoregulation were included. Another inclusion criteria consisted on the actual validation of those models which symbolizes the high level of trust in the researched models.

Another criterion was the sub-domain of the studies which focused only on the indoor space of simulation (bedrooms, office rooms and climate chambers).

The final inclusion criterion, was made by including the human factor, by a wide range of approximations that considered thermal manikins, virtual manikins, and simple objects as heat sources. This last criterion translates the study human presence influence in the indoor environment, which is one of the main goals of the present research.

To sum up, this was the best strategy of research defined to this systematic review.

3 RESULTS AND DISCUSSION

Different types of indoor spaces have been evaluated such as office rooms, climatic chambers, bedrooms and operating rooms.

About 67% of the selected studies had another type of heat sources included in the experimental

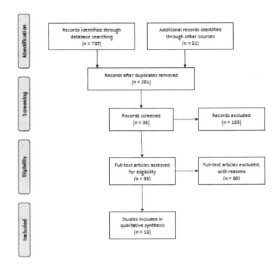

Figure 2. Systematic review stages.

campaign. The remaining 33% had zero heat sources included (Figure 2).

The studies sample consisted in 83% thermal manikins and 17% of real human occupants. Considering now the type of work evaluated, about 67% considered only the seated position, while approximately 8% considered the standing posture and the remaining 25% studied both positions.

According to Portuguese law ("Portaria nº987/93") which is based on the European Framework Directive 89/654/EEC of 30 November 1989, concerning the minimum safety and health requirements for the workplace, the minimum volume per worker is established in 11,50 m³ (10,50 m³ in cases where is secured to have a good ventilation).

Analyzing all the selected work, approximately 18% had a rate of volume per worker until 11.50 m³, 27% had a rate between 11.50 and 23.0 m³ and the remaining 55% with an excessive rate of more than 23 m³ per person.

Also, from the 24 articles obtained, approximately 90% (22 papers) had their experience validated under steady-state conditions which has relevance in the thermodynamics field. The remaining 10% (2 papers) have their experimental study validated under transient and non-uniform conditions.

A presumed limitation to this survey includes the search process itself, due to the exclusion of good documents simply because they didn't respect all the criteria simultaneously, but were exceptionally developed at some topics which were relevant to the present work. Consequently, this may not have allowed the identification of all studies related to this research field, leading the results to more accurate and concrete cases of analysis.

From the studies that investigated ventilation systems, thermal comfort and thermoregulation some important conclusions and citations are mentioned:

"The higher the temperature difference between the surface temperature of the manikin and the air temperature, the faster the airflow in the microclimate" (Voelker et al. 2014). An automatic observation to the previous statement is: It's in summer conditions that there is a need to maximize ventilation, but the temperature differential is higher in winter. So it's a paradox issue.

"Stratum ventilation can be applied in a room with a multiple row of occupants" (Cheng & Lin 2015).

"The Regional-air-conditioning (RACM) produces an airflow circulation cell for satisfying the thermal comfort demands for users, and can potentially be energy-saving" (Huang & Tuan 2010).

"Displacement Ventilation (DV) provides better inhaled air quality than Mixing Ventilation (MV) except in the situation where contaminants are emitted from the floor" (Gao et al. 2007).

"DV is more energy efficient than MV since it is only aimed at conditioning the occupied zone" (Gao et al. 2007).

"Using better insulation and a low-temperature panel radiator is more effective with regard to thermal comfort and energy savings" (Sevilgen & Kilic 2011).

"Intensity of fatigue, headache and difficulty in thinking clearly decreased when subjects worked at slightly lower levels of air temperature and humidity" (Fang et al. 2004).

"The Thermal Displacement Ventilation (TDV) and Under Floor Air Distribution (UFAD) systems had better ventilation performance than the MV system in cooling mode. For heating mode, the TDV and UFAD system created mixing conditions except in the vicinity of the floor" (Lee et al. 2009).

"The desk-mounted fans were able to reduce the convection plumes around the occupant and improved the performance of the single jet PV nozzle by doubling the ventilation effectiveness and improving comfort. They permitted also to achieve a reduced energy saving by up to 13% when compared with conventional mixing ventilation systems" (Makhoul et al. 2013).

"The results have indicated that having both ceiling and floor-based air conditioning systems would allow the flexibility to change supply and exhaust air diffuser locations to optimize thermal and ventilation performance more efficiently in countries that are subject to distinct seasonal changes" (Park & Chang 2014).

"The proposed coaxial personalized ventilation achieved high air quality in the breathing zone demonstrated by a personal exposure effectiveness of 32% at fresh airflow rate of 10L s^{-1} per person. It contributed also to the attainment of temperature differences up to 2°C between the occupant's microenvironment and the rest of the room air leading to consider able energy savings compared to the mixed convection air conditioning" (Makhoul et al. 2015).

"When operating the chair fans, the ventilation effectiveness (23.39%) increased by almost 2.5 times that when chair fans were turned off (9.31%). The fan height and fan flowrate have a dual effect on the thermal comfort and IAQ, where the best IAQ is achieved at different fan configuration than that of thermal comfort. Therefore, optimal height and flow rate are selected to help maintain the best combination of IAQ and thermal comfort. A peak in energy savings is achieved at the best IAQ case (fan height 50 cm (19.69 in.) and a total flow rate 10 L/s [0.35 ft^3/s]) reaching 17% when compared with mixing ventilation" (El-Fil et al. 2016).

"There are several models that can predict human physiological response to heat and cold environment. However, there are only a few models that simulate and predict the temperature distribution all over the body, considering the individual differences and transient conditions" (Guedes et al. 2014).

The main goal was to present a systematic review on the studies that can conduce to a final 3D model capable of predicting the needs of ventilation of a space due to the presence of human beings. This model should maximize its energetic efficiency and satisfaction to workers.

4 CONCLUSIONS

In the present systematic review, 24 papers were included to represent the selected criteria. Only articles that were free for downloading by using the University of Porto federate credentials were included.

The final selected papers included every single criterion mentioned above simultaneously experimentally validated and respected simultaneously the 8 criteria (date, type of document, "full-text" availability, language, mathematical models, experimental part, proper validation and indoor space domain) initially established in consequence of the different 5 types of combinations.

Although several studies examined the significance of thermoregulation models and its correlation with the indoor environment, as well as ventilation systems, which are crucial to investigate the internal comfort conditions by the CFD technique, no systematic review related to this topic was found.

Nevertheless, a wide range of articles that studied different types of ventilation systems was found, which is positive considering its impact on the investigation of internal comfort conditions. Therefore, this systematic review was a novelty that considered the models as essential but also their empirical validation.

To sum up, the results provided a good panoply of papers of high confidence that will help on the development of the 3D CFD model in the future project, by respecting all the criteria simultaneously.

REFERENCES

Cheng, Y.. & Lin, Z.., 2015. Technical feasibility of a stratum-ventilated room for multiple rows of occupants. *Building and Environment*, 94, pp. 580–592.

El-Fil, B., Ghaddar, N. & Ghali, K., 2016. Optimizing performance of ceiling-mounted personalized ventilation system assisted by chair fans: Assessment of thermal comfort and indoor air quality. *Science and Technology for the Built Environment*, 22(4), pp. 412–430.

Eurostat, 2016. *Buildings – Energy Efficiency*. Available at: https://ec.europa.eu/energy/en/topics/energy-efficiency/buildings [Accessed January 13, 2017].

Fang, L. et al., 2004. Impact of indoor air temperature and humidity in an office on perceived air quality, SBS symptoms and performance. *Indoor Air*, 14(7), pp.74–81.

Gao, N.P., Zhang, H. & Niu, J.L., 2007. Investigating indoor air quality and thermal comfort using a numerical thermal manikin. *Indoor and Built Environment*, 16(1), pp. 7–17.

Guedes, J.C., Santos Baptista, J. & de Carvalho, J.M., 2014. 3D human thermoregulation model: Bioheat transfer in tissues and small vessels. In G. Arezes, PM and Baptista, JS and Barroso, MP and Carneiro, P and Cordeiro, P and Costa, N and Melo, RB and Miguel, AS and Perestrelo, ed. *Occupational Safety and Hygiene II*. Guimarães: Taylor & Francis, pp. 669–673.

Huang, K.D. & Tuan, N.A., 2010. Numerical analysis of an air-conditioning energy-saving mechanism. *Building Simulation*, 3(1), pp. 63–73.

Lee, K. et al., 2009. Comparison of Airflow and Contaminant Distributions in Rooms with Traditional Displacement Ventilation and Under-Floor Air Distribution Systems. *ASHRAE Transactions*, 115(2), pp. 306–321.

Makhoul, A., Ghali, K. & Ghaddar, N., 2013. Desk fans for the control of the convection flow around occupants using ceiling mounted personalized ventilation. *Building and Environment*, 59, pp. 336–348.

Makhoul, A., Ghali, K. & Ghaddar, N., 2015. Low-mixing coaxial nozzle for effective personalized ventilation. *Indoor and Built Environment*, 24(2), pp. 225–243.

Myhren, J.A. & Holmberg, S., 2008. Flow patterns and thermal comfort in a room with panel, floor and wall heating. *Energy and Buildings*, 40(4), pp. 524–536.

Park, D.Y. & Chang, S., 2014. Numerical analysis to determine the performance of combined variable ceiling and floor-based air distribution systems in an office room. *Indoor and Built Environment*, 23(7), pp. 971–987.

Sevilgen, G. & Kilic, M., 2011. Numerical analysis of air flow, heat transfer, moisture transport and thermal comfort in a room heated by two-panel radiators. *Energy and Buildings*, 43(1), pp. 137–146.

Voelker, C., Maempel, S. & Kornadt, O., 2014. Measuring the human body's microclimate using a thermal manikin. *Indoor Air*, 24(6), pp. 567–579.

Bacteria bioburden assessment and MRSA colonization of workers and animals from a Portuguese swine production: A case report

E. Ribeiro
Escola Superior de Tecnologia da Saúde de Lisboa, ESTeSL, Instituto Politécnico de Lisboa, Lisboa, Portugal
Environment and Health Research Group, Escola Superior de Tecnologia da Saúde de Lisboa, ESTeSL, Instituto Politécnico de Lisboa, Lisboa, Portugal
Research Center LEAF—Linking Landscape, Environment, Agriculture and Food—Instituto Superior de Agronomia, Universidade de Lisboa, Portugal

A. Pereira, C. Vieira, I. Paulos, M. Marques & T. Swart
Escola Superior de Tecnologia da Saúde de Lisboa, ESTeSL, Instituto Politécnico de Lisboa, Lisboa, Portugal

A. Monteiro
Escola Superior de Tecnologia da Saúde de Lisboa, ESTeSL, Instituto Politécnico de Lisboa, Lisboa, Portugal
Environment and Health Research Group, Escola Superior de Tecnologia da Saúde de Lisboa, ESTeSL, Instituto Politécnico de Lisboa, Lisboa, Portugal

ABSTRACT: Pigs are important reservoir of livestock-associated bacteria, including Methicillin-Resistant *Staphylococcus Aureus* (MRSA), which constitute a professional hazard for workers in direct contact with these animals with increased risk of nasal colonization, potentially associated with subsequent clinical diseases and transference of the infection to others.

Here we performed a bioburden characterization concerning bacterial prevalence and resistance (MRSA) in workers and animals from a Portuguese swine production as a case study.

Air samples were collected through an impaction method. Biological samples were obtained through nasopharyngeal swab procedure. Identification of *S. aureus* was performed trough immunologic tests.

We report an exceedingly high prevalence of total bacteria and *S. aureus* colonization (100%) in workers and animals whereas all of identified strains were MRSA. Additionally, air samples demonstrated high values of total bacterial concentration.

This work raises awareness to the relevance of bioburden monitoring and the requirement to create occupational standards and take effective preventive measures.

1 INTRODUCTION

The World Health Organization (WHO) describes antimicrobial resistance to human pathogens as a global health challenge (World Health Organization (WHO) 2016). Currently, it is acknowledged that the extensive use of antibiotics are the driving force for the worldwide escalation of these microorganisms (Morris & Masterton 2002). For the past decades the amount of large animal-feeding operations (AFOs) including swine, has increased expressively (UEFSA 2001) and a large variety of feed additives and drugs, particularly antibiotics, are approved for use in food-animal agriculture (Bloom 2004). Antibiotics are extensively utilized in the management of animal health, and more recently to growth enhancement and feed efficiency in healthy livestock, which may result in an antibiotic selection pressure responsible for the emergence of resistant strains in these contexts. Although, it is largely assumed that resistant strains such as methicillin-resistant *staphylococcus aureus* (MRSA) originated in humans, the emergence of the first pig-associated strain (ST398) (Armand-Lefevre et al. 2005), which in very few years spread worldwide into diverse livestock species, corroborates this hypothesis.

Currently, animals such as pigs are important reservoir of livestock-associated clones of MRSA (LA-MRSA) which in addition to the animal to animal spread one of the early features of these strains was its ability to transfer from pigs to humans (Barton 2014). Consequently, for workers that spend several daily hours in direct contact with MRSA positive animals, MRSA colonization is a patent and significant professional hazard (Denis et al. 2009; Moodley et al. 2008). MRSA carriers have increased risk for subsequent clinical associated diseases and become a bacterial reservoir with associated high risk to transfer the infection to others, including household members (Hatcher et al. 2016), and to contaminate foods and food surfaces during handling (Jordan et al. 2011). Although MRSA infections are well-known worldwide as a cause of numerous hospitalizations and deaths associated with extremely high mortality rates for invasive infections (Klevens et al. 2007), exposure assessment procedures in occupational environments are not adapted to animal production settings, although swine confinement buildings have been placed among the working environments with the highest bioaerosol (Donham et al. 1989). Thus it is imperative to perform a real scenario characterization concerning bacterial prevalence and resistance in order to avoid health hazardous effects in workers and animals, particularly in swine productions.

In the European context, colonization by LA-MRSA have become, in the past years, an exceedingly debated topic particularly in the context of occupational exposure (Goerge et al. 2015), and in Portugal, a recent study also demonstrates the substantial establishment of ST398-MRSA among healthy pigs in swine farms.

In this work we aimed to perform a bioburden characterization of occupational exposure to bacteria through environmental sampling in swine facilities, microbiota prevalence and antimicrobial resistance, namely MRSA prevalence, in animals and workers of a Portuguese swine production.

2 MATERIALS AND METHODS

Study population: The study included all the pig-farm workers (n = 3), which provided a signed written informed consent before enrolment in the study. The studied animals were in the stalls and had with 3 weeks old (n = 15).

Collection, isolation and microbiological procedures

Biological samples were obtained through nasopharyngeal swab procedure using transport swabs with Stuart media, and immediately transported to the laboratory. In the microbiology laboratory samples were cultured in Columbia agar with 5% sheep blood, for MRSA identification, and incubated for 24 hours at 37°C. Tryptic soy agar (TSA) supplemented with nystatin (0.2%) for mesophilic bacterial population and violet red bile agar (VRBA) for bacteria belonging to the Enterobacteriaceae family (e.g. coliforms—Gram-negative bacteria). TSA and VRBA plates were incubated at 30°C and 35°C for 7 days, respectively. After the incubation period, quantitative colony-forming were obtained, colonies were evaluated based on cultural characteristics and *S. aureus* suspicious colonies isolated and incubated for 24 hours at 37°C. Identification of *S. aureus* was performed trough catalase test and Slidex Staph Kit (Biomerieux ref #73115). MRSA strains were identified through Slidex MRSA detection Test Kit (Biomerieux ref #73117). In this work positive (S. aureus MRSA laboratory collection) and negative (S. aureus ATCC 25923) control strains will be included as positive and negative controls.

Air samples of 50 L were collected through an impaction method with a flow rate of 140 L/min (Millipore air Tester, Millipore, Billerica, MA, USA) onto each media plate. TSA supplemented with nystatin (0.2%) and VRBA were used in order to enhance the selectivity for bacterial populations growth and incubated at 30°C and 35°C for 7 days, respectively. After laboratory processing and incubation of the samples, quantitative colony-forming were obtained (colony-forming units—$CFU.m^{-3}$).

The five sampling sites in gestation, maternity, stalls, fattening and quarantine from the studies swine production were selected based on the higher daily exposure of workers to animals. Outdoor samples were also performed to be used as reference.

3 RESULTS

In all biological samples collected from workers normal commensal flora, namely *staphylococcus spp.* and *streptococcus spp.* was observed. Moreover, sample analysis also demonstrated a frequency of 3 in 3 (100% prevalence) of total bacterial load. One worker showed countless mesophilic bacteria, and no Gram-negative bacteria isolates were detected.

In animals, we also identified 15 in 15 frequencies (100% prevalence) of total bacterial load and two swine samples showed countless mesophilic bacteria. Gram-negative bacteria analysis resulted in a frequency of 11 in 15 (73,3% prevalence) individuals, one animal showed countless coliforms and in four animals no isolates were observed (Table 1).

The frequency analysis of *S. aureus* in workers of the studied swine production demonstrated a

Table 1. Frequency analysis of total bacteria and gram-negative bacteria carriers amongst livestock occupational exposed individuals and animals.

Individuals	Swabs	Total bacteria frequency analysis	Gram-negative bacteria frequency analysis
Workers	3	3 in 3	0 in 3
Piglets	15	15 in 15	11 in 15

Table 2. Frequency analysis of *S.aureus* and MRSA carriers amongst livestock occupational exposed individuals and animals.

Individuals	Swabs	S.aureus frequency analysis	MRSA frequency analysis
Workers	3	3 in 3	3 in 3
Piglets	15	15 in 15	15 in 15

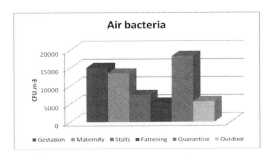

Figure 1. Bacterial load obtained on air samples.

3 in 3 (100% prevalence) colonization occurrence. Regarding the resistance profile, all *S. aureus* strains were resistant to methicillin (MRSA). Additionally, all animals analyzed were also colonized by MRSA. These results are summarized in Table 2.

Total air bacterial load ranged from 5360 CFU.m^{-3} to 18260 CFU.m^{-3} (median 11944 CFU.m^{-3} (Figure 1). Airborne coliforms load only grow up in one sampling site (gestation with 20 CFU.m^{-3}).

4 DISCUSSION

Bacteria bioburden is a key component of bioaerosols which may occur as solid or liquid particles in the air. Prolonged exposure to bioaerosols particularly at workplaces can represent a health hazard (Walser et al. 2015) for workers and for the spread of these microorganisms in the community. For the past years, numerous studies have increased scientific data on occupational exposure to bioaerosols and helped to understand the relationship between exposure and health effects (Ghosh Lal and Srivastava, 2015).

In the context of animal production, it is acknowledged that animals, such as pigs, are important reservoir of livestock-associated bacteria with associated resistances to antibiotics, including LA-MRSA (Armand-Lefevre et al. 2005). Considering that *S. aureus* can colonize the human nostrils via simple inhalation of contaminated air (Masclaux et al. 2013), the direct contact with live animal carriers is possibly the main route of human exposure to bacterial strains such as LA-MRSA. Moreover, LA-MRSA transmission between pig farms mainly occurs through animal trade and to a minor level via humans or livestock transportation (Leibler et al. 2016; Grøntvedt et al. 2016). Nasal LA-MRSA colonization rates in pig farmers is maintained at 59% after time periods with no occupational exposure (holidays) (Köck et al. 2012) indicating that persistent MRSA colonization is expected to be more probable in occupationally exposed individuals (Goerge et al. 2015).

Although there are no national guidelines to impose limit values for airborne bacteria load, some studies (Goyer 2001) were carried out to propose guidelines for eight hours of work indicating 10000 CFU.m^{-3} for total bacteria and 1000 CFU.m^{-3} for Gram-negative bacteria in agricultural and industrial environments.

Here we demonstrated that the majority of the collected air samples presented higher values of total airborne bacterial. The high concentrations of total bacteria instead of airborne coliform could be explained by the fact that 90% of the bacteria isolated from the feces of adult swine are reported as gram positive (Salanitro et al. 1977). The higher concentration of airborne bacteria indoor when compared to the outdoor, suggests that the outdoor air quality can be influenced by the sources of contamination of this activity.

Animal-to-human transmission during farming has been already demonstrated for enterobacteria and enterococci (Armand-Lefevre et al. 2005) and our re-sults suggest that type of transmission can happen, particularly since virtually no protection were used by the workers.

Regarding bacteria identification, we reported concerning high colonization levels of MRSA, both in workers and in animals, as all analyzed individuals were carriers. These levels are exceedingly higher than levels detected in the community for *S.aureus* (31%) and for MRSA (2% - 3%) (Hatcher et al. 2016) and in other Portuguese swine productions (Conceição et al. 2017). Considering that, human MRSA carriers prevalence is higher in intensive antibiotic-using piggeries when compared to antibiotic-free which indicate that antibiotic use is a driver for worker colonization (Rinsky et al. 2013), it would be important to asses antibiotic use in the studied swine production. Moreover, it is also important to notice that occupational exposure to LA-MRSA not only constitutes an important professional hazard but also constitute a relevant risk to individuals that came direct in contact with exposed workers, particularly children that have high colonization prevalence when the worker is a carrier (Hatcher et al. 2016), which represents a public health concern. Additionally, although most of LA-MRSA have been isolated form healthy animals some strains were also isolated from pathological lesions in pigs (Pomba et al. 2009), thus our results may also indicate some concern for animal health.

5 CONCLUSIONS

This work raise the awareness of the urgent need to monitor MRSA strains associated with animal carriers, occupational exposed individuals and potential sources of environmental contamination. Valuable and effective efforts must be made to create occupational health surveillance programs and to determine and regulate the antibiotic selection pressure that is driving the emergence of these strains.

ACKNOWLDEGEMENTS

This work was funded by Concurso Anual para Projetos de Investigação, Desenvolvimento, Inovação e Criação Artística (IDI&CA) do IPL – 2016, BBIOR-Health "Bacterial Bioburden assessment in the context of occupational exposure and animal health of swine productions" Edna Ribeiro— Escola Superior de Tecnologia da Saúde de Lisboa.

REFERENCES

Armand-Lefevre, L., Ruimy, R. & Andremont, A., 2005. Clonal comparison of Staphylococcus from healthy pig farmers, human controls, and pigs. *Emerging Infectious Diseases*, 11(5), pp. 711–714.

Barton, M.D., 2014. Impact of antibiotic use in the swine industry. *Current Opinion in Microbiology*, 19(1), pp. 9–15. Available at: http://dx.doi.org/10.1016/j.mib.2014.05.017.

Bloom, R.A., 2004. Use of veterinary pharmaceuticals in the United States. In *Pharmaceuticals in the Environment*. Springer, pp. 149–154.

Conceição, T., De Lencastre, H. & Aires-De-Sousa, M., 2017. Frequent isolation of methicillin resistant Staphylococcus aureus (MRSA) ST398 among healthy pigs in Portugal. *PLoS ONE*, 12(4), pp. 1–7.

Denis, O. et al., 2009. Methicillin-resistant Staphylococcus aureus ST398 in swine farm personnel, Belgium. *Emerging Infectious Diseases*, 15(7), pp. 1098–1101.

Donham, K. et al., 1989. Environmental and health studies of farm workers in Swedish swine confinement buildings. *British journal of industrial medicine*, 46(1), pp. 31–37.

Ghosh, B., Lal, H. & Srivastava, A., 2015. Review of bioaerosols in indoor environment with special reference to sampling, analysis and control mechanisms. *Environment International*, 85, pp. 254–272. Available at: http://dx.doi.org/10.1016/j.envint.2015.09.018.

Goerge, T. et al., 2015. MRSA colonization and infection among persons with occupational livestock exposure in Europe: Prevalence, preventive options and evidence. *Veterinary Microbiology*. Available at: http://dx.doi.org/10.1016/j.vetmic.2015.10.027.

Goyer, N., 2001. *Les bioaérosols en milieu de travail*,

Grøntvedt, C.A. et al., 2016. Methicillin-Resistant Staphylococcus aureus CC398 in Humans and Pigs in Norway: A "One Health" Perspective on Introduction and Transmission. *Clinical infectious diseases : an official publication of the Infectious Diseases Society of America*, 63, p.ciw552. Available at: http://www.ncbi.nlm.nih.gov/pubmed/27516381.

Hatcher, S.M. et al., 2016. The Prevalence of Antibiotic-Resistant Staphylococcus aureus Nasal Carriage among Industrial Hog Operation Workers, Community Residents, and Children Living in Their Households: North Carolina, USA. *Environmental health perspectives*, (October). Available at: http://www.ncbi.nlm.nih.gov/pubmed/27753429.

Jordan, D. et al., 2011. Carriage of methicillin-resistant Staphylococcus aureus by veterinarians in Australia. *Australian Veterinary Journal*, 89(5), pp. 152–159.

Klevens RM, Morrison MA, Nadle J, Petit S, Gershman K, Ray S, Harrison LH, Lynfield R, Dumyati G, Townes JM, Craig AS, Zell ER, Fosheim GE, McDougal LK, Carey RB, F.S.A.B.C. surveillance (ABCs) M.I., 2007. Invasive methicillin-resistant Staphylococcus aureus infections in the United States. *JAMA*, 298(15), pp. 1763–71.

Köck, R. et al., 2012. Persistence of nasal colonization with livestock-associated methicillin-resistant staphylococcus aureus in pig farmers after holidays from pig exposure. *Applied and Environmental Microbiology*, 78(11), pp. 4046–4047.

Leibler, J.H. et al., 2016. Staphylococcus aureus nasal carriage among beefpacking workers in a Midwestern United States slaughterhouse. *PLoS ONE*, 11(2), pp. 1–11.

Masclaux, F.G. et al., 2013. Concentration of airborne staphylococcus aureus (MRSA and MSSA), total bacteria, and endotoxins in pig farms. *Annals of Occupational Hygiene*, 57(5), pp. 550–557.

Moodley, A. et al., 2008. High risk for nasal carriage of methicillin-resistant Staphylococcus aureus among Danish veterinary practitioners. *Scandinavian Journal of Work, Environment and Health*, 34(2), pp. 151–157.

Morris, A.K. & Masterton, R.G., 2002. Antibiotic resistance surveillance: action for international studies. *The Journal of antimicrobial chemotherapy*, 49(1), pp. 7–10. Available at: http://www.ncbi.nlm.nih.gov/pubmed/11751759.

Pomba, C. et al., 2009. First description of meticillin-resistant Staphylococcus aureus (MRSA) CC30 and CC398 from swine in Portugal. *International Journal of Antimicrobial Agents*, 34(2), pp. 193–194.

Rinsky, J.L. et al., 2013. Livestock-Associated Methicillin and Multidrug Resistant Staphylococcus aureus Is Present among Industrial, Not Antibiotic-Free Livestock Operation Workers in North Carolina. *PLoS ONE*, 8(7), pp. 1–11.

Salanitro, J.P., Blake, I.G. & Muirhead, P.A., 1977. Isolation and identification of fecal bacteria from adult swine. *Applied and Environmental Microbiology*, 33(1), pp. 79–84.

UEFSA, 2001. Development document for the proposed revisions to the national pollutant discharge elimination system regulation and the effluent guidelines for concentrated animal feeding operations. Washington: EPA.

Walser, S.M. et al., 2015. Evaluation of exposure-response relationships for health effects of microbial bioaerosols – A systematic review. *International Journal of Hygiene and Environmental Health*, 218(7), pp. 577–589. Available at: http://dx.doi.org/10.1016/j.ijheh.2015.07.004.

World Health Organization (WHO), 2016. Antimicrobial resistance: fact sheet nº 194. Genebra: WHO. Available at: http://www.who.int/mediacentre/factsheets/fs194/en/ [Accessed June 20, 2012].

Assessment of fatigue through physiological indicators: A short review

J.A. Pérez & J.C. Guedes
FEUP, Faculty of Engineering, University of Porto, Portugal

ABSTRACT: The aim of this review was to obtain relevant information about fatigue assessment through physiological indicators, in order to focus in a future on how it affects psychomotor skills in general and fine motor skills of workers. Four databases (SCOPUS, Science Direct, PubMed and Web of Science) were used to conduct a data search according to crosswords of keywords. Three phases were used: two phases using exclusion criteria (date, type of article, language, other—such as health condition or out-of-range age, duplicates) and one phase using inclusion criteria (objective method of measurement, physiological parameters measurements—HR, oxygen uptake, blood lactate, non-static activity/work). The first phase outcome (125 articles) was reduced to 7 publications considered significant, that used physiological parameters such as HR, oxygen consumption and blood lactate, as well as subjective methods to measure physical exertion, mostly RPE scale. Some studies revealed strong correlations between RPE and both HR and blood lactate, and suggested that the best way to determine physical fatigue is through publications a combination of assessments, not a single variable measure. In the other hand, articles selected have demonstrated the significance of considering, besides physiological parameters, subjective perception of effort during a training process to determine fatigue.

1 INTRODUCTION

1.1 Work fatigue

Work fatigue is a complex but well-known phenomenon, resulting from various factors, often linked to a feeling of exhaustion, lowering of physiological functions, and especially to a decrement in performance and work efficiency (Saito, 1999).

As stated in ISO 6385:2004, is a mental or physical, local or general non-pathological manifestation of excessive strain, completely reversible with rest.

1.2 Fatigue measurement

Despite fatigue cannot be measured itself directly, its consequences and some symptoms are known, or at least physiological as well as psychological indications relevant to these symptoms of fatigue can be measured. In this respect, fatigue (or so called physical exertion) can be understood through changes in physiological functions that include mental or nervous function, autonomic nervous function, endocrinological function, metabolism, etc. These functions are lowered, disturbed or broken down by excessive workload or effort and can be measured by physiological and biochemical techniques.

There can be mentioned a number of indicators which are generally applied in fatigue research: such measures representing cerebral cortical activity level as electroencephalography, channel capacity or perceptual threshold such as flicker fusion frequency, some indications of vegetative functions, biochemical variables relevant to metabolic changes or to endocrine regulation, motor skills and others. Those tests which have been advocated for fatigue measurement can be classified from the viewpoint of methodology into the following categories: questionnaires on subjective feelings of fatigue, psychological tests, neurophysiological tests, biochemical indexes, physiological tests, and autonomic nervous function tests (Saito, 1999). In a further study, physical exertion in work environment is expected to be evaluated through physiological indicators such as body temperature, blood lactate, heart and breath rate, along with movement analysis.

1.1 Fatigue and fine motor skills

There are studies that show, in different areas, that fatigue is in some way related to performance decrease. A study concerning health-care related workers showed that the level of fatigue is directly related to increased number of cognitive errors and decreased psychomotor efficiency and overall task performance (Kahol et al. 2008). Even if, in some cases, cognitive skills are more affected by fatigue than psychomotor skills, impeded cognitive performance may also lead to limited psychomotor proficiency.

Beyond that, it would be interesting to assess, particularly, the effect fatigue has on fine motor skills, above everything in the so called safety-sensitive professions (police officers, firefighters, health care providers, among others) as these occupations need the ability to accomplish tasks that require a combination of fine psychomotor and cognitive decision-making skills to provide safe effective services. Therefore, the decrement of fine motor skills due to fatigue in this kind of jobs will affect not only the workers but the safety of people to whom they provide services.

In this work, it is proposed to determine fatigue through physiological indicators that can lead to making decisions on whether workers are able to keep on performing efficiently and safely. On the other hand, the relationship between fatigue and the decrement of fine motor skills in professionals can be used to create tools that allow redesigning shifts and break hours in safety-sensitive professions.

Keeping this in mind, the main goal of the paper is the search of relevant information about fatigue evaluation mechanisms, with the aim to focus in a future on how it affects psychomotor skills in general and fine motor skills of workers.

2 MATERIAL AND METHODS

2.1 Search strategy

In this review, four databases were used to conduct a data search according to crosswords of keywords: SCOPUS, Science Direct, PubMed and Web of Science. Keywords were categorized in two groups: (A) with only two words, physical fatigue and exertion; and (B) with the words measurement, estimation and assessment. These keywords were selected after an initial search testing, in which the results, at first sight, reflected to be more related to the topic of interest.

In some cases keywords of group (A) where searched in the titles, as results where wide for using the words to search into the entire article. This was the case of SCOPUS and Science Direct databases, as well as the three exertion keywords in Web of Science and PubMed databases.

2.2 Screening

A first phase of screening was carried out establishing criteria, in order to obtain significant results.

The first criteria was "Date": only articles from 2012 to 2017 were considered, except in those cases in which there were few articles or no article at all found in that period, thus the latest five years published that appeared were considered.

The second general criteria used was "Type of article": only articles, articles in press and reviews were considered in this phase, although reviews were not considered later in the final results but as an extra useful complement.

"Language" was the third criterion, excluding all articles but those written in English or Spanish, while the fourth one was "Articles out of topic". As this review is oriented towards ergonomics and it's planned to be included measurement of fatigue of professional hockey players, articles that came from medical sources (mostly journals) about diseases, disorders or just unrelated medical fields were excluded. In other words, all articles that were not from sources that included ergonomics, safety science, sports, sport medicine, sport science, motor skills and/or physiology fields were dismissed.

Lastly, a fifth criterion was named as "Other" to end the first phase of exclusion. In "Other" were included articles referring to diseases or in which participants had any medical condition (therefore considered not healthy) that have not been excluded before for coming from the approved aforementioned sources. Besides, when searching in Web of Science database, it was possible to dismiss some articles that at first sight showed to have a different approach that wouldn't be useful for this research. In the other hand, articles about either young or elder subjects (not in the range of working-age people) were also excluded. All three aspects where included in the "Other" eligibility criterion.

2.3 Duplicates removal

In a second phase, duplicates were excluded by using a title filter to locate repeated articles. Furthermore, papers that are not directly related but serve as extra information were also dismissed (like the ones that don't address fatigue measurements per se but are still related to it in some way).

2.4 Eligibility

In this phase also, now with fewer articles, remaining publications addressing diseases or out-of-range ages were excluded taking into account that these filters where applied before in title articles, and may have been left more with such characteristics that were not filtrated and needed more attention. Finally, it was possible to exclude articles (such as the ones about facial video-based detection of fatigue) that use a methodology that is not worth of attention for this study in which one of the main goals is physiological measurement.

The last phase consisted on the establishment and application of inclusion criteria among the selected articles. First, it was included all articles that use an objective method of measurement. In second place, based on these last results were included only those that measure physiological parameters of interest for this study (HR, oxygen levels, blood lactate). At last, from this latest outcome were selected publications in which the referred activity/work is non-static.

3 RESULTS

3.1 Selection description

By applying this methodology, it was possible to reduce the first phase outcome (125 articles) to 7 significant publications to this work.

In the first phase, considering the six combinations of keywords searched in the four aforementioned data bases, a total of 1.571 articles were rejected by date, 55 by type of article, 16 by language, 543 were out of topic and 172 filtered under the "Other" criteria (Table 1). It should be noted that as this work is health-oriented and physiological measures are involved, it is normal to have an outcome with a considerable number of articles referred to diseases, disorders or in which participants have a medical condition (considered out of topic). Furthermore, in this review were taken into account only the last five years of publications, so it was expected that a notable number of articles were to be excluded initially.

After concluding the first phase of exclusion, 125 articles were identified, of which 53 were duplicates and were dismissed in the second phase, leaving a total of 72 articles.

When considering the criteria established on the second phase of exclusion as shown on Table 2, 24 articles were distinguished.

Table 1. Summary of the first phase of exclusion.

Data base	Total left	Summary of rejected articles				
		Date	TA*	L**	OT***	Other
Scopus	58	565	28	13	168	32
Science Direct	24	435	9	0	234	38
Web of Science	25	296	16	2	99	24
PubMed	18	275	2	1	42	78
Total	125	1571	55	16	543	172

TA*: Type of article; L**: Language; OT***: Out of topic.

Table 2. Summary of the first phase of exclusion.

Filters order	Criteria	Filtered articles	Remaining articles
1	Duplicates	53	72
2	Not directly related	33	39
3	Review paper	1	38
4	Age, disease or disorder	11	27
5	Date	1	26
6	Non-relevant methods	2	24

The third phase, this time of inclusion, was carried out taking these 24 articles as basis. The first thing to consider was whether the publication was experimental or not. As all 24 were experimental, another criterion was used. There were 14 articles that use objective methods of measurement and, of those, were included 10 that measure physiological parameters of interest for this study (HR, oxygen levels, blood lactate).

From this latest outcome were selected publications in which the referred activity/work is non-static (9 articles) and those in which the measurement of physical exertion is involved in their objectives, which left a total of 7 articles relevant to this work.

3.2 Selected papers

At last, seven articles were selected from all the research outcomes, after applying the three phases mentioned above.

Table 3 shows the main characteristics of these articles, including aspects of the people that participate in the experiences, such as age range, mean age, sex and profession, among others. It also shows general features about the trials. It should be noticed that all participants are healthy active people except from one study that also involves non-athletic persons in order to make a comparison.

In the other hand, information about the use of subjective scales of physical exertion measurement is presented, along with the parameters that are measured (objective measurement) in each method of the seven articles. All articles but one use both subjective and objective methods to assess physical exertion. Among the subjective methods used in these publications, almost all use Borg's Scale. In regard to objective methods, 6 of the 7 articles measure heart rate, 4 blood lactate, some of them measure oxygen consumption and movement patterns.

Table 3. Summary of principal characteristics of the selected articles.

Ref.	Measurements Subjective	Measurements Objective	Participants	Research protocol
Hausken & Dyrstad, 2013	Borg's 15-point RPE	Heart rate (HR monitor); accelerometer counts (accelerometer)	35 subjects (22 females), students in academic sports sciences programs.	Four exercise sessions were carried out at the SIS Sports Center at the University of Stavanger, Norway: (a) Zumba (60'); (b) 4 × 4 running (45'); (c) 4 × 4 spinning (45') and (d) pyramid running (45'). Subjects were well nourished and consumed water during each session. Temperature around 22°C, humidity of 45%.
Kakarot & Müller, 2014	CP scale (Heller)	Heart rate (Varioport); chest electrocardiogram data recorded using disposable electrodes	Twenty-nine healthy non-smoking men aged 27–71 years (M = 48, SD = 15) representing three different age groups: 27–41 (10 participants), 42–56 (9 participants) and 57–71 years (10 participants).	Participants took part in 28 minutes of cycling with systematically increased and decreased load as well as in 7 hours of continuous cycling with low to medium exertion, interrupted by brief peak loads at high to very high exertion levels.
Kilpatrick et al., 2016	Borg's CR-10 Scale	Expired gases (metabolic cart); heart rate (HR monitor); blood pressure (by auscultation)	24 recreationally active, healthy students (12 males, 12 females, mean age ± standard deviation [SD] = 22 ± 3 years, mean body mass index [BMI] ± SD = 24 ± 4) from a US university were recruited via email and word of mouth.	Participants completed five exercise trials in the laboratory performed on an electronically braked cycle ergometer.
Kovářová et al., 2015	Borg's Scale	Heart rate (HR monitor); blood lactate (lactometer)	Elite-level short triathlon athletes from the Czech Republic (n = 23, 14 men and 9 women; mean age: 18,7 ± 1,78 years) and non-athletes who were students of the Faculty of Humanities at Charles University (n = 15, 9 men and 6 women; mean age: 19,5 ± 1,84 years).	The data were collected according to a graded exercise test using a bicycle ergometer. The test was performed under standard conditions on a Cyclus 2 laboratory device, which enables the use of an actual bike frame.
Larkin et al., 2014	n.a.	Decision-making performance (video-based test), in-game physical exertion (blood lactate levels) and movement pattern (global positioning system [GPS])	15 Australian football umpires (Mage = 36, s = 13,5 years; Mgames umpired = 235,2, s = 151.3) from a regional Division 1 Australian football competition.	Hypothesis: decision-making performance would be negatively impacted by in-game physical exertion.
Olkoski et al., 2014	Borg's Scale	Oxygen consumption (gas analyser); blood lactate concentration (lactometer); heart rate (HR monitor)	15 healthy women (18–25 yrs). University students who had participated regularly in WEC* for at least six months.	The WEC took place in a laboratory tank, with a cadence of 136 beats per minute (bpm). The protocol was defined taking into account that physiological variables (HR, VO2 and [Lac]) reach a steady state after 140 seconds of submaximal exercise.

(Continued)

Table 3. (Continued).

Ref.	Measurements		Participants	Research protocol
	Subjective	Objective		
Scherr et al., 2013	Borg's Scale	Hart rate (surface ECG and ECG); blood lactate (enzyme-chemically measurement)	2.560 caucasian men and women (1.796 male, 764 female; age range between 13–83 years (mean age: 28 years).	All the participants performed a stepwise incremental exercise test until physical exhaustion (when they were not able to maintain pedal cadence ≥70 rev./min or were not able to run with the given velocity, and the HR was within ±10 bpm of age-predicted HRmax) on either a treadmill or an electromagnetically braked cycle ergometer.

WEC*: water exercises classes.

4 DISCUSSION

Articles selected used physiological parameters such as Heart Rate (HR), oxygen consumption and blood lactate, as well as subjective methods to measure physical exertion, mostly Borg's Rating of Perceived Exertion (RPE) scale. RPE is a widely used and proven valid psycho-physical tool to assess subjective perception of effort during exercise (Borg, 1982). It is well known that in healthy subjects, strong relationships exist between RPE and heart rate during physical activity (1 RPE point is approximately 10 bpm) (Sherr et al., 2013). Hausken & Dyrstad (2013) found that in Zumba exercise session, a very dynamic exercise, subjects had a significant correlation of 0.5–0.6 between Rate of Perceived Exertion (RPE) and percentage of maximum heart rate (%HR max), which encourages to keep on exploring this relationship to determine physical fatigue, despite the fact this relationship was not found in other activities evaluated, as spinning.

Also, Scherr et al. (2013) found a strong relationship for the entire cohort between RPE and blood lactate (even higher than between HR and blood lactate) and between RPE and HR. In their study, individual's subjective overall perception of muscle fatigue and physical stress was assessed alongside objective measures of lactate and HR. These high correlations between RPE and both HR and blood lactate indicate the high precision of the predictive value of these objective measures of intensity (HR or blood lactate) as a function of RPE. What is more, Olkoski et al. (2014) found a correlation between oxygen consumption and both heart rate and blood lactate, which is important as Scherr et al. (2013) demonstrated to be a strong association between HR and blood lactate with RPE; so this could imply there is also a relationship between oxygen consumption and RPE that can be studied and considered in further studies.

Larkin et al. (2014) studied the influence of in-game physical exertion over decision-making performance. The study concluded that physical strain may not affect decision-making performance. This provides useful information for the study, as decision-making is a cognitive skill, and even if, in some cases, cognitive skills are more affected by fatigue than psychomotor skills, impeded cognitive performance may also lead to limited psychomotor proficiency, that is what a further study pretends to approach.

On the other hand, Kovářová et al. (2015) determined that the model of the subjective perception of fatigue does not lend itself to objectivization. They detected through a cubic model, a significant association between heart rate and the Borg Scale of Fatigue score in a non-athlete group (69%) but not in an athlete group (40%). Therefore, they suggest that for non-athletes, the Borg Scale may be used as a tool for the objectivization of physical loading as evaluated by heart rate, but not for elite endurance athletes, for whom the most predictive parameter would the blood lactate level. In any case, the relationship should still be considered, as the study will be oriented to workers in general and not only to elite endurance athletes (in the case of professional athletes who do the activity for a living).

Kakarot & Müller (2014) revealed in their results that, while both measurements are suitable to capture physical strain, HR is not as specific as PE. Although this may be true in some activities, it's considered worth analyzing the relationships in more depth.

To resume, there is information that supports the use of physiological parameters to determine physical fatigue, but always along with the subjective measurement of perceived exertion.

5 CONCLUSIONS

The data reveal that the relationships between RPE and exercise intensity (assessed by blood lactate, and heart rate) are very strong and independent of age, gender, medical history, level of physical activity and exercise modality. An integration of central factors such as HR, oxygen consumption and blood lactate, would explain better the psychophysical variation tan any single physiological variable. This suggests that the best way to determine physical fatigue is through a combination of assessments, not a single variable measure. Nevertheless, articles selected have demonstrated the significance of considering, besides physiological parameters, subjective perception of effort (perceived exertion) during a training process to assess physical fatigue.

This marks a startpoint for the further assessment of fatigue that this work pretends.

REFERENCES

Borg, Gunnar A.V. 1982. Psychophysical bases of perceived exertion. *Medicine and Science in Sports and Exercise* 15(5): 377–381.

Hausken, K. & Dyrstad, S. 2013. Heart rate, accelerometer measurements, experience and rating of perceived exertion in Zumba, interval running, spinning, and pyramid running. *Journal of Exercise Physiology Online* 16(6): 39–50.

Kahol, K., Leyba, M. J., Deka, M., Deka, V., Mayes, S., Smith, M.,... & Panchanathan, S. 2008. Effect of fatigue on psychomotor and cognitive skills. *The American Journal of Surgery* 195(2): 195–204.

Kakarot, N. & Müller, F. 2014. Assessment of physical strain in younger and older subjects using heart rate and scalings of perceived exertion. *Ergonomics* 57(7): 1052–1067.

Kilpatrick, M., Greeley, S. & Ferron, J. 2016. A comparison of the impacts of continuous and interval cycle exercise on perceived exertion. *European Journal of Sport Science* 16(2): 221–228.

Kovářová, L., Pánek, D.; Kovář, K. & Hlinčík, Z. 2015. Relationship between subjectively perceived exertion and objective loading in trained athletes and non-athletes. *Journal of Physical Education and Sport* 15(2): 186–193.

Larkin, P., O'Brien, B., Mesagno, C., Berry, J., Harvey, J. & Spittle, M. 2014. Assessment of decision-making performance and in-game physical exertion of Australian football umpires. *Journal of Sports Sciences* 32(15): 1446–1453.

Olkoski, M., Matheus, S., De Moraes, E., Tusset, D., Dos Santos, L. & Nogueira, J. 2014. Correlation between physiological variables and rate of perceived exertion during a water exercises classes. *Revista Andaluza de Medicina del Deporte* 7(3): 111–114.

Scherr, J., Wolfarth, B., Christle, J., Pressler, A., Wagenpfeil, S. & Halle, M. 2013. Associations between Borg's rating of perceived exertion and physiological measures of exercise intensity. *European Journal of Applied Physiology* 113(1): 147–155.

Saito, K. 1999. *Measurement of fatigue in industries.* Industrial Health: National Institute of Occupational Safety and Health, pp. 134–142.

ISO, B., 6385: 2004 Ergonomic principles in the design of work systems. *European Committee for Standardization.*

Workload mental analysis of the workers in a collective road transport company

T.G. Lima, F.S.M. Gonçalves & R.M.F. Marinho
Federal University of Rio de Janeiro, Rio de Janeiro, Brazil

ABSTRACT: In Brazil, the buses are the main mode of public transportation. This paper looks into the Mental Workload of the workers in a collective road transport company in a city from interior of Rio de Janeiro. It has aimed to measure and compare the Mental Workload in two sectors and then, identify the relation between mental workload with work situation using the Ergonomic Work Analysis and NASA-TLX questionnaires. The results showed that administrative employees have high mental demand index, while the mechanic workshop employees have high index of physical demand. Other point is that both sectors have high effort index, because the company are cutting the number of workers. Finally, it was identified three important elements in EAW with high discomfort index: remuneration, noise and lumbar region. These elements were associated with the mental workload and the results showed low frustration and high performance index in all comparisons, directing the company to investigate theses points more properly.

1 INTRODUCTION

The literature evidences many studies regarding the analysis of mental workload of drivers (De waard, 1996; vGaly & Berthelon, 2017). However, the activities of back office staff such as administrative and mechanics are a matter of importance and deserves to be discussed here.

According to databank from Federation of Passenger Transport Companies of the State of Rio de Janeiro (Fetranspor, 2016), the bus is the main mode populations transport in both municipal and inter-municipal areas. During the 2016, the sector had 83,336 employees in the state of Rio de Janeiro, of which 5,211 were in management area and 8,822 in the maintenance area. These areas are equally responsible for the quality and safety of the travel.

Mental Workload (MWL) is a topic of increasing importance, especially in ergonomics and human factors areas, where it is important to know how MWL impinges on performance (Young et al., 2015) of complex tasks in professional environments (Soria-Oliver et al., 2017).

The mental demand ranges from cognitive aspects, such as attention, concentration, memory, perception and decision making, to emotional aspects, which embrace affections, feelings and motivation towards work (Cardoso & Gontijo, 2012).

The subjective measures are classified as the most used to measure MWL (Cardoso & Gontijo, 2012). According to the authors, the NASA TLX is an important subjective measure, that was developed by Hart and Staveland (1988). NASA-TLX is a multidimensional scale procedure that provides a global workload score based on a weighted average of evaluations on six subscales: Mental, Physical, and Temporal Demands, Frustration, Effort, and Performance (Corrêa, 2003). The main objective of this article is to perform a comparative analysis between two sectors: administrative and mechanic workshop. The second objective is to identify relations between Ergonomic Analysis of Work (EAW) and mental loads indicators according to the data obtained in the NASA-TLX questionnaire.

2 MATERIALS AND METHODS

2.1 Reference material and participants

Corrêa's dissertation (2003) was used as reference for the applied methodology, assimilating the questionnaires with some changes inherent about the company's conditions.

The research established the strategy of capturing two sectors with apparently different requirements, administration and mechanic. Altogether, a sample of 23 administrative staff (88,5%; population: 26) and 17 mechanic workshop workers (81%; population: 21) were studied.

2.2 Instruments

Two questionnaires were applied. The first refers to the Ergonomic Analysis of Work (EAW), that divided the data into three groups: personal, job, occupational biomechanics and environmental conditions, which people answered based on personal information and working conditions; biomechanical and technical conditions. The levels of EAT are comfortable, slight uncomfortable and uncomfortable; and psychosocial, that the people classified different items of organizational conditions between adequate, reasonable and inappropriate. The second questionnaire, NASA-TLX, the person assigned levels on a scale of 0 to 100 to Mental, Physical, and Temporal Demands, Performance, Effort and Frustration.

Mental demand: How much mental and perceptual activity was required? (Low:0; High:100);

Physical demand: How much physical activity was required? (Low:0; High:100);

Effort: How hard did you have to work mentally and physically to accomplish your level of performance? (Low:0, High:100);

Temporal demand: How much did you feel pressured over time during the work? (Low:0; High:100);

Performance: How successful do you think you were in accomplishing the goals of the task set by the high management and by yourself? How satisfied were you with your performance in accomplishing these goals? (Excellent:0; Terrible:100);

Frustration: How insecure, discouraged, irritated stressed and annoyed did you feel during the task? (Low:0; High:100);

NGI: Obtained from the weighted average between the levels of indicated mental loads, corresponding with the predominance reported in the work routine. (Low:0; High:100).

2.3 Procedures

The procedure used for data collection consisted of three stages. At first, the worker was approached individually and was given a brief explanation about the importance of ergonomics at work and the stages of field research. After being accepted to participate, the WEA questionnaire was delivered to the volunteer and, before answer the questions, the researchers explained each item that would be measured on the NASA-TLX questionnaire, asking the worker to think about these items and reply later. After the completion of this phase, which lasted about twenty minutes, the researcher delivered the NASA-TLX questionnaire to the worker, in which, together with the response of the document, an interview was conducted to assimilate the most important points that led to the characterization of the levels described by the official.

2.4 Data analysis

The WEA questionnaires data's were tabulated checking the occurrence levels of each topic asked, being selected the items that, in the point of view of the ergonomic analysis, presented a worrying scenario in both groups that were studied. Then, there was tabulated the data from the NASA-TLX questionnaire, making evident the mental loads indexes on each individual participant of the research. Finally, statistical data were cross-referenced between the averages of the mental loads and the classification of the WEA's highlighted items per person, allowing establishing a line of thought to the problem analysis. In these data crossing, the quartile technique was used, which divides the distribution of the values into 3 groups.

The first quartile (Q1) represents the limit value that is the 25% smaller numerical representativeness of the sample, while the third quartile (Q3) represents the value from which the 25% largest numerical representativeness of the sample. The second quartile (Q2), also known as median, is the value that divides the sample equally in 50% of its numerical representativeness.

3 RESULTS AND DISCUSSION

3.1 Mental loads by sectors

An interesting point to note in Table 1 is the high effort index for both sectors. This behavior occurs because the company was submitted to a large cut

Table 1. Mental loads—sectors.

Sector	ADM				Workshop			
Dimension	Mean	Q1	Q2	Q3	Mean	Q1	Q2	Q3
Mental demand	22.12	16.67	21.33	26.67	12.31	8.00	13.33	16.00
Physical demand	2.23	0.00	0.00	2.00	11.49	6.00	10.00	18.00
Temporal demand	10.20	4.67	10.00	16.00	5.49	2.00	3.33	6.67
Performance	3.22	0.33	2.67	5.67	4.04	1.33	2.67	5.33
Effort	10.12	6.00	10.00	11.33	11.53	8.00	10.67	16.00
Frustration level	5.22	0.00	2.67	8.00	4.47	0.00	1.33	6.67
NGI	53.10	40.33	56.67	66.00	49.33	42.00	46.67	58.67

in the number of employees, causing the accumulation of function to the workers. Thus, it is necessary to make a high effort to meet the current demand of the work and show good results because the country's financial instability supports the possibility of further cuts in the workforce. When analyzing the quartile distribution, it can be stated that the high mean represents a large part of the studied population in both cases, since the scenario of effort levels are above 10 for about 50% of the entire sample.

The rate of time demand is almost doubled for the administrative sector. Based on the interview reports, the workshop staff had the final word on the vehicle liberation. This work behavior can be show in case of problems that would not allow the bus to be liberated and the situation is contacted by another sector that is obliged to provide reposition vehicle. In the administrative area, the predominant activities are create and control the destinations and schedules of bus movements, bureaucratic representation of contracts and vehicles problems, and administration of driver work shifts. Thus, the difference of autonomy between the services in both sectors and the activities carried out characterize the difference in the levels of temporal demand. This picture becomes evident when observing that 50% of the sample of the administrative sector presents levels above the already considered high average 10.20. While the workshop sector, despite averaging 5.49, the third quartile still has close to average values, that is, well below 10.

3.2 *Mental loads comparison with remuneration*

An initial analysis allow verify, for both sectors, the existence of high global NASA indices in groups that do not characterize the remuneration as an appropriate factor. Thus, it can be observed that the workers who are subjected to higher numbers of charges, that is, person who self-charging and is charged more frequently, do not have their due salary recognition.

Observing the Table 2, the idea discussed above can be cognized with the behavior of increasing mental and temporal load acting according to worse salary qualifications in the administrative sector. The reason of this behavior is that the poor Brazilian economic scenario imposed the company to make a cut of employees, delegating more tasks to a single person without a corresponding salary increase.

Thus, is generated a significant increase in the frustration index when compared to the workers who rate as appropriate. It can be stated that the representativeness of high mental demand reaches all the employees when noticing that in first quartile, all employees in the ADM sector are submitted to high values. As for the temporal demand, about 50% of the employees (Q2) already have high valuations.

The Table 3 is an example that shows the scenario commented before. For this, it is enough to analyze the levels of effort, frustration and NGI of the workshop sector when compared to the worst and the best salary qualification. Therefore, based on the workers' own reports during the interview, this scenario not only cause the loss of stimulus to work with the lack of acknowledgment of effort, but also has psychological impacts related to frustration, such as dissatisfaction, stress, insecurity and irritation. In general, this scenario of high NGI representativeness and effort for a large part of the workshop sector is evident when noticing that from the second quartile the sample begins to assume great representativeness in its indices. As for the level of frustration, it remains concentrated with high values in 25% of the analyzed persons who referred the inadequacy of the salary, showing a negative impact on a considerable part of the group of workers studied.

3.3 *Mental loads comparison with noise*

At first, it is interesting to observe that most workers from two sectors (more than 50% of the employees interviewed in each department) agree

Table 2. ADM mental loads—remuneration.

Dimensions	%Mean			25% (Q1)			50% (Q2)			%75(Q3)		
	Inad.	Mod.	Ad.	Inad.	Mod.	Ad.	Inad.	Mod.	Ad.	Inad.	Mod.	Ad.
Mental demand	24.67	22.30	20.00	23.33	17.67	15.00	26.67	21.33	16.67	26.67	25.33	26.67
Physical demand	2.53	0.85	8.48	0.00	0.00	5.67	0.00	0.00	10.00	0.67	0.00	11.00
Temporal demand	11.07	10.91	8.48	10.00	4.67	5.67	10.67	8.00	10.00	16.00	17.00	11.00
Performance	3.07	4.42	1.43	2.67	2.33	0.00	2.67	5.33	0.00	4.00	6.33	1.00
Effort	9.33	9.52	11.62	4.67	5.33	9.00	10.00	8.00	10.00	10.67	11.33	11.33
Frustration level	4.80	8.12	11.62	0.00	2.67	9.00	6.00	6.67	10.00	6.00	11.33	11.33
NGI	55.47	56.12	46.67	53.33	52.33	37.00	56.00	56.67	42.00	68.67	66.33	59.33

Table 3. Mechanic workshop mental loads—remuneration.

Dimensions	%Mean			25% (Q1)			50% (Q2)			75% (Q3)		
	Inad.	Mod.	Ad.	Inad.	Mod.	Ad.	Inad.	Mod.	Ad.	Inad.	Mod.	Ad.
Mental demand	16.22	12.29	10.67	12.33	11.67	8.00	21.33	13.33	8.00	22.67	13.67	13.00
Physical demand	8.22	14.95	9.43	7.00	9.00	2.00	8.00	10.67	4.00	9.33	20.00	15.67
Temporal demand	8.00	4.48	5.43	3.00	2.67	1.67	6.00	3.33	3.33	12.00	6.00	8.33
Performance	2.22	3.62	5.24	0.67	1.67	2.00	1.33	3.33	2.67	3.33	5.33	6.67
Effort	16.67	11.62	9.24	11.67	6.33	7.00	14.00	12.00	10.00	20.33	16.00	10.67
Frustration	15.56	0.29	3.90	6.67	0.00	0.67	6.67	0.00	2.67	20.00	0.00	6.33
NGI	66.89	47.24	43.90	61.33	42.67	32.00	64.00	46.67	44.67	71.00	51.33	51.67

Table 4. ADM mental loads—noise.

Dimensions	%Mean		25% (Q1)		50% (Q2)		75% (Q3)	
	Dist.	Not dist.	Dist.	Not dist.	Dist.	Not dist.	Dist.	Not dist.
Mental demand	25.05	17.56	21.33	16.00	25.33	16.67	30.00	18.67
Physical demand	2.05	2.52	0.00	0.00	0.00	0.00	1.17	2.67
Temporal demand	12.52	6.59	10.00	3.33	12.00	5.33	16.00	10.00
Performance	4.14	1.78	2.67	0.00	4.00	0.00	5.83	0.67
Effort	9.43	11.19	4.83	6.67	10.00	8.00	11.67	10.67
Frustration level	6.90	2.59	0.17	0.00	6.00	2.67	11.33	2.67
NGI	60.10	42.22	56.17	35.33	59.67	38.67	67.33	50.67

about the noise is annoying. In the other hand, the types of noise are different. People who work in an office believe that the conversation noise is more evident, and in the mechanic workshop area, workers said about engines and other facilities noises.

Observing the Table 4, people who works in an office and believe the noise disturb, have higher rates of temporal and mental demand than other who believe the noise not disturb. This behavior is notorious in the quartile analysis, in which the difference in values of these demands increases as it advances in this type of analysis, starting with a contrast of 5 points in the first quartile and reaching differences of up to 11.33 points in the third quartile. Some people spoke that the conversation and phone noise can generate lake of concentration and therefore disrupt the continuity of the activity, generating more frustration.

The Table 5 shows high level of frustration index for mechanic workshop workers who believe the noise is annoying (almost more than 4 times higher than people who agree the noise not disturb). This group also exhibits high performance index, so they have less performance. Comparing the quartile analysis, this conclusion becomes significant when noting that the group that classified the "not disturb" presents value in the third quartile smaller than the representativeness of the first quartile that classified the "disturb". The reason of this behavior is that, although most of these workers are familiarized with the daily noise, they related difficulty in understanding conversations, stress, problems to sleep and lack of concentration, which can influence the performance and frustration levels.

3.4 *Mental loads comparison with lumbar region*

An alarming fact was the high number of people who reported a level of discomfort in the lumbar region, being 59% and 48% in the mechanic workshop and administrative areas, respectively. In the Tables 6 and 7, the impact from these levels can be perceived as increasing frustration in both study groups. The quartile distribution shows that this impact is significantly negative is restricted to about 25% of both sectors studied, since the representativeness of the third quartile presents large difference when compared to the other quartiles.

Considering the mechanic workshop employees who feels uncomfortable in lumbar region, the Table 6 shows a high physical demand, mental demand, temporal demand, frustration and effort index. However, they have the lower performance index, which indicates a high performance behavior. According to the quartile analysis, it can be noted that the average yields good representativeness when evidencing that this behavior begins to

Table 5. Mechanic workshop mental loads—noise.

Dimensions	%Mean Dist.	%Mean Not dist.	25% (Q1) Dist.	25% (Q1) Not dist.	50% (Q2) Dist.	50% (Q2) Not dist.	75% (Q3) Dist.	75% (Q3) Not dist.
Mental demand	11.33	13.00	5.67	8.50	13.33	11.67	15.00	15.33
Physical demand	11.43	11.53	5.00	6.50	8.00	10.33	16.00	16.83
Temporal demand	5.62	5.40	2.67	1.50	5.33	3.33	7.67	6.67
Performance	7.05	1.93	4.33	0.33	6.67	1.67	8.00	2.67
Effort	10.48	12.27	4.67	9.50	9.33	11.33	12.33	16.00
Frustration level	8.19	1.87	0.00	0.00	6.67	0.67	8.67	2.50
NGI	54.10	46.00	45.33	37.00	55.33	45.00	61.67	54.83

Table 6. Mechanic workshop mental loads—lumbar region.

Dimensions	%Mean Comf.	%Mean Slightly	%Mean Uncomf.	25% (Q1) Comf.	25% (Q1) Slightly	25% (Q1) Uncomf.	50% (Q2) Comf.	50% (Q2) Slightly	50% (Q2) Uncomf.	75% (Q3) Comf.	75% (Q3) Slightly	75% (Q3) Uncomf.
Mental demand	11.81	8.57	22.22	9.00	4.67	21.33	13.33	8.00	21.33	13.67	11.67	22.67
Physical demand	12.00	10.10	13.56	5.67	5.00	7.00	10.67	10.00	8.00	15.67	14.67	17.33
Temporal demand	4.86	5.71	6.44	1.67	3.33	0.67	3.33	5.33	1.33	8.00	6.33	9.67
Performance	2.86	6.57	0.89	0.67	3.67	0.67	2.67	5.33	1.33	4.00	7.33	1.33
Effort	9.62	12.76	13.11	5.67	7.67	11.67	10.00	10.67	14.00	13.33	16.67	15.00
Frustration level	1.71	3.43	13.33	0.00	0.00	3.33	1.33	0.00	6.67	2.33	6.67	20.00
NGI	42.86	47.14	69.56	32.00	45.33	65.33	42.00	46.67	66.67	50.33	51.33	72.33

Table 7. ADM mental loads—lumbar region.

Dimensions	%Mean Comf.	%Mean Slightly	%Mean Uncomf.	25% (Q1) Comf.	25% (Q1) Slightly	25% (Q1) Uncomf.	50% (Q2) Comf.	50% (Q2) Slightly	50% (Q2) Uncomf.	75% (Q3) Comf.	75% (Q3) Slightly	75% (Q3) Uncomf.
Mental demand	21.28	22.95	23.17	16.50	18.33	21.33	17.67	24.00	22.33	27.50	28.33	24.17
Physical demand	2.78	2.38	0.33	0.00	0.00	0.00	0.00	0.00	0.00	4.67	0.33	0.33
Temporal demand	9.50	12.00	9.17	3.83	7.67	5.33	10.00	16.00	8.67	12.33	16.00	12.50
Performance	2.50	4.10	3.83	0.00	1.67	3.00	1.00	2.67	4.67	4.33	7.00	5.50
Effort	9.39	13.52	6.33	5.33	10.33	3.50	7.33	12.00	7.33	10.17	15.00	10.17
Frustration level	3.00	5.33	11.67	0.00	0.33	5.50	2.33	2.67	9.67	4.17	9.00	15.83
NGI	48.44	60.29	54.50	35.33	55.67	49.67	55.00	66.67	56.00	57.00	69.33	60.83

be expressed a good representativeness of the sample from the second quartile, that is, in 50% of the sample.

The reason of this behavior is that, based on the workers own reports during the interview, theses workers have to walk a lot, climb stairs and perform crouching activities. Although, the physical demand is lower than the mental demand. It happens also because they have prompt to find defects and fix them, for only later liberate the buses to travel. Thus, they are more sensitive in relation to temporal demand and mental demand and it can interfere in frustration index.

The Table 7 shows that administrative workers who considerate uncomfortable in lumbar region have the highest frustration index and mental demand, and the lowest physical demand and effort index when compared to the workers who rate as comfortable. Observing the quartile distribution, it can be noted that while the mental demand reaches the entire uncomfortable sample with high values, the levels of effort and frustration have a significant contribution to the higher mean in 25% of the people studied. Already, the physical demand has not presented any alarming fact.

The reason of this is that people of this sector stay a long time seated in front of the computer doing activities that require a lot of concentration and mental effort, and consequently lead to discomfort in the spine.

4 CONCLUSIONS

This article has addressed the problem of mental workload of the workers in a Collective Road Transport Company. Our research concluded that the mental loads acting in administrative and workshop sectors are very similar, except for physical and mental demand, where the first is higher in mechanic workshop employees and the second in administrative workers. The comparative analysis to identify relations between Ergonomic Analysis of Work (EAW) and Mental Workload indicators, showed three elements, such as remuneration, noise and lumbar region.

The results were an increased mental demand, time demand, effort and frustration index and reduced performance index when remuneration is not adequate, an increased frustration index and reduced performance when noise causes discomfort, and an increased mental demand index and frustration index when lumbar region causes discomfort.

It was possible to detect the low frustration and high performance index in all comparisons, directing the company to investigate theses points more properly.

Actions and policies to reduce the frustration and improve the performance are recommended, and it is vital to engage the company with more motivated people for best results.

REFERENCES

Cardoso, M.D.S., & Gontijo, L.A. (2012). Evaluation of mental workload and performance measurement: NASA TLX and SWAT. *Gestão & Produção*, 19(4), 873–884.

Corrêa, F.D.P. (2003). Carga mental e ergonomia.

De Waard, D. (1996). *The measurement of drivers' mental workload*. Netherlands: Groningen University, Traffic Research Center.

Federação das Empresas de Transportes de Passageiros do Estado do Rio de Janeiro (FETRANSPOR). 2016. Mobilidade Urbana, setor em números. Available at: https://www.fetranspor.com.br/mobilidade-urbana-setor-em-numeros.

Galy, E., Paxion, J., & Berthelon, C. (2017). Measuring mental workload with the NASA-TLX needs to examine each dimension rather than relying on the global score: an example with driving. *Ergonomics*, 1–11.

Hart, S.G. (2006, October). NASA-task load index (NASA-TLX); 20 years later. In *Proceedings of the human factors and ergonomics society annual meeting* (Vol. 50, No. 9, pp. 904–908). Sage CA: Los Angeles, CA: Sage Publications.

Soria-Oliver, María, López, Jorge S., & Torrano, Fermín. (2017). Relations between mental workload and decision-making in an organizational setting. *Psicologia: Reflexão e Crítica*, 30, 7. Epub May 18, 2017. https://dx.doi.org/10.1186/s41155-017-0061-0.

Young, M.S., Brookhuis, K.A., Wickens, C.D., & Hancock, P.A. (2015). State of science: mental workload in ergonomics. *Ergonomics*, 58(1), 1–17.

Ergonomic analysis of ballistic vests used by police officers in Paraíba State—Brazil

G.A.M. Falcão
Federal University of Pernambuco, PE, Brazil

T.F. Chaves, M.B.F.V. Melo & L.V. Alves
Federal Uuniversity of Paraíba, João Pessoal/PB, Brazil

ABSTRACT: Ergonomics studies the criteria needed to adapt the environment and products to human characteristics. In order to do this, it applies theories, principles, data and methods that can previously preserve human life in aspects related to health, safety, comfort and satisfaction. The ballistic vest, being a protective equipment that is next to the body of the user, causes a great majority of police officers to feel a lot of discomfort, causing worsening of low back pain, since the lumbar segment already suffers constantly with the weight that supports, both of the upper segments of the column and of gravity itself. In this context, understanding that ballistic protection equipment does not present any criteria for the comfort of the policeman, the present work aims to analyze ergonomic aspects of the Baltic vests, seeking to understand the user-vest relationship.

1 INTRODUCTION

Humans wear armor for thousands of years. The ancient tribes, since prehistory, when out hunting, arrested animal skins and plant material around their bodies, while the medieval warriors covered their torsos with metal plates before entering the battle. With the passage of time the armor were became increasingly sophisticated. However, all that changed with the development of new weapons for fighting, and with the appearance of guns and firearms, at the end of the middle ages. These fire guns projectiles at high velocity, giving them enough power to penetrate thin metal layers. You can increase the thick-ness of the materials of traditional armor, but they soon become clumsy and too heavy for one person to wear. Only in the mid-'60 engineers developed a safety vest bullet resistant and reliable, you could dress in a certain way, comfortably (Lopes, 2007).

Currently in Brazil, there is a regrettable reality where there is a daily increase of violence in our country and in the world. The ballistic vest is essential to security professionals, are due to the work environment. Therefore, the kind of police, according to the Art 165 of the consolidation of labor laws (CLT) in Brazil, require personal protective equipment appropriate to the risks which they are exposed. Thus the regulatory standard NR-6 about personal protective equipment claims for vigilantes to work legally with fire and police weapons one of the necessary equipment and mandatory is the ballistic vests for protection of the trunk from the risks of mechanical origin, because the region of the body which is usually achieved with greater constancy.

For the men and women who work in the police, protection against threats like bullets, armor-piercing weapons and fire is crucial. That is why it is necessary the use of protective equipment that provides more and more comfort and efficiency for workers, generating greater security. It is large number of professionals in this area that are targeted by gunshots of firearms, numbers show that 82% of the military police officers killed or injured in services are hit by gunshot in the chest. (DPRF, 2016).

However, today, are frequent questions about the anthropometric measurements of the Brazilian man, especially in relation to those used in the development of products ergonomically correct, when there is a database with reliable and representative of the country measures (Silveira and Silva, 2007).

Therefore by necessity of constant use of this equipment, it is necessary a product that has a perfect interaction with the user. Just like any job, if this interaction is not optimized, the constant inappropriate use, in the case of vests, just getting in the way of body movement, decreasing its usability and comfort, affecting the well-being and health of the worker (Alves, 2011).

Consequently, since the advance of garment production on an industrial scale, it is intended to provide a plausible adequacy to the user's body. For this to happen, it is inevitable that attention is directed to the characteristics of the body, the differences in the forms, in relation to different populations and in between groups of the same population (Heinrich et al., 2008).

The Science study of those responsible for body peculiarities, the anthropometry has a special importance, to the knowledge of the physical dimensions of the man with accuracy, happens to be the fundamental basis for ergonomic applications, because the main goal, when designing products, is to be focused on safety, comfort and satisfaction of your target audience. Soon, the anthropometric data are the main parameters projetct for the dimensioning of ergonomically correct products.

The adaptation of work to the human being has been seen by Ergonomics based on physical media, cognitive, psychosocial and environmental. The organization of human labor has also been considered as one of the elements in the analysis and in ergonomic design, involving, however, broader issues, not limited to the scope of the job or task. To perform a job, the individual is inserted in a broader social context, and the representation that this work has impact socially in your life and well-being, and may lead to situations of greater or lesser wear (Da Silva and Ney, 2006).

In this context, understanding that the ballistic protective equipment do not present any criterion for the comfort of the police officer, the present work aims to analyze ergonomic aspects of Baltic vests, seeking to understand the relation user-vest, with the 5° battalion of military police-BOPE city of Campina Grande, state of Paraiba, Brazil. With some data about the material as well as an entire historical context and evolution in the use of equipment of ballistic protection, look for to understand what needs of cops and functionalities of the vests are being met, as well as understand what are the restrictions for which pass in the use of the artifact. Promptly wanted to see user interactivity with the gadgets and artifacts displayed on the tactical vest.

2 MATERIALS AND METHODS

The methodology consisted in a search at the military police battalion 5-BOPE, to obtaining data were carried out direct observations in the workplace of the officers, as well as structured interviews were applied, assisted by questionnaires with open and closed questions. It was the application of the questionnaire with 40 police officers (all men), as well as an interview with the captain in responsible of the institution. The interviews were conducted with the help of questionnaires to understand the relationship between users and vests, in order to obtain statistical data on police satisfaction regarding ballistic vests.

Respondents were approached to express their opinion regarding the following aspects: Comfort, comfort Degree the weight, heat during use; Size available is adapting to the body measurements of users, size of vest available interferes the handling of arms; Parts of the vest where the measures are uncomfortable; The vest is detrimental to the performance of their military activities; The use of the vest helps to feel fatigued at the end of the work shift; Degree of satisfaction with regard to hygiene/cleaning and conservation of ballistic vests and Police resistance level as for continuous use.

Drawing on the literature review of this work, in which Iida (2005), in congruence with other authors, discusses the "feeling" about the comfort, usability and fatigue in respect of a product must be expressed by the user, in this research we used subjective measures that represent variables categorized nominal (Yes and no) and ordinals in satisfaction, as for example: excellent; good; regular; bad; that best indicate the degree of comfort of product researched the weight, heat, measures, model, fitness activity, among other things.

3 RESULTS AND DISCUSSIONS

According to data collected with the military police, it was observed in the sample under study, a greater prevalence of individuals aged in between 30–35; 35–40 and 40–45 years representing respectively the 22.5%; 35% and 18% of the sample, which totals 75.0%. A minority presented 25–30 years age, the absolute majority of policies interviewed is an adult population, with age between 30 to 45 years, and less expressive, the cash account with cops' youth under 30 years and in another minority with more than 45 years.

Second to the data acquired, it was evidenced that the majority of subjects (all male) has between 11 to 20 years for services rendered to the Corporation, which, undoubtedly, would be a considerable amount of time in speaking of a common worker within an organization. In other words, much of the sample interviewed use the bulletproof vest, there are at least more than 5 years, while the large minority use the equipment of protection for less than five years.

Table 1 below shows the data obtained through inquiries made with respondents. It was possible to observe that with regards to comfort, weight, warmth, hygiene/cleaning and the measures of ballistic vests, most cops frown on the vest, whereas the same bad, regular or bad. And the huge minority considered the EPI as excellent or good. The major

Tabel 1. Reviews of police respondents (%).

Questioning	Excellent	Good	Bad	Regular	Terrible
Comfort	5.0	13.0	37.5	25.0	19.5
Weight	2.5	10.0	42.5	25.0	20.0
Heat	3.0	5.0	17.0	30.0	45.0
Measures	5.0	15.0	42.5	20.0	17.5
Hygiene/ Cleaning	0.0	4.0	18.4	29.0	49.0

complaints regarding the ballistic plates were they cause discomfort, heat and acquire bad odor over time, causing inconvenience to workers. The heat and odor obtaining the highest rates of complaint.

They were also questioned about the adequacy of the size of ballistic vests to your body measurements, and it was found that 77.5%, therefore, the vast majority considers satisfactory. Sought in this matter, between the frontal height measure options, time later, bust width, width of the waist, neckline and caves of ballistic vest, assess which were considered as uncomfortable in the opinion of the users and that, among the respondents indicate the width of the waist and the frontal height as the more uncomfortable with 30 and 27.5% respectively, followed later with height 22.5%. For the case of a sample space, only male great dissatisfaction not observed with respect to measures of the bust. According to respondents, the size of the vests made available to cops do not hinder the handling of arms.

In addition, it was found that the frequency of diagnoses of potentially related health as the use of ballistic vest among the officers. Although with this issue cannot establish a causal relationship between the disease and the use of the vest, tried to establish the user's impression regarding to the potential of the use of the vest to cause any health problems. A minority of 37.5% who responded that they had health problems due to their use of the vest, while 62.5%, the vast majority responded negatively. However, on this question, we can identify that the amount of people that are acquiring health problems caused by the use of the vests is already considered great, proving thus the harm brought by ballistic vests to employees. Where it is concluded that for various aspects the ballistic vests bring fatigue for workers, with a weight of these aspects, generating disruptions in worker's life.

4 CONCLUSION

According to the results obtained, the ergonomic analysis of the ballistic protection vest is little explored by Brazilian specialists who know the subject, but also there is a great lack of information.

The participation of the military police of the State of Paraiba was essential in this work, since the results obtained show that it is necessary to have a readjustment of ergonomic ballistic vest, to make them more compatible with the actual daily needs, not only for protection, but also for the degree of comfort.

The measure of the levels of dissatisfaction on the part of the military police. In relation to the ballistic vest, it was noted that the items that displease the most are: weight, thermal comfort, front height, waist width, height, so I suggest to the manufacturers that with the aid of technological advancement, presenting models of vest that satisfies the degree of comfort of users such as: ballistic vests lighter, less hot, more flexible, noting also the design of Ballistic Panel as: height, width, front waist and chest and cava. Taking into account that a fraction of 37.5% of the officers pointed out the use of ballistic vest as a determinant of health problems, so it is necessary to seek measures to neutralize this result. These variables are very important for the good performance of the policies, avoiding fatigue, loss of mobility and even chronic health problems.

It is concluded that the objectives were achieved by means of the methodology applied, proving what the needs of cops and functionalities of the vests are being satisfied, by means of an ergonomic approach with user participation and by inference their opinion, your experience, your complaints about comfort, usability and fatigue in relation to the product, establishing the ergonomics as favorable element for the design of comfortable and pleasurable products, providing thus positive changes in police scenario through ergonomic principles, giving importance to that profession.

REFERENCES

Alves, H. A. analysis of anthropometric parameters of the head of the brazilian air force's military in the design of ballistic helmets. Brazilian Magazine of Biometrics, p. 472–492, 2011. 1983-ISSN 0823.

Da Silva, E. C. A.; Ney, A. F. V. The organization of work and its transformations, 2006.

DPRF—federal highway police department, 2005. Available in (Http://WWW.dprf.gov.br/coletes) 7/1/2016 Access 10:15 pm.

Heinrich, D. P. Design in Palermo University of Palermo July 2008 ergonomics and Anthropometry applied to clothing—analytical discussion about the impacts on the comfort and quality of the products. 2008.

Iida, I. Ergonomics: design and production. 2nd edition revised and expanded São Paulo: E: Blücher 2005.

Lopes, C. M.; Gonçalves, D. P.; De Melo, F. Ballistic resistance of laminated polymer composites. Signal, v. 75, n. 9, p. 0, 2007.

Silveira, I.; Silva, G. G. Anthropometric measurements and clothing design. 3rd Colloquium of Fashion-Belo Horizonte-cd Anais-2007, 2007.

Facilitators for safety of visually impaired on the displacements in external environments: A systematic review

E.R. Araújo
Faculdade de Engenharia da Universidade do Porto, Porto, Portugal

C. Rodrigues
CITTA—Centro de Investigação do Território, Transportes e Ambiente, Faculdade de Engenharia da Universidade do Porto, Porto, Portugal

L.B. Martins
Universidade Federal de Pernambuco, Recife, Brasil

ABSTRACT: Data supplied by the World Health Organization (WHO) show an increasing number of visually impaired in all the world, in other words, unable to perform a simple displacement for day to day activities with safety and comfort. This study analyzes approaches to support the visually impaired in their mobility in the urban environment. Through a systematic review, following the methodology Prisma, in the data sources: PubMed, Nature.com, and Google Academic. In the studies performed in this review 16 records were included, where it was observed that several assistive technologies are being used to facilitate independent and safe displacement of people with visual impairment, exploring their tactile and auditory capacities. Thus, efforts to improve the performance and soften difficulties of movement of people with visual impairment, which affect their conditions of independence and citizenship, is still a challenge.

1 INTRODUCTION

According to the World Health Organization (WHO) there is a possibility that in 2020 the world reach a value of 225 million people with low sight and 75 million people will be blind, in other words, unable to perform among many other activities, a simple displacement for day-to-day activities.

Walking is a basic necessity exercised daily by all people, and the sight supplies information about the environment, which are cognitively interpreted helping to mold the way pedestrians evolve in the space (Miguel, 2013).

Thus, walking, that is considered a safe, sustainable, accessible and economic form of displacement, is an important factor for an independent life which becomes hampered, when the individual does not present an adequate sight or is in a more critic state of total loss of sight.

Studies have identified that blind people or people with impaired vision are a group vulnerable to risks of falling and slipping, due to cognitive and sensorial deterioration. Besides, the lack of visual information limits their movements and hampers their daily life at work, socially or even in cultural activities (Tange, Takeno, & Hori, 2015).

Equally, for moving safely and independently, visually impaired need mobility skills and direction. This direction requires the interpretation of the location of people and things around them in such a way they can collect and evaluate information for a safe and efficient displacement, with a full extent of possible information designing a mind map of that situation for a well-succeeded displacement (Nunokawa & Ino, 2010).

Studies point a series of approaches to help visually impaired people. Electronic aids (ETA) aimed to acquire information from the environment and improve the mobility of the visually impaired are being increasingly researched and improved (Tapu, Mocanu, & Tapu, 2015; Ghilardi, Macedo, & Manssour, 2016).

Therefore, with these studies, it is intended to synthetize pertinent knowledge and methods used, giving opportunities to developments of projects, new routes of investigation and answering the question: What kind of technologies and materials are being studied to provide more safety and comfort to visually impaired in their walking displacements, in the urban environment?

2 MATERIALS AND METHODS

Systematic searches were performed, according to requirements of the methodology PRISMA, using

Table 1. Aspects approached and concepts.

Item	Concepts
Human march	Walking and running activity of the pedestrian.
Assistive Technology	Systems, services, devices or equipment developed to facilitate daily activities of people with deficiencies, on a safe and comfortable way.
White Cane	They are straight and long, may be foldable. They have aluminum, fiberglass or carbon fiber material.
Mind Map	Corresponds to memorization, organization and representation of information.
Ultrasonic Sensors	Characterized by high performance in the detection of objects or mediation of level with milimetric precision.
ETA (Eletronic Travel Aids)	Devices projected to substitute the system of view by sensors with the objective of obtaining information of the environment.

Table 2. Criteria of inclusion and exclusion.

Criteria		Atributtes	Justification
Problem	Inclusion	Facilitators of mobility of visually impaired.	Focused on the safe and comfortable mobility of the visually impaired.
	Exclusion	Mobility in internal environments.	The study focus on the circulation in the public space.
Intervention	Inclusion	Electronic systems and materials.	Work as environmental perceptors in the place of the sight.
	Exclusion	Recommendations linked to the pathologies	Does not fit in the context
Results	Inclusion	Options with low costs, light and of Easy interaction.	Attends a great percentage of users in the society.
	Exclusion	Options with high costs and difficult interaction.	Does not attend most of the population
Studies	Inclusion	Scientific Studies.	Research of public interest.
	Exclusion	Systematic Reviews.	Secondary Study

resources "Data Base" and "Scientific Magazines" from the SDI-FEUP (Porto University) with the key words: *accessibility, sidewalks, visual impaired, assistive technology, walk, adaptations, displacement, sensory impairment*, permuted in twos. Articles who did not inform the author were not considered either. The electronic search was limited to a period of 10 years, performed from September to October 2017, which articles were raised from indexed publications in PubMed, Nature.com, and Google Academic, presented in the references of articles obtained in the review. Initially, articles dated from 2007 to 2017 were selected, since they contained more updated data for the study, and those from previous years, that were repeated and without access were disregarded. In this ambit, the following concepts were analyzed, according to Table 1.

A first screening was performed based on the titles and abstracts, where the articles selected were examined integrally, with the following criteria: problem, intervention, results and study (Table 2).

These criteria were used to identify what kind of technology and physical elements are being used to improve the circulation of pedestrian in the urban environment providing security and comfort in their activity of displacement.

3 RESULTS AND DISCUSSION

With a total of 18,169 articles selected, four articles were added from quotes in articles collected, totaling 18,173 articles. 1,262 duplicates were discarded, 4,941 for being years before 2008 and 6,640 did not identify the year nor the author. Thus 5,327 studies were screened, being excluded 5,285 by the title, 15 by the abstract, 6 excluded for the complete text and 6 did not have access permitted. Thus, 16 articles had complete text and were analyzed qualitatively, as shown in Picture 1.

The White Cane, known as white cane is the basic equipment and most used by visually impaired, followed by trained dogs to obtain spatial information, and moving with more safety. However, this facility is very limited, as it does not supply detailed information, such as speed, static or dynamic nature, such as distance and time collision, over the detected obstacle (Johnson & Higgins, 2006). Today with the simplicity of monitoring through a white cane, the visually impaired face many problems in the detection of obstacles or fall in platforms (Tange, Takeno, & Hori, 2015).

Strategies and electronic devices for navigation will improve the life of visually impaired users preventing or minimizing risks and lesions and avoiding dangerous expositions in their daily displacements, thereby providing, benefits associated to walking, with a more independent and safe life(Tapu, Mocanu, & Tapu, 2015).

In Tokio (Japan), a group of blind individuals, with earphones and afterwards without the device, participated of a study with the objective

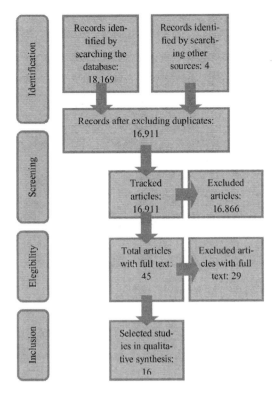

Picture 1. Flowchart of the methodological sequence of the review.

of checking a sensibility of touch produced when using white cane in panels of rubber to stimulate its hardness and comparing the sensation when this touch is produced with the finger. The author emphasizes that the physical characteristics of the white canes have the capacity information of supplying information on the form of vibration to the users, when handled with the technique of two touches, facilitating thus the displacement. The results of the experiment showed that when the cane accompanied of auditory information was used, the sensibility was similar to the touch of a finger in the material to estimate its hardness (Nunokawa & Ino, 2010).

Another element very used in the urban spaces are the tactile floors. Although researches have identified that these floors produce a positive effect when it is intended to direct the way of a visually impaired person, results of other research showed that this kind of pavement reduced the stability of the march (Pluijter, 2015). Thus, their works correlated this kind of floors as potential restriction to the risk of falling. It is also emphasized (Kobayashi, Takashima, Hayashi, & Fu, 2005) that it is interesting a review in the project of the tactile floors nowadays, considering that walking over an irregular surface affects the stability of march also for people with normal sights, specially elderly, wheelchair users and those with strollers, besides a review over its sizing (300 mm), which presents a narrow width to be used by two feet to walk.

The use of guide dogs as auxiliary of locomotion of visually impaired and blind people has a fundamental importance, involving psychological and social questions, in the quality of life of their users, once it makes possible some independence of displacements, it also promotes additional benefits of friendship, companionship, bigger socialization, besides this improvement of self-esteem and bigger confidence. On the other hand, great part of the population who needs this support cannot afford to have one due to the high cost of training and maintenance (Lloyd, La Grow, & stafford, 2016; Wiggett-Barnard & Steel, 2009; Whitmarsh, 2005).

However, researchers have worried increasingly with the development of technologies that facilitate the safe displacement of visually impaired. The Electronic Travel Aid (ETAs), which is a device that allows the feedback between the user and the existing obstacles in their path through the vibrational sound notifications, among others were employed in a research work with help of a tactile screen with answer mechanism, electromagnetic breaks, sensors and a control unit. The detection of obstacles performed along the path taken through six ultrasonic sensors connected to six vibrators put into different directions and the information received through a tactile screen with vibrators that stimulate the hand of the user, according to the location of the obstacles. All this performed with a guide vehicle where the sensors are located. Interestingly is another sensor that when detecting an unevenness the electromagnetic brake is activated avoiding the continuity of the path by the user. All this process of development uses the software LabVIEW (Jeong & Yu, 2016).

In the same way, using the technique of detection of static and dynamic obstacles in internal spaces or urban spaces with the use of sensors and technologies of computational sight, visually impaired people tested a system. The device is attached to the waistline of the user like a belt and the system is formed by a smartphone, ultrasonic sensors, Bluetooth, microcontrollers and earphones that inform through acoustic signals the presence of obstacles that are detected by ultrasonic sensors, cameras and a smartphone. The performance of the system was tested in 21 visually impaired aged between 27 and 67, from whom the youngsters were more audacious while the older ones were afraid of trusting the acoustic signals for their displacement feeling the necessity of combined use with the white cane (Mocanu, Tapu, & Zaharia, 2016).

An experiment in laboratory was developed for people with low sight, simulating crossing a street, using a voice guide and graphic floors that represented 4 private situations such as: gates, exit north, train station and exit south. The results of the experiment emphasize the importance of the use of signalization on the floors, as they justify that the visually impaired tend to walk looking down to the floor that many times is unnoticed by the suspended signalization (Omori, Yanagihara, & Kitagawa, 2015).

A team conducted a transversal study showing that with the help of a white cane blind people had the possibility of distinguishing the contrast of sounds produced by the contact of this cane in different materials and that this feature could be used for directing of these people in their walking displacements, according to *the technique of touch of two points* used. For the authors, the different types of materials could be used in the form of footbridges, and other materials for the pavement would be used working as a base. The tests were performed with two distinct groups. One with 74 people the other with 73. The result of the research shows that the choice of the material to be used in the urban environment must be taken into consideration the background noise, because the more you must use materials that generate levels of higher sounds in high frequency (Secchi, Lauria, & Cellai, 2017).

The Table 3 summarizes the main variables mentioned in the articles, which contributes to make the walking spaces safe for the visually impaired.

The detection of obstacles through sensors connected to a smartphone, more a distance sensor and orientation was tested in an experiment. In this system, the information is sent to the smartphone via Bluetooth. When the obstacle is detected, there is a sound and vibrating alert. (Tange, Takeno, & Hori, 2015).

Considering this context, other work with automatic detection of surface of tactile paving was performed using algorithm of computer combined with decision tree. Thus, it is possible to detect images of tactile surfaces, even before touching them, with a camera connected to the abdomen of the user and a return of audio. 521 images of several pavements with different conditions of illumination were reunited and stored in a form that every pavement detected has associated image. According to the author the system presented a level of 88,48% of precision (Ghilardi, Macedo, & Manssour, 2016).

Based on the proposal approached in the cyber physical system (integrated system that deals with computation, networks and physical processes) a device that is attached to the head of the user was developed. The system uses an ultrasound sensor, a video camera, an intelligent movable device and the aid of a white cane. The premise of this device is to supply the lack of detection of objects suspended above the waistline, not detectable by the white cane (Cheng, 2016).

Table 3. Variables used to help visually impaired.

Variables	Johnson, 2006	Whitmarsh, 2005	Wiggett-Barnard, 2009	Lloyd, 2016	Secchi, 2017	Nunokawa, 2010	Tange, 2015	Jeong, 2016	Omori, 2015	Rheede, 2015	Maidenbaum, 2016	Cheng, 2016	Mocanu, 2016	Ghilardi, 2016	Kobayaski, 2005	Pluijter, 2015
Tactile ground surface									x							
Ultrasonic Sensors						x	x	x					x	x		
Distance Sensor							x									
Camera Video	x										x		x	x	x	
Intelligent moving device								x					x	x	x	x
Informatic application	x											x	x	x	x	
Tactile screen								x								
Glasses											x					
Ear phones										x	x					
Audio Signals																x
Tactile paving						x	x							x	x	x
Auditory track							x	x					x			
White Cane	x	x	x	x	x								x			
Microphones							x									
Guide Vehicle									x							
Index finger									x							
Guide dogs	x	x	x													

4 CONCLUSIONS

The studies showed that there is a great variety of devices and others are being tested to promote a satisfactory performance of help to the visually impaired, for a safe and comfortable displacement. It is seen that there is still a little unsafety for the use of technology assisted by sensors and computers being the white cane yet an element used by most of visually impaired.

It is observed that for a bigger adaptation of this type of supporting the systems must have a better facility of comprehension for a fast and efficient interaction of the users, besides presentation levity and simplicity in its presentation. The ETAs that use sensors and do not use earphones seem to be interesting, once they do not block the auditory via releasing the perception of possible environmental sounds.

It is important to remember that a great part of the population that has visual impairment presents big difficulty of socialization and difficulty to get a job and consequently have low purchasing power. To attend a big percentage of this population it is necessary that the cost for purchasing this equipment with, technological advance have an accessible cost. On the other hand, it will only be available to a small portion of the population.

REFERENCES

Cheng, P.-H. (2016). Wearable ultrasonic guiding device with white cane for the visually impaired: a preliminary verisimilitude experiment. pp. 127–136.

Ghilardi, M., Macedo, R., & Manssour, I. H. (2016). A New Approach for Automatic Detection of Tactile Paving Surfaces in Sidewalks. *Procedia Computer Science, 80*, pp. 662–672.

Jeong, G.-Y., & Yu, K.-H. (2016). Multi-section sensing and vibrotactile perception for walking guide of visually impaired person. *Sensors, 16 (7),1070.*

Johnson, L. A., & Higgins, C. M. (2006). A navigation aid for the blind using tactile-visual sensory substitution. *Conf Proc IEEE Eng Med Biol Soc.*.

Kobayashi, Y., Takashima, T., Hayashi, M., & Fu, H. (2005). Gait Analysis of People Walking on Tactile Ground Surface Indicators. *Neural Systems and Rehabilitation Engineering. IEEE Transactions on 13*(1), pp. 53–59.

Lloyd, J., La Grow, S., & stafford, K. (2016). An Investigation of the Complexities of Successful and Unsuccessful Guide Dog Matching and Partnerships. *3*, pp. Lloyd J, Budge C, La Grow S, Stafford K. An Investigation of the Complexities of Successful and Unsuccessful Guide Dog Matching an114. doi:10.3389/fvets.2016.00114.

Miguel, A. F. (2013). The emergence of design in pedestrian dynamics: Locomotion, self-organization, walking paths and constructal law. *Physics of life reviews, 10*, pp. 168–190.

Mocanu, B., Tapu, R., & Zaharia, T. (2016). When ultrasonic sensors and computer vision join forces for efficient obstacle detection and recognition sensors. *Sensors (Basel)*.

Nunokawa, K., & Ino, S. (2010). An experimental study on target recognition using white canes. *Engineering in Medicine and Biology Society (EMBC), Annual International Conference of the IEEE*.

Omori, K., Yanagihara, T., & Kitagawa, H. (2015). Validation of Mobility of pedestrian with low vision using graphic floor signs and voice guides. *Stud Health Technol Inform*, pp. 398–404.

Pluijter, N. e. (2015). Tactile sidewalk to guide direction of the walk: an assessment of the position of direction and stability of the gait. *Gait & posture*.

Secchi, S., Lauria, A., & Cellai, G. (2017). Acoust wayfinding: a method to measure the acoustic contrast of different paviment materiais for blind people. *Ergonomia aplicada., 58*, pp. 435–445.

Tange, Y., Takeno, S., & Hori, J. (2015). Development of the obstacle detection system combining orientation sensor of smartphone and distance sensor. *Engineering in Medicine and Biology Society (EMBC), 37th Annual International Conference of the IEEE*.

Tapu, R., Mocanu, B., & Tapu, E. (2015). A survey on wearable devices used to assist the visual impaired user navigation in outdoor environments. *Electronics and Telecommunications (ISETC), 11th International Symposium on*.

Visual impaired and blindness. (s.d.). World Health Organization (WHO).

Whitmarsh, L. (2005). The Benefits of Guide Dog Ownership. (T. &. Francis, Ed.) *Visual Impairment Research, 7*.

Wiggett-Barnard, C., & Steel, H. (2009). The Experience of Owning a Guide Dog. (T. & Francia, Ed.) *Disability and Rahabilitation, 30*(14), pp. 1014–1026.

Impacts of accidents with workers in rock processing industries: Short review

André do Couto & J. Santos Baptista
PROA/LABIOMEP/INEGI, Faculty of Engineering, University of Porto, Porto, Portugal

Beda Barkokébas Jr.
Escola Politécnica da Universidade de Pernambuco—POLI, Recife, Brazil

Inaldo da Silva
Instituto Federal de Educação de Pernambuco—IFPE, Recife, Brazil

ABSTRACT: Natural stone production is a major economic sector in many countries around the world. However, are many parameters of difficult control in the materials handling operations that make it one of the most risky industrial activities. Workers in this sector are exposed to significant risks namely in the activities of unloading and storing plates and blocks, cutting and polishing plates, assembling ornamental pieces and loading the final products. This work aims to study the impacts of the accidents with workers in the rock processing industries and the characterization of its causes and effects in this industrial sector. As base of this research was used PRISMA Statement approach. From the 20,436 articles found, 7 matches the research question and were deeply analyzed. According to this research, occupational accidents are the leading causes of death or serious injury and injuries to the upper limbs are an important part of occupational accidents.

1 INTRODUCTION

The production of natural stone is an important sector in many countries of all the continents (Adams, et al., 2013; Estellita, et al., 2010). However, in its production, a large number of carried out operations may, by their nature, may interfere with the safety of workers resulting in accidents what makes this one of the most dangerous industries.

In this industrial sector, the production starts in stone extraction in quarries and they are always associated with incidents and accidents that result in injury, death and property damage (Yarahmadi, et al., 2014). The problem continues in the transformation operations of the big stone blocks into final products, usually for the construction industry and also in this last industry. Both in the quarries and in blocks' processing plants, the workplaces can be open pit, subject to atmospheric conditions and often worsened by high mountain insulation.

There are also services involving the worker's relationship with vehicles and extremely heavy stones. Adding to all this, the main risks are also often related to inadequate safety management and training of workers (Angotzi, et al., 2005).

Within this sector, the most significant branch is rock processing, which presents the final product processing operations where there is always imminence of serious accidents (Melo Neto, et al., 2012). Workers in this sector are exposed to significantly higher risks of multiple and nonfatal injuries than those in construction industry (Adams, et al., 2013). There are effective risks in the unloading and storage of slabs and blocks, in the cutting and polishing of slabs and in assembly the ornamental pieces (task where the most dangerous conditions are present) beyond the loading of final products (Melo Neto, et al., 2012). An analysis of the consequences of severe and mild accidents in Poland has indicated additional attributes such as the type of equipment operation, lack in technical safety measures and characteristics of injury events. The analyzed lesions generally occurred due to the rotary movement of tools or crushing caused by a closing movement, but also a large number of accidents occurred during support actions, such as adjustment of positioning of materials, removal of pollutants and cleaning (Dźwiarek, et al., 2016). Although some of the accidents can be considered as human errors, a poor design of the workplaces and of the protective materials provided, which are the source of many accidents, are the responsibility of the company and not of the worker (Silva, 2015).

All these situations and problems raise a question: what impacts are caused by the occurrence of accidents with workers in rock processing industries?

So, this review aims to collect information to study the impact of accidents involving workers from the rock processing industries and the characterization of its causes and effects in this industrial sector.

2 METHOD

This bibliographic research was supported by PRISMA Statement methodology. The research was developed between September 15 and 21, 2017. The types of resources chosen were data bases and scientific journals. The keywords chosen were: *"rock industry", quarry, marble, granite, industry, accident, crushing and cutting*, which were combined two by two. Were found 20,436 articles and added 1 article identified through independent research. The electronic search resulted in articles from publications indexed in *Annual Reviews, IOP Journals Nature. com, PsycArtcles and PubMed*. With the help of the MSExcel spreadsheet was created a list with all the articles captured in the research. Filters were used to support the work of classifying articles in relation to the inclusion and exclusion criteria. The articles were classified in alphabetical order by authors and evaluated. Articles considered too old, those that were more than twelve years since publication, were discarded for the possibility of containing outdated data on the subject in the area of engineering. There were sorting to separate the repeated articles, in which 6,913 were selected, filtered and removed from the review. The articles were then submitted to the analysis of titles, abstracts and full texts based on four criteria: problem, intervention, results and study. The attributes of inclusion and exclusion of the articles were duly justified. Analyzing the research question, the problem are the accidents in quarries and industries of benefit of rocks and its justification. Thus, environmental and economic impacts were not included because they were not part of the focus of the study. In the selection were included accidents involving workers, as a factor of great importance in the operational activities related to the industries of rock processing; were excluded risks such as air pollution, noise and mechanical vibrations to avoid harming the dimensional purpose of the data in the survey. In the results, because of the intention to quantify the accidents involving workers within the industrial environment, statistical analyzes of accidents were included, but excluding accidents outside the

Table 1. Inclusion and exclusion articles criteria.

Criteria		Atributes	Justificative
Problem	Inclusion	Accidents in rock processing industries	Disorders caused to this industrial sector
	Exclusion	Environmental and economic impacts	Focus on injuries caused to workers
Intervention	Inclusion	Accidents involving workers	Aspects present in all activities of the rock processing industries.
	Exclusion	Air pollution, noise and vibration of machines.	The approach undermines the dimensional purpose of the research activities.
Results	Inclusion	Statistical analysis of accidents	Quantify accidents involving workers.
	Exclusion	Accidents outside the industry	Study focused on the statistics of accidents at work
Studies	Inclusion	Field and laboratory studies	Scientific support for taking decisions
	Exclusion	Sistematic reviews	Secondary study profile with data collection.

industry. Field and laboratory studies were included due to scientific support and opinion studies, as they are also of public interest. Systematic review studies were excluded because they were secondary studies. Table 1 presents the inclusion and exclusion criteria considered in this selection stage.

To the analysis of the included articles, the used spreadsheet presents, in the headings, the particular topics of the study such as: study focus, source of risk exposure, type of machine or used tool, operation/activity, type of accident, part of the body injured and consequence of the accident. Each topic presented items related to the research subject, for later integration of the articles considered relevant. When an item of interest was found in an article, the information was passed to the worksheet by marking an "X" in the appropriate box for the correlation of that item with the analyzed article.

3 RESULTS

Thus, from the 22 complete texts analyzed, 15 were excluded and 7 were selected in the qualitative synthesis because they are relevant to the study of the impact of accidents with workers in the rock processing industries. Although there are serious and frequent accidents in the industrial field of rock processing, only one article was found dealing specifically with risk analysis in this manufacturing sector.

So, this short review presents 7 articles related to the impact of the occurrence of accidents with the involvement of workers in the rock processing industries. When considering the research question prepared to serve as reference to this research, it is clear that the study intends an approach to the industrial sector of rock processing. However, in the course of the research, only one of the analyzed articles was focused on risks in this specific manufacturing sector, and two articles analyzing accidents within quarry environment.

Therefore the results presented in this review also address to the impacts caused by accidents involving workers from industries that use similar equipment to the ones used in the rock processing industries.

Figure 1 Syntheses the entire methodological sequence of this review.

As relevant articles are presented four studies on the subject proposed in the industries of extraction and processing of ornamental stones and three analysis of injuries or damages that occur quite frequently with quarry workers as shown in Table 2.

These articles addresses not only to the accidents exposure levels of these workers, but also

Table 2. Subject matter of the relevant articles.

Research	Adams et al., 2013	Angotzi et al., 2005	Hou et al., 2008	Hua et al., 2013	Melo Neto et al., 2012	Ozçelik et al., 2012	Yarahmadi et al., 2014
Quarries accident	X	X					X
Risk analysis					X		
Orthopedic injuries			X			X	
Traumatic injuries					X		

the most frequent injuries and damages according to the types of the used tools. Comparisons on the analyses of the presented studies show the most frequent accidents in quarries environment, their causes and consequences. It is also checked the body injured parts and the classification of the level of accidents. In a study focused on preliminary risks analysis applied to the finishing process conducted in two rocks processing plants in the metropolitan region of Recife (Brazil) Melo Neto et al., (2012) verified the presence of physical, chemical and accident risks in this production sector. The areas in which the workers are exposed to worse hazardous conditions are those of cutting, grinding and polishing.

Dźwiarek et al., (2016) presented an analysis of 1035 accidents, 341 of them recorded by Poland's National Labour Inspection (PIP) between 2005 and 2011, in order to formulate principles for the application of security techniques by the equipment users.

The accidents analysis indicate additional attributes such as the type of operation, technical measures for the safety and the type of events that cause injuries.

The lesions analysis occurred, usually due to the rotating movement of tools or crushing due to a closing movement (Table 3), but also a large number of accidents during actions of support, as positioning of materials, removing and cleaning of pollutants.

Work-related accidents are the main causes of death or severe disability (Chan, et al., 2009; Ozcelik, et al., 2012). The results of research conducted on the work in marble quarries in the province of Diyarbakir, found that 42.9 percent of the accidents occurred due to breakage of the cutting wire, 17.8% due to the explosion and 3.6% due to the fall from the top of the benches. Of these acci-

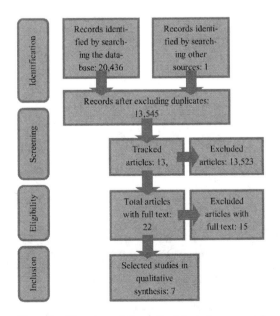

Figure 1. Flowchart of the methodological sequence of the review.

Table 3. Type of harm-causing event.

Executed task	Type of harm-causing event				
	Injury	Injury	Crushing	Cutting	Hitting
Cycle machine	Rotating element	Sharp element	Closing motion	Closing motion	Moving element
Automatic	3	1	8	0	2
Semi-automatic	8	1	18	7	5
Manual operation	186	74	149	51	35

dents, 10.7 percent resulted in death (Yarahmadi, et al., 2014). Other serious accidents were caused by the fall of workers or machinery from the benches, traffic accidents, and electric shock (Statistical center of Iran, 2011; Onder, et al., 2011; Ersoy, 2013; Yarahmadi, et al., 2014). In the area of Massa Carrara, from 1994 to 1998 there were more than 200 accidents per year, of which 10 were too severe or fatal. From 1999 to 2004 in Alta Versilia, there were approximately 20 accidents per year, of which at least 3 were too severe (Angotzi, et al., 2005). Among the serious or fatal accidents, were conferred: fall in height, fall at the same level, projection of stones or materials and projection of equipment parts (Angotzi, et al., 2005). The circumstances stated that work in this industry is a particularly hazardous with respect to the potential for eye injuries, in India, workers expose themselves to significant and multiple risks and non-fatal injuries (Adams, et al., 2013). Adams et al. (2013), accompanied 204 workers from quarries throughout the 6 months and found that the incidence of eye injuries was 18.4% in the group where applied enhanced safety education and 17.8% in the group with standard safety education. It was recorded that 7 (3.4%) of workers (3 in the enhanced group and 4 in the default group) suffered serious injury in eyes while cutting stone process. Upper limbs injuries related to work are the more frequent occurrence what constitute an important proportion between the occupational accidents. The hand performs most of the industrial activities and it is, therefore, the part of the body with more accidents, from 35.3% to 53.1% of occupational injuries. The consequences range from simple skin lacerations to amputations (Ozcelik, et al., 2012). Ozçelik et al. (2012) conducted a retrospective analysis of operated patients between 2005 and 2007 in Turkey and noted that of 4120 upper extremities lesions, 2188 (53.1%) of them were professional injuries. The study analyzed the patterns of injuries resulting from various mechanisms, and found that the two most common mechanisms were crushing-compression and crushing-cut. Accidents involving crushing of hands occurred in greater quantity; totaled 744 (34%), where a

Table 4. Cause of injuries and injury sites.

CAUSE OF INJURIES	N°	%
Traffic accident	78	50,6
Crashing injury	17	11,0
Crushing injury	17	11,0
Penetrating injury	8	5.2
Fall from height	11	7.1
Level fall	17	11,0
INJURY SITES	**N°**	**%**
Major joint or long bone fracture of lower limbs	54	35.1
Foot fracture or soft tissue injury of lower limbs	14	9.1
Major joint or long bone fracture of upper limbs	38	24,7
Hand fracture or soft tissue injury of upper limbs	37	24,0

portion involved workers with stones and marbles. As for the consequences, the two most common types were hamstring injuries + vessels+ nerve and bone amputations. Hou et al. (2008) studied 154 patients with diagnosis of lesions of upper and/or lower limbs, hospitalized in orthopedic and plastic surgery rooms in two University hospitals of Taiwan, between 2004 and 2005 and assessed the injuries resulting from some mechanisms, their causes and consequences (Table 4).

Hua et al. (2013) conducted another study, comprising 561 cases between 2001 and 2010, about the causes of traumatic lesions of the spinal cord. The severity of injury classifications were based on the standards of the American Spinal Injury Association and International Spinal Cord Society (2006), which specify complete injury as the full transition of the spinal cord (total nerve damage below the level of the injury) and incomplete injury due to partial transition of the spinal cord (incomplete damage of nerves below the level of the lesion). Therefore, it was found that 10 (1.8%) of the cases were caused by crushing with 4 incomplete injury consequences (1.5%) and 6 injuries (2.2%). Falls from a height of at least 1 m above

Table 5. Operations performed by the quarry workers.

OPERATIONS	Adams et al., 2013	Angotzi et al., 2005	Melo Neto et al., 2012	Yarahmadi et al., 2014
Stone Cutting	X	X	X	
Drilling			X	
Transport of blocks			X	
Transportation of waste			X	
Sharpening			X	
Polishing in plates			X	
Assembly of ornamental pieces			X	
Diamond wire cutting machines				X

Table 6. Percentages of the lesions studied in the selected articles.

INJURIES	Adams et al., 2013	Angotzi et al., 2005	Hua et al., 2013	Ozçelik et al., 2012	Yarahmadi et al., 2014
Amputation				17%	
Bank-crest decline					3,6%
Complete damage				49,9%	
Crushing				1,8%	
Crushing compression				34%	
Crushing-cut				23,49%	
Cutting wire breaking					42,9%
Death		5,9%			10,7%
Deep crush				16,13%	
Deep-cut				21,52%	
Explosion					17,8%
Eye damage	18%; 9%				
Fall ground level		X	8%		
Fall of height		X	23,9%		
Fracture				10,96%	
Fragments projection	X	X			
Incomplete damage				50,1%	
Nail base daamage				12,56%	
Pieces projection		X			
Severe eye damage	3,4%				
Tendon laceratio + fracture				12,15%	
Tendon + vessel + nerve + fracture				17,41%	

the ground were 134 (23.9%), leading to incomplete injuries 64 (23.9%) and 70 injuries (25.6%). The falls at the same level were 45 (8%), leading to incomplete injuries 25 (9.3%) and 20 injuries (7.3%).

Table 5 shows the operations performed by the quarry workers that were identified in the selected articles and which suggest potential risks for the occurrence of the lesions. The percentages of lesions studied in these articles are presented in Table 6.

4 DISCUSSION AND CONCLUSIONS

The carried analysis of industrial accidents showed that the basic cause of accidents is due to mechanical hazards and, regarding the nature of the activity, the workers in this industry are particularly exposed to risks, generally aggravated by the lack of adequate facilities, organization, healthy and care in their working places. The safety assessment was found to be the major sources of risk, such as traffic accidents, cutting blade or wire rupture and rock fall. Crushes occurrence caused by closing movements and other injuries occurrence due to the rotating movement of tools or in the course of supporting actions such as removing pollutants, correcting the position of the material and cleaning. The high rate of incidents suggests that risk analysis and classification are necessary and unavoidable. Application of additional safety measures should follow the analysis of aspects related to the equipment and their capabilities provided by the design of the workplaces.

REFERENCES

Adams, J. S., Raju, R., Solomon, V., Samuel, P., Dutta, A. K., Rose, J. S., et al. (2013). Increasing compliance with protective eyewear to reduce ocular injuries in stone-quarry workers in Tamil Nadu, India: a pragmatic, cluster randomised trial of a single education session versus an enhanced education package delivered over six months. *Injury*, *44* (1), pp. 118–125.

American Spinal Injury Association and International Spinal Cord Society. (2006). International Standards for Neurological Classification of Spinal Cord Injury. *Chicago, IL*.

Angotzi, G., Bramanti, L., Tavarini, D., Gragnani, M., Cassidoro, L., Moriconi, L., et al. (2005). World at work: marble quarrying in Tuscany. *BMJ Journals*, *62*.

Chan, J. C., Ong, J. C., Avalos, G., Regan, P. J., McCann, J., Groarke, A., et al. (2009). Illness representations in patients with hand injury. *J Plast Reconstr Aesthet Surg*, *62*, pp. 927–932.

Dźwiarek, M., & Latała, A. (2016). Analysis of occupational accidents: prevention through the use of additional technical safety measures for machinery. *International Journal of Occupational Safety and Ergonomics*, 22 (2), pp. 186–192.

Ersoy, M. (2013). The role of occupational safety measures on reducing accidents in marble quarries of Iscehisar region. *Safety Science*, 57, pp. 293–302.

Estellita, L. d., Santos, A. M., Anjos, R. M., Yoshimura, E. M., & Hugo Velasco, A. A. (2010). Analysis and risk estimates to workers of Brazilian granitic industries and sandblasters exposed to respirable crystalline silica and natural radionuclides. *Radiation Measurements*, 45, pp. 196–203.

Hou, W.H., Tsauo, J.-Y., Lin, C.H., Liang, H.W., & Du, C.L. (2008). Worker's compensation and return-to-work following orthopaedic injury to extremities. *J Rehabil Med*, 40 (6), pp. 440–445.

Hua, R., Shi, J., Wang, X., Yang, J., et al. (2013). Analysis of the causes and types of traumatic spinal cord injury based on 561 cases in China from 2001 to 2010. pp. 218–221.

Melo Neto, R. P., & Rabbani, E. R. (2012). Application of preliminary risk analysis at marble finishing plants in Recife's metropolitan área. *Work: IOS Press Contente Library*, 41 (supplement 1), pp. 5853–5855.

Onder, S., Suner, N., & Onder, M. (2011). Investigation of occupational accident occurred at mining sector by using risk assessment decision matrix. *In: Turkey 22th International Mining Congress and Exhibition*, pp. 399–406.

Ozcelik, B., Ertürer, E., Mersa, B., Purisa, H., Sezer, İ., Tunçer, S., et al. (2012). An alternative classification of occupational hand injuries based on etiologic mechanisms: the ECOHI classification. *Turkish Journal of Trauma and Emergency Surgery*, 18 (1), pp. 49–54.

Silva, I. A. (2015). Gerenciamento de Riscos: construindo uma empresa mais segura. *Livro Rápido*, p. 180 p.

Statistical Center of Iran. (2011). Survey results of Iran's active mines in 2011.

Yarahmadi, R., Bagherpour, R., & Khademian, A. (2014). Safety risk assessment of Iran's dimension stone quarries (exploited by diamond wire cutting method). *Safety Science*, 63, pp. 146–150.

Whole-body vibration in mining equipment—a short review

Joana Duarte, M. Luísa Matos, J. Castelo Branco & J. Santos Baptista
PROA/LABIOMEP/INEGI, Faculty of Engineering, University of Porto, Porto, Portugal

ABSTRACT: Whole body vibration is one of the most common occupational hazards in the mining industry due to the use of heavy earth-moving equipment and other machinery. A systematic review was carried in order to determine which equipment contribute the most to the operator's occupational exposure, using for that a combination of seven keywords. The main results found showed that compactors were the equipment with higher acceleration values, followed by blast-hole machines and load-haul-dumpers. Despite the primary results, studies should be performed in order to better understand this issue, including variables such as performed task, equipment velocity, among others.

1 INTRODUCTION

Whole Body Vibration (WBV) can be defined as the "vibration transmitted to a person's entire body via his/her contact with a vibration source" (Derek & Peter, 2005) and it is an important issue for the mining industry, as it occurs in many mining operations (Chaudhary, Bhattacherjee, & Patra, 2015). The vibration phenomenon is related with the Resonant Frequency (RF) which is the speed at which a given object naturally vibrates. Although it is intrinsic to each object, the human body does not have a particulate RF, because the different parts of the body varies in terms of mass and density, leading to diverse vibrating frequencies. Nonetheless, the range between 0.5 and 80 Hz has been suggested as having "significant effects on the human body" (Derek & Peter, 2005). According with the same author, vertical vibration sets the most important RF between 4 and 8 Hz: strong resonances in the neck occur between 3 and 5 Hz, while for the spine is from 4 to 7 Hz.

Heavy equipment such as dozers, scrapers, shovels, haul trucks, loaders and load haul dump vehicles are significant vibrating sources in the mining industry (Chaudhary et al., 2015), which enter the body through the contact with the seat, backrest and via the floor (Kumar, Kumaraswamidhas, & Murthy, 2016).

WBV is also dependent on other parameters such as the vibration magnitude, direction and frequency, duration and distribution of motion within the body (Derek & Peter, 2005) as well as environmental conditions, and factors associated with the machinery itself such vehicle conditions, maintenance, seat type and material, speed, cab layout, position and design (Kumar et al., 2016).

Long term exposure to this type of vibration is a well-known risk factor for degenerative changes in the spine and subsequent back pain (Wolfgang & Burgess-Limerick, 2014). Other consequences can be muscle fatigue, nausea, discomfort, nervous system dysfunction, spinal degeneration, disturbed sleep and gastrointestinal tract problems (Bovenzi & Hulshof, 1998; Derek & Peter, 2005; Tammy Eger et al., 2014).

The purpose of this short review was to analyse which equipment contribute the most to occupational vibration exposure in the mining industry and to understand whether the exposure values were within the range suggested in the Directive 2002/44/EC.

2 MATERIALS AND METHODS

In order to carry out this short review, Preferred Reporting Items for Systematic Reviews and Meta-Analysis (PRISMA Statement) was used.

The research included some of the main engineering databases such as Web of Knowledge (Current Contents, Inspec and Web of Science), Scopus and Academic Search Complete. Scientific journals were screened as well in: Annual Reviews, Directory of Open Access Journals, Elsevier (Science Direct), IEEE Xplore, Emerald and Taylor and Francis. In the occupational health area, PubMed and Medline were also considered.

The keywords set to conduct the study were the combinations "occupational vibration" AND "mining", "occupational vibration" AND "extractive industry", "occupational vibration" AND "mining equipment", "occupational vibration" AND "open cast", "occupational vibration" AND "of the road" as well as "whole body vibration" AND "mining", "whole body vibration" AND "extractive industry", "whole body vibration"

AND "mining equipment", "whole body vibration" AND "open cast", "whole body vibration" AND "of the road". The search fields were "title, abstract, keywords".

As screening criteria were considered only articles and articles in press, published between January 1990 and September 2017 (in order to consider older equipment that is still in use in the mining industry), experimental studies published in journals and written in English.

3 RESULTS

From a total of 1470 identified articles, 252 papers were excluded by date.

The second exclusion criteria applied was the type of paper: only experimental studies were considered, excluding 260 more articles. The paper source was also taken into consideration, leading 22 more articles out of the study. Papers not written in English were also disregarded, which lead to the rejection of 34. Duplicates (223 papers) and works without full-text availability (8 articles) were removed. After reading the title and the abstract, 629 articles that were not in accordance with the proposed objective were excluded.

After this process, 42 papers were considered eligible and were full-text screened in order to determine the whole body vibration produced by various equipment used in the mining industry. The papers included in this study had to perform WBV measures in order to do an accurate analysis; 10 papers met all the criteria, Figure 1.

The 10 papers included in the study mentioned WBV measurement, in the mining industry.

With the exception of one of the considered papers (Leduc, Eger, Godwin, Dickey, & House, 2011) which measured vibration via the feet, acceleration levels were measured on the operator's seat equipment using for that tri-axial accelerometers. From the 10 studies included, only 6 presented the measurements using both A(8) and VDV(8) (T Eger, Stevenson, Boileau, Salmoni, & VibRG (2008); Smets, Eger, & Grenier (2010); Tammy Eger, Stevenson, Grenier, Boileau, & Smets (2011); Chaudhary et al. (2015); Burstrom, Hyvarinen, Johnsen, & Pettersson (2016) and Marin et al. (2017)), 2 of them presented the values from which it was possible to do the calculations (Aye & Heyns, 2011; Mandal & Srivastava, 2010), 1 of them (Leduc et al., 2011) just showed A(8) results and 1 of them (Vanerkar, Kulkarni, Zade, & Kamavisdar, 2008) only referred to VDV(8). In Table 1 the main studies' characteristics are depicted, including sample size and demographic data and evaluated equipment, as well as performed tasks.

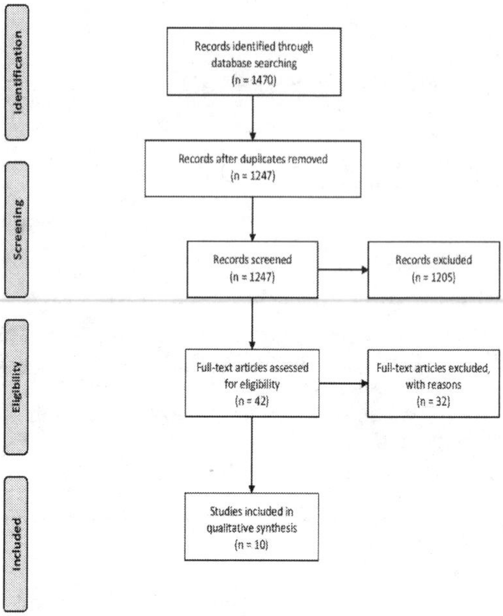

Figure 1. PRISMA Flow Diagram.

T Eger et al. (2008) collected data from three mine sites where the acceleration varied in a range between 0.56 to 0.99 m.s^{-2}, while the VDV varied between 12.38 and 24.67 m.s$^{-1.5}$. LHD F-2 and LHD H-3 displayed dominant x-axis vibrations and lower z-axis vibrations, when compared to the other models.

Vanerkar et al. (2008) evaluated VDV in five different equipment, in two different mines: one bauxite mine and one iron mine, altogether varying in a range between 5.33 and 13.53 m.s$^{-1.75}$.

Mandal & Srivastava (2010) studied the performances of eighteen dumpers during each cycle of operation. According to the authors, the root mean square (RMS) value of acceleration in z-axis measured ranged between 0.644 and 1.82 m.s^{-2}, with a mean value of 1.10 m.s^{-2}. The VDV values varied between 6.05 and 25.13 m.s$^{-1.75}$.

Smets, Eger, & Grenier (2010) measured vibration values for eight different models of haulage trucks, in their different tasks: unloaded travel, loading, loaded travel and dumping, after which they calculated the estimated daily exposure. The results varied between 0.444 and 0.824 m.s^{-2} for A(8) and 8.775 and 16.442 m.s$^{-1.75}$ for VDV.

Aye & Heyns (2011) did a full study of thirty four different equipment (excavators, LHD and other types of equipment such as dozer, drilling machines, graders, etc.) and the 8 hour exposure accelerations calculated from the retrieved data ranged between 0.31 and 1.91 m.s^{-2}. Volvo A35

Table 1. Studied equipment and sample characteristics.

Author (year)	Country	Analysed equipment	Task	Sample (average age in years)	Average experience (in years)
Eger et al. (2008)	Canada	Load Haul Dumper	Loading and driving (with and without load)	7 operators (ages from 38–53 y.o)	Range: 2–20
Vanerkar et al. (2008)	India	Dumper, dozer, drill, shovel, poclain	Driving, loading, drilling	N/A	N/A
Mandal & Srivastava (2010)	India	Dumper	Driving	40 operators (46.8 y.o)	12.33 (SD = 5.86)
Smets et al. (2010)	Canada	Haulage Truck	Driving	7 operators (35.8 y.o)	Range: 1–20
Aye & Heyns (2011)	South Africa	Load Haul Dumper and excavator	Driving	N/A	N/A
Eger et al. (2011)	Canada	Load Haul Dumper	Driving and drilling	17 (ages from 28–53 y.o)	Range: 3–29
Leduc et al. (2011)	Canada	Locomotive, jumbo drill, wooden raise, metal raise and bolter	Driving and drilling	7 operators (36 y.o)	17
Chaudhary et al. (2015)	India	Blast-hole drill machine	Drilling	28 operators (52.64 y.o)	16 (SD = 7.3)
Burström et al. (2016)	Norway	Haul truck, drilling rigs, wheel loader, excavator, dozer, grader and transport car	Driving, drilling, loading	453 operators (N/A)	N/A
Marin et al. (2017)	Colombia	Hydraulic shovel, electric shovel, bull dozer, front loader, wheel dozer, grader, scraper, 190 Ton truck, 240 Ton truck, 320 Ton truck and water truck	Driving and loading	123 operators, data per equipment in order of appearance (38.3, 44.8, 40.8, 43.4, 49.0, 43.9, 44.8, 38.2, 36.6, 32.8, 37.8 y.o)	In order of appearance: 9.9, 19.9, 11.3, 19.3, 22.4, 15, 17.6, 10.8, 12.2, 6, 9.3

LHD, Volvo excavator, CAT excavator and Pit Viper presented higher x-axis vibrations and Bell rear dumper B25D, Bell rear dumper B40D, Bell rear dumper B50D, MAN TGA 33.400 trailer, Dragline, CAT FEL 992G, CAT track dozer, CAT caterpillar, Shantui, Sony grader, Komatsu grader, Bell water, CAT 740 and Landpac compactor presented predominant y-axis vibrations. Every other equipment showed higher z-axis vibrations with the exception of Hitachi excavator LCR with presented the same value both for x and z-axis.

Tammy Eger et al. (2011) measured the vibration in each task (driving loaded, driving unloaded and loading the bucket) for each equipment size. For the small LHD, the acceleration varied between 0.55 and 1.65 m.s^{-2}, while the VDV varied between 9.91 and 40.01 m.s$^{-1.75}$. Concerning big LHD, the vibration values were between 0.70 and 1.79 m.s^{-2} and for VDV the values obtained were between 9.91 and 38.31 m.s$^{-1.75}$. With the exception of large LHD C and D, z-axis vibrations were predominant.

Leduc et al. (2011) measured WBV via the feet on two locomotives, one bolter and one jumbo drill. They also measured the vibration values in to different platforms which were disregarded according to the aim of this study. The accelerations varied between 0.11 and 0.43 m.s^{-2}, with all of the equipment producing highest vibrations for the z-axis.

Chaudhary et al. (2015) only studied the blast-hole drill, leading to a daily RMS acceleration of 1.63 m.s^{-2} with predominant acceleration on the z-axis.

Burstrom, Hyvarinen, Johnsen, & Pettersson (2016) conducted measurements in 95 vehicles in three different mines. The acceleration values ranged between 0.20 and 1.04 m.s^{-2} and 7.3 and 17.3 m.s$^{-1.75}$ for the VDV. With the exception of the wheel loaders and the dozers, every other equipment had predominant vibration on the z-axis.

Marin et al. (2017) analysed the WBV in 38 mining equipment, in which the acceleration ranged between 0.40 and 1.10 m.s^{-2} and VDV from 10.4 to 23.6 m.s$^{-1.75}$.

4 DISCUSSION

The purpose of this paper was to understand which equipment produces highest whole body vibration values potentially harmful for the operator. For the discussion of the main aim, only data concerning daily exposure is going to be considered.

The equipment which produced less vibration was the bolter with a 0.11 m.s^{-2} average and locomotives between 0.36 and 0.43 m.s^{-2} (Leduc et al., 2011).

Haulage truck acceleration varies with its capacity (Smets et al., 2010): 150 ton has vibrations between 0.444 and 0.756 m.s^{-2}, 100 ton trucks produce vibrations on a range between 0.703 and 0.824 m.s^{-2} and smaller trucks have acceleration values varying between 0.693 and 0.777 m.s^{-2}. Marin et al. (2017) also measured acceleration variation concerning three types of truck: 190 ton (0.70 m.s^{-2}), 240 ton (0.64 m.s^{-2}) and 320 ton (0.71 m.s^{-2}); this author learnt that vehicles with slower speeds (meaning less than 3 kilometres per hour) had predominant exposures in the x-axis, vehicles with moderate speeds (between 3–12 kilometres per hour) presented major exposures in the y-axis and vehicles driving at higher speeds displayed predominant exposures in the z-axis. The study linked the average speed reached with the terrain (on the road vs off the road) conditions and performed tasks. Graders acceleration also vary a lot: 0.36 m.s^{-2} (Burstrom et al., 2016) and between 0.52 and 1.17 m.s^{-2}, according to Aye & Heyns (2011). The same author also reported values of 0.53 m.s^{-2} for the drilling machine and 0.62 to 1.04 m.s^{-2} for excavators, dumpers led to higher vibrations between 0.87 and 1.37 m.s^{-2}.

LHD vibrations also vary according to the equipment size: smaller had vibrations between 1.40 and 1.93 m.s^{-2} and larger ranged between 0.55 and 1.79 m.s^{-2} (Tammy Eger et al., 2011). On the study of T. Eger et al. (2008) LHD vibration values were between 0.56 and 0.99 m.s^{-2}.

Blast-hole machine (Marin et al., 2017) had one of the highest reported acceleration values (1.63 m.s^{-2}) however the compactor was the equipment which presented the largest A(8), of 1.91 m.s^{-2}.

What is more, Tammy Eger et al. (2011) concluded that driving with an unloaded bucket resulted in higher levels of vibration exposure and large sized LHD promoted also the exposure to highest vibrations. Proper equipment maintenance may reduce the vibration exposure of the miners (Vanerkar et al., 2008) and regular grading of roads as well as operator's training on how to use the different equipment should also be taken into consideration (Aye & Heyns, 2011).

According to the Directive 2002/44/EC of the European Parliament (2002), the daily exposure limit value with an 8 hour reference should be 1.15 m.s^{-2} and for vibration dose value 21 m.s$^{-1.75}$, and the daily exposure action values stand in 0.5 m.s^{-2} and 9.1 m.s$^{-1.75}$, respectively.

Mandal & Srivastava (2010) dumper tests and Aye & Heyns (2011) LHD tests reported higher vibration values than the recommended for exposure limit and most Tammy Eger et al. (2011) LHD tests, blast-hole drill, wheel dozer and fall outside the range of exposure limit for VDV. Overall, every performed test led to values higher than the recommended concerning the action limits.

5 CONCLUSIONS

Heavy earth machines such as trucks, LHD and compactors are the ones which produce the highest vibrations, mainly of the z-axis type, despite the variability presented in the results. On the other hand, tools such as bolters, do not produce high vibrations. The variability within papers may have occurred due to the different testing conditions: while some of the authors tested real mining conditions, others simulated mining tasks. As it was verified that equipment size and capacity, as well as road conditions also influence the results, in order to do a proper comparison, all studies should have used the same evaluation methodology. On top of that, the authors presented the results in different ways, which made a proper evaluation more difficult.

Further studies should be done in order to better understand how the vibration varies between equipment and a one-by-one task should also be put into consideration for the evaluation process.

ACKNOWLEDGMENTS

This publication has been funded with support from the Portuguese entity *Autoridade para as Condições do Trabalho in the aim of the project Guião para a Avaliação de Riscos na Indústria Extrativa a Céu Aberto*.

REFERENCES

Aye, S. A., & Heyns, P. S. (2011). The evaluation of whole-body vibration in a South African opencast mine. *JOURNAL OF THE SOUTH AFRICAN INSTITUTE OF MINING AND METALLURGY*, *111*(11), 751–757.

Bovenzi, M., & Hulshof, C. T. J. (1998). an Updated Review of Epidemiologic Studies on the Relationship Between Exposure To Whole Body Vibration. *Journal of Sound and Vibration*, *215*(4), 595–611. https://doi.org/10.1006/jsvi.1998.1598.

Burstrom, L., Hyvarinen, V., Johnsen, M., & Pettersson, H. (2016). Exposure to whole-body vibration in opencast mines in the Barents region. *INTERNATIONAL JOURNAL OF CIRCUMPOLAR HEALTH*, *75*. https://doi.org/10.3402/ijch.v75.29373.

Chaudhary, D. K., Bhattacherjee, A., & Patra, A. (2015). Analysis of Whole-Body Vibration Exposure of Drill Machine Operators in Open Pit Iron Ore Mines. *Procedia Earth and Planetary Science*, *11*, 524–530. https://doi.org/10.1016/j.proeps.2015.06.054.

Derek, R., & Peter, A. (2005). Whole-Body Vibration.

Eger, T., Stevenson, J., Boileau, P.-E., Salmoni, A., & VibRG. (2008). Predictions of health risks associated with the operation of load-haul-dump mining vehicles: Part 1-Analysis of whole-body vibration exposure using ISO 2631-1 and ISO-2631-5 standards. *INTERNATIONAL JOURNAL OF INDUSTRIAL ERGONOMICS*, *38*(9-10, SI), 726–738. https://doi.org/10.1016/j.ergon.2007.08.012.

Eger, T., Stevenson, J. M., Grenier, S., Boileau, P.-E., & Smets, M. P. (2011). Influence of vehicle size, haulage capacity and ride control on vibration exposure and predicted health risks for LHD vehicle operators. *JOURNAL OF LOW FREQUENCY NOISE VIBRATION AND ACTIVE CONTROL*, *30*(1), 45–62.

Eger, T., Thompson, A., Leduc, M., Krajnak, K., Goggins, K., Godwin, A., & House, R. (2014). Vibration induced white-feet: Overview and field study of vibration exposure and reported symptoms in workers. *Work*, *47*(1), 101–110. https://doi.org/10.3233/WOR-131692.

Kumar, R.., Kumaraswamidhas, L. A.., & Murthy, V. M. S. R.. (2016). Whole body vibration control and strategies in open cast coal mines: An overview. *International Journal of Control Theory and Applications*, *9*(10), 4519–4525.

Leduc, M., Eger, T., Godwin, A., Dickey, J. P., & House, R. (2011). Examination of vibration characteristics, and reported musculoskeletal discomfort for workers exposed to vibration via the feet. *Journal of Low Frequency Noise, Vibration and Active Control*, 197–206.

Mandal, B. B., & Srivastava, a. K. (2010). Musculoskeletal disorders in dumper operators exposed to whole body vibration at Indian mines. *International Journal of Mining, Reclamation and Environment*, *24*(February 2015), 233–243. https://doi.org/10.1080/17480930903526227.

Marin, L. S., Rodriguez, A. C., Rey-Becerra, E., Piedrahita, H., Barrero, L. H., Dennerlein, J. T., & Johnson, P. W. (2017). Assessment of whole-body vibration exposure in mining earth-moving equipment and other vehicles used in surface mining. *Annals of Work Exposures and Health*, *61*(6), 669–680. https://doi.org/10.1093/annweh/wxx043.

Smets, M. P. H., Eger, T. R., & Grenier, S. G. (2010). Whole-body vibration experienced by haulage truck operators in surface mining operations: A comparison of various analysis methods utilized in the prediction of health risks. *Applied Ergonomics*, *41*(6), 763–770. https://doi.org/https://doi.org/10.1016/j.apergo.2010.01.002.

Vanerkar, A. P., Kulkarni, N. P., Zade, P. D., & Kamavisdar, A. S. (2008). Whole body vibration exposure in heavy earth moving machinery operators of metalliferous mines. *Environmental Monitoring and Assessment*, *143*(1–3), 239–245. https://doi.org/10.1007/s10661-007-9972-z.

Wolfgang, R., & Burgess-Limerick, R. (2014). Whole-body vibration exposure of haul truck drivers at a surface coal mine. *APPLIED ERGONOMICS*, *45*(6), 1700–1704. https://doi.org/10.1016/j.apergo.2014.05.020.

The effect of two training methods on workers' risk perception: A comparative study with metalworking small firms

B.L. Barros & M.A. Rodrigues
Department of Environmental Health, School of Health, Polytechnic of Porto, Porto, Portugal
Research Centre on Environment and Health, School of Health of Polytechnic of Porto (P.Porto), Porto, Portugal

A.R. Dores
CIR, Centro de Investigação em Reabilitação, School of Health, Polytechnic of Porto, Portugal

ABSTRACT: This study aims to evaluate and compare the effect of two training methods on workers' risk perception, when applied in metalworking small firms: an active method, with group discussion, and an expository method, with formal exposure. A sample of 212 workers was divided into three groups: two intervention groups and one group without intervention. A questionnaire was developed and applied before and one month after the training sessions to assess the following dimensions: Perceived susceptibility, Perceived severity, Perceived barriers and Perceived benefits. The results indicate that the training had a positive, but limited, impact on workers' risk perception. Significant differences between both moments were found on Perceived susceptibility in the group where the expository method was applied and in the group submitted to the intervention supported in the active method. No significant differences between both training methods after intervention were observed in this study.

1 INTRODUCTION

Scientific evidences suggest that the workers' unsafe behavior is one of the most important factors responsible for the occurrence of work-related accidents and occupational diseases (Garrett & Teizer, 2009; Hinze et al., 2005; Lingard & Rowlinson, 2005). Occupational Safety & Health (OSH) training is considered one essential tool to improve work conditions and avoid unsafe behaviors, reducing the accident rates and the risk of occupational diseases (Burke et al., 2011; Jacinto et al., 2009; Robson et al., 2010; Stave & Törner, 2007). OSH training has the ability to transmit knowledge on these matters (Aluko et al., 2016; Evanoff et al., 2016; Nielsen et al., 2015), contributing to change workers' risk perception (Evanoff et al., 2016; Williams et al., 2007; Vale, 2015) and the safety commitment (Leiter et al., 2009), as well as behaviors (Zimmer et al., 2017). However, in order to plan and implement an effective intervention in order to promote safe behaviors, it is important to identify the determinants of behavioral change and apply an effective training method that focuses on these determinants (Bryan et al., 2002).

Several models have been developed, namely cognitive models, in order to understand and explain the variables that have a significant impact on the motivation for the adoption of safe behaviors. The Health Belief Model (HBM) is a model that focuses on the cognitive factors that are considered causal mediators of behaviors (Cao et al., 2014; Cheraghi et al., 2014; Mehri et al., 2011). This model has been used to predict behaviors in non-occupational contexts, mainly in the health area. However, some previous studies applied this model in occupational settings with success (see e.g. Cao et al., 2014; Haghighi et al., 2017). When adapted to the OSH field, this model postulates that a worker who feels susceptible to a certain risk, tends to consider the situation more serious. Additionally, the benefits of adopting the target behavior are considered higher than the barriers associated with it, resulting in a higher predisposing to safe behaviors.

Despite the potential of the HBM model to promote safe behaviors in occupational settings, there are not enough evidences regarding the effect of a training method supported in this model, mainly when applied in small-sized firms. Therefore, this study aims to analyze and compare the effect of two OSH training methods in workers' risk perceptions when applied in metalworking small firms, an active method, with group discussion, and an expository method, with formal exposure.

2 METHODOLOGY

2.1 Sample

This study comprised a sample of 212 workers, belonging to 12 small-sized firms from the metalworking industry: 6 companies constituted the intervention group and 6 companies the group without intervention. Three groups were defined for this study: experimental group—EG (subject to the active method with group discussion), active control group—ACG (subject to the expository method with formal exposure) and passive control group—PCG (group without intervention).

Most of workers were males (92.5%) and their mean age was 40.2 years old ($SD = 12.1$ years old).

2.2 Study design and procedures

The present study consisted of four phases: (1) assessment of the training needs through two different approaches, conducting a focus group to identify the most common unsafe behaviors in the sector and a visit by an OSH Practitioner to assess the general conditions of the workplaces, work environment and machines' safety; (2) development of the pedagogical program; (3) training sessions; (4) assessment of the training effect on workers' risk perception, through the application of a questionnaire to the three groups before and after the training sessions (see subsection 2.3.1).

Participants were informed about the purpose of the study. It was explained that their answers would be treated confidentially and only used for the purpose of the present study. The study was approved by the local research ethics committee and complies with the Declaration of Helsinki.

2.3 Instruments

The pedagogical program was designed based on the results of the focus group and the OSH Practitioner visits, and on training materials delivered by the Workplace Safety and Health (WSH) Council (WHS, 2016). It was divided in three parts: (1) introduction to the occupational accidents and diseases in the metalworking industry, legal framework, and employer and workers' duties and responsibilities; (2) risk factors in the metalworking industry: physical, mechanical and ergonomic risk factors, chemical exposure and fire hazards; (3) control measures.

Based on the designed program, a training was conducted through two different methods: (1) an active method supported on the HBM (based on group discussions) and (2) an expository method with formal exposure. In the expository training sessions, theoretical contents were presented using an expositive approach, supported in a power-point format presentation. The training sessions in the discussion group were designed to ensure the same contents taught when applied the expository method. This approach was based on discussions, which were triggered based on images, videos and studies. The planning of this session was based on the HBM, focusing on the following dimensions: Perceived susceptibility, Perceived severity, Perceived barriers and Perceived benefits.

The presentation in power-point was tested with a sample of 10 workers in order to verify contents and its suitability to the reality of the metalworking sector, as well as to the time of the session.

2.3.1 Risk perception analysis

A questionnaire to analyze and compare the effect of the training methods in workers' risk perception was developed. The questionnaire was divided in different parts, being important to this study the following ones: (1) sociodemographic and professional variables, (2) four dimensions to assess risk perception based on the HBM adapted from Haghighi et al. (2017).

Regarding sociodemographic and professional variables, workers were surveyed in relation to age, gender, schooling, seniority in the metalworking sector and in the company, sector, function, seniority in the current function, work shift, employment contract and occupational accidents and diseases.

For the evaluation of risk perception, the dimensions described as determinants of behavior in previous studies were considered (Cao et al., 2014; Cheraghi et al., 2014; Haghighi et al., 2017; Lajunen & Rasanen, 2004): Perceived susceptibility (3 items), Perceived severity (3 items), Perceived barriers (5 items) and Perceived benefits (4 items). These dimensions are based on the HBM described in Haghighi et al. (2017). Authors were asked to authorize the use of the scales to assess the risk perception dimensions under analysis, which was accepted. The items of each dimension were evaluated using a 5-point Likert scale (1 = totally disagree, 5 = totally agree). The validity and reliability of the questionnaire were analyzed in a pilot company with 30 employees.

The questionnaire was applied before and one month after the training sessions. In the control group, the questionnaire was applied in the same periods.

2.4 Data analysis

In order to have agreement in the data analysis, and as the questions of perceived barriers were made in the negative, their answers were reversed.

In a first stage, an exploratory factorial analysis was performed for the subscales of risk perception. To verify its suitability, the Bartltett sphericity test and the Kaiser Meyer Olkin (KMO) measurement were performed. As a method of extraction, and to determine the items underlying the dimensions under study, a principal components analysis was performed, followed by a Varimax rotation, in order to increase the interpretation of the factors. Items with factor loadings higher than 0.4 were selected to define factors, as suggested by Hair et al. (1995).

To evaluate the internal consistency of the questionnaire, the Cronbach's alpha coefficient was determined for each scale.

As a result from this analysis, 8 items from the original scale were deleted: two items from each subscale of risk perception.

Basic descriptive statistics were computed for all variables. Parametric tests were applied, namely t-test for paired samples (used to compare the rankings of each item before and after the training), t-test for independent samples (to compare differences between the two interventions) and Anova (to compare differences between the three types of interventions). A significance level of $\alpha = 5\%$ was considered in the present study. Data analysis procedures were performed using the statistical software package Statistical Package for Social Sciences (IBM SPSS® version 22).

3 RESULTS AND DISCUSSION

3.1 Factor analysis

The data was considered appropriate for the analysis, according to the KMO measure (KMO = 0.73). Bartlett's sphericity test was significant ($p < 0.05$), which indicates a significant correlation between the variables (Hair et al., 1995; Pestana & Gageiro, 2014).

Four factors resulted from the exploratory factorial analysis, which explained 59% of the total variance: Perceived susceptibility, Perceived severity, Perceived barriers and Perceived benefits. These factors are in accordance with the theoretical construct (Haghighi et al., 2017). Table 1 presents the final factorial analysis solution, describing the factor loads from the Exploratory Factorial Analysis and the Cronbach's alpha values for each scale. The obtained Cronbach's alpha were: 0.81 for Perceived susceptibility; 0.68 for Perceived severity; 0.71 for Perceived barriers; 0.83 for Perceived benefits. These results revealed to be appropriate for the analysis, since in all scales the value obtained was higher than 0.60 (Pestana & Gageiro, 2014).

3.2 Analysis of the effect of pedagogical intervention

The effects of a pedagogical intervention on workers' risk perception through the application of two different training methods were analyzed, being the results presented in Table 2. This analysis was

Table 1. Dimensions of risk perception.

Variable	Loading
1. Perceived susceptibility ($\alpha = 0.81$)	
1.1. In the future, it is likely that I have an occupational accident	0.758
1.2. I am at risk for accident while working, even if I regularly comply the safety rules	0.805
1.3. It is very likely for me to have an accident-related injury at my work	0.896
2. Perceived severity ($\alpha = 0.68$)	
2.1. In our company, unsafe working may result in serious health consequences	0.690
2.2. If I do not work safely, my emotional and mental health will be affected	0.654
2.3 Safety equipment use while working will diminish the possible effects of harmful agents	0.667
3. Perceived barriers ($\alpha = 0.71$)	
3.1. Brave and strong men never use personal protective equipment (like helmets, safety gloves and...) while working	0.500
3.2. Sometimes conditions such as heat or harassment resulting safety equipment (e.g. helmets, safety gloves, etc.) hamper me to work safely	0.721
3.3. Sometimes it is necessary to disobey the safety rules at work to increase the production rate	0.723
3.4. In my opinion, occupational accidents depend on the chance of individuals	0.492
3.5. Safe working results in slow progress of the jobs	0.753
4. Perceived benefits ($\alpha = 0.83$)	
4.1. I believe that I can prevent occupational accidents by complying the safety rules	0.447
4.2. In my opinion, all employees should know on how to use personal protective equipment to prevent occupational accidents	0.700
4.3. Using appropriate and safe working methods/instruments while working is necessary to prevent occupational accidents	0.680
4.4. It is necessary to continuously emphasize the safety issues at work to prevent occupational accidents	0.735

considered relevant because several authors have pointed the importance of the training methods applied in safety programs, with particular emphasis in more engaging approaches (Burke et al., 2006, 2011; Hartling et al., 2004; Williams et al., 2007; Rodrigues et al., 2017). However, there is still limited evidence about its real effect, particularly in what regards to risk perception.

No significant differences between the three groups were observed before and after the

Table 2. Comparison of the mean scores of the dimensions of risk perception a before and after the training intervention.

Variable	Group	Before training x(sd)	After training x(sd)	P-value
Perception susceptibility	ACG	10.54 (2.10)	11.37 (1.89)	0.001
	EG	10.47 (3.09)	11.24 (2.23)	0.017
	P-value	0.899	0.754	–
	PCG	9.41 (2.42)	9.58 (2.55)	0.056
	P-value	0.008	0.000	–
Perceived severity	ACG	12.01 (1.32)	12.00 (1.68)	0.952
	EG	12.25 (1.64)	12.49 (1.41)	0.316
	P-value	0.433	0.105	–
	PCG	11.97 (1.67)	11.98 (1.68)	0.657
	P-value	0.582	0.149	–
Perceived barriers	ACG	18.26 (2.95)	18.76 (3.75)	0.349
	EG	18.66 (3.01)	18.72 (3.63)	0.903
	P-value	0.488	0.953	–
	PCG	18.25 (3.36)	18.22 (3.42)	0.703
	P-value	0.720	0.569	–
Perceived benefits	ACG	16.74 (2.11)	16.93 (1.66)	0.443
	EG	16.85 (2.45)	17.21 (1.66)	0.249
	P-value	0.807	0.382	–
	PCG	16.54 (1.73)	16.58 (1.76)	0.320
	P-value	0.641	0.086	–

intervention, except in what regards to the variable Perceived susceptibility ($p<0.05$), which was higher in the active control group; however, it is important to note that the mean values obtained were similar for the three groups.

After the intervention, and in a comprehensive way, it was observed that there was a small improvement in the scores of the risk perception dimensions in both groups that underwent interventions. However, significant differences were only observed for Perceived susceptibility in the active control group and experimental group ($p<0.05$). These results can be related to the growing consciousness of the workers to their exposition to some risks that might cause an occupational accident.

In general, the results of this study showed a limited effect of the training intervention on the variables under analysis. The improvements after the training intervention were small. Additionally, it was expected to achieve a significant improvement in the active method with discussion group, since a more engaging method was applied. Burke et al. (2006) demonstrated that training methods that require greater involvement by workers are more effective in terms of safety performance and knowledge acquisition. However, the improvement found in this study was not significant when comparing the two training methods. Robson et al. (2010) argue that the evidence in favor of the engagement hypothesis is weak and that more evidence should be found. Adams et al. (2013), in a study that aimed to evaluate the effectiveness of two education methods (traditional vs new education paradigm), found that although there was a significant overall decrease in rates of injuries, the difference between the two intervention groups was not significant. Brahm and Singer (2013) have stated that there is no training method better than others and that its effectiveness depends on the needs and characteristics of each company.

4 CONCLUSIONS

This study assessed the effect of training on workers' risk perception, comparing two different training methods. The obtained results showed, in general, a limited but positive effect of the training on workers' risk perceptions. Additional, no significant differences between both training methods were observed.

The results presented in this paper are a part of a study where the effect of the training methods on other dimensions were analysed; only a part of the results is presented. In fact, and despite the importance of the results obtained and presented, there is a need of more research focused on attaining a better understanding of how occupational safety training settings can be made effective in small-sized firms.

REFERENCES

Adams, J. S., Raju, R., Solomon, V., Samuel, P., Dutta, A. K., Rose, J. S., & Tharyan, P. (2013). Increasing compliance with protective eyewear to reduce ocular injuries in stone-quarry workers in Tamil Nadu, India: a pragmatic, cluster randomised trial of a single education session versus an enhanced education package delivered over six months. *Injury*, 44, 118–125.

Aluko, O. O., Adebayo, A. E., Adebisi, T. F., Ewegbemi, M. K., Abidoye, A. T. & Popoola, B. F. (2016). Knowledge, attitudes and perceptions of occupational hazards and safety practices in Nigerian healthcare workers. *BMC Res Notes*, 9, 71.

Brahm, F. & Singer, M. (2013). Is more engaging safety training always better in reducing accidents? Evidence of self-selection from Chilean panel data. *Journal of Safety Research*, 47, 85–92.

Bryan, A., Fisher, J. D. & Fisher, W. A. (2002). Tests of the mediational role of prepara-tory safer sexual behavior in the context of the theory of plannedbehavior. *Health Psychology*, 21, 71–80.

Burke, M., Salvador, R., Smith-Crowe, K., Chan Serafin, S., Smith, A., & Sonesh, S. (2011). The dread factor: How hazards and safety training influence learning and performance. *Journal of Applied Psychology*, 96 (1), 46–70.

Burke, M., Sarply, S., Smith-Crowe, K., Chan-Sherafin, S., Salvador, R. & Islam, G. (2006). Relative effectiveness of worker safety and health training methods. *American Journal of Public Health*, 96 (2), 315–324.

Cao, Z. Chen, Y. & Wang, S. M. (2014). Health belief model based evaluation of school health education programme for injury prevention among high school students in the community context. *BMC Public Health*, 14 (1), 1.

Cheraghi, P., Poorolajal, J., Hazavehi, S. M. M., & Rezapur-Shahkolai, F. (2014). Effect of educating mothers on injury prevention among children aged < 5 years using the Health Belief Model: a randomized controlled trial. *Public Health, 128* (9), 825–830.

Evanoff, B., Dale, A. M., Zeringue, A., Fuchs, M., Gaal, J. Lipscomb, H. J. & Kaskutas, V. (2016). Results of a fall prevention educational intervention for residential construction. *Safety Science*, 89, 301–307.

Garrett, J. W., & Teizer, J. (2009). Human factors analysis classification system relating to human error awareness taxonomy in construction safety. *Journal of Construction Engineering and Management*, 135 (8), 754–763.

Haghighi, M., Taghdisi, M. H., Nadrian, H., Moghaddam, H. R., Mahmoodi, H. & Alimohammadi, I. (2017). Safety Culture Promotion Intervention Program (SCPIP) in an oil refinery factory: An integrated application of Geller and Health Belief Models. *Safety Science*. 93, 76–85.

Hair, J. F., Anderson, R. E., Tatham, R. L. & Black, W. C. (1995). Multivariate Data Analysis with Readings (4th Edition), Prentice-Hall, New Jersey.

Hartling, L., Brison, R. J., Crumley, E. T., Klassen, T. P., & Pickett, W. (2004). A systematic review of interventions to prevent childhood farm injuries. *Pediatrics*, 114, 483–496.

Hinze, J., Huang, X., & Terry, L. (2005). The nature of struck-by accidents. *Journal of Construction Engineering and Management*, 131 (2), 262–268.

Jacinto, C., Canoa, M. & Soares, C. G. (2009). Workplace and organisational factors in accident analysis within the food industry. *Safety Science, 47*, 626–635.

Lajunen, T. & Rasanen, M. (2004). Can social psychological models be used to promote bicycle helmet use among teenagers? A comparison of the Health Belief Model, Theory of Planned Behavior and the Locus of Control. *J Safety Res*, 35 (1), 115–123.

Leiter, M. P., Zanaletti, W. & Argentero, P. (2009). Occupational Risk Perception, Safety Training, and Injury Prevention: Testing a Model in the Italian Priting Industry. *Journal of Occupational Health Psychology*, 14 (1), 1–10.

Lingard, H. & Rowlinson, S. M. (2005). *Occupational Health and Safety in Construction Project Management*. Spon Press.

Mehri, A., Nadrian, H., Morowatisharifabad, M. A. & Akolechi, M. (2011). Determinants of seat belt use among drivers in Sabzevar, Iran: a comparison of theory of planned behavior and health belief model. *Traffic Injury Prevention*, 12(1), 104–109.

Nielsen, K. J., Kines, P., Pedersen, L. M., Andersen, L. P. & Andersen, D. R. (2015). A multi-case study of the implementation of an integrated approach to safety in small enterprises. *Safety Science*, 71, 142–150.

Pestana, M. H. & Gageiro, J. N. (2014). *Análise de dados para ciências sociais. A complementaridade do SPSS.* 6ª Edição, Edições Sílabo. Lisboa.

Robson, L., Stephenson, C., Schulte, P., Amick, B., Chan, S., Bielechy, A., Wang, A., Heidotting, T., Irvin, E., Eggerth, D., Peters, R., Clarke, J., Cullen, K., Boldt, L., Rotunda, C. & Grubb, P. (2010). A systematic review of the effectiveness of training and education for the protection of workers. Report mandated by the Institute for Work & Health (IWH—Canadian Agency) and National Institute for Occupational Safety and Health (NIOSH—US Agency).

Rodrigues, M. A., Vale, C. & Silva, M. V. (2017). Effect of a safety education program on risk perception of vocational students: A comparative study of different intervention methodologies. In Arezes, P.M., Baptista, J.S., Barroso, M.P., Carneiro, P., Cordeiro, P., Costa, N., et al. (Eds). Occupational Safety and Hygiene V. pp. 289–292. London: CRC Press, Taylor & Francis Group. ISBN: 978-1-138-05761-6

Stave, C. & Törner, M. (2007). Exploring the organisational preconditions for occupational accidents in food industry: a qualitative approach. *Safety Science, 45*, 355–371.

Vale, C. (2015). Análise da eficácia de diferentes tipologias de intervenção pedagógica: Programa de educação sobre segurança e saúde no trabalho para futuros jovens trabaladores. Tese de Mestrado em Ambiente, Higiene e Segurança em Meio Escolar. Instituto Politécnico do Porto—Escola Superior de Saúde, Porto. 76 pp.

WHS. (2016). Training Materials. Acedido a 3 de março de 2017, em https://www.wshc.sg/wps/portal/

Williams, W., Purdy, S. C., Storey, L., Nakhla, M. & Boon, G. (2007). Towards more effective methods for changing perceptions of noise in the workplace. *Safety Science*, 45, 431–447.

Zimmer, J., Hartl, S., Standfuß, K., Möhn, T., Bertsche, A., Frontini, R., Neininger, M. P. & Bertsche, T. (2017). Handling of hazardous drugs–Effect of an innovative teaching session for nursing students. *Nurse Education Today*, 49, 72–78.

Diagnoses of the acoustic perceptions of workers for auditory signal design

G. Dahlke & T. Ptak
Poznań University of Technology, Poznań, Poland

ABSTRACT: The article discusses the findings of a preliminary acoustic perception study involving the subjects localizing sound sources. This particular parameter was evaluated by measuring the azimuth angle between the sagittal plane of the subject's body at the level of his/her hearing organ and the straight line passing transversely from the main axis of the subject's body and through the designated center of the sound source. While the study remains consistent with the general principles of psychoacoustic measurements, it is limited by the environment as well as the measuring equipment and tools employed. Patterns have been discovered that are potentially useful in planning further research and developing guidelines for the design of auditory warning signals and managing the acoustic environment in the workplace.

1 INTRODUCTION

1.1 Signal classification

Auditory signals are developed in conformity with the requirements laid down in Polish legislation. The key alert classification criteria have been set out in the Regulation of the Minister of Labor and Social Policy on occupational health and safety (Journal of Laws of 2003, No. 169, Item 1650). The Regulation distinguishes between light, manual, sound and verbal signals (Signalizatory 2013). The final two of these categories can be classified as auditory signals. Specific criteria for the design of auditory alerts are governed by the following standards: PN-EN ISO 7731:2009 (on auditory alerts), PN-EN 60849:2001 (auditory warning systems) and PN-EN 981+A1 (auditory and visual systems for signaling hazards and communicating machine safety information). The above laws and regulations lay down guidelines applying to the design of alert signals.

1.2 Diagnosing auditory signal perception

The perception of auditory stimuli is a logical starting point for the design of the acoustic environment. Such perception refers primarily to the extent to which any desired sounds are audible and the extent to which the sensations experienced by the listener are consistent with signal characteristics. Competence in this area is acquired as listeners gain experience. While, admittedly, ergonomic methodology relies on a personalized approach to every individual, the need to develop design guidelines makes it necessary to prepare border demographic data for a given population. It is therefore essential to understand the mechanism involved and the degree (thresholds and effective levels) of the perception of sounds and their individual components as accurately as possible.

The issues presented in the article fall within the scope of psychoacoustics, which is an empirical approach to the perception of auditory stimuli by man describing relationships between stimuli (acoustic waves) and the responses they elicit. Furthermore, as possible practical applications of such knowledge are explored, psychoacoustics can be classified as one part of ergonomics that can be applied in the design of signals adjusted to reflect the psychophysiological capabilities of workers.

2 METHODOLOGY

2.1 Research objective

This psychoacoustic study was designed to explore selected relationships between physical stimuli and the auditory responses they evoke (Engel 2001; Jorasz 1998; Moore 1999; Ozimek 2002; Radosz 2015; Shinn-Cunningham 2003; Tpia 2008; Wołoszyn 2008; Tytyk 2001, 2011). A wide range of measurement methods can be employed to conduct such research in the most accurate and error-free manner (Dahlke 2014, 2015). Their choice depends, among others, on the research objective and exposure durations. The findings produced relate to responses in a given context and provide an approximate picture of reality. The study is focused on ascertaining the subjects' ability to localize the point in space from which a sound is emanating. The experiment was performed to illustrate the ability of individuals to identify the direction from which the sound was dissipating and gain insights into the patterns behind the spatial perception of sound.

2.2 Research methodology

The proposed methodology relies on multiple presentations of identical signals (i.e. ones having a specific frequency, intensity and sound spectrum) and the measurement of subject responses to each sound. To best conform to the regulatory guidelines governing the design of auditory alerts and the perception differences they describe, which depend on sound frequencies, the procedure was repeated for multiple frequencies (250 Hz, 500 Hz, 1000 Hz, 2000 Hz). The spacing between the frequencies of successive signals corresponded to the octave interval.

The measurements were performed with a view to capturing the perception of the location of a sound source. This particular parameter was evaluated by measuring the azimuth angle between the sagittal plane of the subject's body at the level of his/her hearing organ and the straight line passing transversely from the main axis of the subject's body and through the designated center of the sound source (Figure 1).

As the subject points to the source of the sound by turning directly towards it, differences in sound intensity between the ears and the time differences occuring between the ears while the listener positions himself or herself towards the source of the signal are reduced to the minimum. However, such effects may be helpful in determining signal location, causing differences between the sounds that reach either ear, and giving rise to the so-called rumble effect.

While the study remains consistent with the general principles of psychoacoustic measurements, it is limited by the environment as well as the measuring equipment and tools employed.

The algorithm followed in the study of sound source perception is shown in Figure 2.

The measurements were taken with the following instruments:

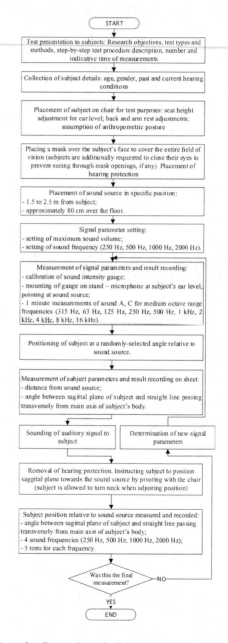

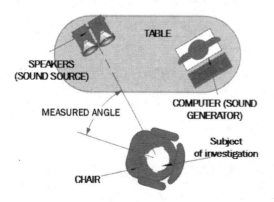

Figure 1. Sound source location identification station designed for conformity with accepted methodology (Ptak 2017).

Figure 2. Research methodology (own research based on: Ptak 2017).

Table 1. Data collected from study subjects (own research based on: Ptak 2017).

No.	Sex	Age	Reported hearing-related health conditions
1	Male	24	None
2	Male	44	Past vestibular neuronitis on the right side of left ear
3	Male	24	None
4	Male	23	None
5	Male	25	None
6	Male	24	None
7	Male	64	None
8	Female	65	Past bacterial infection in right ear canal

- SVAN 971 sound level gauge;
- online sound generator;
- audio source—a set of two speakers connected to a computer with a total power of 120 W and transmission frequencies ranging from 20 Hz to approximately 20 kHz;
- manual goniometer (measuring accuracy of 1°) used for measuring azimuthal angles;
- electronic rangefinder (measuring accuracy: 1 mm).

2.3 *Study sample*

The study comprised eight participants. The data collected is summarized in Table 1. The study group is not sufficiently representative to draw conclusions on a population wider than the subjects broken down by age bracket. However, this is only a preliminary study designed to identify further research needs and the best sample size.

The subjects included individuals who have in the past suffered from severe inner ear conditions.

3 RESULTS

The study provided indications of the distribution of the azimuthal angles at which individual persons turned towards the sound source, broken down by frequency. The sound sources generated an acoustic spectrum whose dominant tones were as selected for the study. At the subject station, these averaged 78.04 dB (with a standard deviation of 4.73 dB). The subjects were off by the biggest margin when attempting to identify the position of the 500 Hz and 2000 Hz sound sources. Significant errors could also be seen in Subject 2. Past health condition were found to impair the perception of high frequency sounds.

The data shown in the graphs (Figures 3, 4, 5, 6, 7, 8, 9, 10) reveal individual differences in the

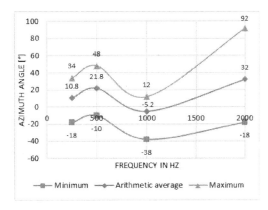

Figure 3. The arithmetic mean curves and the extreme values of azimuthal angles indicated by Subject 1 for the dominant frequency (own research based on: Ptak 2017).

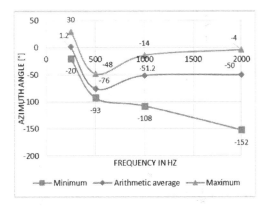

Figure 4. The arithmetic mean curves and the extreme values of azimuthal angles indicated by Subject 2 for the dominant frequency (own research based on: Ptak 2017).

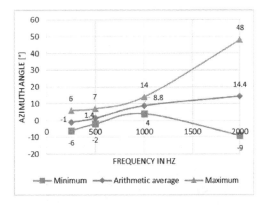

Figure 5. The arithmetic mean curves and the extreme values of azimuthal angles indicated by Subject 3 for the dominant frequency (own research based on: Ptak 2017).

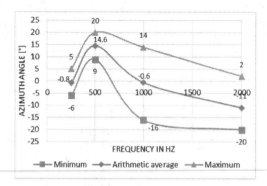

Figure 6. The arithmetic mean curves and the extreme values of azimuthal angles indicated by Subject 4 for the dominant frequency (own research based on: Ptak 2017).

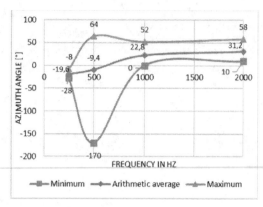

Figure 9. The arithmetic mean curves and the extreme values of azimuthal angles indicated by Subject 7 for the dominant frequency (own research based on: Ptak 2017).

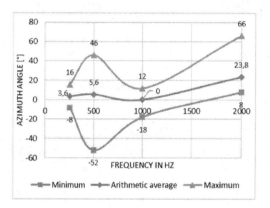

Figure 7. The arithmetic mean curves and the extreme values of azimuthal angles indicated by Subject 5 for the dominant frequency (own research based on: Ptak 2017).

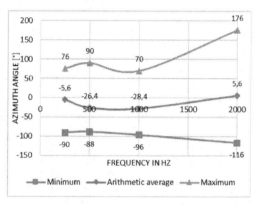

Figure 10. The arithmetic mean curves and the extreme values of azimuthal angles indicated by Subject 8 for the dominant frequency (own research based on: Ptak 2017).

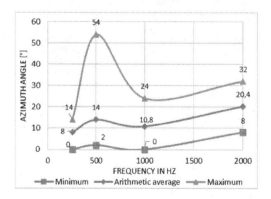

Figure 8. The arithmetic mean curves and the extreme values of azimuthal angles indicated by Subject 6 for the dominant frequency (own research based on: Ptak 2017).

perceptions of the direction from which sounds are emanating. Such perceptions depend mainly on the characteristics of each subject and specifically on subject age, hearing loss, etc. However, a part of the perception process is contingent on the parameters of the signal itself. The study therefore is additionally aimed at demonstrating these sound properties. In the data presented, a positive value of the angle means that the sound source was on the left-hand side of the space demarcated by the sagittal plane of the listener. Negative angle results point to the right-hand side of such space.

An analysis of Spearman correlation coefficients showed a poor correlation between increased performance at 500 Hz and 2000 Hz (the corresponding coefficient being 0.3799).

The standard deviation graph for each individual and each frequency (Figure 11) illustrates a combination of the level and predominance of errors. The highest values for each frequency were obtained by the subjects who were outside of the 23–25 age range. Deterioration has also been

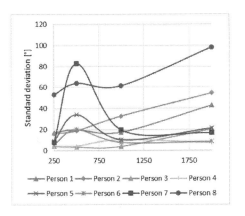

Figure 11. Standard deviations calculated on the basis of the results of individual tests at specific dominant frequencies for successive test subjects (own research based on: Ptak 2017).

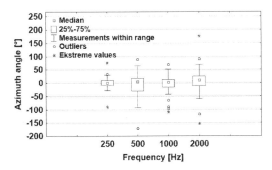

Figure 12. Variability of the azimuth angle for all subjects around the middle point (median). Source: Own research based on: Ptak 2017.

noticed in the 500 Hz and 2000 Hz results for the remaining age ranges (Figure 12).

This pattern has been found in all subjects.

4 CONCLUSIONS

Over the course of the study, the investigators looked at the impact of the dominant signal frequency on the ability to localize sounds. What they noticed was that a change in signal frequency can affect both the accuracy of source indication as well as source identification error rates and severity. When designing signals, it is advisable to reduce the 2 kHz frequency as the dominant component, as it adversely affects perception accuracy while forcing the signal listener to take more time to localize the source. The results for frequencies falling within the 250–1000 Hz range are less error-prone. However, one should bear in mind that the perceptions of signals with a 500 Hz dominant component also showed significant variations. Due to the small sample size, the above conclusions are merely preliminary. They are nevertheless useful for determining sample sizes for future research.

REFERENCES

Dahlke, G. 2015. Exposure to noise of communities residing near slow-traffic public roads. A practical example, [in:] Arezes, P; Baptista, JS; et al., *Proceedings of Sho2015*: International Symposium on Occupational Safety and Hygiene, Guimaraes, Portugal, Pages: 91–93.

Dahlke, G., Grzybowski, W. & Budniak, E. 2014. Analiza poprawności doboru ochronników słuchu. Przykłady praktyczne, w: *Zastosowania ergonomii. Wybrane kierunki badań ergonomicznych w 2014 roku*/Charytonowicz Jerzy [red. naukowa], Wrocław: Wydaw. Polskiego Towarzystwa Ergonomicznego PTErg. Oddział we Wrocławiu, pp. 183–194.

Engel, Z. 2001. Ochrona środowiska przed drganiami i hałasem, Wyd. Naukowe PWN, Warszawa.

Jorasz, U. 1998. Wykłady z psychoakustyki, Wyd. Naukowe UAM, Poznań.

Moore, B. C. J. 1999. Wprowadzenie do psychologii słyszenia, Wyd. Naukowe PWN, Warszawa-Poznań, 1999.

Ozimek, E. 2002. Dźwięk i jego percepcja. Aspekty fizyczne i psychoakustyczne, Wyd. Naukowe PWN, Warszawa-Poznań.

PN-EN ISO 7731:2009 Ergonomics - Danger signals for public and work areas - Auditory danger signals (ISO 7731:2003).

PN-EN 60849:2001 Sound systems for emergency purposes (IEC 60849:1998).

PN-EN 981+A1: 2010 Safety of machinery - System of auditory and visual danger and information signals.

Ptak, T. 2017. Measurement methods in the diagnosis of employees acoustic stimuli perception, diploma thesis under the direction of Grzegorz Dahlke, Poznań University of Technology, Poznań.

Radosz, J., Górski, P. & Młyński, R. 2015. Sygnalizacja akustyczna w środowisku pracy osób niepełnosprawnych, *Bezpieczeństwo Pracy - Nauka i Praktyka*: 2.

Rozporządzenie Ministra Pracy i Polityki Społecznej w sprawie ogólnych przepisów bezpieczeństwa i higieny pracy (Dz.U. 2003 nr 169 poz. 1650).

Shinn-Cunningham, B. 2003. Acoustics and perception of sound in everyday environments, Aisu-Wakamatsu, Japan, Mar. 6–7.

Sygnalizatory świetlne i dźwiękowe, Elektronika Praktyczna 5/2013.

Tipa, R. S. & Baltag, O. I. 2008. Study of the acoustic perception, Rom. Journ. Phys., Vol. 53, Nos. 1–2, P. 183–189, Bucharest.

Tytyk, E. 2001. Projektowanie ergonomiczne, PWN, Warszawa-Poznań.

Tytyk, E. 2011. Inżynieria Ergonomiczna Praktyka, Wyd. Politechniki Poznańskiej, Poznań.

Wołoszyn, P., Jabłoński, M. & Malicki, Ł. 2008. Badanie zdolności percepcji przestrzennej sceny akustycznej u ludzi, Automatyka/Akademia Górniczo-Hutnicza im. Stanisława Staszica w Krakowie, Tom 12, Zeszyt 3, Kraków.

Evacuation study of a high-rise building in fire safety, prescriptive *versus* performance analysis

A.C. Matos & M. Chichorro Gonçalves
Faculty of Engineering, University of Porto, Porto, Portugal

ABSTRACT: The study of fire safety is essential to protect human life and therefore the safeguarding of its heritage and environment. Hence it is very important to do research on this matter. The study of evacuation, particularly in high-rise buildings, due their evident size in altimetry, is keen to guarantee the protection of human life. This document presents the study of evacuation in a building with these characteristics, namely *Antas Tower*, located in Oporto. With the research we present, in this document, we plan to study the evacuation of *Antas Tower* based on a prescriptive analysis in accordance with *FSE* Portuguese legislation, and on a performance analysis using a computational model named *Pathfinder* and the numerical calculation based on *Chapter 2 of Section 4 of the NFPA 2008 Handbook "Calculation Methods for Egress Prediction—Chapter 2"*, doing the comparison with prescriptive analysis.

1 INTRODUCTION

1.1 The importance of fire safety

Nowadays, the study of fire safety in buildings (*FSE*) has been increasing and is now considered a relevant factor, aiming the protection of human life and the defence of the building itself, its content and the environment.

Therefore, among other relevant factors related to this issue, the study of the evacuation of any building, especially in high-rise buildings, where the distance to outside is substantially higher, it is essential to achieve this objective.

When the study focus in existing buildings, the full agreement with *FSE* Portuguese legislation, specifically Decree Law no. 224/2015 (*RJ-SCIE*) and Technical Regulation no. 1532/2008 (*RT-SCIE*) in many cases is difficult to apply, because of their architectural specificities. It is now possible to analyse the evacuation based on performance through computer simulators. That allows the determination of total evacuation time, in certain predefined conditions, allowing all the participants involved in the analysis process a better sustained assessment of the adequacy of the evacuation routes to the number of occupants to be evacuated.

1.2 Objective

The evacuation study of a high-rise building, *Antas Tower* located in Oporto, done in this article, contribute to the knowledge of the implications of the agreement/disagreement between performance analysis and prescriptive in the design result.

2 MATERIAL AND METHOD

2.1 Computer simulation and modeling of egress design

According to the type of building and its characteristics, and the specificity of the calculation results to be obtained, we selected the simulator that is considered the most adequate (based only on those assumptions).

The program chosen for the case study was *Pathfinder*. This is an agent based evacuation simulator, based on the continuous network model and marketed by *Thunderhead Engineering*. This simulator uses two methods of simulation of people movement (Ronchi & Nilsson, 2013), the hydraulic model, SFPE, developed by *Gwynne & Rosenbaum*, based on the calculation of existing means capacity and the agent-based model i.e. the *Reynolds*, based on the study of occupant behaviour.

3 CASE STUDY

3.1 Description of study high-rise building

Antas Tower is a high-rise building, it has 17 floors above the ground floor and 4 below. The 4 subterraneous floors are meant for parking and according to *RJ-SCIE* it is considered high risk. The upper floors are meant for offices, and are considered to have a very high risk, as it is defined in the legislation. The 1st floor has a citizen shop with independent entrance relatively to the parking and office floors.

3.2 Prescriptive analyses

After analysing the building based on the *FSE* Portuguese legislation, there were detected some unconformities that are summarized in Table 1.

Being the largest number of occupants between two consecutive floors, in the present case between the 2nd and 1st floor, equal to 1030, and distributing these occupants by the four existing exit stairs, it is verified that the width for each exit stair should be 2,4 m, and their access doors should also have this dimension, which does not happen. In terms of distance to one escape route, in some places we have 20 m, when the maximum allowed accordingly to the *FSE* Portuguese legislation, is 15 m.

3.3 Fire drill

On the 5th June 2015, at 3 p.m., a fire drill exercise was carried out in the *Antas Tower* (the exception was the citizen shop in the 1st floor), with the collaboration of the Battalion Fire-Fighters (*BSB*) and the Public Security Police (*PSP*). It was verified that the building was considered evacuated after 14 min., and was extinguished 21 min after the fire started. The number of occupants who evacuated was not recorded in the end of the exercise.

3.4 Analyses on performance

Using the Pathfinder simulator, we studied the case study evacuation that we present in Figure 1. Also, several simulations were carried out for later evaluation. In these simulations, the predefined values of maximum velocity attributed by the simulator were maintained for the occupants of a 1.19 m/s in exit access corridors and 1.0 m/s in evacuation through the exit stairs.

3.5 Numerical Calculation using Chapter 2 of Section 4 of NFPA2008 Handbook "Calculation Methods of Egress Prediction"

For the case study, we considered the applicability of *chapter 2 of section 4 of the Handbook standard of NFPA 2* (Proulx, Guylène; L. Bryan, John; F. Fahy, Rita; K. Lathrop, James; J. Fruin, Jonh; Cohn, Bert; J. O'Connor, n.d.), to analyse more concretely the evacuation conditions existing in the building. In this sense, we chose one of the study simulations performed by *Pathfinder 2015*, the evacuation of the 16th to the 4th floor through the possibility of the existence of a refuge zone on 4th floor with 0,9 m, thus allowing a comparison between the results obtained (Figure 2).

Table 1. Prescriptive analysis unconformities.

Simulation	Nonconformity	Comments
1	– Item a) and b) point 1 Article n.º25 of the RT-SCIE	– The exit access corridors do not have any fire protection in relation to the adjacent spaces.
2	– Item b) point 3 and 4 Article n.º56 of the RT-SCIE	– Width of the exit access corridors and exit stairs are less than what t is necessary.
3	– Item a) Point 2 of the Article n.º57 of the RT-SCIE	– Maximum distance to go, one-way is higher than 15 m.
4	– Point 11 of the Article n.º64 of the RT-SCIE	– Minimum width of exit stairs is not fulfilled.
5	– Item a) Point 1 Article n.º68 of the RT-SCIE	– Location and number of refuge areas (not available).
6	– Item a), b) e c) Point 1 Article n.º135 of the RT-SCIE	– Relating to the absence of a smoke control system on the exit access corridors, as well as on the vestibules (floors below the ground floor).
7	– Point 4 Article n.º26 of the RT-SCIE	– Relating to the absence of vestibules for access to exit stairs (floors above the ground floor).

Figure 1. Antas Tower *(Pathfinder 2015)*.

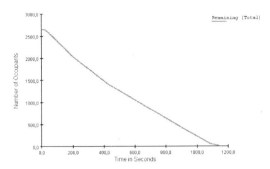

Figure 2. *Pathfinder 2015*, evacuation time (s) according to the evacuated occupants.

4 RESULTS AND DISCUSSION

4.1 Results obtain with Pathfinder 2015

With *Pathfinder 2015*, we obtained the result of 19 min and 19 s, for the evacuation simulation of all floors through the four exit stairs existing on ground floor. Independent simulations were also carried out for the office floors and citizen shop, 1st floor, and the following results were obtained, 14 min 42 s and 4 min 36 s, respectively. Compared with the value obtained in the fire drill indicated in point 3.3, this value was obtained only for the evacuation of the office floors, not including the citizen shop, and it is verified that there is a differential of only 42 s.

It is not possible to make a fully assertive comparison of these two numbers, since we do not know the number of the evacuated occupants during the fire drill exercise. However, the occupants evacuated through the fire drill exercise will been lower than that established in the simulation.

4.2 Results including refuge areas

According to Article no. 68 of the *RT-SCIE*, the high-rise buildings, which are the case of the building under study, should be provided with areas of refuge. Therefore, some simulations were carried out based on the existence of a single zone of refuge to be predicted in 4th floor. Also in order to meet the criteria defined in the article no. 68, it was considered the existence of two refuge areas located of 16th and 4th floor, and its results are shown in Table 2.

According to these results, it was possible to verify that in most situations the strategies with the evacuation priority order from upper floors to lower floors, allow the decrease of the total evacuation time, the explanation for this is when the behaviour refers an evacuation for any existing exits, occupants always choose the closest exit instead the alternatives.

Table 2. Simulations considering refuge areas.

Simulation	Description	Results
1	– Evacuation through any exit, including refuge area provided on 4th floor with 0,9 m.	25 min 26 s
2	– Evacuation of 16th to 4th floor through the refuge area with 0,9 m located on 4th floor, remaining floors through the exits on ground floor.	22 min 53 s
3	– Evacuation through any exit, including a refuge area provided on 4th floor with 1,4 m.	14 min 30 s
4	– Evacuation of 16th to 4th floor through the refuge area with 1,4 m located on 4th floor, remaining floors through the exits on ground floor.	16 min 35 s
5	– Evacuation through any exits, including refuge area provided on 4th floor, with two exits with 0,9 m each.	12 min 40 s
6	– Evacuation of 16th to 4th floor through the refuge area provided in 4th floor, with two exits with 0,9 m each, remaining floors through the exits in ground floor.	12 m 18 s
7	– Evacuation through any exits, including refuge areas with 0,9 m in 16th and 4th floor.	14 min 47 s
8	– Evacuation of the 16th to the 10th floor through the refuge area located on the 16th floor with 0,9 m, evacuation of 9th to 4th floor through the refuge area located on the 4th floor with 0,9 m, remaining floors through the exits on ground floor.	13 m 13 s
9	– Evacuation of 16th to 4th (and only these floors) through the refuge area provided on 4th floor with 0,9 m.	23 min 14 s*

*This simulation was considered also applying the NFPA2008 "Calculation Methods with Egress Prediction—Chapter 2".

It was also verified that, with the inclusion of refuge areas with certain characteristics, for example, considering only the 4th floor as a refuge zone, with two distinct exits with 0,9 m each one, it is possible to do decrease the total evacuation time in 7 min 01 s, which means 36.3% of optimization. It was also possible to verify that it is more advantageous to include two separate exits on 4th floor (refuge zone) with a width 0,9 m, than the inclusion of only one exit with a bigger width, like 1,4 m, the optimization of time for this solution was 1 min 50 s.

4.3 Results Obtain by Numerical Calculation using Chapter 2 of Section 4 of the NFPA2008 Handbook "Calculation Methods of Egress Prediction"

According with the section 3.5, the 9th simulation, shown in Table 2, was also performed by numerical calculation using *section 2, chapter 4 of the NFPA 2008 Handbook "Calculation Methods for Egress Prediction—Chapter 2"*. The results are indicated in Table 3.

With this simulation, it was possible to verify that the result obtained is very similar to the determined using the software *Pathfinder 2015* and the difference is only 45 s. It was also possible to verify that the evacuation time for the emergency stairs V_2 is less than the evacuation time for the emergency stairs V_1. This situation is due to the fact that the access door to the emergency stairs V_2 presents a width of 1.8 m instead of the access door to emergency stairs V_1 which has a width of only 0.9 m. The difference in evacuation times between V_1 and V_2 was 2 min 02 s.

4.4 Results of simulation using lifts in case of fire

Although the Portuguese law does not allow people to use lifts in case of fire, this has been allowed in some countries since the 1980s (Kobes, Helsloot, de Vries, & Post, 2010).

Studies of the WTC 9/11 disaster refer that around 3000 occupants were saved through the use of lifts in the south tower (Kobes et al., 2010).

Based on these factors, simulations including the lifts to evacuation in case of fire were performed. With the results, it was possible to understand if this situation benefits the evacuation in high-rise buildings, especially when we have to evacuate children and people with disabilities. The results are indicated in Table 4.

With the results obtained and indicated in Table 4, it was possible to verify that only up to 12th floor would it be advantageous to evacuate occupants by using existing lifts. When we include the 11th floor, the total time of evacuation exceeds the 19 min 19 s verified for the simulation of total evacuation through the use of only the exit stairs and exit access corridors. As the actual set of occupants between 16th to 12th floors is 538, it is possible to consider that at least 20% of the occupants in this building could be evacuated using the lifts. That percentage can be associated to occupants with disabilities and/or children, which is expected to be lower than the percentage obtained in the simulation. It should be noted that the solution

Table 3. Simulations considering the numerical calculation using chapter 2 of section 4 of the NFPA 2008 handbook "calculation methods for egress prediction".

Floors	tV_1	tV_2
16th Floor – 15th Floor	1 min 26 s	1 min 15 s
15th Floor – 14th Floor	2 min 24 s	2 min 6 s
14th Floor – 13th Floor	3 min 10 s	2 min 42 s
13th Floor – 12th Floor	3 min 44 s	3 min 5 s
12th Floor – 11th Floor	4 min 17 s	3 min 28 s
11th Floor – 10th Floor	4 min 44 s	3 min 45 s
10th Floor – 9th Floor	5 min 18 s	4 min 6 s
9th Floor – 8th Floor	5 min 48 s	4 min 28 s
8th Floor – 7th Floor	6 min 49 s	5 min 18 s
7th Floor – 6th Floor	7 min 19 s	5 min 38 s
6th Floor – 5th Floor	7 min 53 s	6 min 2 s
5th Floor – 4th Floor	8 min 29 s	6 min 27 s
4th Floor – Refuge area	1 min 40 s	–
V1 – Refuge area (4th Floor)	22 min 19 s	–
V2 – Refuge area (4th Floor)	–	20 min 17 s
Total time	23 min 59 s	21 min 57 s

Table 4. Simulations considering the lifts in case of fire.

Simulation	Description	Results
1	– Evacuation of the 16th floor through lifts, remaining floors by exits on ground floor.	16 min 32 s
2	– Evacuation of the 16th and 15th floors through lifts, remaining floors by exits on ground floor.	15 min 36 s
3	– Evacuation of the 16th, 15th and 14th floors through lifts, remaining floors by exits on ground floor.	14 min 45 s
4	– Evacuation of the 16th, 15th, 14th and 13th floors through lifts, remaining floors by exits on ground floor.	15 min 18 s
5	– Evacuation of the 16th, 15th, 14th, 13th and 12th floors through lifts, remaining floors by exits on ground floor.	18 min 44 s
6	– Evacuation of the 16th, 15th, 14th, 13th, 12th and 11th floors through lifts, remaining floors by exits on ground floor.	21 min 41 s

optimized for the use of lifts would correspond to the involvement of the evacuation of the 16th to the 14th floors, which corresponds to 14 m 45 s.

5 CONCLUSIONS

According to this study, it was verified that the result of the evacuation time for the office floors obtained through the *Pathfinder 2015* simulator is very close to the value obtained in the fire drill exercise. However, it was not possible to have access to the number of occupants that were evacuated during the fire drill exercise, so it was not possible to make a more concrete analysis of these results.

Comparing the value obtained for the simulation 9 defined in Table 2 with the *Pathfinder 2015* simulator and with the numerical calculation performed using the methodology defined in *chapter 2 of section 4 of the NFPA 2008 Handbook "Calculation Methods for Egress Prediction—Chapter 2"*, we verified that they are very similar. Based on this comparison and based on the analysis of other articles founded in a systematic review search performed for high-rise buildings, the value of total evacuation time obtained may be considered acceptable. In this sense, and despite the existing non-conformities defined in Table 1, we assume that the building and its present evacuation conditions can be considered adequate for the characteristics that that building presents.

It was also observed that strategies with the evacuation priority order from upper floors to lowers floors and the inclusion of refuge areas can allow an optimization in the total evacuation time of this case study. The use of lifts in case of fire is a situation to be analysed in more detail in future research, analysing all the advantages and disadvantages that this solution entails in order to verify if its implementation and possible legislative changes, can be considered favourable in this type of circumstances.

SYMBOLS AND UNITS

ANPC – National Civil Protection Authority
BSB – Battalion Fire Fighters
NFPA – National Fire Protection Association
PSP – Public Security Police
RJ-SCIE – Legal Regulation of Fire Safety in Buildings (Law Decree 224/2015, October 9th, 2015)
RT-SCIE – Building Fire Safety Technical Regulation (Technical Regulation 1532/2008, December 29th, 2008)
FSE – Facility Safety and Emergency Management
U.T. – Usage-Type
V_1 – Emergency Stair 1
V_2 – Emergency Stair 2
WTC – World Trade Centre

REFERENCES

Decree Law n.o 224/2015 october 9th, first change of Decree Law n.o 220/2008, november 12th. (2015). *Diário Da República/Official Journal of the Portuguese Republic*, 1(198), 8740–8774. https://doi.org/10.1007/s13398-014-0173-7.2.

Ding, Y., Yang, L., Weng, F., Fu, Z., & Rao, P. (2015). Investigation of combined stairs elevators evacuation strategies for high rise buildings based on simulation. *Simulation Modelling Practice and Theory*, 53. https://doi.org/10.1016/j.simpat.2015.01.004.

Jeongin Koo, Yong Seog Kim, Byung-In Kim, & Christensen, K.M. (2013). A comparative study of evacuation strategies for people with disabilities in high-rise building evacuation. *Expert Systems with Applications*, 40(2), 408–417.

Kobes, M., Helsloot, I., de Vries, B., & Post, J.G. (2010). Building safety and human behaviour in fire: A literature review. *Fire Safety Journal*, 45(1), 1–11. https://doi.org/10.1016/j.firesaf.2009.08.005.

Portaria n.o 1532/2008 de 29 de dezembro - Technical regulation of fire safety (RT-SCIE). (2009), 9050–9127.

Proulx, Guylène; L. Bryan, John; F. Fahy, Rita; K. Lathrop, James; J. Fruin, Jonh; Cohn, Bert; J. O'Connor, D. (n.d.). *NFPA Handbook Section 4_20th_2008*.

Ronchi, E. (Lund U., & Nilsson, D. (Lund U. (2013). Assessment of Total Evacuation Systems for Tall Buildings, (January), 45.

Sagun, A., Bouchlaghem, D., & Anumba, C.J. (2011). Computer simulations vs. building guidance to enhance evacuation performance of buildings during emergency events. *Simulation Modelling Practice and Theory*, 19(3), 1007–1019. https://doi.org/http://dx.doi.org/10.1016/j.simpat.2010.12.001.

Thompson, P., Nilsson, D., Boyce, K., & McGrath, D. (2015). Evacuation models are running out of time. *Fire Safety Journal*, 78, 251–261. https://doi.org/http://dx.doi.org/10.1016/j.firesaf.2015.09.004.

Vermuyten, H., Beliën, J., De Boeck, L., Reniers, G., & Wauters, T. (2016). A review of optimisation models for pedestrian evacuation and design problems. *Safety Science*, 87, 167–178. https://doi.org/http://dx.doi.org/10.1016/j.ssci.2016.04.001.

Wu, G.-Y., & Huang, H.-C. (2015). Modeling the emergency evacuation of the high rise building based on the control volume model. *Safety Science*, 73. https://doi.org/10.1016/j.ssci.2014.11.012.

Yuan, J.P., Fang, Z., Wang, Y.C., Lo, S.M., & Wang, P. (2009). Integrated network approach of evacuation simulation for large complex buildings. *Fire Safety Journal*, 44(2), 266–275. https://doi.org/10.1016/j.firesaf.2008.07.004.

Radiological characterization of the occupational exposure in hydrotherapy spa treatments

M.L. Dinis & A.S. Silva
CERENA—Polo FEUP—Centre for Natural Resources and the Environment, FEUP—Faculty of Engineering, University of Porto, Porto, Portugal
PROA/LABIOMEP—Research Laboratory on Prevention of Occupational and Environmental Risk, Faculty of Engineering, University of Porto, Porto, Portugal

ABSTRACT: This work aims to perform a radiological characterization of the occupational exposure in 16 Portuguese thermal establishments based on indoor radon dosimetry. The effective doses received by workers due to radon inhalation were estimated and external gamma dose rates were measured through continuous periods. The dose assessment was estimated based on a deterministic approach while radon risk for indoor exposure was assessed on a probabilistic basis. The total effective dose ranged between 4 and 31 mSv/y. As expected, the results indicate that radon inhalation presents higher risk than gamma radiation exposure and, therefore, inhalation is considered as the main exposure pathway. In particular, two thermal establishments, TE5 and TE14, present considerable higher risk: 10^{-2} and 10^{-3}, respectively. For external exposure, all thermal establishments present a risk with a magnitude of 10^{-6}. From the obtained results, it is possible to identify the situations with higher risk where an urgent action must be addressed.

1 INTRODUCTION

The majority of exposure to ionizing radiation comes from natural sources. The exposure to radon gas and its progeny contribute with more than 50% of the total dose from natural radiation sources and it is recognized as the most important cause of lung cancer incidence except for smoking (UNSCEAR, 2000).

The World Health Organization (WHO, 2009) recommends that countries adopt a reference level of 100 Bq/m^3 for indoor radon concentration but if this level cannot be implemented under the prevailing country-specific conditions, it is recommended that the reference level should not exceed 300 Bq/m^3.

The International Commission on Radiological Protection (ICRP) has revised the reference level for radon gas in dwellings and other buildings with high occupancy rates by 300 Bq/m^3 (ICRP, 2010).

In the European Directive 96/29/EURATOM, it is recommended that spa therapy be considered as a professional activity of enhanced natural radiation exposure due, in large part, to the inhalation of radon released from thermal waters. The same Directive (repealed by Directive 2013/59/EURATOM) refers that each Member State shall identify, by means of survey or any other adequate mean, the work activities where a significant increase in the exposure due to natural radiation sources may occur. These include, in particular, hydrotherapy spa treatments facilities (thermal establishments), where thoron/radon daughters or gamma radiation may be present.

This work aims to develop a methodology for the radiological characterization of the occupational exposure in 16 Portuguese thermal establishments, based on indoor radon dosimetry. An assessment of the effective dose was performed and the resulting carcinogenic risk was estimated based on measurements of indoor radon concentrations. A deterministic approach was used for dose assessment while a probabilistic approach was adopted to estimate the resulting risk with the Monte Carlo method.

2 MATERIALS AND METHODS

2.1 *Occupational exposure and carcinogenic risk*

The recent European Directive 2013/59 EURATOM stipulates a limit on effective dose for occupational exposure of 20 mSv in any single year. Below this dose limit, the principle of optimization requires that any radiation exposure should be kept as low as reasonably achievable (ALARA). When the annual dose limit is exceeded, the regulatory body can permit this exposure by considering the individual case and/or imposing work conditions and dose restrictions for the successive years, providing that the annual dose over any five consecutive years does not exceed 20 mSv.

The US Environmental Protection Agency (EPA, 1997) evaluates the risk due to radiation exposure as the carcinogenic slope factor, representing the lifetime excess total cancer risk per unit intake or exposure. The cancer slope factor represents the slope of the dose-response curve, at very low concentrations, thus quantifying the cancer inducing potential; the unit is the inverse of a dose. The product of the cancer slope factor by the dose received estimates the risk for a member of the critical group. The risk represents the probability of cancer inducing by this particular exposure, in excess relatively to the background risk, known as the incremental lifetime risk.

A radiological risk assessment is an estimate of the probability of a fatal cancer over the lifetime of an exposed individual while a radiological dose assessment calculates the amount of radiation energy that might be absorbed by a potentially exposed individual in a specific exposure (EPA, 1997).

The acceptable risk is generally defined as 10^{-6} for the general public and 10^{-5} for workers. This means that an additional one case of cancer is accepted for populations of 1 million or 100 000, respectively. A risk level of 1 in a million, or 1 in one hundred thousand, also implies a likelihood that up to one person out of one million (or 100 000) equally exposed people would develop cancer if exposed continuously (24 h/d) to a specific radiation dose over 70 years (average lifetime). This value is in excess to the normal background number of cancer originated by multiple and indeterminate causes, respectively 200 000 or 20 000 (EPA, 1997).

In what concerns to the most type of radiation, the relationship between dose and development of cancer is well characterized for high doses, however, for lower doses, it is not well defined. In these cases, risks estimates are obtained by extrapolating data available for high dose exposures, typically based on linear/no-threshold model (LNT) that assumes there is no level below which radiological doses are safe (Shah et al., 2012).

The exposure scenario adopted in this study considers both internal and external exposure for estimating the dose and the associated risk of the occupational exposure in water treatments therapeutics at 16 Portuguese thermal establishments. For the critical group it was considered an average adult worker exposed during 8 hour/day, 5 days/week and 49 weeks/year for the last 20 years. For the exposure frequency, it was considered no job or task rotation of the workers (Silva et al., 2016; Dinis et al., 2017).

2.2 Description of the study

The International Basic Safety Standards for Protection against Ionizing Radiation (IAEA, 2003) states that in cases where individual monitoring is inappropriate, inadequate or not feasible, the occupational exposure of the worker shall be assessed based on the results of monitoring of the workplace and information on the locations and durations of exposure of the workers.

For the purpose of this study, 16 Portuguese thermal establishment (out of 38 active) agreed to participate in the assessment of the occupational exposure in their facilities under a confidentially agreement. Therefore, the concentration of radon was measured at different workplaces of each one of these establishments. The chosen locations included treatment rooms, pools and some access spaces where workers remain during treatment sessions. The assessment was carried out between November 2013 and September 2015, during two different periods: spring/summer and autumn/winter. Gamma dose rates were also measured at the same locations. The assessment was taken under normal activities and operating conditions (Silva et al., 2016; Silva & Dinis, 2016).

2.3 Indoor radon concentrations and gamma dose rates measurements

Radon concentration measurements were performed with CR-39 nuclear track detectors placed in each room at approximately 2 meters from the floor and for an exposure period of 45 days. After this period, the detectors were removed and stored individually in sealed containers to prevent any contamination from others sources. The analysis was performed in the Natural Radioactivity Laboratory of the University of Coimbra. Gamma dose rates were hourly registered with a Geiger counter (Gamma Scout) for the same period of time, 45 days, (Dinis et al., 2017).

2.4 Effective dose assessment

A deterministic methodology was adopted to calculate the effective dose due to radon inhalation using the expression given by UNSCEAR (2000) where all inputs parameters are defined as a single fixed value (Eq. 1):

$$H_{int} = C_{Rn} \cdot F \cdot E \cdot (DCF) \quad (1)$$

H_{int} is the annual effective dose (mSv/y), C_{Rn}, (Bq/m^3) is the average radon concentration measured at 2 m from the ground, F is the equilibrium factor between radon and its progeny, E (hours/year) is the occupancy factor and DCF (mSv/y per Bq/m^3) is the dose coefficient factor (UNSCEAR, 2006).

The equilibrium factor for radon decay products (F) is a measure of the degree of radioactive equilibrium between radon and its short-lived radioactive decay products representing the fraction of potential alpha decay energy of the short-lived radon decay products, compared to secular equilibrium. Due to the high variably of radon equilibrium factor, it is common to find a wide range of values in literature: 0.4–0.6 (Kojima, 1996; Koo et al., 2001; IAEA, 2003; Chen & Marro, 2011). A value of 0.51 was adopted for (F) in this study (Dinis et al., 2017).

For the exposure frequency (E) it was considered 1960 hours per year (Dinis & Fiúza, 2012; Dinis et al., 2017).

The Dose Coefficient Factor (DCF) or radon Equilibrium Equivalent Concentration (EEC) represents the conversion of potential alpha energy exposure ($Bq\,h/m^3$) to effective dose equivalent (nSv). A value of 9 nSv/h per Bq/m^3 (9×10^{-6} mSv/y per Bq/m^3) was adopted from UNSCEAR (2006).

The external doses (H_{Ext}, mSv/y) result from the measured gamma dose rates (D_γ, mSv/d) adjusted for the considered exposure frequency (Eq. 2):

$$H_{ext} = D_\gamma \cdot E \quad (2)$$

2.5 *Risk assessment*

The annual risk incurred to the workers both by internal exposure (radon inhalation) and by external exposure (gamma radiation) was assessed with a probabilistic risk-based approach. First, the annual risk from the internal exposure due to radon inhalation (R_{Rn}) was calculated with the indoor average radon concentrations (C_{Rn}, Bq/m^3); the inhalation rate (BR, m^3/d); the exposure frequency (E, d/y); and the radon cancer slope factor (SCF, Risk/Bq) (Eq. 3):

$$R_{Rn} = C_{Rn} \cdot F \cdot BR \cdot E \cdot (SCF) \quad (3)$$

The indoor radon data can be described by a log-normal distribution (UNSCEAR, 2006).

For the breathing rate (BR) a log-normal was also adopted with a mean and standard deviation of 16.45 and 4.69 m^3/d, respectively (Beals, 1996; Dinis & Fiúza, 2012; EPA, 2016; Dinis et al., 2017).

A triangular distribution was adopted for the indoor exposure frequency: 180 d/year (minimum), 307 d/year (maximum), and 245 d/year (most probable) (EPA, 2016).

For long-term variation of radon equilibrium factor a log-normal distribution was adjusted to the data with a mean and standard deviation of 0.51 ± 0.12, respectively (Kojima, 1986).

The cancer risk for external exposure due to gamma dose (the health risk from a given radiation dose) (R_{ext}) was obtained with Eq. 4:

$$R_{ext} = H_{ext} \cdot (DRC) \quad (4)$$

where (DRC) is the Dose-to-Risk Conversion Factor for Cancer Mortality (fatal cancer risk per Sievert). The ICRP (1990) considers a value of 0.04 Sv^{-1} for workers and 0.05 Sv^{-1} for the public.

The considered probabilistic distributions were used in the risk calculation for both internal and external exposure; the Monte Carlo method was used to generate the distributions for the input parameters as well as the output distribution of the resulting risk, developed in Matlab® code (Dinis et al., 2017).

3 RESULTS AND DISCUSSION

3.1 *Radon concentration*

The indoor radon concentrations registered in the studied Thermal Establishments (TE) are presented in Table 1 (Silva et al., 2016).

In average, the values of radon concentration are widely distributed, between 166 and 3124 Bq/m^3; the standard deviation is relatively high due to the high variability of the sampling locations within each thermal establishment, in addition to their individual geographical location which is conditioned by local and/or regional geology settings (Minda et al., 2009; Mihci et al., 2010).

Almost all establishments (97%) reported indoor radon levels higher than 100 Bq/m^3. In 69% of the establishments the radon concentration, in average, exceeded the EU reference level of 300 Bq/m^3 and 56% of the thermal establishments presented indoor radon concentrations above 400 Bq/m^3, the national action level.

Radon concentration measurements are just a part of the information needed to conduct a

Table 1. Indoor radon concentration (Bq/m^3).

TE N°	μ ± σ (Bq/m^3)	Min. (Bq/m^3)	Max. (Bq/m^3)
1	1189 ± 1232	238	3479
2	569 ± 116	422	707
3	364 ± 210	152	724
4	428 ± 70	274	502
5	3124 ± 1212	1912	4335
6	1256 ± 430	878	2181
7	1047 ± 564	366	1681
8	354 ± 7	347	361
9	262 ± 107	143	376
10	233 ± 90	121	406
11	480 ± 255	209	1079
12	1122 ± 403	813	1692
13	187 ± 136	73	498
14	2090 ± 553	1130	2873
15	166 ± 46	93	235
16	288 ± 100	172	467

radiological assessment (Guedes et al., 1999; Wang et al., 1999; Ioannides et al. 2000). For radiological protection purpose, radon progeny has a very important contribution to the internal exposure and should also be measured, however, most of the available data for radon progeny are extrapolated from the radon concentration by using equilibrium factors not generally known due to the complexity in determining radon equilibrium factor (UNSCEAR, 2000).

3.2 Dose assessment

The annual effective dose received by the workers due to radon inhalation ranged between 1 and 28 mSv/y; for the gamma radiation dose, the results are presented in Table 2 (Dinis et al., 2017).

The total effective dose was obtained by combining the doses from the internal exposure with the doses from the external exposure; the results ranged between 4 and 31 mSv/y showing that the external irradiation due to gamma radiation is negligible when compared to the internal irradiation caused by radon inhalation. In addition, for the cases where the effective dose is above 6 mSv/y, workers are classified as 'category A' (i.e., subject to individual monitoring and medical surveillance).

3.3 Risk assessment

The probabilistic distributions of the input parameters, generated by the Monte Carlo method, are presented in Figure 1. As all distributions have the same shape, only one set of distributions is presented here: Radon Equilibrium Factor (F); Expo-

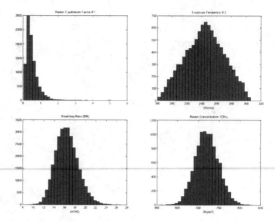

Figure 1. Probabilistic distributions of the input parameters.

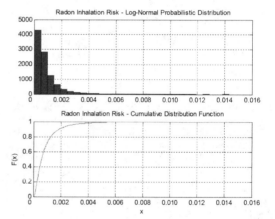

Figure 2. Radon inhalation risk: Log-normal probabilistic distribution and cumulative function.

sure Frequency (E, d/year); Breathing Rate (BR, m³/d) and Radon Concentration (CRn, Bq/m³) (Dinis et al., 2017).

The probabilistic distributions for the calculated risk are presented in Figure 2 for internal exposure (radon inhalation) and in Figure 3 for external exposure (gamma radiation) (Dinis et al., 2017).

A summary of the results from the probabilistic approach is also presented in Table 3. The central tendency and the high-end estimate (mean μ, median Md and 95th percentile) were used for the characterization of the individual risk (Dinis et al., 2017).

The results showed that inhalation is the main exposure pathway as radon inhalation presents a considerable higher risk than external radiation. The exposure at TE5 and TE14 presented the highest inhalation risk: 0.0025 (50th percentile)

Table 2. Gamma dose rates (mSv/y).

TE N°	μ ± σ (mSv/y)	Min. (mSv/y)	Max. (mSv/y)
1	2.73 ± 2.40	1.30	4.95
2	3.05 ± 0.32	2.73	3.59
3	4.24 ± 0.92	2.24	7.98
4	2.54 ± 1.73	2.18	2.88
5	2.79 ± 0.18	1.91	6.50
6	4.29 ± 0.22	1.29	4.29
7	3.47 ± 0.40	3.24	3.71
8	2.51 ± 2.12	1.16	3.88
9	3.76 ± 0.35	3.65	3.87
10	3.67 ± 0.46	2.51	4.83
11	2.48 ± 0.08	2.04	2.83
12	3.56 ± 0.20	2.89	5.52
13	2.57 ± 1.80	2.54	2.60
14	3.00 ± 0.08	2.91	3.08
15	3.67 ± 2.40	2.45	9.08
16	3.38 ± 0.32	2.50	5.02

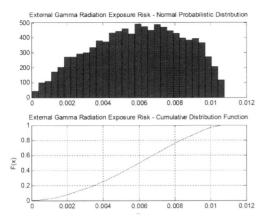

Figure 3. External gamma radiation exposure risk—normal probabilistic distribution and cumulative probability functions.

Table 3. Annual risk from radon inhalation and external exposure (Dinis et al., 2017).

TE N°	Annual risk Radon inhalation			Annual risk External exposure		
	μ	Md	95th	μ	Md	95th
1	1×10^{-3}	1×10^{-3}	4×10^{-3}	5×10^{-6}	5×10^{-6}	8×10^{-6}
2	7×10^{-4}	5×10^{-4}	2×10^{-3}	6×10^{-6}	6×10^{-6}	9×10^{-6}
3	4×10^{-4}	3×10^{-4}	1×10^{-3}	8×10^{-6}	8×10^{-6}	1×10^{-5}
4	5×10^{-4}	3×10^{-4}	1×10^{-3}	5×10^{-6}	5×10^{-6}	8×10^{-6}
5	4×10^{-3}	2×10^{-3}	1×10^{-2}	5×10^{-6}	5×10^{-6}	8×10^{-6}
6	1×10^{-3}	1×10^{-3}	4×10^{-3}	8×10^{-6}	8×10^{-6}	1×10^{-5}
7	1×10^{-3}	9×10^{-4}	3×10^{-3}	6×10^{-6}	7×10^{-6}	1×10^{-5}
8	4×10^{-4}	3×10^{-4}	1×10^{-3}	5×10^{-6}	5×10^{-6}	8×10^{-6}
9	3×10^{-4}	2×10^{-4}	9×10^{-4}	7×10^{-6}	7×10^{-6}	1×10^{-5}
10	3×10^{-4}	2×10^{-4}	8×10^{-4}	7×10^{-6}	7×10^{-6}	1×10^{-5}
11	6×10^{-4}	4×10^{-4}	2×10^{-3}	5×10^{-6}	5×10^{-6}	8×10^{-6}
12	1×10^{-3}	9×10^{-4}	4×10^{-3}	6×10^{-6}	7×10^{-6}	1×10^{-5}
13	2×10^{-4}	1×10^{-4}	7×10^{-4}	5×10^{-6}	5×10^{-6}	8×10^{-6}
14	2×10^{-3}	2×10^{-3}	7×10^{-3}	5×10^{-6}	6×10^{-6}	9×10^{-6}
15	2×10^{-4}	1×10^{-4}	6×10^{-4}	7×10^{-6}	7×10^{-6}	1×10^{-5}
16	3×10^{-4}	2×10^{-4}	1×10^{-3}	6×10^{-6}	6×10^{-6}	1×10^{-5}

and 0.0017 (50th percentile), respectively; 0.0108 (95th percentile) and 0.0071 (95th percentile), respectively. For the gamma radiation exposure, in all cases the risk presents a magnitude of 10^{-6} with higher values in TE3, TE6, TE9, TE10 and TE15.

The key contributors to variability or uncertainty in the predicted exposures or risk estimates can be identified by the Monte Carlo method and used to compare the risks concerned to different management options, or to invest in research with the greatest impact on risk estimate uncertainty.

4 CONCLUSIONS

The results from this study showed that several reference levels (indoor radon levels, effective dose) were exceed. In 9 establishments (out of 16) the values of the indoor radon concentrations were above 400 Bq/m^3. Approximately 80% of the total measurements of indoor radon concentration exceeded the EU reference level of 300 Bq/m^3.

In what concerns to the effective dose, there are 11 establishments with values above 6 mSv/y, the reference level of "an existing exposure situation" and in these cases, the exposure should be classified as "a planned exposure situation" and actions should be taken (Directive 2013/59/EURATOM).

From the results obtained in Table 3, is possible to identify the situations with higher risk where an urgent action must be addressed. Such places must be subjected to special and more detailed investigation, taking into account the whole mobility of radon and progeny in the ambient air.

The probabilistic estimates of risk should be presented as a supplement of the deterministic approach; these may be used to answer scenario specific questions and to facilitate risk communication while a probabilistic approach may be used to provide quantitative estimates of the uncertainty and variability providing clear and concise information in individual exposures and risks.

REFERENCES

Beals, J.A.J., Funk, L.M., Fountain, R., Sedman, R., 1996. Quantifying the Distribution of Inhalation Exposure in Human Populations: Distribution of Minute Volumes in Adults and Children, Environ Health Perspect., V. 104, N° 9.

Chen, J., Marro, L., 2011. Assessment of radon equilibrium factor from distribution parameters of simultaneous radon and radon progeny measurements, Radiat Environ Biophys 50:597–601, doi: 10.1007/s00411-011-0375-8.

Dinis, M.L., Fiúza A., 2012. Occupational exposure during remediation works at a uranium tailings pile, J. Environ. Radioact., V. 119, 63–69, doi: 10.1016/j.jenvrad.2012.08.009.

Dinis, M.L., Silva, A.S., Pereira, A.J.S.C., Fiúza, A. (2017). Radiological Characterization of Selected Thermal Centres Based on Indoor Radon Dosimetry, Proceedings of the International Conference Waste Management Symposium 2017, Phoenix, USA, March 05–09, 2017.

EPA, 1997. Health Effects Assessment Summary Tables, Radionuclide Carcinogenicity Slope Factors: HEAST.

EPA, 2016. Exposure Factors Handbook: 2011 Edition, EPA/600/R-090/052F, 09/2011.

Guedes, S., Hadler, J.C., Iunes, P.J., Navia, L.M.S., Neman, R.S., Paulo, S.R., Rodrigues, V.C., Souza, W.F., Tello, S.C.A., Zúñiga, A., 1999. Indoor radon and radon daughters survey at Campinas-Brazil using CR-39 first results, Radiat. Meas., 31, 287–290.

IAEA, 2003. Radiation Protection against Radon in Workplaces other than Mines, IAEA Safety Reports Series, No.33.

ICRP, 1990. Recommendations of the International Commission on Radiological Protection, ICRP Publication 60, Ann. ICRP 21 (1–3).

ICRP, 2010. Lung Cancer Risk from Radon and Progeny and Statement on radon, Publication 115 Ann, ICRP 40(1).

Ioannides, K.G., Stamoulis, K.C., Papachristodoulou, C.A., 2000. A survey of 222Rn concentrations in dwellings of the town of Metsovo in north-western Greece. Health Phys., 79 (6), 697–702.

Kojima, H., 1996. The equilibrium factor between radon and its daughters in the lower atmosphere, Environ. Int, V. 22, S187-S192, PI1 SOl60-4120(96)00107–9.

Koo, Y., Yip, Y., Ho, Y., Nikezic, D. Yu, N., 2001. Experimental study of track density distribution on LR115 detector and deposition fraction of ^{218}Po in diffusion chamber, Nucl. Instr. Meth. Phys. Res, V. 491, 470–473.

Mihci, M., Buyuksarac, A., Aydemir, A., Celebi, N., 2010. Indoor and outdoor Radon concentration measurements in Sivas, Turkey, in comparison with geological setting. J. Environ. Radioact., V. 101, 952–957.

Minda, M., Tóth, Gy., Horváth, I., Barnet, I., Hámori, K., Tóth, E., 2009. Indoor radon mapping and its relation to geology in Hungary. Environ. Geol., V. 57, 601–609.

Shah, D.J., Sachs, R.K., & Wilson, D.J., 2012. Radiation-induced cancer: a modern view. Br J Radiol, 85(1020), e1166–e1173, doi: 10.1259/bjr/25026140.

Silva, A., Dinis, M.L., Pereira, A.J.S.C., 2016. Assessment of indoor radon levels in Portuguese thermal spas, Radioprotection, V. 51, 249–254, doi: 10.1051/radiopro/2016077.

Silva, A.S., Dinis, M.L., 2016. Measurements of indoor radon and total gamma dose rate in Portuguese thermal spas, in Occupational Safety and Hygiene IV, 485–489, Taylor & Francis, doi: 10.1201/b21172-93.

UNSCEAR, 2000. Sources and effect of ionizing radiation, United Nation Scientific Committee of the Effect Atomic, Radiation Report on The General Assembly, New York.

UNSCEAR, 2006. Report to the General Assembly, with Scientific Annexes, Annex E, Sources-to-effects assessment for radon in homes and workplaces, ISBN 978-92-1-142274-0.

Wang, Y., Ju, C., Stark, A.D., Teresi, N., 1999. Radon mitigation survey among New York State residents living in high radon homes, Health Phys., 77 (4), 403–409.

WHO, 2009. Handbook on indoor radon: a public health perspective, ISBN 978 92 4 154767 3.

Labor claims and certification in occupational health and safety management

Fernanda Daniela Dionísio, Nélson Costa & Celina Pinto Leão
ALGORITMI Centre, School of Engineering, University of Minho, Guimarães, Portugal

ABSTRACT: The main objective of this work is to evaluate the influence of the certification in safety and health management in workplaces in the accidents rates. To achieve this objective, a mixed research methodology was chosen: document analysis and questionnaire used as data collection techniques. The bibliographic review was carried out on the various aspects of Occupational Health and Safety (OHS) Certification and the questionnaire for the risk evaluation in OHS allowing work accidents and professional diseases identification. The developed questionnaire was distributed to both workers from companies with External OHS Service and from companies with Internal OHS Service, in a total of 74 questionnaires (60 and 14 respondents, respectively). Also, data on occupational accidents were used to gather information on OHS Services. The results of the study point out to an existing relationship between the type of OHS practiced/implemented in companies and the compliance with the statutory requirements of OHS, which may intervene directly in the workplace accidents. After analysing the studied variables, recommendations are proposed, aiming to an improvement of the OHS certifications.

1 INTRODUCTION

Work accidents are a problem at a national, European and global level, with serious economic and social consequences (EASHW 2002). The number of accidents, including fatal accidents, results in costs for countries, for organizations and with serious consequences for the health and well-being of the population, being a labor accident one of the heaviest suppliers that currently affects the working population (Alves, 2012). In view of the above, it is increasingly important that managers consider safety based on behavior, by identifying and monitoring the behavior in all the activities in the workplace (Pecillo, 2016). Despite the implementation of many safety and health management practices at work, know as OHSMS (Occupational Health and Safety Management System), occupational accidents still occur. However, they occur because there is a risk, and this risk arises from a set of working conditions that influence the behaviour of each worker (Pecillo, 2016). With proper control of the work conditions and therefore the associated risk, it may be possible to prevent accidents, regardless of the individual factors of each case (Garcia-Herrero et al., 2012).

In recent years, investigations on accidents tended to be more on an organizational level/causes than on the technical and human causes (Okoh & Haugen, 2013). Despite of many years of research on the performance of OHSMS, there is still no conclusive evidence of its effectiveness (Podgórski, 2015). There are, of course, reports showing the strong correlation of its implementation with safety and health indicators (Yoon et al., 2013). In order to reduce the number of accidents at work, the International Labor Organization, ILO, (ILO, 2004) recommends the implementation of an OHSMS in companies to manage the risks, and also to propose a rule to guide OHSMS' deploying (Health, 2001). The OHSMS are management tools that contribute to an efficient performance improvement of companies in relation to OHS. To compliance with legislation, increase of productivity, reduce of accidents, credibility with the public and growing awareness of the safety and health of the organization's employees and partners is necessary (Oliveira, Oliveira, & Almeida, 2010). The H&S certification companies have a key role, since they must alert the business of the risks involved, evaluate and implement a culture of security. Preventing such risks starts with involving the company's own workers (Yuan and Wang, 2012).

To have a successful OHSMS, it is necessary that the organization has an established safety culture, i.e., it is essential to know in what situation is the safety culture maturity in the company, for the formulation of changes in plans, where needed, to thereby obtain the success of this system (Mara & Marino, 2011). The OHSMS should take an approach of the kind Plan-Do-Check-Act (PDCA), commonly known as continuous improvement cycle, which is based on planning, implementing, evaluating and updating in a correct and/or preventive way so that the organization can systematically obtain better results in relation to the OHS indicators (Health, 2001). Implement and certify a management system for OSH is a strategic issue for companies. The need for companies to demonstrate commitment through certification, either by image issues, competitive advantage or customer, pressure has been assumed as an irreversible process of modern times.

Preserving OHS of workers is one of the basic components and complementary to ensure full quality assurance and production processes competitiveness. Concerns about OHS are now being treated in the form of management system, using the standard OHSAS 18001 Occupational Health and Safety Assessment Series (OHSAS 18001, 2007) structured in the PDCA method (Plan-Do-Check-Act), i.e., a systematic and organized way to pursue continuous improvement.

The relevance and importance of the development of the study comes from the need to study the variation of work accidents with certification in OHSMS. This is a possible and effective way to know what are the relationship between accidents at work in the companies that hired an external entity responsible for ensuring that the OHSMS system is implemented, practiced and meets all legal requirements. Finally, we intend to propose improvements in the OHSMS implementation method in order to delegate preventive principles and possible improvements also in terms of work accidents.

2 MATERIAL AND METHODS

To achieve the objectives of this study, two tools were used in order to collect data: (1) a survey by questionnaire, with questions adapted to the business reality under study. Both companies' workers with Internal Service OHS and External OHS services were the respondents; (2) data on occupational accidents (maintained the confidentiality and anonymity of the information provided by workers and businesses). In the following sub-sections, the characterization of the sample, tools brief descriptions and the analyze performed.

2.1 Sample

A total of 74 answers were received, of which 52.7% were from female workers (aged 19 to 55 years with an average of 33 years, SD = 9). With respect to male workers (47.3%), they were aged 19 to 62 years (SD = 10). With regard to Health and Safety at Work services, there are respondents who work in companies with common OHS Services and workers from companies exclusively with OHS Internal Services. Of companies with external services, 35% have some responsibility for OHS workers, and 45% have no one responsible for this area in the company.

Most respondents have secondary education (55%), followed by those with higher education (28%), and bachelor's degree (22%). 67.6% have an individual contract of employment and have been in the company for more than 2 years. Concerning the professional category of the respondents, 15% are embroiderers, and the remainder being from various categories, from administrative staff, electricians, locksmiths, hairdressers, agents of authority, senior technicians, veterinarians, among others.

2.2 Tools used

The analyses of the labor claims and certification in Occupational Health and Safety Management was performed by a questionnaire and based on data on occupational accidents collected from a database of a company providing OSH services.

2.2.1 Questionnaire

The questionnaire addressed various aspects of the OHS, from topics related to their involvement in the company's security issues, workers posture and conscience before most delicate situations on OHS and on accidents at work, and occupational diseases. The questionnaire includes two main parts. After the characterization of the respondents (age, gender, professional activity, number of years working at the company, number of years at the actual professional activity), each respondent evaluates the risk in OHS and identifies work accidents and professional diseases. Some questions were of "Yes"/"No" answers, others were assessed using a 5-point Likert scale of importance (each item ranged from "1 = Never" to "5 = Always").

The questionnaire had an expected duration of completion of approximately 15 minutes, and was applied during a work pause of the workers. The questionnaire was distributed to the workers and was filled in with the support of the authors of this study (as an interview by questionnaire), thereby eliminating respondents confusion and avoiding or reducing the subjectivity inherent. The chosen firms were those that have responded and expressed willingness for contributed. The questionnaires were distributed in the municipality of Fafe (Portugal) directly to officials known to the authors. The study was conducted from January to September 2017.

2.2.2 Database

The company's database was another tool used to obtain information regarding OHS Services information. In order to have access to the company's accident records and communicated by the companies that contracted their services, it was necessary to overcome some bureaucratic questions. First, a proposal was presented to the company responsible and then a formal request through a letter of authorization guaranteeing anonymity. Some difficulties were encountered during this process: the waiting time to obtain the data and the lack of some essential data to further study, for example, number of employees of the company's distressed and number of working hours.

2.2.3 *Data analysis*

A descriptive statistical analysis was performed in order to make a characterization of the sample and to identify possible trends and special behaviors. Subsequently statistical tests were performed, including the chi-square test (χ^2), with a 5% significance level. The results were compared to those purchased through the literature review. Thus, it was possible to check whether there is an association between OHS mode practiced in business and compliance with legal requirements of OHS, which could be directly involved in work accidents.

3 PRESENTATION AND DISCUSSION OF RESULTS

3.1 *Perception of workers obtained from the questionnaires*

Workers, from companies with external services, and have someone responsible for OHS, defined that the responsibility, regarding the OHS issues, should not only be the hoster company' responsibility to perform. Usually this is performed in collaboration with the clerk dealing with OHS issues. On the contrary, in the companies where it was considered the Internal Service, (14 situations), there is at least one SST engineer. Of respondents who said that there is no responsible for OHS in the company, 70.6% attaches importance to the existence of a responsible worker, while 29.4% say it is important not to appoint an employee to perform these duties.

With regard to external services, about 40% of respondents in each of these services say that security is not the top priority of management. In order to perform tasks that jeopardize the safety and modality of OHS applied in the company and the availability to talk about safety in the company and to reflect on safety during the work, there is a significant association between the quantity of work and the modality of OHS mode practiced in the company.

Workers who are most familiar with OSH measures before they are put into practice are the External Services mode. 43% of respondents from the Internal Service replied that they rarely know the security measures before they are put into practice. There is a perception that the Internal Service has fewer OHS inspections. With regard to compliance with safety standards, in the opinion of both External Service and Internal Service workers, there are sometimes the application of safety standards.

The perception of respondents on the introduction of updated safety standards is similar to the two modes, where the Internal Service, 14% say always be updated, compared to 20% of the External Service. It is in the form of external services that are less valued compliance with safety rules, 25% said rarely or never, while the Internal Service 7% said rarely. The Internal Service stands out with 36% of responses on how to always value the compliance with safety rules.

According to the perception of workers, the safety equipment is more often available to respondents the Internal Service and, in these companies, they value the safety of people and equipment, and ideas of workers on OHS were the most requested, there is also featured in this mode in response that the boss cares more about the safety compared to the modality of the External Services. Respondents from companies with External Services are working fewer times safely, ie, 14% said they rarely or never work safely. The modality of the Internal Service stands out with 43% of respondents claiming to always work safely and is also where there are fewer respondents who have suffered accidents, 7%, represented by only one worker, in a 14 this modality universe in the company, compared with 25% of respondents from companies with External OHS services. On the issue of accidents and these are discussed and learn with them, in the form of Internal Services, 14% of respondents replied that rarely discusses the incident and learn from it, as in the form of external services, were obtained 21% of responses so and 3% still said they never discussed the accident and never learn from it. The Internal Service, 43% of respondents said that in their companies is always discussed the accident and learn with him, while in companies with external services the percentage drops to 33%.

Respondents of the Internal Service are the ones that give more importance to OHS services, represented by 50% of responses, while respondents External Service have 40% of responses. About 36% of the two methods consider the important security services. Respondents from companies with external services are those with largest number of responses when we analyze the category of indifference of these services, with 20% of respondents, compared to 7% of respondents' answers the Internal Service. The number of responses is similar to the two modalities in respect of respondents who consider unimportant security services. For 40% of respondents, the purpose of the implementation of OHSMS is to reduce risks of accidents and occupational diseases, 26% claims to be to improve working conditions, 19% is of the opinion is to create a culture towards continuous improvement the organization, 11% that only serves to comply with the legislation and only 3% has the idea of being to increase the motivation of employees and 1% to improve the image of the organization.

On the other hand, the aspects where that the companies with External Services stood out for the positive, while the companies with the Internal Service function were in the participation of the employees in the process of identification of hazards and evaluation of risks, in the knowledge of the measures of OHS by part of the employees, before they are put into practice, in the sense of safety, in the valuation of the employee as a worker, in the surveys of security companies and in the occurrence of accidents in recent times, where, in this case, all employees are considered part of the company where the respondent works.

In summary, Table 1 represents the issues that were seen more positively by the majority of respondents from companies with Internal OHS Services and External OHS Services modality.

3.2 *Analysis of accident data provided by the company for the provision of External OHS services*

In this work, it was still possible to access the claims register for about 5800 companies. To register we found that, from 2001 to 2016 there are records in 1875 company claims, however, there are two situations that do not have the year of the accident. In 2011 there were 207 accidents (11%) and an increase in the remaining years, by 2015, 21%. In 2016 had decreased to 290 accidents, i.e. to 16%. In order to try to understand this decrease, the data obtained in this study were compared with the national scene and found that at national level also decreased quite accidents at work 2015 to 2016, regarding to serious accidents (ACT 2017). According to an ACT Inspector General interview, great efforts have been made to improve the safety situation, which explains the current decrease in the number of accidents in recent years (Pinto, 2017).

To make the most detailed study of accidents at work, communications were deemed to relate to 2016, when there was a 290 occurrences record, the majority of male gender cases, 69%. For the place of occurrence of the accident, the majority, 73% is within the company premises (excluding the transport means), 20% occurred outside the company premises (excluding the transport means) 4% was *in intinere*, and 3% in transport means. With regard to the Economic Activity Code companies of the victims where there is highest accident in manufacturing, 112 cases and public administration and defense; Social Security, with 87 cases. In companies with mode Internal OHS services, 50% of victims are male gender and 50% of females. Already in companies where OHS is the company's responsibility for providing these services, 69% of victims in 2016 are male gender and only 31% are of the female gender. Analyzing the question of the location of the lesion and mode of OHS services, about 33% of companies with Internal Service was injured in the trunk, the other in the lower limbs (excluding feet) and other hands. With regard to companies with external service areas of the lesions are more diversified, 26% correspond to the hands, 18% relate to the lower limbs (excluding feet), 14% higher members (except hands), 12% to the trunk, 11% to feet, 8% multiple injuries, 6% head (except eyes) and 5% to eyes.

Table 1. Summary positive results addressed without consultation, obtained from companies with External Health Service and with Internal Health Service.

Internal service	External service
Requesting tasks that do not put OSH at risk	Participation in the process of hazard identification and risk assessment
Clarification of doubts in training actions	Knowledge of OSH measures before they are put into practice
Control of safety standards	Feeling safe
Security standards update	Appreciation as a worker
Enhancement of compliance with safety rules	Surveys of security companies
Availability of Safety Equipment	Occurrence of accidents in recent times in the company
Importance of the safety of people and equipment	
Discussion and learning with work accidents	
Work safely	
Work without Security to quickly accomplish the task	
Workers involved in safety issues	
Chief concern on security issues	
Importance of OSH services	
Occurrence of work-related accidents by respondents	

4 CONCLUSION

The certification of Occupational Health and Safety Management System promotes the development of a safer and healthier work environments for workers, allows the organization to identify and control labor risks consistently, ensures compliance with legislation, and demonstrates that is clearly focused on the continuous improvement of occupational health and safety conditions throughout the organization and is an essential tool for improving working conditions and reducing occupational accidents and diseases. Transforming organizational safety culture poses a challenge, since it is necessary to break the paradigm, where the issue is no longer solely the responsibility of the security professional, to be the responsibility of the entire organization, and competence must be seen as the center of the action with regard to OHS issues.

After analysis of the data, it was verified that the companies aim to comply with regulatory norms and other pertinent legal requirements, with significant differences in some aspects of OHS in companies with Internal Health Service and in companies with External Health Service. The employees of companies with internal service stand out from those of companies with external OHS service in the following aspects: clarification of doubts in training actions, valorization of compliance with safety rules, availability of safety equipment, importance of safety of people and safety, involvement in OHS issues, concern of the head in OHS issues, importance of OHS services, and OHS respondents suffered fewer accidents at work than respondents from the External Services modality.

In this study, in addition to analyzing the data related to the questionnaire, data on labor claims, provided by an outsourced OHS service provider, were also analyzed. Through the analysis of these data, it was verified that from 2011 to 2015 there was an increase in accident rates, which decreased significantly, from 21% in 2015 to 16% in 2016. Regarding the place of accident occurrence, a majority (73%) is inside the company's premises (excluded in means of transportation), 20% occurred outside the company's premises (excluded in means of transport), 4% was *in intinere*, and 3% in the means of transportation. There is greater loss in manufacturing, 112 cases and in public administration and defense; Social security registered 87 cases. In the course of this study, a set of limitations was identified, the time that the worker had to make available to fill out the questionnaire, since, although it appeared simple to complete, it was exceeded for the time provided, since the respondents were the response options were: always, often, sometimes, rarely and never, and respondents reported that, for example, the barrier between the response sometimes and rarely is very so they kept thinking longer than originally foreseen for this issue.

Another limitation was related to the acquirement of data provided by a company providing External OHS Services, related to labor claims.

After this work, it is considered that there will be a benefit in the sharing of the analysis of their results, which may give the workers the perception that they are effectively heard and considered, and also that their participation and active intervention as agents of change are important within the scope of OHS.

As a proposal for improvement in OHS Certification, it is essential that there be a preventive and proactive attitude of the organization, from the companies responsible for OHS, to employers and workers, assuming all responsibilities and working together to achieve the same objective: continuous improvement in OHS, where the safety, comfort and well-being of the organization are assured, systematically and systematically evaluating the facilities and the work stations in all aspects, being essential the exchange of impressions among all.

REFERENCES

ACT. (2017). Acidentes de Trabalho Graves. 29 de Setembro. Retrieved from http://www.act.gov.pt/(pt-PT)/CentroInformacao/Estatistica/Paginas/AcidentesdeTrabalhoGraves.aspx.

Alves, A.M.R.C.D. (2012). Análise de acidentes de trabalho numa indústria metalomecânica. Escola Superior de Ciências Empresariais. Retrieved from: http://comum.rcaap.pt/handle/123456789/4305.

EASHW. (2002). European Agency for Safety and Health at Work. (E. Communities, Ed.). https://doi.org/10.16373/j.cnki.ahr.150049.

García-Herrero, S., Mariscal, M.A., García-rodríguez, J., & Ritzel, D.O. (2012). Working conditions, psychological / physical symptoms and occupational accidents. Bayesian network models. Safety Science, 50, 1760–1762. https://doi.org/10.1016/j.ssci.2012.04.005.

Health, O. (2001). African Newsletter, 11(3).

Mara, M., & Marinho, D.O. (2011). Cultura e gestão da segurança no trabalho: uma proposta de modelo, 205–220.

OHSAS 18001:2007, Detailed Guide. Retrieved from: https://www.nsai.ie/getattachment/Our-Services/Certificati on/Management-Systems/OHSAS-18001/MD-19-02-Rev-4—OHSAS-18001-Occupational-Health-and-Safety.pdf.aspx.

Okoh, P., & Haugen, S. (2013). Maintenance-related major accidents: Classification of causes and case study. Journal of Loss Prevention in the Process Industries, 26(6), 1060–1070. https://doi.org/10.1016/j.jlp.2013.04.002.

Oliveira, J.O., Oliveira, A.B., & Almeida, R.A. (2010). Gestão da segurança e saúde no trabalho em empresas produtoras de baterias automotivas: um estudo para identificar boas práticas. Produção, 20(3), 481–490. https://doi.org/10.1590/S0103-65132010005000029.

Pecillo, M. (2016). The resilience engineering concept in enterprises with and without occupational safety and health management systems. Safety Science, 82, 190–198. https://doi.org/10.1016/j.ssci.2015.09.017.

Pinto, L. (2017). Acidentes: Sinistralidade laboral diminuiu. Construção continua a liderar as estatísticas.

Newspaper article, Público. Retrieved from https://www.publico.pt/2017/04/06/sociedade/noticia/sinistralidade-laboral-diminuiu-construcao-continua-a-liderar-as-estatisticas-1767880.

Podgórski, D. (2015). Measuring operational performance of OHS management system - A demonstration of AHP-based selection of leading key performance indicators. Safety Science, 73, 146–166. https://doi.org/10.1016/j.ssci.2014.11.018.

Yoon, S.J., Lin, H.K., Chen, G., Yi, S., Choi, J., & Rui, Z. (2013). Effect of occupational health and safety management system on work-related accident rate and differences of occupational health and safety management system awareness between managers in South Korea's construction industry. Safety and Health at Work, 4(4), 201–209. https://doi.org/10.1016/j.shaw.2013.10.002.

Yuan, X., & Wang, K. (2012). Study on safety management of small and medium-sized enterprises based on BBS. Procedia Engineering, 45, 208–213. https://doi.org/10.1016/j.proeng.2012.08.145.

Ergonomic risk measurement in prioritizing corrective actions at workstations

W. Czernecka & A. Górny
Poznan University of Technology, Poznan, Poland

ABSTRACT: One of the most important steps in the management of work safety is to diagnose dangerous workplace-related irregularities and to implement corrective actions that prevent recurrence of hazards for workers. This article describes the relationship between ergonomic risk measurement at workstations and the ability to diagnose dangerous work items and the opportunity to propose appropriate corrective actions. Therefore, in this paper the problem of decision making based on measuring ergonomic risk and particular diagnosis of musculoskeletal loads was described. The study also presents the types of measures that can be used to measure ergonomic risk with an indication of what type of load they can measure. The article also outlines how to use the ergonomic risk measurement results to plan and implement corrective actions on the example of different workplaces. These activities can be undertaken on a technical or organizational levels and can also help to shape employees' awareness of workplace hazards. The importance of this type of activity is also emphasized in order to ensure safe and comfortable workplaces, which should be part of the work safety management in enterprises.

1 INTRODUCTION

Ergonomics is a science that deals with designing and adaptation of workplaces to human capabilities. Practical ergonomics due to its methodology reduces work accidents and by focusing on human reliability factors ensures high level of human performance (Butlewski et al. 2014). In one of its branches focuses on the diagnosis of musculoskeletal disorders of employees, which are often consequences of way the work is performed. Musculoskeletal disorders are main problem in in various industries (Lasota & Hankiewicz 2016a). There are many methods of measuring ergonomic risk at workplaces. Their use is based on analysis of many parameters – the pace of work, the position of the individual segments of the worker's body during work, the repetition of the employee's movements, the load on the worker during lifting and carrying etc. It is important to choose the appropriate tool for the type of workstation and performed work movements (Kale & Vyavahare 2016). An ergonomic diagnosis carried out by such methods leads to a report which shows the work items that cause the greatest discomfort and the musculoskeletal disorders of the workers (Bryska & Herma 2011). The ergonomic risk (distinguished usually as small, medium, requiring immediate actions) is also assessed.

Results of ergonomic risk measurements at the workplace allow to carry out corrective actions that promote optimal working conditions and contribute to the safety of workers. These actions are also an important element of the workplace safety management system, which assumes, among other things, organizing, planning, implementing and evaluating activities to continuously improve safety and working conditions (Taderera 2012, Górny 2014, 2017).

2 THE PROBLEM OF ERGONOMIC RISK MEASUREMENT

An ergonomic diagnosis is a set of activities that must be performed to define and evaluate the relationships that exist between a person, a technical object and the task being performed. The purpose of such a diagnosis is to estimate the ergonomic quality of the system. The assessment of ergonomic risk factors leads to verifying whether a worker-machine system can cause an accident or ailments in employees. In some cases it is also possible to determine the likelihood of anomalies occurring at the workstations (Lubaś 2010).

The effectiveness of the ergonomic risk management process depends to a great extent on defining of worker-maschine interaction, defining of worker exposure to risk factors, and the actions taken to eliminate or reduce them (Burgess-Limerick et al. 2009).

The process of ergonomic risk management can be graphically represented (Fig. 1).

In the assessment of ergonomic risk it is possible to diagnose mental, physical, organizational

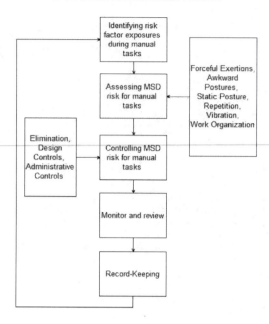

Figure 1. Ergonomic risk management proccess (based on Burgess-Limerick et al. 2009).

and individual factors that may affect the occurrence of musculoskeletal disorders (Lasota & Hankiewicz 2006b). It is also often possible to determine the level of impact of a given factor on a human during work. Results of ergonomic risk assessments are of interest not only to ergonomists but also to employees, physicians, epidemiologists, employers and supervisors. Increasingly, it is noted that ergonomic research may be the basis for undertaking risk reduction activities at workplaces (David 2005, Górny 2014).

3 CHARACTERISTICS OF SELECTED ERGONOMIC RISK MEASURES

There are many tools for measuring ergonomic risk at workstations. They can be classified into four groups:

a. checklists – based on the observation, they give the possibility of describing the task in terms of duration and frequency of exposure of the employee to dangerous factors. They also allow to identify potential factors that can cause a worker's disorder.
b. observation-based methods – they allow to evaluate the posture of the employee at work according to the load category and strength used by the employee.
c. direct measurement methods – they allow to assess ergonomic risk by means of measurements carried out in the course of performing tasks by the employee at the workplace. The appropriate measuring equipment is used, which is most often worn by an employee during work.
d. applications for ergonomic assessment – their performance is based on observations and the use of artificial intelligence components (Ndukeabasi 2013).

There are many attempts of improvement the process of ergonomic risk assessment by using different scanning methodology (Butlewski et al. 2017) or based on performance levels (Dahlke et al. 2016). But most of methods are requiring assessment of performed tasks and motions therefore are time consuming.

Ergonomic analysis tools can be used singly or in groups for one assessment. Frequently used methods are:

a. Washington State Ergonomic and MSD Risk Assessment Checklist – a checklist for evaluating the employee's posture with the division of body segments into four parts: knee, neck and shoulder, low back, hand and wrist. The time of keeping body posture during operation is checked (http://personal.health.usf.edu).
b. REBA – the tool proposed by Hignett and McAtamney in 2000. It allows to quickly evaluate the posture of the employee's body segments, it is also possible to take into account the load and the static or dynamic postures. With this tool it is possible to assess the ergonomic risk, which can be indicated in five levels – from negligable to very high (Fig. 2) (Madani & Dababneh 2016).
c. RULA – the method for ergonomic evaluation of upper body segments. It includes the number of work movements, strength, static workload. The ergonomic risk assessment is presented in four categories, from acceptable to very high (McAtamney & Corlett 1993).
d. OWAS – the method developed in Finland at a steel company. It allows to identify musculoskeletal disorders in employees. It is possible to assess the position of the body parts of workers and to assess the ergonomic risk, on the basis of which it is possible to propose appropriate corrective actions (Beheshti et al. 2016).
e. NIOSH Revised Lifting Equation – this method consists in determining the permissible weight of lifting by workers during a work shift that does not cause musculoskeletal pain. The lifting index is also calculated, so it is possible to

estimate ergonomic risk (http://ergo-plus.com/niosh-lifting-equation-single-task/).

f. Strain Index – the method that was developed on the basis of knowledge in physiology, biomechanics and epidemiology. It refers to the area of the wrist. This indicator is calculated as the product of 6 variables describing the postures of the employee's body. Evaluation by this method can be considered as close to quantitative (http:// archiwum.ciop.pl/53956).

g. Jack – a software developed by Siemens Corporation. It gives the opportunity to model the posture of the employee while working and to make a simulation. Jack includes work tools and workplace environment. Thanks to the detailed report, an ergonomic assessment is possible (Ramadhan et al. 2017).

h. AnyBody – a software that allows, among others, to advanced simulations of: individual muscle forces, metabolism, antagonistic muscle action, elastic energy in tendons (https://www.anybodytech.com/).

i. Humos – it is full human body model that was developed by a consortium of European car companies, universities, research institutes and software vendors. It is possible to distinguish Humos-1 (a human model of a male in a driving position) and Humos-2 (with the positioning tool to put the human models in a proper pose) (Kunwoo 2006).

j. Univ. of Michigan's 3DSSPP - 3DSSPP software allows to set static loads for tasks such as lifts, presses, pulls and pushes. The program allows to simulate the work, with data on postures, strength parameters and anthropometry of male and female (https://c4e.engin.umich.edu/tools-services/3dsspp-software/).

k. Auburn Engineers ERGO Job Analyzer – the tool that allows to quantify dozens of risk factors and to evaluate the likelihood of ergonomics illness/injuries at work (http://www.ergopage.com/ About_Us.html).

The use of any of the aforementioned tools leads to the presentation of an ergonomic risk report. The results may be presented in a score scale, graph or simulation, and their highest values may be obeyed to prioritize in the area of proposed corrective actions. These actions should be taken immediately if there is a high risk to workers, and it is also advisable to determine how to monitor future risks.

4 ERGONOMIC RISK MEASUREMENT IN PRIORITIZING CORRECTIVE ACTIONS AT WORKSTATIONS

On the basis of the ergonomic risk assessment, it is possible to propose corrective actions, with a division into urgent and possible to be introduced at a later time. There are many examples of such an approach to ergonomic risk (Table 1).

Based on the characteristics of individual ergonomic risk measurement tools, the authors decided to review the artices in the above-mentioned table. The results of the scientific studies presented in the public databases were analyzed and 5 of them met the criteria assumed at the beginning: the possibility of unambiguous identification of the assessment area, the possibility of distinguishing ergonomic risk assessment methods and the possibility of referencing the results of the reviewed research to the specificity of the tool used. Methodology in this article is therefore a form of prisma analysis.

Ergonomic risk assessment tools can greatly facilitate the introduction of corrective actions aimed at reducing the musculoskeletal disorders in workers. The choice of tools for the ergonomic assessment should also take into account their limitations and weaknesses. These can, among other things, relate to:

a. some of the methods do not take into account the position of the lower limbs and the angle of bend of the knee (e.g. RULA, OCRA),
b. some of the methods do not take into account the position of hands and wrists (e.g. NIOSH Lifting Equation, OWAS),
c. some of the tools do not include the force of gripping and pinching (e.g, NIOSH Lifting Equation, RULA, REBA).
d. some of the tools do not include the force of pushing and pulling (e.g, NIOSH Lifting Equation, RULA),
e. some methods do not allow to evaluate vibration and stress contact (e.g. NIOSH Lifting Equation, RULA, REBA, OWAS) (https://www.iwh.on.ca/ system/files/documents/msd_prevention_toolbox 3c_2007.pdf).

If the evaluation of parameters that are not included in the selected methods is relevant from the point of view of the analysis and the possibility

REBA Score	Risk Level	Action
1	Negligable	None necessary
2 - 3	Low	May be necessary
4 - 7	Medium	Necessary
8 - 10	High	Necessary soon
11 - 15	Very High	Necessary now

Figure 2. Levels of ergonomic risk – REBA (ergo.human.cornell.edu/CUErgoTools/REBA%206.xls).

Table 1. Review of corrective actions taken on the basis of an ergonomic risk assessment.

Area of ergonomic assessment	Applied tool	Actions proposed and references
Quality control in the production company	Questionnaire, checklist	Adjusting the work space to operators, extending the machine feeder, changing the location of the loading boxes and their height, using a rotary chair, providing adequate lighting and hearing protection (Wachowiak & Kujawińska 2013).
Repair workstation	REBA by Ergo Intelligence Upper Extremity Assessment Tools v. 1.8	Equip the employee with appropriate pneumatic tools to reduce the wrist movement and work platform allowing the hands to work below the shoulder line (Krasoń & Mączewska 2017).
The ceramic industry	Couto's checklist	Implementation of gymnastics in order to minimize and prevent musculoskeletal problems, training of workers for the development of an ergonomic education, carrying out periodic evaluation of the working methods (Bastos et al. 2016).
Automobile industry	Questionnaire	Taking actions to continually improve lean manufacturing principles taking into account the shaping of ergonomic working conditions (Rao & Niraj 2016).
Welding operator	RULA, REBA, OWAS, Discomfort Assessment Checklist	Applying ergonomic guidelines to companies in various industries, promoting employees' ergonomic awareness by giving them the opportunity to propose changes, conducting periodic ergonomic assessments to ensure continuous workplace changes (Mahendra et al. 2016).

of proposing corrective measures, additional analyzes should be completed.

Correct analysis of the ergonomic risk and working out ideas for corrective action on this basis do not guarantee the success of the project. Consideration should also be given to barriers to the application of ergonomic solutions in engineering design. Such barriers may include:

a. financial and commercial factors,
b. changes in products, rotation of employees, other initiatives of the company,
c. lack of adequate planning,
d. lack of time to incorporate ergonomics factors into the design,
e. lack of knowledge of the ergonomics issues among the engineers,
f. discrepancies between expectations and the effect of planned and implemented ergonomic solutions,
g. irrational behaviours and conflicts of interest during design,
h. restrictions related to the need to meet complex legal requirements (Pinder 2015).

Therefore, when planning corrective actions based on ergonomic analyzes, attention should be paid not only to the priority of actions but also to the real possibilities of their designing and implementation.

5 CONCLUSIONS

The results of an ergonomic risk assessment can be used to plan and implement corrective actions in companies, which is part of the work safety management. Posture at work recognized as the most incorrect according to the methods of analysis (highest scores) should be prioritized – actions must be taken as soon as possible, as workplace postures contribute to the rapid development of musculoskeletal disorders in workers. However, the limitations of the methods used should be also taken into account. When there is a need to evaluate more parameters than the tool predicts, it is important to extend the evaluation by additional methods. It is also important to take into account the barriers to implementing ergonomic solutions in companies. The proposed actions should therefore be appropriate to the results of the assessment, but also realistic to be carried out in accordance with the company's financial, technical and organizational capabilities.

The above-presented result of the analysis of documented actions taken on the basis of ergonomic evaluations suggests that it is possible to plan them after an earlier analysis of similar cases presented in scientific articles. However, it is important to initially define the assumptions of the analysis, thanks to which it will be possible to select publications relevant to the selection of ergonomic assessment methods adapted to the given industry.

REFERENCES

Ahmad A m Ramadhan, A., Alotaibi, A., Kasap, S. 2017. "Ergonomic Weightlifting" Competition by Using JACK, *Proceedings of the International Conference on Industrial Engineering and Operations Management, Rabat, Morocco, April 11–13*: 2474–2478.

Beheshti, M.H., Firoozi, A., Alinaghi Langari, A.A., Poursadeghiyan, M. 2016. Risk assessment of musculoskeletal disorders by OVAKO Working posture Analysis System OWAS and evaluate the effect of ergonomic training on posture of farmers. *JOHE* 4(3): 132.

Burgess-Limerick, R., Steiner, L., Torma-Krajewski, J. 2009. Ergonomics Processes: Implementation Guide and Tools for the Mining Industry: 7–19, Pittsburgh: Department Of Health And Human Services Centers for Disease Control and Prevention National Institute for Occupational Safety and Health, Pittsburgh Research Laboratory.

Butlewski, M., et al. 2014. Design methods of reducing human error in practice. *Safety and Reliability: Methodology and Applications*, Proceedings of the European Safety and Reliability Conference ESREL.

Butlewski, M., Sławińska, M. & Niedźwiecki, M. 2017. 3D Laser Models for the Ergonomic Assessment of the Working Environment. In: *Advances in Social & Occupational Ergonomics*: 15–23, Springer International Publishing.

Byrska, K., Herma, S. 2011. Kształtowanie jakości ergonomicznej w kontekście projektowania antropotechnicznych stanowisk pracy. In: K. Bzdyra (ed.), Wydawnictwo Uczelniane Politechniki Koszalińskiej.

Dahlke, G., Drzewiecka, M. Butlewski, M. 2016. Influence of work processes on variations in the parameters of complex reactions to stimuli. In: P.M. Arezes, et al. (eds.), *Occupational Safety and Hygiene, SHO 2016*: 58–60. Guimarães: Portuguese Society of Occupational Safety and Hygiene (SPOSHO).

David, G. 2005. Ergonomic methods for assessing exposure to risk factors for work-related musculoskeletal disorders. *Occupational Medicine* 55: 190–191.

ergo-plus.com/niosh-lifting-equation-single-task/, 25.10.2017.

ergo.human.cornell.edu/CUErgoTools/REBA%206.xls, 25.10.2017.

Górny, A. 2014. Human Factor and Ergonomics in Essential Requirements for the Operation of Technical Equipment. In: C. Stephanidis (ed.), *Posters Extended Abstracts: International Conference, HCI International 2014*, CCIS, 435: 449–454.

Górny, A. 2017. The use of working environment factors as criteria in assessing the capacity to carry out processes. In: G. Oancea, M.V. Drăgoi (eds.), *MATEC Web of Conferences*, 94: 04011.

Kale, P., Vyavahare, R. 2016. Ergonomic Analysis Tools: A Review. *Int. J. of Current Engineering and Technology* 6(4): 1271–1272.

Krasoń, P., Mączewska, A. 2017. Organizacja stanowiska naprawczego z uwzględnieniem zasad ergonomii, *Konferencja Innowacje w Zarządzaniu i Inżynierii Produkcji*: Opole.

Kujawińska, A, Wachowiak, F. 2013. Wpływ wybranych czynników ergonomicznych na skuteczność kontroli jakości. *Inżynieria Maszyn* 18(1): 51–60.

Kunwoo, L. 2006. CAD System for Human-Centered Design. *Computer-Aided Design & Applications* 3(5): 617.

Lasota, A.M., Hankiewicz, K. 2016a. Assessment of risk to work-related musculoskeletal disorders of upper limbs at welding stations, In: P.M. Arezes, et al. (eds.), *Occupational Safety and Hygiene, SHO 2016*: 138–140. Guimarães: Portuguese Society of Occupational Safety and Hygiene (SPOSHO).

Lasota, A.M., Hankiewicz, K. 2016b. Evaluation of ergonomic risk in the production line of frozen food products. *Proceedings of the 2016 International Conference on Economics and Management Innovations*, 57: 272–278.

Lubaś, P. 2010. Diagnoza ergonomicznych czynników ryzyka, online: https://szczecin.pip.gov.pl/pl/f/v/6351/Diagnoza% 20 ergonomicznych%20czynnikow%20 ryzyka.pdf, 23.10.2017.

Mahendra, K., Thimmana, A., Virupaksha, H. 2016. Ergonomic analysis of welding operator postures. *Int. J. of Mechanical And Production Engineering* 4(6): 9–21.

Moreira Caland Bastos, F., Orsano de Sousa, P., Cavalcanti Albuquerque Neto, H. 2016. Ergonomic Diagnosis of a Ceramic Industry: A Case Study Using Couto's Checklist. *Int. J. Social Science and Humanity* 6(10): 794–797.

Ndukeabasi, I. 2013. *A framework for ergonomic assessment of residential construction tasks*, A Thesis submitted to the Faculty of Graduate Studies and Research in partial fulfilment of the requirements for the degree of Doctor of Philosophy In Construction Engineering and Management Department of Civil and Environmental Engineering, University of Alberta.

Niraj, M., Rao, S. 2016. A case study on implementing lean ergonomic manufacturing systems (LEMS) in an automobile industry, *IOP Conference Series: Materials Science and Engineering*: 2–8.

Pinder, A. 2015. Literature review: Barriers to the application of Ergonomics/Human Factors in engineering design, Prepared by the Health and Safety Laboratory for the Health and Safety Executive, RR1006 Research Report: 4–5.

Taderera, H. 2012. Occupational Health and Safety Management Systems: Institutional and Regulatory Frameworks in Zimbabwe, *Int. J. of Human Resource Studies*, 2(4): 100.

http://archiwum.ciop.pl/53956, 25.10.2017.

http://personal.health.usf.edu/tbernard/HollowHills/WISHA_Checklist_20.pdf, 25.10.2017.

http://www.ergopage.com/About_Us.html, 28.10.2017.

https://c4e.engin.umich.edu/tools-services/3dsspp-software/, 28.10.2017.

https://www.anybodytech.com/, 26.10.2017.

https://www.iwh.on.ca/system/files/documents/msd_prevention_toolbox_3c_2007.pdf, 28.10.2017.

Main mitigation measures—occupational exposure to radon in thermal spas

A.S. Silva & M.L. Dinis
CERENA-Polo FEUP—Centre for Natural Resources and the Environment, Faculty of Engineering, University of Porto, Porto, Portugal
PROA/LABIOMEP—Research Laboratory on Prevention of Occupational and Environmental Risk, Faculty of Engineering, University of Porto, Porto, Portugal

ABSTRACT: The exposure to natural sources of ionizing radiation is the most significant contribution to the annual dose received by the general population. The risk of cancer due to radon exposure is increased for smokers The purpose of this work was to perform a review on occupational exposure to radon considering mitigation measures that may be efficiently adopted to decrease the levels of indoor radon in thermal spas. All architectural and constructive solutions that reduce building contact with soil, prevent radon infiltration, and facilitate diffusion into the atmosphere contribute to reducing the risk of reaching hazardous concentrations within the building. It is therefore concluded that all efforts to reduce radon infiltration from the soil are based on active or passive ventilation techniques, mainly soil below ground level, increased soil tightness in contact with the ground and in the increase of the ventilation of the spaces of housing.

1 INTRODUCTION

The exposure to natural sources of ionizing radiation is the most significant contribution to the annual dose received by the general population. On average, 80% of the annual dose is due to naturally occurring terrestrial and cosmic radiation sources. The largest natural source of radiation to human exposure is the radon gas and its progeny, which can contribute with more than 50% of the total dose from natural sources (UNSCEAR, 2000).

The radioactive elements present, for the most part, in natural radiation sources are ^{238}U (uranium-238), ^{232}Th (thorium-232) and ^{40}K (potassium-40).

Uranium can be found naturally in the environment in rocks and soils or even dissolved in groundwater in the form of three isotopes: ^{234}U, ^{235}U and ^{238}U, however their proportion in the earth's crust is very different: 0.0054%, 0.7204% and 99.2742%, respectively, therefore the ^{238}U chain is the one of greatest concern (Silva, 2015).

Radon has been identified by the World Health Organization (WHO) as the second leading cause of lung cancer after tobacco smoke (UNSCEAR, 2000; OMS, 2007). It is responsible for an estimated 21 000 deaths from lung cancer annually and 2 900 of these deaths occur among people who have never smoked (EPA, 2009). The risk of cancer due to radon exposure is increased for smokers, as the radiation emitted by tobacco synergizes when in the presence of radon gas (Darby, 2005; Al Zoughool et al., 2009; EPA, 2013; Erdogan et al., 2013; Silva & Dinis, 2016; Silva et al., 2016).

Radon may be also present in water, and in the case of groundwater passing through uranium-bearing soils and uranium-rich rocks (e.g. uranium-rich granites and pegmatites), commonly contains significant radon concentrations. In these cases, if groundwater is used as drinking water, people may be exposed both through water consumption and also by the inhalation of radon released from water into air (UNSCEAR, 2006).

In thermal spas groundwater with diverse chemical compositions is usually used in therapeutic treatments, which may contain additionally significant amounts of natural radionuclides (Silva et al., 2013, Müllerova, et al., 2016; Walczak, et al., 2016). Radon concentrations in indoor air of hydrotherapy treatments rooms can lead to intense exposure both for users and workers. However, this will result in a short-term impact for users but in a long-term for workers who have longer and continuous exposure periods. In this way, this additional exposure may become extremely significant to this workers group and radon occupational exposure should be addressed: first, by assessing radon concentration and exposure dose; then, by implementing efficient measures to mitigate radon exposure and lastly, by monitoring to control and correct, if necessary, the adopted measures.

The purpose of this work was to perform a review on how to decrease the exposure to radon considering the mitigation measures that may be efficiently adopted to reduce the levels of indoor radon in thermal spas.

2 METHODS

A systematic review was conducted in multiple databases with Metalib tool using a combination of a pair of key-words: "radon" and "spas" refined with "exposure" or "mitigation measures" in the subject research field. The methodology adopted in this work focused on a few issues: i) identify main mitigation measures that decrease the radon levels in indoor air; ii) identify the main measures that can be adopted in thermal spas to minimize occupational exposure to radon; and iii) identify the main measures that can be addressed to workers in order to decrease the exposure.

Fifty-seven articles were selected according to the identification of mitigation measures to reduce levels of indoor radon concentration from 1993 to 2017. Nineteen articles were selected from the following databases for review: Web of Sicence (8) and Scopus (5). Others sources were also consulted such as databases from the WHO and EPA. The criteria for the selection of the articles were: a) date of the article; ii) the articles address the issues described above.

3 RESULTS AND DISCUSSION

For the present study, the articles and documents selected in the research were: Campos et al., 2010; Louro et al., 2010; Vazquez et al., 2011; Nikolov et al., 2012; Steck, 2012; Erdogan et al., 2013; Pressyanov, 2013; Antão, 2014; Valmari et al., 2014; Ziane et al., 2014; Darby et al., 2015; Palacios, 2015; Silva et al, 2015; Djamil, 2016; Müllerova, 2016; Shahrokhi et al., 2016; Silva et al., 2016a; Silva et al., 2016b, Walczak et al., 2016 and Silva, 2015. Documents form international organizations were also consulted: EPA, 1993; EPA, 1997; WHO, 2009 and ICRP, 2010.

3.1 *Measures to reduce indoor radon concentration in general*

The radon concentration inside buildings depends on several factors, among which the relationship between buildings and the soil and the potential for radon emanation of such soil. All architectural and constructive solutions that reduce building contact with soil, prevent radon infiltration and facilitate radon diffusion into the atmosphere, contribute to reduce the risk of accumulating high radon concentration within the building (Silva, 2015).

In the case of buildings built in past times in areas of risk of high radon concentration, the available techniques that aim at the urban and architectural rehabilitation of these buildings are sufficient, as a general rule, to take into account some constructive details which improve the ventilation conditions of the building, in particular, the space between the lower floor and the ground, by placing a radon-impermeable barrier on the walls and floors in contact with the ground so that there is no radon infiltration into the interior (Silva, 2015).

Only in special cases of buildings situated on sites with high radon exhalation, the use of active ventilation systems (with mechanical ventilation) or passive ventilation (using natural ventilation) will be justified by suctioning between the ground floor and the ground. The cost of installing these systems, when foreseen in projects, is very low, being easily included in the total budget of the work, with many public health advantages for the future users of the buildings (Silva, 2015).

In what concerns to the techniques available to reduce the risk of radon infiltration into buildings to be constructed (new buildings), they are similar to those used in the rehabilitation of existing buildings, but as they can be designed before the execution on site, are more economic and more efficient due to a better planning.

It is therefore concluded that all efforts to reduce radon infiltration from the ground are mostly based on active or passive ventilation techniques, (EPA, 1997). In case of already existing indoor radon accumulation, ventilation is one of the most important mitigation measures to reduce levels of radon concentration in indoor air.

3.2 *Residents*

The minimisation of radon concentrations in new buildings is therefore of great importance. This can be achieved, for example, through the enforcement of building regulations and the education of construction professionals. With regard to radon mitigation in existing buildings, synergies with the ongoing renewal of the building stock should be exploited (Palacios et al., 2015).

According to WHO (2009) most prevention strategies address steps to limit soil gas infiltration due to air pressure differences between the soil and the indoor occupied space. Radon prevention strategies should consider the specific mix of construction practices, radon sources, and transport mechanisms in the region or country, in order to be cost-effective.

For example, Active Soil Depressurization (ASD) (Figure 1) is simple to install and provides greater radon reduction compared to Passive Soil Depressurization (PSD) systems (EPA 1993).

On the other hand, Passive Soil Depressurization (PSD) may also be used in new constructions. This technique is similar to active soil depressurization (ASD), with the following exceptions (Figure 2): i) the effectiveness of PSD depends on the thermal buoyancy of air in the vent pipe and its ability to slightly depressurize the soil under the

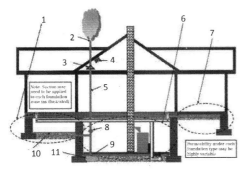

Legend: 1 – Crawlspace; 2 – System discharge; 3 – Fan; 4 – Electrical supply; 5 – Vent pipe; 6 – Basement foundation; 7 – Slab on ground foundation; 8 – System indicator; 9 – Suction point; 10 – Crawlspace membrane; 11 – Suction pit (if no drain tile).

Figure 1. Active soil depressurization for radon control in new constructions (WHO, 2009).

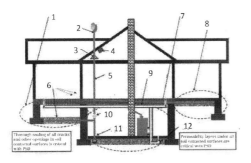

Legend: 1 – Crawlspace foundation; 2 – System discharge; 3 – Future fan location (if needed); 4 – Electrical supply; 5 – Vent pipe; 6 – Membrane (under each foundation type); 7 – Basement foundation; 8 – Slab on ground foundation; 9 – Air sealed ducts; 10 – Future system indicator (if fan needed); 11 – Suction point; 12 – Avoid subslab ducts.

Figure 2. Passive soil depressurization for radon control in new constructions (WHO, 2009).

dwelling; ii) the elements of the building that are in contact with the soil must be sealed to prevent soil gas infiltration; iii) since air pressure differences are so small between the vent pipe and the occupied area, the only way to monitor system performance is via periodic or continuous radon monitoring (WHO, 2009; Vazquez et al., 2011; Steck, 2013).

Several authors point out that one of the mitigation measures to reduce levels of indoor air radon concentration is to improve the ventilation conditions of the dwelling, namely at the aeration level through the daily opening of windows and doors (Figure 3) (WHO 2009; Campos et al., 2010; Louro et al., 2010; Nikolov et al., 2012; Pressyanov, 2013; Antão, 2014; Valmari et al., 2014; Ziane et al., 2014; Shahrokhi et al., 2016).

Antão (2014) reports that where possible, insulation measures for radon exhalation (painting, sealing of fractures, etc.) should be carried out (Figure 4).

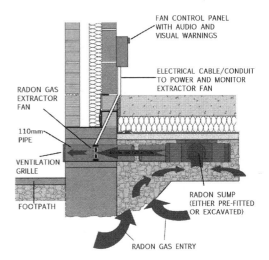

Figure 3. Ventilation system (ORBV, 2017).

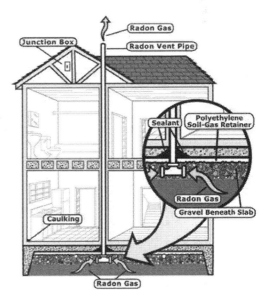

Figure 4. Insulation measures for radon exhalation (Radon Resistant New Construction (RRNC) General Information, 2010).

3.3 Case-study of the Portuguese thermal spas

Silva (2015) in a study on occupational exposure to radon in thermal spas carried out in Portugal between 2011 and 2015, proposed several mitigation measures to be implemented in the studied cases, according to the results obtained for radon concentration and occupational exposure doses (Silva et al., 2015; Silva et al., 2016a; Silva et al., 2016b). In this study, very high values of radon concentration were registered both in the indoor air and water: 73–4335 Bq/m³ and 76–6949 Bq/L, respectively, in thermal spas and 68–4051 Bq/m³ for indoor radon concentration in workers' dwellings, revealing concerning exposure doses: between 2.05 and 31.85 mSv/y for thermal spas and between 0.49 and 29.17 mSv/y for workers' dwellings. Nevertheless, it is stressed (Silva, 2015) that the complexity of the relationship between radon indoor occupational exposure in this working environment and the potential health effects resulting from this exposure should be considered with caution when interpreting these values (e.g. seasonality of thermal spas or the variety of factors that can affect indoor radon concentration).

In what concerns to legal requirements, many countries adopted 200 Bq/m³ as an action level which means that at this value mitigation measures should be taken to reduce indoor radon levels. The WHO recently recommended that the concentration of indoor radon should not exceed 100 Bq/m³ or 300 Bq/m³ in exceptions cases, if the above indicated cannot be achieved (WHO, 2009). The International Commission on Radiological Protection (ICRP) has revised the reference level for radon gas in dwellings and other buildings with high occupancy rates to the public by 300 Bq/m³ (ICRP, 2010). In the Directive 2013/59/EURATOM) it is stated that indoor radon concentration should be kept below 300 Bq/m³. The EU Action Level for water (1000 Bq/L) does not apply to mineral waters as it is considered that they are not consumed on a regular basis and therefore no reference value has been established for radon concentration in thermal or mineral waters yet.

In Portugal, there is a long tradition in the use of natural mineral waters for medicinal purposes (thermalism) (Silva, 2015), as there are several mineral-medicinal thermal sources in the country. The potential to radon exposure may be intensified if these are located in regions of high level of natural radiation, which seems to be the case. Therefore, from Silva (2015), a set of general recommendations can be derived, taking into account the results obtained for radon exposure in thermal spas:

- Implement a system of monitoring and radiological protection of workers in thermal spas whenever justified;
- Acquire and provide workers with personal protective equipment, in particular masks, to be used in thermal spas, whenever justified, thereby optimizing exposure;
- Mandatory rotation of the tasks, thus optimizing the exposure;
- Make aware workers to the importance of using the protection measures proposed by the managers of thermal spas, such as the use of mask.
- Improve the ventilation conditions of the thermal spas buildings, where appropriate, in particular equip all workplaces with an efficient and, if possible, air-conditioned ventilation system. Ventilation must be performed from the lower floors.
- Ensure compliance with the legislation, in particular at the level of radiation monitoring and the levels laid down in legal diplomas for the concentration of radon in indoor air (Silva, 2015; Djamil, B., 2016).

Regarding the workers behaviour from thermal spas, the first mitigation measures to be implemented are using, whenever justified, the proper individual protective equipment provided by the employer (e.g. masks).

4 CONCLUSIONS

This study allows to conclude that there are several measures that may be implemented in the different places and can reduce the concentration and occupational exposure to radon in thermal spas: the used building systems, the ventilation and air-renewal systems are the most important factors that will influence the levels of indoor radon concentration.

The results from the case-study of the Portuguese thermal spas also showed some variations related with the occupation of the space and ventilation. According to the results and the reference levels from the Directive 2013/59/EURATOM, some recommendations are proposed in order to optimize the workers' exposure (improve the ventilation conditions, implement a system of surveillance, monitoring and radiation protection) as well as some mitigation measures to decrease radon levels within workers' dwellings (improve the ventilation conditions and reduce the flux from the subsoil).

REFERENCES

Al Zoughool, M., Krewski, D. (2009). Health effects of radon: a review of the literature. Int J Radiat Biol. 85(1): 57–69, doi: 10.1080/09553000802635054.

Antão, A.M. (2014). Assessment of radon concentrations Inside a Hight School Building in Guarda (Portugal): Legislation implications and mitigation measures proposed. Uranium, Environment and Public Health, 8: 7–12, doi: 10.1016/j.proeps.2014.05.003.

Campos, M.P., Pecequilo, B.R.S., Mazzilli, B.P. (2010). ^{222}Rn and 212Pb exposures at a Brazilian SPA. Radiation Protection Dosimetry 141(2): 210–214, doi:10.1093/rpd/ncq167.

Darby, S., Hill, D., Auvinen, A., Barros-Dios, J.M., Baysson, H., Bochicchio, F., Deo, H., Falk, R., Forastiere, F., Hakama, M., Heid, I., Kreinbrock, L., Kreuzer, M., Lagarde, F., Mäkeläinen, I., Muirhead, C., Oberaigner, W., Pershagen, G., Ruano-Ravina, A., Ruosteenoja, E., Rosario, A.S., Tirmarche, M., Tomásek, L., Whitley, E., Wichmann, H.E., Doll, R.,

(2005). Radon in homes and risk of lung cancer: collaborative analysis of individual data from 13 European case-control studies. BMJ 330(7845): 223–227.

Djamil, B. (2016). Indoor radon mitigation in South Korea. International Journal of Applied Engineering Research, 11: 8521–8523. ISSN 0973-4564.

EPA (1993). Radon Reduction Techniques for Existing Detached Houses: Technical Guidance (Third Edition) for Active Soil Depressurization. UPEPA Publication 625-R-93-011, Washington D.C.

EPA (1997). United States Environmental Protection Agency, Health Effects Assessment 14 Summary Tables, Radionuclide Carcinogenicity Slope Factors: HEAST, www.epa.gov/radiation/heast/docs/heast_ug_0401.pdf.

EPA (2009). United States Environmental Protection Agency, Health Risk of Radon. www.epa.gov/radon/health-risk-radon | January 2017.

EPA (2013). United States Environmental Protection Agency, Consumer's Guide to Radon Reduction, How to fix your home. URL:http://www.epa.gov/radon, EPA 402/K-10/005 | March 2013.

Erdogan, M., Ozdemir, F., Eren, N. (2013). Measurements of radon concentration levels in thermal waters in the region of Konya, Turkey. Isotopes in Environmental and Health Studies 49(4): 567–574. doi:10.1080/10256016.2013.815182.

ICRP, (2010). "Lung Cancer Risk from Radon and Progeny and Statement on radon", ICRP Publication 115 Ann, 40(1).

Louro, A., Belchior, A., Cunha, G., Gil, O.M., Peralta, L. (2010). Riscos para a saúde humana da exposição ao radão habitacional: um caso de estudo na região da Guarda. Décimas Quartas Jornadas Portuguesas de Proteção Contra Radiações. Complexo Interdisciplinar do Instituto Superior Técnico de Lisboa, pp.6.

Müllerova, M., Mazur, J., Blahušiak, P., Grzadziel, D., Holý, Kovács, T., Kozak, K., Csordás, A., Neznal, Martin, Neznal, Matej, Shahrokhi, A. (2016). Indoor radon activity concentration in thermal spas: the comparison of three types of passive radon detectors. J. Radional Nucl Chem: 310, 1077–1084. DOI 10.1007/s10967-016-4961-8.

Nikolov, J., Todorovic, N., Pantic, T.P., Forkapic, S., Mrdja, D., Bikit, I., Krmar, M., Veskovic, M. (2012). Exposure to radon in the radon spa Niška Banja, Serbia. Radiat. Meas 47(6): 443–450, doi:10.1016/j.radmeas.2012.04.006.

OMS—Organização Mundial da Saúde (2007), Radon and cancer. Fact Sheet Nr. 291. WHO, Geneve.

Palacios, M., Barazza, F., Murith, C., Ryf, S. (2015). Implementation of the new international standards in swiss legislation on rádon protection in dwellings. Radiat. Prot., 164:1–2, April 2015.

Pressyanov, D.S. (2013), Use of polycarbonate materials of high rádon absorption ability for measuring radon. Romanian J. Phys.: 58-S221-S229.

Shahrokhi, A., Nagy, E., Csordas, A., Somlai, J., Kovacs, T. (2016). Distribution of indoor rádon concentrations between selected Hungarian termal baths. Nukleonika, 61: 333–336, doi: 10.1515/nuka-2016-0055.

Silva, A.S. & Dinis, M.L. (2016). Measurements of indoor radon and total gamma dose rate in Portuguese thermal spas. In: Occupational Safety and Hygiene IV (P. Arezes, J.S. Baptista, M. Barroso, P. Carneiro, P. Cordeiro, N. Costa, R. Melo, A.S. Miguel, G. Perestrelo, Eds.), pp. 485–489. Taylor & Francis, London.

Silva, A.S. (2015). Exposição Ocupacional ao Radão em Estabelecimentos Termais. Tese de Doutoramento em Segurança e Saúde Ocupacionais, FEUP, Porto.

Silva, A.S., Dinis, M.L., Diogo, M.T. (2013). Occupational Exposure to Radon in Thermal Spas, Book chapter in: Occupational Safety and Hygiene, Eds. P. Arezes, J.S. Baptista, M. Barroso, P. Carneiro, P. Cordeiro, N. Costa, R. Melo, A.S. Miguel, G. Perestrelo, pp. 273–277. ISBN: 9781138000476, London: Taylor & Francis.

Silva, A.S., Dinis, M.L., Pereira, A.J.S.C. (2016). Assessment of indoor radon levels in Portuguese thermal spas. Radioprotection, pp. 249–254, doi: 10.1051/radiopro/2016077.

Silva, A.S., Dinis, M.L., Pereira, A.J.S.C., Fiúza, A. (2015). Radon levels in Portuguese thermal spas. Proceedings of the: "Third International Conference on Radiation and Application in Various Fields of Research, RAD2015", Budva, Montenegro, June 08–12, 2015.

Silva, A.S., Dinis, M.L., Pereira, A.J.S.C., Fiúza, A. (2016a) Radon levels in Portuguese thermal spas. RAD Journal, Radiation and Applications in Physics, Chemistry, Biology, Medical Sciences, Engineering and Environmental Sciences, 76–80.

Silva, A.S., Dinis, M.L. (2016b). Measurements of indoor radon and total gamma dose rate in Portuguese thermal spas. Occupational Safety and Hygiene IV, pp. 485–489, London: Taylor & Francis, doi: 10.1201/b21172–93.

Steck, D.J. (2012). The effectiveness of mitigation for reducing radon risk in single family Minnesota homes. J. Environ. Radioact., 3: 214–248, doi: 10.1097/HP.0b013e318250c37a.

Torgal, F. Pacheco (2013). Gás radão: um perigoso contaminante do ar interior das habitações. Maquinaria, Edição 228, Fevereiro de 2013.

UNSCEAR (2000). United Nations Scientific Committee on Effects of Atomic Radiation, Sources and Effects of Ionizing Radiation. United Nations, New York, United Sales publication E.00.IX.3.

UNSCEAR (2006). Effects of Ionizing Radiation: UNSCEAR 2006 Report: Volume II.

Valmari, T., Arvela, H., Reisbacka, H., Holgren, O. (2014). Radon measurement and mitigation activity in Finland. Radiat. Prot. Dosim, 160: 18–21.

Vazquez, B.F., Adan, M.O., Poncela, L.S.Q., Fernandez, C.S., Merino, I.F (2011). Experimental study of effectiveness of four radon mitigation solutions, based on underground depressurization, tested in prototype housing built in a high radon area in Spain. J. Environ. Radioact., 102: 378–385, doi: 10.1016/j.jenvrad.2011.02.006.

Walczak, K., Olszewski, J. Zmyslony, M. (2016). Estimate of radon exposure in geothermal SPAs in Poland. Int J Occup Environ Med, 29(1):161–6.

WHO (2009). Handbook on Indoor Radon, A Public Health Perspective, ISBN 978 92 4 154767 3. http://apps.who.int/iris/bitstream/10665/44149/1/9789241547673_eng.pdf.

Ziane, M.A., Lounis-Mokrani, Z., Allab, M. (2014). Exposure to indoor radon and natural gamma radiation in some workplaces at Algiers, Algeria. Radiat. Prot. Dosim, 160(1–3): 128–133, doi:10.1093/rpd/ncu058.

Safety performance and its measurement: An empirical study concerning leading and lagging indicators

A. Job & I.S. Silva
School of Psychology, University of Minho, Braga, Portugal

ABSTRACT: When it comes to Safety Performance measurement, the literature has identified two types of indicators, the "leading" and the "lagging". This study aims to identify if there's any correlation between two "leading" indicators and the number of accidents reported (i.e. lagging indicator). The sample is composed by 174 participants who had management/supervisor responsibilities. Most of them had more than 40 years old, were college educated and almost half worked in medium-size to large companies in the industrial sector. The results obtained don't allow us to establish a significant link between the indicators considered.

1 INTRODUCTION

In 2015, private industry employers reported approximately 2.9 million non-fatal workplace injuries in the United Stated (US) (Bureau of Labor Statistics, 2016). Moreover, in the European Union (EU), the Eurostat (2016) reports a number close to 3.2 million non-fatal accidents in 2014, and lastly, in Portugal, a report from the *Autoridade para as Condições de Trabalho* (2017) shows a total of 264 non-fatal accidents and 138 fatal accidents in 2016.

These global statistics are important and we should analyze them, trace their evolution and take them under consideration. Although at an organizational level, how much does this statistic help us to improve safety performance?

Some authors (e.g. Hubbard, 2004; Sgourou et al. 2010) argue that the traditional approach to evaluate the safety performance of an organization relies on the measurement and analysis of data related to incidents (such as accident frequency, accident costs and severity rates). Despite being more easily understood by employees and managers (Cooper and Philips, 2004), this kind of indicators (usually called "lagging" or "retrospective") are often criticized because "*they tell us that there is a problem, but they do not tell us what the problem is*" (Carder and Ragan, 2004, p. 158). Furthermore, Wu et al. (2015) says that, because of this retrospective and performance-based metric sometimes companies wouldn't report minor incidents so they can "maintain" their safety performance.

Petersen (2000) says that when these problems became obvious, a different measure was created by safety professionals—the audit. Although, the same author refers that using this method as a valid measure of excellence is questionable unless the audit has passed some rigorous tests and its largely correlated with a lagging indicator. If not, then some caution is suggested.

Having this as a consideration, some authors (e.g. Grabowski et al., 2007; Manuele, 2009; Wu et al., 2015) started endorsing the Prospective Safety Performance Evaluation methodology (PSPE). This method is characterized by its proactive nature and considers not only safety management activities (i.e. training, hazard identification), but also employee activities (i.e. observable safe behaviors), supervisor activities (i.e. communication related to safety) and management activities (i.e. management commitment, polities) (Wu et al. 2015). Moreover, Hinze et al. (2013) considers this method as a "*set of selected measures that describe the level of effectiveness of the safety process (…) when one or more of these measures suggest that some aspect of the safety process is weak or weakening, interventions can be implemented to improve the safety process*" (p. 24) and therefore, reduce the possibilities of an accident. This kind of indicators are usually called "leading".

Despite this, Dyreborg (2009) in their research reached the conclusion that the PSPE is still under investigation regarding its inherent terminology, scientific theory and practice. Wu et al. (2015) fulfil this idea by referring that even though the large quantity of studies regarding risk assessment, occupational accidents and safety practices, there's still not a reliable methodology defined for the analysis and evaluation of this concept.

1.1 *The present study*

With this need already identified, this study aims to compare two measures of safety performance that fall into the "leading" category (i.e. Compliance with Safety Behaviour [CSB] from Hayes et al., 1998; and Safety Performance measure [SPM] developed by Fernández-Muñiz, Montes-Peón & Vázquez-Ordás, 2007), to a "lagging" measure (i.e. number of accidents last year). Petersen (2000) argued that these surveys should query hourly employees', because it's their perception that reflects the key reality. Notwithstanding, in this study we tried do "capture" the supervisor's and management's perspective. This is not only justified by the fact that this study is a complement of a larger study that used this audience as a target, but also by the fact that the management perspective is often overlooked, even though they play a major role in safety matters.

Lastly, the following hypotheses are established as main objectives of this study. H1 - *"The CSB is significantly correlated with the SPM"*; H2 - *"The CSB isn't significantly correlated with the number of accidents reported last year"*; H3 - *"The SPM isn't significantly correlated with the number of accidents reported last year"*.

To test the hypotheses formulated, it's important to notice that both instruments were adapted, not only to the Portuguese language, but also to a different target audience that they were developed for, so an adaptation is necessary and was made in this study.

2 METHOD

2.1 Sample

The sample consisted in 174 participants that had management/supervisor responsibilities. Most of them, had more than 40 years old and were college educated and almost half of them worked in medium-size to large companies in the industrial sector (see Table 1). More information regarding the sample can be consulted in Table 1.

2.2 Procedures

The first step was to request the authorization to use both instruments to their authors. After their approval, we started to translate the instruments to Portuguese and adapt them to our target audience. We also used the back-translation method to make sure that there were no mistakes in our original translation. This back-translation was assured by a bilingual (English-Portuguese) individual, familiarized with safety subjects. Next, we carried out a pre-test with three volunteers that had extensive knowledge regarding the occupational safety theme. This pre-test revealed that the instrument was ready to use leading us to the sample collection phase.

The sample collection was made via online and paper method. The former method resulted in 222

Table 1. Sample description (N = 174).

Variables	n (%)	%
Gender		
Male	109	62.6
Female	64	36.8
Age		
Under 25 yrs old	3	1.7
Between 26 and 40 yrs old	82	47.1
41 yrs old or above	87	50.0
Qualifications*		
Level 2	2	1.1
Level 3	3	1.7
Level 4	11	6.3
Level 5 (High School)	48	27.6
Level 6 or above (College)	110	63.2
Role		
CEO	39	22.4
Manager	41	23.6
Middle Management	51	29.3
Supervisor	22	14.9
Other	9	5.2
Sector		
Industrial	86	49.4
Construction	14	8.0
Services	64	36.8
Outro	8	4.6
Size (number of workers)		
Micro (<10)	30	17.2
Small (10–50)	50	28.7
Medium (51–250)	42	24.1
Large (>250)	52	29.9

Note: The values shown in the table in some cases may not correspond to the N due to missing values in some variables.
*Qualifications shown under the European Qualifications Framework (EQF) classification.

accesses to our link and 42 valid responses (18.92% response rate). The latter method resulted in 220 delivered questionnaires and 132 valid responses (60% response rate).

Finally, we proceeded to the statistical analysis of the data using the Statistical Packaged for the Social Sciences (SPSS), version 21. We started with an exploratory factorial analysis using a Varimax rotation to identify the measure's structure, and analyse its construct validity, followed by a reliability analysis. Afterwards, we conducted Pearson correlations between them.

2.3 Measures

As previously mentioned, for this study we used three instruments.

The first was the Compliance with Safety Behaviours developed by Hayes et al. (1998) and consists of 11 items, each one reflecting a safe or unsafe behaviour. Respondents are asked to indicate how often this behaviour happens in their organization through a Likert scale that varies from 1 ("Never") to 5 ("Always"). The score is obtained by averaging the responses after recoding the three negative items present in the scale. Higher scores indicate greater compliance with safety work behaviours. Also, we must note that the original instrument was developed to be applied on a different target audience (i.e. employees) and one item-example is "*Keep my work area clean*". Due to this, we've made some changes so it could be applied to managers. This version of the instrument can be consulted at Table 2.

Table 2. CSB's rotated component matrix results.

Items	F1	F2	F3
1. I encourage my work team to overlook safety procedures in order to get the job done more quickly.	0.05	**0.82**	−0.29
2. I advise my work team to follow all safety procedures regardless of the situation I am in. (*)	**0.52**	**0.51**	0.23
3. The workers under my supervision handle all situations as if there is a possibility of having an accident. (*)	0.55	−0.18	0.24
4. I encourage my team to wear safety equipment required by practice.	**0.84**	0.15	−0.03
5. The workers under my supervision keep their work area clean	0.50	0.04	**0.58**
6. I encourage my team to adopt safety behaviours.	**0.81**	0.21	0.28
7. The workers under my supervision keep the work equipment in safe and working condition.	**0.72**	−0.00	0.37
8. My work team takes shortcuts to safe working behaviours in order to get the job done faster.	0.07	**0.80**	0.31
9. Workers do not follow safety rules that I think are unnecessary.	−0.03	**0.74**	0.33
10. When my team finds out about a security problem, they report it to me immediately.	0.17	0.14	**0.84**
11. I correct safety problems to ensure accidents will not occur.	0.34	0.22	**0.63**
Total variance explained (%)	38.7	17.1	9.2

Note: F = Factor; (*) Items not considered in subsequent analysis.

The second instrument used is the Safety Performance Measure developed by Fernández-Muñiz et al. (2007). It's composed by a unidimensional structure of 4 items and the respondents must answer through a 5-point Likert scale, ranging from 1 ("Extremely Dissatisfied") to 5 ("Extremely Satisfied"). This instrument, in its original version presented a Cronbach's Alfa of .74 and it doesn't have any item formulated in the negative. Higher scores indicate greater satisfaction with the safety performance of the employees. Since this instrument in its original form was already developed and suitable to be applied to management participants, we didn't need to make any changes.

Finally, the third and last instrument consisted in a brief item asking participants the following: "*number of accidents that my team reported to me last year*". Although workplace injuries and near misses are relevant lagging indicators (e.g., Li et al., 2013; Probst et al., 2013), the latter wasn't accounted for in this study. Most of the time, this data is based in self-reports (Wachter and Yorio, 2014) which my produce biased results.

3 PRESENTATION AND DISCUSSION OF THE RESULTS

3.1 *Measures' psychometric properties*

Regarding the CSB, our exploratory factorial analysis revealed a KMO value (Kaiser-Meyer-Olkin measure of sample adequacy) of .82, which Field (2009) considers as adequate. Additionally, Barlett's Test of Sphericity indicated that there was a sufficient item-correlation to proceed our analysis ($\chi^2(77) = 703,03$, $p < .000$).

We had no information regarding the original structure of this instrument, although in this study a tridimensional structure was identified. Following the recommendations of Costello and Osborne (2005), we've decided to use a load of >.50 as a criterion to retain items and choose not to consider item number 2 (see Table 2) in subsequent analysis due to the cross loading verified. In terms of reliability, Field (2009) considers that a Cronbach's Alfa superior to .7 is acceptable. In our study, the F1 presented a value of .71, although if we removed the item number 3 (see Table 2), it would rise to .78, so we've decided to also drop this item in future analysis. The Cronbach's alfa for F2 and F3 was .74 and .72 respectively, and the disregarding of any item in these factors wouldn't improve this value.

The exploratory analysis of the second measure (SPM) revealed a KMO value of .79 and a significant Barlett's test of Sphericity ($\chi^2(6) = 212,48$, $p < .000$) which allowed us to proceed with the analysis. The results revealed a unidimensional structure as expected, with great factor loadings

(Table 3). In terms of reliability, the analysis revealed a Cronbach's Alfa of .80.

Lastly, regarding the "lagging" indicator (i.e. number of accidents), we've made a descriptive analysis revealing a $\bar{x} = 3.10$ (SD = 9.32).

3.2 Measures' association analysis

In this section, we present the results of the analysis conducted to test the association between the leading indicators and the association of these indicators with the lagging indicator. As stated in Table 4, the results show a positive and statistically significant correlation between every factor of the CSB with the SPM, with r values varying between .22 and .41, revealing that the leading indicators are correlated, supporting our first hypothesis (H1). Our second hypothesis (H2) was also supported by data, revealing that none of the three CSB factors correlate with the lagging indicator (i.e. Accidents). Finally, the third hypothesis (H3) of our study which stated that the SPM didn't had a statistically significant correlation with the number of accidents, is rejected. These results are discussed in the next section.

3.3 Discussion

Hollnagel et al. (2006) state that the lagging indicators of safety performance are the most commonly used by organizations. Although, the contemporary view on safety stresses the need to proactively evaluate and manage their safety activities, challenging organizations to anticipate vulnerabilities rather than merely react (Hollnagel et al., 2006; Reiman and Pietikainen, 2012).

Table 3. SPM's saturated factor weights.

Items	Factor
Personal injuries	**0.787**
Material damage	**0.828**
Employees' motivation	**0.799**
Absenteeism/Lost time	**0.769**
Total variance explained (%)	63.4

Table 4. Pearson correlation between measures.

Measure	1.	2.	3.	4.	5.
1. F1 – CSB	–	**0.20***	**0.59****	**0.31****	0.08
2. F2 – CSB		–	**0.26****	**0.22****	–0.01
3. F3 – CSB			–	**0.41****	–0.01
4. F1 – SPM				–	**–0.16***
5. Accidents					–

*$p < 0.05$; **$p < 0.01$.

The main objective of this research was to compare lead and lag measures of safety performance. Our first hypothesis (H1), which stated that the leading measures were significantly correlated was confirmed. Even though we recommend interpreting these results with caution due to some sample limitations, this correlation reveals, as expected, some consistency between the leading indicators. Cooke and Rohleder (2006) advocate that most of the times there's a lack of a multidisciplinary input to understand safety performance, leading to an inconsideration of all relevant factors that influence safety. Despite this, Sinelnikov, Inouye & Kerper (2015) state that the literature regarding leading indicators is "*a compilation of thoughts, opinions, case studies and some empirical research (…) their nature and utility remains murky*" (p. 240). Having this as consideration, the instruments used in this research may lack the multidisciplinary input, but the fact that they correlate significantly, helps us to take share some light in the "murkiness" referred by Sinelnikov et al. (2015).

Some authors (Hopkins, 2009; Lingard et al., 2017; Petersen, 2000) argue that the use of incident rates is problematic and should be avoided. This happens due to underreporting to obtain rewards for accidents-free periods (Pedersen et al., 2012). In this line of thoughts, our H2 stated that the CSB wasn't significantly correlated with the lagging indicator. The results show that there isn't a significant correlation between indicators. Taking into account that we obtained some consistency between leading indicators, this may mean that our research supports the problematic view on the accident measures. Anyhow, these results should be interpreted carefully.

Finally, the third hypothesis affirmed that the SPM wasn't significantly correlated with the lagging indicator. The results obtained reveal a negative and statistically significant correlation leading us to the rejection of H3, meaning that the greater the score obtained in SPM, the less the number of accidents reported. Even though the problematic with the lagging indicator, Dyreborg (2009) defends that it would be desirable to find that the incident rate decreases as the safety activities (and consequently the safety performance) increase. Therefore, we identify some disagreement between authors regarding the use of this measure. Notwithstanding, the correlation between measures is considered weak by Ratner (2009) which doesn't allow us to establish a steady link between them.

This difference in the results obtained may be due differences amongst the leading indicators used. Some authors (i.e. Hinze et al., 2013; Hopkins, 2009; Webb, 2009) do recommend that we should focus on using leading indicators, most fail to specify which ones should we really use due to the lack of empirical research in this field and the "murkiness" referred by Sinelnikov et al. (2015). Not only that, as acknowledged by Hinze et al. (2013), leading indicators is for the most part and emerging technology and its definition, usage and effectiveness needs more investigation. Notwithstanding, some authors (Grabowski et al, 2007; Zwetsloot, 2009) have connected leading indicators of safety performance to the concept of safety climate and culture, proposing these concepts as leading indicators, although this still doesn't solve the "measurement" problem.

The present study has some limitations that we must point out. First, our lagging measure was based in the management's perception and not on official records. This emphasizes the underreporting problem that we've mentioned before. Haight and Thomas (2003) refer that in the absence of real data (i.e. records) the representative effectiveness measurement is questionable. Also, the leading measures are also prone to social desirability, which can lead to biased results. Secondly, our sample isn't representative of the population, therefore the generalization of this result is not possible and the results should be interpreted with caution.

3.4 Conclusions

The main objective of this study was to compare the two leading indicators with the number of work related accidents reported to the manager or supervisor. We choose only to present the results obtained and compare them to the literature, and not to take position on the discussion about the problematic (or not) use of the lagging indicator. Although instead of focusing in this discussion, our recommendation to future lines of investigation is that it should focus more on developing and consolidating the concept, definition and utility of leading measures, so we can move towards a more solid ground in terms of safety performance measurement.

REFERENCES

Autoridade para as Condições de Trabalho. 2017. Acidentes de trabalho graves. Available at: http://www.act.gov.pt/(pt-PT)/CentroInformacao/Estatistica/Paginas/AcidentesdeTrabalhoGraves.aspx.

Bureau of Labor Statistics. 2016. Employer-reported workplace injuries and illnesses. US Department of Labor. Available at https://www.bls.gov/news.release/pdf/osh.pdf

Carder, B., & Ragan, P. 2003. A survey-based system for safety measurement and improvement. *Journal of Safety Research* 34(2): 157–165.

Cooke, D., & Rohleder, T. 2006. Learning from incidents: from normal accidents to high reliability. *System Dynamics Review* 22: 213–239.

Cooper, M., & Phillips, R. 2004. Exploratory analysis of the safety climate and safety behavior relationship. *Journal of Safety Research* 35(5): 497–512.

Costello, A. B. & Osborne, J. W. (2005). Best practices in exploratory factor analysis: four recommendations for getting the most from your analysis. *Practical Assessment Research & Evaluation, 10(7)*: 1–9.

Dyreborg, J. 2009. The causal relation between lead and lag indicators. *Safety Science* 47: 474–475.

Eurostat. 2016. Accidents at work statistics. Available at http://ec.europa.eu/eurostat/statistics-explained/index.php/Accidents_at_work_statistics

Fernández-Muñiz, B., Montes-Peón, J., & Vázquez-Ordás, C. 2007. Safety culture: Analysis of the causal relationships between its key dimensions. *Journal of Safety Research* 38: 627–641.

Field, A. (2009). *Discovering statistics using SPSS*. London: SAGE Publications, Ltd.

Grabowski, M., Ayyalasomayajula, P., Merrick, J., Harrald, J.R., & Roberts, K. 2007. Leading indicators of safety in virtual organizations. *Safety Science* 45(10): 1013–1043.

Haight, J. & Thomas, R. 2003. Intervention effectiveness research: A review of the literature on leading indicators. *Chemical Health & Safety*: 21–25.

Hayes, B., Perander, J., Smecko, T. & Trask, J. 1998. Measuring perceptions of workplace safety: Development and validation of the work safety scale. *Journal of Safety Research* 29(3): 145–161.

Hinze, J., Thurman, S., & Wehle, A. 2013. Leading indicators of construction safety performance. *Safety Science* 51: 23–28.

Hollnagel, E., Woods, D., & Leveson, N. 2006. *Resilience engineering: Concepts and precepts*. Aldershot: Ashgate Publishing.

Hopkins, A. 2009. Thinking about process safety indicators. *Safety Science* 47(4): 460–465.

Hubbard, E. 2004. *The diversity scorecard: Evaluating the impact of diversity on organizational Performance*. Burlington: Elsevier Inc.

Li, F., Jiang, L., Yao, X., & Li, Y. 2013. Job demands, job resources and safety outcomes: the roles of emotional exhaustion and safety compliance. *Accident Analysis and Prevention* 51: 243–251.

Lingard, H., Hallowell, M., Salas, R., & Pirzadeh, P. 2017. Leading or lagging? Temporal analysis of safety indicators on a large infrastructure construction project. *Safety Science* 91: 206–220.

Manuele, F. 2009. Leading and lagging indicators. *Professional Safety* 54(12): 28–33.

Pedersen, L., Nielsen, K., & Kines, P. 2012. Realistic evaluation as a new way to design and evaluate occupational safety interventions. *Safety Science* 50(1): 48–54.

Petersen, D. 2000. Safety management 2000: Our strengths & weaknesses. *Professional Safety* 45(1): 16–19.

Probst, T., Graso, M., Estrada, A., & Greer, S. 2013. Consideration of future safety consequences: A new predictor of employee safety. *Accidents Analysis and Prevention* 55: 123–134.

Ratner, B. 2009. The correlation coefficient: Its values range between +1/–1, or do they? *Journal of Targeting, Measurement and Analysis for Marketing* 17: 139–142.

Reiman, T., & Pietikainen. 2012. Leading indicators of system safety—Monitoring and driving the organizational safety potential. *Safety Science* 50: 1993–2000.

Sgourou, E., Katsakiori, P., Goutsos, S., & Manatakis, E. 2010. Assessment of selected safety performance evaluation methods in regards to their conceptual, methodological and practical characteristics. *Safety Science* 48(8): 1019–1025.

Sinelnikov, S., Inouye, J., & Kerper, S. 2015. Using leading indicators to measure occupational health and safety performance. *Safety Science* 72: 240–248.

Watcher, J., & Yorio, P. 2014. A system of safety management practices and worker engagement for reducing and preventing accidents: An empirical and theoretical investigation. *Accident Analysis and Prevention* 68: 117–130.

Webb, P. 2009. Process safety performance indicators: A contribution to the debate. *Safety Science* 47: 502–507.

Wu, X., Liu, Q., Skibiniewski, M., & Wang, Y. 2015. Prospective safety performance evaluation on construction sites. *Accident Analysis and Prevention* 78: 58–72.

Zwetsloot, G. 2009. Prospects and limitations of process safety performance indicators. *Safety Science* 47: 495–497.

Comparison between methods of assessing the risk of musculoskeletal disorders on upper limb extremities: A study in manual assembly work

S. Costa, P. Carneiro & N. Costa
ALGORITMI Centre, School of Engineering, University of Minho, Guimarães, Portugal

ABSTRACT: The upper limbs extremities are more vulnerable to the development of musculoskeletal disorders, intrinsically related to professional manual assembly. There are several methods that assess the risk of work-related musculoskeletal disorders on upper limbs, yet there are few to assess their extremities, specifically the joints of the fingers. This study aims to compare and evaluate the adequacy and sufficiency of ergonomic analysis methods in assessing the risk of work-related musculoskeletal disorders of the upper limbs extremities, in manual assembly work of a medical device. ART and OCRA checklist methods were selected, and the results compared with the results of the Strain Index method. There were verified statistically significant differences between the three methods applied ($p < 0.001$), concluding that, according to this study, methods that better assess the risk are, in descending order, the Strain Index, OCRA checklist and ART.

1 INTRODUCTION

According to EU-OSHA (2010), Work-Related Musculoskeletal Disorders (WMSD), have become the most common form of occupational disease worldwide and represent one of the main concerns regarding Occupational Safety and Health (ESENER, 2010).

The socioeconomic impact of the WMSD on several branches of economic activity (Eurostat, 2010), often characterized by a Taylorist/Fordist organization system, whose inherent risk factors are recognized by several studies as categorical for the emergence of WMSD.

According to the 5th European Working Conditions Survey, in 2010, exposure to repetitive hand or arm movements is by far the most prevalent risk, with 63% reporting perform repetitive hand or arm movements at least a quarter of the time (Eurofound, 2012). In Portugal, according to the same survey, posture-related risks are the most prevalent, being the fourth most prevalent European country (Eurofound, 2012).

Musculoskeletal disorders affect different parts of the human body. More specifically, musculoskeletal disorders of upper limbs extremities are common and result in large costs (Garg et al., 2012). According to Xu et al. (2012), among the various segments of the human body, the upper limbs extremities are the most vulnerable to the development of WMSD and, according to the European Foundation for Living Conditions, is a significant problem in the workplace with regard to incidence and costs. This development of WMSD in upper limbs extremities presents relevant prevalence in some professional activities such as manual assembly (Wang et al., 2014). Assembly is a typical activity in a variety of industries, implying repetitive movements, with little opportunity for employees to change their postures (Wang et al., 2014).

There are several methods that classify the risk of WMSD of the upper limbs, however, there are few that assess the risk of WMSD of upper limbs extremities, more specifically of the hands, wrists and joints of the fingers. There are still few studies comparing the risk levels obtained by the various methods (Chiasson et al., 2012; Jones & Kumar, 2007; Kjellberga et al., 2015), and there is a lack of knowledge and guidance on what the methods are more effective and valid (Kjellberga, et al., 2015).

The present study aims to compare and evaluate the adequacy and sufficiency of ergonomic analysis methods in the evaluation of the risk of WMSD of upper limbs extremities on manual assembly of medical devices.

2 METHODOLOGY

The present research is a cross-sectional exploratory study (Saunders et al., 2012) based on the prior definition of the research guiding questions, followed by the observation of the reality of the problem in a real work context and qualitative and quantitative data collection. In order to meet the objectives, the study was carried out according to the following steps: (1) Bibliographic review on the issue; (2) Survey of existing methods that incorporate in their evaluation the risk of developing WMSD in the upper limbs extremities, and manual assembly work characteristics, through an exhaustive bibliographical research, through the application of the PRISMA methodology (Moher et al., 2009) to select the methods of ergonomic analysis to be applied in the study; (3) Selection of the study sample, which performed a detailed manual assembly, with demanding use of the upper limbs extremities, in the productive process; and a significant number of workers performing manual assembling (sample ≥ 30 workers to ensure the possibility of using parametric tests) and identification of the main risk factors of WMSD; (4) Application of an adapted Musculoskeletal Nordic questionnaire to characterize the study sample and to evaluate the prevalence of WMSD associated symptomatology; (5) Application of the selected ergonomic analysis methods; (6) Analysis, comparison and treatment of the data, using: statistical tests appropriate to the nature of the data; using tools such as LiteCAD® version 2.0.0.109 to obtain the angles performed with the upper limbs; Excel®, Microsoft Office 2013 for the treatment of the most common descriptive statistics and IBS SPSS 22® - Statistical Package for the Sciences for performing statistical tests.

3 RESULTS AND DISCUSSION

3.1 *Selection of methods of ergonomic analysis and their characterization*

As criteria for the selection of the methods of ergonomic analysis to be applied in the present study were considered those that incorporated in their evaluation more risk factors for evaluation of the risk of developing musculoskeletal injuries of the upper limbs extremities in manual assembly activities. Through the application of the PRISMA methodology (Moher et al., 2009), the methods selected for ergonomic analysis of the workstations under study were the ART (Assessment of Repetitive Tasks of the Upper Limbs (the ART tool, 2010) and the OCRA Checklist,—Occupational Repetitive Actions (Colombini et al., 2013) with 25 and 20 risk factors incorporated in their evaluation, respectively.

3.2 *Selection of workstations and study sample*

A company that manufactures medical devices in Portugal was selected. The company under study has two manual assembly lines, namely: manual assembly and automated assembly line with imposed cadence.

It was possible to obtain the information that the most critical workstations are those of automated assembly line due to the following factors: high repetitiveness of tasks; high precision and thoroughness of tasks; cadence imposed by the machine; every 4,8/5,0 seconds, on average, a medical device is produced; high work rate; existence of complaints reported by employees; high absenteeism; evidence of occupational diseases. In view of the above, the sample of the present study refers to the employees of the company that carry out automated assembly line activity with imposed cadence. The study sample is made up of 46 workers, representing 79% of the study population and 64% of the workforce of the company.

3.3 *Characterization of workstation*

In the automated assembly line, 17 work stations (n = 17) were evaluated, for the production of the medical device DISPMED, all of which were manually assembled, and a representative line was selected from all lines on the automated assembly line, according to the information given by the company. All the workstations (WS) are in the seated position, without any type of displacement. In general, in all workstations, actions performed are related to manipulate small components and execute collages, grips, inserts, various types of grips/grips with hands and fingers, among others, with continued use of the upper limbs extremities cadence imposed by the automated assembly line. The total working time is 460 minutes/day with a maximum working time with no breaks of 155 minutes.

3.4 *Characterization of the study sample and results of application of the questionnaire*

The questionnaire was applied to 45 workers (n = 45), 100% female, working in the automated assembly line, at fixed time.

The anatomical regions with the highest reported pain were, in descending order, the lower back region (96%), the cervical region (93%), joint of right fingers (71%), the right shoulder (69%), joint of left fingers (62%), right wrist (58%), right hand (56%) and left wrist (51%). It is important to emphasize that, in a general way, the right limb presented more pain reported by the workers, which may be intrinsically linked to the demands of the tasks, since there is a greater request of the right upper limb in the majority of WS evaluated, related to a higher number of technical actions performed by WS when compared to the left limb.

Particularly the anatomical regions evaluated, after the lower back and cervical region, there was a great percentage of symptomatology reported at the level of the finger joints.

3.5 Results of application of the selected ergonomic analysis methods—ART and OCRA Checklist

The ART and OCRA Checklist methods were applied to the 17 workstations, independently for each arm. In order to perform an easy and correct comparison of the results obtained, once the final WMSD risk assessment levels are different in both methods, a standardization of the levels was created in order to categorize the risks and compare the scores. This same methodology has already been used in other studies comparing methods of ergonomic analysis such as Chiasson et al. (2012), Jones & Kumar (2010), Kjellberga et al. (2015), Lavatellia et al. (2012), Paulsen, et al. (2015), Serranheira & Uva (2010) and Serranheira & Uva (2009). The categorization of risk levels is shown in Table 1.

In view of the above, the results shown in Graphs 1 and 2 were obtained by standardizing the risk classification.

The analysis of the graphs shows that, for the left and right upper limbs, the risk levels of WMSD development vary between low and medium, and in case of the ART method the risk index of the WS (15 WS) is mostly low, unlike the OCRA Checklist method for which the majority of the WS present medium risk (15 WS). The results obtained were not expected, taking into account the characteristics of the workstations.

The Friedman test was performed in order to verify if there are significant differences between the methods. For both upper limbs (left and right) it can be stated with 99.9% confidence that there are significant differences between methods X^2 (1, N = 17) = 13,000, p <0.001.

According to Lavatellia et al., (2012) there are no clear references in the literature on the repetitiveness/frequency of work, which they sustain when a task is considered "repetitive". In general, it is considered that there is repetitiveness in a working situation when it is recognized that identical movements are performed more than 2 to 4 times per minute (Lavatellia et al., 2012). Taking into account this premise all the workstations evaluated in the present study are repetitive, since the cycle times of WS vary between 4 and 9 seconds and the number of movements/minute, between 6 and 14. However, the ART and OCRA Checklist only assign score values to the frequency, when 11 movements/minute and 22.5 movements/minute, respectively, are performed for each task.

The force risk factor is similarly assessed between methods, duration and a descriptive scale are considered in the evaluation of strength. But, none of the methods takes into account factors such as grip strength, namely taking into account the size, orientation, location and weight of the object/piece to be manipulated. However, studies such as Hwang et al. (2010), Lee and Jung (2014) and Lee & Jung (2015) consider these factors in the application of force and posture adopted by the hand. It should also be mentioned that the type of grip with the "tips"/"extremities of the fingers" are generally more susceptible to cause musculoskeletal disorders than those performed with the palm (Pires, 2011), since the movements are performed by small and weaker muscle groups (Pastre, 2001).

With respect to posture, the two methods do not evaluate the same anatomical regions of the upper limbs. In the case of defining the type of work posture, the OCRA Checklist turns out to be more specific, since it takes into account the angles performed with the anatomical regions evaluated, also entered with the "repeating stereotypes" factor, that is, evaluating the presence of identical gestures (or movements) at the shoulders, elbows, hands and/or time.

Another difference is in the final classification of the risk. The ART method enters with the individual score of each anatomical region, and

Table 1. Standardization of levels in order to categorize risks and compare scores.

ART	OCRA Checklist	Risk Classification WMSD
≤11	≤7,5	Low risk worksation
12 a 21	7,6 a 14	Medium risk workstation
≥22	≥14,1	High risk workstation

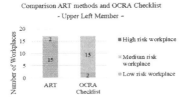

Graph 1. Comparison between the ART method and OCRA Checklist—Upper left limb.

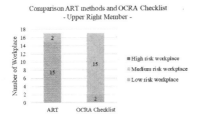

Graph 2. Comparison between the ART method and OCRA Checklist—Upper right limb.

the OCRA Checklist only enters with the highest value of the scores obtained in the various anatomical regions adding the value obtained in the "repeatability stereotypes" factor. In relation to the hands and fingers, the posture is only evaluated by the type of handgrip performed, namely, forceps, hook, open and normal, all of these postures being at the same evaluation level, not considering flexion, extension, abduction or adduction of the fingers in each one. That is, according to the methods, the risk of developing a disease is the same if a standard handle, a hook handle, an open handle with a finger or a tweezing handle is used. Only the duration of the handle influences the increase of the risk, however it should be noted that "fingertip"/"finger grip" handles are generally more susceptible to musculoskeletal disorders than those performed with the palm of the hand (Pires, 2011). It was also possible to verify, that the posture adopted by the fingers (flexion, extension, abduction or adduction movements) is not considered, in any method, in assessing the risk of WMSD in the upper limbs. However, Brunnstom (1997) cited by Pastre (2001) defines amplitudes of articular movements of the upper limb (wrist, fingers and thumb), as well as the ranges of average normal movement, with respect to movements of flexion, extension, abduction and adduction.

A new study opportunity was found, namely, to compare the results obtained with the ART and OCRA Checklist method with one of the most used observational methods (according to Pascual & Naqvi, 2008), the Strain Index method by Moore & Garg (1995), being a specific method of evaluation of the upper limbs extremities.

3.6 Results and discussion of the application of the Strain Index (SI) method

In order to compare the results of this method with ART and OCRA Checklist, similarly to what was previously done, a standardization of the levels was created, according to Table 2.

In view of the above, the results shown in Graphs 3 and 4 were obtained by standardizing the risk classification.

By the analysis of the graphs we conclude that the results obtained with the SI method

Table 2. Uniformization of levels in order to categorize risks and compare scores.

SI	ART	OCRA Checklist	Risk Classification WMSD
≤2,9	≤11	≤7,5	Low risk worksation
3 a 6,9	12 a 21	7,6 a 14	Medium risk workstation
≥ 7	≥ 22	≥ 14,1	High risk workstation

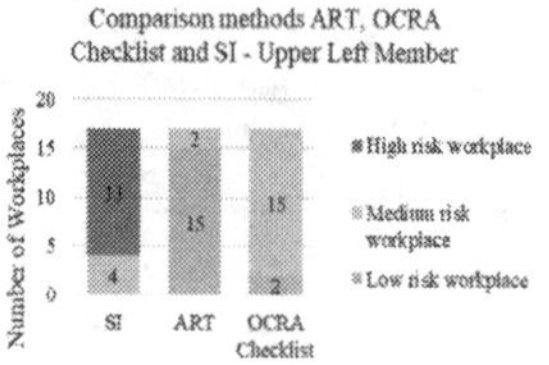

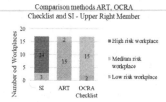

Graph 3. Comparison between the ART method, OCRA Checklist and SI—Upper left limb.

Graph 4. Comparison between the ART method, OCRA Checklist and SI—Upper right limb.

are quite different from those obtained with the other methods, with the SI being obtained a more predictable risk index according to the requirements and characteristics of the WS's, approach with fewer risk factors in their assessment. However, once you enter more scales of evaluation by risk factor, sensitivity increases in the evaluation.

With regard to "efforts per minute"/Repeatability/Frequency (MR), the evaluation with the SI method is very different compared to the other two methods. There is a great disparity between the three methods, namely: (1) ART—assigns punctuation from 11 moves/minute; (2) OCRA Checklist—assigns punctuation from 22.3 movements/minute; (3) Strain Index—assigns score from 4 moves/minute. Taking into account that, according to several studies such as Lavatellia et al. (2012), it is considered that there is repetitiveness in a work situation whenever the same movements performed more than two to four times per minute are recognized so, the most appropriate method to evaluate the "repetitiveness/frequency" risk factor is the Strain Index.

The Friedman test was carried out in order to verify if there were significant differences between the methods. For the left upper limb it can be stated with 99.9% confidence that there are significant differences between methods $X^2 (2, N = 17) = 30,000$, $p < 0.001$. For the right upper limb, it is also possible to affirm with 99.9% confidence that there are significant differences between methods $X^2 (2, N = 17) = 30.525$, $p < 0.001$.

In a study by Kjellberga et al. the ART, OCRA and SI methods (also with three levels of risk, low, medium and high), similar levels of risk were obtained between the ART and OCRA methods, but the SI method also obtained a higher percentage of high-risk tasks. A study by Murgia et al. (2012) who applied the OCRA Checklist method and SI also identified, in the tasks evaluated, a higher level of risk with the SI than with OCRA.

According to this study, in terms of sensitivity and adequacy of assessment of the risk of developing WMSD in the upper limbs extremities for manual assembly activity, the most appropriate methods are, in descending order, the Strain Index, OCRA Checklist and ART. However, all of them present many limitations regarding the ergonomic analysis of the upper limbs extremities.

4 CONCLUSION AND FUTURE PROSPECTS

The methods applied in the present study only consider the fingers with regard to some types of handle, and only the duration of the handle effects the increase of the risk, classifying similarly several types of handle. The methods do not evaluate finger kinesiology, hierarchizing their risk according to the health damage caused to each finger movement.

It was found that there were statistically significant differences between the 3 methods applied ($p < 0.001$), and it was concluded that, according to the present study, the sensitivity and adequacy of the evaluation of the risk of WMSD development in the upper limbs extremities for manual assembly activity, the most appropriate methods of ergonomic analysis are, in descending order, the Strain Index method, OCRA Checklist and ART.

Considering all the results of the present study, it is therefore urgently necessary to develop research to assess the risk of developing WMSD in the upper limbs extremities, in particular for the creation of the method, including of the joints of the fingers, having, for repetitive tasks, that they will have to be duly investigated and validated. Some suggestions for creating a new method may include evaluating: (1) strength indicators (blisters/callus on the fingers and white tip of the fingers); (2) finger positioning, evaluating finger kinesiology (3) evaluate other types of handles with the fingers and hands, and the increase in risk should be hierarchized, not only according to the duration of the handle, but also according to the damage to health caused by each one, since, for example, fingertip/finger grips are generally more likely to cause musculoskeletal disorders than those performed with the palm (Pires, 2011).

REFERENCES

Chiasson, M., Imbeau, D., Aubry, K., & Delisle, A. (2012). Comparing the results of eight methods used to evaluate risk factors associated with musculoskeletal disorders. *International Journal of Industrial Ergonomics 42*, 478–488.

Colombini, D., Occhipinti, E., & Álvarez-Casado, E. (2013). The revised OCRA Checklist method—updated version. Barcelona—Spain. *Editorial Factors Humans.*

ESENER. (2010). European Survey of Enterprises on New and Emerging Risks: Managing safety and health at work. Luxembourg: *European Survey of Enterprises on New and Emerging Risks.* European Agency for Safety and Health at Work.

EU-OSHA. (2010). OSH in figures: Work-related musculoskeletal disorders in the EU Office of the European Union. Luxembourg: *Facts and figures Publications.*

Eurofound. (2012). Fifth European Working Conditions Survey. Luxembourg: *Publications Office of the European Union.*

EUROSTAT. (2010). Health and safety at work in Europe (1999–2007). Belgium: *Publications Office of the European Union.*

Garg, A., Hegmann, K.T., Wertsch, J.J., Kapellusch, J., Thiese, M.S., Bloswic, D.K., Holubkov, R. (2012). The WISTAH hand study: A prospective cohort study of distal upper extremity musculoskeletal disorders. *BMC Musculoskeletal Disorders.*

Hwang, J., Kong, Y., & Jung, M. (2010). Posture evaluations of tethering and loose housing systems in dairy farms. *Applied Ergonomics 42*, 1–8.

Jones, T., & Kumar, S. (2010). Comparison of Ergonomic Risk Assessment Output in Four Sawmill Jobs. *International Journal of Occupational Safety and Ergonomics Vol. 16*, No. 1, 105–111.

Kjellberga, K., Lindbergc, P., Nymana, T., Palme, P., Rhenb, I.M., Eliassona, K., Forsmana, M. (2015). Comparisons of six observational methods for risk assessment of repetitive work-results from a consensus assessment. *Proceedings 19th Triennial Congress of the IEA*, Melbourne 9–14.

Lavatellia, I., Schaubb, K., & Caragnanoc, G. (2012). Correlations in between EAWS and OCRA Index concerning the repetitive loads of the upper limbs in automobile manufacturing industries. *Work 41*, 4436–4444.

Lee, K.-S., & Jung, M.-C. (2014). Ergonomic Evaluation of Biomechanical Hand Function. *Safety and Health at Work*, 1–9.

Lee, K.-S., & Jung, M.-C. (2015). Investigation of hand postures in manufacturing industries according to hand and object properties. *International Journal of Industrial Ergonomics 46*, 98–104.

Moher, D., Liberati, A., Tetzlaff, J., Altman, D., & PRISMA, G. (2009). Preferred Reporting Items for Systematic Reviews and Meta-Analyses: The PRISMA Statement. *Annals of Internal Medicine 151*, 264–269.

Moore, J.S., & Garg, A. (1995). The strain index: A proposed method to analyze jobs for risk of distal upper extremity disorders. *American Industrial Hygiene Association Journal*, 56(5), 443.

Murgia, L., Rosecrance, J., Gallu, T., & Paulsen, R. (2012). Risk evaluation of upper extremity musculoskeletal disorders among cheese processing workers: A comparison of exposure assessment techniques. *International Conference RAGUSA SHWA Ragusa—Italy "Safety Health and Welfare in Agrofood Agricultural and Forest Systems"*, 49–55.

Pascual, S.A., & Naqvi, S. (2008). An Investigation of Ergonomics Analysis Tools Used in Industry in the Identification of Work-Related Musculoskeletal Disorders. *International Journal of Occupational Safety and Ergonomics (JOSE)*, Vol. 14, No. 2, 237–245.

Pastre, T.M. (2001). Análise do Estilo de Trabalho em Montagem de Precisão. Trabalho de Conclusão do Curso de Mestrado Profissionalizante em Engenharia como requisito parcial à obtenção do título de mestre em Engenharia—modalidade profissionalizante—*Ênfase Ergonomia*.

Paulsen, R., Gallu, T., Gilkey, D., Reiser II, R., Murgia, L., & Rosecrance, J. (2015). The inter-rater reliability of Strain Index and OCRA Checklist task assessments in cheese processing. *Applied Ergonomics 51*, 199–204.

Pires, L.E. (2011). Contributo para a validacão de uma estratégia de diagnóstico do risco de LMLT: empresas de triagem de resíduos orgânicos. Mestrado em Segurança e Higiene no Trabalho - *Projeto de Investigação, Escola Nacional de Saúde Pública*—Universidade Nova de Lisboa.

Saunders, M., Lewis, P., & Thornhill, A. (2012). Research Methods for Business Students. *Financial Times Prentice-Hall*, 6th Ed.

Serranheira, F., & Uva, A. (2009). Avaliação do risco de Lesões Músculo-Esqueléticas: será que estamos a avaliar o que queremos avaliar? *Saúde & Trabalho*. 7, 69–88.

Serranheira, F., & Uva, A. (2010). LER/DORT: que métodos de avaliação do risco? *Revista Brasileira de Saúde Ocupacional*, São Paulo, 35 (122), 314–326.

Wang, H., Hwang, J., Lee, K.-S., Kwag, J.-S., Jang, J.-S., & Jung, M.-C. (2014). Upper Body and Finger Posture Evaluations at an Electric Iron Assembly Plant. *Human Factors and Ergonomics in Manufacturing & Service Industries 24* (2), 161–171.

Xu, Z., Ko, J., Cochran, D.J., & Jung, M.-C. (2012). Design of assembly lines with the concurrent consideration of productivity and upper extremity musculoskeletal disorders using linear models. *Computers & Industrial Engineering* 62, 431–441.

Stress misconduct and reduced ability to express dissent: A study on a sample of students at the University of Siena

O. Parlangeli, P. Palmitesta, M. Bracci & E. Marchigiani
Department of Social Political and Cognitive Science, University of Siena, Italy

P.M. Liston
*Centre for Innovative Human System School of Psychology, Trinity College Dublin,
The University of Dublin, Ireland*

ABSTRACT: The safety of socio-technical systems depends on a multitude of factors, including stress levels and compliance with relevant ethical principles. A recent study has identified a relationship between stress and unethical behaviour in specific areas such as scientific research (Parlangeli et al. 2017a, b). This paper reports a study conducted on students at the University of Siena in order to examine whether a relationship between stress and ethical behaviour can be identified in student populations. The results indicate a clear relationship between stress levels and the perception of the frequency of unethical behaviours. Of particular interest are the data suggesting how stress levels can negatively impact respondents' abilities to manifest dissent or disagreement, thus depriving people who experience high stress levels of the ability to contribute to the construction of workplaces characterised by honesty and integrity.

1 INTRODUCTION

Stress can be a precursor to low levels of safety in socio-technical systems in many ways (Clarke & Cooper 2004). First of all because it affects our ability to recognise that we are facing danger, but also because it alters our ability to solve moral problems (Jones 1991; Selart & Johansen 2011; Parlangeli et al. 2017a, b). Behaving ethically requires commitment, energy and self-control (De Wall et al. 2008), thus when high levels of stress are experienced it is likely that people will be less able to behave in altruistic and/or prosocial ways. Stress therefore appears to influence health and well-being both directly, but also indirectly, through our "unethical" choices, which, in turn, may impact system safety levels. Research in the third-level education sector demonstrates that inappropriate student behaviour such as the commission of plagiarism, copying during examinations, and other similar behaviours are becoming increasingly prevalent (Carroll 2005; Jones 2011; Kokkinaki et al. 2015), and further it points to a situation in which as student dishonesty increases so the perceived severity of such behaviour decreases (Murdock et al. 2001). Online plagiarism and cyber cheating moreover does not seem to constitute a new set of actions but rather they closely replicate and consolidate already existing behaviour (Selweyn 2008).

If education and training institutions prove to be failing to properly transmit fundamental ethical principles to students, it is likely that this will have consequences on moral attitudes and behavioural misconduct that will likely find their way to future work environments. These consequences are all the more likely if fundamental ethical principles are not adequately inculcated in third level educational establishments which educate and train the workforce comprising professionals of the highest calibre. As such, the study herein explores the link between the perception of stress and inappropriate behaviour in a sample of university students.

2 THE STUDY

To investigate the possible relationship between the perception of stress and the commission of unethical behaviour in education and training contexts, all students enrolled at the University of Siena were contacted by e-mail and requested to complete a questionnaire. The purpose of the questionnaire was to examine whether an increase in the perception of stress is related to an increase in ethical violations. The questionnaire also sought to understand whether, if such a relationship were identified, specific types of ethical violations were differentially affected by stress.

3 THE SAMPLE

The University of Siena student population returned 1030 questionnaires, 727 female respondents and 303 males with an average age of 23.54 (23.46 years for females and 23.74 for males). Respondents were citizens of many nations but the majority were Italian (N = 998). The majority were students enrolled on three-year degree courses (n = 494) and masters courses (n = 427). The others were enrolled in one-cycle degree (n = 100) or doctoral degree programs (n = 3). 276 respondents were enrolled in courses related to the faculty of economics, jurisprudence and political sciences, 258 in the faculty of humanities, history and philosophy of arts, 258 in the faculty of biomedical and health sciences and 215 in the area of experimental sciences.

4 TOOLS AND PROCEDURE

The data were collected through a specifically formulated questionnaire and made available online for the respondents to complete. The questionnaire consisted of 21 questions and is divided into 3 sections. The first section gathered demographic information on the sample; the second measured the degree of perceived stress through the use of Perceived Stress Scale, PSS-4 (Warttig et al. 2013); the third section was structured to gather information on (a) the ethical behaviour of students in relation to teaching staff; and (b) the ethical behaviour of students in relation to other students; and finally (c) the ethical behaviour of teaching staff towards the students. A question was also included about the frequency of students' misconduct in general. For each of the investigated behaviours the respondent judgments were collected relating to the frequency, severity and manifestation of dissent on a 5-point Likert scale.

5 RESULTS

The results related to the perception of stress obtained from the PSS-4 indicate a non-negligible level of stress, equal to 8.99 (the maximum score is equal to 16) with no significant differences between the faculties in which the respondents study. In a previous study conducted with 70 professors at the University of Siena this value amounted to 7.24 (Parlangeli et al. 2017b) and this reveals a significant difference (p<.000) with the sample of students considered here. On the basis of these data it seems therefore feasible to assume a higher level of perceived stress in student populations than in teaching staff (see Figure 1).

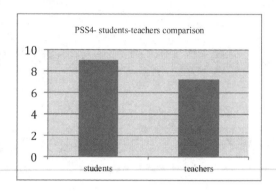

Figure 1. Differences in stress levels measured using the PSS-4 relative to a sample of students and a sample of teaching staff.

Figure 2 shows the differences in the students' judgments about teaching staff behaviour towards students, about student behaviour towards teaching staff and about students' behaviour towards other students. In particular, the first graph shows the frequency ratings (top), the second shows gravity ratings (centre) and the third shows ratings of the propensity to express dissent in response to unethical behaviours (bottom).

Considered in their entirety, these results provide a glimpse into a context in which unethical behaviours are affected by the difference of roles between teachers and students. Indeed, it can be seen that the most common inappropriate behaviours are those performed by students in relation to other students. Furthermore, although the judgments of severity are all quite harsh, it is possible to highlight a decrease in the degree of severity when inappropriate behaviours are committed by students with regards to teaching staff. Also, manifestations of dissent highlight this difference between the roles: as evident from Figure 2, they appear to be much more pronounced when they are expressed as a result of inappropriate behaviour performed by students in relation to other students.

Significant correlations emerged between perceived stress levels (PSS-4) and some unethical behaviours, and it appears that rising levels of stress correspond to an increase in the frequency and severity assessments of unethical behaviours when these are committed by teaching staff in relation to students (respectively: $r = .169$; $p<.000$; $r = .103$; $p<.001$). Also highlighted was a decrease in the willingness of students to express their dissent in high stress contexts when teaching staff behave inappropriately ($r = -.100$; $p<.001$). Which seems to suggest that when students are more stressed they perceive teachers as more misleading but seem less capable of expressing their disapproval.

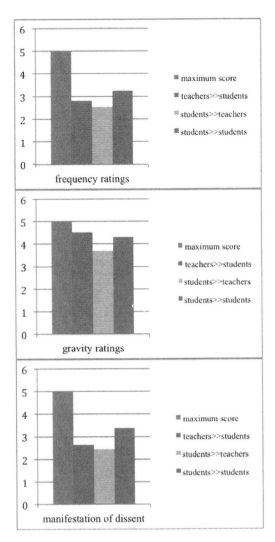

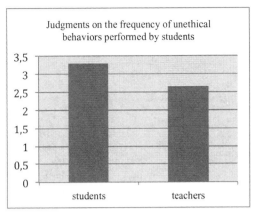

Figure 3. Judgments on the frequency of unethical behaviours performed by students in general.

Figure 2. The graph on the top frequency ratings for unethical behaviour. The centre graph shows the ratings of gravity. The third graph illustrates manifestations of dissent.

The frequency of students' inappropriate behaviour towards other students correlates positively with the perception of stress. With an increase in stress the perceived frequency of misbehaviour increases and it becomes easier to evaluate ones' peers behaviour as inappropriate (r = .157, p<.000).

In general, there is a difference between the judgments of teaching staff (Parlangeli et al. 2017b) and students: in almost all the judgements reported it is clear that students perceive inappropriate behaviour less frequently and less severely than teaching staff. Students also report being less likely to manifest their dissent when they witness unethical behaviour. In the case of assessments on the frequency of misbehaviour committed by the student population in general, students reported values higher than those of teaching staff (students = 3.31, faculty = 2.66, p<.000) (Fig. 3).

There are no significant differences, however, regarding assessments of the frequency of teaching staff's inappropriate behaviour towards students (teachers = 2.61, students = 2.80; p>.150). Overall, with regard to teaching staff, students perceive a less ethical world, but they also evaluate it with less severity and are less inclined to address the inappropriate behaviours that they perceive.

6 CONCLUSION

Increased levels of stress seem to have an impact on the perception of frequency judgments of inappropriate behaviours. In addition, in stressful situations, the ability to manifest dissent when confronted with violations of ethical behaviour codes seems to have greater chances of expression only amongst peers. This, amongst other things, can lead us to assume that working environments, particularly those that are hierarchical (Bagnara et al. 2008), can also inhibit the evolution of working styles based on open and ethical collaboration. The results obtained on the relationship between stress, ethical behaviour and the willingness to express dissent seem particularly relevant in that they relate to a sample of young people who, in the near future, will find themselves in leadership positions in working environments (Hayes & Introna 2005), apparently without being adequately equipped with the tools needed to react to the realities of modern working environments in an honest, ethical and appropriate way.

REFERENCES

Bagnara, S., Parlangeli, O., & Tartaglia, R. 2010. Are hospitals becoming high reliability organizations?. *Applied Ergonomics* 41(5): 713–718.

Carroll, J. 2007. A handbook for deterring plagiarism in higher education, 2nd edn. Oxford: Oxford Brookes University, Oxford Centre for Staff and Learning Development.

Clarke, S. & Cooper, C.L. 2004. Managing the risk of workplace stress: Health and safety hazards. London: Routledge.

DeWall, C.N., Baumeister, R.F., Gailliot, M.T., & Maner J.K. 2008. Depletion makes the heart grow less helpful: Helping as a function of self-regulatory energy and genetic relatedness. *Personality and Social Psychology Bulletin* 34(12): 1653–1662.

Hayes, N. & Introna, L.D. 2005. Cultural values, plagiarism and fairness: when plagiarism gets in the way of learning. *Journal of Ethics and Behaviour* 15(3): 213–223.

Jones, D.L.R. 2011. Academic Dishonesty: Are More Students Cheating? *Business Communication Quarterly* 74(2): 141–150.

Jones, T. 1991. Ethical Decision Making by Individuals in Organizations: An Issue-Contingent Model. *The Academy of Management Review* 16(2): 366–395.

Kokkinaki, A., Demoliou, C., & Lakovidou, M. 2015. Students' perceptions of plagiarism of relevant policies in Cyprus. *International Journal for Educational Integrity* 11(3): 1–11.

Murdock, T.B., Hale, N.M. & Weber, M.J. 2001. Predictors of cheating among early adolescents: Academic and social motivations. *Contemporary Educational Psychology* (26): 96–115.

Parlangeli O., Bracci M., Guidi S., Marchigiani, E., & Liston, P.M. 2017a. Stress and Unethical Behaviour in the Research of the Untenured University Researchers; *Conference proceedings of the ICERI, 10th International Conference of Education Research and Innovation, Seville, Spain 16–19 November 2017: 1905:1911.*

Parlangeli, O., Palmitesta, P., Bracci, M., Caratozzolo, M.C., Liston, P.M. & Marchigiani, E. 2017b. Stress and Perceptions of Unethical Behaviour in Academia. *Conference Proceedings of the ICERI, 10th International Conference of Education Research and Innovation, Seville, Spain 16–19 November 2017: 1912–1919.*

Selart, M. & Johansen, S.T. 2011. Ethical Decision Making in Organizations: The Role of Leadership Stress. *Journal of Business Ethics* 99(2): 129–143.

Selwyn, Neil. 2008. A Safe haven for Misbehaving? An Investigation of Online Misbehavior among University students. Social Science Computer Review 26(4): 446–465.

Warttig, S.L., Forshaw, M.J., South, J. & White, A.K. 2013. New, normative, English—sample data for the Short Form Perceived Stress Scale (PSS-4). *Journal Of Health Psychology* 18: 1617–1628.

Wearable technology for occupational risk assessment: Potential avenues for applications

I. Pavón & L. Sigcha
Instrumentation and Applied Acoustics Research Group (I2A2), ETSI Industriales, Universidad Politécnica de Madrid, Madrid, Spain

P. Arezes & N. Costa
The Ergonomics and Human Factors Research Group (E&HFg), Universidade do Minho (UMinho), Guimarães, Portugal

G. Arcas & J.M. López
Instrumentation and Applied Acoustics Research Group (I2A2), ETSI Industriales, Universidad Politécnica de Madrid, Madrid, Spain

ABSTRACT: This paper analyzes the opportunities offered by wearable technology for occupational risk assessment. A wide range of electronic devices known as "wearables" include sensors that provide interesting features with the potential to be used in workers' monitoring, sending alert signs (e.g., visual, sound, haptic) and training. In this paper, it has been identified the primary applications in which wearables are being used for risk assessment (motion and activity detection, recognition of work-related musculoskeletal disorders, fall detection, evaluation of exposure to different physical agents, evaluation of exposure to chemical agents, and location of potential hazards). Also, it has been identified the opportunities for improvement in the future using the potential of emerging technologies.

1 INTRODUCTION

The latest estimate on occupational accidents published by International Labour Organization (ILO) in September 2017, shows that 2.78 million workers die each year from occupational accidents and occupational diseases. Occupational mortality represents 5% of the world total. The majority of work-related mortality comes from occupational diseases (2.4 million) 86.3% of the total. The remaining 13.7% is due to occupational accidents. Work-related injuries and illnesses result in the loss of 3.9% of world GDP, with an annual cost of 2680 billion €. In the case of the EU, the cost of work-related injuries and injuries accounts for 3.3% of GDP, 476 billion € per year. Occupational accidents and its costs can be significantly reduced by establishing and practicing good occupational health and safety policies (Hämäläinen et al. 2017).

There is an increasing number of technological solutions that use a preventative approach (versus a reactive model), making possible to detect situations of occupational risk before they occur. Using these advanced technologies it is possible to reduce occupational risk and its costs. (Ding et al. 2013)

Information and Communications Technologies (ICT) can be used to monitor the well-being, progress, and quality of life of someone over a period of time. Devices like smartphones, tablets or wearables (e.g., smartwatches, smart wristbands) are used to manage specific health problems. This technology continues to evolve rapidly, the devices now can communicate with each other taking advantage of the omnipresent internet, and cloud computing services. The combination of the Internet of Things (IoT), Big Data (collection and analysis of large amounts of data) and the smart working environments (SWE) provides an opportunity for monitoring activities conducted by the worker, machinery, and tools (EU OSHA 2017, Li 2018).

Holistic solutions need to be used in managing Occupational Risk Prevention (ORP). Areas for developing and improving include analyzing the inputs and outputs of the system, analyzing the users involved and their needs, improving efficiency and reducing costs, accelerating the risks detection processes and integrating technologies that are currently under development.

Additionally, the resulting synergies should also be considered in managing ORP, such as prevention enhancement, accident and illness reduction, productivity improvement, costs reduction, availability and sharing of information, decision making in quasi-real-time, integration with the concepts of digitization of industry (Industry 4.0).

Devices called wearables can be carried on the body or integrated into clothing or other accessories. Wearable's sensors are used to monitor and track the wearer's health, identify the medical needs, and for fitness. Wearables are also used in areas such as entertainment, education, finance, and music (Sazonov et al. 2014, Wright 2014, Page 2015).

There are some initiatives for the use of wearable devices in industrial operations, mainly to the monitoring process and measuring health. Advantages such as portability, energy autonomy, and discrete use have been exploited in several solutions, creating a new paradigm for risk assessment with information shared real-time (Negi et al. 2011, Chan et al. 2012, Kim et al. 2013, Peppoloni 2016, Bieber et al. 2012, Daponte et al. 2013).

The intelligent wearable devices have processing, data storage, and communication capabilities, which gives them the potential to be a part of integrated systems. Thus, wearables can be used in the field of industrial safety (Podgorski et al. 2017, Bernal et al. 2017, Li 2018).

The aim of this article is to summarize the current state and the main applications of wearable technology related to the ORP. Also, it is presented the opportunities generated by the wearable technology and emerging technologies to improve the current risk management systems. Finally, the foremost challenges and future perspectives of the weareable technology related to ORP will be discussed.

2 WEARABLE TECHNOLOGY FOR OCCUPATIONAL RISK ASSESSMENT

The authors searched for published literature using a range of related keywords such as wearable, risk prevention, occupational safety, work monitoring and, risk assessment. The literature was selected according to its relevance with the ORP and with the date of publication after the year 2000.

Based on the literature review have been found numerous solutions that use wearable devices for the management of different occupational hazards. These solutions include motion and activity detection, recognition of work related musculoskeletal disorders, fall detection, evaluation of exposure to different physical agents (e.g., vibrations, noise, thermal environment, and light), exposure to chemical agents, and location of potential hazards.

Recent developments in the areas of environmental intelligence (AmI), the IoT and Cyber-Physical Systems (CPS) have created various solutions for occupational health and safety using smart personal protection devices or smart equipment (Podgorski et al. 2017). These developments emerged from the standardization of low-cost sensors called MEMS (Micro-Electro-Mechanical Systems).

2.1 Wearables for motion and activity detection

One of the primary applications of the wearables is detecting activity and movement. Different studies have used specific wearable devices and commercial devices for their variety of sensors (mainly inertial sensors e.g., accelerometers or gyroscopes) (Yang et al. 2010, Labrador et al. 2013, Mortazavi et al. 2015, Gjoreski et al. 2016).

Also, smartwatches and smartphones have been used simultaneously for the automatic recognition of activities. In (Ramos et al. 2016) the use of two intelligent devices with MEMS inertial sensors shows an increase the accuracy of physical activity recognition. The results show that wearables are a viable alternative to automatic recognition using commercial devices.

For occupational safety and health of construction workers, in (Lee et al. 2017) using a commercial wearable device with several sensors intended for physiological and activity monitoring have made it possible to perform personal measurement on health and well-being of workers. Showing that is possible to identify relationships between health variables to improve the performance, productivity and the safety of workers.

2.2 Wearables for detection of work-related musculoskeletal disorders

For detection of musculoskeletal disorders in the industry, some solutions have been developed mainly using inertial sensors.

In (Peppoloni et al. 2016) a wearable device that analyzes muscle strains and posture have been tested. The usefulness of this device was compared with the traditional method of observational inspection, showing that this type of system can be used in a complete work shift in a discreet way.

Similarly, in (Valero et al. 2017) the analysis of construction workers' postures has been carried out using a network of wireless sensors that identify postures and detect those that may be potentially dangerous. The results suggest that by using this approach, there is an improvement in the productivity and performance of workers employing favorable job positions.

In (Nath et al. 2017, Yan et al. 2017) a similar approach is followed, but using the sensors of several smartphones placed with adapters, in one case on the arms and waist as well on back and neck of construction workers. Showing that it is possible to identify possible ergonomic risks related to the work using two sensors. The systems are presented as a low-cost and efficient solution to prevent works from dangerous positions.

2.3 Wearables for fall detection

Detection of falls has been one of the relevant research topics in recent years. In (Pannurat et al. 2014) a compilatory study of fall detection and prevention systems was conducted using wearable (with inertial sensors) and environmental sensors. The study shows that wearables have some advantages over the environmental sensor systems such as better mobility, easy installation, and have a higher area of use and coverage.

The use of smartphones for fall detection is a growing topic to investigate since these devices have the necessary hardware for fall detection. In the studies of (Habib et al. 2014, Casilari et al. 2015) are present solutions for falls detection and prevention by using smartphones and Android devices.

Although smartphones can be considered as an excellent alternative to the dedicated systems, they also present technical challenges in regards to the quality of sensors, power autonomy, and identification of the best smartphone's location on the body.

2.4 Wearables for assessment of exposure to physical agents

Another foremost application for wearable devices is the assessment of exposure to physical agents. One of the relevant topics is the vibration exposure, which can be divided into hand-arm vibration (HAV) and whole body vibrations (WBV).

In (Tarabini et al. 2012) the advantages and disadvantages of the use of accelerometers for the measurement of HAV and WBV were evaluated. This paper concludes that it is possible to design systems that include MEMS accelerometers inside any hand tool, in the operator interface, or inside the seats of automobile structures, tractors, and trucks.

2.4.1 *Hand Arm Vibration (HAV)*

Experiences in the development of HAV meters were carried out by (Morello et al. 2010, Aiello et al. 2012) in which portable devices with MEMS accelerometers were used to measure HAV. These studies are presented as a new approach to safety and risk management, performing in real-time.

Then, in (Liu et al. 2013) a smartphone with a metering application was used as a wearable on the hand. The smartphone was secured to a crafted glove that was placed on a hand, showing a versatile and low-cost solution for risks assessment, train and bring awareness to workers.

Regard the use of smartwatches, in (Pavón et al. 2016) an HAV measurement system based on a commercial smartwatch was developed. Showing that this kind of system can be used in specific vibration risk management tasks due to the operating system restrictions. This study identifies improvement opportunities and presents the advantages (portability, continuous measurement, and cost reduction) of wearable systems over standard measuring devices.

Then, in (Matthies et al. 2016) a system to determinate the exposure to HAV based on classification algorithms were developed using the built-in microphone and accelerometer data. The results indicate some deviation when compared to a reference measuring device. However, using this approach demonstrates that such systems are a viable solution for risk management and monitoring of workers, overcoming the limitations of the device or sensor.

2.4.2 *Whole Body Vibrations (WBV)*

Smartphones with built-in MEMS accelerometers have been used to measure WBV. In (Cutini et al. 2014) a software application that measures the exposure to vibrations was developed in an Android phone, the system shows deviations in the measurements but presents a simple and low-cost system for carrying out the initial risk assessment and training.

In (Wolfgang et al. 2014 a, b) a WBV metering system was developed in a multimedia device (iPod) and compared with a gold standard vibration analyzer. The results of their experiments suggest that measurements made with the iPod can be used to measure full-body vibrations with minimal error, reducing the costs and complexity to manage this risk.

2.4.3 *Noise*

In recent years smartphones and mobile applications have been used as a measurement tool for environmental noise and occupational noise assessments. The accuracy of these systems using the built-in microphone have been analyzed in (Kardous et al. 2014, Kardous et al. 2015, Nast et al. 2015, Murphy 2016), the results show variation in the measurements made with these systems.

Then, solutions using smartphones with a high-quality external microphone have been tested by (Kardous et al. 2016), in which conclude that the calibration can make the system a useful tool for qualified professionals. Methods for calibration of these systems have been proposed in (Dumoulin 2013, Zhu et al. 2015).

2.4.4 Occupational thermal environment

In (Pancardo et al. 2015) several sensors (accelerometer, humidity, and temperature) from a smartphone were used to evaluate occupational heat stress produced by a hot work environment, the results suggest that this type of system is more efficient, discreet and less costly than using standard methods. As well it allows workers to take care of their health based on objective information and warnings. Similar to other studies, it concludes that including new sensors and devices, such as wristbands can offer new opportunities for preventing and monitoring health.

2.4.5 Light

An initiative regarding exposure to physical agents was made by (Cerqueira et al. 2017) where the accuracy of smartphones is reported using different applications for the measurement of light exposure with a focus on occupational health. Showing that smartphones have limitations and variability and may not be considered useful in occupational lighting assessments.

2.5 Wearables for assessment of exposure to chemical agents

A study made by (Negi et al. 2011) evaluated the exposure of chemical agents through the use of an ad-hoc wearable device with sensors for measuring hydrocarbons, total acids, humidity, and temperature. This system can monitor and detect gases (toxic hydrocarbons and acid environments) and can be applied to occupational and environmental health. This study sets the paradigm of continuous monitoring in real-time and shows that it is possible to use a smartphone as a user interface for the sensor device.

2.6 Wearables for location of potential hazards (moving machinery)

Another approach to safety was presented by (Ferreira et al. 2017) in which wearable devices were placed on worker safety vests and arms. The devices were used as an alarm (visual and sound) to alert maintenance workers near train tracks the proximity of a train. This study identified the challenges that need to be addressed to present a plausible technological security solution for railway personnel involved in maintenance work, in which includes the wireless communication and usability related to worker's perceptions of the alerts.

3 DISCUSSION & CONCLUSIONS

Based on the literature review, a growing number of publications that use wearable devices for occupational risk assessment have been identified. The solutions use both commercial and specific wearable devices.

In the analyzed studies, it has been identified the use of devices developed specifically for particular risk assessment tasks as well as a growing number of solutions that use smart devices mainly smartphones and smartwatches, but some of these solutions only use one or more sensors and do not take advantage of their communication capabilities.

Several applications of this kind of technology for occupational risk assessment have been identified, among which are:

- Motion and activity detection.
- Detection of risk factors related to the musculoskeletal system.
- Fall detection and prevention.
- Exposure to physical and chemical agents (noise, vibration, gases, light).
- Location of potential hazards (moving machinery).

Most of this wearable solutions are focused on monitoring and alert (visual or haptic feedback), but with the support of other technologies or devices (e.g., smartphones or tablets) they could make use of their characteristics for making a complete system for management, training, and workers' awareness.

Despite the fact that the adoption of wearable devices is beginning and that many innovative solutions have been presented for the management of specific occupational risks, the integration of this solutions is not yet complete due to the lack of a common framework for development and future needs. A framework can make use of the features and advantages provided by the information and communications technologies and data analysis (based for example on the Big Data paradigm).

Also, the accuracy and reliability of these systems must be taken into account when they are used as a measurement system. At present, the built-in sensors of some intelligent devices are less precise than precision equipment, which makes difficult to perform a correct risk assessment. However, these devices could be used to obtain measurements with better representativeness because they can be worn in a continuous manner over the time.

In the future, it is expected to use the IoT approach in sensors, tools, and workers to improve the management of occupational safety. The first experiences in the use of wearables as integral management systems have been detected (Podgorski et al. 2017, Bernal et al. 2017, Li 2018), in which include the use of ad-hock wearable devices or personal protective equipment with sensors. Showing the need for innovative solutions that integrate monitoring, management, and workers training. Taking into account the diversity of risks that can be found in the workspace, as well as the real training and education needs of workers in different industries.

REFERENCES

Bernal, G., Colombo, S., Al Ai Baky, M. & Casalegno, F. 2017, "Safety: Designing IoT and Wearable Systems for Industrial Safety through a User Centered Design Approach", *Proceedings of the 10th International Conference on Pervasive Technologies Related to Assistive Environments* ACM, 163.

Casilari, E., Luque, R. & Moron, M. 2015, "Analysis of Android Device-Based Solutions for Fall Detection", Sensors, vol. 15, no. 8, 17827–17894.

Cerqueira, D., Carvalho, F. & Melo, R.B. 2017, "Is It Smart to Use smartphones to measure illuminance for occupational Health and Safety Purposes?" *International Conference on Applied Human Factors and Ergonomics* Springer, 258.

Chan, M., Estève, D., Fourniols, J., Escriba, C. & Campo, E. 2012, "Smart wearable systems: Current status and future challenges", *Artificial Intelligence in Medicine*, vol. 56, no. 3, 137–156.

Cutini, M. & Bisaglia, C. 2014, "Whole body vibration monitoring using a smartphone", Ageng 2014 Zurich; International Conference of Agricultural (EurAgEng), 8.

Daponte, P., De Vito, L., Picariello, F. & Riccio, M. 2013, "State of the art and future developments of measurement applications on smartphones", *Measurement: Journal of the International Measurement Confederation*, vol. 46, no. 9, 3291–3307.

Ding, L.Y., Zhou, C., Deng, Q.X., Luo, H.B., Ye, X.W., Ni, Y.Q. & Guo, P. 2013, "Real-time safety early warning system for cross passage construction in Yangtze Riverbed Metro Tunnel based on the internet of things", *Automation in Construction*, vol. 36, 25–37.

Dumoulin, R. & Voix, J. 2013, "Calibration of smartphone-based devices for noise exposure monitoring: Method, implementation, and uncertainties of measurement", *The Journal of the Acoustical Society of America*, vol. 133, no. 5, 3317–3317.

EU OSHA 2017, Monitoring technology: the 21st century's Pursuit of well-being?

Ferreira, B, et al., C. 2017, "Wearable computing for railway environments: proposal and evaluation of a safety solution", *IET Intelligent transport systems*, vol. 11, no. 6, 319–325.

Gjoreski, M., Gjoreski, H., Lustrek, M. & Gams, M. 2016, "How Accurately Can Your Wrist Device Recognize Daily Activities and Detect Falls?", *Sensors*, vol. 16, no. 6, 800.

Habib, M., Mohktar, M., Kamaruzzaman, S., Lim, K., Pin, T. & Ibrahim, F. 2014, "Smartphone-Based Solutions for Fall Detection and Prevention: Challenges and Open Issues," *Sensors*, vol. 14, no. 4, 7181–7208.

Hämäläinen, P., Takala, J. & Kiat, T.B. 2017, *Global estimates of occupational Accidents and work-related illnesses 2017*, Workplace Safety and Health Institute.

Kardous, C. & Shaw, P. 2014, "Evaluation of smartphone sound measurement applications", *Journal of the Acoustical Society of America*, vol. 135, no. 4, EL186-EL192.

Kardous, C.A. & Celestina, M. 2015, "Use of smartphone sound measurement apps for occupational noise assessments", *Journal of the Acoustical Society of America*, vol. 137, no. 4, 2292.

Kardous, C.A. & Shaw, P.B. 2016, "Evaluation of smartphone sound measurement applications (apps) using external microphones—A follow-up study", *The Journal of the Acoustical Society of America*, vol. 140, no. 4, EL327-EL333.

Kim, S. & Nussbaum, M. 2013, "Performance evaluation of a wearable inertial motion capture system for capturing physical exposures during manual material handling tasks", *Ergonomics*, vol. 56, no. 2, 314–326.

Kim, K. & Shin, D. 2015, "An acceptance model for smartwatches Implications for the adoption of future wearable technology", *Internet Research*, vol. 25, no. 4, 527–541.

Labrador, M.A. & Lara Yejas, O.D. 2013, *Human activity recognition: using wearable sensors and smartphones*, 1st EDN, Taylor & Francis, Boca Raton.

Lee, W., Lin, K., Seto, E. & Migliaccio, G. 2017, "Wearable sensors for monitoring on-duty and off-duty worker physiological status and activities in construction", *Automation in Construction*, vol. 83, 341–353.

Li, R.Y.M. 2018, "Smart Working Environments Using the Internet of Things and Construction Site Safety" in *An Economic Analysis on Automated Construction Safety* 137–153.

Liu, B. & Koc, A.B. 2013, "Hand-Arm Vibration Measurements and Analysis Using Smartphones", *2013 Kansas City, Missouri, July 21-July 24, 2013* American Society of Agricultural and Biological Engineers, 1.

Matthies, D., Bieber, G. & Kaulbars, U. 2016, "AGIS: automated tool detection & hand-arm vibration estimation using an unmodified smartwatch", ACM, 1.

Mortazavi, B., Nemati, E., VanderWall, K., Flores-Rodriguez, H., Cai, J., Lucier, J., Naeim, A. & Sarrafzadeh, M. 2015, "Can Smartwatches Replace Smartphones for Posture Tracking?", *Sensors*, vol. 15, no. 10, 26783–26800.

Murphy, E. & King, E. 2016, "Testing the accuracy of smartphones and sound level meter applications for measuring environmental noise", *Applied Acoustics*, vol. 106, 16–22.

Nast, D., Speer, W. & Le Prell, C. 2014, "Sound level measurements using smartphone "apps": Useful or inaccurate?" *NOISE & HEALTH*, vol. 16, no. 72, 251–256.

Nath, N., Akhavian, R. & Behzadan, A. 2017, "Ergonomic analysis of construction worker's body postures using wearable mobile sensors", *Applied Ergonomics*, vol. 62, 107–117.

Negi, I., et al., 2011, "Novel monitor paradigm for real-time exposure assessment", *Journal of Exposure Science and Environmental Epidemiology*, vol. 21, no. 4, 419–426.

Page, T. 2015, "A Forecast of the Adoption of Wearable Technology", *International Journal of Technology Diffusion*, vol. 6, no. 2, 12–29.

Pancardo, P., Acosta, F., Hernandez-Nolasco, J., Wister, M. & Lopez-de-Ipina, D. 2015, "Real-Time Personalized Monitoring to Estimate Occupational Heat Stress in Ambient Assisted Working", *Sensors,* vol. 15, no. 7, 16956–16980.

Pannurat, N., Thiemjarus, S. & Nantajeewarawat, E. 2014, "Automatic Fall Monitoring: A Review", *Sensors,* vol. 14, no. 7, 12900–12936.

Peppoloni, L., Filippeschi, A., Ruffaldi, E. & Avizzano, C. 2016, "(WMSDs issue) A novel wearable system for the online assessment of risk for biomechanical load in repetitive efforts", *International Journal of Industrial Ergonomics,* vol. 52, 1–11.

Podgorski, D., Majchrzycka, K., Dabrowska, A., Gralewicz, G. & Okrasa, M. 2017, "Towards a conceptual framework of OSH risk management in smart working environments based on smart PPE, ambient intelligence and the Internet of Things technologies", *International Journal of Occupational Safety and Ergonomics,* vol. 23, no. 1, 1–20.

Ramos, F., Moreira, A., Costa, A., Rolim, R., Almeida, H. & Perkusich, A. 2016, "Combining smartphone and smartwatch sensor data in activity recognition approaches: An experimental evaluation", 267.

Sazonov, E., Neuman, M.R. 2014, Wearable sensors: fundamentals, implementation and applications, Academic Press.

Tarabini, M., Saggin, B., Scaccabarozzi, D. & Moschioni, G. 2012, "The potential of micro-electro-mechanical accelerometers in human vibration measurements", *Journal of Sound and Vibration,* vol. 331, no. 2, 487–499.

Valero, E., Sivanathan, A., Bosche, F. & Abdel-Wahab, M. 2017, "Analysis of construction trade worker body motions using a wearable and wireless motion sensor network", *Automation in Construction,* vol. 83, 48–55.

Wolfgang, R. & Burgess-Limerick, R. 2014, "Using Consumer Electronic Devices to Estimate Whole-Body Vibration Exposure", *Journal of Occupational and Environmental Hygiene,* vol. 11, no. 6, D77-D81.

Wolfgang, R., Di Corleto, L. & Burgess-Limerick, R. 2014, "Can an iPod Touch Be Used to Assess Whole-Body Vibration Associated with Mining Equipment?", *Annals of Occupational Hygiene,* vol. 58, no. 9, 1200–1204.

Wright, R. & Keith, L. 2014, "Wearable Technology: If the Tech Fits, Wear It", *Journal of Electronic Resources in Medical Libraries,* vol. 11, no. 4, 204–216.

Yan, X., Li, H., Li, A. & Zhang, H. 2017, "Wearable IMU-based real-time motion warning system for construction workers' musculoskeletal disorders prevention", *Automation in Construction,* vol. 74, 2–11.

Yang, C. & Hsu, Y. 2010, "A review of accelerometry-based wearable motion detectors for physical activity monitoring", *Sensors,* vol. 10, no. 8, 7772–7788.

Zhu, Y., Li, J., Liu, L. & Tham, C. 2015; 2016, "iCal: Intervention-free Calibration for Measuring Noise with Smartphones", 85.

Occupational Safety and Hygiene VI – Arezes et al. (Eds)
© *2018 Taylor & Francis Group, London, ISBN 978-1-138-54203-7*

Maturity level of hearing conservation programs in metallurgical companies and the workers noise perception

I.C. Wictor & A.A. de P. Xavier
Federal Technological University of Paraná, Ponta Grossa, Brazil

ABSTRACT: This research aimed to evaluate the influence of level maturity about hearing conservation programs on the workers risk perception. This study evaluated five metallurgical companies with a sample of 243 workers exposed to sound pressure levels above the action level in national legislation 85 dB (A). Based on the bibliographic review, two questionnaires were used to evaluate qualitative variables. The first questionnaire was developed and applied to companies to evaluate maturity levels in hearing conservation programs. Subsequently a questionnaire was applied to the worker considering the individual noise perception; It was concluded that the different levels of maturity in hearing conservation program presented no significant differences in the workers perception, however, there has been an increased risk perception in the different levels of exposure to noise.

1 INTRODUCTION

The Hearing Conservation Program (HCP) is a set of technical and administrative measures aimed at protecting workers' health when exposed to occupational noise.

An important aspect when considering the structuring and implementation of a Hearing Conservation Program, is to assume that professional hearing loss is totally preventable (Arezes, 2002; Bies and Hansen, p 163, 2009).

The hearing conservation program is not only to provide ear protection systems to exposed people, but to seek solutions directly at the source, to eliminate or reduce noise to the maximum using all resources, administration and engineering.

The main objective of this work is to evaluate the hearing conservation programs in metallurgical companies and the workers perception. As none of the surveyed companies have a HCP implemented, but each company adopts differentiated measures for hearing protection of its workers, this work sought to evaluate a level of maturity to make the comparisons.

1.1 *Hearing conservation programs in metallurgical companies*

Companies that present noise levels above 85 dB (A) for an exposure of 8 hours daily (NHO, 2002), should take measures to promote hearing conservation of its employees.

However, companies have a set of measures aimed at worker protection, but measures are not provided in a hearing conservation planning. The measures are included in Occupational Health and Safety planning, which performs measurements of industrial noise through the worker's daily exposure level, included in the Environmental Risk Prevention Program.

According NR-9 (MTE, 2011) establishes the obligation to elaborate and implement the Environmental Risk Prevention Program-which according to the Standard considers the physical agents (noise, vibration, pressure, temperature, and radiation), chemical and biological, depending on their nature, concentration or intensity, can cause harm to the worker's health.

Even though there is no hearing conservation program—HCP in place, the measures adopted by the companies are part of a program planning that can be included in a HCP, so that companies use this information to improve their methods and planning to obtain positive results.

1.2 *Maturity level of hearing conservation program*

As no researched company has a hearing conservation program, the best way to evaluate them would be to establish values for each action that are part of an HCP, to differentiate them through a maturity level in project management.

According Andersen & Jessen (2002) in the real world, there will be no companies that have reached the maximum level of maturity, and none will reach, so it makes sense to talk about maturity levels and make an effort to measure or characterize the maturity of the organization.

The concept of maturity indicates that there may be differences in the development of a level of capacity between one and another company.

The notion of stages indicates that maturity develops over time. However, the use of steps is not the only model to present different stages of maturity.

2 MATERIALS AND METHODS

2.1 Data collection procedures

To develop an analysis about the hearing conservation programs and workers risk perception 243 workers from four medium-sized and one large metallurgical enterprises in Parana State-Brazil were interviewed.

The data collection was performed in four steps:

1. Collection of noise levels (informed by company);
2. Selection of the workers (which were working in environment with level up than 85 dB (A);
3. Employees questionnaire (first, general data were collected: age, sector of working, service time in the same sector, after, main questionnaire about noise perception;
4. Company questionnaire.

To select the sample study, some inclusion and exclusion criteria were applied: Inclusion—agree to participate voluntary in the survey; be working in the same sector at least 1 year. Exclusion—working in environment with levels below 85 dB (A), according Brazilian regulatory limits, which limit value is 85 dB (A) for eight hours, and the maximum dose of 100% were used. In sequence, data collected were analyzed.

This study was approved by the National Commission of Ethics in Research (CONEP) under the number 53661315.4.0000.5547. All selected workers in this research signed a voluntary commitment agreement.

2.2 Interviewees and questionnaire

The questions about risk perception were made using a Likert scale questionnaire, this method was developed by Arezes (2002), and the ranking perception answer used ranges from 1 = no risk to 5 = too much risk, and, 1 = totally disagree to 5 = totally agree.

The questionnaire applied to workers provides four questions that assess the individual perception of risk treated as follows: sources of risk; knowledge about noise; perception of self-efficacy; and perception of protection.

Through the questionnaire applied to the companies was determined the level of maturity of each of the companies surveyed, exclusively considering the actions of the companies for the hearing conservation of their employees.

The questionnaire consists of 50 questions, the questions have general and specific approach, closed questions (yes or no) answers and multiple-choice answers. Each issue was classified among the five levels of the OPM3 model that is based on the PMBoK concept, which observes the processes involved in project management.

3 RESULTS

3.1 Interview questions and answers

First, the noise level was obtained in all sectors of the companies surveyed. Only workers exposed to noise levels above 85 dB (A) participated in the survey.

The data presented in Table 1 show the environmental noise in the percentage of workers are exposed.

We can observe that, in company (A) and (D), noise levels are very high and there are many workers exposed in this environment. In the other hand, company (E) has lower noise levels comparing with the other.

Table 2 presents the general data of the companies surveyed, as number of employees, average age and time service in the same sector.

Subsequently a questionnaire was applied to evaluate the actions of companies in the auditory conservation of workers.

Figure 1 presents a diagram of the main points of the questionnaire applied as companies.

3.2 Workers noise perception

The questionnaire applied to the employee provides four (4) questions that assess the individual perception of the risk treated below. After data collection, the reliability analysis of the questions was performed with all 243 respondents.

Table 1. Percentage of workers exposed to noise levels exceeding 85 dB (A).

Noise levels dB (A)	A	B	C	D	E
85–86	10%	0%	24%	0%	68%
87–89	28%	40%	48%	16%	26%
90–92	26%	58%	25%	84%	6%
93–98	36%	.2%	3%	0%	0%
Sample	**58**	**43**	**89**	**19**	**34**

Table 2. General data of the companies surveyed.

Company	Employees number	Sample number	Age average	Sd	Service time	Sd	HPD ear plug	HPD earmuffs
A	385	58	40,8	11,1	8,9	6,08	79%	21%
B	80	43	37,6	12,5	7,9	6,8	93%	7%
C	720	89	38,9	11,5	7,1	7,3	87%	13%
D	165	19	31,5	8,4	4,3	4,5	100%	0%
E	220	34	34,5	9,7	4,6	5,3	56%	44%

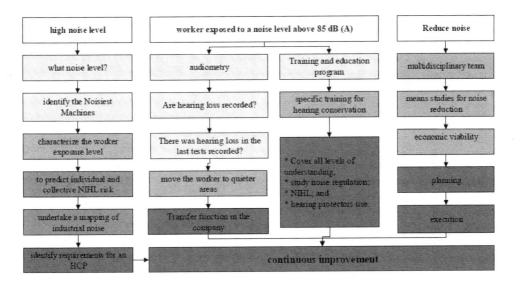

Figure 1. Basic diagram of an HCP questions for companies.

The worker responded according to his or her perception, on a severity or severity scale, expressing the opinion for each item described for sources of risk: "What type of risk do you think is associated with each of the following situations?" in between 1 – no risk; 2 – little risk; 3 – no opinion; 4 – some risk; and 5 – too much risk.

Knowledge about noise was intended to determine the level of knowledge about occupational exposure to noise by the following expression: "to what extent do you agree with the following statements": responses varied in the likert scale to 1 – totally disagree; 2 – partially disagree; 3 – no opinion; 4 – partially agree; 5 – totally agree.

The evaluation of the individual perception of the worker with regard to self-efficacy, the questions assess the opinion of the respondents regarding the use of ear protectors and the perception of protection and capacity of use in an adequate way in the workplace.

The perception of protection assessment sought to assess the ways in which workers protect themselves from exposure to noise, and how they feel protected.

Table 3. Average points for each question evaluated.

Questions	Amplitude possible [Min.; Max.]	Amplitude obtained [Min.; Max.]	Score (Company A to E) Average
WORKERS PERCEPTION			
Sources of risk	[7; 35]	[11; 35]	26,9
Knowledge	[4; 20]	[9; 20]	19,3
Self—efficacy	[7; 30]	[10; 35]	28,6
Protection	[5; 25]	[8; 25]	18

Table 3 shows the maximum amplitude of possible points for each question and the amplitude obtained, followed by the average obtained among the 5 companies surveyed.

The results on the perception of noise by the workers found in Table 5 were obtained by the media of points in each question for each company surveyed.

3.3 Maturity level characterization at Surveyed Companies

Through the questionnaire applied to the companies was determined the level of maturity of each of the companies surveyed, exclusively considering the actions of the companies for the auditory conservation of their employees.

Each question was classified among the five levels of the OPM3 model that is based on the PMBoK concept, which observes the processes involved in project management: initialization; planning; execution; control; and continuous improvement. The HCP maturity level uses the MMGP reference (Prado, 2004), with the same level of adherence proposed in the model.

The maturity model for Hearing Conservation Programs can also be described as a Project, as it encompasses all the processes required for deployment, management and improvement.

For this maturity model, all levels are interconnected. Thus, as levels become higher, it means that the company has a greater number of actions and control and of greater significance within the HCP.

The sums obtained through the questionnaire applied in the companies were tabulated. For each positive question a score was assigned, which depended on the level that was distributed. The scores are as follows:

– Level 1—Initialization: 1 point per positive response;
– Level 2—Planning: 2 points per positive response;
– Level 3—Execution: 4 points per positive response;
– Level 4—Control: 8 points per positive response;
– Level 5—Continuous improvement: 10 points per positive response.

Table 2 presents the scoring results of each company for actions within the auditory conservation program.

To evaluate the level of maturity from the data in Table 2, the model of Prado (2004) was used, which suggests a level of adherence for each level that the company scored, this model uses the concept of percentage of adhesion that should be used in conjunction with the level of maturity to understand at what level the sector of the company is.

It is called the Adherence Percentage a certain level obtained through a percentage in the maturity evaluation test.

At level 2 company A, company B and company E obtained 100%, which means complete adherence. Both companies' C and D achieved 86% which results in a good adhesion level.

For level 3 the results differ, where company C obtained 90% considered good at the adhesion level, followed by companies' A and B with 60%, in this result the two companies reached the minimum score to enter the good level. The companies' D and E obtained a percentage lower than 60%.

3.4 Relationship between maturity levels and risk perception

Every company has a way of applying worker protection methods. The measures vary from one company to another, according to a set of factors, but mainly with the culture of company management.

Therefore, we evaluated each of the companies surveyed through a questionnaire with the safety and hygiene department, on the measures and commitments that the company has in relation to hearing conservation.

With the sum of the questions, each company has a number that represents the points accumulated to its level of maturity. Table 4 presents a relationship between the score of each company and the employee perception score in each company for each item of individual perception.

Company A and company C had a higher level of maturity in HCP, while D had the lowest level

Table 4. Score obtained by each company on HCP.

Levels Maturity	A	B	C	D	E
Level 1	13	12	11	14	12
Level 2	18	16	12	12	14
Level 3	24	24	36	20	20
Level 4	24	*	24	*	8
Level 5	*	*	*	*	*
Sample	79	52	83	46	54

Table 5. Maturity levels of hearing conservation program and mean perceptions.

	Company A	Company B	Company C	Company D	Company E
MATURITY	79	52	83	46	54
Sources of risk	27,6	26,3	26,7	29,9	25,3
Knowledge about Noise	19,2	19,4	19,1	19,4	19,4
Perception of Self efficacy	29,8	27,6	27,8	28,3	30
Perception of protection	17,8	16,7	17,7	18,5	17,8

among the companies surveyed. However, it is important to note that the perception of sources of risk and the perception of protection are greater in company D, with less maturity.

It is possible to perceive that the perception has no relation with the level of maturity, because in the company with higher level of maturity (A and C) the perception is not significantly different from the companies with lower level of maturity.

It is important to note that workers 'perceptions may be more related to the work environment in which they are exposed than the companies' PCA maturity levels, some authors (Rabinowitz et al., 2007; Morata et al., 2005; Ahmed, 2012) also reached the same conclusions.

Therefore, it is important to analyze the branch of activity in which the research is carried out, many sectors, because they present characteristics of high noise levels, workers, regardless of the actions of the company, may have a higher perception of risk than workers in other sectors. This positioning is reinforced by (Rabinowitz et al., 2007), who observed workers in the aluminum industry with higher noise exposure at the workplace and had less hearing loss than co-workers in less noisy areas.

(Bockstael, De Bruyne et al., 2013) concluded that companies that use more rigid policies the noise risk perceive by the worker is higher than those that are more flexible, but notes that the perception of risk of exposure to noise by workers also varies among companies, and infer that noise exposure is a more determinant factor in the perception of risk than corporate policy.

4 CONCLUSIONS

Understanding workers' perceptions, the safety culture of a workplace, and attitudes are important factors in assessing safety needs. The perception about the risk addressed through the questions, seek mainly to understand the dimension of the perception of risk and to relate among the companies surveyed.

As the authors mentioned, the results of this research suggest that the worker's perception regarding the individual's perception of risk, perception of effects, safety culture and risk behavior is not related to maturity levels in hearing conservation programs in the different metallurgical companies, however, it is observed that the perception varies between companies, and the level of exposure to occupational noise is strongly linked to workers' perception.

The perception of the work environment can be related to the culture of the company and how much it invests in hearing conservation of the workers. At this point, it can influence, therefore, the companies that presented higher level of maturity (company A and company C) registered a higher perception index for the working environment.

REFERENCES

Ahmed, H.O., et al. 2001. Occupational noise exposure and hearing loss of workers in two plants in eastern Saudi Arabia. *Annals of Occupational Hygiene* 45 (5): 371–380.

Ahmed, H.O. 2012. Noise exposure, awareness, practice and noise annoyance among steel workers in United Arab Emirates. *Open Public Health Journal* 5: 28–35.

Andersen, E.S.; Jessen, S.A 2002. Project maturity in organization. *International Journal of Project Management*, N. 21, p. 457–461.

Arezes P.M. F.M. 2002. Percepção do Risco de Exposição Ocupacional ao Ruído. 240 f. *Tese (doutoramento em Engenharia de Produção) Escola de Engenharia da Universidade do Minho.*

Arezes, P.M.; Miguel A.S. 2005. Hearing protection use in industry: The role of risk perception. *Safety Science* 43(4): 253–267.

Bockstael, A., et al. 2013. Hearing protection in industry: Companies' policy and workers' perception. *International Journal of Industrial Ergonomics* 43(6): 512–517.

Byrne, D.C., et al. 2011. Relationship between comfort and attenuation measurements for two types of earplugs. *Noise and Health* 13(51): 86–92.

Hunashal, R.B.; Patil, Y.B. 2012. Assessment of Noise Pollution Indices in the City of Kolhapur, *India. Procedia – Social and Behavioral Sciences*, v. 37, n. 0, p. 448–457.

Ministério do Trabalho e Emprego. 2011. *Norma Regulamentadora nº 9*. Brasília Ministério do Trabalho e Emprego.

Morata, T.C., et al. 2005. Working in noise with a hearing loss: Perceptions from workers, supervisors, and hearing conservation program managers. *Ear and Hearing* 26(6): 529–545.

PMI, Project Management Institute, Inc. 2013. *A Guide to the Project Management Body of Knowledge (PMBOK)*. 5ed. Pennsylvania: Newtown Square.

Prado, D. 2004. *Gerenciamento de Programas e Projetos nas Organizações*. 3 ed. Nova Lima: Minas Gerais.

Rabinowitz, P.M., et al. 2007. Do ambient noise exposure levels predict hearing loss in a modern industrial cohort? *Occupational Environment Medicine* 64:53–59.

Pitfalls of measuring illuminance with smartphones

R.B. Melo & F. Carvalho
Laboratório de Ergonomia, Faculdade de Motricidade Humana and Centro de Investigação em Arquitetura, Urbanismo e Design, Faculdade de Arquitetura, Universidade de Lisboa, Portugal

D. Cerqueira
Laboratório de Ergonomia, Faculdade de Motricidade Humana, Universidade de Lisboa, Portugal

ABSTRACT: The development of many applications and the incorporation of different sensors in smartphones potentiate their use in different contexts, including the Occupational Safety & Health domain. Like any other measurement device, mobile phones must comply with accuracy and precision requirements. Therefore, a set of nine smartphones and 14 applications were tested against a lux meter, over four illuminance levels (300 lx, 500 lx, 750 lx and 1000 lx) and three correlated color temperatures (2700 K, 4000 K and 6500 K). Statistical significant differences were registered between the values displayed by the mobile phones and those of the lux meter, suggesting that smartphones are not suitable for measuring illuminance and decide on the adequacy of lighting in occupational settings.

1 INTRODUCTION

Smartphones have gone through an amazing evolution since their appearance and nowadays allow us to do things that previously required stand-alone hardware to accomplish, namely measuring. In recent years, different sensors have been incorporated in these mobile phones and many applications have been developed (Daponte et al. 2013). Yet, accuracy and precision, which are crucial assets of measuring instruments, should be a concern to those relying on smartphones to do measurements.

Designing occupational settings for visual performance necessarily demands lighting design in accordance with international standards (CEN 2011; ISO 2002) and must be concerned with visual comfort, the workers' visual impairments and age, as well as with the nature of the tasks and the workplace characteristics. Recommendations are based in photometric measures such as illuminance, luminance, Correlated Color Temperature (CCT), color rendering index and unified glare rating (UGR). Nevertheless, the most used measure to evaluate lighting is illuminance, which has to be often measured to check on lighting's adequacy. The lux meter is the traditional dedicated equipment worldwide recognized as the most appropriated to accomplish this task (Goodman 2009).

Due to their ubiquity, portability and social acceptance, we can find smartphones being used in areas such as education, business, health, fitness, transportation, and social life. A huge number of applications facilitate the users to manage prescriptions, provide price comparisons, promote alternative treatment options, etc. (Sarwar & Soomro 2013). Within the Occupational Safety & Health (OSH) domain, the options cover measurement of noise and vibration as well as of illuminance levels.

Recently published studies have analyzed the potential of smartphones to measure noise (Ibekwe et al. 2016; Kardous & Shaw 2014; Kardous & Shaw 2016; Kardous & Shaw 2015; Murphy & King 2016), and vibration (McGlothlin et al. 2015; McGlothlin et al. 2014). Applications designed to measure noise have been extendedly analyzed for their accuracy and improvements have been made recently by NIOSH (Roberts et al. 2016). As for lighting measurements, new apps have been developed (Kim et al. 2016; Gutierrez-Martinez et al. 2017) but only two studies dealt with accuracy issues (Goldschmidt 2016; Gutierrez-Martinez et al. 2017).

There are three possibilities to measure lighting with a mobile phone: a) relying on the ambient light sensor; b) using the digital camera; c) with external sensors (Gutierrez-Martinez et al. 2017). The ambient light sensor is used to adjust the screen's brightness based on the surroundings' lighting aiming at saving battery and optimizing the screen's readability. The functionality of smartphone's digital cameras has been extended beyond taking pictures allowing them to be used in areas such as telemedicine, microscopy, and biology, among others. External sensors are sold separately representing an additional cost.

Gathering the recent advances made on smartphones with their portability, connectivity and price, makes it wise to think of them as fairly accurate tools to measure lighting. To check on the veracity of this assumption, the authors completed a study to evaluate smartphones' performance while measuring illuminance in different conditions of illuminance and CCT.

2 MATERIALS AND METHODS

2.1 Hardware and software

College students were asked to volunteer their mobile phones to be tested. Nine mobile phones from different manufacturers were gathered: three from Apple (iPhone 5S, 6S and 6 Plus), three from Samsung (Galaxy A3 2015, Galaxy A3 2016, and Galaxy SIII), one from Huawei (P8 Lite), one from Vodafone (Ultra 6) and one from Nokia (Lumia 520). All these devices allowed measuring illuminance levels, either through their camera or their ambient light sensor. All cameras were used in the automatic mode.

Bearing in mind that this was a preliminary and exploratory study, camera and light sensor quality differences among smartphones were not considered as independent variables. Attention was directed towards the operating system.

Fourteen applications dedicated to measure illuminance were selected and covered three main operating systems: six were Android-based, four were suitable for iOS only and the last four were directed to Windows. The inclusion of applications in the study was restricted to those:

- Able to measure and report illuminance values expressed in lux;
- Available for download free of charge;
- Highly rated by previous users.

Calibration was possible in five of them but only worked in four Android-based.

2.2 Experimental design

Different lighting variables were controlled in a black chamber to ensure that all smartphones and applications would be subjected to the exact same type and amount of light. Two compact fluorescent lamps (Correlated Color Temperature – CCT: 2700 K and 6500 K) and a LED lamp (CCT: 4000 K) were used as light sources. A light fixture mounted on an adjustable arm allowed creating different illuminance levels (300 lx, 500 lx, 750 lx and 1000 lx), either moving it away from the measurement device or towards it.

A digital lux meter from Hagner (model EC1-X) was used to measure illuminance before each smartphone/app test. Both the lux meter and the mobile phones were placed successively at the same position on a platform located axially below the light fixture.

Apps requiring calibration were subjected to the process using an illuminance level of 100 lx, previously checked with the lux meter.

2.3 Data analysis

Smartphone illuminance readings were compared to those of the lux meter by means of equation (1).

$$Rel\ dif = \left(\frac{E_{app} - E_{lux}}{E_{lux}} \right) \times 100 \qquad (1)$$

where $Rel\ dif$ = Relative difference; E_{app} = reading from the smartphone; and E_{lux} = reading from the lux meter.

Perfect agreement between the readings of the mobile phone and those of the lux meter would produce a zero value for $Rel\ dif$, meaning that a particular smartphone/app combination is accurate. Positive values represent overestimation of illuminance, whereas negative ones are associated with underestimation.

Version 24 of the Statistical Package for the Social Sciences (SPSS), from IBM, was chosen for data analysis and a 0.05 significance level was used to reject the null hypothesis in all tests. Due to sample size, only non-parametric tests were applied (Kruskal-Wallis, Mann-Whitney, Friedman and Wilcoxon).

3 RESULTS

Figure 1 compares smartphone/app readings with those obtained with the lux meter for the illuminance reference levels of 300 lx, 500 lx, 750 lx and 1000 lx.

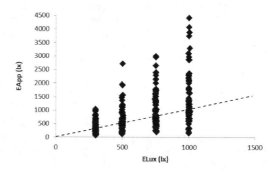

Figure 1. Illuminance values: readings from smartphones (EApp) *versus* readings from the lux meter (ELux).

The lack of agreement between the smartphone/app and the lux meter outputs stands out and a closer look reveals that dispersion increases as the established illuminance levels increase.

Scatterplots of *Rel dif* across illuminance and CCT values are presented in Figures 2 and 3, respectively.

The *Rel dif* mean value across the reference illuminance levels was around 25%. However, while some smartphone/app combinations displayed illuminance values below the reference levels for as much as 85%, others exceeded the gold standard for as much as 440% (Figure 2).

The analysis of smartphone/app performances across different CCT also revealed significant dispersion and different responses (Figure 3). Nevertheless, best *Rel dif* mean was obtained at 4000 K.

It can be noticed in Figure 4 that Windows-based apps appear to more precise than the other ones, as dispersion is much less pronounced. On the other hand, readings from most iOS' apps tend to be lower than those of the lux meter.

Figure 5 shows that smartphones presented different accuracy levels when measuring illuminance ($p < 0.001$). Nokia's Lumia 520 model

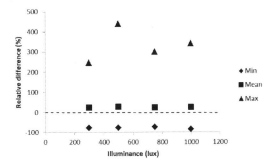

Figure 2. Relative differences obtained for illuminance levels of 300 lx, 500 lx, 750 lx and 1000 lx.

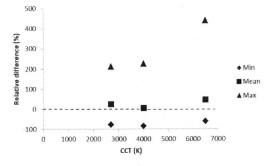

Figure 3. Relative differences obtained for correlated color temperatures of 2700 K, 4000 K and 6500 K.

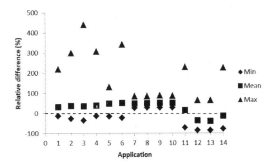

Figure 4. Relative differences obtained for each application: (1–6) Android-based; (7–10) Windows-based; (11–14) iOS-based.

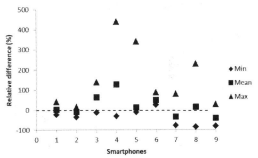

Figure 5. Relative differences obtained with different smartphones: 1) Vodafone Ultra 6; 2) Samsung Galaxy A3 2016; 3) Samsung Galaxy SIII; 4) Huawei P8 Lite; 5) Samsung Galaxy A3 2015; 6) Nokia Lumia 520; 7) iPhone 5S; 8) iPhone 6 Plus; 9) iPhone 6S.

systematically overestimated the illuminance values, whereas iPhone models 5S and 6S underestimated this particular photometric measure (*Rel dif* mean values of –34% and –41%, respectively). The best *Rel dif* mean values were achieved for two Android-based phones: Vodafone Ultra 6 (2.27%) and Samsung Galaxy A3 2016 (–8.53%).

4 DISCUSSION

A similar study dealing with the accuracy level of smartphones used to measure illuminance was found (Goldschmidt 2016) but it was impossible to match the mobile phone models and the apps used in both studies. Applications are added and removed from app stores on a daily basis and updates are frequent. Nevertheless, our results reveal much less accurate smartphone/app combinations.

Despite the fact of presenting only the results obtained with a LED lamp for 100 lx, 500 lx and 1000 lx, Goldschmidt (2016) confirms that iPhones tend to display values well below the reference ones.

Curiously, in our study only Windows-based apps did not present statistically significant different *Rel dif* values (p = 0.94) but displayed values above the reference ones. A single phone (Nokia Lumia 520) was tested with these four apps, which may explain this fact.

On the other hand, both studies found that some apps do not have any influence on the value displayed. It appears that the only function played by the app is to display the value. This happened particularly with those relying on the ambient light sensor to measure illuminance instead of the camera. It seems that the fact of being located inside the device makes the light sensor able to measure only direct light, while the lux meter also measures indirect light (Gutierrez-Martinez et al. 2017). Nevertheless, using the smartphone's camera also has some drawbacks because the illuminance measured takes into account the distance of the device from the surface where the measure is being taken.

Moreover, we have found that the same phone/app combination shows different response abilities across both the illuminance (p < 0.001) and CCT (p < 0.001) ranges, which means that the *Rel dif* values are not the same for different environmental conditions. These results are also in accordance with those of Goldschmidt (2016) but not with the ones from Gutierrez-Martinez et al (2017).

Calibration is imperative among measuring devices. However, it was found that not all illuminance apps allow this procedure to be accomplished. Surprisingly, whenever this option was available, results were not much improved and *Rel dif* mean values ranged between 31% and 51%. Goldschmidt (2016) achieved similar results concerning the calibration impact on the measurements' accuracy. The fact that some apps come with the calibration function sounds professional but it may mislead the user, and may have detrimental effects on workers' visual comfort, performance and health when it comes to control lighting in occupational settings. On the contrary, Gutierrez-Martinez et al (2017) concluded that calibration did improve the smartphones ability to measure illuminance. However, these researchers only tested two calibrated mobile phones and compared the obtained results with those of a different device. Moreover, it is clear that the calibration process always relies on a lux meter.

5 CONCLUSIONS

Incorporating new sensors and recently developed apps in smartphones increases their potential. Using these devices to measure illuminance, instead of a dedicated lux meter, sounds quite attractive but not before ensuring they provide acceptable accuracy levels.

A set of smartphones and apps never analyzed before were tested for their accuracy while measuring illuminance, under controlled conditions. CCT was also taken into consideration.

Results were not encouraging as a wide-range of deviations from the "gold standard" was found, and phone/app combinations did not reveal a definite/predictable pattern for readings as a function of illuminance or CCT.

It does not sound smart to use smartphones to measure illuminance, particularly relying on both the tested models and apps.

Further developments of this study should consider exploring paid apps; extend the sample of mobile phones and use of external devices to filter light and overcome the drawbacks of using the camera.

The camera settings' adjustment (ISO, shutter speed and aperture) is another relevant issue to be analyzed in the future, instead of operating it exclusively in the automatic mode.

REFERENCES

CEN, 2011. EN 12464-1:2011 Light and lighting - Lighting of work places Part 1: Indoor work places.

Daponte, P. et al., 2013. State of the art and future developments of measurement applications on smartphones. *Measurement: Journal of the International Measurement Confederation*, 46(9), pp. 3291–3307.

Goldschmidt, J., 2016. Luxmeter App versus measuring device: Are smartphones suitable for measuring illuminance? *DIAL*. Available at: https://www.dial.de/en/blog/article/luxmeter-app-versus-measuring-deviceare-smartphones-suitable-for-measuring-illuminance/.

Goodman, T., 2009. Measurement and specification of lighting: A look at the future. *Lighting Research and Technology*, 41(3), pp. 229–243.

Gutierrez-Martinez, J.-M. et al., 2017. Smartphones as a Light Measurement Tool: Case of Study. *Applied Sciences*, 7(6), p. 616.

Ibekwe, T.S. et al., 2016. Evaluation of Mobile smartphones App as a screening tool for Environmental Noise Monitoring. *Journal of Occupational & Environmental Hygiene*, 13(2), pp. D31–D36.

ISO, 2002. ISO 8995-1:2002(en) Lighting of work places — Part 1: Indoor.

Kardous, C.A. & Shaw, P.B., 2015. Do Sound Meter Apps Measure Noise Levels Accurately? *Sound and Vibration*, (July), pp. 10–13.

Kardous, C.A. & Shaw, P.B., 2014. Evaluation of smartphone sound measurement applications. *The Journal of the Acoustical Society of America*, 135(4), p. EL186–92.

Kardous, C.A. & Shaw, P.B., 2016. Evaluation of smartphone sound measurement applications (apps) using external microfones - A follow-up study. *The Journal of the Acoustical Society of America*, 140(4), p. EL327–33.

Kim, Y.-S., Kwon, S.-Y. & Lim, J.-H., 2016. Implementation of Light Quality Evaluation System using Smartphone. *International Journal of Bio-Science and Bio-Technology*, 8(3), pp. 259–270.

McGlothlin, J., Burgess-Limerick, R. & Lynas, D., 2015. An iOS Application for Evaluating Whole-body Vibration Within a Workplace Risk Management Process. *Journal of Occupational and Environmental Hygiene*, 12(7), pp. D137–142.

McGlothlin, J., Wolfgang, R. & Burgess-Limerick, R., 2014. Ergonomics: Using Consumer Electronica Devices to Estimate Whole-Body Vibration Exposure. *Journal of Occupational & Environmental Hygiene*, 11(6), pp. D77–D81.

Murphy, E. & King, E.A., 2016. Testing the accuracy of smartphones and sound level meter applications for measuring environmental noise. *Applied Acoustics*, 106, pp. 16–22.

Roberts, B., Kardous, C. & Neitzel, R.L., 2016. Improving the Accuracy of Smart Devices to Measure Noise Exposure. *Journal of Occupational & Environmental Hygiene*, 13(11), pp. 840–846.

Sarwar, M. & Soomro, T.R., 2013. Impact of Smartphones on Society. *European Journal of Scientific Research*, 98(2), pp. 216–226.

Effects of thermal (dis)comfort on student's performance: Some evidences based on a study performed in a secondary school

M. Talaia
DFIS—CIDTFF—University of Aveiro, Aveiro, Portugal

M. Silva
DEP—University of Aveiro, Aveiro, Portugal

L. Teixeira
DEGEIT—IEETA—GOVCOPP—University of Aveiro, Aveiro, Portugal

ABSTRACT: In modern society people spend much of their time indoors of building—home, work office/place or schools. Particularly in the case of students, they spend about 80% of their daytime inside of schools, having as theirs main mission the learning of the subjects that are taught in the class. For this mission, the physical and psychological well-being of the student is very important, and the thermal environment of the classroom has an impact on the thermal comfort, which in turn can influence the mission referred above. The present study aims to analyse the influence that the thermal environment can have on the learning process of the students. It was performed in a secondary school, located in central region of Portugal. The obtained results showed that from the thermo-hygrometric data records it is possible to predict the type of thermal environment of a classroom and that the thermal environment and the thermal insulation of the clothing influence the well-being, the thermal sensation and the learning process of the students.

1 INTRODUCTION

Nowadays there has been a number of evaluations in schools conducted by qualified entities, such as the Organization for Economic Co-operation and Development (OECD). This type of evaluations has given little importance to the aspects related to the thermal environment, a determinant factor for the well-being of the students, evaluated in terms of thermal comfort.

The American Society of Heating and Air Conditioning (ASHRAE) defines thermal comfort as the state of mind in which the individual expresses satisfaction with the thermal environment (ASHRAE 55, 2010; ISO 7730, 2006). This definition implies a certain degree of subjectivity and presupposes the analysis of two aspects: physical aspects inherent to the thermal environment and subjective aspects inherent to the individual's state of mind (Morgado, Teixeira, & Talaia, 2015; Nico, Liuzzi, & Stefanizzi, 2015; Talaia, Meles, & Teixeira, 2013; Teixeira, Talaia, & Morgado, 2014).

Thermal comfort is obtained when an individual is in a condition of energetic balance with the surrounding thermal environment. For example, Corgnati, Filippi and Viazzo (2007) and Wargocki and Wyon, (2017) have shown that in conditions of extreme thermal discomfort, the individuals may exhibit several physical and psychological effects, such as fatigue and decreased mental capacity.

There are already some studies that have been performed, demonstrating the relationship between thermal comfort and worker productivity and/or student performance (Corgnati et al., 2007; Dias Pereira, Raimondo, Corgnati, & Gameiro da Silva, 2014; Mishra & Ramgopal, 2015; Morgado, Talaia, & Teixeira, 2015; Nico et al., 2015; Wargocki & Wyon, 2013, 2017). Although the results of these experimental activities have not led to definitive conclusions, they have shown a tendency on the sensation of discomfort in individuals, provided by hot or cold environments, to reduce intellectual and and/or productive performance.

Krüger and Dumke (2001) have shown, based on laboratory and field studies, a relationship between thermal comfort and an individual's performance. In this relationship, an important component is the thermal insulation of clothing, that represents the resistance to the transfer of energy as heat between the surface of the skin and clothing. The value of the thermal insulation of clothing depends on the properties and characteristics of the materials used in the confection of fabrics (Lazzarotto, 2007). ISO 7730 (2006) indicates, for some clothing, the typical accepted values for the thermal insulation of clothing, in units of *clo*.

Currently, when we talk about cognitive development (usually associated with teaching and learning processes) we must consider important aspects related to motivation, social relations, environmental stimulus (where the surroundings in which the thermal environment is part of), among others.

Typically, most school buildings have standardized building systems, without taking into account the specific characteristics of the type of climate in each region. This leads to certain school buildings not meeting the requirements of thermal comfort. Certainly, these conditions have a negative impact on the motivation, concentration and evaluation of the students. As such, it is necessary that in a school architecture it be taken into account the needs of thermal comfort, in order to provide a pleasant environment that favours teaching and learning process.

The study of Rebelo et al. (2008) revealed that the thermal environment in classrooms is one of the factors that conditions the learning process at any level of education. These authors also showed the strong influence of solar radiation throughout the day in the comfort conditions in classrooms.

Additionally, Talaia and Silva (2016) showed that the type of thermal environment of a classroom influences the results of a student's evaluation with a reduction of the evaluation results of about 3.9% for each °C referring to the sum of the temperature with the temperature of the humid thermometer.

In fact, and according to Nico et al. (2015), studies related with the evaluation of thermal comfort becomes more important when the aim is to maximize the productivity of the worker or the performance of a student in the learning and teaching process, as is the objective of the present work.

The present study aims to analyse the influence that the thermal environment can have on the learning process of the students.

2 MATERIALS AND METHODS

This study was performed in a secondary school, located in central region of Portugal. The methodology used was a research-action, based on mixed method for data collection – quantitative and qualitative.

This section briefly presents the methodology used, explaining the procedures that were followed to collect data, the experiment design and the methods and indexes used in data processing.

2.1 Data collection, sample and experiment design

In this study, a set of data from various sources was used, namely: (i) data from the results of the students' evaluations in the classroom; (ii) thermal sensation data registered by students in a colour scale; (iii) thermo-hygrometric data registered using the measuring instrument, and (iv) data on the type of clothing worn by the students.

Data collection took place in a classroom environment during a period of time considered cold, and involved 41 students from two classes of the 8th year of the secondary school. Regarding the gender of the sample, 18 female and 23 male students participated in this study. Regarding the age, 39 students were 13 years old and 2 were 14 years old. Students will be identified in this study by Ai, with i ranging from 1 to 41 (i = 1 to 41).

Particularly in the context of this sample, performing the activities in the classroom in written support, commonly referred as questions-problem, is often the basis for the organization of the students' knowledge. The questions-problem can be used, for example, to consolidate learning and synthesize content.

In the scope of this research, students participated in different activities, through the application of a set of questions-problem.

These questions were about contents that were being taught and aimed to evaluate the development of skills related to the acquisition of knowledge on the contents taught in the class where the experience was performed. The questions-problem was applied whenever it was considered appropriate, and the students had no prior knowledge of its application.

Additionally, within the classroom experience, each student recorded their thermal sensation on a thermal colour scale, selecting the position that best corresponded to the thermal comfort they felt.

Figure 1 shows the colour scale used in this study.

Regarding the thermo-hygrometric data, it was registered using the measuring instrument Center 317- temperature humidity meter, and allowed the application of the EsConTer thermal index (Talaia and Simões, 2009) to predict the thermal sensation inside the classroom.

Finally, and taking into account the method of observation, data on the type of clothing used by the students was also collected, in order to calculate the *clo*.

2.2 Method and indexes used

The EsConTer index, developed by Talaia and Simões (2009) and confirmed in different applications by Morgado, et al. (2017) and Morgado,

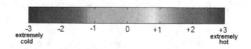

Figure 1. Thermal sensation scale (Talaia and Rodrigues, 2008).

et al., (2015) determines a value of the seventh scale of thermal sensation in the range of -3 (very cold thermal environment) to +3 (very hot thermal environment). This index is calculated by the Expression (1), where T represents the air temperature (°C) and Tw the temperature of the humid thermometer (°C).

$$EsConTer = -3.75 + 0.103(T + Tw) \quad (1)$$

Regarding the clothing insulation (Icl), it can be estimated directly from the data presented in available tables for typical combinations of clothes (the values are for static thermal insulation) or indirectly, by summation of the partial insulation values for each item of clothing. One clothing unit (1 *clo*) is defined as the thermal resistance offered by each 0.155 m² of the worn clothing set, when the covered skin cools 1 °C due to the heat transfer rate of 1 W to the outer surface. One *clo* is equal to 0.155 m².°C/W or 0.155 m².K/W.

For a sedentary activity (classroom of a school for example) with a metabolism of 70 W/m² or 1.2 *met*, one metabolic unit (1 *met*) is equal 58.2 W/m².

The thermal insulation of clothing (*TiClo*) index is determined, in this work, by the Expression (2) and the result is present in units of *clo*.

$$TiClo = -0.7418 EsConTer - 0.3250 + 0.0764 Tw \quad (2)$$

The clothing is characterized by its thermal resistance, Icl, in the m².K/W units, which gives the clothing a thermal resistance between the body and the environment, representing a barrier for heat exchanges by convection. To each type of clothing corresponds the thermal resistance index Icl (*clo*). The thermal resistance of clothing is expressed in *clo* and relates to the protective power of clothing (ISO 7730, 2006).

The heat transfer between the skin and the outer surface of the clothing is quite complex, involving internal convection and radiation processes, at the places where the air passes, and conducted through the clothing itself (Rodrigues, 1978).

$$Icl = \sum Icli \quad (3)$$

In this work the Expression (3), where *Icli* represents the thermal resistance for each item of clothing *i*, was applied for the calculation of the thermal resistance of the clothing.

3 RESULTS AND DISCUSSION

The results partially presented in this paper were achieved from an experimental study performed on a secondary school located in centre of Portugal.

The prediction models used, EsConTer [*Expression (1)*] and *TiClo* [*Expression (2)*], show that the data follows an adjustment line by applying the least squares method with a Pearson correlation coefficient of 0.9730. Also, the slope of the fit line with a Pearson correlation coefficient of 0.8212 shows that there is excellent agreement between the thermal sensation registered and the thermal insulation of clothing used.

Regarding the influence of thermal comfort on student performance, Figure 2 shows the results obtained from the evaluation of eight students on a specific question of students' assessment (Evaluation#5).

The obtained results, represented (i) by columns in percentage for 8 students (A1–A8) and for the evaluation#5, (ii) by the straight blue line with the predicted thermal sensation in the classroom for the thermo-hygrometric conditions recorded when the EsConTer index is applied; and (iii) by the red line with the value of the actual thermal sensation of each student are present in Figure 2.

As expected, the results show that students have different thermal sensations.

Additionally, the results show that for a thermal sensation below –0.5 (limit considered for thermal comfort), the results of the evaluations are usually lower (negative results). It should be noted that it is possible to observe that the values of the thermal sensation registered by the students are in agreement with the predicted thermal sensation when using the EsConTer thermal sensation index. Student clothing is a factor to consider through the insulation of the used clothing.

Figure 3 shows the results obtained by the student A4 in five different evaluation activities.

The results seem to be influenced by the thermal sensation registered by the student (red line). For values greater than –0.5 and in the thermal sensation range [–0.5; +0.5] the results are positive.

From this experience, it is possible to observe that when the thermal sensation values recorded

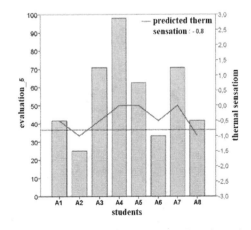

Figure 2. Results obtained by eight students for evaluation #5.

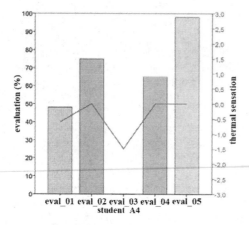

Figure 3. Results obtained by the one student A4 in five different evaluations.

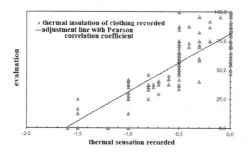

Figure 4. Influence of students' thermal sensation on evaluation.

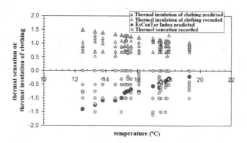

Figure 5. Influence of air temperature on thermal sensation and thermal insulation of students' clothing.

by the students are in the thermal comfort range, the results obtained by the evaluations of the questions-problem are usually positive. If the thermal sensation registers values lower than –0.5, the observed evaluation results are less than 50%.

Thus, and taking into account the observed results, it is possible to conclude that the thermal sensation of the environment can influence the students' results in terms of performance, since thermal comfort or discomfort has an impact on the concentration, and, consequently, on the student's knowledge learning process.

This conclusion is in line with the results of Wargocki and Wyon (2017) that concluded that thermal discomfort causes distraction and thus further negative effects on cognition and, consequently, on student performance.

Figure 4 shows how the thermal sensation experienced by the students influences the evaluation results. The triangles represent the data recorded in this experience. The adjustment line, with a Pearson correlation coefficient of 0.8087, suggests the expected trend, i.e., when the classroom's thermal environment approaches a cold sensation of temperature, the results of the evaluation tend to worsen.

Figure 5 shows how the thermal sensation and thermal insulation of clothing, recorded by students and predicted through the prediction formula of EsConTer index and *TiClo* index, are influenced by the classroom air temperature.

In the Figure 5, the green colour triangles represent the thermal insulation of the clothing predicted by the application of the *TiClo* index, while the yellow triangles represent the thermal insulation of the clothing used for the temperatures recorded in the classroom. The dark blue circles represent the expected thermal sensation by application of the EsConTer index and the light blue circles represent the thermal sensation registered by the students in the thermal colour scale.

With the results shows in Figure 5 it is possible to conclude that the students, in general, used a clothing thermal insulation lower than expected. This can be justified as the thermal environment is considered cold and the clothing used by the students is not adequate to balance the insufficiency of thermal insulation of the classroom due to the influence of the atmospheric conditions outside the building.

Additionally, Figure 5 shows that the students registered in the thermal colour scale a thermal sensation with a lot of variability, which confirms the subjectivity of a thermal environment against the thermal sensation that each individual registers. On the other hand, it can be verified, as one would expect, that when the thermal environment approaches a comfortable level, the thermal insulation of the clothing decreases, suggesting a "lighter" clothing.

4 CONCLUSIONS

This study showed that it is possible to understand the thermal sensations of comfort and discomfort of students in the classroom and the way these sensations condition the learning process. The learning process is affected by the thermo-hygrometric conditions of the environment surrounding the students, as, considering the results of this study, when the values of the thermal sensation registered by the students are lower than –0,5 of the seventh scale of colours, the results obtained from the knowledge evaluation tend to be less than 50%.

Additionally, the results showed that the methods used in this work are important tools to evaluate how situations of thermal discomfort can condition the learning process in a thermal environment considered cold.

In the current climate changes phenomenon, where global warming is a proven reality, studies of this nature are important in the teaching and learning process. It is therefore hoped, that this study may help in the definition of strategy measures for the improvement of the thermal environments in classrooms, as well as contributing to the demonstration of how the results of students can be influenced by the thermal environment of a classroom where the teaching and learning occurs.

The main limitation of this work is the lack of representativeness, because it was applied in a single school. Thus, as future work, it is intend to extend this study to other schools, in order to understand this phenomenon in other contexts.

REFERENCES

ASHRAE 55. (2010). *Thermal Environmental Conditions For Human Occupancy*. Atlanta, Georgia: American Society of Heating, Refrigerating and Air-Conditioning Engineers.

Corgnati, S.P., Filippi, M., & Viazzo, S. (2007). Perception of the thermal environment in high school and university classrooms: Subjective preferences and thermal comfort. *Building and Environment*, *42*(2), 951–959.

Dias Pereira, L., Raimondo, D., Corgnati, S.P., & Gameiro da Silva, M. (2014). Assessment of indoor air quality and thermal comfort in Portuguese secondary classrooms: Methodology and results. *Building and Environment*, *81*, 69–80. http://doi.org/10.1016/j.buildenv.2014.06.008

ISO 7730. (2006). Ergonomics of thermal environment-Analytical determination and interpretation of thermal comfort using calculation of PMV and PPD indeces and local thermal comfort criteria. In *International Standards Organization*.

Krüger, E.L. e Dumke, E.M. (2001). Avaliação integrada da vila tecnológica de Curitiba. *Tuiuti Ciência e Cultura*, 25(3):63–82. Lazzarotto, N. (2007). Adequação do modelo PMV na avaliação do conforto térmico de crianças do Ensino Fundamental de Ijuí-RS. *Dissertação de Mestrado em Engenharia Civil*, Brasil.

Lazzarotto, N. (2007). Adequação do modelo PMV na avaliação do conforto térmico de crianças do Ensino Fundamental de Ijuí-RS. *Dissertação de Mestrado em Engenharia Civil*, Brasil.

Mishra, A.K., & Ramgopal, M. (2015). A comparison of student performance between conditioned and naturally ventilated classrooms. *Building and Environment*, *84*, 181–188.

Morgado, M., Talaia, M., & Teixeira, L. (2017). A new simplified model for evaluating thermal environment and thermal sensation: An approach to avoid occupational disorders. *International Journal of Industrial Ergonomics*, *60*, 3–13.

Morgado, M., Teixeira, L., & Talaia, M. (2015). Creating PRODUCTIVE Workers in Industrial Context from the Definition of Thermal Comfort. *International Journal of Industrial Engineering and Management*, 6(2), 75–84.

Nico, M.A., Liuzzi, S., & Stefanizzi, P. (2015). Evaluation of thermal comfort in university classrooms through objective approach and subjective preference analysis. *Applied Ergonomics*, *48*, 111–120.

Rebelo, A., Santos Baptista, J. e Diogo, M.T. (2008). Caracterização das Condições de Conforto Térmico na FEUP. Proceedings CLME'2008 II CEM. 5º Congresso Luso-Moçambicano de Engenharia e 2º Congresso de Engenharia de Moçambique: Maputo. Porto: Edições INEGI.

Rodrigues B (1978). A Bioclimatologia e a Produtividade Laboral. *Revista do Instituto Nacional Meteorologia Geofísica* 1(1), 9–51.

Talaia, M. e Rodrigues, F. (2008). Conforto e stress térmico: uma avaliação em ambiente laboral. Em J.F.S. Gomes et al., Proceedings CLME'2008 II CEM. 5º Congresso Luso-Moçambicano de Engenharia e 2º Congresso de Engenharia de Moçambique: Maputo. Porto: Edições INEGI.

Talaia, M. e Silva, M. (2016). O ambiente térmico de uma sala de aula influencia os resultados de avaliação de um estudante. *Revista de Formación e Innovación Educativa Universitaria*. 9(2):67–76.

Talaia, M. e Simões, H. (2009). EsConTer: um índice de avaliação de ambiente térmico. *V Congreso Cubano de Meteorología* (pp. 1612–1626). Somet-Cuba, Sociedad de Meteorología de Cuba.

Talaia, M., Meles, B., & Teixeira, L. (2013). Evaluation of the Thermal Comfort in Workplaces- a Study in the Metalworking Industry. In P. Arezes, J.S. Baptista, M.P. Barroso, P. Carneiro, P. Cordeiro, N. Costa, … G. Perestrelo (Eds.), *Occupational Safety and Hygiene* (pp. 473–477). London: Taylor & Francis Group.

Teixeira, L., Talaia, M., & Morgado, M. (2014). Evaluation of indoor thermal environment of a manufacturing industry. In Occupational Safety and Hygiene II—Selected Extended and Revised Contributions from the International Symposium Occupational Safety and Hygiene, SHO 2014.

Wargocki, P., & Wyon, D.P. (2013). Providing better thermal and air quality conditions in school classrooms would be cost-effective. *Building and Environment*, *59*, 581–589.

Wargocki, P., & Wyon, D.P. (2017). Ten questions concerning thermal and indoor air quality effects on the performance of office work and schoolwork. *Building and Environment*, *112*, 359–366.

Usability of an open ERP System in a manufacturing company: An ergonomic perspective

Krzysztof Hankiewicz
Poznan University of Technology, Poland

K.R. Kumara Jayathilaka
University of Bern, Switzerland

ABSTRACT: Nowadays, Information Technology (IT) plays an essential role in industries and organizations, but decision makers underestimate the usability of IT systems. This paper concerns the usability of an Open ERP system implemented in a manufacturing company. Moreover, the paper reflects on Open ERP incorporation of key business functions, business processes and implementation issues by using the System Usability Scale. Suggestions and considerations are made after analyzing the current Open ERP system and business process of the organization.

1 INTRODUCTION

An ERP (Enterprise resource planning) system is a software solution that integrates various functional spheres in an organization – a link through the entire supply chain, aimed at adopting the best industry and management practices for providing the right product at the right place and at the right time at minimum cost. The system consists of integrated multi module applications and common database. Nowadays, ERP systems produce gigantic be-hoof to enterprise, successfully meeting their business goals effectively and efficiently. However, the system is widely criticized by end users and decision makers due to its complexity and adoptability (Wesson, Singh, 2009; Ysnhong, 2009).

Adoptability and complexity issues arise due to the integration of many modules with large amount of data into one system. This leads to insufficient usability which leads to deficient set of common goals of an organization (Ray 2014). Particularly insufficient usability emerging mainly in user interface. Contrarily, common business process mapping in to specific organization could lead to unsuccessfulness.

A functional and basic user interface needs to be consistent and predictable when selecting interface elements. The end user needs to be aware and familiar with those elements that are potentially needed to improve efficiency, satisfaction and successful task completion. Extensively, the interface should incorporate easiness of use, easiness of learning and support task in process completion. Detecting and identifying usability, specifically in an ERP system, is quite challenging due to the lack of suitable criteria and a limited amount of research (Sadiq, Pirhonen, 2014; Wesson, Singh 2009).

In the process of developing information systems not all ergonomic aspects are taken into consideration. Usually ergonomic requirements, which the software fulfills, are concerned only with the user interface. Designers should remember about other ergonomic requirements, especially about utilitarian features of the whole system, which determine users' work conditions (Bishu et al., 2001; ISO 9241-11; Hankiewicz, 2005; Hankiewicz, 2012).

2 METHODOLOGY

2.1 *Research context*

Usability assessment can be performed in a few different ways—empirically, conceptually or analytically—by using various inspection methods. This research study is based on a combination of basic and applied research category related on usability. Identifying criteria has been done by merging various criteria and methods by mixing and gathering. Particularly ISO 9241, Singh and Wesson's usability mapping, Nielsen's ten heuristics and SUS (System Usability Scale) were very helpful in developing criteria needed to carry out tests in the field of research interest (Holzinger, 2005, Rauer, 2011; Nielsen, 1995; Nielsen, 2001). Current industry research is carried out based on surveys, end user interviews, industry expert evaluations, questionnaires and observations. These methods revealed valuable information on ERP system usability which can improve the usability of a specified system. The revealed information can be used, by persons who need it to consider the development phase (Wesson, Singh, 2009; Bevan, Macleod, 1994).

The main goal of this research is to identify why the Open ERP system failed to implement in a manufacturing company. The study includes an empirical evaluation of an Open ERP system by a qualitative and quantitative data analysis of the end users responses in a usability questionnaire. Five main usability criteria were used to analyze the current situation of the system and recorded on the System Usability Scale sheet. The study revealed where the system needed to improve on usability and collaborative work between end users and developers.

2.2 Usability inspection methods and criteria of evaluation

The users, tasks and environment are characteristic of the context of usability. Inspection methods include heuristic evaluation, cognitive walkthrough, formal usability inspection, pluralistic usability walkthrough, feature inspection, consistency inspection and standards inspection (Garrett, 2011; Hollingsed, Novick, 2007). The heuristics evaluation method, introduced by Jakob Nielsen together with Rolf Molich in the early 90's, is mainly used by usability experts and is a comparatively cheap method to evaluate the system interface. The 10 most general principles for interaction design—'heuristics' are (Hollingsed, Novick, 2007; Wesson, Singh, 2009):

– visibility of system status;
– match between system and the real world;
– user control and freedom;
– consistency and standards;
– error prevention;
– recognition rather than recall;
– flexibility and efficiency of use;
– aesthetic and minimalist design;
– help users recognize, diagnose and recover from errors;
– help and documentation.

Cognitive walkthrough is a more detailed procedure which can implemented during each development phase. It does focus mostly on user oriented goals and user performance evaluated by a particular system. Formal usability inspection methods include a review by the developer's peers of a user's specific task performance; this method has a six step procedure which includes a combination of heuristic evaluation and cognitive walkthrough. Pluralistic usability walkthrough incorporates with all kinds of users, product developers, production team, usability experts discussing each dialogue element of the system. Feature inspection lists a sequence of features used to accomplish a specific task, checks sequences and steps, familiarizes the user with steps, checks whether the user needed to have extensive knowledge or experience with the regard of interface. Consistency inspection is performed by developers who use the same design for multiple projects. Standard inspections are conducted by industry experts to check whether the interfaces meet the required standards (Baker et al., 2002 Hollingsed, Novick, 2007; Nielsen, 1994).

ISO 9241 defines the effectiveness, efficiency and satisfaction of a specific user achieving specific goals in a particular environment. Effectiveness: the accuracy and completeness with which specified users can achieve specified goals in particular environments. Efficiency: the resources expended in relation to the accuracy and completeness of goals achieved. Satisfaction: the comfort and acceptability of the work system to its users and other people affected by its use.

Singh and Wesson's (2009) extrapolated usability criteria mapping reveals that most usability issues in an ERP system are connected with (Fig. 1):

– navigation;
– learnability;
– task support;
– presentation (input and output);
– customization.

2.3 System Usability Scale

System Usability Scale (SUS) designed by John Brooke (2013) provides a "quick and dirty", reliable tool for measuring the usability of a particular system. It consists of a 10 item questionnaire with five response options for respondents: from "strongly agree" to "strongly disagree". The tool has the capacity to evaluate a wide variety of products, systems or services Participants are asked to score the following items (Leavitt, 2016):

– I think that I would like to use this system frequently.
– I found the system unnecessarily complex.
– I thought the system was easy to use.
– I think that I would need the support of a technical person to be able to use this system.
– I found the various functions in this system were well integrated.
– I thought there was too much inconsistency in this system.
– I would imagine that most people would learn to use this system very quickly.
– I found the system very cumbersome to use.

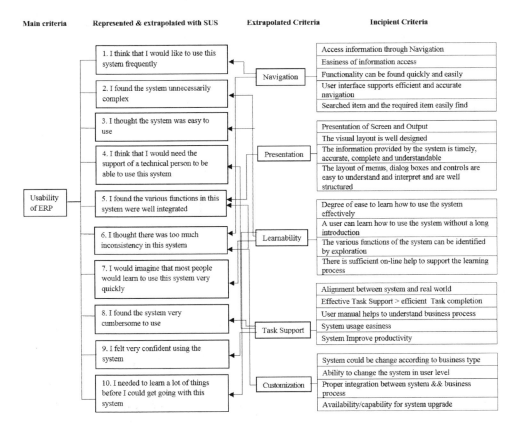

Figure 1. Hierarchical structure of usability criteria (elaborated on the basis of Brooke 2013; ISO 9241-11; Prussak & Hankiewicz, 2007; Rauer, 2011; Wesson & Singh, 2009).

– I felt very confident using the system.
– I needed to learn a lot of things before I could get going with this system.

The SUS scoring system is relatively complex, however, the methodology could be used even by a non-expert to evaluate their system. The participant's scores for each question are converted to a new number, added together and then multiplied by 2.5 to convert the original scores of 0–40 to 0–100. Though the scores are 0–100, these are not percentages and should be considered only in terms of their percentile ranking (Brooke, 2013; Leavitt, 2016).

An assessment of ERP system usability is quite challenging to perform due to limited research and resources on the topic. Although past researches concerned some cases related to inter-dependencies between.

The usability estimation inter-depends on a four level hierarchical structure which includes main criterion, criteria represented and extrapolated with SUS, extrapolated criteria, and incipient criteria. In order to obtain main criterion, extrapolated criteria focused on SUS (System Usability Scale) ten questions in various perspective ways. Finaly extrapolated criteria based on incipient criteria as elementary criteria.

The analytical questionnaire-based evaluation method was conducted based on ISO 9241, Singh and Wesson's usability mapping, Nielsen's ten heuristics and SUS (System usability scale) due to cost-effectiveness. This enables the evaluation of ERP system usability issues without a number of experts and laboratory space.

Process of estimation takes place in the following steps:

– identify the incipient criteria based on ISO 9241, Singh and Wesson's usability mapping, Nielsen's ten heuristics and SUS (System usability scale),
– obtain responses to ten SUS questions with the user's personal perspective on the system,
– estimate results based on user responses and review through the criteria set.

The SUS questions are divided into the following five main categories: navigation, presentation, task support, learnability and customization. Each question represents at least one category and some represent more than one (Wesson, Singh, 2009).

Process of estimation takes place in the following steps:

- identify the incipient criteria based on ISO 9241, Singh and Wesson's usability mapping, Nielsen's ten heuristics and SUS (System usability scale),
- obtain responses to ten SUS questions with the user's personal perspective on the system,
- estimate results based on user responses and review through the criteria set.

The SUS questions are divided into the following five main categories: navigation, presentation, task support, learnability and customization. Each question represents at least one category and some represent more than one (Wesson, Singh, 2009).

3 RESULTS

The estimation of criteria is based on users' responses of their personal perspective and on SUS score of the system. This section highlights the potential usability issues based on the above-mentioned factors.

The following 23 usability problems were identified:

1. Error message window popup during business operations
2. User credential issues frequently appearing
3. Users panicking while using the system
4. System language automatically changing
5. Error message notification not providing information about error
6. Errors cannot be solved at the user level
7. Selection of company geographic location not functioning
8. Reference counter errors
9. Country based tax system difficult to handle
10. Lack of information providing while printing documents
11. Internal search result is complicated
12. Character searching unavailable
13. Customer cannot categorize information by country
14. Lack of information related to business process and operation
15. Guiding information not available for operations
16. Proper navigation not available for next step
17. Sequence of steps on process not available
18. Process output not easily visible
19. Completion of task is not efficient
20. Users feel training is necessary
21. Operating manual is necessary to complete tasks
22. Unable to master the system within a short time period
23. Limited customization of the system

While conducting personal interview mapping with criteria, the above usability problems were identified. The feedback of 24 end users from different countries is shown as three values of central tendency (mean, median, and mode) in Table 1.

An analysis of the results from Table 1 revealed that after calculating the mean values for each question and multiplying by 2.5 the SUS score is 48. Based on user responses, the most essential criteria of usability models were "customization" and "task support".

The least essential criteria of the usability model were "navigation", "presentation" and "learnability" (Fig. 2).

The results of the "customization" and "task support" criteria show that end users are struggling to customize their system according to their needs. The alignment between the system and the real world contains significant differences in areas such as effective task completion and effective task support. The results of the "navigation", "presentation"

Table 1. Central tendency values based on SUS Questionnaires.

	SUS Questionnaires (Represented and extrapolated with other criteria)	Mean	Median	Mode
1	I think that I would like to use this system frequently	2.91	3	3.0
2	I found the system unnecessarily complex	3.37	3	3.0
3	I thought the system was easy to use	4.20	4	4.5
4	I think that I would need the support of a technical person to be able to use this system	3.20	3	4.0
5	I found the various functions in this system were well integrated	3.80	4	4.0
6	I thought there was too much inconsistency in this system	2.30	2	3.0
7	I would imagine that most people would learn to use this system very quickly	2.45	2	2.0
8	I found the system very cumbersome to use	3.50	4	4.0
9	I felt very confident using the system	1.80	2	2.0
10	I needed to learn a lot of things before I could get going with this system	3.62	4	3.5

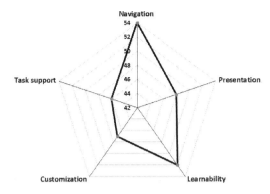

Figure 2. Criteria result based on SUS.

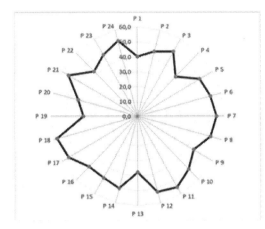

Figure 3. Results for research participants.

and "learnability" criteria show that the process of identifying and accessing information in the system has some issues but is generally successful. The visual layout design represents a significant level of success. The degree of effort required to learn and use the system shows that it is less successful in that regard. Participants' responses (Fig. 3) on the user based result score vary between 35 and 60. The values show that the system's usability is rated as acceptable.

4 CONCLUSION

ERP systems are designed to integrate the many different functions of an enterprise, and due to their complexity often suffer from usability issues which lead to a failure in achieving the company's common goals. This research study was conducted by mapping and mixing many different types of usability criteria and it reflects a new perception of an ERP system. A qualitative and quantitative data analysis of open-ended responses to personal questions and interview was performed. The results of the usability evaluation of an Open ERP system in this manufacturing company revealed that the overall and aggregated quantitative and qualitative usability criteria of "presentation", "navigation" and "learnability" were rated as generally positive. The best result was obtained for "navigation", which may indicate that developers focus on this component the most. However, to be successful improvement is needed for "presentation" component. "Customization" and "Task support" need to be significantly improved by the system's developers, and the designers need to consider agile development, where usability testing occurs at the very early stages of development.

ACKNOWLEDGEMENT

The support and facilitation of this research study by the company's end users from various locations is hereby acknowledged.

REFERENCES

Baker, K., Greenberg, S. & Gutwin, C. 2002. Empirical Development of a Heuristic Evaluation Methodology for Shared Workspace Groupware, Proceedings of the 2002 ACM conference on Computer supported cooperative work: 96–105.

Bevan, N. & Macleod M. 1994. Usability measurement in context, Behavior and Information Technology, 13: 132–145.

Bishu, R.R., Kleiner, B.M. & Drury, C.G. 2001. Ergonomic Concerns in Enterprise Resource Planning (ERP) Systems and Its Implementations. In J.P.T. Mo et al. (eds.), *Global Engineering, Manufacturing and Enterprise Networks*, New York: 146–155.

Brooke, J. 2013. SUS—Retrospective. *Journal of Usability Studies*, Vol. 8, Issue 2, February 2013: 29–40.

Garrett, J.J. 2011. *The Elements of User Experience: User-Centered Design for the Web and Beyond* (Second Edition), Berkeley, California, USA.

Hankiewicz, K. 2005. Ergonomic Profile of Computerized Management Information Systems. In CAES'2005 International Conference of Computer-Aided Ergonomics, Human Factors and Safety Kosice.

Hankiewicz, K. 2012. Ergonomic Characteristic of Software for Enterprise Management Systems. In P. Vink (ed.), *Advances in Social and Organizational Factors*: 279–287.

Hollingsed, T. & Novick, D.G. 2007. Usability Inspection Methods after 15 Years of Research and Practice, *SIGDOC'07*, October 22–24, 2007, El Paso, Texas, USA: 249–255.

Holzinger, A. 2005. Usability engineering methods for software developers, *Communications of the ACM*, January 2005/Vol. 48, No. 1: 71–74.

ISO 9241-11:1998. Ergonomic requirements for office work with visual display terminals (VDTs) — Part 11: Guidance on usability, ISO—International Organization for Standardization (www.iso.org). Retrieved from https://www.iso.org.

Leavitt, M.O. & Shneiderman, B. *Research-Based Web Design & Usability Guidelines*, U.S. Department of Health and Human Services (HHS) and the U.S. General Services Administration (GSA), Washington DC, USA.

Nielsen, J. 1994. *Usability Inspection Methods*, Conference Companion CHI 94, Boston Massachusetts, USA.

Nielsen, J. 1995. January 1, 1995, *10 Usability Heuristics for User Interface Design*, https://www.nngroup.com/articles/ /ten-usability-heuristics/.

Nielsen, J. 2001. January 21, 2001, *Usability Metrics*, https://www.nngroup.com/articles/usability-metrics/.

Prussak, W. & Hankiewicz, K. 2007. Quality in use evaluation of business websites. In L. Pacholski & S. Trzcieliński (eds), *Ergonomics in Contemporary Enterprise*, Poznań: 84–91.

Rauer, M. 2011. *Quantitative Usablility-Analysen mit der System Usability Scale (SUS)* [https://blog.seibert-media.net/blog/2011/04/11/usablility-analysen-system-usability-scale-sus/] Access on 25/9/2016.

Ray, R. 2014. *Enterprise resource planning*, McGrow Hill Education Private Limited, New Delhi.

Sadiq, M. & Pirhonen, A. 2014. Usability of ERP Error Messages. *International Journal of Computer and Information Technology* Volume 03 – Issue 05, September 2014: 883–893.

Singh, A. & Wesson, J., 2009. Evaluation Criteria for Assessing the Usability of ERP Systems, SAICSIT'09,12–14 October 2009, Riverside, Vanderbijlpark, South Africa: 87–95.

Ysnhong Z. 2009. ERP Implementation Process Analysis on the key success factor, International forum on information technology and application.

Postural adjustment for balance in asymmetric work. A practical example

G. Dahlke & K. Turkiewicz
Poznań University of Technology, Poznań, Poland

ABSTRACT: In the article, the authors present the part of findings of research on changes in loads brought to bear on the musculoskeletal system during the operation of various cameras models. The study programme includes measurements of the positions of musculoskeletal system segments during the operation of cameras (taken with electronic goniometers and torsiometers) as well as static and dynamic posturographic measurements taken to identify shifts in camera operator foot pressure against the ground and ascertain the centre of gravity of camera operator bodies. To maintain balance, the operators compensate the asymmetrical positioning of cameras (on their right shoulders) by, *inter alia*, laterally bending their trunks. Other postural adjustments include changes in the pressure exerted by the feet. The asymmetry of both the loads and camera operator postures may become habitual, increasing the risk of musculoskeletal system degeneration (Gnat et al. 2006).

1 INTRODUCTION

Any dynamic or static work involving the placement of segments of the musculoskeletal system in postures that result in varying degrees of asymmetry requires continuous postural adjustments to maintain body balance. During each activity, the position of a worker's musculoskeletal system depends not only on the layout of the work area, the range of motions performed and muscle fatigue but also on the body's propensity to maintain balance. Because the balance system (the vestibular system, muscle and joint receptors, internal organs and vision (Silbernagl 1994; Traczyk 2016)) can be affected by multiple stimuli, workers assume a wide range of postures leading to overloads and ultimately to the development of health conditions. In poor lighting (or wherever reliance on vision in the performance of work is greatly diminished), the role played by vision in maintaining body balance is significantly reduced. Similarly, once the body cools down, the performance of proprioceptors becomes impaired. The vestibular system, in its turn, becomes stimulated by loud noise (Damijan et al. 2012). A short interval of static work allows the centre of gravity to become more stable through trunk alignment. When work is dynamic, motions of the limbs become more frequent to restore balance (Dahlke et al. 2014, 2015, 2016). This phenomenon is investigated in a study involving camera operators. The authors enrol all camera operators employed in a local television station, all of whom have had many years' experience (the authors' study is part of a dissertation (Turkiewicz 2013). Camera operation involves the use of a camera lens, which largely precludes the reliance on vision to maintain balance. Asymmetrical loads and the asymmetrical postures assumed at work increase the risk of developing postural problems and exacerbate musculoskeletal conditions. By measuring foot pressures against the ground, the authors examine the impact of changes in loads and the body postures assumed at work on balance compensation and on the aggravation of postural asymmetries. The study additionally helps identify the causes of employee-reported symptoms.

2 METHODS

During the study, measurements and assessments are performed of:

- work postures (photographs in the sagittal and frontal plane—the measurements are made with the use of an electro-goniometer from Biometrics Ltd.) (Figure 1); the positions of musculoskeletal system segments are measured to analyse their acceptability under the ISO 11226 standard;
- changes in the position of the body's centre of gravity using an electronic pedometer.

Foot pressure against the ground and shifts of the centre of gravity are measured with a pedometer (from the company Medicapterus). Its 2304

Figure 1. An example of an upper-limb measurement using goniometers and torsiometers attached to worker skin (Source: Turkiewicz 2013).

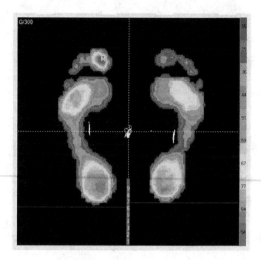

Figure 2. A Win-Pod (software) screen view is used to operate the pedometer. The colours show the foot pressure exerted against the ground. The white curves in the centre and the two curves near the centres of the feet represent shifts in the centre of gravity (Dahlke, Drzewiecka, Stasiuk-Piekarska, 2014).

(48 × 48) tensometric sensors are used to obtain readings at the frequency of 200 Hz to show changes in foot pressure against the ground (the data are captured by a computer programme).

The pedometer also allows measurements of the centre of gravity of the feet as well as the whole body (Figure 2).

The study involved three camera operators (Table 1). The subjects used a variety of equipment depending on the tasks performed. Postural adjustments for balance were observed with the heaviest of the cameras, i.e. Sony DSR-250, weighing 4.4 kg.

For each subject, six measurements are taken using a WIN-POD pedometer under the following conditions:

- **Measurement 1**: Static measurement (a momentary measurement for the preliminary identification of the distribution of foot pressure on the ground) without a camera; Eyes open; Gaze focused on proximate object.
- **Measurement 2**: Posturographic measurement (lasting 30 seconds) without a camera; Eyes open; Gaze focused on proximate object.
- **Measurement 3**: Posturographic measurement without a camera; Eyes closed.
- **Measurement 4**: Static measurement with camera on right shoulder; Left eye closed—work with camera.
- **Measurement 5**: Posturographic measurement with camera on right shoulder; Eyes open—work with camera.
- **Measurement 6**: Posturographic measurement with camera on right shoulder; Left eye closed.

Table 1. Subject details (Source: Turkiewicz 2013).

Subject designation	Year of birth	Weight [kg]	Height [cm]
Subject 1	1988	75	192
Subject 2	1986	65	175
Subject 3	1978	111	178

All of the above measurements were taken in a television studio.

3 RESULTS AND DISCUSSION

Measurements 1 and 4 were omitted in the analysis of results but are used for the preliminary observations of patterns in camera operator behaviours. Only those posturographic measurements that lasted 30 seconds were selected for such analysis.

The operators who placed their cameras on the right arm compensated for the asymmetry caused by the extra weight of the camera (about 4.4 kg). To that end, they adjusted the positions of their upper limbs by shifting their body weight leftward by way of bending their spine (Figure 3). Despite the lateral bend of the torso, posturographic measurements indicate that the greatest load is applied to the right lower limbs of camera operators (Figures 4 and 10).

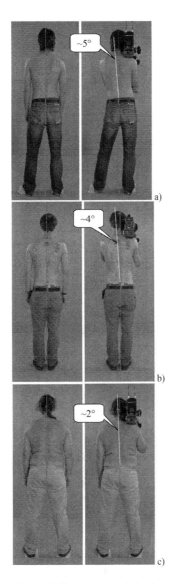

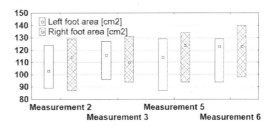

Figure 4. Distribution of the foot surface area of camera operators during posturographic measurements (Source: own research).

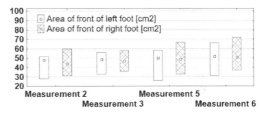

Figure 5. Distribution of changes in the area of the anterior parts of the feet of the camera operators during posturographic measurements (Source: own research).

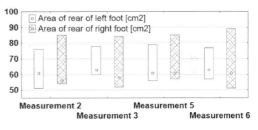

Figure 6. Distribution of changes in the surface area of the posterior part of the feet of the camera operators during posturographic measurements (Source: own research).

Figure 3. Postural adjustments made by the subjects by bending their torso laterally (a – Subject 1, b – Subject 2, c – Subject 3). Shown on the right are the bending angles during camera operation (Source: own work based on Turkiewicz 2013).

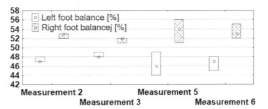

Figure 7. Percentage distribution of pressure on left and right feet of the camera operators during posturographic measurements (Source: own research).

This tendency becomes even more pronounced once the operators pick up their cameras. This results from, among other factors, the rightward (outward) extension of the left lower limb to provide additional support for the camera. A load shift can be observed towards the anterior part of the right foot and towards the posterior part of the left foot (Figures 5, 6, 8 and 9).

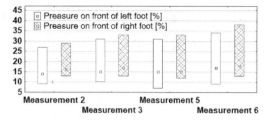

Figure 8. Percentage distribution of pressure on front of left and front of right feet of the camera operators during posturographic measurements (Source: own research).

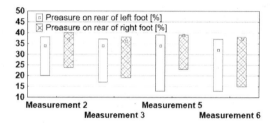

Figure 9. Percentage distribution of pressures on left and right feet of the camera operators during posturographic measurements (Source: own research).

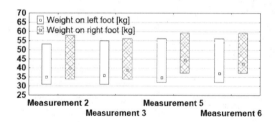

Figure 10. Distribution of pressures, in kg, on left and right feet of the camera operators during posturographic measurements (including the extra weight of the camera) (Source: own research).

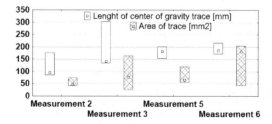

Figure 11. Distribution of the length of the lines and the area within the lines plotted by the centres of gravity of the camera operators during posturographic measurements (Source: own research).

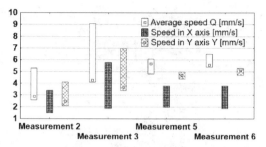

Figure 12. Distribution of the speed of shifts of the centres of gravity of the camera operators during posturographic measurements (total speed – Q and axial speed: X – lateral, Y – anterior-posterior) (Source: own research).

As the camera operators were observed at work, the authors identified the postures discouraged by the ISO 11226 standard by using the Working Method Audit worksheet (Horst, Dahlke 2002; Horst 2004) (Table 2 provides a list of discouraged postures identified during the examination of camera operators). Many inappropriate static positions were observed during prolonged work and movement. Photographs showing lateral bending angles revealed that such angles are considerably smaller for heavier individuals who appear to require lesser postural adjustments for balance. The aforementioned ISO standard defines a static posture as one that requires holding a musculoskeletal system segment still for 4 or more seconds. The standard also indicates the maximum holding time of four minutes for conditionally accepted postures. Hence, for the postures shown in Table 2, work cycles should not exceed 4 minutes.

When operating a camera, changes were also observed in other parameters:

– greater area of right foot due to extra weight added by camera (Figure 4);
– relative increase in weight on the front of feet (Figures 5 and 6 and Figures 8 and 9) coinciding with greater changes in weight distribution between rear and front (especially for the right foot) (Figure 13);
– increased weight on the right foot (Figures 7 and 10);
– shorter lines plotted by the centre of gravity but larger areas within such lines (Figure 11) – for measurement 3, deviation from the vertical is greater with the eyes closed suggesting the significance of vision for maintaining balance. When the eyes remain open (measurement 4), the area within the lines plotted by the centres of gravity is substantially smaller thanks to the stabilisation of balance enabled by the other eye;
– slower deviations from the vertical (Figure 12) caused by a focus on the filmed area and the

Table 2. Characteristics of postures discouraged by the ISO 11226 standard that are adopted during the operation of the Sony DSR-250 camera, as identified by the authors during the study (Turkiewicz 2013).

No.	Posture assumed during work
1	2
1	**Static postures**
1.1	**Trunk:**
1.1.1	Worker leaning the trunk laterally at an angle greater than 10°
1.1.2	Worker leaning unsupported trunk backward
1.2	**Arms:**
1.2.1	Worker holding unsupported arm extended forward at angles ranging from 20° to 60°
1.2.2	Worker holding unsupported arm extended laterally at angles ranging from 20° to 60°
1.3	**Forearms and hands:**
1.3.1	Worker holding right arm bent outward to either the right or the left to the maximum range of motion (at an angle greater than 20°)
1.4	**Head and neck:**
1.4.1	Worker holding the neck bent laterally at an angle greater than 10°

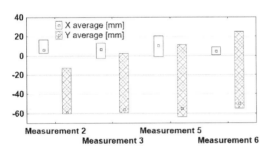

Figure 13. Distribution of mean deviations of the centre of gravity of the camera operator's bodies during posturographic measurements (axial: X – lateral, Y – anterior-posterior) (Source: own research).

effort to ensure image quality. Increased tension was observed in the muscles responsible for maintaining a stable position with the camera;
– greater deviation along the Y axis (instability and anterior-posterior shifting) and, in measurement 6, reduced shifts along the X axis (lateral) (Figure 13).

When operating a camera with the left eye closed, camera operators sought to maintain a stable posture. However, having to look through a lens does not provide the same stability that is attainable with the eyes open. The result is an increased area plotted by the movement of the centre of gravity. In practice, this translates into the stabilisation and higher quality of the camera recorded image. In both static and posturographic measurements, the subjects are observed to transfer their centres of gravity towards the heels. This response is very common in people wearing high-heeled footwear or living a predominantly sedentary lifestyle (own research).

4 CONCLUSIONS

In the case of static work, the prolonged holding of asymmetric postures may lead to changes precipitated by overloads. Asymmetry may also become habitual and accelerate degeneration (Gnat et al. 2006). Dynamic work, and especially work involving a big range of motions and the handling of heavy weights also requires the balancing of the body and the associated compensatory adjustments. Particularly detrimental to the musculoskeletal system in such work are high repetitiveness of motions and strong forces impacting the musculoskeletal system (Błaszczyk et al. 1997; Fidelus 1972).

The majority of unnatural postures in the study were found to be assumed when operating the SONY DSR 250 camera. This particular model was used more frequently than any other. It was both the largest and the heaviest of the devices examined.

The camera forced the operators to bend the trunk laterally at an angle of more than 10° and keep it tilted backwards without support. These postures were induced by the weight of the camera, which at times exceeds 5 kg with accessories. The operators balance their bodies laterally as well as backwards and forwards to reduce the pressure exerted by the load.

The shape of the camera, the placement of its controls and the manner in which it is used force operators to lift an unsupported arm forward at an angle of 20° to 60° and bend their right hand outward to the maximum range of motion.

To see the display and preview the material they are recording, the workers are forced to bend their necks sideways at an angle of more than 10°.

The prolonged muscular tension during work reduces blood circulation to a level below normal, consequently causing fatigue. It is therefore advisable to use tripods that relieve the musculoskeletal system. Nevertheless, while relatively easy to handle in a studio, tripods present a greater challenge in the field.

REFERENCES

Błaszczyk J. & Lowe D.L. & Hansen P.D. 1997. Age - related differences in performance of stereotype arm movements: Movement and Posture Interaction, *Acta Neurobiol. Exp.*, 57, 49–57.

Dahlke G., Drzewiecka M. & Butlewski M. 2016. Influence of work processes on variations in the parameters of complex reactions to stimuli, [in:] Arezes, P; Baptista, JS; et al., *Proceedings Book of Sho2016: International Symposium on Occupational Safety and Hygiene*, Guimaraes, Portugal, 23–24 March 2016, Pages: 58–60.

Dahlke, G. & Drzewiecka, M. & Stasiuk-Piekarska, A. 2014. Pozasłuchowy wpływ elektrowni wiatrowych na człowieka (Non-auditory impacts of wind farms on humans). Logistyka 5/2014: 290–300.

Dahlke, G. & Drzewiecka, M. 2015. Wind farm impact on the human balance system. In Arezes, P; Baptista, JS; et al., Proceedings of Sho2015: *International Symposium on Occupational Safety and Hygiene*. Guimaraes, Portugal: 94–96.

Damijan, Z. & Iwański, D. & Dahlke, G. 2012. Investigation of the vibroacoustic climate inside the buses Solaris Urbino 12 used in public transport systems. In: Acta Physica Polonica A., Vol. 121, No. 1-A: 32–37.

Damijan, Z. & Uhryński, A. 2012, The influence of general low frequency vibration on posture stability. In: Acta Physica Polonica A., Vol. 121, No. 1-A:28–31.

Fidelus K. 1972. Teorie procesu sterowania ruchami człowieka. Sport Wyczynowy, 2, 1–8.

Gnat, R., Saulicz, E., Kuszewski, M. 2006, Identification of the causes of functional pelvis asymmetry, [in:] Obciążenie układu ruchu. Przyczyny i skutki. Oficyna Wydawnicza Politechniki Wrocławskiej, Wrocław.

Horst, W., Dahlke, G. 2002, Ergonomics risk factors identification audit, [in:] Occupational risk in didactic, learning and training of ergonomics, work safety and labour protection; Monograph. - Poznań: Institute of Management Engineering, Poznań University of Technology, pp. 115–132.

Horst W. 2004. Ryzyko zawodowe na stanowisku pracy. Część I Ergonomiczne czynniki ryzyka. Wyd. Politechniki Poznańskie, Poznań.

http://biometricsltd.com.

ISO 11226:2000 Ergonomics - Evaluation of static working postures.

PN-EN 1005-5:2007 Safety of machinery - Human physical performance - Part 5: Risk assessment for repetitive handling at high frequency.

Silbernagl S. & Despopoulos A. 1994. Kieszonkowy atlas fizjologii, Wydawnictwo Lekarskie PZWL, Warszawa.

Traczyk W.Z. 2016. Fizjologia człowieka w zarysie, PZWL, Warszawa.

Turkiewicz K. 2013. The ergonomic risk factors in camera operator workstation, Diploma thesis under the direction of Grzegorz Dahlke, Poznań University of Technology, Poznań.

Working conditions and occupational risks in the physiotherapist's activity

L.S. Costa
ESTeSC—Coimbra Health School, Instituto Politécnico de Coimbra, Coimbra, Portugal

M. Santos
Faculdade de Psicologia e Ciências da Educação, Universidade do Porto, Porto, Portugal

ABSTRACT: The assessment of working conditions is crucial to determine occupational risks. Physiotherapists during their work activity are increasingly faced with specialized and prolonged interventions. The aim of this study was to determine how physiotherapists characterize their working conditions and what risks they identify. The results show that physiotherapists recognize problems in their work conditions and characteristics. They are exposed to situations that generate discomfort, namely, emotional demands, painful postures and a complex, unrecognized and overwhelming activity. They mainly find risks related to dealing with the patients, to physical constraints and with the pace of work.

1 INTRODUCTION

Work is a process (individual or collective) of transformation and with a purpose. In this way, working is not just about producing or transforming nature. Work(ing) is gestures, know-how, mobilize capacities, reflect, feel, interpret and react to situations (Dejours, 2006). The influence that work can have on life and health can be seen, e.g., in the negative contribution of unemployment to well-being and the positive influence of stable employment. But for the protection of workers' health, are essential the quality and feasibility of working conditions, which are linked to the activity conception, work organization and interpersonal relationships at work (Iavicoli, 2014). Risks are not just, more or less, predictable potential hazards. At work they shape many forms that are associated with the society's evolution where such work is performed and are an intrinsic dimension of the work activity (Echternacht, 2008, Thébaud-Mony, 2010).

The evaluation and understanding of the conditions under which work is done are central, because even if the work is well designed, its organization is strict, the instructions, procedures and rules are clear it is difficult to reach the quality if only what is prescribed is scrupulously respected. Actually, common work situations are permeated by unexpected events, incidents, malfunctions, organizational incoherence and unforeseen occurrences whether coming from materials, equipment and instruments, or other workers, leaders, subordinates, team, suppliers and customers (Dejours, 2004; Gollac & Volkoff, 2000).

As such, working conditions are experienced in a unique way by each worker. Its repercussions and influences, its contribution to the life in work and outside of the work depend on the contexts, of the ways that each worker, by his or her will, treads in the field of work, but also in the way in which he or she evaluates these conditions, how they perceive and what they do to deal with them. This involves considering diversity and variability in the assessment of working conditions, as well as the need to understand a set of factors that, mutually and at different levels, interact at distinguished moments in life. These factors require a careful and differentiated approach that does not neglect the need to integrate into the history of each individual the conditions in which he works (Volkoff, Molinié & Jolivet, 2000 as cited in Barros-Duarte & Lacomblez, 2006).

The provision of health care is increasingly an important pillar of societies, even by the population aging and the presence of new and chronic pathologies. Thus, health professionals are increasingly called towards a specialized, but also, extended intervention. Health care are provided in multiple contexts, with different actors (professionals, users, relatives) from different backgrounds and owners of different cultures (Rios, 2008).

Physiotherapists are health professionals whose central intervention consists in treating, enable and physically rehabilitating people of all age groups, with different kind of dysfunctions, including those resulting from aging and chronicity. In addition, the intervention of these professionals, although carried out within a more or less extensive team of health professionals, is fulfilled in a direct relationship with patients and their families, without the intervention of any other professional (Associação Portuguesa de Fisioterapeutas, 2010).

It is a professional group whose work has been little analyzed and the research carried out, focusing on health issues, deal almost exclusively with the musculoskeletal injuries and the relationships between them and the physical component of their work. Not only, but also for these reasons is a working population that is worth being analyzed with regard to the conditions and characteristics of their work.

The present study aimed to determine how physiotherapists characterize their working conditions and what professional risks they identify.

2 METHODOLOGY

Concerning the purposes outlined, the survey was used as a data collecting method, as it allows the gathering of information about different behaviors related to individuals present or past, and the access to descriptions of behaviors that for several reasons are not observable (Ghiglione & Matalon, 1997). A self-reported survey was chosen, since it has the advantage of less reactivity by the respondents and less interference of the researcher (Alferes, 1997).

2.1 Participants and data collection procedures

Physiotherapists from different providing health care institutions (public and private) in Portugal were asked. A convenience and non-random sampling process was used (Marôco, 2011). The final sample comprises 249 physiotherapists.

2.2 The tool and its application

The INSAT2010 (Health and Work Survey) of Barros-Duarte and Cunha (2010) was used as a data collection instrument. This survey encompasses indicators from different disciplinary areas, focuses on the lived experience of the workers, and incorporates a broad health conception, including subjective dimensions. It promotes a statistic of prudence, which avoids deterministic analysis, favoring a perspective that allows the understanding of the facts, rather than an explanation of them (Barros-Duarte & Cunha, 2010).

INSAT is organized into seven sets of questions or axes, which allow, among others: the characterization of the sociodemographic and professional aspects of the worker; the report of its work concerning is nature, type of contract, working hours and shifts; the evaluation of the conditions and characteristics of work that the workers perceive to be exposed and, if so, the degree of discomfort they feel (the discomfort is expressed through a Likert scale, from: 1- very high discomfort; 2-high discomfort; 3-discomfort, 4-low discomfort, 5-no discomfort) and describe some living conditions outside of work and the work life balance. In the last two axes, it is possible to obtain information about the perception of health and well-being, as well as the relation established with the work. For this study, was used the group of questions concentrates on work constraints and their effects on workers, namely: environmental and physical constraints; organizational (pace and autonomy) constraints; relationship (colleagues and customers) constraints, and work characteristics. We ensured the confidentiality on the data collected by maintaining the anonymity of the participants.

3 RESULTS

3.1 Sample sociodemographic and professional characterization

Most physiotherapists are female (76.7%). They are mostly unmarried (51.4%) with less than 40 years of age (mean 34 ± 7.8 years) and have worked in the public sector (50.6%) for less than 10 years (59.8%).

The physiotherapists work in the continental territory and in the two autonomous regions of Portugal. With reference to the Nomenclature of Territorial Units for Statistical Purposes (NUTS III), the places where they work are located in 21 of the 30 sub-regions of Portugal. The workplace of 53.4% of physiotherapists is made up more than 249 workers. 205 (82.3%) stated to be employees and 43 (17.3%) self-employed workers. Almost seventy percent of physiotherapists (173) work with a permanent employment contract and 71 (28.5%) with a fixed-term or temporary one.

3.2 Exposure to work conditions and characteristics and perceived discomfort

Prior to the data statistical analysis, we proceeded to reverse the Likert scale discomfort values. This reversal allowed that the highest value of the scale corresponded to the greater degree of discomfort.

3.2.1 Environmental and physical constraints

Regarding the physical environment (Table 1), physiotherapists consider them to be more exposed to hot/cold conditions (46.6%) and to biological agents (38.2%).

Table 1. Exposure to environmental and physical constraints and perceived discomfort.

Working conditions	Exposure		Discomfort	
	n	%	Mean	Sd
Physical environment				
Harmful Noise	73	29,3	3,22	0,79
Heat/cold	116	46,6	3,16	0,78
Biological agents	95	38,2	3,22	1,14
Physical constraints				
Repetitive gestures	169	67,9	3,12	0,92
Accurate and meticulous gestures	145	58,2	2,72	0,98
Painful postures	202	81,1	3,51	0,96
Intense physical strain	184	73,9	3,59	0,98
Standing for a long period of time	130	52,2	3,40	1,00
Standing with displacement	160	64,3	2,94	0,98

Table 2. Exposure to work pace, autonomy and initiative constraints and perceived discomfort.

Working conditions	Exposure		Discomfort	
	n	%	Mean	Sd
Work pace constraints				
Follow imposed pace	46	18,5	3,02	1,05
Relying on the work of co-workers	63	25,3	2,82	1,09
Relying on clients/users direct orders	93	37,3	2,25	0,90
Production norms or strict deadlines	95	38,2	2,78	1,12
Constantly to adapt to method or tools changes	87	34,9	2,28	1,01
Having to do several things at the same time	183	73,5	2,79	1,00
Frequent interruptions	141	56,6	2,83	0,97
Having to hurry up	168	67,5	3,24	1,00
Solve sudden situations/problems without help	152	61,0	2,67	1,04
"Skip" or shorten meals or not even having a break	138	55,4	3,24	1,07
Exceed normal schedule	119	47,8	2,96	1,07
Autonomy and initiative constraints				
Not having the freedom to decide how to do work	35	14,1	3,39	1,32
Not being able to change the order of the tasks to do	43	17,3	3,56	1,36
Not being able to choose breaks	33	13,3	2,94	1,27
Not being able to change a hard working schedule	55	22,1	3,09	1,28

Still in Table 1 are the situations that physically can be source of constraints. Physiotherapists reported greater exposure to painful postures (81.1%), intense physical strain (73.9%) and repetitive gestures (67.9%). On average are postures and physical efforts that are considered more uncomfortable.

3.2.2 *Work pace, autonomy and initiative constraints*

With regard to the work pace/rhythm (Table 2), the physiotherapists' exposure was mainly due to the fact that they had to do several things at the same time (73.5%), to hurry (67.5%) and to solve sudden situations or problems without help (61.0%). On average, the biggest discomfort is when "Skip" or shorten meals or not even having a break and having to hurry up.

In the same table, it can be seen that from the situations related to Autonomy and Initiative, physiotherapists consider that they are more exposed to having to comply with a strict working schedule (22.1%) and to obey a predefined order of accomplishment of tasks (17.3%). It is also in these situations that they consider themselves more disturbed.

3.2.3 *Work relationship and dealing with the public constraints*

Needing help (59.8%), verbal aggression (30.5%), opinions not always taken into consideration (28.5%) and intimidation (28.1%) are the situations that physiotherapists believe to be more exposed. The average discomfort is greater for intimidation and for the impossibility of expressing themselves. When dealing with the public, the most reported situations are those of having to respond to the difficulties or suffering of other people (88.0%), suffer the demands of the public (85.5%) and deal with situations of tension in the relationship (74.7%). Verbal and physical aggression are the conditions with the highest average amount of discomfort (Table 3).

3.2.4 *Work characteristics*

Most physiotherapists work in the presence of others and in their work, they have to learn new things, conditions that do not disturb them. They are also exposed to an activity that will be difficult to perform when they are 60 years old (79.9%), which involves overwhelming moments (75.1%) and that is little recognized by the leaders (50.6%). On average, the discomfort is greater when work has conditions that undermine dignity and when physiotherapists ponder their work at age 60 (Table 4).

Table 3. Exposure to work relationship and dealing with the public constraints and perceived discomfort.

Working conditions	Exposure		Discomfort	
	n	%	Mean	Sd
Work relationships				
Frequently needing help but not always getting it	149	59,8	2,18	1,14
Not having my opinion always taken into consideration	71	28,5	3,55	1,26
Impossible to express myself	42	16,9	3,88	1,09
Verbal aggression	76	30,5	3,51	1,20
Physical aggression	46	18,5	3,28	1,29
Sexual harassment	26	10,4	3,80	1,44
Intimidation	70	28,1	3,97	1,06
Age discrimination	26	10,4	3,41	1,24
Dealing with the public				
Direct contact with the public	244	98,0	1,30	0,67
Suffer the demands of the public	213	85,5	2,56	0,97
Deal with situations of tension in the relationship with the public	186	74,7	2,82	0,97
Verbal aggression from the public	161	64,7	3,10	1,09
Physical aggression from the public	90	36,1	3,05	1,09
Having to respond to the difficulties or suffering of others	219	88,0	2,56	1,13

Table 4. Exposure to work characteristics and perceived discomfort.

Work characteristics	Exposure		Discomfort	
	n	%	Mean	Sd
Always in the presence of others	222	89,6	1,32	0,64
Always learning new things	211	84,7	1,19	0,49
Monotonous	27	10,8	3,35	0,84
Excessively varied	70	28,1	1,84	0,98
Little creative	33	13,3	2,79	1,24
Very complex	101	40,6	1,85	0,87
With overwhelming moments	187	75,1	2,91	0,99
Ergonomically little organized	118	47,4	3,34	0,97
Where adequate equipment/instruments are lacking	139	55,8	3,48	0,96
With inadequate facilities	91	36,5	3,48	1,00
Not usually recognised by my colleagues	65	26,1	3,52	0,97
Not usually recognized by the leaders	126	50,6	3,97	0,89
With conditions that undermine dignity	16	6,4	4,53	1,00
Difficult to perform at age 60	199	79,9	4,10	0,97
Where I generally feel exploited	115	46,2	4,04	0,95
I wish my children would not do it	83	33,3	3,67	1,36
With which I am little satisfied	81	32,5	4,01	1,05

4 DISCUSSION

The aim of this study was to determine how physiotherapists characterized their working conditions and to identify risks associated with these conditions. The results presented show the perception that these professionals have regarding the conditions and characteristics of their work.

It is found that physiotherapists identify adverse working conditions concerning their physical working environment, namely the temperature and the contact with biologic agents. As such conditions cause them discomfort, they can be considered of risk. In fact, physiotherapists can work in several public or private professional institutions. In the public sector, they work in Hospitals (with or without hospitalization) and Health Care Centers/Family Health Units. In private institutions, the practice of physiotherapy is carried out both on behalf of others and on its own account. Also can take place in the private sector (clinics convened with the National Health Service or other health subsystems) or unconventional. Therefore, the diversity of places and work situations can justify environmental conditions that are not always optimized. On the other hand the hospital and health care environments often implies a greater exposure to pathogenic material.

Physiotherapists claim to have to adopt painful postures, to make intense physical strains and repetitive gestures. These physical constraints, in addition to being justified by the characteristics of the patients and the treatments to be carried out, are in line with the results of the latest European Survey on Working Conditions, which concludes that exposure to physical risks shows no significant decrease and some risks (painful and tiring postures and repetitive movements) shows an increase (Eurofound, 2012).

To carry out their work most physiotherapists need to do several things at the same time, to hurry up and solve sudden problems without help.

They are often interrupted, need to "skip" or shorten meals and do not take breaks. Thus, they are confronted with a high work pace, a situation that can be identified as a risky condition. Autonomy issues do not appear to be risk to most of these workers, although for some there are adverse working conditions with regard to having to carry out a hard working schedule and performing tasks whose order of implementation is predefined.

For the category work relationships and dealing with the public, we also identify constraints to which physiotherapists are exposed and which can create occupational risks. Actually they frequently need help but not always get it, which helps to understand the intense work rhythm and the excessive physical strain. They are exposed to verbal aggression, their opinion is not considered and they feel intimidation in the relationship with colleagues and leaders. The contact with the customers and their relatives or companions premise that they have to answer to difficulties or suffering, to bear their demands and to deal with stiffness situations. Like most health professionals, the physiotherapist deals with a set of more or less common circumstances, such as the pressure of having to work face-to-face with the public and the responsibility for people. These circumstances are intensified by the professional involvement with patients with disabilities, serious injuries and physical pain. So, their interactions with the customers are completely different from those that other professional groups, outside the health area, have with the public with which they relate. These interactions, their characteristics, as well as the characteristics of their work, give rise to demanding working conditions (Sauter & Murphy, 1995). Research results show that health professionals are exposed, among others to lack of preparation and capacity, role overloading, extended hours of work, teamwork conflicts, difficulties in work/family balance, and insufficient human and material resources (Camelo, 2006).

Physiotherapists characterize their work activity as being complex, difficult to accomplish at age 60, with overwhelming moments, where adequate equipment/instruments are lacking and which is little recognized by the leaders. These constraints may contribute to the risks identified above. Indeed a high work pace, little time for breaks, lack of help, dealing with users and their suffering, verbal aggressions, e.g., are certainly associated with the characterization they do of their work.

Thus, physiotherapists recognize that they are exposed to constraints in all the categories of work conditions and characteristics that have been established. Towards a better identification of risk factors to which these professionals are exposed, another data analysis was also carried out. Therefore only items with a percentage of exposure equal to or greater than half of the inquired physiotherapists were considered in each category.

It was found that in all categories there are items with these amount of exposure, except for physical environment and autonomy and initiative. Dealing with the public and physical constraints are the categories where there is the highest percentage of items marked by more than half of physiotherapists. The work relationship category is the one that show a smaller percentage. The data, concerning the discomfort that the physiotherapists perceive with the situations to which they are exposed, allow to confirm this analysis. All the constraints to which more than half of the physiotherapists are exposed are stated as causing discomfort.

Therefore, it seems possible to say that physiotherapists are subject, mainly and above all, to risks related to Dealing with the Public, to Physical Constraints, to Work Characteristics and to Pace of Work Constraints. Actually, these results converge with a professional activity that has as one of its objectives the physical recovery of people with different kind of problems and with different degrees of complexity. They fit with permanent contact with patients with difficult, severe and often very difficult clinical conditions. Since in clinical settings, patients have difficulty caring for themselves, are insecure and stressed it can be assumed the existence of a degree of demand and care that will tend towards the existence of tension and anxiety at the relational level.

5 CONCLUSIONS

Physiotherapists characterize their working conditions as having several constraints. These conditions lead to an exposure to disturbing situations, essentially when dealing with the public (users and their families), their demands, difficulties or suffering and when they have to get over stressful situations. Also, physical constraints lead to risks through painful postures, strong efforts and repetitive gestures. Furthermore, they have a high pace of work, have to deal with emotional demands and with a complex, unrecognized and overwhelming work. Then in the physiotherapist's activity the risks associated with their working conditions are mostly related to dealing with the public and to physical constraints.

REFERENCES

Alferes, V. (1997). *Investigação científica em Psicologia: Teoria e prática.* Coimbra: Almedina.
Associação Portuguesa de Fisioterapeutas. (2005). *Normas de Boas Práticas para a prestação de serviços de*

Fisioterapia. Retrived from http://www.apfisio.pt/Ficheiros/N_B_Praticas.pdf.

Barros-Duarte, C., & Cunha, L., (2010). INSAT2010 – Inquérito Saúde e Trabalho: outras questões, novas relações, *Laboreal*, 6(2), 19–26, http://laboreal.up.pt/revista/artigo.php?id=48u56oTV6582234;5252:5:5292.

Barros-Duarte, C., & Lacomblez, M. (2006). Saúde no trabalhoediscriçãodasrelaçõessociais.*Laboreal*,2(2),82–92. http://laboreal.up.pt/revista/artigo.php?id=37t45nSU547112278541446881.

Camelo, S. (2006). Riscos Psicossociais Relacionados ao Estresse no Trabalho das Equipes de Saúde da Família e Estratégias de Gerenciamento (Tese de Doutoramento não publicada). Escola de Enfermagem de Ribeirão Preto, São Paulo, Brasil.

Dejours, C. (2004). Subjetividade, trabalho e ação. *Revista Produção*, 14(3), 27–34.

Dejours, C. (2006). Subjectivity, work, and action. *Critical Horizons*, 7(1), 45–62. doi:10.1163/15685160677930 8161.

Echternacht, E. (2008). Atividade humana e gestão da saúde no trabalho: Elementos para a reflexão a partir da abordagem ergológica. *Laboreal*, IV(1), 46–55. http://laboreal.up.pt/revista/artigo.php?id=48u56o TV65822343965929; 38;2.

Eurofound. (2012). *Fifth European Working Conditions Survey*. Luxembourg: Publications Office of the European Union.

Ghiglione, R., & Matalon, B. (1997). *O inquérito, teoria e prática* (3ª ed.). Oeiras: Celta Editora.

Gollac, M., & Volkoff, S. 2000. *Les conditions de travail*. Paris: Editions La Découverte.

Iavicoli, S. (2014). Les facteurs de risques psychosociaux dans le monde changeant du travail. In L. Lerouge (Dir.), *Approche interdisciplinaire des risques psychosociaux au travail* (pp. 10–21). Toulouse: Octares Editions.

Marôco, J. (2011). *Análise estatística com o SPSS Statistics* (5ª Ed.). Pero Pinheiro: Report Number, Lda.

Rios, I. 2008. Humanização e Ambiente de Trabalho na Visão de Profissionais da Saúde. *Saúde e Sociedade*, 17(4), 151–160.

Sauter, S., & Murphy, L. (1995). *Organizational risk factors for job stresse*. New York: John Wiley & Sons.

Thébaud-Mony, A. (2010). Riscos. *Laboreal*, VI(1), 72–73. http://laboreal.up.pt/revista/artigo.php?id=37t 45nSU54711238:7626984121.

Influencing factors in sustainable production planning and controlling from an ergonomic perspective

Maximilian Zarte
Faculdade de Ciencias e Tecnologia, Universidade Nova de Lisboa, Portugal

Agnes Pechmann
Department of Mechanical Engineering, University of Applied Sciences Emden/Leer, Germany

Isabel L. Nunes
Faculdade de Ciencias e Tecnologia, UNIDEMI, Universidade Nova de Lisboa, Portugal

ABSTRACT: The need for sustainable development has been widely recognized by governments and enterprises and has become a hot topic of various disciplines including ergonomics and industrial engineering. Initiatives in this domain will be more successful if they follow an integrated approach. For instance, the conventional production planning and controlling (a methodology used in industrial engineering) considers resources, such as material, labor and production capacity at its respective costs, but often neglects the role of ergonomic factors. The goal of this paper is to provide an overview of ergonomic factors which influence sustainable production planning and controlling in manufacturing enterprises. These influencing factors are discussed based on a root cause analysis. A root cause diagram is shown illustrating the relationships between sustainable production planning and controlling and ergonomic factors.

1 INTRODUCTION

The Brundtland Commission defined sustainability as "development that meets the needs of the present without compromising the ability of future generations to meet their own needs" (WCED 1987). Moreover, sustainability is bonded by the triple bottom line which balances economic goals, environment cleanness, and social responsibility (Elkington 1997). The need for sustainable development has been widely recognized by governments and companies and has become a hot topic of various disciplines even in the field of ergonomics (Radjiyev et al. 2015). Ergonomics is "the scientific discipline concerned with the understanding of the interactions among humans and other elements of a system, and the profession that applies theory, principles, data and methods to design in order to optimize human wellbeing and overall system performance" (IEA 2000). Obviously, this definition already matches two dimensions of sustainability: social responsibility (human well-being), and economic success (system performance). Moreover, Thatcher (2013) investigates the relationship between ergonomics and the environmental dimension of sustainability. This author discusses the term "green ergonomics" and presents ideas on the design of low resource systems and products, design of green jobs, and design of processes and products for behavior change to meet sustainable goals.

The change of behavior plays an important role in sustainable development (Hanson 2013). Particularly, high polluting and hazard-prone industries changed their behavior to make their practices more sustainable (Horberry et al. 2013). Professionals working within these industries can help integrate an ergonomic approach to tackle sustainable concerns. Thatcher et al. (2017) argue strongly that ergonomics should be an integral part of finding solutions to the current global challenges for sustainability. Moreover, the authors argue that ergonomics needs to go a step further to create transdisciplinary approaches that combine theory and practice from different disciplines. Sutherland et al. (2016) reviewed efforts to integrate ergonomics in the companies' Production Planning and Controlling (PPC) process and identified a lack of social data and standardized approaches (Sutherland et al. 2016). PPC considers resources, such as material, labor and production capacity at its respective costs, but neglects the role of environmental aspects and ergonomic factors (Pechmann & Zarte 2017).

The lack of consideration for ergonomic factors in PPC is also recognized in the study of Giret et al. (2015). These authors suggest addressing ergonomics in PPC in future research to provide a more global analysis guided by the principles of sustainability.

The goal of this paper is to provide an overview on the benefits of an integrated sustainable PPC process for manufacturing companies, including an Ergonomics perspective.

For this purpose, the initial stage was identifying the groups of stakeholder that are affected by PPC. Then, an analysis of an existing set of indicators for Corporate Social Responsibility (CSR) reporting was made to select the ones that are relevant to each stakeholder group, bearing in mind that CSR indicator sets provide a comprehensive overview on indicators, which can be used to measure the sustainable performance of products and processes. On a subsequent stage, a group of influencing factors for sustainable PPC was identified for each relevant CSR indicator integrating an Ergonomics perspective. Some of these key influencing factors are presented below in a root cause diagram that shows the relationship between sustainable PPC and ergonomic factors contributing to well-being and performance.

After this introduction, Section 2 presents materials and methods, which were used for the analysis. Section 3 discusses the results of the analysis summarized on a root cause diagram. Section 4 offers the main conclusions of the paper, followed by references.

2 MATERIALS AND METHODS

A root cause analysis was performed with the purpose of identifying ergonomic factors contributing for a sustainable PPC. In fact, the goal of the root cause analysis methodology is to reveal key relationships for a specific topic among various main variables, and possible causes, providing additional insight into process behavior (Wilson et al. 1993).

The root cause analysis is structured in three steps:

1. Determination of the specific topic
2. Definition of main variables related to the specific topic
3. Identification of causes which influence the main variables

In the present analysis, the specific topic is 'sustainable PPC' integrating also an Ergonomics' perspective, as discussed in the introduction.

For the definition of the related main variables, an approach based on the stakeholders for sustainability in production was adopted. A set of stakeholder types was defined on the study of Sutherland et al. (2016), where stakeholders were categorized in the following six groups:

– Owners
– Costumers
– Suppliers
– Employees
– Local Community
– Global Community

These groups are further described in the Section 3.

To identify the causes related to the six stakeholder groups, CSR reporting indicators were analyzed. CSR reporting is an organizational report on the sustainable performance of countries, organizations, or companies (GRI 2017). The Global Reporting Initiative (GRI) is an international, independent organization that, since 1997, helps businesses, governments and other organizations to understand and communicate the impact of business on critical sustainability issues, such as climate change, human rights, corruption and many others. The published GRI standards for sustainable reporting are widely used and accepted in industry (GRI 2017), and used in the current analysis.

To evaluate the CSR indicators, a set of criteria was defined to select ergonomic factors to integrate in a sustainable PPC process:

– The factor must have a predictable qualitative and quantitative impact on production planning;
– The factor must have a qualitative and quantitative impact on production control;
– The factor must be relevant regarding sustainable manufacturing and ergonomics
– The indicator must be understandable and easy to interpret by the company.

Applicable indicators for PPC are identified through an evaluation of the GRI indicators according the defined criteria.

3 RESULTS

In the following subsections, the above-mentioned stakeholder groups are briefly described, and the related ergonomic factors are presented. An overview of the results is presented in Figure 1 as a root cause diagram.

Because of the high number of indicators, a presentation of all evaluated indicators according the criteria is not possible. Therefore, the following subsections contain only the resulting applicable indicators.

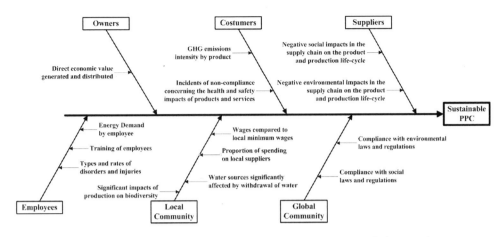

Figure 1. Root cause diagram for a sustainable PPC process integrating an Ergonomics' perspective.

3.1 Owners

Company owners are individuals with a financial investment in the business and responsible for the strategic orientation of the company (Sutherland et al. 2016).

For this stakeholder group, the GRI indicator 201-1 (Direct economic value generated and distributed) is applicable according to the defined criteria. Owners are interested in the economic value creation of the company is required for a long-term survival on the market. The lack of economic value precludes sustainability actions, such as employees' fair wages and investment on production environmental optimization. A positive economic value creation of a manufacturing system must always be the goal of sustainable PPC process.

3.2 Costumers

In general terms, a costumer can be viewed as any end-user of a product, service, or process (Sutherland et al. 2016).

For this stakeholder group, the GRI indicators 305 (Emissions), and 416-2 (Incidents of non-compliance concerning the health and safety impacts of products and services) are applicable according to the defined criteria. Through the demonstration of social responsible practices, and the meeting of sustainable social and environmental expectations from the civil society, new markets can be accessed, with a positive impact to the company wealth. Social and environmental expectations can be set as limits or goals for sustainable PPC.

3.3 Suppliers

A supplier provides goods or services to the company (Sutherland et al. 2016).

For this stakeholder group, the GRI indicators 308-2 (Negative environmental impacts in the supply chain and actions taken), and 414-2 (Negative social impacts in the supply chain and actions taken) are applicable according to the defined criteria. The supplier has an indirect impact on the outcomes of a sustainable PPC process. The selection of adequate suppliers and the influencing of their behaviors, can help in mitigating any potential or real social and environmental negative impacts along the supply chain, and on the product and production life-cycle (Nunes 2012; Ustailieva et al. 2012).

3.4 Employees

An employee is an individual who provides their skills to a firm, usually in exchange for a monetary wage (Sutherland et al. 2016).

For this stakeholder group, the GRI indicators 302-1 (Energy consumption within the organization), 403-2 (Types of injury and rates of injury, occupational diseases, lost days, and absenteeism, and number of work-related fatalities), and 404-1 (Average hours of training per year per employee) are applicable according to the defined criteria. By monitoring employees, their behavior can be evaluated and then shaped to increase the production efficiency (corresponding to the controlling stage of the PPC process). For example, the energy required to manufacture a product can vary depending on the employees' skills' level. The employees' energy consumption while performing their tasks can be considered in production planning to increase the energy efficiency of the production (corresponding to the planning stage of the PPC process). For instance, a continuously training program can contribute to increase the employees' skills and awareness, resulting in increased production efficiency and health and safety (e.g., by avoiding work-related disorders, injuries, down-time on production and employees turn-over).

3.5 Local community

A local community is a spatially-related group of individuals utilizing a shared resource base within which a company enterprise exists (Sutherland et al. 2016).

For this stakeholder group the GRI, indicators 202-1 (Ratios of standard entry level wage by gender compared to local minimum wage), 204-1 (Proportion of spending on local suppliers), 303-2 (Water sources significantly affected by withdrawal of water), and 304-2 (Significant impacts of activities, products, and services on biodiversity) are applicable according to the defined criteria. Through the PPC controlling stage, the negative social and environmental impacts on the local environment can be measured and mitigated. Thus, the integration of a comprehensive ergonomic approach to the sustainable PPC process contributes to the social and economic wealth of the local community. The benchmarking with local competitor can be used regarding social and environmental impacts to set goals for production planning. By demonstrating social responsibility and supporting the local community, companies' image and penetration in local markets can be improved.

3.6 Global community

A global community is a community outside the boundaries of local community, e.g. state, national, and international entities (Sutherland et al. 2016).

For this stakeholder group, the GRI indicators 203-2 (Significant indirect economic impacts), 307-1 (Non-compliance with environmental laws and regulations), and 419-1 (Non-compliance with laws and regulations in the social and economic area) are applicable according to the defined criteria. National and international laws and regulations on social and environmental matters (namely on working conditions, resource usage, and waste disposal) should carefully be observed by companies in their integrated and sustainable PPC process. Through a proactive strategy, companies stay ahead of the entry on force of new laws and regulations, which can result in market advantages regarding competition.

4 CONCLUSIONS

The paper provides an overview on the key topic of a sustainable PPC process for manufacturing companies, referring also the benefits of integrating ergonomic factors in this process. This was done based on an analysis of the existing set of indicators for CSR reporting provided by the GRI.

For the analysis of the key topic, the related main variables were defined (i.e., the stakeholders for sustainability in production), and the main factors (including some ergonomic factors) influencing these main variables were identified and illustrated on a root cause diagram. To identify applicable influencing factors, the GRI indicators were evaluated against a set of criteria for selection of applicable ergonomic-related indicators for PPC. The identified ergonomic factors should be considered in the planning and controlling processes to improve the sustainability performance of manufacturing.

The findings of this analysis suggest that, in order to achieve high CSR standards through PPC companies should:

i. provide training to their employees, developing new qualifications at the work place;
ii. benchmark with competitors;
iii. influence suppliers;
iv. engage with local community;
v. anticipate new laws and regulation from global community, and requirements from costumers; and
vi. set comprehensive goals and limits in their PPC processes, including an Ergonomics perspective.

The current results can be used as basis to support and improve the outcome of the PPC process discussions with company managers, experts and employees' representatives. The integration of the Ergonomics' perspective ensures that the sustainable PPC process, is guided by principles of social responsibility (human well-being), and economic success (system performance).

ACKNOWLEDGMENT

The study was funded by project UID/EMS/00667/2013.

REFERENCES

Elkington, J. 1997. *Cannibals with forks: The triple bottom line of 21st century business.* Oxford: Capstone.
Giret, A., Trentesaux, D., & Prabhu, V. 2015. Sustainability in manufacturing operations scheduling: A state of the art review. *Journal of Manufacturing Systems*, 37: 126–140.
GRI 2017. GRI Standards. Retrieved from https://www.globalreporting.org/standards, Last Access 01.11.2017.
Hanson, M.A. 2013. Green ergonomics: challenges and opportunities. *Ergonomics*, 56(3), 399–408.

Horberry, T., Burgess-Limerick, R., & Fuller, R. 2013. The contributions of human factors and ergonomics to a sustainable minerals industry. *Ergonomics*, *56*(3): 556–564.

IEA 2000. IEA Executive Defines Ergonomics. *On-Site Newsletter for the IEA 2000/HFES 2000 Congress.*

Nunes, I. L. 2012. The nexus between OSH and subcontracting. *Work (Reading, Mass.)*, *41 Suppl 1:* 3062–3068.

Pechmann, A., & Zarte, M. 2017. Procedure for Generating a Basis for PPC Systems to Schedule the Production Considering Energy Demand and Available Renewable Energy. *Procedia CIRP*, *64*: 393–398.

Radjiyev, A., Qiu, H., Xiong, S., & Nam, K. 2015. Ergonomics and sustainable development in the past two decades (1992–2011): Research trends and how ergonomics can contribute to sustainable development. *Applied ergonomics*, *46 Pt A // 46:* 67–75.

Sutherland, J. W., Richter, J. S., Hutchins, M. J., Dornfeld, D., Dzombak, R., Mangold, J., & Friemann, F. 2016. The role of manufacturing in affecting the social dimension of sustainability. *CIRP Annals—Manufacturing Technology*, *65*(2): 689–712.

Thatcher, A. 2013. Green ergonomics: definition and scope. *Ergonomics*, *56*(3), 389–398.

Thatcher, A., Waterson, P., Todd, A., & Moray, N. 2017. State of Science: ergonomics and global issues. *Ergonomics*: 1–33.

Ustailieva, E., Eeckelaert, L., & Nunes, I. L. 2012. Promoting occupational safety and health through the supply chain: Literature Review. *European Agency for Safety and Health at Work*. Retrieved from https://osha.europa.eu/en/publications/literature_reviews/promoting-occupational-safety-and-health-through-the-supply-chain/view.

WCED 1987. Report of the World Commission on Environment and Development: Our Common Future.

Wilson, P. F., Dell, L. D., & Anderson, G. F. 1993. Root cause analysis: A tool for total quality management. Milwaukee, Wis: ASQC Quality Press.

Proof of concept—work accident costs analysis tool

Catarina Correia & J. Santos Baptista
PROA/LABIOMEP/INEGI, Faculty of Engineering, University of Porto, Porto, Portugal

Paulo A.A. Oliveira
CIICESI—School of Management and Technology—Polytechnic of Porto, Porto, Portugal

ABSTRACT: The occupational safety and health system has become increasingly visible at workplace. This article presents a proof concept of a computer tool, which allows the analysis of costs associated with work accidents. To achieve this goal, a computer support tool was developed as an aid to its analysis. At this point a brief study of the accident cost is attached. The tool was implemented on MSExcel and tested using data from heath sector. A brief analysis of general labor costs, in a discriminatory manner was done. It was concluded that the tool is extremely useful as it allows to evaluate the costs associated with labor accidents and compares them with the implementation costs and amounts charged by the insurers.

1 INTRODUCTION

1.1 Occupational safety and health management system

Occupational safety and health is an area of expertise dedicated to studying and developing techniques and technologies focused at the prevention of accidents and occupational diseases. Its main objective is to improve and monitor all working conditions.

Therefore, an Occupational Safety and Health Management Systems (OSH-MS) aims to provide a method for assessing and improving issues related to prevention of accidents, incidents and occupational diseases in the workplace (International Labor Organization, 2011).

The most impactful approach to occupational accidents and diseases is to determine their cost and, consequently, to calculate the savings that may be associated with their non-occurrence. So, in a very pragmatic point of view, a first step involves the distinction between direct and indirect costs, which can be defined as follows:

– Direct costs—can be identified and assigned directly to each type of product/equipment/accident/illness. That means no apportionment is required to assign the cost to the identified object.
– Indirect costs—cannot be attributed directly to each type of product/equipment/accident/illness. Indirect costs are allocated through the use of pre-determined pricing criteria and related, for example, with indirect labor, energy costs, hours/used equipments, etc.

This is a possible distinction, since there is no literary consensus on the distinction between these two components (Jallon et al., 2011). So, indirect costs are much more difficult to measure than direct costs (Thiede & Thiede, 2015). The prevention of accidents at work presents themselves as an effective solution in reducing costs related to poor working conditions. In this way, it makes sense to invest in prevention (Jallon et al., 2011).

1.2 Work accident concept

According to portuguese law, "It is an accident that occurs in the place and time of work and, directly or indirectly produces bodily injury, functional disturbance or diseases resulting in a reduction in the ability to work capacity or to obtain a salary or in death of the worker."

For the purposes of this Book Chapter, workplace means any place where the worker is or had to be in virtue of his or her work and in which is directly or indirectly subject to the control of the employer.

'Working time, beyond the normal working time' means the time that precedes the work, in preparatory or work-related acts, and the time in preparatory or work-related acts after working time, under normal or forced conditions of work (art 8.º, Lei n.º 98/2009). From the point of view of the authors identified and selected from the literature review, safety must be intertwined with the concept of management not only when an occupational accident occurs, but always taking preventive measures. Not only from the point of view of safety, but also from the point of view of business, it is recognized the importance of work accident analysis and the need to address this issue in a more holistic way (Veltri et al., 2013). As is well known, an occupational accident or disease can cause a significant disruption in the stability of the work environment where it occurs (Jallon et al., 2011).

1.3 Objectives

When the subject to be addressed is safety at work, the goal is to promote safety and health of workers to reduce absenteeism and prevent the occurrence of accidents and occupational diseases.

As such, some organizations choose to implement an occupational safety and health management system. This system can addresses subjects like implementation/certification of a health and safety standards like OHSAS 18001 (2007), with risk assessment and work accident analysis and, many times, the implementation/certification of an integrated management system (IMS).

The discussion that rises is not the relevance of implementing an occupational safety and health management system in an organization. The core of the problem is certainly if it is economically feasible, when examined strictly at the level of economic balance for the company to implement it or not. Thus, there are factors such as the costs associated with work accidents that must be analyzed.

Considering the importance of this problem to the companies and trying to give a contribution to its solution, the goal of this work was to develop a tool dedicated to the analysis and recording of occupational accidents, assessing the cost associated with them in a method that could be usable in any organization inside health sector. With that knowledge is possible to know the prevention costs associated with the safety and health system and to evaluate the results from the different methods of analysis.

The different tasks can be summarized as follows:

- Develop a computer support for the automatic calculation of accidents costs;
- Collect data that allows testing the tool;
- Analysis of accident costs in global terms and by categories supported by data collection;
- Present a forecast of the results to be obtained.

2 MATERIALS AND METHODS

The starting point for the tool development was the definition of all variables to be considered in the research. These were defined bearing in mind a previous bibliographic review.

With this focus, an informatic support tool based on Microsoft Excel was developed, in order to simplify and assist the organization itself in the analysis of accident data. The tool was developed to be as intuitive and effective as possible and to deliver results in real time. Some fields are prepared to answer of pre-established questions.

The informatics support tool is divided into separate tabs, each one dedicated to a specific topic:

- Information about the injured;
- Data relating to the accident investigation;
- Data associated with accident costs;
- Each occurrence is treated individually and is related to a predefined typology and trying to collect the detailed information of each occurrence.

In addition to all this information, an analysis was done considering the focus of incidence, being possible to carry out the analysis by all the categories that prove to be significant.

3 RESULTS

The tool was tested, with data from health sector. Exclusively, for an analysis being the following criteria: classification of the accident and profession.

The data were obtained from a thesis (Castro, 2010). The coding is according to EUROSTAT.

The aim is to prove the benefits of the tool itself, focusing on the analysis of results from its use.

For formatting purposes, the tool is divided into 3 parts and 20 accidents are presented as well as the total and average value of each category evaluated. It is only showed an analysis based on profession (Tables 1, 2 and 3).

Seven accidents of each profession were randomly selected to evaluate the influence of this parameter on its average cost.

According the Figure 1, can be observed that, considering the same number of accidents under the same conditions, these costs will vary greatly according to the professional category.

Analyzing Figure 1, can be observed that accident costs are very different according the profession. Since it is a test to the tool, it is rather pertinent to demonstrate the type of available information and its organization from the tool. This enables a data layout in which it is easier and faster to understand.

As the analysis for the assessment according to the classification and the type of profession has been demonstrated, other relevant parameters can be also assessed.

Table 1. Working accident analysis—health sector data (part 1).

N°	Classification	Number of days hospitalized	Professional category	Salary €/hour	Health (€)	Hospitalar costs Hospitalization N° days	Hospitalar costs Hospitalization Value (€)	Rehab (€)	Pharmaceutical (€)
1	1 polytrauma	1	AMM	3,41	30	1	50,00	—	—
2	1 polytrauma	1	AD	4,55	30	1	50,00	—	—
3	1 polytrauma	1	ENF	14,20	30	1	50,00	—	—
4	1 polytrauma	1	FARM	9,09	30	1	50,00	—	—
5	1 polytrauma	1	MED	28,41	30	1	50,00	—	—
6	1 polytrauma	1	OP	3,98	30	1	50,00	—	—
7	1 polytrauma	1	TEC	9,09	30	1	50,00	—	—
8	2 fracture	2	AMM	3,41	30	2	100,00	—	—
9	2 fracture	2	AD	4,55	30	2	100,00	—	—
10	2 fracture	2	ENF	14,20	30	2	100,00	—	—
11	2 fracture	2	FARM	9,09	30	2	100,00	—	—
12	2 fracture	2	MED	28,41	30	2	100,00	—	—
13	2 fracture	2	OP	3,98	30	2	100,00	—	—
14	2 fracture	2	TEC	9,09	30	2	100,00	—	—
15	3 dislocation/sprain	3	AMM	3,41	30	3	150,00	—	—
16	3 dislocation/sprain	3	AD	4,55	30	3	150,00	—	—
17	3 dislocation/sprain	3	ENF	14,20	30	3	150,00	—	—
18	3 dislocation/sprain	3	FARM	9,09	30	3	150,00	—	—
19	3 dislocation/sprain	3	MED	28,41	30	3	150,00	—	—
20	3 dislocation/sprain	3	OP	3,98	30	3	150,00	—	—
Total 161	—	2268	—	1672,73	4830,00	1988	99400,00	—	—
Average 81	—	14,09	—	10,39	30,00	13	617,39	—	—

AMM—Medical Action Assistant; AD—Administrative Assistant; ENF—Nurse; FARM—Pharmaceutical; MED—Doctor; OP—Operational Assistant; TEC—Technical Assistant;

Table 2. Work accident analysis tool, using data from hospital sector (part 2).

N°	Compensation costs														
	ATD*						PTD*					APD*		PPD*	
	Death	N° days	Base salary (€/day)	Salary L98/2009 (€)	Salary DL503/99 (€)	Cost (€)	N° days	Base salary (€/day)	%	Salary 98/2009 (€)		%	Cost (€)	%	Cost (€)
1	0	1	23,33	16,33	23,33	16,33	1	23,33	0,01	0,16		0,01	0,16	0,00	0,00
2	0	1	31,11	21,78	31,11	21,78	1	31,11	0,01	0,22		0,01	0,22	0,00	0,00
3	0	1	97,22	68,06	97,22	68,06	1	97,22	0,01	0,68		0,01	0,68	0,00	0,00
4	0	1	62,22	43,56	62,22	43,56	1	62,22	0,01	0,44		0,01	0,44	0,00	0,00
5	0	1	194,44	136,11	194,44	136,11	1	194,44	0,01	1,36		0,01	1,36	0,00	0,00
6	0	1	27,22	19,06	27,22	19,06	1	27,22	0,01	0,19		0,01	0,19	0,00	0,00
7	0	2	62,22	87,11	124,44	87,11	2	62,22	0,01	0,87		0,01	0,44	0,00	0,00
8	0	2	23,33	32,67	46,67	32,67	2	23,33	0,02	0,65		0,02	0,33	0,00	0,00
9	0	2	31,11	43,56	62,22	43,56	2	31,11	0,02	0,87		0,02	0,44	0,00	0,00
10	0	2	97,22	136,11	194,44	136,11	2	97,22	0,02	2,72		0,02	1,36	0,00	0,00
11	0	2	62,22	87,11	124,44	87,11	2	62,22	0,02	1,74		0,02	0,87	0,00	0,00
12	0	2	194,44	272,22	388,89	272,22	2	194,44	0,02	5,44		0,02	2,72	0,00	0,00
13	0	2	27,22	38,11	54,44	38,11	2	27,22	0,02	0,76		0,02	0,38	0,00	0,00
14	0	3	62,22	130,67	186,67	130,67	2	62,22	0,02	2,61		0,02	0,87	0,00	0,00
15	0	3	23,33	49,00	70,00	49,00	3	23,33	0,03	1,47		0,03	0,49	0,00	0,00
16	0	3	31,11	65,33	93,33	65,33	3	31,11	0,03	1,96		0,03	0,65	0,00	0,00
17	0	3	97,22	204,17	291,67	204,17	3	97,22	0,03	6,13		0,03	2,04	0,00	0,00
18	0	3	62,22	130,67	186,67	130,67	3	62,22	0,03	3,92		0,03	1,31	0,00	0,00
19	0	3	194,44	408,33	583,33	408,33	33	194,44	0,03	12,25		0,03	4,08	0,00	0,00
20	0	3	27,22	57,17	81,67	57,17	3	27,22	0,03	1,72		0,03	0,57	0,00	0,00
Total 161	—	—	11448,89	99960,00	142800,00	99960,00	—	11448,89	—	16301,54		—	989,58	—	0,00
Average 81	—	—	71,11	620,87	886,96	620,87	—	71,11	0,12	101,25		0,12	6,15	—	0,00

* ATD—Absolute Temporary Disability; PTD—Partial Temporary Disability; APD—Absolute Permanent Disability; PPD—Partial Permanent Disability.

Table 3. Work accident analysis tool, using data from hospital sector (part 3).

| | Production losses | | | | | | | Replacement of the victim (€) | | Other costs (€) | | | | | |
| | Victim | | Other workers | | | | | | | Accident investigation costs (€) | | | | Material damages | Total cost (€) |
N°	Hours	€	Hours	€	Hours	€	Pay check		N° hours Int.	Cost Int.	N° hours Ext.	Cost Ext. (€)		
1	3	10,23	2	6,82	1	15,63	27,27		1	9,09	1	10,80	0,00	176,49
2	3	13,64	2	9,09	1	15,63	36,36		1	9,09	1	10,80	0,00	196,82
3	3	42,61	2	28,41	1	15,63	113,64		1	9,09	1	10,80	0,00	369,59
4	3	27,27	2	18,18	1	15,63	72,73		1	9,09	1	10,80	0,00	278,12
5	3	85,23	2	56,82	1	15,63	227,27		1	9,09	1	10,80	0,00	623,67
6	3	11,93	2	7,95	1	15,63	31,82		1	9,09	1	10,80	0,00	186,66
7	3	27,27	2	18,18	1	15,63	145,45		1	9,09	1	10,80	0,00	394,84
8	3	10,23	2	6,82	1	15,63	54,55		2	18,18	2	21,60	0,00	290,64
9	3	13,64	2	9,09	1	15,63	72,73		2	18,18	2	21,60	0,00	325,72
10	3	42,61	2	28,41	1	15,63	227,27		2	18,18	2	21,60	0,00	623,89
11	3	27,27	2	18,18	1	15,63	145,45		2	18,18	2	21,60	0,00	466,04
12	3	85,23	2	56,82	1	15,63	454,55		2	18,18	2	21,60	0,00	1062,38
13	3	11,93	2	7,95	1	15,63	63,64		2	18,18	2	21,60	0,00	308,18
14	3	27,27	2	18,18	1	15,63	218,18		2	18,18	2	21,60	0,00	583,19
15	3	10,23	2	6,82	1	15,63	81,82		3	27,27	3	32,40	0,00	405,12
16	3	13,64	2	9,09	1	15,63	109,09		3	27,27	3	32,40	0,00	455,06
17	3	42,61	2	28,41	1	15,63	340,91		3	27,27	3	32,40	0,00	879,56
18	3	27,27	2	18,18	1	15,63	218,18		3	27,27	3	32,40	0,00	654,82
19	3	85,23	2	56,82	1	15,63	681,82		3	27,27	3	32,40	0,00	1503,83
20	3	11,93	2	7,95	1	15,63	95,45		3	27,27	3	32,40	0,00	430,09
Total 161	—	5 018,18	—	3 345,45	—	2 515,63	166 909,09		—	18 070,92	—	21 470,40	—	438 810,79
Average 81	—	31,17	—	20,78	—	15,63	1 036,70		—	112,24	—	133,36	—	2 725,53

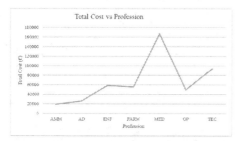

Figure 1. Graphic illustrating the variety of the total cost of accidents according to profession.

4 DISCUSSION

The tool is set in a way that allows a constant update of the information letting the results to be disposed immediately.

The data structure is somehow adjustable from organization to organization with the creation of a computer support tool to aid in the whole process. The tool can be adapted to any type of working scope, being just necessary to make some changes. The tool is also intuitive and easy to understand.

The main advantage of this tool is the information that can be extracted from its use, starting from the analysis of the inserted data. Since, several analyzes can be carried out by category, by profession, total costs, costs by sector. This tool is a good complement to the organizations, since they have standards to fulfill, so an analysis of accidents is an added value.

This tool, not only allows the organizations to know the incidence of accidents, but also the actual costs associated with them, in order to find out where there are higher or lower and act appropriately.

To the previous advantages, it is added that this tool gives to the organization the knowledge of the real cost of their own accidents, and not exclusively the value presented by the insurance company. So, it is possible to compare these two values, which allows a correct evaluation of the real situation. Another advantage from a management point of view, therefore, should be emphasized in having discriminated data, since it allows acting in a more adequate and effective way in accidents prevention.

Summarizing, the main advantage of the tool is that it allows a complete and very detailed analysis in real time.

The interconnection between all the information makes possible a wide range of data that can be analyzed, and which allows useful conclusions to the organization. In addition, the analysis allows not only the total costs but also provides the amount spent on each work accident.

The discrimination of all data type, allows an analysis of the different parameters by adding useful and valuable information to the prioritization of the intervention. Knowing the origin of the accidents, allows to limit the intervention with measures more appropriate to the critical areas and, therefore, with better results.

5 CONCLUSIONS

The first conclusion concerns the issue—costs and accidents at work—that usually is not sufficiently analyzed by the organizations and is generally treated only by the insurance company.

It is concluded that this tool is useful for its purpose and adds value to any organization that decides to use it and can be adapted to all other sectors.

When assessing the incidence of accidents by nature and type of injury is perceptible where to act to reduce the number of accidents.

The tool has proved to be very useful, not only for presenting results in real time but also for allowing a complete analysis regarding the work accidents problem in the company.

6 FUTURE WORKS

The first future perspective, which is quite obvious, is to use the created tool in real context (business, companies) since it adds value to the analysis of accident costs. Improve their adaptability to the different business contexts is also an important target.

Another possibility of improvement can be added: considering the analysis performed on the incidence of accidents, was evidenced that the focus must be given to a bigger emphasis on prevention issues.

REFERENCES

Castro; M. J. (2010) Investimento na Prevenção vs Custos do Acidente. Tese de Mestrado em Engenharia em Segurança e Higiene no Trabalho. Faculdade de Engenharia da Universidade do Porto. Porto.

International Labor Organization (2011). Dia Mundial da Segurança e Saúde no Trabalho. Organização Internacional do Trabalho, sistema de gestão da segurança e saúde no trabalho: um instrumento para uma melhoria contínua. Retrieved from: http://www.dnpst.eu/uploads/relatorios/relatorio_oit_2011_miolo.pdf.

Jallon, R., Imbeau, D., & Marcellis-warin, N. De. (2011). Development of an indirect-cost calculation model suitable for workplace use. Journal of Safety Research, 42(3), 149–164. https://doi.org/10.1016/j.jsr.2011.05.006.

Lei n.º 98/2009, de 4 de setembro, Diário da República, 1.ª série – N.º 172 – 4 de Setembro de 2009. Assembleia da República. Lisboa.

Thiede, I., & Thiede, M. (2015). Quantifying the costs and benefits of occupational health and safety interventions at a Bangladesh shipbuilding company, 127–136. https://doi.org/10.1179/2049396714Y.0000000100.

Veltri, A., Pagell, M., Johnston, D., Tompa, E., Robson, L., Amick, B. C., … Macdonald, S. (2013). Understanding safety in the context of business operations: An exploratory study using case studies. Safety Science, 55, 119–134. https://doi.org/10.1016/j.ssci.2012.12.008.

Wrist-hand work-related musculoskeletal disorders in a dairy factory: Incidence, prevalence and comparison between different methods for disease validation

Ana Raposo
PROA/LABIOMEP/INEGI, Faculty of Engineering, University of Porto, Porto, Portugal

Renato Pinho
Faculty of Medicine, University of Porto, Portugal

J. Santos Baptista
PROA/LABIOMEP/INEGI, Faculty of Engineering, University of Porto, Porto, Portugal

J. Torres Costa
PROA/LABIOMEP/INEGI, Faculty of Medicine, University of Porto, Porto, Portugal

ABSTRACT: The involvement of wrist and hand in working postures is very common in work environments like food manufacturing. Risk factors increasing due to repetitive tasks and long working hours with few breaks. The aim of this study is to analyze the prevalence and incidence of wrist/hand Work-Related Musculoskeletal Disorders (WRMSD), as well the wrist/hand disease assessment methods, in a dairy factory. The company has 620 employees and data were collected among the 166 workers belonging to the cheese sector. Two evaluations were done and a total of 134 respondents (80,7%) were evaluated in both. Wrist/hand WRMSD prevalence rate was assessed by self-reported symptoms, clinical evaluation and imaging, in both evaluation periods, varying between 18% and 45%. The incidence rate from the first to the second evaluation varied between 7% and 23%, depending on the sensitivity and specificity of each evaluation method's criteria. The results of this study emphasize the importance of joint evaluation of outcomes of different assessment methods.

1 INTRODUCTION

The Italian physician Bernardo Ramazzini, among office clerks, described the first case reported of wrist/hand problems work-related, in 1713. Ramazzini observed that the main causes of these disorders were highly repetitive work, static working postures and stress (Buckle 1997). These problems are still present nowadays. According the fifth European Working Conditions Survey, about 60% of Europeans make repetitive hand and arm movements (Living 2014)\. In fact, those risk factors are associated with the main wrist/hand musculoskeletal disorders such as Carpal Tunnel Syndrome (CTS), tenosynovitis, tendinitis, and De Quervain's disease (Armstrong et al. 1982; Armstrong et al. 1986; Silverstein et al. 1986; Atterbury et al. 1996; Melchior et al. 2006; Ricco and Signorelli 2017).

Wrist/hand disorders account for about 55% of all WRMSD (Kong et al. 2006) and are the leading cause of absenteeism, loss of productivity, injury and illness work-related disability (Alexopoulos et al. 2006, Kaka et al. 2016). It is estimated that injuries in wrist/hand leads to a 13 days median, away from work, accounting around 12% of lost time (Barr et al. 2004; Kim et al. 2015; Sutton et al. 2016). Even so, the wrist/hand musculoskeletal disorders aren't recognized as occupational diseases in the majority of EU members. Only tendinitis of the wrist and hand is recognized in DK, F, IRL and L.

Reported incidence of upper limb WRMSD including wrist/hand disorders increased slightly in south Europe between 2007 and 2016 presenting values around 15% (Stocks et al. 2015, BLS 2016).

The association of the musculoskeletal disorders and work possibly have a temporal cause-effect relation (Barr et al. 2004) but prevention and the early diagnosis are not valorised. This is due to the absence of information able to identify economic sectors and working conditions that increase risk factors, but also because there is no a complete knowledge of the true "size" of the problem (Costa et al. 2015). In fact, in the last decades, various studies found strong evidence that some jobs requiring highly repetition, handling heavy loads, forceful wrist/hand exertions, awkward posture and vibration, increases risk for wrist/hand WRMSD (Putz-Anderson et al. 1997; Carey and Gallwey 2005; Akesson et al. 2012). However, most of these studies are of poor quality, without rigorous reporting, both at the level of results and at the level of exposure assessment (Gerr, Letz et al. 1991).

The lack of globally accepted criteria for the diagnosis of wrist/hand diseases makes the studies difficult to compare. There are evident limitations in establishing causality between methodologies applied in numerous studies (Armstrong et al. 1993).

The aim of the present study is investigate the prevalence and incidence and of work-related wrist/hand musculoskeletal disorders in a dairy products factory as well the wrist/hand diseases assessment methods.

2 MATERIAL AND METHODS

This study was developed in a food factory specialized in dairy products, in the cheese sector, which activity started in 2008. The company has 620 employees and the data were collected between 2010 and 2014 among all the 166 cheese sector workers. A total of 134 respondents (corresponding to a rate of participation of 80.7%) participated in the two evaluation periods defined for this study: one in 2011/2012 and another in 2013/2014. Data were collected from the application of the Nordic Questionnaire survey on musculoskeletal symptoms (Kuorinka et al., 1987) validated for Portuguese population (Serranheira & Santos, 2003) with assisted response, from the clinical orthopedic examination and from the upper limb ultrasonography.

A group of 401 workers with non-repetitive work was also included as a control group, but at this group only the Nordic Questionnaire, survey was applied. Statistics were performed using SPSS 22.

3 RESULTS

3.1 General characteristics and working profile of participants

From the 134 participants, 41 (30.6%) were male and 93 (69.4%) female, aging from 20 to 61 years old. Respect to working profile, 80% of the participants worked less than 8 years in the factory, and 94% of workers are right-handed. Demographic characteristics of participants and control group are summarized in the Tables 1 and 2.

Table 1. Demographic characteristics of participants and control group.

Cheese factory workers	
Variable	Mean ± standard deviation
Age	32.42 ± 7.6 years old
Height	1.65 ± 0.09 m
Weight	67.63 ± 13.5 kg
BMI*	25.99 ± 7.63 kg/m^2
Control Group	
Variable	Mean ± standard deviation
Age	30.00 ± 9.8 years old
Height	1.71 ± 0.09 m
Weight	68.22 ± 12.6 kg
BMI*	23.22 ± 3.30 kg/m^2

*Body Mass Index.

Table 2. General characteristics of participants.

Variable	Percentage (%)
Smokers	28,0
Non smokers	72,0
Age Groups (Years)	
19–30	53.0
31–40	23.1
41–50	18.7
51–60	5.2
Work Stations	
Supply of raw materials; Pressing and Moulding.	18.7
Salting and Maturing	20.1
Packing	39.6
Others	21.6

3.2 Prevalence of wrist/hand WRMSD disorders

In the first evaluation the total symptoms prevalence of wrist/hand WRMSD was of 23.9% and the prevalence in the second evaluation was 18.7%. In terms of gender women reported more wrist/hand musculoskeletal symptoms than men (93,7% vs 6,3%). Men do not perform tasks with as many repetitive hand/wrist movements as women do. Regarding the clinical orthopaedic examination the prevalence was 41,0% and 44,8%, respectively for first and second evaluation. Considering only the results of ultrasonography, done in both

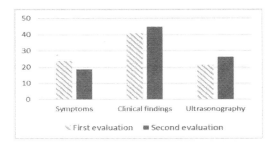

Figure 1. Prevalence in both evaluation periods (cheese sector).

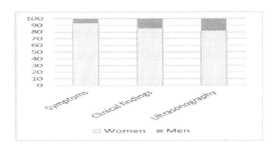

Figure 2. Prevalence by gender – first evaluation (cheese sector).

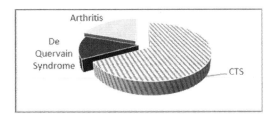

Figure 3. Type of shoulder disorders encountered in the study.

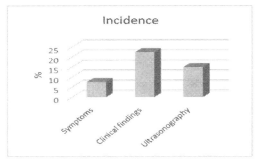

Figure 4. Incidence over the follow-up period.

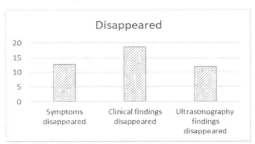

Figure 5. Percentage of cases that disappeared, by the evaluation methods of the study.

evaluation periods, the prevalence was of 21,4% and 26,2% respectively. Graphic results are presented in Figures 1 and 2.

In the control group, was found a shoulder WRMSD prevalence of 8,2%. As shown in Figure 3, the major disorder diagnosed by the orthopaedist was carpal tunnel syndrome (74,5%). De Quervain syndrome (14,5%) and Arthritis (21,8%) are among others. On the other side, many workers presented simultaneously CTS and De Quervain syndrome (14,5%) or CTS and Arthritis (5,5%)

3.3 Incidence of wrist/hand WRMSD disorders

Over the follow-up period, there were 10 new cases of wrist/hand symptoms, 30 new findings by the orthopaedist and 20 new abnormalities found in ultrasonography, giving rise to a period incidence rate of 7,5%, 22,4% and 14,9%, respectively (Figure 4).

Furthermore, the cases that disappeared were accounted, considering the three evaluation methods of the study. Thus, 17 respondents became asymptomatic (12,7%), 25 clinical findings disappeared (18.7%) and 16 ultrasonography abnormalities became normal (11,9%) as showed in Figure 5.

Considering ultrasound in particular, a statistical analysis was made. So, can be observed that 9,0% of the findings achieved in first evaluation period remain in the second evaluation and of these 16,7% are exactly the same, while 83,3% mostly change from unilateral to bilateral abnormalities.

3.4 Validation methods of wrist/hand WRMSD disorders

Finally, in order to analyze the possible presence of work-related wrist/hand MSD, suggestive shoulder disease indicators were used, based on the clinical examination, and upper limb ultrasonography. The establishment of validation methods for wrist/hand WRMSD, is based in groups representing positive response in the considered methods.

Table 3. Group's percentage representing positive response of wrist/hand disease.

Groups	% (2011/2012)	% (2013/2014)
A: Signs of disease in one of the methods	30,6	32,8
B: Signs of disease in two of the methods	18,7	18,7
C: Signs of disease in three of the methods	6,0	6,7
D: No signs of disease in any method	44,8	44,8

Therefore, three evaluation methods and four groups were considered. The evaluation methods were: symptoms, clinical examination and upper limb ultrasonography. The groups were named as A, B, C and D, which represents positive response to wrist/hand disease in one, two, three, or in any of the methods, respectively (Table 3).

4 DISCUSSION

The schedule of a dairy factory processing industry worker involves a series of events such as suppling raw material, moulding, salting, moulding and packing. These activities involve the frequent use of forceful exertions, repetitive motions, rapid work pace, and non-natural body postures sustained over a long period. So, the expected prevalence and incidence of wrist/hand WRMSD disorders were high, considering a population highly exposed to repetitive work (Fredrisson et al. 1999; Thomsen et al. 2007; Denbeigh et al. 2013). In fact, wrist/hand WRMSD prevalence rates achieved by self-reported symptoms, clinical evaluation and imaging, in both evaluation periods, varied between 18,7% and 44,8%, and the results are concordant from the first to the second evaluation, in the three methods applied. This result is consistent with other studies in manufacturing industries, reporting wrist/hand WRMSD prevalence of 15–40%. (Werner et al. 1997; Thomsen et al. 2007; Klussmann et al. 2008; Kebede and Tafese 2014; Hembecker, et al. 2017).

In this study the prevalence increase when the assessment was made by clinical methods. This can be explained by the increase of specificity or by the the underestimated symptoms perceived by workers and/or decision not to report those symptoms in the survey.

Ultrasonography is the most commonly requested examination in musculoskeletal shoulder disorder diagnosis due of absence of radiation exposure, availability, low cost and patient friendly examination. So, it was selected method for wrist/hand imaging in this study (Gupta and Robinson 2015). As this method can detect a variety of asymptomatic wrist/hand abnormalities, and detect several "abnormalities" that may be clinically unrelated to the patient's complaints (Lento and Primack 2008), it would be expected a high prevalence value of the three methods used in the present study. However the ultrasonography prevalence values were smaller than expected, when compared with clinical findings probably because it considers only the wrist image.

In terms of gender the prevalence of self-reported WMSD was higher for women than for men, which agrees with described in literature (Rocha et al. 2005; Sutton et al. 2016). This can be explained because the proportion of female population is substantially higher than male, and also because women may be more likely to express pain and symptoms comparing with men or even the shorter muscles in women (Fagarasanu and Kumar 2003).

To access with more certainty the effect of work conditions in the development of wrist/hand disorders, a control group was included in the study, without exposure to the same risk factors than cheese factory workers and the symptoms prevalence of this group (8,2%), was lower than cheese factory workers symptoms prevalence as expected. The prevalence value achieved, agree with the prevalence wrist/hand injuries reported in others studies in no manufacturing industries, like third sector jobs (Alrowayeh et al. 2010, d'Errico et al. 2010).

Regarding the most common disorder encountered, carpal tunnel syndrome (CTS) appeared to be the most prevalent, followed by arthritis and De Quervain syndrome. These findings agree with several studies reporting severe impairment associated with CTS disorder (Sutton et al. 2016). Others studies confirm that repetitive work is associated with wrist/hand arthritis (Dillon et al. 2002).

The risk of development of wrist/hand WRMSD, is measure by the incidence rate. Over the follow-up period, incidence value was between 7% and 23%, depending of the considered method. These values are in agreement with some literature which report incidence rates between 2% and 20% (Kivekas et al. 1994; Gell et al. 2005; Ricco and Signorelli 2017).

Despite the large number of studies of upper limb disorders, there are only a few that describe the outcomes assessed by more than one or two methods (surveys mainly) and relates them together.

One of the objectives was to define methodologies for the validation of MSD of the wrist/hand, and the defined groups enabled to link the three methods information. So, to use criteria with high degree of sensitivity, leading to the detection of the disease in earlier stages of development, the results of group A should be taken into account. In turn, if the criteria is having a high degree of specificity, results of group C should be considered.

5 CONCLUSIONS

The obtained results are consistent with some and discrepant with other similar studies.

The considerable heterogeneity between wrist/hand WRMSD studies settings as well the adoption of a widely different methodologies, exposure assessment and statistical approaches make comparison and interpretation of outcomes difficult.

This study highlighted the importance of joining evaluation of different assessment methods. The use of only one assessment method could explain, in part, the large differences in prevalence rates observed in the literature.

This study reaffirm the importance of establishing appropriate and consensual diagnosis criteria that allow a better comparison between results. Nonetheless, more research is required in understanding the relationships between different assessment methods, around the cost-effectiveness of the different strategies and at the prevention or treatment field

Some limitations could be discussed however, including the participants number, that should be higher or the study design provided no information on events during the two years of follow-up (it would be better if there were more examinations along the time). That may be a limitation, given the rapid evolution of these disorders.

The high prevalence and incidence rates found in the present study, underlined, the importance of having WMSDs education and prevention programs, as well including the need of an improved medical routine surveillance, as well better ergonomics in the workplaces.

REFERENCES

Akesson, I., Balogh, I., & Hansson, A. (2012). Physical workload in neck, shoulders and wrists/hands in dental hygienists during a work-day. Applied Erg., 43(4), 803–811. doi:10.1016/j.apergo.2011.12.001.

Alexopoulos, E., Tanagra, D., Konstantinou, E. & Burdorf, A. (2006). Musculoskeletal disorders in shipyard industry: prevalence, health care use, and absenteeism. BMC Musc Dis, 7, 88. doi:10.1186/1471-2474-7-88.

Alrowayeh, H. N., Alshatti, T. A., Aljadi, S. H., Fares, M., Alshamire, M. M., & Alwazan, S. S. (2010). Prevalence, characteristics, and impacts of work-related musculoskeletal disorders: a survey among physical therapists in the State of Kuwait. BMC Musc Dis., 11, 116. doi:10.1186/1471-2474-11-116.

Armstrong, T. J., Buckle, P., Fine, L. J., Hagberg, M., Jonsson, B., Kilbom, A., . . . Viikari-Juntura, E. R. (1993). A conceptual model for work-related neck and upper-limb musculoskeletal disorders. Scandinavian Journal of Work, Environment & Health, 73–84.

Armstrong, T. J., Foulke, J. A., Joseph, B. S., & Goldstein, S. A. (1982). Investigation of cumulative trauma disorders in a poultry processing plant. The American Industrial Hygiene Ass. J., 43(2), 103–116.

Armstrong, J., Radwin, R. G., Hansen, D. J., & Kennedy, K. W. (1986). Repetitive trauma disorders: job evaluation and design. Human Fact., 28(3), 325–336.

Atterbury, R., Limke, J., Lemasters, G., Li, Y., Forrester, C., Stinson, R., & Applegate, H. (1996). Nested case-control study of hand and wrist work-related musculoskeletal disorders in carpenters. Am. J. of Ind. Med. 30(6), 695–701. Doi:10.1002/(SICI)1097-0274 (199612)30:6<695::AID-AJIM5>3.0.CO;2-Q.

Barr, A., Barbe, M., & Clark, B. (2004). Work-related musculoskeletal disorders of the hand and wrist: Epidemiology, pathophysiology, and sensorimotor changes. J. of Orth. & Sp. Phl Ther., 34(10), 610–627.

BLS. (2016). Nonfatal Occupational Injuries and Illnesses Requiring Days Away From Work, 2015. Econ Releases. From https://www.bls.gov/news.release/osh2.t04.htm.

Buckle, P. (1997). Fortnightly review: work factors and upper limb disorders. Bmj, 315(7119), 1360–1363.

Carey, E., & Gallwey, T. (2005). Wrist discomfort levels for combined movements at constant force and repetition rate. Erg, 48(2), 171–186. doi:10.1080/001401304 10001714760.

Costa, J. Torres da, Baptista, J. Santos & Vaz, M. (2015). Incidence and prevalence of upper-limb work related musculoskeletal disorders: A systematic review. 51(4), 635. doi:10.3233/WOR-152032.

d'Errico, A., Caputo, P., Falcone, U., Fubini, L., Gilardi, L., Mamo, C., Coffano, E. (2010). Risk factors for upper extremity musculoskeletal symptoms among call center employees. J Occ Health, 52(2), 115–124.

Denbeigh, K., Slot, T. R., & Dumas, G. A. (2013). Wrist postures and forces in tree planters during three tree unloading conditions. Erg., 56(10), 1599–1607. doi:10.1080/00140139.2013.824615.

Dillon, C., Petersen, M., & Tanaka, S. (2002). Self-reported hand and wrist arthritis and occupation: Data from the US National Health Interview Survey-Occupational Health Supplement. Am. J. of Industrial Medicine, 42(4), 318–327. doi:10.1002/ajim.10117.

Fagarasanu, M., & Kumar, S. (2003). Shoulder Muscu loskeletal Disorders in Industrial and Office Work. 7(1), 1.

Fredriksson, K., Alfredsson, L., Koster, M., Thorbjornsson, C. B., Toomingas, A., Torgen, M., & Kilbom, A. (1999). Risk factors for neck and upper limb disorders: results from 24 years of follow up. Occup Environ Med, 56(1), 59–66.

Gell, N., Werner, R., Franzblau, A., Ulin, S. & Armstrong, T. (2005). A longitudinal study of industrial and clerical workers: incidence of carpal tunnel syndrome and assessment of risk factors. J Occup Rehabil, 15(1), 47–55.

Gerr, F., Letz, R., & Landrigan, P. J. (1991). Upper-extremity musculoskeletal disorders of occupational origin. Ann. review of public health, 12(1), 543–566.

Gupta, H., & Robinson, P. (2015). Normal shoulder ultrasound: Anatomy and technique. Sem in Musc Rad, 19(3), 203–211. doi:10.1055/s-0035-1549315.

Hembecker, P., Konrath, A., & E. A, D. (2017). Investigation of musculoskeletal symptoms in a manufacturing company in Brazil: a cross-sectional study. Braz J Phys Ther, 21(3), 175–183. doi:10.1016/j.bjpt.2017.03.014.

Kaka, B., Idowu, O., Fawole, H., Adeniyi, A., Ogwumike, O., & Toryila, M. (2016). An Analysis of Work-Related Musculoskeletal Disorders Among Butchers in Kano Metropolis, Nigeria. Saf Health Work, 7(3), 218–224. doi:10.1016/j.shaw.2016.01.001.

Kebede Deyyas, W., & Tafese, A. (2014). Environmental and organizational factors associated with elbow/forearm and hand/wrist disorder among sewing machine operators of garment industry in Ethiopia. J Env. Publ Health, 732731. doi:10.1155/2014/732731.

Kim, J., Suh, B. S., Kim, S. G., Kim, W. S., Shon, Y. I., & Son, H. S. (2015). Risk factors of work-related upper extremity musculoskeletal disorders in male cameramen. Ann Occup Environ Med, 27(1), 5. doi:10.1186/s40557-014-0052-x.

Kivekas, J., Riihimaki, H., Husman, K., Hanninen, K., Harkonen, H., Kuusela, T., ... Zitting, A. J. (1994). Seven-year follow-up of white-finger symptoms and radiographic wrist findings in lumberjacks and referents. Scand J Work Env. Health, 20(2), 101–106.

Klussmann, A., Gebhardt, H., Liebers, F., & Rieger, M. A. (2008). Musculoskeletal symptoms of the upper extremities and the neck: a cross-sectional study on prevalence and symptom-predicting factors at visual display terminal (VDT) workstations. BMC Musc. Disord, 9, 96. doi:10.1186/1471-2474-9-96.

Kong, Y., Jang, H., & Freivalds, A. (2006). Wrist and tendon dynamics as contributory risk factors in work-related musculoskeletal disorders. Human Fact. and Erg in Manuf, 16(1), 83–105. doi:10.1002/hfm.20043.

Kuorinka, I., Jonsson, B., Kilbom, A., Vinterberg, H., Biering-Sørensen, F., Andersson, G., & Jørgensen9, K. (1987). Standardised Nordic questionnaires for the analysis of musculoskeletal symptoms.18(3),233–237.

Lento, P. H., & Primack, S. (2008). Advances and utility of diagnostic ultrasound in musculoskeletal medicine. Current Rev. in Musculosk. Medicine, 1(1), 24–31. doi:10.1007/s12178-007-9002-3.

Living, E. F. f. t. I. o., & Conditions, W. (2014). Changes over time–First findings from the fifth European Working Conditions Survey: Publications Office of the European Union.

Melchior, M., Roquelaure, Y., Evanoff, B., Chastang, J. F., Ha, C., Imbernon, E., Leclerc, A. (2006). Why are manual workers at high risk of upper limb disorders? The role of physical work factors in a random sample of workers in France (the Pays de la Loire study). Occ Env Med, 63(11), 754–761. doi:10.1136/oem.2005.025122.

Putz-Anderson, V., Bernard, B., Burt, S., Cole, L., Fairfield-Estill, C., Fine, L., Hurrell Jr, J. (1997). Musculoskeletal disorders and workplace factors. NIOSH, 104.

Ricco, M., & Signorelli, C. (2017). Personal and occupational risk factors for carpal tunnel syndrome in meat processing industry workers in Northern Italy. Med Pr, 68(2), 199–209. doi:10.13075/mp.5893.00605.

Rocha, L. E., Glina, D. M., Marinho Mde, F., & Nakasato, D. (2005). Risk factors for musculoskeletal symptoms among call center operators of a bank in Sao Paulo, Brazil. Ind Health, 43(4), 637–646.

Serranheira, F., Pereira, M., & Santos, C. (2003). Auto-referência de sintomas de lesões músculo-esqueléticas ligadas ao trabalho LMELT numa grande empresa em Portugal. Rev port de saúde pública, 21(2), 37–47.

Silverstein, B. A., Fine, L. J., & Armstrong, T. J. (1986). Hand wrist cumulative trauma disorders in industry. Occ. and Env. Medicine, 43(11), 779–784.

Stocks, S., McNamee, R., van der Molen, H., Paris, C., Urban, P., Campo, G., Agius, R. (2015). Trends in incidence of occupational asthma, contact dermatitis, noise-induced hearing loss, carpal tunnel syndrome and upper limb musculoskeletal disorders in European countries from 2000 to 2012. Occ. and Env. Med., 72(4), 294–303. doi:10.1136/oemed-2014-102534.

Sutton, D., Gross, D. P., Cote, P., Randhawa, K., Yu, H., Wong, J. J., Taylor-Vaisey, A. (2016). Multimodal care for the management of musculoskeletal disorders of the elbow, forearm, wrist and hand: a systematic review by the Ontario Protocol for Traffic Injury Management (OPTIMa) Collaboration. Chiropr Man Therap, 24, 8. doi:10.1186/s12998-016-0089-8.

Thomsen, J., Mikkelsen, S., Andersen, J., Fallentin, N., Loft, I., Frost, P., Overgaard, E. (2007). Risk factors for hand-wrist disorders in repetitive work. Occ Env Med, 64(8), 527–533. doi:10.1136/oem.2005.021170.

Werner, R., Franzblau, A., Albers, J., & Armstrong, T. (1997). Influence of body mass index and work activity on the prevalence of median mononeuropathy at the wrist. Occ Env Med, 54(4), 268–271.

Work ability of teachers in Brazilian private and public universities

A.J.A. Silva & T.G. Lima
Federal University of Rio de Janeiro, Macaé, Rio de Janeiro, Brazil

ABSTRACT: Over the years, the number of teacher's removals from their activities for health problems have grown in Brazil. In a survey conducted in 2010 with 2685 teachers from state of São Paulo, 27% reported that they had to stay away from work for health reasons and 18% claimed to have depression. This article uses the Work Ability Index (W.A.I) to compare the work ability between teachers from Brazilian one private and one public universities and to relate this index with specific demographic variables. The study was conducted with 53 Engineering teachers. Mann-Whitney Test and Tukey Test were used as statistical tests to certify the sample studied. Results demonstrated that the private university index is similar the public university and oldest people feel more capable about teaching in public university, for example. Studies about teacher ability of work must increase to improve its health and life quality at work.

1 INTRODUCTION

1.1 The work in higher education sector in Brazil

The education is a right that must be assigned to all citizen. The Higher Education in Brazil has in its history a direct development relation around political periods experienced in the country. The graduation courses were formed in accordance with the necessity and localization, aiming at the formation of strategic professionals according to the moment lived by the country. From 1808, the first courses that were offered were Engineering and Medicine. The Higher Education institutions are categorized in public institution and private institution. To Stallivieri (2007, p. 6), this classification occurs according to "forms of financing with which each of the models seeks to survive in the scenario of higher education." To ensure the rules, rights and duties of higher education institutions in Brazil, in 1996 it was formalized the text of the Law of Guidelines and Bases of National Education (Law number 9394/1996).

1.1.1 The public university

According to Chaui (2003, p. 5), the Public University is defined as "a social institution founded on the public recognition of its legitimacy and its attributions, in a principle of differentiation, which gives it autonomy before other social institutions, a structured by regulations, rules, norms and values of recognition and legitimacy internal to it."

In Brazil, the first university to emerge was the University of Rio de Janeiro in 1920, according to Stallivieri (2007), currently known as the Federal University of Rio de Janeiro (UFRJ). From its foundation, another 20 federal universities were initiated within the period of 1930 to 1964, marked historically by the Industrial Revolution and Military Dictatorship, respectively.

In the period from 2003 to 2014, the public institutions suffered restructuring in their processes. It resulted in a significant increase in access to higher education "raising the number of municipalities attended by federal universities of 114 to 289 municipalities, which represented a growth of 153%" (Brazil, 2014). Also, there was an increase in students enrolled in public universities is also expressive: while at 2001 the rate was 16.1% of the citizens between 18 and 24 years, at 2015 this rate increased to 34.6%.

1.1.1.1 The public server: Teacher

The teacher of the public university exercises its position in public educational units and is governed by the laws and norms that relate to their duties. Currently, the teacher of the public university needs, in addition to the execution of its formal activities, "be a specialist technician, competent researcher and scientist, excellent teacher and administrator for managing projects and guiding groups" (Pessoa, 2014, p. 4).

To work in a public institution, the teacher needs to pass through qualification and can perform its duties in two types of regime: 40 hours weekly with exclusive dedication, 40 hours weekly without exclusive dedication or 20 hours weekly.

1.1.2 The private university

Private institutions are "created by accreditation with the Ministry of Education and are maintained and administered by a physical or legal person in private law, may have or not for profit" (Brazil, 2007).

Access to private institutions can happen through payment of tuition by students or by granting scholarships.

In Brazil, as a complement to admission, the federal government has funding programs and scholarships that offer the student the opportunity to study in private universities.

1.1.2.1 *The employee of private network: Teacher*
The teacher of private university has no specific features to be able to teach in this type of institution. Universities establish their own internal requirements for hiring.

They are governed by a Collective Convention. This convention establishes the rights, duties and rules of the teaching profession.

According to a survey conducted in 2015 by the National Institute of Educational Studies and Research Anísio Teixeira, the number of teachers who work in private institutions is around 70 thousand professionals.

Bispo et al (2014, p. 2), affirm "most of the university teachers do not assume their teaching identity and face it as a form of salary complementation, as the title of teacher, alone, suggests a lesser identity."

Camargo (2012) ensures that exists a relation between the function of the teacher in the private institution with a productive process, where it defines teaching as a production process that aims to generate skilled labor force, being represented by students in the process of learning.

1.2 *The health of the worker*

Occupational Health characterizes the actual state of the conditions that the worker must perform his duties in his employment.

In a survey conducted by Union of Professors of the Official Teaching of the State of São Paulo in 2010 with 2685 teachers from state of São Paulo, 27% of respondents reported that they had to stay away from their activities for health reasons, 29% complained of muscle aches and 18% claimed to have depression (Oliveira, 2010).

Servilha & Correia (2014), carried out a study with university professors in Campinas and with a sample of 112 teachers, 81.2% claimed to possess vocal fatigue, 79.4% had an effort to speak and 60.7% stated that they had hoarseness due to the teaching practice.

Therefore, "establishing relations between the multiple stressors agents and the perception of stress in university teachers can help the understanding on which aspects should be considered in control of these situations" (Camargo et al, 2013, p. 590).

1.3 *Capacity for work*

Capacity for work represents the capacity that the worker possesses to accomplish his labor activities, considering his physical and mental conditions. This characteristic is influenced by aspects such as age, health status and working conditions.

For Martinez (2006), the determinant factors of work capacity are: demographic aspects, lifestyles, health, education and competence and work.

For Martinez et al (2010), obtaining knowledge about the factors that contribute directly to the determination of the capacity for the professional's work contributes in an effective way to maintain this capacity.

"The cost of actions to promote the ability to work to prevent diseases and accidents is lower when compared to the cost of treatments." [Bergström (1998) *in* Martinez (2010, p. 1556)].

1.4.1 *The Work Ability Index (W.A.I)*
The Work Ability Index (W.A.I) is a predictive value instrument that enables the evaluation and prior identification of changes and obtaining information that assists in the direction of preventative measures related to the activities executed [Tuomi et al (2005) *in* Menegon (2011)].

Established through a questionnaire applied to employees, the index is determined and the level of capacity for the work of these employees is known.

Table 1. Scores by dimension of the Work Ability Index questionnaire.

Item	Pointing
1. Capacity for current work	0–10
2. Capacity for work in relation to labor requirements	0–10
3. Current number of diseases diagnosed by physician	1–7
4. Estimated loss of capacity for work due to diseases	1–6
5. Absences to work on grounds of illness in the past 12 months	1–5
6. Prognosis on the capacity for work for the next 2 years	1, 4, 7
7. Mental Resources	1–4

Table 2. Work Ability Index.

Points	Work Ability Index
7–27	Low
28–36	Moderate
37–43	Good
44–49	Excellent

According to Meira (2004), the W.A.I questionnaire is established through the responses of workers in questions related to the demands of work, health status and physical, mental and social capacities, considering the worker's self-assessment of his or her condition and previous documentation of diseases diagnosed by doctors or licenses obtained because of illness and is composed by seven dimensions.

The weighted sum of the points assigned to each of these dimensions defines the total index score. The items have minimum and maximum scores and the equivalence of their values are weighted according to the specific characteristics of the activity performed at work.

After calculating the total score, the index is divided into four categories:

Thus, in this tool, "the worker's own concept of his ability to work is as important as the evaluation of specialists" [Tuomi et al (2010) *in* Moreira (2013, p. 23)].

2 METHODOLOGY

The study presented in this article was carried out in two universities located in two municipalities of the state of Rio de Janeiro: one public, established in the municipality of Macaé and one private, in the municipality of Rio das Ostras.

The research was constituted of 53 Engineering teachers – 42 from public university (population, 55) and 11 from private university (population, 27) – through application of the questionnaire of the Work Ability Index. The rate of response was of 64,6%.

The Tukey Test and Mann-Whitney Test were used to confirm the obtained rate may be considered.

The questionnaire was available online and it was applied for data collection between October 23rd and November 29th, 2017.

3 RESULTS

The Mann-Whitney Test and Tukey Test were applied at Minitab software to data analysis. With the application of Mann-Whitney Test the medians of the private and public universities are 43.500 and 41.500, respectively. The test was performed with a confidence level of 95% and presented statistical W = 837.5. The null hypothesis was similar medians and non-null hypothesis was different medians. As the result showed significance of the test with p-value equal to 0.2753, the null hypothesis is not rejected. This means that the test presents fundamentals that the samples studied are of populations that have similar medians. And the Tukey Test shows the averages of the Work Ability of the private and public universities are respectively 41.9833 and 40.6739, obtaining as a difference of 1.31. This difference is considered small, then the test shows that both universities have indications of owning similar averages. The null hypothesis was similar averages and non-null hypothesis was different averages. The test also presented p-value equal to 0.248, which means that the null hypothesis is not rejected. In this way, the p-value contributes as another evidence of the similarity of the universities averages. The test was performed with a confidence level of 95%. As medians and averages were established with similarities, the sample of 53 teachers was validated with the use of both tests.

According this present study, the comparison of Work Ability Index with demographic variable are:

Analyzes must be applied relating the Work Ability Index to each demographic information extracted from the interviewees of each university model.

According to Work Ability Index result, it is possible to notice interesting information. The first important data is related to Work Ability Index of public university and private university: the score of public university is 41,2 and of private university is 41,7.

The index when is related to gender it's observed that female teachers from public university have Work Ability Index higher than female teachers from private university and male teachers index from private university is higher than the index from male teachers from public university. The age relation shows that in both of universities the youngest teachers feel more capable when it compared with the oldest teachers, however only in private university it's possible to observe an index from youngest teachers that may be classified as an "Excellent" Work Ability Index.

Another interesting result is the lower index found in teachers from private university with higher level of education. Who feels more capable are those who have not yet completed their doctorate's degree in private university and in public university. The index from teachers who feel more capable is classified as "Excellent" index.

Table 3. Work Ability Index of each University model.

University model	Work Ability Index	Class*	S.D.**
Public University	41,2	Good	4,3
Private University	41,7	Good	3,7

*Class = Classification.
**SD = Standard Deviation.

Table 4. Work Ability Index by gender.

Gender	Work Ability Index average	
	Public	Private
Female	42,1	41,6
Male	41,0	43,0

Table 5. Work Ability Index of each age group.

Age Group	Work Ability Index average	
	Public	Private
From 20 to 29 years old	41,9	44,0
From 30 to 39 years old	41,6	41,8
Over 40 years old	40,2	38,5

Table 6. Work Ability Index by teacher qualification.

Teacher Qualification	Work Ability Index average	
	Public	Private
Complete Doctorate	40,7	38,0
Incomplete Doctorate	42,8	45,0
Complete Master	39,8	41,8
Incomplete Master	41,5	41,8

Table 7. Work Ability Index by function time.

Period	Work Ability Index average	
	Public	Private
Less 1 year	45,5	–
From 1 to 5 years old	41,4	41,3
From 5 to 10 years old	42,2	42,0
Over 10 years old	39,2	45,0

Table 8. Work Ability Index by salary range.

Salary range	Work Ability Index average	
	Public	Private
R$ 1.200,00 to R$ 2.099,99*	–	43,5
R$ 2.100,00 to R$ 3.299,99	39,3	40,3
R$ 3.300,00 to R$ 4.799,99	35,0	41,0
R$ 4.800,00 to R$ 6.299,99	44,1	38,0
R$ 6.300,00 to R$ 7.799,99	44,3	–
R$ 7.800,00 or more	40,0	45,0

*R$ = $ 0,30712 in 10/31/2017.

The index related to the function time presents that youngest as teachers in public university and those who have more than 10 years teaching in private university have an index that must be classified as "Excellent" Work Ability Index.

And the index when is related to the salary range shows as result that the highest index in public university is in a salary range of R$ 4.800,00 to R$ 7.799,99 and not in the largest possible range, that is over R$ 7.800,00. But in private university it is possible to observe that salary has a direct impact in capacity of work: the high average is in a range of R$ 7.800,00 or more. However, a big index can be observed on teachers who receive the lower salaries. Possibly, it is necessary to enlarge the analysis of this criterion, that is, to correlate the higher rates of salary with other variables, such as: teaching time in the current university, type of contract, level training or even total teaching time.

4 CONCLUSIONS

The research allowed the observation of the factors that have a significant impact or not on the capacity for the work of the professors of the public and private universities of two municipalities of the state of Rio de Janeiro.

The instrument for evaluating the work capacity has allowed to analyze comparatively the work ability of teachers from two universities, being one a public university and another a private university, both from higher education. The data showed that both exhibit "good" rating of the ability to work.

It was noted that, by correlating the indexes with the demographic variables, in the private university, professionals who have incomplete doctorate, which have more than 10 years of teaching, and that earn more than R$ 7.800,00 have an excellent capacity index for the job. In the private institution, it was not observed in the average teachers with moderate indices of work capacity.

In relation to the public university, by correlating the indexes with the demographic variables, it was realized that teachers with less than one year of teaching and with a salary range from R$ 4.800,00 to R$ 7.799,99 have an excellent average work ability level. In the public institution, it was noted in the average that teachers with moderate indexes of work capacity, are those who receive between R$ 3.300,00 to R$ 4.799,99.

It is noted that the factors contributing to an "excellent" level of capacity for the work are different between public and private institutions, and the salary range can contribute to reducing the work ability index at the public university.

The studies in this area bring important results for a continuous monitoring of the capacity of the teacher as they are exposed to several stressors, using tools to ascertain the levels of physical and emotional health are relevant so that these professionals, whose mission is to educate people, always feel good when performing their professional functions.

REFERENCES

Bergström M.; Kaleva, S.; Koskinen, K. SMEs: towards better work ability. Työterveiset [journal on line]. 1998 [cited 2007 Jan 10];1(Spec N):4–7.

Bispo, F. C. S.; Junior, A. B. S.; Dos Santos Junior, A. B. The Higher Education Teacher: Educator or Service Provider? XI Symposium on Excellence in Management and Technology (SEGeT), 2014.

Brazil. Justice ministry. Secretariat of Economic Law. Booklet of Private Institutions of Higher Education. Brasília, DF, 2007.

Brazil. Law n. 9394, December 20, 1996. Law of the Guidelines and Bases of National Education. Brasília, 1996.

Brazil. Ministry of Education. Secretary of Higher Education. The democratization and expansion of higher education in the country: 2003–2014. Social Balance Sheet 2003–2014. Brasília, 2014.

Camargo, E. M.; Oliveira, M. P.; Rodriguez-Añez, C. R.; Hino, A. A. F.; Reis, R. S. Perceived stress, health-related behaviors and working conditions of university teachers. Bibliographic databases Psicol. Argum, v. 31, N. 75, p. 589–597, 2013.

Camargo, L. F. F. The condition of the private higher education teacher: structural characteristics of the teaching activity and the processes of transformation in the labor relations. 2012. Doctoral Thesis. University of Sao Paulo.

Chaui, M. The public university from a new perspective. Brazilian journal of education, v. 24, p. 5–15, 2003.

Martinez, M. C. Study of the factors associated with the capacity to work in workers of the Electric Sector. 2006. Thesis. Thesis (Doctorate in Public Health) – Faculty of Public Health, University of São Paulo, São Paulo.

Martinez, M. C.; Latorre, M. R. D. O.; Fischer, F. M. Ability to work: literature review. Science & Collective Health, v. 15, n. suppl 1, p. 1553–1561, 2010.

Meira, L. F. Ability to Work, Risk Factors for Cardiovascular Diseases and Working Conditions of Workers of a Metal-Mechanical Industry of Curitiba/PR—Post-Graduation Program in Mechanical Engineering, Technology Sector, Federal University of Paraná, 2004.

Menegon, Fabrício Augusto. Structural assembly activity of aircraft and factors associated with the ability to work and fatigue. 2011. Doctorate Thesis. São Paulo University.

Moreira, Patrícia Santos Vieira et al. Application of the capacity index for the work in the nursing team: descriptive study. 2013.

National Institute of Educational Studies and Research Anísio Teixeira. Statistical Synopsis of higher Education 2015. Brasília: Inep, 2016.

Oliveira, Leandro Romani de. The health of the teachers. Union of the Teachers of the Official Education of the State of São Paulo, 2010.

Pessoa, C. E. Q. Precarious of Teaching Work at the Brazilian Public University. CONEDU Magazine. V. 1, 2014.

Servilha, E. A. M.; Correia, J. M. Correlations between conditions of the environment, work organization, vocal symptoms self-referenced by university teachers and speech-language assessment. Communication Disorders. v. 26, n. 3, 2014.

Stallivieri, L. The Brazilian Higher Education System—Characteristics, Trends and Perspectives. University of Caxias do Sul (UCS), Rio Grande do Sul, 2007.

Tuomi, K.; Ilmarinen, J.; Jahkola, A.; Katajarinne, L.; Tulkki, A. Work Ability Index. São Carlos: EdUFSCar, 2005.

Tuomi, K.; Ilmarinen, J.; Jahkola, A.; Katajarinne, L.; Tulkki, A. Work Ability Index. Translated by Frida Marina Fischer (coord), São Carlos: EdUFSCAR, 2010. 59p.

Maintenance manual for equipment on construction site

Cristina Reis
Universidade de Trás-os-Montes e Alto Douro, Vila Real, Portugal

Carlos Oliveira
Instituto Politécnico de Viana do Castelo, Viana do Castelo, Portugal

M.A. Araújo Mieiro
PERITS Engenharia, Braga, Portugal

Paula Braga
Universidade de Trás-os-Montes e Alto Douro, Vila Real, Portugal

J.F. Silva
Instituto Politécnico de Viana do Castelo, Viana do Castelo, Portugal

José António F.O. Correia
INEGI—FEUP, Portugal

ABSTRACT: In this work the main differences between the machines, tools and equipment of construction site in the civil construction are presented, and to do it so, on this research work is presented a practical example of the types of maintenance required. Given the diversity of machines and equipment, it would not be possible in a job of this type to approach them all, so it was decided to choose one. It is emphasized that it was decided to speak in detail about the cranes, since it is one of the most used equipment in the most diverse works.

1 INTRODUCTION

In this work, the types of revisions to which they are subject will be discussed and who is responsible for their safety and verification.

With this work we hope to be able to explain the differences between these three concepts (tool, machine and equipment), since some doubts arise in the classification of the same in the construction site. Many questions arise, for example, "Is a crane a machine? Or is it an equipment, and why is it not considered a tool?" It is these types of questions that are intended to be clarified in this paper.

This issue raises many questions as there is no concrete definition for the objects that are considered machines, equipment and tools. This is a problem that does not have a clear solution, since there is nothing in concrete that says, for example, that a backhoe is a machine and not a tool or it is a machine and not an equipment, and this can generate some confusion when you need to name this type of object. This fault can be seen in the definition of equipment, that is, for example, when it is said that an equipment "is a tool". It is for these reasons that there should be a universal classification on which machines, equipment or tools are classified.

In this work, it is intended to make known the TSCMI, as an entity that certifies machines, in this case of civil construction, in Portugal.

2 CRANES

2.1 *What is a crane?*

The crane is an equipment whose function is to lift and move loads, by means of a hook suspended by a cable and its transport several in a radius of different meters, at all levels and in all the directions. The crane consists of a metal tower, horizontal component and motors (which can be lifting, rotating, distributing or translating).

2.2 *Operation principal of a crane*

To work properly, the worker need to have some knowledge regarding balance rules. If the worker have a 10 tons load on 2 meters on the left, he must also have another 10 tons load to 2 meters on the right, according to Figure 1.

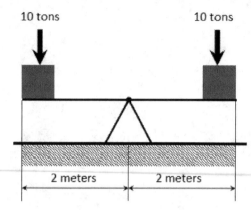

Figure 1. Balanced loads.

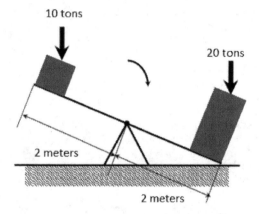

Figure 2. Tipping caused by unbalanced loads.

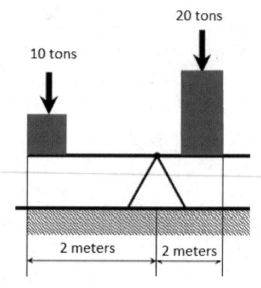

Figure 3. Different balanced loads.

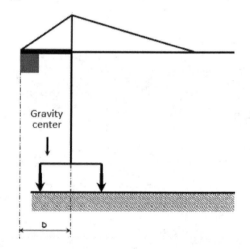

Figure 4. Center of gravity without loads.

If the distances to the joint do not change, and if a load of 20 tons is placed to the right, an imbalance and a respective tipping will occur, as seen in Figure 2.

In this case, in order to obtain equilibrium, the distance from the heavier load side must be shortened (Figure 3), i.e.:

– Right moment: 20 t × 1 m = 20 t.m
– Left moment: 10 t × 2 m = 20 t.m

When there is no load on the crane, i.e., maximum unladen imbalance, the gravity center of the frame must remain within the frame area (as can be seen in Figure 4).

When a load is place in maximum unbalance, the mass center should be as in the previous situation, i.e. in the chassis.

In order to obtain the maximum crane moment, the following expression must be applied:

M = Product of mobile load weight × distance D

The stability of the crane depends in part on this moment, therefore, it must remain constant with respect to one point and, at each position of the car, a load value will not be exceeded, in order not to cause the crane to be tipped. This procedure is called the load curve, as shown in Figure 6 by way of example.

2.3 *Assembly/disassembly*

The most important when working on the works is safety and for this, it is necessary to have continuous prevention by the responsible people, to avoid accidents. This condition is the fundamental element in order to succeed in any productive activity involving a modern company. There are several articles that can be found reporting accidents involving cranes, for example at 5th January

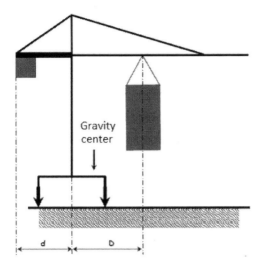

Figure 5. Gravity center with load.

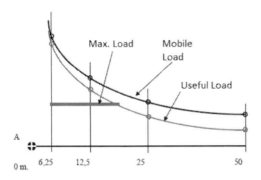

Figure 6. Load curve.

2012, it is reported an accident in Braga (Portugal) according "Jornal de Notícias", in which the victim passed away. After a little research, many reports of these accidents are easily reported and a question remains: "Why are they still going on?".

There are some criteria to be observed before starting to build a crane, for example, given the type of needed crane:

– Action radius (spear length)
– Maximum load/End load
– Service height under the hook

One of the most important aspects is the place of the crane implementation, therefore:

– Avoid embankments, landslides and fragility land areas (such as plumbing);
– Ensure the mandatory separation of voltage cables.

Construction site preparation:

– Preparation of assemblies with the necessary dimensions and strength.

– Guarantee conditions of access to the site in order to stabilize the auto-crane and park of semi-trailers.
– Ensure the electrical power necessary to feed the crane.

Crane characteristics and techniques (to be supplied by the manufacturer):

– Crossings;
– Electric power crane;
– Rated power;
– Starting power.

Preparation of test loads:

– Load at the tip;
– Maximum load;
– High speed.

2.4 Interdicted maneuvers

It is necessary to be careful when working with a crane, and there are aspects to take into account to ensure not only the safety of the people involved in the construction, but it is also important to be careful with it, such as the crane fallen case that happened during the stadium 2014 European Football construction. There are preventive measures such as:

– Cranes must not be operated in strong winds or rains;
– Ensure that no person is in the cargo handling area before any movement and that there are no tools or obstacles to obstruct the rails;
– Do not place hands/feet under load;
– It is forbidden to exceed the maximum load capacity established in the equipment or to handle loads that are poorly packed or poorly balanced;
– Avoid making sudden movements and maneuvering the loads smoothly;
– Do not abandon the load handling controls, leaving the loads suspended.

3 INSPECTION TYPES

Inspections are the main purpose of the job, because it needs a special careful when using a crane and failure to comply with safety rules can lead to serious injury or even death.

Crane maneuvers should only be carried out by responsible persons with adequate technical and practical training and proven experience. In addition to training it is essential that these workers have good physical and health conditions, proven by the work doctor's examinations and do not suffer from vertigo.

It should be noted that there are legal requirements that must obey, such as:

Table 1. Checklist example.

Verification	Verifications types	Y	N
1	Access to the crane tower is inaccessible to outsiders	X	
2	The crane gear systems are in operation		X
3	The metal elements of the crane are in good condition	X	
4	Checking the screws	X	
5	Supports verification	X	

– perform the end-of-assembly check;
– complete the cargo test form;
– comply with the rules of Decree-Law no. 50/2005.

In this decree you can find several articles that are important to take into account, for example, Article 6, entitled "Verification of work equipment". The most important parts of Article 6 are highlighted here, for example, "...the employer must verify them after installation or assembly in a new location...", "The employer must carry out periodic checks...", among others. Table 1 presents a verification check-list:

4 TSCMI

Designated as "Technological Support Center for Metalworking Industry" (TSCMI) is a private non-profit public institution. It was created in 1986 and resulted from the association of interests, industrial companies and their associations with public organisms.

Its areas of intervention are:

– Technical and Technological Support to Companies of the Metal-Mechanical Sector and to the Related or Complementary Sectors;
– Promotion of the Improvement of the Quality of Products and Industrial Processes;
– Technical and Technological Training of Company Personnel;
– Disclosure of Technical and Technological Information;
– Realization of Essays and Analysis of the Characterization of Raw Materials, Products and Equipment;
– Sectoral Diagnoses implementation, aiming at the Identification of Priority Actions;
– Quality Management System implementation;
– Environmental Management System implementation;
– Diagnostics and Environmental Studies;
– Coatings and Surface Treatment;
– Calibrations, Measurements and Standardizations;
– Diagnostics and Quality Audits;
– CE Type Examination—Machinery Directive;
– Integration of Safety in Machinery and Equipment;
– Those corresponding to the notified Sector Standardization Organism;
– Health and Safety Services organization;
– Diagnostics and Industrial Safety Studies.

5 CONCLUSIONS

A tool is a instrument, device, or physical or intellectual mechanism used by workers from the most diverse areas, to perform a certain task. Initially, this term was used to designate objects of iron or other type of material (like plastic, wood) for domestic or industrial purposes. Some types of utensils can be used as weapons (where hammer and knife are included). On the other hand, some weapons such as explosives, can be used as tools.

One of the objectives of this research work was to try to get a clearer idea of the type of inspections that must be done on the equipment and how to operate safely, using equal balanced loads for example, and of the standards that must be complied with. It was also hoped that some knowledge about the principle of crane operation, as well as assembly and disassembly, and the precautions that have to be taken to operate the crane have been conveyed.

REFERENCES

http://www.catim.pt/Catim/princ_apresenta.html.
http://www.portugalventures.pt/en/directories/access-point/catim-centro-de-apoio-tecnologico-a-industria-metalomecanica.html.
http://www.segurancaonline.com/fotos/gca/dl50_2005_1307528343.pdf.
http://pt.wikipedia.org/wiki/Ferramenta.
http://pt.wikipedia.org/wiki/Equipamento.
http://pt.wikipedia.org/wiki/M%C3%A1quina.
http://www.estig.ipbeja.pt/~rasmi/seminarios/1_ciclo/SEGURANCA_EM_GRUAS_TORRE_1_1.pdf.
http://pt.wikipedia.org/wiki/Betoneira.
http://pt.scribd.com/doc/22902030/RAC-EQUIPAMENTOS-MOVEIS.
http://www.jn.pt/paginainicial/pais/concelho.aspx?Distrito=Braga&Concelho=Braga&Option=Interior&content_id=2222512.
http://www.protecao.com.br/noticiasdetalhe/JyjjJ9ja/pagina=1.
http://www.maisfutebol.iol.pt/mundial-2014-acidente-grave-dois-mortos/52962a44e4b00828d4088dae.html.
http://negocios.maiadigital.pt/hst/sector_actividade/armazenagem/riscos_armazenagem/riscos_medidas_prev.

Ergonomic risk management of pruning with chainsaw in the olive sector

M.C. Pardo-Ferreira
Universidad de Málaga, Spain

A. Zambrana-Ruíz
Universidad Carlos III de Madrid, Spain

J.A. Carrillo-Castrillo
Universidad de Sevilla, Spain

J.C. Rubio-Romero
Universidad de Málaga, Spain

ABSTRACT: Manage ergonomic hazards and issues at the work is essential. For this, the first step is to assess the risk to know what is its magnitude and define the preferred lines of action. In the case of the Musculoskeletal disorders, they are one of the most common work-related ailments. In Agriculture, 48.2% of workers adopted painful or fatiguing postures, lift or move heavy loads or apply important forces. Thus, the present study is proposed to study the ergonomic risks of pruning with chainsaw in the olive sector. Thus, the ergonomic evaluation methods OWAS, RULA and REBA were used. The results indicate that pruning activities with chainsaw in the olive sector are risky ergonomically and require specific actions that minimize ergonomic risk.

1 INTRODUCTION

Musculoskeletal Disorders (MSDs) are one of the most common work-related ailments. Throughout Europe they affect millions of workers and cost employers billions of euros. Tackling MSDs helps improve the lives of workers, but it also makes good business sense. According to the latest Working Conditions survey by the National Institute for Safety and Hygiene at Work in Spain (INSHT), musculoskeletal disorders are one of the main complaints of workers. In fact, 77.5% of workers feel some discomfort that they attribute to postures and efforts derived from the work they do. Among the most frequent complaints are those located in the lower back area, neck/neck and upper area of the back. Specifically, in Agriculture, 48.2% of workers adopt painful or fatiguing postures, lift or move heavy loads or apply important forces.

On the other hand, Spain ranks first in the world in surface and production of olive oil. The Spanish production represents approximately 60% of the production of the EU and 45% of the world (Ministerio de Agricultura y Pesca, Alimentación y Medio Ambiente, 2017). Andalusia produces 80% of the production in Spain. Specifically, the province of Jaén produces 50% of the production of Andalusia, only this province concentrates 18% of the world production (La guía del Aceite de Oliva, 2013).

Based on this, the present study was proposed to analyze the task of pruning with chainsaw in the olive sector with the objective of analyzing MSD for the chainsaw operator. For this, different methods of ergonomic evaluation, presented below, were used and the results obtained were compared.

1.1 Olive tree pruning

Pruning is an activity that takes place at the olive tree at the end of the olive campaign. Depending on when the campaign ends, it can be done in the months of February, March and April at the latest.

Its main objective is to prepare the olive for production, that is, replace old branches with the purpose of growing new branches and eliminate branches that interfere with others or make it harder to collect the olive. The new form of harvesting of olives also affected the pruning, since in the past it was trying to get a tall and very voluminous tree, whereas now a lower and more accessible tree is trying to get.

1.2 Rapid Upper Limb Assessment (RULA)

This method was developed by McAtamney & Corlett (1993) in the Institute for Occupational Ergonomics with the objective of to investigate the exposure of individual workers to risk factors associated with work related upper limb disorders. Thus, the method assessments individual postures adopted and uses diagrams of body postures and three scoring tables to provide evaluation of exposure to risk factors such as numbers of movements; static muscle work; force; work postures determined by the equipments and furniture and time worked without a break (McAtamney & Corlett, 1993). This method is very widespread and is widely used today.

1.3 Rapid Entire Body Assessment (REBA)

According to Hignett & McAtamney (2000), a team of ergonomists, physiotherapists, occupational therapists and nurses collected and individually coded over 600 postural examples to produce a new tool incorporating dynamic and static postural loading factors, human load interface (coupling), and a new concept of a gravity-assisted upper limb position, this tool was REBA method.

In the same way as the previous method, the use of the REBA method is widespread. The main difference between the RULA and the REBA is that the REBA includes in the evaluation of the lower extremities. Another novelty regarding the RULA method is the consideration of the existence of sudden changes in posture or unstable postures, and if the posture of the arms is maintained in favor of gravity (Diego-Mas, J.A., 2015a).

1.4 Ovako Working Analysis System (OWAS)

The OWAS method was developed by Karhu et al. (1977) and is a practical method for identifying and evaluating poor working postures. The method consists of two parts. The first is an observational technique for evaluating working postures. The second part of the method is a set of criteria for the redesign of working methods and places. The criteria are based on evaluations made by experienced workers and ergonomics experts (Karhu et al., 1977).

Unlike other methods of postural evaluation such as RULA or REBA, which value individual postures, OWAS is characterized by its ability to assess globally all the postures adopted during the performance of the task. However, OWAS provides less accurate assessments than the previous ones. (Diego-Mas, J.A., 2015b).

2 METHODOLOGY

2.1 Place

Field measurements were developed on March 22 and 23, 2016. The measurements were collected in an olive grove located in Torreperogil, a village in the province of Jaén in Andalucía (Spain). The plantation is a young olive tree about 30 or 40 years old, with bushes of 3 to 4 feet. Due to a propitious age, an unbeatable situation and the fertility of the land, this farm is in full production.

2.2 Participants

To evaluate the work task of pruning with chainsaw, three workers participated as volunteers in the field measurements. Each of these workers had more than 20 years of experience in the execution of this type of work tasks. The characteristics of each of them are included in the Table 1.

2.3 Materials

This study has been developed using a hardware and software solution for capturing objective 3D motion data (iSen 2.20). The system uses inertial sensors that provide real-time orientations, that is, angles as well as angular velocities and accelerations. Thus, four sensors (STT-IBS), have been used to perform the fieldwork. They communicate wirelessly using Bluetooth up to 50m for outdoor tracking. Thus, they easily transferred data to the laptop used (Thosiba Satélite Pro C50) in real time. Also, a table was used, as shown in Figure 2, in order to have an adequate surface to locate the equipment.

In addition, all monitored work tasks were recorded with a mobile phone (Nokia Lumia) in order to be analyzed later. To prune the olive trees the chainsaw (Echo CS-4200) was used as work equipment as well as all the necessary PPE.

Once the data has been collected, a postural analysis has been developed using the data collected applied different ergonomic assessment methods such as OWAS, RULA and REBA, for each selected posture. For this purpose an online software has been used (Diego-Mas, 2006).

Table 1. Characteristics of volunteer workers.

Participant	Sex	Age Years	Height cm	Weight kg	Hand preference handed
Worker 1	Male	51	171	90	right
Worker 2	Male	73	162	68	right
Worker 3	Male	49	175	93	right

2.4 Methods

In order to achieve the objective proposed in this study four stages have been established and are described below.

Stage 1: Analysis of the activity

The different steps and tasks developed to carry out the pruning activity have been studied accurately. The pruning activity can be divided into two tasks, the one of cutting with a chainsaw and the one of removing the cut branches outside the olive. In this case, only the first task, in which the chainsaw is used, will be studied. This is due to the operators that cut with the chainsaw usually are different to the operators that perform the removal of branches. Thus, these operators are less qualified and less experienced than those who handle the chainsaw. In this way, the different steps carried out by the operator who uses the chainsaw are summarized in Figure 1.

Stage 2: Field measurement

Once the task to be studied has been defined, the data collection or measurements are started. Firstly, the table is assembled with the equipment and the sensors are paired with the laptop. Secondly, the sensors are placed to the worker according to the selected measurement protocol. In this study, three different protocols were used to collect all the necessary data in successive measurements:

1. Shoulder, Arm, Full Right
2. Shoulder, Arm, Full Left
3. Head, neck

The sensors placed according to protocol 2 are observed in the left part of Figure 2. The position of the sensors in protocol 1 would be the same but in the right arm. In the right part of Figure 2 the location of the sensors in protocol 3 is observed. It is very important that the location of the sensors be performed properly, otherwise the data collected will not be valid.

Thirdly, the data record begins and the worker must mark the initial position according to the chosen protocol. Finally, the pruning of the olive tree with a chainsaw starts.

Additionally, the measurement is recorded in video to facilitate the analysis of the data and to select the most critical positions of the activity.

In total, at least two researchers are needed for field measurement, one to record and another to monitor the activity on the laptop.

Stage 3: Selection of posture to evaluate

Two criteria have been used to decide which postures would be the ones selected to be evaluated with RULA and REBA. The first criterion was focused on evaluating postures applying the OWAS method. All the postures that were identified as non-hazardous were reviewed by the researchers to determine if they could pose a risk to the worker. Those positions in which there were doubts about whether they were hazardous or not, despite being classified as non-dangerous, were selected. In this way, it was intended to clarify its level of real risk through other evaluations applying RULA and REBA methods.

Thus, the OWAS method was applied to 17 postures for worker 1, 12 for worker 2 and 16 for worker 3. Finally, three postures were selected for each worker.

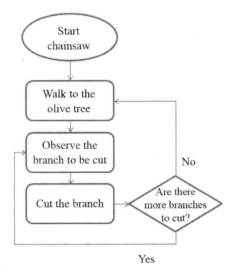

Figure 1. Pruning process performed by the chainsaw operator.

Figure 2. Images of the field work done.

Table 2. Results of the OWAS method.

Level of risk	Number of postures by level of risk		
	Worker 1	Worker 2	Worker 3
1	11	8	9
2	2	1	6
3	4	3	1
4	0	0	0
Total	17	12	16

Table 3. Results of the OWAS method.

	Posture 1	Posture 2	Posture 3
Worker 1			
RULA	Score: 4	Score: 4	Score: 4
REBA	Score: 2	Score: 2	Score: 2
Worker 2			
RULA	Score: 4	Score: 4	Score: 4
REBA	Score: 2	Score: 2	Score right side: 3 Score left side: 2
Worker 3			
RULA	Score: 4	Score: 4	Score: 4
REBA	Score: 2	Score: 2	Score: 3

Stage 4: Application of ergonomic methods
There are many methods that can be used to perform a postural evaluation of the different activities. According to the Manual of Ergonomics applied to the Prevention of Occupational Risks by Llorca et al. (2015), when movements of extension, abduction, flexion less than 60° or abduction of the arm and movements of extension or flexion less than 40° of the neck are adopted, an evaluation should be made with level II methods, that is, RULA, OWAS and REBA. In addition, other authors such as Leszckynski & Pyzia (2012) have used the RULA and REBA methods to evaluate the work of chainsaw operators for forestry. Based on this, the different selected positions have been evaluated applying these methods.

3 RESULTS

The results obtained for the activity of pruning with chainsaw in the OWAS method are included in the Table 2.

Thus, 64.70% of the postures of the worker 1, 66.66% of the postures of the worker 2 and 56.25% of the postures of the worker 3 present a level of risk 1, that is, they are not dangerous. The 23.52% of the postures of the worker 1 and worker 2 have a level of risk 3. However, only the 6.25% of the postures of the worker 3 have this level of risk. According to the results obtained in OWAS method, 60% of the postures evaluated have no risk of causing ergonomic damage to the worker.

The results obtained for RULA method are summarized in Table 3. The scores are the same both on the right side of the body and on the left side. It is highlighted that all evaluated positions obtained a level of risk 4 for the RULA method. This is the maximum score, therefore, they are postures that cause ergonomic damages and that require immediate action. This reveals the risk of the pruning task from the ergonomic point of view.

The results for the REBA method are also included in Table 3. It was obtained that 83.33% of the evaluated postures have a level of risk 2 according to the REBA method, that is, the action is necessary. In addition, the rest of the positions have a level of risk 3. This is at a higher risk level than the previous one, which indicates that action is necessary as soon as possible. For the three positions selected for each worker in 88.88% of the cases the score on the left side is equal to the score on the right side. Only in position 3 of worker 2 these scores are different.

4 DISCUSSION

The postures that OWAS method evaluates as non-hazardous when analyzed by applying RULA and REBA methods have been classified as dangerous and the need to carry out actions in this regard has been established. Therefore, OWAS is a much less accurate method than RULA and REBA methods.

It is important to notice that in 77.77% of the cases evaluated in the pruning task it is observed that for a value of the RULA method of risk level 4 for REBA is a level of risk 2. Hence the results seem to indicate that RULA is more strict than REBA, but both are coincident in that it is necessary to perform a performance or redesign of the task. It is necessary to consider that RULA method presents 4 levels of risk they range from 1 to 4. However, REBA method presents 5 levels of risk ranging from 0 to 4. Another factor that may have influenced this difference in risk level is the fact that the global scores of groups A and B increase very differently in RULA and REBA. In RULA method the load ranges are stricter than in REBA method and 2 points are added to both body parts. However, in REBA method the chainsaw load is less than 5 kg and there is a good grip so the scores of groups A and B respectively are not increased at any point. This difference in the scales must be taken into account when interpreting the results properly.

5 CONCLUSIONS

It is very important to analyze the consequences that could cause the MSDs in the life of a worker, because when this type of injury appears it is very likely that it could be repeated and many of these workers suffer these injuries throughout their working life. To avoid that, it is essential to manage ergonomic hazards and issues at the work. The first step should be to assess the ergonomic risks to which the worker is exposed. Next, the actions that must be carried out to manage this risk by eliminating it or reducing it to the maximum have to be defined. According to the results obtained, it can be concluded that pruning activities with chainsaw in the olive sector are risky ergonomically and require specific actions that minimize ergonomic risk. Thus, among the preventive measures that should be carried out in order to prevent MSDs could be highlighted:

– Perform stretches before the task.
– Train workers in pruning techniques to use the appropriate postures.

- Encourage short breaks to prevent fatigue.
- Due to the eventuality of the pruning, at least an annual informative talk is recommended in order to remember the preventive measures to the workers when they start the pruning tasks that year.
- Use properly PPEs such as chainsaw protective trousers or hearing protection.
- Prevent workers from performing their tasks alone, it is better that there are always two workers working in parallel in different lines of olive trees, never in the same tree. This facilitate prompt assistance in case of emergency.

REFERENCES

Diego-Mas, J. (2006). Ergonautas—la ergonomía online. Available at: https://www.ergonautas.upv.es/ [Accessed 4 Nov. 2017].

Diego-Mas, J.A, (2015a). Evaluación postural mediante el método REBA. Ergonautas, Universidad Politécnica de Valencia. Available at: http://www.ergonautas.upv.es/metodos/reba/reba-ayuda.php [Accessed 4 Nov. 2017].

Diego-Mas, J.A., (2015b). Evaluación postural mediante el método OWAS. Ergonautas, Universidad Politécnica de Valencia. Available at: http://www.ergonautas.upv.es/metodos/owas/owas-ayuda.php [Accessed 4 Nov. 2017].

European Agency for Safety and Health at Work (2017). Musculoskeletal disorders—Safety and health at work—EU-OSHA. Available at: https://osha.europa.eu/en/themes/musculoskeletal-disorders [Accessed 4 Nov. 2017].

Hignett, S., & McAtamney, L. (2000). Rapid entire body assessment (REBA). Applied ergonomics, 31(2), 201–205.

Karhu, O., Kansi, P., & Kuorinka, I. (1977). Correcting working postures in industry: a practical method for analysis. Applied ergonomics, 8(4), 199–201.

La guía del Aceite de Oliva (2013). Zonas Productoras. Available at: http://www.guiadeaceite.com/zonasproductoras/ [Accessed 4 Nov. 2017].

Leszckynski, K. & Pyzia, P., (2012). Analysis of musculoskeletal disorders during chainsaw work using REBA and RULA methods. Nauka Przyroda Technologie, 6(3), 3–10.

Llorca Rubio, J.L., Llorca Pellicer, L, y Llorca Pellicer, M., (2015). Manual de ergonomía aplicada a la prevención de riesgos laborales. Madrid: Pirámide.

McAtamney, L., & Corlett, E.N. (1993). RULA: a survey method for the investigation of workrelated upper limb disorders. Applied ergonomics, 24(2), 91–99.

Ministerio de Agricultura y Pesca, Alimentación y Medio Ambiente (2017). Aceite de oliva-Aceite de oliva y aceituna de mesa—Producciones agrícolas—Agricultura Available at: http://www.mapama.gob.es/es/agricultura/temas/producciones-agricolas/aceite-oliva-y-aceituna-mesa/aceite.aspx [Accessed 4 Nov. 2017].

What kind of lower limb musculoskeletal disorders can be associated with bus driving?

M. Cvetković
Faculty of Occupational Safety, University of Niš, Niš, Serbia

M.L. Dinis
CERENA—Polo FEUP—Centre for Natural Resources and the Environment, Faculty of Engineering, University of Porto, Porto, Portugal

E. Stojiljković
Faculty of Occupational Safety, University of Niš, Niš, Serbia

A.M. Fiuza
CERENA—Polo FEUP—Centre for Natural Resources and the Environment, Faculty of Engineering, University of Porto, Porto, Portugal

ABSTRACT: Prolonged sitting position (more than 2 hours) with lower limbs static postures may lead to a great variety of consequences, such as venous disorders, cardiovascular diseases, pain after sitting, muscles cramps among others. To accomplish the aim of this study, PRISMA Statement Methodology was used. Thereby 13 Scientific Journals and 11 Index—Database were scanned, through where 5361 works were found, of which 17 were included in this paper. Questionnaires were the main source of data collected on lower limbs musculoskeletal disorders occurrence in city bus drivers. Concerning the aforementioned disorders, neither the laboratory research nor the practical one has been studied well enough. Potential symptoms of musculoskeletal disorders of bus drivers' lower limbs could be: venous disorders, joint edema, lower limbs muscle pain, changes in knee bone density, knee meniscus and cartilage damage as well as and osteoarthritis, among others.

1 INTRODUCTION

While working, workers'/operators' body can stay in awkward postures, which can induce a great variety of disorders, known as work-related musculoskeletal disorders (Grace P. Y. Szeto & Lam, 2007; Okunribido & Lewis, 2010). Prolonged standing/sitting, climbing up a ladder or steps, carrying a heavy non-ergonomic burden (both in the short and in the long distance) or occasional kneeling when performing working tasks can cause a variety of risk factors related to musculoskeletal disorders of workers' lower limbs (Okunribido & Lewis, 2010).

In case of professional drivers, after a long awkward sitting position and required movements by the working process, the most frequent disorders occur in the form of back, neck and upper extremities pain (Tamrin, Yokoyama, Aziz, & Maeda, 2014; Yasobant, Chandran, & Reddy, 2015). Analysis methods for body posture (one example of such is given in references (Tamrin et al., 2014)) may be beneficial to the designing of a corresponding seat, which has to meet basic standards, relating to support of soft tissue, pelvic, back and lower limbs of the driver, and so that it, when sitting, does not cause discomfort.

Whole-body vibration, non-ergonomic work, individual factors (for instance age, worker gender, height and weight, body mass index, health status), as well prolonged sitting and processes required by the workplace may endanger and consequently increase the risk of musculoskeletal disorders (Grace P. Y. Szeto & Lam, 2007; Tamrin et al., 2014; Yasobant et al., 2015). Information on the occurrence of bus drivers' musculoskeletal disorders and corresponding symptoms (Tamrin et al., 2014) can be found, where it was emphasized that 57% of drivers had symptoms of pain in some parts of the body.

Furthermore, in the same study (Tamrin et al., 2014) it can be found that the percentage of these symptoms of bus drivers' musculoskeletal disorders is different in the right lower limb (24%) than in the left lower limb (21%). The aforementioned data may depend on various factors, one of which is the vehicle manoeuvring, i.e. the use of an automatic transmission. Prolonged sitting with

Table 1. Lower limb disorders conditions that have been associated with occupational populations (Okunribido & Lewis, 2010).

Variable	LLD conditions according to the regions of the lower extremity		
	Hip/thigh	Knee/lower leg	Ankle/foot
Overuse Injuries	Osteoarthritis (OA)	Beat knee/Hyperkeratosis	Achilles tendonitis
	Piriformis syndrome	Bursitis	Blisters
	Trochanteritis	Meniscal lesions	Foot corns
	Hamstring strains	Osteoarthritis (OA)	Halux valgus (bunions)
	Sacroiliac pain	Patellofemoral pain syndrome	Hammer toes
		Pre-patellar tendonitis	Pes traverse planus
		Shin splints	Plantar fasciitis
		Infra-patellar tendonitis	Sprained ankle
		Stress fractures	Stress fractures
			Varicose veins
			Venous disorders
Acute Injuries		Meniscal tear	Ankle fractures
			Metatarsal fractures

a static posture of the lower limbs (more than 2 hours (Okunribido & Lewis, 2010)) can cause a variety of consequences such as venous disorders (Łastowiecka-moras, 2017; Souza, Franzblau, & Werner, 2005), cardiovascular diseases (Thosar & Health, 2014), lower limbs edema (Ema et al., 2015; Wrona, Jöckel, Pannier, Bock, & Hoffmann, 2015), the pain after sitting, cramps muscles and the similar (Wrona et al., 2015).

Moreover, the excessive use of the lower limbs can cause damages in some areas of workers' feet/ankle, knee and hip. Professional bus drivers spend a lot of years, driving from 6 to 8 hours per day, in the same posture with, in fact, the same exact movements, both on lower limbs and body; the consequences of their excessive use are expected. A possible condition of the lower limbs after an acute injury and excessive use is shown in the paper (Okunribido & Lewis, 2010), and in Table 1.

Thus, there are potential risks that awkward body posture during a long drive and prolonged sitting may induce musculoskeletal disorders in the region of the hip, knee and ankle/foot. As an example, the pain in the knee of the taxi drivers was mentioned, where, as a testing method, The Nordic musculoskeletal questionnaire was applied and the obtained data was crossed with the time spent driving (Chen et al., 2004).

The main objective of this short systematic review is to present the results of the lower limbs musculoskeletal disorders which can be found in the bus drivers. Methodological evaluation for symptoms and syndromes of the musculoskeletal disorders, as well as the risk factors that cause this problem, are also displayed. On that basis, a question emerges: "Do driving and prolonged sitting induce musculoskeletal disorders on bus driver's lower limbs?". Systematic literature review and subsequent analysis of the works put forward in this paper will further benefit a long-term objective in the selection of methodologies and equipment for the determination of lower limbs disorders in the bus drivers.

2 MATERIALS AND METHODS

For carrying out this study, PRISMA Statement Methodology was used (see Figure 1), which includes the application of search methodology in four steps (analysis of available literature, classification, selection and analysis of selected papers). It is relevant that when applying the PRISMA methodology we use appropriate keywords and selection criteria as well as exclusion criteria.

This research was conducted in 13 Scientific Journal editors: ACM Digital Library, ACS Journals, AIP Journals, ASME Digital Library, CE Database (ASCE), Geological Society of America, Highwire Press, IEEE Xplore, IEEE Xplore, IOP Journals, nature.com, Royal Society of Chemistry, Science-Direct, Scitation, SIAM, and 11 Index—Databases: Academic Search Complete, Current Contents, Inspec, MEDLINE (EBSCO), PubMed, Science & Technology Proceedings, ScienceDirect, SCOPUS, SourceOECD, TRIS Online, Web of Science.

The keywords combination were used: "Musculoskeletal disorders" + "Lower limb"; "Musculoskeletal disorders" + "Lower limb" + "Risk factors"; "Musculoskeletal disorders" + "Lower limb" + "Sit-ting"; "Musculoskeletal disorders" + "Lower limb" + "Occupation"; "Musculoskeletal disorders" + "Lower limb" + "Driver"; "Lower limb" + "Occupation" + "Sitting"; "Lower limb" + "Occupation" + "Driver"; "Lower limb" + "Risk factors" + "Driver"; "Lower limb" + "Disease" + "Sitting"; "Lower limb" + "Driver" + "Sitting".

Firstly, the paper's research was limited to the period from January 2008 to June 2017, where 5361 articles were found, to which duplicates were removed (4496); afterwards, from the article's references, some other older papers were also included. At this point, exclusion criteria had to be defined and all papers had to fulfill all of the following phases: 1) the papers should be in written in English, 2) title and abstract should be in accordance with the main goal of the study and 3) papers to which content could be accessed. After analyzing the results, 72 papers were considered eligible.

These articles were then full-text screened in order to determine which papers had data describing symptoms, syndromes, lower limbs disorders during a prolonged sitting or driving, as well as the risk factors that might be caused or aggravated by the condition of the bus driver's lower limbs, so it came up to a total of 17 studies.

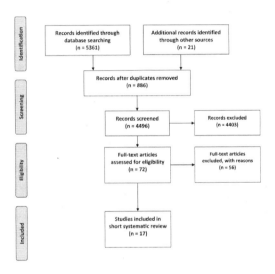

Figure 1. PRISMA flow diagram.

3 RESULTS

After analyzing the papers, 17 of them were included, 4 of which are presented in Table 2 and 8 more in Table 3.

The results of the lower limbs musculoskeletal disorders occurrence in city bus drivers have been obtained through questionnaires. Moreover, neither the laboratory research nor the one carried out in practice, of the aforesaid lower limbs disorders in bus drivers, have been studied enough.

Personal characteristics and gender differences of drivers are one of the key elements of musculoskeletal disorders of lower limbs. It should also be emphasized that the presence of both genders as professional bus drivers was about the same in the first 5 years of work experience, but then the difference is much bigger (Grace P. Y. Szeto & Lam, 2007; Messing, Stock, & France Tissot, 2009; Tamrin et al., 2014; Wrona et al., 2015).

The differences in height, weight and other personal characteristics are expected. Grace, Szeto and Lam (2007) state, in their research, that data after 5 years spent in an awkward posture, i.e. in a prolonged sitting when driving, shows that the pain in the thigh and knee occurs. As a conclusion, they also stated that this is mainly a chronic disease and that more than 50% (both genders) of drivers reported this type of discomfort.

Messing (2009) includes data related to the period of 12 months spent on a variety of jobs, where they are represented in percentage and directly compared with both genders. The presence of pain in the knee is higher in men (9.8%) than in women (7.3%), whereas in women, chronic pain in the hips, feet and ankles is on average greater for 1.5% (Messing, Tissot, & Stock, 2008; Okunribido & Lewis, 2010).

Different types of the seats demand different postures of the bus driver's and (Kyung &

Table 2. Odds Ratios (ORs) and 95% Confidence Intervals (CIs) for Prevalence of lower limb pain.

Author	Country	Hip Risk factors		Knee risk factors		Ankle/foot Risk factors		Lower leg risk factors		Lower limb risk factors		P value
		OR	CI	OR	CI	OR	CI	OR	CI	OR	CI	
Messing (2009)	Canada	N. A.	N. A.	N. A.	N. A.	N. A.	N. A.	1.20*	0.65–2.22*	N. A.	N. A.	N. A.
Wrona (2015)	Germany	N. A.	N. A.	N. A.	N. A.	N. A.	N. A.	N. A.	N. A.	1.8**	(1.6–2.2)**	N. A.
Jiu-Chiaun Chen (2004)	USA	N. A.	N. A.	1.99***	(1.00–3.98)***	N. A.	N. A.	N. A.	N. A.	N. A.	N. A.	N. A.
Szeto (2007)	China	N. A.	N. A.	0.50****	(0.25–0.98)****	N. A.	N. A.	N. A.	N. A.	N. A.	N. A.	0.044

OR—Odds Ratio; CI—Confidence Interval (95%); *Sitting in a constrained posture; **chronic venous disease—Any symptom; ***Total driving per day (6–8 h/day); N.A. – not available.

Table 3. Lower limb prevalence of musculoskeletal disorders in different areas and postures of workers.

Author	Questionnaire	Validity	Population (n)	Sample (n)	Hip/thigh	Knee	lower leg	Ankle/foot	Position	
Negahban et al. (2014)	LLFS		304	304	9.9	68.1**		22.0	St	Si
Messing et al., (2009)	Nordic	*	5434	5434	4.3	9.8	5.3	8.2	St	Si
Tamrin et al., (2014)	Nordic	*	1181	896	19.9	27.5		28.9	Bd	
Reilly et al., (2000)	Questionnaire		82	82		27.8			Bd	
Messing et al., (2009)	Nordic	*	7757	4534	3.3	6.9	2.5	4.2	Si	
Yasobant et al., (2015)	Nordic	*	280	280		6		4	Bd	
Chen et al., (2004)	Nordic	*	1355	1242		17			Td	
Szeto et al., (2007)	Nordic	*	481	404		35***	18	17	Bd	

LLFS–Persian lower extremity functional scale; Nordic–Nordic questionnaire on musculoskeletal symptoms; Questionnaire–non–specified questionnaire; ** – Knee and lower leg; *** – knee and thigh; St–standing; Si–sitting; Dr–driving; Td–Taxi driving; Bd–Bus driving.

Nussbaum, 2008) presented results which shows that left side of the thigh makes bigger pressure on the driver seat. This information can be interesting for further research.

Different positions angles of lower limb joint system, with the additional risk (prolonged sitting, static position of the left lower limb with a slight outer rotation, the whole-body vibration, etc.), are leading to the occurrence of pain in the knee. Reilly, Muir and Doherty (2000) carried out an analysis of pain prevalence in the driver's knee, where 27.8% of 82 patients reported pain in the knee. Pain in the knee depends on the time period spent while driving.

Table 2 presents risk factors associated with significant lower limb pain and presented through Odds ratios and confidence intervals in correlation with prolonged sitting or driving.

The research conducted by Chen (2004) (within the period of one year) on taxi drivers, showed that the occurrence of pain in the knee was 11%, 17%, 19% and 22% subject to the driving time (≤6; 6–8, 8–10; and 10≥ hours).

4 DISCUSSION

An interesting observation can be seen in Łastowiecka (2017) article wherein he introduced a data sheet of venous disorders in which the results of the left lower limb differ from the ones of the right lower limb (the research carried out by using Doppler ultrasound device).

Furthermore, postural edema is also connected with the static and prolonged sitting, which is expected in a city bus, where the driver spends several hours a day driving, and it affects the appearance of swelling and deformation of the joints. Studies researching musculoskeletal injuries on bus drivers report lower limbs disorders, in particular, knee disorders (Chen et al., 2004; Grace P. Y. Szeto & Lam, 2007). However, it should be noted that the relation of pain in the knee to the length of driving is significant. The research by Chen (2004) states, indicates the presence of the knee bone density disorder, as well as cartilage damage of the same after prolonged sitting when driving, is longer than 4 hours/a day. The reason of this disorder can be re-repeating the leg movement when driving and other negative factors as well (straining knee postures with the static position of the left lower limb while driving a bus with automatic gear shift, vibrating floor with the consequence of the whole-body vibrations, inadequate/restrained space and similar).

Besides the previously mentioned disorders of lower limbs both in Table 1 and in the discussion, it is also added a Lower Extremity Functional Scale whose research was conducted by (Negahban et al., 2014). It is very important that the expected range of disorders be expanded so that for future analysis of symptoms and syndromes there will be specific data on musculoskeletal disorders of lower limbs.

At the lower functional extremity (Negahban et al., 2014) lists the types of musculoskeletal disorders of lower limbs such as: degenerative joint disease, muscle strain, bursitis, ligament sprain, patellofemoral pain syndrome, tendonitis, meniscal injury, ligament and meniscal injury, plantar faciatis or heel pain, metatarsalgia, and nonspecific sprain/strain. The mentioned Lower Extremity Functional Scale is a reliable, consistent and important tool for presenting and determining the functionality of the lower extremities, and as such, the presented musculoskeletal disorders of lower limbs may be considered.

Table 3 displays prevalence of musculoskeletal disorders considering different positions for drivers, as well as workers who must endure awkward postures specifically in lower limbs

5 CONCLUSIONS

Research based on the use of the questionnaire presents a difference when making the conclusion about the prevalence of the pain on lower limbs. Table 3 clearly presented the usage of invalidated questionnaires presents bigger percentage in lower limbs prevalence than validate questionnaires. However, a significant percentage refers to the knee pain that occurs upon prolonged exposure to prolonged sitting during driving (≥ 4 hours/day) (Chen et al., 2004; Grace P. Y. Szeto & Lam, 2007; Souza et al., 2005).

Awkward posture, length of driving and sitting induce disorders in the hips, knees, ankles and feet. Potential symptoms of musculoskeletal disorders of bus drivers' lower limbs could be venous disorders, joint edema, lower limbs muscle pain, changes in knee bone density, knee meniscus and cartilage damage and osteoarthritis. In addition to the above, additional potential hazards may include degenerative joint disease, muscle strain, ligament sprain, tendonitis, ligament and meniscal injury, plantar fasciitis or heel pain, metatarsalgia, and nonspecific sprain/strain. The risk factors that may affect or deteriorate the above disorders are occupational (physical) risk factors, personal (and demographic) risk factors and psychosocial risk factors.

ACKNOWLEDGMENTS

This publication has been funded with support from the European Commission under the Erasmus Mundus project Green-Tech-WB: Smart and Green technologies for innovative and sustainable societies in Western Balkans (551984-EM-1-2014-1-ES-ERA MUNDUS-EMA2).

REFERENCES

Chen, J., Dennerlein, J.T., Shih, T., Chen, C., Cheng, Y., & Wushou, P. (2004). Knee Pain and Driving Duration : A Secondary Analysis of the Taxi Drivers ' Health Study, 94(4), 575–581.

Ema, C., Belczak, Q., Maria, J., Godoy, P., Seidel, A.C., Ramos, R.N., ... Caffaro, R.A. (2015). Influence of prevalent occupational position during working day on occupational lower limb edema, 14(2), 153–160.

Grace P.Y. Szeto, & Lam, P. (2007). Work-related Musculoskeletal Disorders in Urban Bus Drivers of, (July). https://doi.org/10.1007/s10926-007-9070-7.

Kyung, G., & Nussbaum, M.A. (2008). Driver sitting comfort and discomfort (part II): Relationships with and prediction from interface pressure, 38, 526–538. https://doi.org/10.1016/j.ergon.2007.08.011.

Kyung, G., Nussbaum, M.A., Babski-reeves, K.L., Kyung, G., Nussbaum, M.A., & Babski-reeves, K.L. (2010). Enhancing digital driver models: Identification of distinct postural strategies used by drivers, 139(May). https://doi.org/10.1080/00140130903414460.

Łastowiecka-moras, E. (2017). How posture influences venous blood flow in the lower limbs : results of a study using photoplethysmography, 3548(May). https://doi.org/10.1080/10803548.2016.1256938.

Messing, K., Stock, S.R., & France Tissot. (2009). Should studies of risk factors for musculoskeletal disorders be stratified by gender ? Lessons from the 1998 Quebec Health and ... Should studies of risk factors for musculoskeletal disorders be stratified, (May 2017). https://doi.org/10.5271/sjweh.1310.

Messing, K., Tissot, F., & Stock, S. (2008). Distal Lower-Extremity Pain and Work Postures in the Quebec Population, 98(4), 705–713. https://doi.org/10.2105/AJPH.2006.099317.

Negahban, H., Hessam, M., Tabatabaei, S., Salehi, R., Soheil Mansour Sohani, A., & Mehravar, M. (2014). Reliability and validity of the Persian lower extremity functional scale (LEFS) in a heterogeneous sample of outpatients with, (November). https://doi.org/10.3109/09638288.2013.775361.

Okunribido, O., & Lewis, D. (2010). Work-related lower limb musculoskeletal disorders—A review of the literature, (August).

Reilly, S.C.O., Muir, K.R., & Doherty, M. (2000). Occupation and knee pain : a community study, 78–81. https://doi.org/10.1053/joca.1999.0274.

Souza, J.C.D., Franzblau, A., & Werner, R.A. (2005). Review of Epidemiologic Studies on Occupational Factors and Lower Extremity Musculoskeletal and Vascular Disorders and Symptoms, 15(2). https://doi.org/10.1007/s10926-005-1215-y.

Tamrin, S.B.M., Yokoyama, K., Aziz, N., & Maeda, S. (2014). Association of Risk Factors with Musculoskeletal Disorders among Male Commercial Bus Drivers in Malaysia, 24(1991), 369–385. https://doi.org/10.1002/hfm.

Thosar, S., & Health, O. (2014). Effect of Prolonged Sitting and Breaks in Sitting Time on Endothelial Function, (August). https://doi.org/10.1249/MSS.0000000000000479.

Tissot, F., Messing, K., & Stock, S. (2005). Standing, sitting and associated working conditions in the Quebec population in 1998 Standing, sitting and associated working conditions in the Quebec population in 1998, *139*(May). https://doi.org/10.1080/001401305123313 26799.

Wrona, M., Jöckel, K., Pannier, F., Bock, E., & Hoffmann, B. (2015). Association of Venous Disorders with Leg Symptoms : Results from the Bonn Vein Study 1. *European Journal of Vascular & Endovascular Surgery*, *50*(3), 360–367. https://doi.org/10.1016/j.ejvs.2015.05.013.

Yasobant, S., Chandran, M., & Reddy, E.M. (2015). Are Bus Drivers at an Increased Risk for Developing Musculoskeletal Disorders ? An Ergonomic Risk Assessment Study. https://doi.org/10.4172/2165-7556.S3-011.

Lean thinking practices in ergonomics in industrial sector

M. Gibson
Saturn Ergonomics Consulting, Auburn AL, USA

B. Mrugalska
Faculty of Engineering Management, Poznan University of Technology, Poznan, Poland

ABSTRACT: The paper presents the integration of "Lean Thinking" into the application of industrial ergonomics. When properly focused, ergonomics can be a powerful tool for achieving Lean objectives such as improving process efficiency, increasing productivity, and reducing quality defects. As ergonomics is applied to reduce injury risk it can simultaneously identify and reduce movement or motion-related wastes. By thoughtfully merging Lean and ergonomics, organizations can better achieve production process excellence and improved customer satisfaction. This research is based on project case studies within organizations where the principles were applied. It refers to wastes such as motion, waiting, inventory and defects. In details the examples are described to show how to reduce lean-related wastes.

1 INTRODUCTION

Over the last few decades the increase of customer' expectations has led to high levels of competition between companies, internationalization and economic conjuncture. To cope with these challenges a number of new management methods and tools were introduced. For example, lean systems were introduced to combine continuous improvement and unify lead-time, cost and quality as key performance indicators (Kokareva et al. 2014).

The concept of "lean thinking" derives from the Toyota Production System (TPS). It was developed within Toyota over the second half of the twentieth century, however, it was coined in the U.S. in the 1990s. Its boost was visible after the publication of "The machine that changed the world" (Womack et al. 1990) and "Lean thinking: banish waste and create wealth in your corporation" (Womack & Jones, 1996). According to these authors it is defined as: "...less of everything compared with mass production—half the human effort in the factory, half the manufacturing space, half the investment in tools, half the engineering hours to develop a new product in half the time ... keeping far less than half the needed inventory on site, results in many fewer defects, and produces a greater and ever growing variety of products" (Womack et al., 2007). It refers to standardizing processes within workplaces to minimize redundancies and improve value (Wyrwicka & Mrugalska, 2017).

Nowadays Lean is applied with increasing frequency in multiple manufacturing industries including automobile, automotive, metallurgical, pharmaceutical, coal and mining companies (Chowdary & George, 2011; Grzybowska & Gajsek, 2012; Koukoulaki, 2014, Santos et al. 2015).

As it can be noticed, lean concentrates on the production and quality related interests of manufacturers and their customers; whereas ergonomics typically viewed as focusing on the human, ensuring that job demands do not exceed the capabilities (physical, cognitive, etc.) of the worker. Ergonomics allows understanding the interactions among humans and other elements of a system (Górny, 2015), to optimize human well-being and overall system performance (Naranjo-Flores & Ramírez-Cárdenas, 2014).

This paper shows how lean thinking and ergonomics can be merged in the industrial workplace to better complement each other. This is demonstrated through practical examples of the application of ergonomics to reduce lean-related wastes such as motion, waiting, inventory and defects.

2 LEAN THINKING AND ERGONOMICS PRINCIPLES

The effective implementation of management practices often depends not only on technology but also people, processes, structure and culture. However, not all organizations can or even should implement the same practices (Mrugalska & Wyrwicka, 2017).

Lean thinking relies on "streamlined work process, reduced inventory, no backlog, maximizing throughput, and eliminating bureaucracy" (Ikovenko & Bradley, 2005). It focuses on such three areas as:

- visibility
- velocity
- value.

Visibility, also called transparency or visual control, makes sure that problems become immediately (or very quickly) known to all stakeholders so that corrections and adjustments can be made.

Velocity, defined as flow, refers to the speed of process delivery. The faster an order is realized, a product built, or a service provided the less it costs to make the customer happy. Rapidly completing actions provides flexibility and improved responsiveness to customer demand. And reduction of process lead time allows rapid response to new orders or other changes defined by the customer.

Value is the third core principle. Work should only be performed if it creates customer value. Therefore, value must be precisely defined in terms of a specific product in reference to the needed capabilities at defined prices through a dialogue (Charron et al., 2015). To satisfy the customer, work should be performed (and value created) in the simplest and most efficient way.

Lean achieves these three objectives—visibility, velocity, and value—by implementing tools and techniques: Value Stream mapping, Quick Changeover/Setup Reduction, Single Minute Exchange of Dies (SMED), Kaizen, Flow Manufacturing, Visual Workplace/5S Good Housekeeping, Total Productive Maintenance (TPM) and Pull/Kanban Systems. In reference to production practices they also encompass: bottleneck removal (production smoothing), cellular manufacturing, competitive benchmarking, continuous improvement programs, cross-functional work force, cycle time reductions, focused factory production, Just-in-Time/continuous flow production, lot size reductions, maintenance optimization, new process equipment/technologies, planning and scheduling strategies, preventive maintenance, process capability measurements, quality management programs, quick changeover techniques, reengineered production process, safety improvement programs, self-directed work teams, total quality management (Shah & Ward, 2003). These tools create a system so they contribute to the elimination of a particular type of waste (Table 1) and they should be applied together (Langstrand 2012; Wyrwicka & Mrugalska, 2015).

Lean and ergonomics share common ground in their historical beginnings. Early lean pioneer and author, Shigeo Shingo, wrote that his "thinking is based on Frederick Taylor's analytical philosophy" and "deeply colored by Frank Gilbreth's exhaustive pursuit of goals and the single best method" when

Table 1. 7 + 2 wastes of lean.

Type of waste	Description
Transport	Moving people, products & information
Inventory	Storing parts, pieces, documentation ahead of requirements
Motion	Bending, turning, reaching, lifting
Waiting	For parts, information, instructions, equipment
Over production	Making more than is immediately required
Over processing	Tighter tolerances or higher grade materials than are necessary
Defects	Rework, scrap, incorrect documentation
Non-utilized talent	Under-utilizing capabilities, delegating tasks with inadequate training
Resources	Failure to make efficient use of electricity, gas or water

evaluating manual labor. Both Taylor and Gilbert practiced what today might be termed *performance ergonomics* … leveraging human physical and cognitive capabilities to maximize productive work output. It was clear, from his own words, that Shigeo Shingo was influenced heavily by Taylor and Gilbreth (Shingo, 1987).

In addition to his interests in scientific job management, Taylor sought to increase productivity by preventing the onset of physical and cognitive fatigue. He specifically described what today would be termed "localized muscle fatigue". Emerging new ergonomics fatigue evaluation models such as the RCRA, Recommended Cumulative Recovery Allowance (Gibson & Potvin, 2016) provide answers to the type of question posed by Taylor over 100 years ago … *How much rest & recovery time is necessary to prevent the onset of muscle fatigue?* Fatigue evaluation models such as the RCRA are yet another example of common ground between Lean and ergonomics. It is not a stretch to recognize human fatigue (both physical and cognitive) as a potential lean-related waste.

Ergonomics ensures that the physical and mental demands of work systems (which includes equipment, materials, tools, interfaces, environment, etc.) is within human capabilities. Ergonomics is perceived as a system approach, being design driven and focusing on human performance and well-being (Dul et al., 2012). Over the years, this concept has been influenced by industrial engineering what led to the promotion of working smarter, not harder, elimination of waste, and considering

Table 2. Ergonomics and lean (Industrial NewsB, 2015).

Integration	Outcomes
Fitting the task to the person	Improved morale
Human capability design	Reduction of absenteeism
The best work methods	Reduction of turnover
Minimizing injuries and their costs	Enhanced corporate culture/climate
Improving productivity & quality	Increased ownership in the processes by the people actually doing the work

the economic impacts (Konz & Johnson, 2004). It found its practical applications in prevention of operator fatigue and stress leading to potential work-related musculoskeletal and neurovascular disorders. Its key principles for workplace design include (Dul & Weerdmeester, 2008; Konz & Johnson, 2004; Walder et al., 2004):

– avoiding prolonged, static postures
– keeping joints in neutral position
– locating work, parts, tools, and controls at optimal anthropometric locations
– providing adjustable workstations and a variety of tool sizes
– if appropriate, providing adjustable seating, arm rests, back rests, and foot rests
– utilizing feet and legs, in addition to hands and arms
– using gravity to limit the energy expenditure
– conserving momentum in body motions
– providing strategic location (in the *power zone*) for lifting, lowering, and releasing loads
– respecting a broad variety of workers with their size, strength, and cognitive abilities
– providing more frequent short breaks than a single long one.

Many of them can be applied to redesign of work and its standardization.

Ergonomics is a vital tool to help achieve the goals of Lean. This can be realized through elimination of waste such as unnecessary motions and reduction of mistakes leading to improvements in quality (Weber, 2001). Table 2 shows the integration of ergonomics and lean and their outcomes.

3 CASE STUDIES

3.1 *Motion*

Many ergonomic risks involve movement of the human body resulting in awkward postures and positions. Common examples include extended reaches to obtain parts from a bin located at the back of a worktable, and forward bending to access parts from a large container. Activities such as these, when performed on a repetitive basis, may result in localized muscle fatigue and if performed over an extended period lead to the onset of discomfort and subsequently work-related musculoskeletal disorders.

Ergonomics typically looks at these awkward postures from the standpoint of presenting increased injury risks. Lean typically quantifies the time required to complete these motions in standard work calculations. But there typically is a disconnect. The ergonomist typically fails to quantify the time value of the awkward posture, while the Lean practitioner fails to identify the injury-risk component in the standard work calculations. Both are byproducts of the same thing ... human movement.

Lean practitioners are familiar with calculating the time value associated with human movement ... such as walking. Common time estimates for walking are 0.2 s for every 12" (or 30.5 cm) traveled, or 0.5 s for every step taken by the worker (Table 3).

A commonly used Lean tool is the Spaghetti Diagram. The Spaghetti Diagram provides a visual model to capture/quantify human movement associated with walking. See Figure 1 below.

Combining the logic of Table 1 values with the Spaghetti Diagram, it can be readily calculated that the time value of movement is 5.6 seconds (28 ft. × 0.2 s = 5.6 s). This logic can be extended to the motion of reaching. See Figure 2, Forward Reach for Box.

Reaching to the far side of the work surface would require more time than if the box were positioned closer to the front edge of the work surface. Table 4 provides approximate time values

Table 3. Walking motion—waste estimator.

Walk distance	Motion waste
Per Foot (1 ft. or 30.5 cm)	0.2 s
Per Step	0.5 s

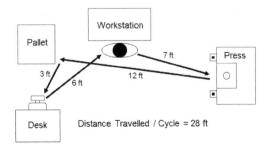

Figure 1. Spaghetti Diagram.

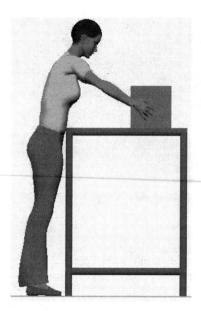

Figure 2. Forward Reach for Box.

Table 4. Reaching motion—waste values.

Reach distance	Motion waste
9–15"	0.2 sec.
15–21"	0.4 sec.
21–27"	0.6 sec.
27–33"	0.8 sec.

for the motion-waste due to reaching. These predetermined movement times have their origin in MTM, Methods-Time Measurement data. MTM values are statistically compiled, predetermined/synthesized time values that reflect commonality in standard movement time estimates for common tasks such as reaching, irrelevant of individual variations as size, gender, age, etc. This motion-waste can be thought of a "time penalty" (additional time required compared to complete the motion).

It should be noticed that if the reach is <9", there is no reach/time penalty. The farther the reach, the more time it takes (the higher the "motion waste"). Example, if we reduce 5 reaches within the 27–33" range to the 9–15" range; then the motion waste savings is: $5 \times (0.8\ sec - 0.2\ sec)$, or $5 \times 0.6\ sec = 3\ sec$.

This method effectively translates the ergonomic risk factor (reaching) into time. Reducing reach saves time. And if the task occurs at a "bottleneck" station, this reach reduction translates into a productivity increase in addition to reducing the potential injury risk. Reach reduction benefits both ergonomics and Lean.

3.2 Waiting

When a line is not balanced, one or more stations may outpace the main line. When this occurs, the employee at the station outpacing the line may become idle much of the time. Idleness can turn into impatience, and if the worker chooses to continue producing units, resulting is the creation of excessive (WIP) work-in-progress.

The excessive work-in-process results in a large quantity of parts residing on the surface of the worktable, resulting in the employee elevating the arm and reaching ... presenting ergonomic-related injury risk to the shoulder. From the Lean perspective, if it were discovered that a component was defective; then it would be necessary to re-work the entire table of subassemblies. Lean considers this type of re-work as the waste of *Correction*.

In this example, the station should utilize the pull principle, and produce the quantity needed, when it is needed, rather than overproduce. It could be argued that the Figure 3 example results in another form of Lean waste ... *Overproduction*. This is yet another example of how the objectives of Lean and ergonomics are closely intertwined.

3.3 Inventory

Tenets of Lean are pull production and the subsequent benefit of smaller batch sizes. Smaller batch sizes typically benefit ergonomics as well. Smaller containers of parts & materials are lighter weight, and easier to position than larger containers. Once again, the objectives of Lean and ergonomics are closely intertwined (Figure 4).

3.4 Defects

Analyzing a warehouse picking operation, it was found that 85% of defects were related to picking the incorrect item type or the incorrect count. The Tote

Figure 3. Waste due to waiting.

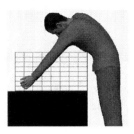

Figure 4. Smaller container (right) requires less reaching & bending than the larger container (left).

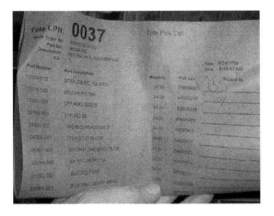

Figure 5. Tote Pick List on red paper with black font.

Figure 6. Contrast between letters.

Pick List used by employees performing this work was on red paper with black font (Figure 5).

Also of interest:

- Area lighting levels had recently been reduced
- Majority of employees were 40+ years of age (at this age, people begin to need more light).

Figure 6 presents a practical guideline related to background and font color.

The red Pick Sheet with the black font is classified as "Poor" according to the above criteria. In this example, redesign of the pick sheet (black font on a white or yellow background), and potentially increasing area lighting levels, should improve picking accuracy. This is an example of how vision ergonomics has the potential to not only reduce eye strain but to achieve the Lean objective of reducing defects. In this example, defects (wrong items or wrong count of items in tote) results in unsatisfied customers if undiscovered, or results in re-work if discovered in subsequent packing or inspection.

4 CONCLUSIONS

Effective ergonomics is a necessary part of any sustainable organization. And correctly implemented, lean thinking requires effective utilization of ergonomics. The successful merger of Lean thinking and ergonomics may involve typical activities such as the redesign of work, standardizing work, and reduction or elimination of MSD risk factors. But to go beyond the current state of ergonomics being viewed as mostly an injury prevention tool (not fully complimenting Lean), it is necessary to better understand the commonality between ergonomics and Lean ... historically, and in current application.

Ergonomics and Lean can be more effectively integrated by focusing on the inherent common objectives—such as reduction or elimination of non-value activities such as reaching, increasing efficiency of manual materials handling, etc. This requires shifting from the focus of ergonomics being an injury prevention tool, to viewing ergonomics as a valuable means to increase productivity and operational efficiency. This objective can only be achieved through training and leadership articulation of a clear message of how ergonomics compliments the objectives of Lean ... as the Case Study examples provided in this paper demonstrate.

REFERENCES

Charron, R., Harrington, H.J., Voehl, F., & Wiggin, H. 2015. The lean management system handbook. CRC Press: Boca Raton.

Chowdary, B.V., & George, D. 2011. Improvement of manufacturing operations at a pharmaceutical company: a lean manufacturing approach. *Journal of Manufacturing Technology Management*, 23(1), 56–75.

dos Santos, Z.G., Vieira, L., & Balbinotti, G. 2015. Lean manufacturing and ergonomic working conditions in the automotive industry. *Procedia Manufacturing*, 3, 5947–5954.

Dul, J., & Weerdmeester, B. 2008. *Ergonomics for Beginners. A Quick Reference Guide*. Third Edition. CRC Press: Boca Raton.

Dul, J., Bruder, R., Buckle, P., Carayon, P., Falzon, P., Marras, W.S., & van der Doelen, B. 2012. A strategy for human factors/ergonomics: developing the discipline and profession. *Ergonomics*, 55 (4), 377–395.

Gibson, M., & Potvin, J.R. 2016. An equation to calculate the recommended cumulative rest allowance across multiple subtasks, *Assoc. of Canadian Ergonomists (ACE) Conference*, Niagara Falls, ON, October.

Górny, A. 2015. Man as internal customer for working environment improvements. *Procedia Manufacturing*, 3, 4700–4707.

Grzybowska K., & Gajdzik B. 2012. Optimisation of equipment setup processes in enterprises. *Journal Metalurgija*, 51(4), 563–566.

Ikovenko, S., & Bradley, J. 2005. TRIZ as lean thinking tool. *The Triz Journal*. Retrieved from: https://triz-journal.com/triz-lean-thinking-tool/.

Industrial NewsB. 2015. Ergonomics & lean manufacturing. Retrieved from: http://www.industrialnews.in/blogs/3305-ergonomics-lean-manufacturing.html.

Kokareva, V.V., Malyhin, A.N., & Smelov, V.G. 2014. The organization of engineering shop "lean" management system. *Applied Mechanics and Materials*, 682, 555–560.

Konz S., & Johnson S. 2004. *Work Design: Occupational Ergonomics*. 6th Edition, Arizona: Holcomb Hathaway.

Koukoulaki, T. 2014. The impact of lean production on musculoskeletal and psychosocial risks: An examination of sociotechnical trends over 20 years. *Applied Ergonomics*, 45(2), 198–212.

Langstrand, J. 2012. *Exploring organizational translation—a case study of changes toward lean production*. Linköping Studies in Science and Technology, Dissertation No. 1422, Linköping University, Sweden.

Mrugalska B., & Wyrwicka M.K. 2017. Towards lean production in Industry 4.0. *Procedia Engineering*, 182, 466–473.

Naranjo-Flores A.A., & Ramírez-Cárdenas E. 2014. Human factors and ergonomics for lean manufacturing applications. In: García-Alcaraz J., Maldonado-Macías A., Cortes-Robles G. (eds.) *Lean manufacturing in the developing world*. Springer, pp. 281–299.

Shah R., & Ward P. 2003. Lean manufacturing: context, practice bundles, and performance. *Journal of Operations Management*, 21, 129–149.

Shingo, S. 1987. *The Sayings of Shigeo Shingo*. Massachusetts: Productivity Press.

Walder J., Karlin, J., & Kerk C. 2007. Integrated Lean Thinking & Ergonomics: Utilizing Material Handling Assist Device Solutions for a Productive Workplace. An MHIA white paper. South Dakota. Retrieved from: http://www.mhi.org/downloads/industrygroups/lmps/whitepapers/Integrating_Lean_Thinking.pdf.

Weber, A. 2001. Ergonomics: issues & trends. Retrieved from: https://www.assemblymag.com/articles/84114-ergonomics-issues-trends.

Womack J., Jones D., & Ross D. 1990. *The machine that changed the world*. New York: Macmillan.

Wyrwicka M., & Mrugalska B. 2015. *Barriers to eliminating waste in production system*. Proceedings of the 6th International conference on engineering, project, and production management, 2–4 September 2015, Gold Coast, Australia, pp. 354–363.

Wyrwicka M.K., & Mrugalska B., 2017. Mirages of lean manufacturing in practice. *Procedia Engineering*, 182, 780–785.

Occupational activity on blood pressure profile

J. Pereira, R. Souza & H. Simões
ESTeSC—Coimbra Health School, Instituto Poltécnico de Coimbra, Coimbra, Portugal

ABSTRACT: Hypertension (HT) is an important risk factor for cardiovascular disease, which can suffer alterations resulting of lifestyle and working conditions of individuals. It has been verified in some workers, variations in blood pressure due to professional features and stress faced at work.

Evaluate, in working individuals, the influence of working conditions (stress and risk factors) and mild physical exercise on blood pressure.

The sample consisted of 19 subjects aged between 28 and 59 years, of which 15 were female (78.9%) and 4 males (21.1%) who practiced their professional activity during the period of data collection. A considerable percentage shows overweight (38.9%), and a small percentage (16.7%), history of hypertension. Some of the individuals were practitioners of physical exercise (31.6%) and a small percentage had smoking habits (26.3%).

During the class period, there was an average increase of 2.44 mmHg in Systolic Blood Pressure (SBP) and an average increase of 4.37 mmHg in Diastolic Blood Pressure (DBP), by considering the sample included in this study, DBP undergoes a greater change when the individual is exposed to the stress of their profession.

There is a slight rise in systolic and diastolic blood pressure (mean) during the work period compared to the previous time.

1 INTRODUCTION

Cardiovascular diseases are a leading cause of morbidity and mortality in the world[1].

Hypertension (HT) in turn is a major cardiovascular risk factor, and can lead to serious consequences in organs such as the heart, brain, kidneys and blood vessels. It can be characterized by a chronic disease with slow and silent evolution leading to late detection in most cases.[1]

Some aspects such as heredity, age, race, obesity, stress, sedentary lifestyle, alcohol, gender, anticoncecionais and high sodium intake are part of the main risk factors for HT[1].

The health condition is critical to life quality and working capacity of people, so that increasingly more researchers are concerned about the interplay between these factors[6].

Regarding education, there has been a large aggravation of teachers health, often associated with the characteristics and conditions of work[6]. Some of most frequent health complications among teachers may be: psychological disorders (like *stress*, depression, mental exhaustion and burnout syndrome), work related ergonomic requirements (as musculoskeletal symptoms and voice disorders) and other general problems. The observed lesions affect both the physical and psychological health and may compromise the performance and working capacity. Some data also indicate that for such professions as the education sector, where the workload is high, there is a limitation with regard to the control of activities and little social support[6].

Previous studies have shown that job strain tends to raise Blood Pressure (BP), and this is verified mostly in men in relation to women[2].

Work stress was also associated to a BP raise during sleeping hours, leading to the possibility that work stress can impact the circadian rhythm of blood pressure. An abnormal variation or a "non-dipping" of night time blood pressure is a risk factor for cardiovascular diseases, leading to serious damage in end-organs and a higher cardiovascular morbidity and mortality[4].

By Ambulatory Blood Pressure Monitoring (ABPM), we can obtain measurements which will allow to observe the variation of BP and heart rate throughout the day for teachers. Through this method, is possible to acquire more reliable estimates of blood pressure (due to observer errors and the increased number of reads compared to measurements of casual blood pressure)[2], thus allowing, achieve the proposed objective to investigate the influence that job characteristics trigger in the health conditions of teachers, particularly the BP behavior of in teachers that are within and outside of their teaching activity. We also studied the effect of moderate exercise (walking) on blood pressure and heart rate, in particular, whether they normalize.

This study has a sample of teachers who practice their profession in Infanta Dona Maria High School, having completed the ABPM in the same school.

2 MATERIAL AND METHODS

The present study had a duration of 2 academic years, having been initiated in October 2012 and ended in June 2014.

The data collection was taken in Infanta Dona Maria High School through questionnaires and results obtained from ABPM, having a duration of two months (during the month of February and April 2014).

The sample consisted of 19 teachers of both genders (inclusion criteria), who voluntarily participated in the examination of ABPM during the sample collection.

The Ambulatory blood pressure monitoring was performed noninvasively, has been used an apparatus of the brand Meditech ABPM-04, 2007/411043 series, which followed the following protocol: 6 am to 22 pm—measurements 20/20 minutes; from 22 h to 6 h—measurements 30/30 minutes. Was informed to all teachers, that they should conduct a walk of around 30 minutes, at least 1 log BP measurement during the same period is required.

Once completed the sampling, all data were treated in a database, with the knowledge and acceptance of those surveyed (tacit consent), having fulfilled all ethical questions relative to study type, where anonymity and confidentiality respected data, as well as curricular and academic interest only.

3 RESULTS

The sample consisted of 19 subjects aged between 28 and 59 years, of which 15 were female (78.9%) and 4 males (21.1%).

According to Graph 1, more than half the population have a family history of hypertension (73.7%) and a considerable percentage shows overweight (38.9%), low physical activity (52.6%) and cholesterol (36.8%). Still has a percentage of 16.7% of hypertensive patients and smokers 26.3%. Diabetes was a factor in low prevalence in this sample, with a percentage of 5.3%.

The whole sample (100%) presented with normal mean values of SBP, PP and HR, and a large part of normal DBP (94.74%).

For the hypotheses, we carried out a correlation analysis of the data and the appropriate statistical tests were applied, yielding the following results:

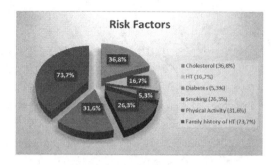

Graph 1. Sample distribution according to risk factors.

For the first hypothesis (H1), we intended to study whether during lective teaching activity in the classroom, the blood pressure values tend to be higher when compared to times before and after that activity.

The mean values of SBP and DBP were compared during the teaching activity in the classroom, as well as the mean values of SBP and DBP recorded two hours before and after that moment.

By applying ANOVA for repeated measures was observed that the mean SBP at the time of teaching activity is slightly higher (2.44 mmHg) compared to the preceding moment, however, this difference is not statistically significant (F = 0.377, df = 2, p = 0.693) (Table 1).

The same happened when we applied the same statistical test for PAD, with an increase of 4.37 mmHg, without statistical significance (F = 2.767, df = 2, p = 0.097) (Table 2).

We tried to analyze if the values of SBP and DBP during class, vary with gender. Student's t-test for independent samples (Table 3) was applied.

It was found that there were no statistically significant differences in SBP between genders (t = −0.938, df = 17, p = 0.361), although males have registered slightly higher values than females (Graph 2).

By applying the same statistical test, it was also found that males recorded DBP (during class) slightly higher than females, there was statistically significant difference (t = −1.253, df = 17, p = 0.227) (Graph 3).

In an attempt to satisfy a fourth hypothesis, we sought to determine whether Pulse Pressure (PP), prevails in hypertensive individuals, ie, we intended to test whether hypertensive individuals also had elevated values of PP during the daytime and nighttime periods, proving that these individuals have a higher cardiovascular risk.

As can be seen in Table 4, only 3 individuals are hypertensive.

Thus, given the size of the groups with and without hypertension, despite the average pulse

Table 1. Average SBP before, during and after the class period.

Mean SBP	N	$\bar{x}$	s
2 hours before class period	19	121,1116	10,624
During the class period	19	123,5511	9,922
2 hours after class period	17	120,7665	8,891

Table 2. Average DBP before, during and after the class period.

Mean DBP	N	$\bar{x}$	s
2 hours before class period	19	76,369	7,468
During class period	19	80,742	7,633
2 hours after class period	17	76,942	6,490

Table 3. Relationship between mean SBP and DBP (during class) with female and male gender.

Gender		N	$\bar{x}$	s	t	Sig. (2-tailed)
SBP (during class)	Female	15	122,445	10,481	–0,938	0,361
	Male	4	127,700	7,000		
DBP (during class)	Female	15	79,627	7,907	–1,253	0,227
	Male	4	84,925	5,315		

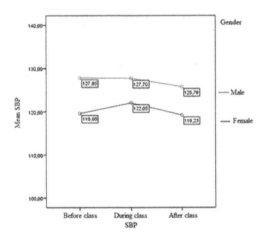

Graph 2. Influence of gender on average SBP before, during and after the teaching activity in the classroom.

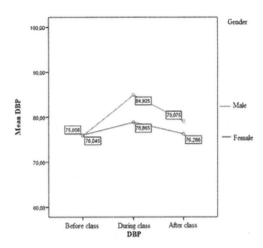

Graph 3. Influence of gender on average DBP before, during and after the teaching activity in the classroom.

Table 4. Relationship between HTA and PP.

	HT	N	$\bar{x}$	s
PP	No	15	42,73	4,667
	Yes	3	42,33	4,933

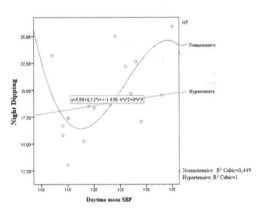

Graph 4. Relationship between HT and nocturnal dipping in SBP.

pressure to be similar in both, you can not test the hypothesis of equality of means.

Ultimately, we sought to determine whether hypertensive individuals exhibit a nocturnal blood pressure profile with high cardiovascular risk, ie, we sought to assess whether hypertensive individuals exhibit an abnormal BP night dipping.

Through the analysis of curvilinear regression (cubic), was obtained Graph 4, allowing to observe the relationship between the mean systolic blood pressure and nocturnal dipping.

Through Graph 4, it is observed that systolic nocturnal dipping has a curvilinear relationship with the daytime systolic blood pressure.

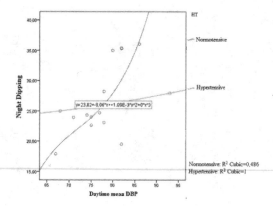

Graph 5. Relationship between HT and nocturnal dipping in DBP.

Thus, for non-hypertensive individuals, when the SBP variation ranges up to ±117 mmHg, there is a nocturnal dipping less marked than from that value.

The value of nocturnal dipping increases, the greater the average daytime of SBP.

In hypertensive individuals, the value of the nocturnal dipping remains straight without great variation between night and day.

By applying the same statistical test, was obtained Graph 4, which relates the average diastolic blood pressure and nocturnal dipping.

According to Graph 5, it is observed that in non-hypertensive, the higher the mean daytime DBP, the greater the night dipping.

In hypertensive individuals, the value of the nocturnal dipping as the SBP, remains straight without severe changes between night and day (Graph 5).

4 DISCUSSION

HT is considered an important risk factor for cardiovascular diseases and is closely related to habits and lifestyles of individuals. To these factors is joined work or other activities, which also influence BP values and can alter them (Schnall, Schwartz, Landsbergis, Warren, & Pickering, 1992).

Thus, had as main objective to study the behavior of blood pressure in school and extracurricular environment by Ambulatory Blood Pressure Monitoring (ABPM) in high school teachers.

Were included in this sample, high school teachers of both genders, who participated voluntarily in the study and that carried out teaching activity during the collection period.

In our investigation, it was observed that BP vales (SBP/DBP) increased during the classes in relation to other periods considered.

To Steptoe (2000), the teaching profession is considered a stressful occupation due to the demands of students and parents along with low financial rewards. Although these results are not statistically significant, agree with studies from Vrijkotte, van Doornen, & de Geus (2000), P. Schnall, Schwartz, Landsbergis, Warren, & Pickering (1992) and G.Cesana et al. (1996), showing that SBP and DBP are increased during working hours due to stressful situations to which they were subjected.

In the study by Schnall P., Schwartz, Landsbergis, Warren, & Pickering, (1992), it was observed that the work strain has a greater effect on blood pressure with increasing age, and their exposure have little or no effect on individuals who stand at around the age range of 30/40 years of age. This may explain, in part, the low significance in this study, since the average age of teachers is around 48 years old.

We sought to determine whether gender influences the variation in blood pressure during the teaching activity, it was examined that there are no statistically significant differences in SBP between genders, although males have registered slightly higher values than females.

According to Tobe et al. (2005), certain factors such as age, stress at work and gender are significant variables to higher blood pressure, and also stated that work stress was positively associated with higher levels of blood pressure in men and more recently women.

In an attempt to satisfy a fourth hypothesis, we attempted to analyze whether hypertensive individuals also had high values of Pulse Pressure (PP) during the day and night periods, proving that these individuals have a higher cardiovascular risk.

In our work, only 3 individuals in this sample were hypertensive. It was noted that the mean pulse pressure was similar in subjects with and without hypertension (HTA), possibly given the fact of these subjects were medicated. However, given the size of these 2 groups was not possible to test the hypothesis of equality of means.

5 CONCLUSIONS

In this study could be noted that the teaching profession affects the blood pressure values when they perform their duties and that some variables such as age, sex and physical exercise, can lead to alterations of those values.

Throughout the work, there were some limitations, the main being related to the small sample size, not allowing a consistent result on the different assumptions and the lack of studies that considered the influence of blood pressure with the teaching profession. In the questionnaire given to individuals, responses should have been better quantified and qualified in the sense that responses were obtained only through the individual, without any analysis or quantitative clinical information.

It is than possible to conclude that exposure to stress at work can actually cause changes in blood pressure of an worker individual and may even cause a base BP gained by years of work, higher than the individual would have with less years of career.

REFERENCES

[1] Carvalho, M.V., Siqueira, L.B., Sousa, A.L., & Jardim, P.B. (2013). A Influência da Hipertensão Arterial na Qualidade de Vida.

[2] Cesana G, Ferrario M, Sega R, Milesi C, De Vito G, Mancia G, Zanchetti A. (1996). Job strain and ambulatory blood pressure levels in a population-based employed sample of men in northern Italy. *Scand J Work Environ Health*, pp. 294–305.

[3] Leary, A.C., Donnan, P.T., MacDonald, T.M., & Murphy, M.B. (October de 2000). The Influence of Physical Activity on the Variability of Ambulatory Blood Pressure. *AJH*, pp. 1067–1073.

[4] Lin-Bo Fan, M.D., Blumenthal, A.J., PhD, Hinderliter, A.L., & Sherwood, A. (2013). The effect of job strain on nighttime blood pressure dipping among men and women with high blood pressure.

[5] Ribeiro, F., Grubert, C.S., Campbell, Mendes, G., Arsa, G., Moreira, S.R., ... Simões, H.G. (18 de October de 2011). Exercise lowers blood pressure in university professors during subsequent teaching and sleeping hours. pp. 711–716.

[6] Santos, M.N., & Marques, A.C. (2013). Condições de saúde, estilo de vida e características de trabalho de professores de uma cidade do sul do Brasil.

[7] Schnall, P., Schwartz, J., Landsbergis, P., Warren, K., & Pickering, T. (May de 1992). Relation between job strain, alcohol, and ambulatory blood pressure. *Hypertension*, pp. 488–494.

[8] Steptoe, A. (2000). Stress, social support and cardiovascular activity over the working day. *International Journal of Psychophysiology*, pp. 299–308.

[9] Tobe, S.W., Kiss, A., Szalai, J.P., Perkins, N., Tsigoulis, M., & Baker, B. (August de 2005). Impact of Job and Marital Strain on Ambulatory Blood Pressure. *AJH*, pp. 1046–1051.

[10] Vrijkotte, T.G., van Doornen, L.J., & de Geus, E.J. (April de 2000). Effects of Work Stress on Ambulatory Blood Pressure, Heart Rate, and Heart Rate Variability. *Hypertension*, pp. 880–886.

Reusing single-use medical devices, where are we standing now?

L.B. Naves & M.J. Abreu
2C2T-Centre for Textile Science and Technology, Department of Textile Engineering, University of Minho, Azurém, Guimarães, Portugal

ABSTRACT: Reprocessing single-use devices has attracted interest on the medical environment over the last decades. The reprocessing technique was sought in order to reduce the cost of purchasing the new medical device, which can achieve almost double of the price of the reprocessed product.

This paper seeks to assay the situation of reprocessing single-use medical devices worldwide and used as methodology a literature review, aiming the reuse of medical devices that were firstly designed for single use only, but has become, more and more, effective on its reprocessing procedure. We also show the regulation, the countries worldwide which allows this procedure, the classification of these device and also the most important issue concerning the re-utilization of medical devices and different reprocessing process, minimizing the risk of gram positive and negative bacteria, avoid cross-contamination, Hepatitis B (HBV) and C (HCV) virus, and also Human Immunodeficiency Virus (HIV).

Keywords: Hepatitis A and B; HIV; reusing; reprocessing; single-use medical device

1 INTRODUCTION

The European Union (EU) and the United States (US), together, represent two of the most important world markets for medical equipment and devices. When using these medical devices on the patients, they present vastly different approaches for approving new and reused devices.

In the United States, is necessary the PMA (Premarket approval) approval by the FDA (Food and Drug Administration) for reprocessing medical devices, which is responsible for ensuring the effectiveness and safety of the medical device. The PMA may include several requirements as bench and animal tests. There is also regulation called GMPs (Good Manufacturing Practices) which is responsible for determining some rules as labeling and packaging that describe how the reprocessed devices should be used, and the facilities that have done the processing. The range of the device of GMP's can be from non-invasive products as a bandage to invasive supplies as artificial hearts and cardiac catheters [U. S. G. A. Office, 2003].

The Single-Use Device (SUD) is a disposable device labeled for single use and not intend to be reused in another patient nor to be reprocessed. The main goal of reprocessing these devices is to decrease the environmental pollution and save money, by amortizing the cost of a single-use item between two or, even more, times, can be financially rewarding, the reprocessing cost is less than that of its acquisition [Belkin, 2001].

The practice of reusing and reprocessing Single Use Devices (SUDs) should not be done, unless the staff members, administrators, and perioperative members are prompt to assume any risk and responsibility for the performance and effectiveness of this reused single-use medical device [Issues, 1996]. The initial step that must be done at the hospital which intends to use the reprocessed medical device is to establish a committee that includes several departments with distinguished activities as financial, risk management, infection control, administration and central processing.

It is mandatory for hospitals to have a third-party as reprocessor "A business establishment, separate from the user facility and the device manufacturer, one of whose primary businesses is to reprocess single-use/disposable medical devices" [Statement, 2006] which will be responsible for the reprocessing procedure to establish a testing protocol in order to verify and ascertain the effectiveness of the sterilizing, disinfection and cleaning. When an external reprocessing company is chosen, the contractual arrangement of the service should be of full responsibility of the user facility's in order to achieve and ascertain the quality of the service provided. The user facilities must report only serious injury and deaths, notwithstanding, manufacturers are required to report more extensive and may be necessary to provide supplemental information surrounding the medical device.

In the United States, before any newly medical device go to the market, is necessary the approval of the US Food and Drug Administration (FDA). The FDA will be responsible for establishing the regulation for this new medical device, requiring further information that the device can be cleaned, sterilized and disinfected effectively. Once that the device is intended for reuse, some aspects such as user design, material design, physical design and total system design must be considered during the design processes [AAMI, 1988].

A huge concern must be taken into consideration when reusing medical devices that has been in touch with body fluids or human blood in order to provide adequate protection for health professionals and patients against several viral infections as hepatitis B (HBV) and C (HCV) viruses and Human Immunodeficiency Virus (HIV). This device requires completely reprocessing in terms of hygiene, safety, reliability and functionality [Tessarolo et al, 2006], regulation of original equipment manufacturers, ethics and patient informed consent. The use of reprocessed SUDs could increase risks to the patients when compared to the use of a new single-used device if the safety requirements established are not followed and state that available evidence indicate that some SUDs can be reused and reprocessed safely on others patients.

2 REPROCESSING MEDICAL DEVICES

According to Alfa and Castillo [2004], they reported types of single-use devices that were reprocessed as followed:

– BTCA, biopsy, laparoscopic, blood pressure, endoscopic, cardiovascular, respiratory, pulse oximeter, angiographic catheter, electrode recording catheter, tissue saturation oximeters, compression, hemodialyzers, surgical burrs and saw.

In the same study, further, they describe the areas within health care facilities in which reprocessing of SUDs is performed, such as neurology, obstetrics (gynecology), cardiovascular, orthopedics, respiratory, gastroenterology (urology) and surgery.

2.1 Classification of medical devices in the USA

This device can be classified into three different class.

2.1.1 Class I medical device
Is a class I medical device, those which the device is neither life-sustaining nor life-supporting, it does not present a potential illness or injury for patients and, the use of this type of device is not substantially important in preventing impairment of human health [FDA, 2016] or low-risk items such as orthopedic saw blades and bandages, requiring least regulations [Dunn, 2002].

2.1.2 Class II medical device
According to the AORN – Association of periOperative Registered Nurses. Guidance Statement, this is a medical device for which the control by itself does not provide effectiveness and reasonable assurance of device safety, however, there is enough information regarding special controls to provide the assurance of performance standards, patient registers, and premarket surveillance, it represents medium risk devices, can include blood pressure cuffs and urethral catheter.

2.1.3 Class III medical device
The use and application of class III medical device must include valid scientific research and report evidence demonstrating the effectiveness and the safety of the original or reprocessed device. This type of device is life-sustaining or life-supporting, it presents a potential risk of injury or illness, and is of substantial importance in preventing impairment of human health [FDA, 2016]. Present significant risk factors for the patients, these devices can be heart valves, intra-aortic balloon angioplasty catheter, and percutaneous transluminal coronary angioplasty catheters. All class III devices must apply for PMA.

2.2 Classification of medical devices in the European Union (EU)

The use of Single-Use Medical Devices (SUMD) has considerably increased in hospitals in particular to reduce the risks of cross contamination between patients. Definition and classification of single-use medical devices emerge in Council Directive 93/42/EC concerning to medical devices, adopted on 14 June 1993 and amended by Directive 2007/47/EC as "a device intended to be used once only for a single patient".

Essentially, all devices fall into four basic categories:

– Non-invasive devices;
– Invasive medical devices;
– Active medical devices;
– Special Rules (including contraceptive, disinfectant, and radiological diagnostic medical devices).

Devices are further segmented into the classes noted below:

– Class I—Provided non-sterile or do not have a measuring function (low risk);
– Class I—Provided sterile and/or have a measuring function (low/medium risk);
– Class IIa- (medium risk);

- Class IIb- (medium/high risk);
- Class III- (high risk).

Council Directive 93/42/EEC amended by Directive 2007/47/EC not defines reprocessing but impose the need to ensure that the reprocessing of medical devices does not constitute a danger to the safety or health of patients by clarifying the definition of "single use" and the establishment of labeling and instructions for use uniformly [Abreu, 2015].

2.3 Cleaning and decontamination protocols

It is important to note, that some reprocessors suggest that is necessary to establish a maximum number of times that a device can be reprocessed, however, there is no published recommendation or universal standards for limits of reprocessing procedures. Different types of device respond differently to reprocessing process.

In 2007, Tessarolo and co-workers, have reported in their study four different protocols for cleaning and decontamination of the medical devices, using some effective HIV disinfectants as phenolic and chlorine, however, these disinfectants have some limitations as residues which are highly toxic and fixative effect on plasma proteins respectively. As shown in the following Table 1:

In the same study, Tessarolo et al. state that the protocol 1 has an effective efficiency of new polyphenols against bacterias Gram-negative and Gram-positive, in addition to that, has been also been reported the HIV disinfection using the surfactant components of the polyphenolic emulsion combined with a single chemical treatment [Bloomfield et al., 1989].

When dealing with phenolic derivates, is necessary to remove all toxic residues. By using this chemical derivate is possible to achieve a better and effective tuberculocides and fungicides. The phenolic components must be carried with precaution, once that it can be absorbed by the polymeric part of the device which is been cleaned, it may also provoke modification of the polymeric parts used on the medical device by interaction and penetration of phenolic components with the polymeric matrix. If the toxic residues are not completed eliminated, it can be released to the patient via bloodstream while he or she, is undergoing a specific treatment utilizing reused medical device [Rutala, 1990].

2.4 Labeling

Labeling single-use medical device, that is intended to be reused is very important, by doing so, is necessary information of the devices and this may include package inserts, requiring the location and the name of the manufacturer, and any printed information for the device's intended use [Donowa, 2002]. In the United States, the FDA intends to make immediate enforcement action against third-party reprocessors or hospitals, that fail to submit administratively poorly or incompletely organized paperwork.

3 THE USE OF REPROCESSED SINGLE-USE MEDICAL DEVICES WORLDWIDE

Due to financial constraints, the reuse of single-use device is a necessity in many of semi or developing countries of Asia, Africa, Central and South America, and Eastern Europe, due to a shortage of financial resource and medical supplies. When the SUDs are reprocessed in-house, the cost of the device can be less than 10% of a new one, reprocessed by third-party the savings are expected to be around 50% of the cost of a new device [U. S. G. A. Office, 2003].

3.1 Asia

This procedure is widespread [Ahuja et al, 2000], particularly for injection needles. Third-party reprocessors do not offer their services in Asia, the reuse is conducted in an unregulated—manner at the user facility level. According to the Association of Medical Device Reprocessors, in India, the hospital routinely reuse SUDs, the process is not regulated by the government, however, private hospitals may have guidelines regarding SUD reuse [AMDR, 2010]. A survey from 2003 found that in Japan, 80–90% of hospitals reused SUDs, though, reprocessing is not currently regulated in Japan.

3.2 United States

All the medical device must be approved by the FDA, it is reprocessed by hospital facilities and

Table 1. Different experimental protocols for decontamination of medical devices [Tessaralo, 2007].

Cleaning and decontamination protocols	
Protocol	Treatment
1	Polyphenolic emulsion
2	Polyphenolic emulsion followed by enzymatic solution
3	Chlorine solution followed by enzymatic solution
4	Enzymatic solution followed by chlorine solution

third-party. SUDs have some statutory requirements made by FDA, including requirements for labeling, quality system regulation, registration and listing, medical device reporting, premarket notification and approval, medical device tracking, correction and removal [Statement, 2006].

3.3 Canada

The Advisory Committee on Health Service of Canada indicate that the reusing procedure occurs around 40% across the country [CADTH, 2015], gastroenterology, cardiovascular and surgery are the specialties which reuse most the SUDs.

In the US and Canada, is very common the reuse of invasive SUDs as sphincterotomes, electrophysiology catheter and biopsy forceps [8].

Dunn [2002], stated *"Hospital clearly owe a duty to the patient to provide proper and adequate facilities and equipment so as to reasonably ensure the patient's safety. This duty extends to the maintenance of these facilities and equipment, including handling, testing, inspecting, sterilizing, and reusing disposables. If patients are uninformed about the use of recyclable items and the product fails or the patient develops a postoperative infection or other adverse reaction, the hospital could face liability for breach of duty to provide a safe facility".*

3.4 New Zealand and Australia

All reprocessors of the medical device (hospital, third-party and OEM) must follow several requirements for safety as regulated by the Therapeutic Goods Administration (TGA) [n.d.]. The government of Australia does not endorse the reuse of SUDs if these products are used is necessary to have the patient consent [Qian, 2002].

3.5 Africa

Unfortunately, in many countries in Africa, the lack of resource, including the distribution channels, there is a necessity of reusing SUD that has not been sterilized, this may include the reuse of rubber gloves, needles, and syringes [AMDR, 2010].

3.6 Middle east

Is common to reuse SUDs in Arabic countries, the most used product is a cardiac catheter, there is an absence of a regulatory framework. In the other hand, in Israel, the country does not have specific regulation for the reprocessing SUDs, all medical device must be registered with the Ministry of Health (MOH), if the device is approved by the U.S. FDA, then, there will be no necessarily to perform further testing requirement, resulting in a lawfully marketed product in the country. The Kingdom of Saudi Arabia, has done Interim Regulation regarding SUDs, for devices that may be marketed in Saudi Arabia, The Saudi Food and Drug Authority (SFDA) published [Campbell, 2012]: *"Comply with the relevant regulatory requirements applicable to one or more of the jurisdiction of Australia, Canada, Japan, the USA and the EU/EFTA, and additionally with provisions specific to the KSA concerning labeling and condition of supply and/or use".*

3.7 European Union (EU)

The EU does not have a single policy regarding the use of reprocessed SUDs, some countries are permitted the use of this procedure as Germany, Portugal, Sweden, Holland, Belgium, some countries do not recommend this technique: Italy Czech Republic, Austria, United Kingdom, Ireland, Spain. In France, this technique is forbidden [Abreu, 2015]. In 2010, the European Commission highlighted the risks of unregulated reprocessing in a report, based on it the European Parliament is considering legislation for SUDs reuse by the EU Medical Device Regulation [EC, 2010]. Reprocessing should be performed according to the instructions provided by the manufacturer and must follow technical specifications that ensure the conformity of the device at the time of its use towards the essential requirements of the apply Directive. It rises a difficult situation that single-use medical devices manufacturer instructions don't inform about the reprocessing procedures, once that it's an only once single use product [Abreu, 2015].

4 CONCLUSION

In conclusion, we can state that reuse single-use device has attracted interest in many countries around the world. Especially in the developing countries, this practice has gained incredible interest in order to save money and minimize the environmental impact.

It is wildly accepted the reuse of these devices once that the regulations are followed. New regulations and safety requirements have been made either for hospital facilities, health care or third-party reprocessors.

The patients must be informed about the procedure of reusing any reprocessed medical device, and therefore, the facilities such hospital should have their consent. In addition to that is very important to ascertain that the reuse of this device will not compromise the patient healthy, will not cause any cross infection through gram positive or gram negative bacteria's, are completely free of contamination of hepatitis B (HBV) and C (HCV) virus, and also human immunodeficiency virus (HIV), which may be a hazard for patient undergoing any kind of treatment and having low self-immunity.

ACKNOWLEDGMENT

This work was financed by FEDER funds through the Competitivity Factors Operational Program—COMPETE and by national funds through FCT – Foundation for Science and Technology within the scope of the project POCI-01-0145-FEDER-007136.

REFERENCES

AAMI, "Human Factors Engineering Guidelines and Preferred Practices for the Design of Medical Devices," *Assoc. Adv. Med. Instrum.*, pp. 35–77, 1988.

Abreu, I. and Abreu, M. J. "Reprocessing Single-Use Devices," *Legislative Issues in the EU*, European Medical Hygiene. February, pp. 36–39, 2015.

A. G. D. of H. T. G. Administration, "Statement by the TGA on regulations for sterilisation of single-use devices."

Ahuja, V. and Tandon, K., Rakesh, "WORKING PARTY REPORT: CARE OF ENDOSCOPES Survey of gastrointestinal endoscope disinfection and accessory reprocessing practices in the Asia—Pacific region," *Sci. Med. Delhi, New*, pp. 78–81, 2000.

Alfa, M. J. and Castillo, J., "Impact of FDA policy change on the reuse of single-use medical vices in Michigan hospitals," *Am. J. Infect. Control*, vol. 32, no. 6, pp. 337–341, 2004.

Association of Medical Device Reprocessors, "AMDR Summary: International Regulation of 'Single-Use' Medical Device Reprocessing," no. 202, pp. 1–6, 2010.

Belkin, N. L., "Reuse of single-use devices," *AORN*, vol. 74, no. 1, p. 2001, 2001.

Bloomfield, S. F. and Miller, E. A., "A comparison of hypochlorite and phenolic disinfectants for disinfection of clean and soiled surfaces and blood spillages," *J. Hosp. Infect.*, vol. 13, no. 3, pp. 231–239, 1989.

CADTH, "Reprocessing of Single-Use Medical Devices_ A 2015 Update _ CADTH."

Campbell, G., "Medical Devices Iterim Regulation," no. 1, pp. 756–767, 2012.

Donawa, M., "FDA user fee and Modernisation Act," *Med. Device Technol.*, vol. 13, no. 10, pp. 27–29, 2002.

Dunn, d., "Reprocessing Single-use Devices—Regulatory Roles," *AORN J.*, vol. 76, no. 1, pp. 98–127, 2002.

European Commission, "Proposal for a REGULATION OF THE EUROPEAN PARLIAMENT AND OF THE COUNCIL on medical devices, and amending Directive 2001/83/EC, Regulation (EC) No 178/2002 and Regulation (EC) No 1223/2009".

Food and Drug Administration, "Recalls, Corrections and Removals (Devices)" 2016.

Issues, C., "Biologic test indicators; classification of medical devices; scope disinfection; reused single-use medical devices," vol. 64, no. 6, pp. 963–964, 1996.

Qian, Z. and Castañeda, W. R. "Can Labeled Single-Use Devices Be Reused? An Old Question in the New Era," *J. Vasc. Interv. Radiol.*, vol. 13, no. 12, pp. 1183–1186, 2002.

Rutala, W. A. "APIC guideline for selection and use of disinfectants," *AJIC Am. J. Infect. Control*, vol. 18, no. 2, pp. 99–117, 1990.

Statement, G., "AORN guidance statement: Reuse of single-use devices.," *AORN J.*, vol. 84, no. 5, pp. 876–884, 2006.

Tessarolo F., Caola I., Fedel, M., Stacchiotti, A., Caciagli P., Guarrera, G. M., Motta, A. and Nollo G., "Different experimental protocols for decontamination affect the cleaning of medical devices. A preliminary electron microscopy analysis," *J. Hosp. Infect.*, vol. 65, no. 4, pp. 326–333, 2007.

U. S. G. A. Office, "Report to Congressional Requesters," *Area*, no. July 2003.

Active workstations to improve job performance: A short review

Sara Maheronnaghsh & Mário Vaz
Research Laboratory on Prevention of Occupational and Environmental Risks (PROA), Faculty of Engineering, University of Porto, Portugal

Joana Santos
Scientific Area of Environmental Health, Research Centre on Health and Environment (CISA), School of Health of Polytechnic Institute of Porto (ESS.PPorto), Portugal

ABSTRACT: Modern changes in communication, transportation, domestic-entertainment technologies and workplaces are related to a substantial increase of sedentary living. This review was performed to summarize and analyze the studies investigating the impact of decreasing work sedentary time and implementing active workstations to improve work performance. This review was based on relevant articles published in Scopus, Medline, and Web of science from 2007 until 2017. The results of this brief review indicate statistically significant relationships between several interventions for increasing physical activity and employee work performance. Type of intervention and implementation procedure were identified as factors that can determine effects on the job performance. Some studies have shown interventions with positive influence on productivity, others suggest that it is feasible to implement a multi-component intervention such as Stand Up desk with high fidelity with no perceived decrease in productivity. In conclusion, more field studies are necessary to determine work performance and acceptance of dynamic workstations in a real work environment.

1 INTRODUCTION

When people is sit still for too long without sufficient active breaks, this is called a sedentary behavior. In fact, the prevalence of sedentary occupations or light intensity activities has increased from approximately 50% to more than 80% during the past five decades (Pronk, 2015). This change in occupational energy expenditure has been associated with a decrease of 100 calories per day, which, in turn, is purported to account for as much as 80% of the average increase in body weight among working men and women during this period (Church et al., 2011). Boyle and colleagues (2011) conducted a population-based case-control study of colorectal cancer in Western Australia in 2005–2007 and found that long-term sedentary work may increase the risk of distal colon cancer and rectal cancer (tumors that develop in the large intestine). Sedentary behavior and physical inactivity have independent health risks; even with regular exercise, people are still exposed to increased health risks if they have a sedentary lifestyle (Groenesteijn, Commissaris, Van Den Berg-Zwetsloot, & Hiemstra-Van Mastrigt, 2016).

Sitting less than 8h/day and meeting the physical activity recommendation of the World Health Organization (WHO) independently protected against all-cause mortality. WHO (2010) recommends 150 minutes of moderate-intensity aerobic physical activity (PA) per week (e.g. walking), or 75 minutes of vigorous-intensity aerobic PA per week (e.g. jogging), for adults between 18 and 64 years old. Center for Disease Control and Prevention (CDC) (2008) recommended 2.5 h of moderate intensity exercise and some muscle strengthening training per week—in total about 3.5 h per week of purposeful exercise. The average person sleeps 8.5 h/day and, consequently, 15.5 h/day or 108.5 h/week waking. This means that even for a person who meets the exercise guidelines, there are 105 h/week in which a person is not purposefully exercise or sleeping. During this period of time the activity level could be considered as sedentary (i.e., standing in line, sitting, or lying down) ranging between 1–1.5 met or as light activity (i.e., moving around, or leisurely walking at <3 mph (mile per hour) which vary between 1.6–2.9 met. Decreasing sedentary time and increasing light activity during the waking hours may be an important factor to control body weight regulation and chronic disease risk (Dutta, Koepp, Stovitz, Levine, & Pereira, 2014).

People with sedentary jobs (e.g. working on the computer) spend about 5 hours to 8 hours sitting during the work day (Genevieve N. Healy et al., 2013; McCrady & Levine, 2009). Sedentary work has been associated with several adverse health outcomes such as increased risk of cardiovascular disease, obesity, depression, Type II diabetes, musculoskeletal disorders and premature mortality (Gao et al. 2016).

The need to reduce the sedentary behaviors in workers is urgent. Thus, introducing periods of physical activity during working hours, with active workstation, is considered a good intervention to reduce sedentary work (Poochada and Chaiklieng 2015). These consists of adapting the workstation with devices, such as treadmill (see Figure 1) or cycle ergometer, that workers use while working normally at their desks. This will increase the physical activity and improve the health of employees (Bergman, Boraxbekk, Wennberg, Sorlin, & Olsson, 2015). The sit-to-stand workstation (see Figure 1) is another intervention that could be implemented in workplace with a good efficacy to reduce sedentary time (sitting) (Probst et al., 2013; Tudor-Locke et al., 2014).

In general, it has been proven that the introduction of physical activity during working hours is beneficial to the worker and organization, increasing productivity (von Thiele Schwarz & Hasson, 2011). Emerging evidence indicates positive effects of physical activity on work performance, fitness program participation on absenteeism, and vigorous physical activity on sick leave (Proper and van Mechelen, 2008).

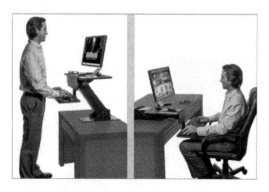

Figure 1. Examples of active workstations (treadmill and sit-to-stand, respectively).

Work performance have proven difficult to measure. Absenteeism could be used as to measure it. Researchers agree that performance has to be considered as a multi-dimensional concept. On the most basic level one can distinguish between a process aspect (i.e., behavioral) and an outcome aspect of performance (Sonnentag, Volmer, and Spychala 2008).

Therefore, the aim of this study was to explore the different types of intervention to increase physical activity in the workplaces and how these interventions affect work performance.

2 MATERIALS AND METHOD

The search was focused on literature pertaining to the effect of active workstations on quality of life and job performance.

The search strategy consisted of a comprehensive search that could locate the widest spectrum of articles published from 2007 until 2017 and was performed in selected electronic databases, namely: Scopus, Medline, and Web of science. Based on the electronic database used, the search terms were as follows: "active workplace", "sit stand desk", "treadmill", "work performance", "physical activity", "sedentary work" and "quality of life".

After importing all results, we screened with respect to title, abstract and key words. The exclusion criteria were as follows: duplicate printings and non English/Portuguese language. Only articles reporting quantitative or qualitative assessment or analysis of the influence of active workstations were included. One systematic review was also included in this study.

3 RESULTS AND DISCUSSION

The search strategy yielded a total of 40 citations before duplicates removal. After checking the duplicates, 38 papers were considered for screening. After application of the eligibility criteria while considering the full text, a total of 7 studies (see Table 1) were considered for the final analysis.

Therefore, we want to develop a conceptual framework that focuses on the impact of physical activity on work performance.

Ben-Ner et al. (2014) conducted a 12 month-long experiment in a financial services company on 43 employees (sedentary work), to understand how the availability of treadmill workstations affects employees' physical activity and work performance. Overall, performance was assessed with online detailed quarterly questionnaire concerning work, life and health. They found that, in general, quality and quantity of work performance and the interactions with coworkers improved as a result of adoption of treadmill workstations. Bernaards, Proper & Hildebrant (2007) investigated associations among three modifiable risk

Table 1. Overview of results—effects on job performance.

Ref.	Type of intervention	Effect on JP
Pronk et al. (2004)	Exercise	Increase
Carr et al. (2016)	Multi-component intervention: • elliptical machine • Organizational-level	Increase
Ben-Ner et al. (2014)	Treadmill workstations	Increase
Chau et al. (2016)	Sit-stand desks	Increase
Commissaris et al. (2014)	Three dynamic workstations: • Treadmill • Elliptical trainer • Bicycle ergometer	Objective: Increase Subjective: Decrease
Neuhaus et al. (2014)	Multi-component intervention: • height-adjustable workstations, • Organizational-level, • Individual-level.	No decrease
Bernaards et al. (2007)	Physical fitness and exercise	No association

JP: Job Performance.

factors as physical activity, cardiorespiratory fitness, body mass index and work productivity and sickness absence in computer workers. Productivity and sickness absence were assessed with Health and Performance Questionnaire (HPQ). The effects of physical activity on worker performance were less clear-cut. No association between self-reported physical fitness and work performance was found. Pronk et al. (2004) conducted a study that also tested the association between lifestyle-related modifiable health risks and work performance. In this study, dependent variables included number of work loss days, quantity and quality of work performed, overall job performance, extra effort exerted, and interpersonal relationships. To collect the data, they used a traditional Health Risk Assessment (HRA) survey and expanded the content to include questions related to work performance. Data were obtained from 683 workers and the results indicated that higher levels of physical activity increased the quality of work performed and overall job performance and decreased severe obesity related to a higher number of work loss days. Researchers concluded that lifestyle-related modifiable health risk factors had significantly positive impact on employee work performance.

Coulson et al. (2008) used a mixed methods design combined a randomized cross-over trial with concurrent focus groups. Three workplaces (two private companies, one public service organization) were purposefully selected for their provision of on-site exercise facilities, size (>250 employees) and large proportion of sedentary occupations. Two mood diary questionnaires were distributed to employees exercising on-site only. A 15-item work performance grid was completed at day-ends. Three on-site focus groups were held concurrently to explore performance-related topics. They found that employees self-rated job performance and mood were higher on days they exercised in the company gym than on days they did not.

Neuhaus et al. (2014) systematically summarizes the evidence for activity-permissive workstations on sedentary time, health-risk biomarkers, work performance and feasibility indicators in office workplaces. They identified studies with acceptability measures reported predominantly positive feedback. Findings suggested that activity-permissive workstations can be effective to reduce occupational sedentary time, without compromising work performance.

Commissari et al. (2014) evaluated short term task performance while working on three dynamic workstations: a treadmill, an elliptical trainer, a bicycle ergometer and a conventional standing workstation. A standard sitting workstation served as control condition. Fifteen Dutch adults performed five standardized office tasks in an laboratory setting. Both objective and perceived work performance were measured. With the exception of high precision mouse tasks, short term work performance was not affected by working on a dynamic or a standing workstation. In this study they clarified that standing and dynamic (or active) workstations offer the possibility to increase physical activity and reduce sedentary time at the workplace during daily office work. They considered performance of standard office tasks (reading and correcting, re-typing a presented text and cognitive tasks) that was hardly affected while using a standing or a dynamic workstation. However, a computer task that requires fine motor actions of the hands (mouse pointing and clicking) was affected by the movements at a dynamic workstation. Despite, generally, an equal objective performance on dynamic workstations, participants perceived that their performance deteriorated.

Carr et al. (2016) tested an integrated intervention on occupational sedentary/physical activity behaviors, cardiometabolic disease biomarkers, musculoskeletal discomfort, and work productivity. They estimated workplace costs of health problems in terms of reduced job performance, sickness absence, and work-related accidents/injuries. Participants received an ergonomic workstation optimization (elliptical machine, active Life Trainer) intervention and three e-mails/week promoting rest breaks and posture variation. Other group of participants received the intervention plus access to a seated activity permissive workstation. No significant intervention effects were observed for any work productivity items measured on the Health and Work Performance Questionnaire.

As a result they found that increased occupational physical activity and greater activity permissive workstation adherence was associated with improved health and work productivity outcomes.

Some studies (see Table 1) have shown interventions with positive influence on productivity (Ben-Ner et al. 2014; Carr et al. 2016; Chau et al. 2016; Pronk et al. 2004). Neuhaus et al. (2014) suggested that it is feasible to implement a multi-component intervention such as was used in Stand Up desk with high fidelity, no perceived decrease in productivity.

Bernaards et al. (2007) didn't found any association between self-reported physical fitness and work performance. Commissaris et al. (2014) found positive results on objective measures of performance with dynamic workstations, however participants perceived that their performance deteriorated. The results of this brief review indicate statistically significant relationships between several interventions for increasing physical activity and employee work performance. A low-to-high degree of association has been reported in the different literatures between active workstations and performance.

4 CONCLUSION

The results of this review demonstrate that providing interventions for increasing physical activity in various workplaces have different impacts on productivity. More research in real work environment are necessary to determine work performance and acceptance of dynamic workstations. To the best of our knowledge, using different interventions simultaneously (organizational measures and design good active workstations), could be the most effective strategy to increase physical activity in the workplaces.

REFERENCES

Ben-Ner, Avner et al. 2014. "Treadmill Workstations: The Effects of Walking While Working on Physical Activity and Work Performance." *PLoS ONE* 9(2).

Bernaards, Claire M, Karin I Proper, and Vincent H Hildebrandt. 2007. "Physical Activity, Cardiorespiratory Fitness, and Body Mass Index in Relationship to Work Productivity and Sickness Absence in Computer Workers with Preexisting Neck and Upper Limb Symptoms." *Journal of occupational and environmental medicine* 49(6): 633–40.

Boyle, Terry, Lin Fritschi, Jane Heyworth, and Fiona Bull. 2011. "Long-Term Sedentary Work and the Risk of Subsite-Specific Colorectal Cancer." *American journal of epidemiology* 173(10): 1183–91.

Church, T S et al. 2011. "Trends over 5 Decades in U.S. Occupation-Related Physical Activity and Their Associations with Obesity." *PLoS One* 6. http://dx.doi.org/10.1371/journal.pone.0019657.

Commissaris, Dianne A C M et al. 2014. "Effects of a Standing and Three Dynamic Workstations on Computer Task Performance and Cognitive Function Tests." *Applied Ergonomics* 45(6): 1570–78. http://www.sciencedirect.com/science/article/pii/S0003687014000891.

Coulson, J C, J McKenna, and M Field. 2008. "Exercising at Work and Self-Reported Work Performance." *International Journal of Workplace Health Management* 1(3): 176–97. https://www.scopus.com/inward/record.uri?eid=2-s2.0-84986021398&doi=10.1108%2F17538350810926534&partnerID=40&md5=d162cf0459f0bd289e3bf9ecb9cdf51e.

Dutta, Nirjhar et al. 2014. "Using Sit-Stand Workstations to Decrease Sedentary Time in Office Workers: A Randomized Crossover Trial." *International Journal of Environmental Research and Public Health* 11(7): 6653–65. http://www.mdpi.com/1660-4601/11/7/6653.

Gao, Ying, Nina Nevala, Neil J Cronin, and Taija Finni. 2016. "Effects of Environmental Intervention on Sedentary Time, Musculoskeletal Comfort and Work Ability in Office Workers." *European journal of sport science* 16(6): 747–54.

Groenesteijn, L., D. A. C. M. Commissaris, M. Van Den Berg-Zwetsloot, and S. Hiemstra-Van Mastrigt. 2016. "Effects of Dynamic Workstation Oxidesk on Acceptance, Physical Activity, Mental Fitness and Work Performance." *Work* 54(4): 773–78.

Neuhaus, M et al. 2014. "Reducing Occupational Sedentary Time: A Systematic Review and Meta-Analysis of Evidence on Activity-Permissive Workstations." *Obes Rev* 15. http://dx.doi.org/10.1111/obr.12201.

Poochada, W, and S Chaiklieng. 2015. "Ergonomic Risk Assessment among Call Center Workers." *Procedia Manufacturing* 3: 4613–20. https://www.scopus.com/inward/record.uri?eid=2-s2.0-85009962397&doi=10.1016%2Fj.promfg.2015.07.543&partnerID=40&md5=14a5705e9aec78ebc5e0033700ca80ed.

Probst, K., Lindlbauer, D., Perteneder, F., Haller, M., Schwartz, B., & Schrempf, A. 2013. Exploring the Use of Distributed Multiple Monitors within an Activity-Promoting Sit-and-Stand Office Workspace. In Kotze, P and Lindgaard, G and Wesson, J and Winckler, M (Ed.), Human-computer Interaction—Interact 8119: 476–493.

Pronk, Nicolaas P et al. 2004. "The Association between Work Performance and Physical Activity, Cardiorespiratory Fitness, and Obesity." *Journal of occupational and environmental medicine* 46(1): 19–25.

Pronk. 2015. "Fitness of the US Workforce." *Annual review of public health* 36: 131–49.

Proper, K., & Van Mechelen, W. 2008. Effectiveness and economic impact of worksite interventions to promote physical activity and healthy diet. Dalian, China. Retrieved from http://www.who.int/dietphysicalactivity/Proper_K.pdf?ua=1.

Sonnentag, Sabine, Judith Volmer, and Anne Spychala. 2008. "Job Performance." *The SAGE Handbook of Organizational Behavior* 1: 427–47.

Tompa, Emile. 2002. "The Impact of Health on Productivity: Empirical Evidence and Policy Implications." *The Review of Economic Performace*: 181–202.

Tudor-Locke, C., Hendrick, C. A., Duet, M. T., Swift, D. L., Schuna Jr., J. M., Martin, C. K., ... Church, T. S. 2014. Implementation and adherence issues in a workplace treadmill desk intervention. Applied Physiology Nutrition and Metabolism, 39(10). http://doi.org/10.1139/apnm-2013-0435.

Case study of risk assessment in single family housing

Cristina Reis, Vera Gomes & Paula Braga
Universidade de Trás-os-Montes e Alto Douro, Vila Real, Portugal

J.F. Silva
Instituto Politécnico de Viana do Castelo, Viana do Castelo, Portugal

José António F.O. Correia
INEGI—FEUP, Portugal

Carlos Oliveira
Instituto Politécnico de Viana do Castelo, Viana do Castelo, Portugal

ABSTRACT: In civil construction, prevention is done by identifying and eliminating and/or controlling the accidents causes.
Risk is present everywhere, so it's essential to prevent that what is a possibility may become a reality. Accidents always bring great damage to companies: humans; social rights; materials; image and confidence on the part of the insurance companies and banking entities. The builder should have in this document an aid in controlling the risks in order to reduce the above losses. Fewer accidents contribute to increased productivity and production quality. The management of worker safety is a matter of great importance to the developer, as is cost management.
The present work has as background the study of the problematic related to the area of the security in the construction work. Although in recent years investing, more and more, in this topic has been a reality it has not yet been seen significant improvements in the construction sector.
At a first stage of the investigation daily visit to works of the company Augusto and Afonso da Cunha Ld was made. During these visits, the nonconformities found were collected and later analyzed in detail. From this analysis 25 variables were defined, in terms of importance in the occurrence of accident risk.
In a second phase of the study, the two-variable variables were cross-checked, which allowed the analysis and interpretation of the dependence relationship between them. By comparison with the mathematical model, a matrix was elaborated where the variables that, when conjugated increase the risk of accident, were observed.

1 INTRODUCTION

1.1 *Considerations*

As far as safety is concerned, the civil construction sector is one of the poorest sectors, with a high rate of accidents, serious injuries and deaths.

For socio-cultural nature reasons and despite the positive evolution of the most recent years, Portuguese society, at all levels, still has a strong desire to take an active and rational attitude towards accidents and, as a consequence, to relegate the prevention of occupational risks to a second plan.

Safety in construction has long been a subject that was not always treated with due importance.

Since it is an economic sector in which work has a strong manual component, where the worker's predisposition to perform the tasks directly influences the performance and is dependent on the capacities and conditions of the employees, all actions to improve the work conditions can be positively reflected in productivity.

Reducing occupational accidents is therefore an essential objective, the pursuit of which requires the involvement of all actors, in particular SMEs, the self-employed, the social partners, public authorities, insurers, social security institutions and labor inspection services.

It is therefore necessary to embark on a path towards the improvement of physical and psychic working conditions, and it was in this sense that this work was developed.

It is in this perspective that the accomplishment of this work may become advantageous, insofar as it can perceive the true fatality problem, evaluate it and seek solutions/improvement, based on a comparative and in-depth approach to occupational accident data.

The purpose of this research was to survey non-conformities found in the works, analyze them, identify the dependence relation of the variables that are the source of the risk of accidents and evaluate the possible consequences.

2 METHODOLOGY

This work consisted on the analysis of the implementation of safety at work making daily visits and data collection that allowed examination of the work situation, relative to safety.

The duration of the data collection period was a constraint in this study, since it was limited to less than three months and it was not possible to analyze the evolution of the work.

The existing mathematical model, on which this study was based, is an exhaustive work of surveying and analyzing the processes of work accidents, being the comparison of this study with the same limited model, since there were no accidents.

This study was carried out only in a small, family-run company with few workers, which confined our inquiries to the same small number of workers.

2.1 Data collection

For the elaboration of this study four works were visited and, as it might be expected, in all of them problems were found related to the health and safety of workers.

In order to achieve this purpose, documents were created that better identify the anomalies. Fifty-six non-conformities were filed against the applicable legislation, and descriptions of the preventive actions to be carried out were drawn up.

In order to obtain an overall situation of the works with regard to safety, it was decided to submit work-visit reports. These contain information on situations that merited immediate and/or short-term intervention at the level of:

✓ Safety papers
✓ Site organization
✓ Collective protection
✓ Individual protection
✓ Work equipments
✓ Excavations
✓ Earth movements
✓ Cranes
✓ Scaffolding
✓ Stairs
✓ Electrical installation and power tools of the shipyard
✓ Pillars, beams and slabs
✓ Roof works

Following the collection and analysis of non-conformities, and in order to understand how accidents can occur, work records were created, which were subdivided into twenty-six variables, referring to the worker, work and means, and used equipment:

✓ Gravity
✓ Hour
✓ Day
✓ Season
✓ Material factor
✓ Consequences
✓ Collective protection
✓ Individual protection
✓ Job
✓ Age
✓ Equipment malfunction
✓ Company size
✓ Administrative region
✓ Nationality
✓ Preventive measures
✓ Site organization
✓ PSS
✓ Security personnel
✓ Possible consequences
✓ Work schedule
✓ Job situation
✓ Task
✓ Sub-task
✓ Kind of job
✓ Type of work

3 DATA ANALYSIS

Some examples of the characterization of each variable and its accounting are presented below.

After crossing two variables and comparing with the existing mathematical model it was necessary to compile the results into a matrix.

It is essential to present proposals for improving safety conditions in construction so that all the actors know what the main risk factors are, and if a combination of variables is to be found on the job, the actors may be alerted in order to avoid a possible accident.

The obtained results gave us a better understanding of situations where the risk of accidents is higher and which must be avoid.

Through the matrix it is possible to observe which factors have not affected the severity of the accident risk, such as: the size of the company, the region, the nationality, the existence or not of PSS and safety personnel, working hours and the employment situation, the type of work and the duration of the work.

Hour	Accidents Risk
8h00-10h00	20
10h00-12h00	13
12h00-14h00	3
14h00-17h00	14
17h00-20h00	6
Total	56

Figure 1. Quantification of the hour variable.

Gravity	Accidents Risk
Not Serious	13
Serious	21
Very Serious	22
Total	56

Figure 2. Quantification of gravity variable.

Possible Consequences	Accidents Risk
Rollover	1
Fall of objects	15
Fall of height	36
Level drop	3
Burial	1
Total	56

Figure 3. Quantification of possible consequences.

The variables that most affected possible accidents are the variables directly related to the construction, such as: the preventive measures infringed the lack of individual and collective protection equipment, the task and the sub-task to be executed.

4 CONCLUSION

With the realization of this study more knowledge was acquired of the construction safety situation. Recommendations may now be proposed to improve safety.

It has been verified throughout this work that, in terms of security, there is still a lot of work ahead, both in the situation of workers and at the level of employers.

With the obtained results it is possible to better understand which variables cross each other to increase the probability of occurrence of accidents.

Following this work it was verified that the workers do not have the necessary training for the execution of their tasks, in security. Whether due to ignorance or unfamiliarity, workers continue to use neither personal protective equipment nor collective protection equipment.

Employers continue to fall into safety and modernization of means; obsolete, incomplete and dangerous equipment for the safety and health of workers continue to exist. They do not train workers for the tasks they will perform and for the equipment they will use.

Workers are still tied to experience, with high levels of self-confidence that put them at risk.

Based on this research, safety coordinators, safety technicians, employers, and construction managers are advised on a series of suggestions for the improvement of their professional activities in order to provide critical awareness and training. The following recommendations are suggested:

✓ To analyze the risks to which the workers are exposed, in the execution of the main tasks in the civil construction works
✓ To raise the real costs of accidents occurring in the works, in order to motivate and raise the awareness of the management of the companies through them,
✓ Carry out the safest methods and the appropriate equipment to carry out the main tasks,
✓ Carry out a detailed analysis of the accidents that occurred and through these a profile of the greatest needs in terms of safety,
✓ Contact the workers themselves to identify the main risks to which they are subject,
✓ Research incentive methods for workers who can sensitize workers to the need to use personal and collective protective equipment and provide training in how to use them.

REFERENCES

Cristina Madureira dos Reis (2008). Melhoria da Eficácia dos planos de segurança na redução dos acidentes na construção. Dissertação de doutoramento apresentada para a obtenção do grau de Doutor Engenharia Civil, pela Faculdade de Engenharia da Universidade do Porto.

Dos Reis, Cristina (2000) - Análise da Implementação dos Planos de Segurança em obra—Faculdade de Engenharia da Universidade do Porto, Porto.

Ministério da Segurança Social e do Trabalho—Decreto-lei n.º 273/2003, DR—I Série A, n.º 251, de 29 de Outubro de 2003, referentes as condições de segurança no trabalho em estaleiros temporários ou móveis que vem revogar o decreto-lei n.º 155/95 de 1 de Julho de 1995.

Reis, Cristina, Soeiro Alfredo (2005) – Economia dos Acidentes na construção—Simulação e Análise, Estudo de Alguns Casos, ISHST, Lisboa.

Vera Fernandes Gomes Vera Dulce (2010). Avaliação em obra dos riscos de acordo com o risco provável de acidente em habitações unifamiliares. Dissertação apresentada para a obtenção do grau de Mestre em Engenharia Civil, pela Universidade de Trás-os-Montes e Alto Douro.

Evaluation of the luminous performance of a public university library in the Northeast of Brazil

A.R.M.V. Silva, M.N. Almeida, J.I.F. Machado, B.C.C.B. Carvalho & H.C. Albuquerque Neto
Department of Industrial Engineering, Federal University of Piauí, Brazil

ABSTRACT: The environmental comfort of a place has a great influence on the performance of occupants when they carry out their activities, besides contributing to their well-being; it is basically divided into thermal, acoustic and luminous comfort. Thus, this work has as main objective to evaluate the levels of illuminance to which occupants are submitted in the reading areas of the Central Library of the Federal University of Piauí, in the city of Teresina. Its methodology is of a basic and descriptive nature, with a mixed approach, being framed as a case study. For that, one instrument were used: the Lux Meter, and with the aid of R × 64 2.15.0® software, Statistic and Microsoft Excel®, statistical tests were performed to evaluate the behaviour of the illuminance variable amidst the shifts, days and reading areas and their fulfilment to NBR 8995-1, a surface response was elaborated to verify the distribution of the illumination. From the study, it was observed that, in general, the library is not in compliance with the current norm, and there was no significant change in the average level of illuminance between the shifts and days of measurement, except for the reading areas chosen for analysis. It was also verified that the reading area on the second floor has a better luminous performance in comparison to the others. In this way, it is advisable to repair the electrical installations of this library, besides the placement of more luminaires near the windows and replacement of the lamps by the LED ones.

1 INTRODUCTION

The comfort sensation is essential for the humans to have pleasure and well-being in accomplishing their activities. This sensation is related to psychophysiological factors and may interfere in the behaviour of individual, who reacts differently to each external stimulus (Segatin & Maia, 2007).

Corroborating with the authors, Coutinho Filho et al. (2010) explain that a system's favourable environmental conditions are important for the performance, protection and well-being of occupants. These good environmental conditions directly reflect upon the productivity and quality of the activity carried out. In view of that, so that occupants feel fine in their environment, they need to enjoy a favourable environmental comfort situation. Therefore, the place's environmental comfort conditions directly influence the occupant's performance in carrying out leisure and labour activities, especially, in learning-oriented facilities, as is the case of libraries.

For Ochoa; Araújo & Sattle (2012), in order to meet a satisfactory environmental performance, a series of variables should be involved in the study of the environment, amongst them the acoustic, thermal and luminous comfort variables. Thereby, the environmental comfort of a place can be understood as sufficiency of the variables involved to the use of man, thus, thermal, acoustic and luminous conditions must be taken into account (NERBAS, 2009).

Concerning an educational environment, some of these variables take over the others as regards importance level when referring to the reading activity, e.g. the luminous variable. Such fact was confirmed by Samani (2012) who researched and verified that luminous comfort in learning environments improves and increases the occupant's performance and productivity.

For Lambert; Dutra & Pereira (2014), the determining factor for efficiently developing activities are the quality and the quantity of illumination, as it is through them that the visual perception of spaces and the object of attention is attained, defining it as visual comfort. According to Couto (2007), quality of illumination refers to luminance and the quantity is related to illuminance of the environment. For that reason, this work aims to evaluate the illuminance levels to which occupants of a higher education library are submitted in its reading areas.

Brazil establishes the minimal illuminance levels to be reached in function of the visual activity type as per NBR 8995-1, in effect as of 2013, replacing NBR 5413 of 1992 due to a large time without any revision.

Poor luminous ambience demands a larger effort from an individual's sight, who may exhibit visual strain and cephalalgia (headaches) as an immediate consequence (Queiroz et al. 2010). In addition, the author comments that, in the long term, depending on the task exposure time, may have his/her visual capacity reduced, although this does not cause permanent sight damages. Iida and Guimarães (2016) add that visual strain causes eye rash and watery eyes, with that the individual starts to blink more frequently, making the sight blurred or even causing diplopia.

For Mendonça (2016), in the activities demanding visual effort, as is the case of reading, low or excessive luminosity may lead to this strain condition, in addition to causing irritability and the consequent yield decrease. Therefore, a study that deals with this aspect of environmental comfort in an educational setting has an important impact not only on the performance of tasks, but also on life quality of those enjoying this setting.

Furthermore, such research is justified by the fact that the library was established in 1973, hence the architectural design took into account the luminous standards from that time. However, NBR 8995-1 took effect in April 2013, and there is no research regarding it. Thus, this luminous study is pertinent and it may contribute to increase the current knowledge context on the theme.

2 METHODOLOGY

2.1 *General aspects of data collection*

Initially, the study's variables were divided into two dimensions: the first is the dependent variable and the second refers to the independent ones, which can vary the illuminance levels, shown in Table 1.

In Table 1, the level of the Illuminance variable was considered as the dependent one, expressed in lux by the pointer, while the remainder ones represent the study's independent variables that may likely vary the illuminance level.

Additionally, as data collection instruments for the quantitative data survey referring to levels of illuminance, the LDR-22 model handheld digital lux Meter was used, duly gauged, in addition to the aid from a Microsoft Excel® spreadsheet.

Accordingly, the illuminance measurement points of the reading areas were attained in the library, resulting in 101 points. At each point, three measurements were accomplished, so that the average of such measurements could be established afterwards. This procedure occurred during the three shifts (morning, afternoon and evening) for 3 days, adding up 2727 measurements implying 909 observations.

Table 1. Dimensions of the variables and pointers.

Dimension 1 – Illuminance values observed of the illuminance level with the aid of a lux meter, expressed in lux

Variable	Pointer
Illuminance level	Lux

Dimension 2 – Variations in the illuminance levels amongst areas, shifts and days

Variables	Pointers
Area	Upper or Lower
Shift	Morning, Afternoon or Evening
Day	1, 2 or 3

With that, these values were compared to the ones recommended by NBR 8995-1; however, before this measurement, a pilot test was carried out for knowing the equipment's technical specifications.

2.2 *Treatment and analysis of data*

The data collected were systematically organized for carrying out a detailed analysis, then there was tabulation of observations collected referring to levels of illuminance with the aid of electronic spreadsheets. After this tabulation and the organization, the data were compared to the ones established by NBR 8995-1, as mentioned before.

Afterwards, the data were imported to the R × 64 2.15.0® software, allowing the performance of the basic statistics calculations (central tendency and dispersion measurements) for each shift, reading area and measurement day, as well as a general analysis that considers all points indistinguishably of area, shift or day. Also on the R × 64 2.15.0® software, boxplot graphs were elaborated aiming to have a better visualization of the dependent variable's behaviour (illuminance) per shift, area and measurement day.

As a supplement for the above-mentioned tools, some statistical tests were used. Firstly, the normality of observations of the dependent variable was verified by means of the histogram and then the Shapiro-Wilk and Lilliefors test was used, in which both have as alternatives the null hypothesis (H_0): normal distribution, and the alternative hypothesis (H_1): non-normal, in which for a p-value less than 0.05, that is, for 5% significance, the null hypothesis is rejected, not exhibiting a normal probability distribution.

The use of the Wilcoxon non-parametric test was also necessary, which considers as a null hypothesis (H_0): the means are equal, and alternative hypothesis (H_1): the means are different; the test statistics is the p-value that, in case it is higher than 0.05, then one cannot reject the null hypothesis.

3 RESULTS AND DISCUSSION

From the tabulation of observations, the statistical data analyses were started; for them, three perspectives of the illuminance level variation were regarded, due to the independent variables, the area, the shift and the measured day. In Table 2, there are the central tendency and dispersion measurements of the dependent variable (illuminance) for each independent variable previously exhibited, and, lastly, a general analysis that considers all points indistinguishably of area, shift or day.

With Table 2, a better visualization can be attained of the behaviour of illuminance levels, and it is possible to verify that none of the means have values higher than or equal to 500 lux, which is the required value for library reading areas as per NBR 8995-1.

Another important issue to be mentioned referring to Table 2 is that in the maximum values found only areas 1 and 2 did not reach values within the recommended limit by the standard and, thus, these areas did not have points above the recommended limit, being totally outside the required standard. Contrariwise, the other variables (shifts, area 3 and days) had points within the tolerance limit of the current standard. Important to point out is that all values of this column in different shifts and days belong to measurements of area 3.

As for the standard deviations, they were shown similar amongst shifts, days and the general, whereas for the areas there was larger disparity, especially between area 3 and areas 1 and 2, being this the first indication of a more significant variation amongst the levels of illuminance measured in distinct areas.

For a better visualization of these measurements, box plot graphs were created that show, more clearly, the variations of the illuminances performances per shift, area and day, according to Figures 1–3.

Observing the box plot graphs of the independent variables Shift and Day, represented by Figure 1 and Figure 2, respectively, one can note that they do not show a large variation in the mean of values observed of illuminance and, additionally, a concentration of observations is seen around the value 200 lux.

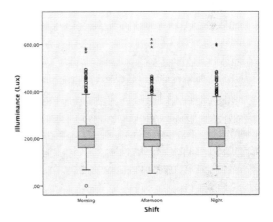

Figure 1. Illuminance × Shift graph.

Table 2. Central tendency and dispersion measurements.

Variables	Illuminance values (lux)		
	Minimal values	Maximum values	Standard deviation
Shift 1	67.8	583.4	92.30
Shift 2	52.5	621.3	90.65
Shift 3	70.3	600.3	90.39
Area 1	92.8	**345.2**	58.98
Area 2	86.1	**303.6**	44.83
Area 3	52.5	621.3	128.55
Day 1	52.5	621.3	91.21
Day 2	70.3	598.7	91.72
Day 3	69.1	604.7	91.40
GENERAL	52.5	621.3	91.02

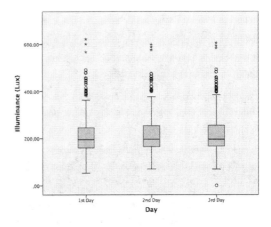

Figure 2. Illuminance × Day graph.

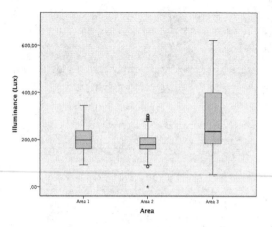

Figure 3. Illuminance × Area graph.

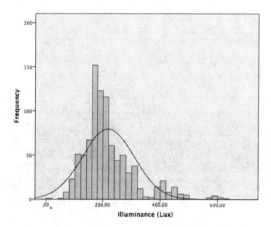

Figure 4. Illuminance histogram.

Table 3. Shapiro-Wilk and Lilliefors test.

Variables	P_{values} – Shapiro-Wilk Test	P_{values} – Lilliefors Test
Shift 1	$1.287 \times e^{-14}$	$<2.2 \times e^{-16}$
Shift 2	$5.594 \times e^{-14}$	$2.873 \times e^{-16}$
Shift 3	$6.659 \times e^{-15}$	$<2.2 \times e^{-16}$
Area 1	$1.353 \times e^{-5}$	0.006918
Area 2	$3.51 \times e^{-6}$	$3.851 \times e^{-9}$
Area 3	$3.239 \times e^{-7}$	$1.479 \times e^{-12}$
Day 1	$1.814 \times e^{-14}$	$3.226 \times e^{-15}$
Day 2	$1.174 \times e^{-14}$	$<2.2 \times e^{-16}$
Day 3	$2.621 \times e^{-14}$	$<2.2 \times e^{-16}$

However, the graph of the Area variable (Fig. 3) presents indications of a variation in Illuminance, especially between area 3 and the remainder ones (1 and 2), evidencing, once more, the indication of variation amongst the areas already shown by the standard deviation in Table 2. Nevertheless, this indication should be verified by means of statistical tests to prove the exposition herein.

For such, the normality of observations of the dependent variable was verified by means of the illuminance histogram (Fig. 4) and then the Shapiro-Wilk and Lilliefors test was used, shown in Table 3.

With the aid of Figure 4, the majority of samples are found within the intervals referring to the values of 150 to 250 lux, and also noteworthy is that there are only very few values higher than 500 lux, required by NBR 8995-1. It is important to point out that these values belong to the same point of the upper area. It is also possible to suppose that the profile of observations is not similar to the profile of a normal probability distribution, as the observations would have to be supposedly equally divided around a mean, which in this case is not well defined. Still, for proving this indication the Shapiro-Wilk and Lilliefors normality tests, shown in Table 3, were applied.

As seen, for the above-mentioned tests the alternatives of hypotheses are thus regarded: null hypothesis (H_0): normal distribution and alternative hypothesis (H_1): non-normal. Hence, through Table 3, one can observe that the results of the Shapiro-Wilk and Lilliefors tests for the three independent variables (shift, day and area) showed a p-value less than 0.05, i.e., for 5% significance the null hypothesis can be rejected, therefore, the illuminance measured does not exhibit a normal probability distribution, confirming thus the indication suggested by the observation of Figure 4 of the histogram graph.

In view of that, it is common to perform the Wilcoxon non-parametric test to confirm whether there is or not a variation in the illuminance mean due to the three independent variables (shift, day and area). Therefore, it was applied to validate the indication of variation of the means shown in Table 2, with the column of standard deviation, and in the analysis of the box plot graphs of Figures 1–3, the test results are exhibited in Table 4.

Since, for this test to be considered as a null hypothesis (H_0): the means must be equal, and the alternative hypothesis (H_1): the as means must be different, through Table 3, one can observe that the Wilcoxon test results for the variables Shift and Day showed a p-value higher than 0.05, that is, for 5% significance, one can infer that the null hypothesis cannot be rejected, and the means are equal. Therefore, only for the Area variable the means can be considered distinct, which also confirms the difference found in the box plot graph of Figure 4, as the result for this independent variable exhibited

Table 4. Wilcoxon Test.

Wilcoxon Test	P_{values}
Shift 1 – Shift 2	0.8558
Shift 2 – Shift 3	0.9191
Shift 3 – Shift 1	0.7702
Area 1 – Area 2	$4.981 \times e^{-5}$
Area 2 – Area 3	$<2.2 \times e^{-16}$
Area 3 – Area 1	$7.531 \times e^{-12}$
Day 1 – Day 2	0.5485
Day 2 – Day 3	0.9104
Day 3 – Day 1	0.449

a p-value less than 0.05, where the null hypothesis is rejected in which there is variation of the means for 5% significance.

Such evidence shows that daylighting may not be properly utilized, once there is no significant difference between measurements performed in the morning and the afternoon shifts and the measurements in the evening, which can be related both to the library's architectural design and the shape of the individual study desks, which are booths.

4 CONCLUSION

With the measurements, the majority of the illuminance levels measured in the points are from 150 to 250 lux, the overall mean of illuminance is 217.5 lux, and by comparing it to NBR 8995-1, the illuminance levels of points are, in a general manner, lower than the minimal limit of 500 lux recommended by the standard. With only one point, in area 3, which attained values higher than or equal to those of the standard in all shifts and days of measurements.

The statistical tests also revealed that only among the areas there was variation in the illuminance levels, and area 3 was the one to vary the most as regards the others. As for the shifts and days of measurement, there was no significant variation.

Also concerning the three reading areas analysed, area 3 (upper) presented a larger mean of the illuminance level, when compared to the rest. Such fact can be justified by the correct proximity from the lamps to the desks, the quantity of lamps operating and their uniform distribution. Therefore, there should be a decrease of the height between the lamps and the reader's visual field in areas 1 and 2 (lower), for a better luminous performance in these areas. Moreover, the library needs a periodic maintenance of installations of the luminaires in the three areas studied.

REFERENCES

Associação Brasileira de Normas Técnicas (ABNT). 1992. *NBR 5413: Iluminância de interiores*. Rio de Janeiro: Brasil.
Associação Brasileira de Normas Técnicas (ABNT). 2013. *NBR 8995-1: Iluminação de ambientes de trabalho - parte 1: Interior*. Rio de Janeiro: Brasil.
Coutinho Filho, E. F. et al. 2010. Avaliação do conforto ambiental em uma escola municipal de João Pessoa. In: *IX Encontro Nacional de Extensão*. Anais... João Pessoa.
Couto, H. A. 2007. *Ergonomia Aplicada ao Trabalho: conteúdo básico: guia prático*. Belo Horizonte: Ergo Ltda. 272p.
Iida, i.; Guimarães, L. B. M. 2016. Ergonomia: Projeto e Produção. (3. ed.) São Paulo: Blucher Ltda., 850p.
Lambert, R.; Dutra, L. & Pereira, F.O.R. 2014. *Eficiência Energética na Arquitetura*. (3 ed.) São Paulo: Eletrobras/Procel.
Mendonça, S. E. A. *Efeitos fisiológicos da iluminação*. Disponível em: <https://www.google.com.br/url?sa=t&rct=j&q=&esrc=s&souce=web&cd=2&cad=rja&uact=8&ved=0ahUKEwi2wP7G-6_QAhXHTZAKHQpVB5gQFggmMAE&url=http%3A%2F%2Fwww.sinal.org.br%2Fdownload%2FInsalubridade%2Filuminacao.doc&usg=AFQjCNElOvA4AutEhEP25K5LYGaYkkvi8Q&sig2=PjZbETz_cwHrbY5U_cOEvw&bvm=bv.139250283,d.Y2I> Acesso em: 20 set. 2016.
Nerbas, P. F. 2009. *Estudo arquitetônico para gestores imobiliários*. Curitiba: IESDE Brasil S.A.,
Ochoa, J. H.; Araújo, D. L. & Sattler, M. A. Análise do conforto ambiental em salas de aula: comparação entre dados técnicos e a percepção do usuário. *Ambiente Construído*. Porto Alegre, v. 12, n. 1, p. 91–114, jan./mar. 2012.
Queiroz; M. T. A. et al. Estudo de caso: Impactos da iluminação inadequada em área de internação hospitalar. In: *VII Simpósio de Excelência em Gestão e Tecnologia*, 2010, Rio de Janeiro. Anais... Rio de Janeiro, p. 1–12, 2010.
Samani, S. A. 2012. The impact of indoor lighting on students' learning performance in learning environments: A knowledge internalization perspective. *International Journal of Business and Social Science*. Malaysia, v. 3, n. 24, p. 127–136.
Segatin, B. G. O. & Maia, E. M. F. L. 2007. *Estresse vivenciado pelos profissionais que trabalham na saúde*. 44p. Monografia do curso de pós-graduação em saúde da família, Instituto de Ensino Superior, INESUL, Londrina.

Thermal comfort assessment of orthopaedic health professionals in an operating room

Nelson Rodrigues, A.S. Miguel, Senhorinha Teixeira & Constantino Fernandes
University of Minho, Portugal

J. Santos Baptista
PROA/LABIOMEP/INEGI, Faculty of Engineering, University of Porto, Porto, Portugal

ABSTRACT: The present work aims to study thermal sensation of orthopaedic health professionals during the execution of their activities in an operating room. The development of a questionnaire allowed the knowledge of the personal factors and thermal sensation of those professionals when exposed to their work environment. Considering the data similarities, four different classes were created: "Surgeon", "Surgeon's Assistant", "Instrumentalist" and "Nurses and Auxiliaries". Among the considered classes, the study focuses in two: "Surgeons" and "Nurses and Auxiliaries" which presented considerable differentiation. The first class presented a thermal sensation directed towards a hotter thermal sensation, while the latter on was directed to a colder thermal sensation. Additionally, it was verified that the surgeons present a greater intolerance towards thermal environment variations when compared to the nurses and auxiliaries.

1 INTRODUCTION

In industrialised countries, people spend most of their time indoors where they develop a wide variety of tasks (Höppe & Martinac, 1998). This amount of time pushes people to develop a comfortable indoor environment to increase their health and well-being.

According to Parsons (2014), a comfortable environment is also crucial to increase people's professional performance. Among the indoor buildings, healthcare facilities, namely operating rooms, present one of the most complex and demanding environments (Balaras, Dascalaki, & Gaglia, 2007). The first role of an operating room is to provide a clean environment, protecting the patient and the surgical staff from infectious particles. The control of the indoor environment varies according to the room requirements and local legislation (Decreto Regulamentar N.º 63/94, 1994). However, due to the restrict conditions, it is necessary the implementation of HVAC systems that guarantee a minimum of air changes per hour, and that is capable to maintain a low temperature, generally not higher than 24°C (ASHRAE Standard 170P, 2006). While focusing on the biological protection, the environment can lose quality regarding the thermal comfort, and should also be taken into account (Van Gaever, Jacobs, Diltoer, Peeters, & Vanlanduit, 2014). Assessing the indoor quality of an operating room is the first step to guarantee that thermal comfort is not neglected. A study performed by Balaras et al. (2007), in 20 operating rooms at Greece, showed that the majority of the rooms did not presented the conditions stipulated on the norms, presenting a poor control of the thermal conditions, between other problems.

In addition to the increased productivity, well-being and health, a comfortable staff is less prone to errors. This is especially important in the case of operating rooms where an error can jeopardise the patient's life. Thermal comfort assessment is commonly achieved using predictive methods such as the Fanger index that make use of the thermal neutrality concept (Djongyang, Tchinda, & Njomo, 2010; Fanger, 1970; Khodakarami & Nasrollahi, 2012). This is achieved when the heat balance between the human body and the environment is zero. However, the use of Fanger index has been showing incoherencies in the hospital dynamics (De Giuli, Zecchin, Salmaso, Corain, & De Carli, 2013). The direct methods, on the other hand, are more precise since they directly measure people's perception of the thermal environment and account for the cases where the thermal preference is different from thermal neutrality (Miguel, 2014; Parsons, 2014). On the downside, they require a high quantity of data to account for everyone's opinion. For a feasible data acquisition, authors generally use questionnaires to obtain a sample of the population in the study. The present study also adopted the use of questionnaires to address the professionals' thermal sensation, alongside with their environment acceptability. For a more precise determination of the thermal sensation, the study was conducted while the professionals were developing their activities and assisting them during the questionnaires completion.

2 METHODOLOGY

2.1 Questionnaires creation

In the present work, the subjective assessment of thermal comfort was achieved through the application of an assisted questionnaire. The questions used were based on the questionnaire proposed by Parsons (2014). Through pervious works (Rodrigues, 2011), irrelevant questions and of difficult comprehension were eliminated, assuring the creation of a questionnaire that was fully understood by the population in study. Towards limiting the data variables, this study was developed in the same orthopaedic operating room, located in a hospital of the northern region of Portugal. All the participants were medical professionals or assistants developing their activities during surgeries. The first questions aim to identify individual parameters that influence thermal comfort in a small scale, such as gender, age, height and weight.

The questionnaires were also used to obtain the personal data about the activity the professional performed (profession), as well as the clothing insulation. Through observation, it was determined that the tasks developed by a professional with the same profession were similar, which lead to the approximation that professionals with the same profession had the same metabolic rate, grouping them together. The inquired profession was directly asked in the questionnaire, while the actual metabolic rate was determined through observation of the developed tasks, using the method B, described in ISO 8996:2004. This method assigns a metabolic rate value to different workloads and tasks. However, since the developed workload and tasks during the surgeries were variable, the final value for the metabolic rate was calculated through a time-weighted average, of all the different workloads.

A similar methodology was used to determine the clothing insulation through the usage of tabled values, described in ISO 7730:2005. The different garments used by the health professionals were assessed through the questionnaires and the final insulation value corresponded to the sum of each individual insulation value. After assessing the main personal factors, the questionnaires aimed to assess the individual's thermal sensation, starting by identifying his location in the room, between two regions with different environmental conditions.

The first region consisted in the surrounding area of the surgical room with lower air velocity and without the surgical focus influence. In contrast, the second region consisted in the central area of the surgical room, where the surgeries took place, with higher air velocity and under the surgical focus influence. Having the location in mind, the following questions targeted the individual's thermal sensation regarding the time before, during and after the execution of their activities. For this purpose, the ASHRAE 7 points scale for the thermal sensation. The scale comprises values from −3 (cold) to +3 (hot) with 0 being the thermal neutrality. To complement the obtained data and better understand the professional's thermal sensation, each response was followed by the question of how the individual would like to feel, i.e.: "colder", "no change" or "hotter". Additionally to the information about the thermal sensation, the last questions focused on finding sources of local discomfort, as well as the affected body part.

2.2 Questionnaires application

The application of the questionnaires was carried out by the researchers, in an orthopaedic surgical room, on the health professionals and assistants. To minimise the intrusion in the professional's work and to avoid any danger to the patient, all the questionnaires were applied at the end of the surgeries. Furthermore, to ensure that all the questions were clear, the researchers were always present and assisted in the replies.

3 RESULTS

3.1 Sample characterisation

The questionnaires were applied in 8 different surgeries and a total sample of 69 replies were obtained. From the total replies, 38 individuals were males (55.1%) and 31 individuals were females (44.9%). The age range varied from 24 to 66 years old with a median of 43 years old, an average of 43.78 years old and a standard deviation of 13.96 years old. The questionnaires assessed a total of 7 activities, defined as "Surgeon", "Surgeon's Assistant", "Instrumentalist", "Anaesthetist", "Nurse", "Auxiliary" and "Technician". However, due to similarities in the developed activities, they were grouped into 4 distinct activities: "Surgeon", "Surgeon's Assistant", "Instrumentalist" and "Nurses and Auxiliaries". Due to time restrictions, the calculation of the metabolic rate was performed

Table 1. Approximated metabolic rate for the different activities' groups.

	Surgeon	Surgeon's assistant	Instrumentalist	Nurses and auxiliaries
Metabolic rate [met]	2.29	2.43	2.07	1.8
Clothing insulation [clo]	0.82	0.91	0.78	0.53

Table 2. Clothing insulation calculated for the different activities' groups, respective standard error and number of cases.

	Surgeon	Surgeon's assistant	Instrumentalist	Nurses and auxiliaries
Clothing insulation [clo]	0.82	0.91	0.78	0.53
Standard error	0.03	0.01	0.04	0.03
Number of cases	13	4	8	44

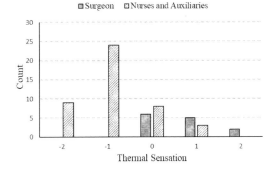

Figure 1. Thermal sensation responses for the classes surgeons and nurses and auxiliaries.

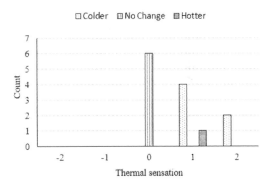

Figure 2. Environment's judgment for the class "Surgeons", according to the thermal sensation response.

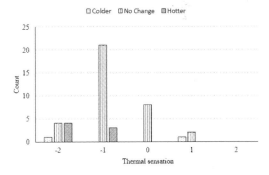

Figure 3. Environment's judgment for the class "Nurses and Auxiliaries", according to the thermal sensation response.

for a sample orthopaedic surgery with 60 minutes duration. Despite this, this study englobed different orthopaedic surgeries. The time-averaged metabolic rate for the different activities, assessed using tabled approximated values, is represented in Table 1.

The clothing insulation calculated using the data collected about the different cloths used by the professionals is presented in Table 2.

Although data regarding thermal sensation for before, during and after the execution of activities was collected, the focus of this article is directed at during the execution of the activities. To better interpret the thermal sensation, the collected data are also expressed by the different activities. Figure 1 represents the thermal sensation response for the classes "Surgeons" and "Nurses and Auxiliaries".

The latter class, Nurses and Auxiliaries, comprised most of the results with 64% of the total replies and the class Surgeons 19%. The average thermal response and standard deviation were, respectively, 0.38 ± 0.96 and −0.98 ± 0.79. Regarding the location in the operating room, the class "Surgeons" occupied the central area, while the class "Nurses and Auxiliaries" occupied the area surrounding the patient.

In order to better understand the thermal sensation distribution, the environment's judgment is also represented. Figure 2 represents the environment's judgment by the class "Surgeons", according to the thermal sensation response, with a population of 13 inquiries.

Similarly, Figure 3 represents the environment's judgment by the class "Nurses and Auxiliaries", according to the thermal sensation response, with a population of 44 inquires.

4 RESULTS AND DISCUSSION

The different classes had their own metabolic rate with the surgeon's assistant presenting the higher value. However, since the sample for this class was composed of only 4 elements, the statistical significance is not as high when compared to the "Surgeons" class with 13 elements. Nevertheless, it is possible to observe a considerable difference between the first and the fourth class, where the latter presented a lower metabolic rate. The same tendency is verified for the clothing insulation between the first and last class, with the surgeons presenting a higher level of protection and consequently thermal insulation. This result shows the tendency for a higher thermal sensation regarding the class "Surgeons" since both, a high metabolic rate and high thermal insulation contribute to a hotter sensation.

This result is later verified in terms of thermal sensation as seen in Figure 1, where the surgeons replied with a neutral thermal sensation or towards a hot environment. Contrasting with the previous result, however following the personal variables tendency, the class "Nurses and Auxiliaries" replied with a thermal sensation mostly directed at colder sensations. The most frequent reply was "slightly cold" (−1) with only three replies at "slightly hot" (+1). For their advantage, this class was mostly located in the area surrounding the patient and the effects of the ventilation was not as strong as for the class "Surgeons". In both cases, however, the extremes of the scale, "very cold" (−3) and "very hot" (+3) were not selected. The difference between both classes was accessed with a Mann-Whitney test for a 0.05 significance level. This result is in accordance with other authors' studies that also show significant differences between classes of healthcare professionals (Dascalaki, Gaglia, Balaras, & Lagoudi, 2009; Zwolińska & Bogdan, 2012).

Considering the environment's judgment for the class "Surgeons" there is a visible tendency to search for thermal neutrality. As seen in Figure 2, when the thermal sensation is "Neutral" (0), all the individuals replied that they would not change how they felt. However, when the thermal sensation is "Slightly hot" (+1) or "Hot" (+2), the individuals replied that they would like to feel colder. In terms of room location, this class has the advantage of being located directly under the ventilation, which provides a greater cooling effect. In contrast, the class "Nurses and Auxiliaries" did not present the same tendency towards thermal neutrality. In the latter case, the environmental judgment followed a similar tendency as of the thermal sensation, with the reply "No change" being the most common. The only exception is in the "Cold" (−2) thermal sensation where the judgment hotter has the same number of responses. The greatest difference in the number of responses is in the "Slightly cold" (−1) thermal sensation. This preference showed that for the class "Nurses and Auxiliaries" the comfort zone was wider, and a colder environment was accepted as comfortable alongside the thermal neutrality. This behaviour is more common for classes that develop physical demanding tasks, where a colder environment helps the body to cool off. However, the result obtained in this class points towards a different cause, indicating that the individuals adapted to their working conditions, accepting a colder environment as comfortable.

5 CONCLUSIONS

The present study considered the thermal comfort of health professionals and auxiliaries during the execution of their activities. The first step towards the objective was the creation of the questionnaires for data acquisition. This demonstrated to be a proper tool in the thermal environment study, allowing assessing the personal variables that influence thermal sensation, as well as the professional's individual vote. The assistance provided while completing the questionnaire proved to be essential during the data acquisition.

Considering the professional's classes, namely, the classes "Surgeons" and "Nurses and Auxiliaries", it was possible to identify the existence of differentiation. The results demonstrated that the first class expressed a thermal sensation towards hot, while the latter class expressed a thermal sensation directed towards cold, being in accordance with the determined personal variables for metabolic rate and cloth insulation. In terms of the operating room location, both cases were favoured. The class "Surgeons", having a hotter thermal sensation, was located directly under the ventilation, which provided a stronger cooling effect. In contrast, the class "Nurses and Auxiliaries" was located in the surrounding area, being less affected by its cooling effect.

In addition to the thermal sensation, the professionals replied about their satisfaction with the thermal environment. In this situation, the class "Surgeons" demonstrated their will in maintaining a neutral thermal sensation, indicating that this class was less prone to accept thermal variations, namely, towards hot.

On the other hand, the class "Nurses and Auxiliaries" demonstrated to have a higher acceptance of their thermal environment, despite expressing to perceive the thermal environment mostly as cold. This acceptance was demonstrated by their choice of not wanting to change the thermal environment, whether they had a neutral of slightly cold thermal sensation.

For a further verification of the results obtained, a grater sample should be considered, as well as widening the study to operating rooms of different hospitals. Additional research should also address the variability present in the orthopaedic surgeries, focusing the study only in one type of surgery.

REFERENCES

ASHRAE Standard 170P. (2006). *Ventilation of Health Care Facilities*. Atlanta, GA: American Society for Heating, Refrigerating and Air-Conditioning Engineers Inc.

Balaras, C.A., Dascalaki, E., & Gaglia, A. (2007). HVAC and indoor thermal conditions in hospital operating rooms. *Energy and Buildings*, *39*(4), 454–470. http://doi.org/10.1016/j.enbuild.2006.09.004.

Dascalaki, E.G., Gaglia, A.G., Balaras, C.A., & Lagoudi, A. (2009). Indoor environmental quality in Hellenic hospital operating rooms. *Energy and Buildings*, *41*(5), 551–560. http://doi.org/10.1016/j.enbuild.2008.11.023.

De Giuli, V., Zecchin, R., Salmaso, L., Corain, L., & De Carli, M. (2013). Measured and perceived indoor environmental quality: Padua Hospital case study. *Building and Environment*, *59*, 211–226. http://doi.org/10.1016/j.buildenv.2012.08.021.

Decreto Regulamentar N.º 63/94. (1994). Requisitos relativos a instalações, organização e funcionamento das unidades privadas. Diário da República. Retrieved from https://www.ers.pt/uploads/document/file/203/16.pdf.

Djongyang, N., Tchinda, R., & Njomo, D. (2010). Thermal comfort: A review paper. *Renewable and Sustainable Energy Reviews*, *14*(9), 2626–2640. http://doi.org/10.1016/j.rser.2010.07.040.

Fanger, P.O. (1970). *Thermal Comfort: Analysis and applications in environmental engineering*. New York: Mcgraw-hill.

Höppe, P., & Martinac, I. (1998). Indoor climate and air quality. Review of current and future topics in the field of ISB study group 10. *International Journal of Biometeorology*, *42*(1), 1–7. Retrieved from http://www.ncbi.nlm.nih.gov/pubmed/9780844.

ISO 7730. (2005). Ergonomics of the thermal environment—Analytical determination and interpretation of thermal comfort using calculation of the PMV and PPD indices and local thermal comfort criteria. Geneva: International Organization for Standardization.

ISO 8996. (2004). Ergonomics of the thermal environment—Determination of metabolic rate.

Khodakarami, J., & Nasrollahi, N. (2012). Thermal comfort in hospitals—A literature review. *Renewable and Sustainable Energy Reviews*, *16*(6), 4071–4077. http://doi.org/10.1016/j.rser.2012.03.054.

Miguel, A.S. (2014). *Manual de Higiene e Segurança do Trabalho* (13th ed.). Porto Editora.

Parsons, K.C. (2014). *Human Thermal Environments: The Effects of Hot, Moderate, and Cold Environments on Human Health, Comfort, and Performance* (3rd ed.). London: Taylor & Francis. Retrieved from https://www.crcpress.com/product/isbn/9781466595996.

Rodrigues, N.J.O. (2011). *Caracterização do ambiente térmico ocupacional em salas de operação – construção de um modelo CFD*. University of Minho.

Van Gaever, R., Jacobs, V. a., Diltoer, M., Peeters, L., & Vanlanduit, S. (2014). Thermal comfort of the surgical staff in the operating room. *Building and Environment*, *81*, 37–41. http://doi.org/10.1016/j.buildenv.2014.05.036.

Zwolińska, M., & Bogdan, A. (2012). Impact of the medical clothing on the thermal stress of surgeons. *Applied Ergonomics*, *43*(6), 1096–1104. http://doi.org/10.1016/j.apergo.2012.03.011.

Study of the impact of investments in occupational safety and accident prevention in the construction sector

Rui Pais
PROSPECTIVA, Projects, Services, Studies, SA, Lisboa, Portugal

Paulo A.A. Oliveira
CIICESI—School of Management and Technology, Polytechnic of Porto, Portugal

ABSTRACT: Worldwide, construction workers are three times more likely to be killed and twice as likely to be injured as workers in other areas. The costs of these accidents are enormous both for the individual, for the employer and for society and may account for a considerable proportion of the contract price. Portugal is one of the countries of the European Union that has one of the highest labor loss rates, with a higher incidence in the Civil Construction sector. The importance of this sector to the Portuguese economy is very well known, since it is a direct source of employment of workers. Its activity moves several sectors, which is why it is considered one of the driving forces of the national economy, not only because of its specific weight in wealth creation but also in employment, taking into account the Its obvious multiplier effect, and is therefore a fundamental activity for the growth of the economy. Therefore it is urgent to change this paradigm, in favor of occupational safety and prevention. This is a sector with a low-skilled workforce, presenting a great precariousness and job instability, offering low salaries and demanding high levels of income. It is also characterized by a large and constant displacement of personnel from site to site, who are also required a frequent change of place and job within each work. It also requires a great deal of versatility on the part of workers in that it has a huge diversity of activities and professions.

1 INTRODUCTION

In terms of the incidence of work accidents in the EU-28, between micro, SME (Small and Medium Enterprises) and large companies, it is verified that in the first per thousand workers there are 36 serious and fatal work accidents and in the second only 2 accidents (Commission of the European Communities n.º 2003/361/CE). This means that the probability of an accident in the euro area (EU-28) is about 18 times higher in micro, SME than in large companies.

Investment in Occupational Safety and Accident Prevention in the Construction Sector, whether in the field of training, protections, among others, will result in short and medium term, in productive gains, reduction of accident rates, improvement of corporate image, valuation of "life", decrease of income or work stoppage due to the occurrence of accidents.

Construction has a set of very specific and unique characteristics, which are associated with a strong precariousness and labor turnover, plus the general practice of subcontracting. These peculiarities give it a unique character among the different sectors of economic activity and transform it into one of the ones that present a higher accident rate, especially with regard to fatal accidents. For this reason, it deserves careful attention (Oliveira, 2011).

It is recalled that in the construction industry, the construction process does not take place around the machine, as a static logic of factory scope, but rather as a function of the dynamics of the project that is carried out. It follows that prevention must be developed according to its own methodologies that accompany the dynamics and particularities of projects and construction processes (Pais, 2016).

It is important to invest in the various components (training, technical, human, etc.) of all workers, in order to assess the impact on the reduction of the accident rates.

In civil construction, one of the great pillars of prevention is training. Given the great mobility and turnover of workers, it can be seen on the ground that most of the operational workers "only" have OSH training, when they start an activity or a work, in fact "only" a meeting, often of little time, where the risks associated with the tasks to be performed are transmitted.

Being the Construction sector, possessing a wide set of characteristics and specificities of different nature, which determine a different intervention action of the generality of the sectors of economic activity, that is, they are basic factors for the development of its own OSH management model In Construction, and this should diverge from the generic form currently seen (Oliveira, 2011).

This is the sector which contributes most to the high economic costs to the community at large as result of accidents at work and occupational diseases. What justifies, a growing responsibility in the action of all the actors, throughout the constructive process.

2 MATERIALS AND METHODS

The main objective of the present study was to try to understand the relationship between the value of investment in prevention/protection versus the number of incidents/accidents at work and versus sinistrality indices of companies/works. This work had a methodological basis based on three fundamental bibliographical research components: Scientific component—Consultation and analysis of several articles and scientific papers with peer review, as basis for framing and orientation of the theme and proposed objectives; Legal component—Support and support in various legal and normative documents, both historical and current, in order to support the development of this work in the guidelines issued by the bodies that supervise and supervise the law and working conditions, both at the national level (ACT; DGS; among others), either international (OIT; AESST; AISS; among others); and Technical component—Research and study of several statistical models, of treatment and data analysis.

The bibliographic research components were complemented with a statistical analysis of the behavior of the work accident in Construction that met the requirements identified as factors common to several organizations.

For this purpose, a data survey was carried out, with the latter as a base instrument for two questionnaires, which were submitted to a pre-test. The data were treated for statistical purposes in the present research work.

The questionnaires carried out in a real context of execution of two works (Figures 1 and 2), that moved several sectors, upstream and downstream of its production chain. In both situations, the questionnaires were applied to a sample of 109 companies and 416 workers, in a total universe of 134 companies and 852 workers who participated in the entire production process.

Notwithstanding the insistence, the sample resulted from the kindness and cooperation of the respondents who agreed to participate in the study.

Figure 1. Construction of large sewage treatment plant.

Figure 2. Construction of drainage networks for sewage and water supply.

The questionnaire was applied to the companies based on the following requirements: companies with economic activity in the field of Civil Construction and Public Works, with different dimensions (staff, annual turnover, qualification, safety and medical services at work, investment in OSH and training provided to workers in OSH), within the period under study; and existence of accident records within the study period.

The application of the questionnaire to the workers was carried out based on the following requirements: collection of personal data of the worker (professional category, age, gender, nationality, schooling); collection of professional data of the worker (possessing CAP for the professional activity, professional situation, time practiced, time of activity in the company, time of experience in the current function, time of experience in the construction sector, family background in construction and data on OSH training) within the study period; collection of information indicating how the accident occurred, the circumstances in which it occurred, and how the injuries occurred (the event is divided into three sequences: specific physical activity, deviation, contact—mode of injury, and agents associated materials) within the study period; and collection of information regarding the nature and severity of the injuries and consequences of the accident (part of the injured body, type of injury and number of days lost) within the study period.

The reference period is defined between January 2013 and December 2014, and may in some cases be broader and/or shorter due to the data provided by the entities.

The sectoral scope refers to the activities of section and subsection F of the CAE Revision 3, namely in the Civil Construction and Public Works sector, which covers companies with a geographical location in the national territory (continental Portugal).

Thus, the questionnaire covered the sample selected, with the necessary guarantees of confidentiality of the data collected, with the intention of minimizing any deviation from the results.

The questionnaires were a research instrument consisting of a set of written questions, with objective knowledge of the views, interests, situations lived, and were applied through the personal interview method, guaranteeing the anonymity of the interviewees. Responses have been treated confidentially and will only present general results of the study, without any information that could lead to the specific identification of the participant.

3 RESULTS AND DISCUSSION

All of the following data are the result of an investigation of a sample, based on two questionnaires, carried out in a real context in the execution of two public works that have operated in various sectors upstream and down-stream of their production. In both situations, the questionnaires were applied to a sample of 109 companies and 416 workers, in a total universe of 134 companies and 852 workers who participated in the entire production process, during the period of analysis of this research study (2013–2014). Regarding the number of hours of OSH training, it was verified that, except for occasional cases of technicians attending a TSST and/or recycling course in both years of study, the only contact with training in the area was given in sections on-site. On the ground, it can be seen that the workers of temporary work companies are more at risk. It is observed that about 94% of workers had up to 4 hours of OSH training (Tables 1 and 2).

A natural and obvious situation results in the idea that all workers feel a need for more OSH training. In the field, they say that given the technological evolution, of materials and constructive solutions, it is essential to know the inherent risks, as well as the knowledge of preventive measures.

Despite the range of values mentioned in the survey, on the ground it is clear that investment in OSH, not only in training, but especially in local conditions, including collective protection, is small (Tables 3 and 4). Many workers have IPE's unused *"but they use it because they have it"*.

Table 1. Number of hours of OSH training (2013).

OSH Training (h) Year 2013	Frequency	Percentage
0	40	9,62
4	348	83,65
5	2	0,48
8	4	0,96
10	11	2,64
15	6	1,44
20	1	0,24
25	2	0,48
250	2	0,48
TOTAL	416	100,00

Table 2. Number of hours of OSH training (2014).

OSH Training (h) Year 2014	Frequency	Percentage
0	3	0,72
3	1	0,24
4	381	91,59
5	3	0,72
8	5	1,20
10	12	2,88
15	7	1,68
20	1	0,24
25	3	0,72
TOTAL	416	100,00

In the analysis of the number of hours of training per worker, it is observed that around 4 to 5% of companies comply with the stipulated in the Labor Code. (Table 5) They minister at least 35 hours per worker.

As a result of the investigations carried out, 20 sample workers had at least 1 work accident during the study period. More than half of the accidents occurred in the afternoon. In annual terms, it is verified that the month of June was where there were more accidents. 25% of the injured workers were on temporary employment contracts. 50% of the injured were carpenters, and servants and masons accounted for 20% of the victims. The age of the victims is predominant until the age of 40, 50% are between 31 and 40 years old. Until the age of 40, 75% of accidents occurred, all of them Portuguese nationals.

Next, the characterization of the sample in relation to the companies is presented. 95% are SME (Small and Medium Enterprises) (Commission of the European Communities n.º 2003/361/CE) (Tables 6 and 7).

Table 3. Value of the company's investment in OSH (2013).

Company Investment in Safety and Hygiene at Work		
Value	Frequency 2013	Percentage
[1; 100 000]	107	98,17
[100 001; 200 000]	2	1,83
TOTAL	109	100

Table 4. Value of the company's investment in OSH (2014).

Company Investment in Safety and Hygiene at Work		
Value	Frequency 2014	Percentage
[1; 100 000]	107	98,17
[100 001; 200 000]	2	1,83
TOTAL	109	100

Table 5. Hours of training given by the company per worker (2013–2014).

Training/Worker (h) in OSH		
Value	Frequency 2013	Frequency 2014
[1;20]	104	105
[21;40]	5	4

Table 6. Number of employees per company (2013).

Number of Employees		
Type of Company	Frequency 2013	Percentage
[1;4]	3	2,75
[5;9]	14	12,84
[10;19]	28	25,69
[20;49]	39	35,78
[50;99]	10	9,17
[100;249]	10	9,17
[250;499]	1	0,92
[500;999]	–	0
[1000;+∞]	4	3,67
TOTAL	109	100,00

From the analysis to the data of the sample, it was verified that the companies in general, does not have services of safety and hygiene at work, as well as medicine at work. The modality of organization is external in its almost totality in both situations. In the analysis of the accident rates, the sample shows that there was no fatal accident in the sample and non-fatal accidents recorded a decrease of 9.45%.

Table 7. Number of employees per company (2014).

Number of Employees		
Type of Company	Frequency 2014	Percentage
[1;4]	3	2,75
[5;9]	12	11,01
[10;19]	31	28,44
[20;49]	38	34,86
[50;99]	11	10,09
[100;249]	9	8,26
[250;499]	–	0
[500;999]	1	0,92
[1000;+∞]	4	3,67
TOTAL	109	100,00

As a result of the statistical treatment, the following data were verified:

It is the intermediate age classes (namely the 30–39 years-old class) who have the most work-related accidents. It should be noted that these age classes, both the sample and the global population, are also the most representative.

The intermediate classes have the most training, namely 30 to 34 years and 40 to 44 years. It is also verified that the classes with more than 50 years are the ones that present less frequency of formation in OSH which, related to the one mentioned in the previous point, can indicate that the collaborators with less formation in OSH could be more exposed to the occurrence of accidents.

If we associate 'Time of Professional Experience in the Construction Sector' and 'OSH training', it is verified that there is a greater discrepancy of employees with and without training in the older classes. This means that although there is still a marked shortage of OSH training for all employees, there is a considerable difference in the older classes between those who did not have training and those who did.

After analyzing the intersection between 'Company investment value in OSH and 'Sinistrality Indices', it is verified that the 'Sinistrality Indices' is lower where there is a higher value of the company's investment in OSH.

4 CONCLUSIONS

In the particular case of the sample, the companies investment in prevention, especially in training, is not significant for the change in any of the accident rates. This may be due to the fact that there is a clear trend towards divestment of firms as a result of the current economic climate and over time as well as the lack of data needed for the construction of the accident rates in the period under study.

It was found that instruments for an overall policy on occupational safety and health were designed to meet the need to promote the approximation of european standards on accidents at work and occupational diseases and also to achieve the overall objective of constant reduction and consolidated labor accident rates (Decree-law 273/2003 of 29/10).

To achieve these objectives, it is essential to invest in a culture of prevention, to facilitate the consultation and active participation of workers in the process of improving work organization, to adopt measures aimed at improving well-being at work, with a view to adapting work to man and their compatibility with family life, introduce factors of control and development of workers physical and mental health and promote health surveillance.

The study shows that the vast majority of workers do not have training in OSH. They only have in many cases only the training of reception on the job.

From the data collected in the surveys, it should be noted that the total number of workers in the sample who reported OSH training provided was insufficient.

As shown by the data, there is a higher prevalence of OSH training in employees with less seniority, and is even more important, in the technical staff.

For the treatment of data, it is verified that the employees who have been in the position for more years have fewer literary qualifications and less training in OSH and, consequently, a greater number of occupational accidents. This reflects, according to the individual factors referred to by the HSE (1999), the importance that the school qualifications and levels of acquired competences possess with respect to the labor accidents.

One of the relevant situations, verified on the ground, is related to temporary work agencies.

The most relevant explanatory variable in the explanation of the accident rates (Incidence and Severity) is the unemployment rate. Viewed in isolation from each other, but together with the variable unemployment rate, the variables participation rate in training actions and training costs per worker are also important, as evidenced by statistical reliability. Interestingly, these two variables act in the opposite direction, that is, an increase in the participation rate in training actions leads to an increase in the two accident rates, and an increase in training costs per worker leads to an increase in Index of severity.

These facts do not mean that the investment in Prevention and OSH should not be done. What it translates is that, as it is done, not only does it not lead to expected results, but even leads them to the opposite direction, to what would be intended.

In economic terms, over the period 2013–2014, what the average company spends per year in Prevention and OSH is more than what it saves with the possible decrease in work accidents, which only reinforces the conclusion of inefficiency of investment in Prevention and OSH.

Due to the results obtained, it is urgently necessary to place a question: what is so poor about the Prevention and Safety in the Construction sector, in the face of such huge investments, which are often called for by successive legislation in order to make it more demanding?

The easiest, and most dangerous, conclusion (because it becomes demagogic and reductive) would be: investment in Prevention and OSH does not pay off. This conclusion would not go to the root of the causes, nor would it even throw clues. And as the purpose of any research work, when, when answering certain questions, others arise, it will be necessary to try to outline new hypotheses of investigation and not to draw conclusions early. Because a more serious hypothesis is to say that investment, as it is done at the moment, is not compensatory. That is, it is not concerned what is spent on Prevention and OSH, but how it is spent or declared to be spent.

These results are highly worrisome and once again allow us to strongly question the objectives and quality of the training provided in companies, as well as the legal model that is defined in this area. It is once again proven that training is not enough, but the most fundamental is its adequacy to the reality of each company, with the risk here proven of the complete lack of return on investment made under these conditions.

REFERENCES

MESS—Ministério do Emprego e da Segurança Social: Código do Trabalho – Law n.º 7/2009, of 12/02, actualizada pela Lei n.º 28/2016, de 23/08.

MESS—Ministério do Emprego e da Segurança Social: Diário da República – Série A, Decree-law nº 273/03 of 29/10, transposição da Directiva Europeia nº 92/57/CEE relativa às Prescrições Mínimas de Segurança e Saúde a aplicar nos Estaleiros Temporários ou Móveis, (2003).

Oliveira, P.A.A.: Modelo de Análise da Sinistralidade Laboral versus Investimento em Prevenção, para o Setor da Construção. Tese doutoral em HSST, departamento de Ciências Biomédicas, Universidad de León, pp. 1–219 (2011).

Pais, R.M.S.: A Tendência Evolutiva da Relação entre as Empresas/Obras e Trabalhadores no âmbito da Segurança e Saúde Ocupacional versus Sinistralidade Laboral do Setor da Construção Civil, aplicável às Empresas Portuguesas—Estudo de Caso. Tese doutoral em HSST, departamento de Ciências Biomédicas, Universidad de León, (2016).

Recommendation of the Commission the European Communities n.º 2003/361/CE de 6 de Maio, relativa à definição de micro, pequenas e médias empresas, (artigo 2.º do anexo da Recomendação, 2003).

Lean construction and safety

A.M. Reis, B.V. Zeglin & L.L.G. Vergara
UFSC, Florianópolis, SC, Brazil

ABSTRACT: Lean Construction is a philosophy that seeks both to maximize value and minimize waste in the Construction Industry. Work safety, in turn, aims to implement inspection and prevention measures to improve protection to employees. This article aims to identify how Lean Construction practices impacts work safety through the perspectives of workers in the construction industry. Following a qualitative research approach, 15 semi-structured interviews were carried out with professionals with an academic background. It was mandatory that all interviewees were applying Lean Construction principles to construction management, and were responsible to guarantee safety within the construction site. Interviews went then through a rigorous process of content analysis. In this process, three topics of study were categorized as, Planning, Visual Management, and Organization and Cleaning. The results drawn from the careful consideration of this data seems to indicate that work-site safety was positively impacted by the adoption of Lean Construction practices.

1 INTRODUCTION

Construction tasks are dynamic and complex, the economic demand to increase productivity may lead to increasing exposure to new risks, accidents, and injury rates (Ghosh & Young-Corbett, 2009). The expression "Lean Construction" was first created by Koskela (1992). As previously demonstrated by the experience of the manufacturing industry, the author propounded that the adoption of a new philosophy on production would be a fundamental paradigm shift for the construction industry.

The basic principles of the "Lean Construction" are the elimination of tasks that do not add value to the final product, and the search for the most efficient conversion flow. In general, it is understood that Lean Construction extends the objectives of a lean production system (maximize value and minimize waste) to the Construction Industry. Considering that accidents generate administrative costs and productivity losses, the safety of the construction workers would be a natural result of Lean Construction.

Thereby, this article aims to identify the perspectives of Florianopolis' civil construction professionals on how lean construction affects safety.

2 METHODOLOGY

This article is characterized as a qualitative research, and, by the nature of its objective, it is an exploratory research. As proposed by Guerra (2006), content analysis was used to treat the data gathered by 15 to 20 interviews. A total of 15 professionals working in the civil construction industry were recruited. All workers must already have monitored tasks –at different levels of management– and must have been responsible for controlling and maintaining safety at the work-site. Semi-structured interviews were conducted in order to understand how –according to their experience and opinion –lean construction impacts safety. The content analysis method is a suitable and flexible way of collecting and communicating ideas and patterns or themes that arise during an interview. Interviews were recorded and transcribed verbatim in order to organize and prepare the data for analysis. The transcribed copy has been read several times for a good understanding of the general ideas and to identify the crucial ideas in all the interviews, related to the study's objective, as suggested by Creswell et al. (2006). Codes were assigned to words, phrases, and sections within the relevant data for the research question. These codes were then categorized into relevant topics (planning, visual management, and organization and cleaning), which were confirmed with the literature.

3 RESULTS

3.1 *Interviewee's profile*

Interviews were conducted with 15 construction professionals, with different levels of work-site experience, whose organizations are involved in lean construction practices. The sample includes Civil Production Engineers (I1, I6, I7, I10), Civil

Table 1. Interviewee's profile.

Interviewee	Work experience	Role
I1	27 years	Civil Production Engineer
I2	6 years	Safety Technician
I3	8 years	Building Technician
I4	17 years	Building Technician
I5	19 years	Civil Engineer
I6	1 year	Civil Production Engineer
I7	5 years	Civil Production Engineer
I8	8 years	Civil Engineer
I9	8 years	Civil Engineer
I10	2 years	Civil Production Engineer
I11	7 years	Architect
I12	2 years	Civil Engineer
I13	2 years	Civil Engineer
I14	3 years	Civil Engineer
I15	2 years	Civil Engineer

Engineers (I5, I8, I9, I12, I13, I14, I15), Building Technicians (I3, I4), one Safety Technician (I2) and one Architect (I11). The level of experience in construction sites differs among the interviewees, as shown in Table 1.

3.2 Safety and lean construction

To avoid losses linked to human factors, such as work accidents, which are waste and non-value-added events in any kind of production system, these human factors should be part of the evaluation system used in the process. I6 says that the *"Lean"* is about respect for people: *"If something affects safety, you must stop anything that discourages or puts people at risk."* According to Ogunbiyi et al. (2014), safety biases social, economic and environmental sustainability. From the social point of view, the implementation of lean thinking in construction tends to improve the workplace, ensuring the health and well-being of the workforce with better layouts, well-organized workstations and reducing the exposure to safety risks.

3.2.1 Planning
It is possible to conclude from the interviewees' opinion that lean construction generates positive impacts on safety since safety management is strictly linked to the issue of planning and controlling the general production of civil construction. I15 summed up the idea saying that: *"Construction sites with well-defined processes avoid overlaps of activities, team accumulation and generation of waste. The combination of these three things is the major cause of accidents within a construction site. Therefore, performing better-planned constructions reduces the risk within a construction site."* I9 believes that planning minimizes the accident risks and helps in physical ergonomics, he says that *"[…] if we create processes and methods, a logical sequence of work steps and a routine of material transportation, with all that well outlined and transparent, we are avoiding possible accidents, accidents with the open gate or with the transportation of a certain type of material, or carrying a bag of cement."* Yet, I1 exemplifies this idea, considering the importance of collective protections: *"The more you plan, the less you have to change later on because a safe construction is the one on that the collective protection is efficient. Therefore, when you plan, you alter less this equipment. You put it up once and it lasts longer. So, a lean construction, well planned, is safer, without a doubt."* I2 illustrates this idea attesting that *"for example, there is a protection on a balcony and you need to run the marble or the waterproofing, you need to carry out a series of services that require you to take off the guard-rails and leave it without protection. So, having that planned is easier, because you already have that and will not need to take it off to do the work. There were situations when we had to perform certain types of services, and we would ask ourselves, 'how am I going to get there?', 'how am I going to build a scaffold where there is not enough space for a scaffold?', 'there is no room for the work seat!' So, if there is a plan to execute all these steps, everything is easier."*

Another issue raised by the interviewees that relates to safety on lean construction is about the Last Planner system, a philosophy that seeks to minimize uncertainties and variabilities in processes through the implantation of new levels of planning, such as short and medium terms plans, in order to increase performance and productivity, reverberating primarily on deadline accomplishments (Ballard & Howell, 1994).

If applied correctly and implemented in all its elements, it is believed that it can bring success to lean construction, because, besides the advantages related to productivity, there are opportunities to include safety issues in the system management.

According to I6, "When planning the construction, we have the balance line, which is the macro planning. Then a medium term, around three months, with the restrictions and so on. And then you will control weekly the 'PPC' (Percent Plan Complete) with these three levels of planning there could be a safety indicator and long, medium and short-term safety strategies." I7 adds that "[…] safety is a restriction, as well as material, labor, space, all those restrictions of the Last Planner. Safety also plays a restriction to start a service. So, I see that safety is aligned with lean philosophy, they must walk together. Obviously, each case is a case, each service is a service, but safety has to be within it."

Saurin et al. (2001) research have considered managing safety through production planning and control. In their exploratory study, they found that some lean production concepts and methods, such as the Last Planner Method, which has been used for production planning and control, can be easily extended to safety planning. The study suggested an integrated safety planning and control model with four essential functions: anticipating safety resources that are necessary to control risks, identifying and controlling risks originated in production planning decisions, evaluating safety performance (based on both proactive and reactive indicators), and enabling workers to identify risks and make suggestions to control them.

Awada et al. (2016) carried out interviews with professionals on the reduction of work-related accidents in the construction industry in Lebanon. All professionals interviewed were engaged in construction projects with Last Planner System's implementation. As a result, it was concluded that effective integration between the Last Planner system and safety planning can improve safety conditions in construction projects. 70% of the interviewees claimed that the involvement of workers in the planning phase, as well as their ongoing training, helps to reduce construction accidents. In addition, the interviewees stated that the use of a weekly work plan in the construction phase is of great importance.

The study conducted by Leino & Elfying (2011) proposed an integrated implementation of the Last Planner system measures with a safety program. A qualitative and quantitative investigation was carried out during 2007–2010, through meetings, discussions, and data collection. The authors show that the Last Planner system provides a fertile platform for the involvement of the workforce since the worker has experience and knowledge about the risks and obstacles of the tasks and can identify construction works risks and their respective preventive measures.

3.2.2 Visual management

Visualization is an effective lean tool in notifying workers about standards of duration, quality, and safety, as well as to motivate them through relatively simple and inexpensive improvements to the immediate area around the working field. According to I6, *"Lean [system] works hard on the issue of visual management, tries to work to make things quite visual, and this applies to the planning of the work, and also in any procedure within the working area, from individual protection equipment, collective protection equipment, areas that you can or cannot circulate, construction site layout."* I3 and I4 also said that *"visual control"* is a lean tool to promote safety.

Providing risk forecasts with detailed information visually can contribute to better weekly planning and make each worker more aware of the specific risks they face. To provide risk predictions with visual information can contribute to a better weekly planning and make each worker more watchful to the specific risks they face (Sacks et al., 2009).

Nakagawa (2014), another author that discusses the importance on visualization as a concept of Lean Construction on construction sites, says that when visualization is not implemented, the workers tend to become indifferent to other activities. This creates waste, especially on projects with many activities and workers. Visualizing the objectives, progress and minimum requirements for safety, quality and environmental control can encourage them to pay attention to other activities and overall project progress, thereby reducing waste. Allowing all workers to visualize the goals in a shared way encourages communication, improves team motivation and engagement.

3.2.3 Organization and cleaning

A question raised by four of the interviewees refers to the importance of organizing and cleaning the construction site. For I5, *"to operationalize the worker, so that he can develop his task in less time, with the required quality and without material waste, he must have a clean and organized environment, with all the necessary protection and equipment, precisely for him to focus on executing the work without risks."*

For I8, "[...] as far as safety at the site is concerned, improvement is visible, starting with cleaning. We dispose of all rubbish and material separately. Each material has a compartment for separation and subsequent disposal. Every execution is more efficient and with fewer accidents."

I12 believes that "[...] the impacts that lean construction can have on safety only tend to be positive" and explains: "if lean construction tends to decrease waste production, it organizes better waste management, it means that the workers will be less vulnerable to risks within the working site, because there will be less waste such as nails, lumber, things that cause risks to the workmen. And those that still exist won't be spread around the construction site decreasing accidents."

In I11's opinion, "lean construction helps with safety because it brings a lot of cleanliness to the construction site." Besides the cleanliness, I11 points to the organization of the construction site, saying that "the materials must go to the site at the moment that they are going to be used. One thing that happens in our construction is the release of the materials by kits, so this helps with safety because you do not have materials on the site unnecessarily, which makes it more organized."

The organization of the construction site was outlined in Zhang et al. (2015) study, through the visualization of potentially congested workspaces. The study developed a method that contributes to the improvement of the safety performance from a BIM (Building Information Modeling) platform, used to detect potential workspace conflicts between work teams or between material lifting equipment. The method developed by these authors can support project collaborators, such as engineers, planners, construction site supervisors, and workers. In this way, it improves the basis on how decisions are made regarding the safety of the construction site, as well as its potential impact on a productive and unobstructed environment.

4 CONCLUSION

This article identified the perspectives of construction professionals from Florianópolis, SC, Brazil, on how lean construction impacts safety.

The perspectives provided by the professionals working in the sector –whether by their systemic vision, applied knowledge in management, or work-site safety experience– contributed with highly qualified interviews, and the generation of an invaluable database, which was categorized in themes: planning, visual management, and organization and cleaning. The planning was pointed as a crucial point to mitigate the risks in construction sites, always adapting the most appropriate strategies to each reality. Integration of production principles into safety planning would help in overcoming the inherent shortcomings of construction safety planning. To include safety management into the Last Planner System could improve safety performance in the same way that the system improves production performance. Visual management, in turn, is one of the main lean strategies to promote the engagement and sharing of information among workers, makes them pay attention to all the procedures performed within the site, as well as being aware of the use of personal protective equipment. In addition, lean construction practices corroborate with the organization and cleanliness of the construction site, which makes construction workers less exposed to risks.

Thus, the three categorized themes point out that the correct approach given to safety at the site has positive implications for achieving the goals of lean construction. The exploitation of this synergy, although characterized as a gradual change, adds value throughout the entire production chain, especially so that workers have a better life quality within the work environment.

Increasing safety is critical in the context of improving productivity and efficiency in the construction industry. Thus, the development of valid safety performance metrics is a significant first step towards improving safety.

For a further understanding of the subject, it is suggested to carry out empirical research to test the hypothesis that lean construction is positive for safety at construction sites, as well as the measurement of performance indicators applied to lean construction and safety.

REFERENCES

Awada, M.A. & Lakkis, B.S. & Doughan, A.R. & Hamzeh, F.R. 2016. Influence of lean concepts on safety in the Lebanese construction industry. In *Proc. 24th ann. conf. of the int'l. group for lean construction, Boston, USA*, (11): 63–72.

Ballard, G. & Howell, G. 1994. Implementing lean construction: stabilizing work flow. *Lean Construction*, 101–110.

Creswell, J.W. & Shope, R. & Plano Clark, V L. & Green, D.O. 2006. How interpretive qualitative research extends mixed methods research. *Research in the Schools*, 13(1): 1–11.

Ghosh, S. & Young-Corbett, D. 2009. Intersection between lean construction and safety research: a review of the literature. *Industrial Engineering Research Conference, Florida, 2 June 2009*.

Guerra, I.C. 2006. *Pesquisa qualitativa e análise de conteúdo: sentidos e formas de uso*. Lucerna: Princípia.

Koskela, L. 1992. *Application of the new production philosophy to construction* 72. Stanford, CA: Stanford University.

Leino, A. & Elfving, J. 2011. Last Planner and zero accidents program integration-workforce involvement perspective. In *Proceedings of the 19th annual conference of the international group for lean construction, Lima, Peru*: 622–632.

Nakagawa, Y. 2005. Importance of standard operating procedure documents and visualization to implement lean construction. In *Proceedings of the 13th annual conference of the international group for lean construction. Sydney, Australia*.

Ogunbiyi, O. & Goulding, J.S. & Oladapo, A. 2014. An empirical study of the impact of lean construction techniques on sustainable construction in the UK. *Construction innovation*, 14(1): 88–107.

Sacks, R. & Treckmann, M. & Rozenfeld, O. 2009. Visualization of work flow to support lean construction. *Journal of Construction Engineering and Management*, 135(12): 1307–1315.

Saurin, T.A. & Formoso, C.T. & Guimaraes, L.B.M. 2001. Integrating Safety into Production Planning and Control Process: An Exploratory Study. In *Proceedings of the 9th annual conference of the international group for lean construction, Singapore*.

Zhang, S. & Teizer, J. & Pradhananga, N. & Eastman, C.M. 2015. Workforce location tracking to model, visualize and analyze workspace requirements in building information models for construction safety planning. *Automation in Construction* (60): 74–86.

Analysis of the noise level in a university restaurant: A case study

A.D.S. Oliveira, H.C. Albuquerque Neto, M.N. Almeida, A.R.M.V. Silva &
B.C.C.B. Carvalho
Department of Industrial Engineering, Federal University of Piauí, Brazil

ABSTRACT: The present study had to analyse the noise level which workers are exposed in a University Restaurant (UR) responsible for the process of preparing meals of a Federal Higher Education Institution (FHEI). For this the methods used consisted of punctual measurements followed by dosymmetries to portray the acoustic conditions of the studied environment, using tools of descriptive and inferential statistics for the realization of the data analysis. The results showed that, taking in consideration the NR-15 (1990) which establishes at 85 dB(A) the noise exposure limit for 8 hours of work, the University Restaurant presented values above the permitted limit with exposure levels ranging from 83,4 dB (A) to 94,4 dB (A), taking into account all groups of analysed workers. In front of that, it was concluded that the vast majority of workers are in a insalubrity situation in relation to noise, without taking into consideration the attenuation provided by the Individual Protection Equipment.

1 INTRODUCTION

Noise has been considered an important agent of occupational hazard, considering that the workers exposed may present beyond hearing complaints, psychological problems, digestives, communication, sleep and physiological alterations (ARAÚJO, 2010). In addition to these issues, Silva et al (2016) affirm that the exposition to the intense noise is an occupational threat and causes problems that overcome the hearing loss and affect the life quality of the people who are exposed.

It is in this context that is inserted food service companies, in which the restaurants are evidenced due to this sector employs a large number of people in the world (TO; CHUNG, 2014) playing an important role in terms of economy (ARAÚJO, 2010). The working process within a restaurant requires that the food preparation activities, in most cases, be developed under different environmental conditions, such as the presence of excessive noise (TEIXEIRA et al, 2015). This condition besides affecting the operational performance may increase the risk to the worker, taking in consideration the susceptibility of each individual, such as the type of noise and the exposure time (SASAKI, 2008).

Thereby, the present article aims to analyse the noise level in which the workers are exposed in a University Restaurant (UR) responsible to the meal preparation process of a Federal Higher Education Institution (FHEI).

2 METHODOLOGICAL PROCEDURES

2.1 Technical procedures for measuring the noise levels

For the execution of the noise level measurements of the university restaurant was adopted a dosimeter with model decibelimeter function DOS-600, made in accordance to the IEC 61672-1 type 2, IEC 61252, IEC 60651 type 2, IEC 60804 type 2, ANSI S1.25 type 2 standarts, beyond being in agreement to the brazilian standarts NR-15 and NHO-01.

According to the standart ISO 1996-1 (2003), the initial use of the equipment in the sound pressure levels measurements will be on time, and comprising a period of 5 minutes each. Subsequently, a more comprehensive measurement is made with 8 hours of measurement, representing the worker's working day. It is emphasized that during the rest period of the workers, the device was paused, resuming the measurement with the return of the worker.

Finally, it is clarified that the above-mentioned equipment was calibrated and gauged by a duly accredited certification laboratory following ISO/IEC 17025 (2005), which establishes general requirements for the competence of testing and calibration laboratories. This calibration is valid for one (1) year taking effect during the data collection.

2.2 Treatment and data analyse

Through all the measurements collected, descriptive statistics are used in the elaboration of tables that show the evolution of the level of attenuation and exposure of the noise during the time of collection. Then, non-parametric tests were performed between the values obtained by collecting the punctual measurements (5 minutes) and by the continuous measurements (hours of the workday) in order to verify if there are statistically significant differences between the collected data.

For this, the Kruskall-Wallis test is used, which is the non-parametric alternative when it is desired to decide whether random and independent samples come from a given population or from different populations (MELLO; GUIMARÃES, 2015). The study by Nascimento (2017) highlights the two options offered by the test:

- If the null hypothesis is maintained (p-value > 0.05) it means that there are no statistically significant differences between the distributions of the samples under analysis;
- If the null hypothesis is rejected (p-value < 0.05) it indicates that there are statistically significant differences between the distributions of the samples under analysis.

This test was performed using software SPSS version 23, a statistical analysis software where it will be fed with the data collected by the measuring equipment.

3 RESULTS ANALYSIS

3.1 *The productive process description of the university restaurant*

The University Restaurant in study is a reference in the sector within the state of Piauí producing about 5,500 (five thousand five hundred) meals per day. The restaurant has several employees among them those who are subjected to the conditions of work with occupational noise are: cooks, kitchen assistants, cleaning assistants and stockman.

The cooks are responsible for the cooking itself, exercising a more fixed function, is the best part of the time in the general kitchen however there are still times when they assist in different parts of cooking, depending on the hours. Kitchen assistants are not a fixed activity performing functions according to a demand of the moment, among them are the cutting of meat and vegetables, sanitation, distribution of meals received, storage, hygiene and washing utensils and distribution. The cleaning assitant is responsible for the general cleaning of the restaurant in its entirety and the stockman is in charge of receiving, conferring and storing the inputs for production. After the explication about the operation of the research space, the following subsection will deal with the results of the punctual measurements which aim to provide an initial perception about the audiometric conditions of the environment.

3.2 *Punctual measurements results*

Punctual measurements were taken following the proposed methodology in order to verify if workers are exposed to noise levels above the 85 dB (A) limit established by NR-15 for an 8-hour working day.

As punctual measurements, they were made in some subsectors of the restaurant, in different shifts for each day, in a spot close to the employees' work station to portray with fidelity. The choice of subsectors based on audiometric conditions at sites with different sources of noise that are not express noise sources were not targeted by the research such as an office room, typically quiet environment. The major values found are organized in Table 1.

From the values found in Table 1, it can be seen that the tray washing room is the noisiest space with a sound pressure level of 110.9 dB(A), which is quite high compared to that established by the Regulatory Standard 15 (NR-15) setting in 85 dB (A) as the exposure limit for 8 hours of work. This same violation of the standard limit, in a lower form, was found in the Kitchen, with 102.4 dB(A), and in the salting room, with 85.8 dB(A). The refectory also had high noise levels mainly due to the large conversational intensity on the spot and the handling of trays and cutlery.

With the results obtained from the punctual measurements and taking in consideration that most of the workers perform functions in the different locations analyzed throughout the work day it is possible to infer that these are in a insalubrity situation if there is no protective measure to mitigate noise. In order to reinforce this statement dosimetry were performed.

Table 1. Punctual measurements results in the week 1 and 2.

Physical space	Week 1		Week 2	
	Day 1 (dB)	Day 2 (dB)	Day 1 (dB)	Day 2 (dB)
Vegetables preparation	85,8	84,1	84,8	83,2
Meat preparation	84,2	81,5	86,1	80,5
Tray washing	107,8	110,1	107,2	110,9
Receiving	79,8	85,2	82,1	90,2
General kitchen	90,8	100,2	102,4	96,8
Refectory	82,1	87,8	91,4	89,2

3.3 Dosimetry results

Thirteen dosimetry were performed in different UR workers, comprising the dates of 05/23/2017 to 06/14/2017. Each dosimetry consisted in 8 hours daily and during the time of the worker rest the dosimetry was paused, being restored with the return of this one to the work. It should be emphasized that dosimetry is not necessary in all employees, with the division into homogeneous groups sufficient with the evaluation of one or more workers.

The restaurant has 44 employees directly involved in the production. The division into homogeneous groups which according to Fundacentro (2001) are groups composed of workers who exhibit the same characteristics of exposure. It was done in the following way: 8 cooks, 4 storageman, 28 kitchen assistants and 4 cleaners. The number of dosimetry per homogeneous group is given in Table 2.

Most of the dosimetry were carried out with kitchen assistants since they represent the homogeneous group that comprises the majority of the employees involved in the production, the days referring to the kitchen assistant were days 1; 2; 5; 6; 7; 9 and 11. However, some dosimetry were made specifically with certain workers because they represent the other groups, in the cooks case the dosimetry days were 3; 4; 8 and 10, and the cleaning responsible and stockman the dosimetry was made only in 1 day, day 12 and day 13 for each respectively. With this division it is possible to portray the audiometric situation of all employees of the studied environment. It was established by the result of the punctual measurements that at some point during his working day the limit value of 85 dB was exceeded.

Regarding the inferential analysis of the measurements, initially it was sought to verify if there are statistically significant differences between the days of dosimetry measurements. Following what had already been explained, thirteen (13) days of dosimetry were performed however the last day presents a distinct time which makes the inferential comparison unfeasible. For the analysis of the remaining 12 days the non-parametric Kruskal-Wallis test was used which compares more than two independent samples.

Furthermore, it is also comparable the measurements shifts, videlicet, the results obtained by the morning shift and by the evening shift. For this, the Mann-Whitney test was used which follows the two same options evidenced by the Kruskal-Wallis test. Therefore, it is verified that there is a statistically significant difference between the days and the shifts of the measurements (p-value = 0.00). This result gives the premise that there is a variation of noise during the workday that is not equalized between days and shifts, in other words, there is no noise variation pattern. In front of that, even though the noise is intermittent its variation is unstable during the development of the collaborators work.

In addition, as shown in Table 2, it is possible to associate the days of measurement and the collaborators positions. Considering that the kitchen assistant represents seven (7) days and the cooks (4) days, Kruskal-Wallis and Mann-Whitney non-parametric tests can also be performed for days and shifts respectively.

It was found that there is a statistically significant difference between the days and shifts of the measurements made in the kitchen assistant (p-value = 0.00). This result reflects the overall result found previously in which there is a great variability of the noise level for this collaborator. Regarding the cooks results there was also a statistically significant difference between the days of measurements (p-value = 0.01) however the same does not occur in the two different shifts (p-value = 0.058). With this, it can be said that there is no difference if the shift is morning or evening due to the noise variation does not change so much.

3.4 Exposition level and noise attenuation

The dosimetry equipment offers a range of ready-to-use information at the end of the measurement. The level of exposure (dB) is the level at workers are exposed and it should not exceed the legal limit of 85 dB because it would represent an unhealthy situation as elucidated by NR-15 (BRAZIL, 1990). If the value found exceeds the limit of 85 dB one must verify the level of attenuation necessary to preserve the health of the workers and to characterize the unhealthiness. For the purposes of the attenuation level the NR-09 standard (BRASIL, 1990) is considered which establishes action level as the value above which preventive actions should be initiated in order to minimize the probability of exceeding the limits exposure. Therefore, they are shown in Table 3.

Table 2. Number of dosimetry by homogeneous group.

Homogeneous group	Number of dosimetry	Days
Kitchen assistant	7	1; 2; 5; 6; 7; 9; 11
Cooks	4	3; 4; 8; 10
Cleaning responsible	1	12
Stockman	1	13
Total dosimetry	13	

Table 3. Dosymetry results summary.

Days	Role	Dose (%)	Exposition level (dB)	Attenuation level (dB)
Day 1	Kitchen assistant	289,00	92,6	12,6
Day 2	Kitchen assistant	330,29	93,4	13,4
Day 3	Cooks	374,67	94,4	14,4
Day 4	Cooks	330,62	93,4	13,4
Day 5	Kitchen assistant	300,25	92,8	12,8
Day 6	Kitchen assistant	117,24	85,9	5,9
Day 7	Kitchen assistant	131,48	86,8	6,8
Day 8	Cooks	312,60	93,1	13,1
Day 9	Kitchen assistant	338,79	93,6	13,6
Day 10	Cooks	378,02	94,4	14,4
Day 11	Kitchen assistant	143,33	87,4	7,4
Day 12	Cleaning responsible	120,67	86,3	6,3
Day 13	Stockman	79,92	83,4	3,4

From the analysis of the results demonstrated in Table 3 it was verified that every day except for day 13 they showed higher levels than the established limit of 85 db. Regarding the attenuation results, it was possible to observe the highest obtained value of 14.4 dB(A). Thus, with the guarantee of conformation with the action level of 80 dB(A) by the NR-9, and also the limit of 85 dB(A) given by the NR-15, it is necessary to attenuate the noise reaching the worker's hearing aid by 14.4 dB(A).

The form of attenuation chosen by the company was the distribution of hearing protection equipment which, according to the manufacturer, has an attenuation level of 18 dB(A) sufficient to deconfigure the insalubrity situation caused by the excessive noise and to preserve the workers health. However is not able to infer that the problem caused by noise is completely remedied by that and some preventive measures must be taken to definitely protect employees' health and performance.

3.5 *Improvement suggestions*

As noted in subsection 3.4 it was possible to conclude that the employees of the university restaurant in study are exposed to high noise levels being necessary outline some measures aiming to minimise the degree of harmfulness that noise can offer.

The measures of noise control are summarized in the performance in the source, in the propagation and on the man, being the recommendation to act in the source, then in the propagation and finally in the man. Some measures of control at the source may be the acquisition of quieter equipment if the restaurant has sufficient resources or periodic maintenance on the noise emitting machines, cooker hood and trays washing machine, reducing the sound pressure level emitted by worn parts or with malfunction. If action at the source is not possible, it is possible to use acoustic barriers in the machines, thus acting in the propagation way of sound where the noise level that reaches the worker's audition pavilion is attenuated by this last action.

The university restaurant uses as a control measure the atuaction on the man through Individual Protection Equipment (IPE). This is due to the distribution of atrial protrusions that offer an attenuation level necessary to discourage the insalubrity caused by noise but it should be emphasized that such protectors must be well maintained and within the validity period besides the need of its use throughout the working day to actually offer the attenuation described by the manufacturer. In this way, a more rigorous inspection is suggested to ensure the correct IPE's use as well as lectures aiming to make workers aware of the health hazards that the noise causes and the importance of the use of hearing protectors to avoid such harm.

4 FINAL CONSIDERATIONS

The aim of the present article was to analyse the noise level in a University Restaurant in the state of Piauí/Brazil due to the production activity configuration occurs in this type of environment with different noise sources such as hoods and the constant noise impact generated by the handling of the various cooking utensils.

The measurements consisted of punctual calibrations at first followed by dosimetry which allowed the analysis of a larger number of data since it consisted by the sound pressure collect levels every 10 seconds in a period of 8 hours, making it possible to obtain a more reliable level of exposition with the real situation in which the workers are bound.

Exposure level values for the 13 days of measurements gave values above the limit established by NR-15, except the 13th day which presented the value of 83.4 dB(A). The maximum value recorded was for the 3rd day and the 10th day both presenting a value of 94.4 dB(A), a very high level of exposure, which according to the No. 1 of NR-15 (1990), would only allow a 2-hour work day daily. One main point of observation is that in these two days the monitored employees were cooks, demonstrating the higher degree of exposure of this workers group. By calculating the level of attenuation it was found that the IPE offered by the institution is sufficient to descharacterize the insalubrity but the ideal would be a priority action at the noise source and then in the propagation middle and in the last case on the man. The limitations that provoke this direct action on man are mainly from economic issues since alternatives of the noise control at the source and in the propagation middle usually are more expensive.

In front of the results the article development is important to portray the acoustic situation inside a University Restaurant evidencing the occupational risk that the workers are exposed in relation to noise. Employers should be aware of the productivity loss caused by noise and, above all, the health damage it can bring to their employees. Finally, some limitations can be illustrated in the present research design. Moreover, it is also important to draw up suggestions for future researches, in order to the limitations and the suggestions contribute to the scientific knowledge that guides the theme.

Due to the short time available, the authors did not have sufficient time to perform a larger number of dosimetry during different months. This would allow a better understanding of the acoustic situation of the environment as well as identify possible differences between levels of noise exposure in different periods of the month or even of the year.

The second limitation found was the impossibility of carrying out the study involving the other two University Restaurant headquarters of the Federal Higher Education Institution (FHEI) in question. This would make it possible to verify possible differences in the level of exposure between the headquarters and later to identify the causes for this situation.

One suggestion for future researches is to evaluate not only the environment noise levels but also the perception of exposed individuals through a qualitative analysis using questionnaires resulting in a more comprehensive problem view and consequently increases the knowledge about the subject in a general perspective. Another suggestion is an analysis about the noise attenuation conditions seeing that for the ear protectors to attenuate the level that is specified by the manufacturer they have to be in good conditions and its use has to be correct. In addition to verifying the feasibility of measures of noise control at the source or in the propagation means.

ACKNOWLEDGEMENTS

The authors thank the managers and workers of the University Restaurant who allowed and supported the elaboration of this article.

REFERENCES

Araújo, E.M. G de. *Análise da organização e das condições de trabalho em uma Unidade de Alimentação e Nutrição em relação ao desempenho e à satisfação do trabalho: um estudo de caso*. 2010, 100. Dissertação (Mestrado em Sistema de Gestão). Universidade Federal Fluminense, Niterói.

Brasil. *Ministério do Trabalho. Norma Regulamentadora n.º 15*. Brasília, 1990. Disponível em: <http://www.mtps.gov.br/images/Documentos/SST/NR/NR15/NR15-ANEXO15.pdf.>. Acesso em: 22 jun. 2016.

Brasil. *Ministério do Trabalho. Norma Regulamentadora n.º 09*. Brasília, 1990. Disponível em <http://sislex.previdencia.gov.br/paginas/05/mtb/9.htm.>. Acesso em: 20 nov. 2016.

Fundacentro. *Norma de higiene ocupacional: Procedimento técnico*. Avaliação da exposição ocupacional ao ruído. 2001.

ISO 1996-1. *Ergonomics of the thermal environment – Instruments for measuring physical quantities*. Genève: International Organization for Standardization. 2003.

ISO/IEC 17025. *Requisitos Gerais para Competência de Laboratórios de Ensaio e Calibração*. Rio de Janeiro: International Organization of Standardization, 2005.

Mello, F.M.; Guimarães, R.C. *Métodos Estatísticos para o Ensino e a Investigação nas Ciências da Saúde*. Lisboa: Sílabo, 2015.

Nascimento, R.A.M. *Avaliação do nível de ruído em uma marmoraria: um estudo de caso na cidade de Teresina-PI*. 2017, 55 f. Trabalho de Conclusão de Curso I (Graduação em Engenharia de Produção) Universidade Federal do Piauí, UFPI, Teresina.

Sasaki, K.P.B. *Relações entre adoecimento, fatores de risco e desenvolvimento seguro do trabalho entre trabalhadores de duas unidades de alimentação hospitalares*. 2008. 95f. Dissertação (Mestrado em Nutrição) – Universidade de Brasília, UNB, Brasília.

Silva, M.S. et al. Percepção do ruído ocupacional e perda auditiva em estudantes de Odontologia. *Revista da ABENO*, v. 16, n. 2, p. 16–24, 2016.

Teixeira, S.A. et al. Investigação dos riscos ambientais e ergonômicos em restaurantes privados de um município do Piauí–Brasil. *Revinter Revista de Toxicologia, Risco Ambiental e Sociedade*, v. 8, n. 1, 2015.

To, W.M.; Chung, A.W.L. Noise in restaurants: Levels and mathematical model. *Noise & Health, Hong Kong*, v. 16, n. 13, p. 368–373, nov. 2014.

The influence of noise on the perceptions of discomfort, stress, and annoyance

R. Monteiro & M.A. Rodrigues
Department of Environmental Health, School of Health, Polytechnic of Porto, Porto, Portugal
Research Centre on Environment and Health, School of Health of Polytechnic of Porto, Porto, Portugal

D. Tomé
Department of Audiology, School of Health, Polytechnic of Porto, Porto, Portugal
CIR, Centro de Investigação em Reabilitação, School of Health, Polytechnic of Porto, Porto, Portugal

ABSTRACT: The aim of this study was to investigate the effect of different noise conditions on subjects' perceptions of discomfort, stress and annoyance: standard condition (C1), environmental noise without alert sounds (C2) and environmental noise with alert sounds (C3). An experiment was designed in order to simulate the noise normally prevailing in a real working context. Fifteen undergraduate students were included (20 to 23 yrs; all female). The noise levels were fixed at 45 ± 0.3 dB(A) (C1), 60 ± 0.4 dB(A) (C2) and 68 ± 0.4 dB(A) (C3). The influence of noise on subjects' discomfort, stress and annoyance perceptions was assessed with Visual Analogue Scales. The results demonstrated that subjects experienced higher levels of discomfort, stress and annoyance when exposed to more adverse noise conditions. In the three conditions, significant positive correlations were found between discomfort and stress levels. In C3, significant positive correlations were observed between both annoyance and discomfort with stress levels.

1 INTRODUCTION

Occupational noise is considered one of the most important risk factors in almost all working settings (see e.g., Concha-Barrientos et al., 2004; Planeau, 2005; Ismail, 2011; Jahncke et al., 2011; Rodrigues et al., 2014). The exposure to high levels of noise has been related to several adverse effects on physical health and overall psychological well-being (Stansfeld & Matheson, 2003; Basner et al., 2014; Rodrigues et al., 2014; Rodrigues et al., 2015). The most well-known effect is the Noise Induced Hearing Loss (NIHL) (EU-OSHA, 2005; Nelson et al., 2005; Kirchner et al, 2012; Basner et al., 2014). In fact, most of the current literature about workers' exposure to occupational noise focuses on the risk NIHL. However, psychological effects should not be overlooked, because of its role to workers' performance and well-being. Among them, particular attention has been given to stress (Haines et al., 2001; EU-OSHA, 2005; Fouladi et al., 2012) discomfort and annoyance (Stansfeld & Matheson, 2003; Basner et al., 2014).

Annoyance reactions are one of the most important psychological effects related to noise exposure. It has influence on daily activities, feelings, thoughts, or rest, and brings negative responses, such as anger, displeasure, discomfort, distress (Guski, 1993; Guski et al., 1999; Leventhall, 2004; Basner et al., 2014). In fact, stress and discomfort responses are frequently related to annoyance reactions (Babisch, 2003; Stansfeld & Matheson, 2003; Rylander, 2004; Miyakawa et al., 2006). These psychological effects are of particular relevance in occupational settings, since they can limit workers' performance in tasks that require divided attention, working memory, retrieval of information from memory, and decision-making (LeBlanc, 2009).

The degree of annoyance, discomfort and stress are influenced by several factors, such as: noise characteristics; individual susceptibility; type of activity; time of exposure; attitude towards the source of noise; individual noise sensitivity; perceived control (Waye et al., 2002; Morrison et al., 2003; Bluhm et al., 2004; Björk et al., 2006; Schütte et al., 2007; Portela et al. 2013; Hasfeldt et al., 2014; Skagerstrand et al., 2017). Sound pressure levels are frequently described in the literature as one of the most important factors related to these symptoms. According to Bluhm et al. (2004) and Skagerstrand et al. (2017), the prevalence of annoyance increases with increasing noise levels. In the study conducted by Hasfeldt et al. (2014), the results showed that when the subjects perceived noise levels as very high, they experienced the noise as annoying and stressful. Other studies have shown increased stress is related to noise exposure during work (Haines et al., 2001; Waye et al., 2002; Fouladi et al., 2012). However, despite the importance of the loudness, other noise characteristics can be related to these symptoms, such as the frequency response and the noise type. High frequency noise is more irritating and disturbing compared to the low frequency noise (Stansfeld & Matheson, 2003; Nassiri et al., 2015). Intermittent noise was also found to be more disruptive than continuous noise (Taylor et al., 2004; Nassiri et al., 2013).

Individual susceptibility is also frequently emphasized in the literature. For example, Portela et al. (2013) found that the feeling of discomfort was notably higher in younger workers than in the older ones, as well as in workers with longer working time.

Because research on this topic is still scarce, the present study aims to analyze effect of different noise conditions on subjects' perceptions of discomfort, stress, and annoyance. To this end, an experiment in a lab was designed and three noise conditions tested: standard condition (C1), environmental noise without alert sounds (C2) and environmental noise with alert sounds (C3). During the study, the following hypotheses were tested:

Hypothesis 1: Subjects' perceptions of discomfort, stress, and annoyance are higher under more adverse noise conditions.

Hypothesis 2: Annoying sound perceptions are positively correlated with the discomfort and stress levels felt by the participants.

2 METHODOLOGY

2.1 *Participants*

Fifteen female undergraduate students from a higher education school were part of this study. The participants were aged between 20 and 23 (M = 21.6; SD = 0.8).

Inclusion criteria for this study was normal hearing, lack of visual disorders, non-smoking, lack of sleep disorders in the past 24 hours and good mental health. This information was obtained through questionnaires.

Before beginning the experimental procedure, all participants signed an informed consent form. The study was conducted in accordance with the Helsinki Declaration.

2.2 *Experimental design*

An experimental procedure that involved 3 acoustic conditions was designed: (1) Standard condition 45 ± 0.3 dB(A), (2) Environmental noise without alert sounds 60 ± 0.4 dB(A), (3) Environmental noise with alert sounds 68 ± 0.4 dB(A).

These conditions were designed considering the noise exposure levels determined in a food establishment, without risk of NIHL (Daily exposure lower than 80 dB(A), NIOSH, 1998, Directive 2003/10/EC and OSHA, s.d.).

A laboratory was adapted to create a simulated environment. A desk, which allowed each participant to be seated during the tests, was placed in the room. Four speakers, two on each side, were placed around the desk, projecting recorded environmental noise and producing the other alert sounds. The noise and alert sounds stimuli were controlled by Otometrics software and the Madsen Astera audiometer equipment.

A visual analogue scale was used to measure the discomfort, stress and annoying sound perceptions. All the tests were completed by the subjects under the same noise conditions. Sound pressure levels were monitored throughout each section with a CESVA SC310 Sound Level Meter. A variation of 0.5 dB(A) during each trial in relation to the predefined sound pressure levels was considered acceptable for this study.

All the experiments were done during the morning, ensuring that the subjects had slept at least 7 hours the night before. Each experimental procedure had, on average, a duration of 30 minutes per subject and during this period, the subjects were asked to perform tasks that required attention and memory.

2.3 *Visual Analogue Scales (VAS)*

Visual analogue scales (VAS) were used in order to estimate the subjects' perceptions about discomfort, stress and annoyance in the 3 acoustic conditions, in the same way as used in previous studies (Lundquist et al., 2000; Lesage et al., 2012; Sjödin, 2017). The VAS consisted of a line with 100 mm in length, labelled at each end as "Not at all…" at the left end, and "Extremely…" at the right. Subjects were asked to mark across the line the point that indicated the level of discomfort, stress and annoying sound perception that they were feeling. The distance of the mark from the left line was taken as the score.

2.4 *Data treatment*

Comparisons of the means were made through the three-factor ANOVA test. Correlation analyses were made using Pearson's and Spearman's correlation coefficients. Spearman's correlation was used in the cases were normality was violated.

All statistical analyses were conducted using statistical software package IBM Statistical Package for the Social Sciences (SPSS Statistics) version 23. The significance level used was $p < 0.05$.

3 RESULTS AND DISCUSSION

Subjects' perceptions about discomfort, stress and annoyance were assessed. The results are presented in Tables 1–3. According to the obtained results, subjects' perceptions of discomfort, stress, and annoyance are higher under more adverse noise conditions ($p < 0.01$). In C3 were observed higher levels of discomfort (4.6 ± 1.9), stress (4.7 ± 1.9) and annoying sound perceptions (5.5 ± 2.0). These levels can be classified as moderate, according the scale applied in this study. These results were expected since in C3 the sound pressure levels were higher. This louder sound, together with the intermittent noise, made this noise condition the most adverse one.

Table 1. Descriptive statistics of discomfort levels in the 3 conditions of noise.

	Discomfort level		
	Mean	SD	*p*-value
C1	1.4	1.5	
C2	3.2	2.0	0.00
C3	4.6	1.9	

Note: C1 – Standard condition; C2 – Environmental noise without alert sounds; C3 – Environmental noise with alert sounds.

Table 2. Descriptive statistics of stress levels in the 3 conditions of noise.

	Stress level		
	Mean	SD	p-value
C1	1.9	1.1	
C2	3.1	1.7	0.00
C3	4.7	1.9	

Note: C1 – Standard condition; C2 – Environmental noise without alert sounds; C3 – Environmental noise with alert sounds.

Table 3. Descriptive statistics of annoyance levels in the 3 conditions of noise.

	Annoyance		
	Mean	SD	*p*-value
C1	0.1	0.3	
C2	2.9	1.2	0.00
C3	5.5	2.0	

Note: C1 – Standard condition; C2 – Environmental noise without alert sounds; C3 – Environmental noise with alert sounds.

These results were expected, since other studies also observed that noise exposure was associated with higher levels of perceived stress, discomfort and annoyance (Haines et al., 2001; Morrison et al. 2003; Björk et al, 2006; Muzet, 2007; Ljungberg & Neely, 2007; Fouladi et al., 2012; Sjödin et al., 2012; Portela et al., 2013; Hasfeldt et al., 2014; Skagerstrand et al., 2017). According to Björk et al. (2006) and Muzet (2007), when a sound exceeds a certain level the sound is perceived as loud and/or annoying. In the studies conducted by Hasfeldt et al. (2014) and Skagerstrand et al. (2017) the results showed that the subjects experienced the noise as annoying, disruptive and stressful when the sound pressure levels were higher.

It is important to notice that the sound pressure levels used in this study were considerably lower than the ones used in other studies where similar results were obtained (see e.g. Fouladi et al., 2012; Sjödin et al., 2012; Portela et al., 2013). This was already noticed by Holmberg et al. (1995). According the authors, even when the levels of noise exposure are lowest, they may cause annoyance and discomfort (Holmberg et al., 1995).

In the present study, subjects' perception on discomfort, stress and annoyance was higher when the sound pressure levels were about 68 dB(A) with intermittent noise. The obtained results suggest the relevance of the intermittent noise to the psychological health effects. In fact, intermittent noise is more disruptive than continuous noise (Joseph et al., 2000; Nassiri et al., 2013; Nassiri et al., 2015). Intermittent noise patterns can result in decreased performance more than continuous noise, because under these conditions, subjects are more likely to disturb their attention (Taylor et al., 2004; Nassiri et al., 2013; Nassiri et al., 2015). As a result, other psychological health effects can be enhanced, in particular annoyance due to the difficulty in complete the tasks. Therefore, it is not surprising that a great number of subjects report complaints related to intermittent noise patterns (van Dijk et al., 1987).

Table 4 presents the results about the correlations between the different variables. In the three conditions (C1-C3), significant positive correlations were found between discomfort and stress levels ($p < 0.01$). As the discomfort level increased, the stress level also increased and vice-versa. Positive correlations were also observed for C3 between annoyance and discomfort ($r = 0.801$; $p < 0.01$), as well as between annoyance and stress levels ($r = 0.560$; $p < 0.01$). According to these results, in this noise condition, while the annoying sound perception increased, the discomfort and stress levels also increased. These results suggest that relationship between annoyance reactions with stress and discomfort levels (Babisch, 2003; Stansfeld & Matheson, 2003; Rylander, 2004; Miyakawa et al.,

Table 4. Correlations between discomfort and stress levels and annoying sound perception with the three conditions.

		(1)	(2)	(3)
C1	(1) Discomfort level	–	0.847**	0.425
	(2) Stress level	0.847**	–	0.150
	(3) Annoyance level	0.425	0.150	–
C2	(1) Discomfort level	–	0.811**	0.186
	(2) Stress level	0.811**	–	0.004
	(3) Annoyance level	0.186	0.004	–
C3	(1) Discomfort level	–	0.518*	0.801**
	(2) Stress level	0.518*	–	0.560*
	(3) Annoyance level	0.801**	0.560*	–

Note: $*p < 0.05$; $**p < 0.01$; C1 – Standard condition; C2 – Environmental noise without alert sounds; C3 – Environmental noise with alert sounds.

2006) is mediated by sound levels and type. Similar results were found by Waye et al. (2002) and Sjödin et al. (2012), where the authors observed that only under high noise levels, annoyance was associated with higher stress levels.

4 CONCLUSIONS

Results of this study showed that noise levels without risk for NIHL can have an important influence on subjects' perceptions about stress, discomfort and annoyance. It was observed that they experienced higher levels of discomfort, stress and annoying sound perceptions when the sound pressure levels increased. Moreover, significant correlations between discomfort and stress levels in the three conditions and between annoyance and discomfort and stress levels in the more adverse noise condition were observed.

In this study, the workers' noise exposure in a food establishment was simulated. This occupational context was selected because workers deal with attentional and short-term memory tasks, being the influence of intermittent noise due to the existing sound signals particular important. In view of this, results from this study suggest the importance to create alternative solutions in food establishments, were alarm signs are regular and several mistakes with the clients' requests observed everyday.

The results were limited to the conditions of this experiment, i.e., a relatively small sample, the participants were all female and students. Additionally, it is important to note that the sound pressure level in condition C1, 45 dB(A), although the room was quieter than the others, ventilation could not be turned off as it would leave the subjects uncomfortable.

REFERENCES

Babisch, W. 2003. Stress hormones in the research on cardiovascular effects of noise. *Noise & Health* 5(18): 1–11.

Basner, M., Babisch, W., Davis, A., Brink, M., Clark, C., Janssen, S. & Stansfeld, S. 2014. Auditory and non-auditory effects of noise on health. *The Lancet* 383: 1325–32.

Björk, J., Ardö, J., Stroh, E., Lövkvist, H., Östergren, P.O. & Albin, M. 2006. Road traffic noise in southern Sweden and its relation to annoyance, disturbance of daily activities and health. *Scandinavian Journal of Work, Environmental & Health* 32: 392–401.

Bluhm, G., Nordling. E. & Berglind, N. 2004. Road traffic noise and Annoyance - an increasing environmental health problem. *Noise & Health* 6: 43–9.

Concha-Barrientos, M., Campbell-Lendrum, D., Steenland, K. 2004. Occupational noise: assessing the burden of disease from work-related hearing impairment at national and local levels. In: *Environmental Burden of Disease Series, No. 9*. Geneva: WHO.

Directive 2003/10/EC of the European Parliament and of the Council on the minimum health and safety requirements regarding the exposure of workers to the risks arising from physical agents (noise). Journal Official n° L 042, 6 of February.

EU-OSHA. 2005. *The impact of noise at work*. Factsheet 57. Accessed on 1 of February of 2017, available at: https://osha.europa.eu/en/publications/factsheets/57/view.

Fouladi, D.B., Nassiri, P., Monazzam, E.M., Farahani, S., Hassanzadeh, G. & Hoseini, M. 2012. Industrial noise exposure and salivar cortisol in blue colar industrial workers. *Noise & Health* 14 (59): 184–189.

Guski, R. 1993. Personal and social variables as co-determinants of noise annoyance. *Noise & Health* 3: 45–56.

Guski, R., Flescher-Suhr, U. & Schuemer, R. 1999. The concept of noise annoyance: How international experts see it. *Journal of Sound and Vibration* 223: 513–27.

Haines, M.M., Stansfeld, S.A., Job, S.R.F., Berglund, B. & Head, J. 2001. A follow-up study effects of chronic aircraft noise exposure on child stress responses and cognition. *International Journal of Epidemiology* 30(4): 839–845.

Hasfeldt, D., Terkildsen, H., Toft, P. & Birkelund, R. 2014. Patients' perception of noise in the operating room – A descriptive and analytic cross-sectional study. *Journal of PeriAnesthesia Nursing* 29(5): 410–417.

Holmberg, K., Landström, U. & Nordström, B. 1995. Annoyance and discomfort during exposure to high-frequency noise from an ultrasonic washer. *Perceptual and Motor Skills* 81(3): 819–827.

Ismail, A.R. 2011. Multiple linear regressions of environmental factors toward discrete human performance. In *Proceedings of the 4th international conference on environmental and geological science and engineering* (EG '11), Barcelona, 15–17 September 2011.

Jahncke, H., Hygge, S., Halin, N., Green, A.M. & Dimberg, K. 2011. Open-plan office noise: Cognitive performance and restoration. *Environmental Psychology* 31: 373–382.

Joseph, C., Aravindakshan, B. & Vyawahare, M.K. 2000. Non-auditory effects of noise: Psychological Task Performance. *Indian Journal of Aerospace Medecine* 44(1): 49–56.

Kirchner, D.B., Everson, E., Dobie, R.A., Rabinowitz, P., Crawford, J., Kopke, R. & Hudson, T.W. 2012. Occupational noise-induced hearing loss: ACOEM task force on occupational hearing loss. *Journal of Occupational & Environmental Medicine* 54(1): 106–108.

LeBlanc, V.R. 2009. The effects of acute stress on performance: implications for health professions education. *Academic Medicine* 84(10): 25–33.

Lesage, F.-X., Berjot, S. & Deschamps, F. 2012. Clinical stress assessment using a visual analogue scale. *Occupational Medicine* 62(8): 600–605.

Leventhall, H.G. 2004. Low frequency noise and annoyance. *Noise & Health* 6: 59–72.

Ljungberg, J.K. & Neely, G. 2007. Stress, subjective experience and cognitive performance during exposure to noise and vibration. *Journal of Environmental Psychology* 27: 44–54.

Lundquist, P., Holmberg, K. & Landstrom, U. 2000. Annoyance and effects on work from environmental noise at school. *Noise & Health* 2: 39–46.

Miyakawa, M., Matsui, T., Kishikawa, H., Murayama, R., Uchiyama, I., Itoh, T. & Yoshida, T. 2006. Salivary chromogranin A as a measure of stress response to noise. *Noise & Health* 8(32): 108–113.

Morrison, W.E., Haas, E.C., Shaffner, D.H., Garrett, E.S. & Fackler, J.C. 2003. Noise, stress and annoyance in a pediatric intensive care unit. *Critical Care Medicine* 31(1): 113–119.

Muzet, A. 2007. Environmental noise, sleep and health. *Sleep Medicine Reviews* 11: 135–142.

Nassiri, P., Monazam, M., Dehaghi, B.F., Abadi, L.I.G., Zakerian, S.A. & Azam, K. 2013. The effect of noise on human performance: A clinical trial. *International Journal of Occupational and Environmental Medicine* 4(2): 87–95.

Nassiri, P., Monazzam, M.R., Asghari, M., Zakerian, S.A., Dehghan, S.F., Behzad, F. & Azam, K. 2015. The interactive effect of industrial noise type, level and frequency characteristics on occupational skills. *Performance Enhancement & Health* 3: 61–65.

Nelson, D.I., Nelson, R.Y., Barrientos, M.C. & Fingerhut, M. 2005. The global burden of occupational noise-induced hearing loss. *American Journal of Industrial Medicine* 48(6): 446–458.

NIOSH. 1998. *Criteria for a recommended standard: Occupational noise exposure – Revised criteria*. Ohio: NIOSH.

OSHA. s.d. *The Occupational Noise Exposure Standard: 29 CFR 1910.95*. Accessed on 10 of December of 2017, available at: https://www.osha.gov/pls/oshaweb/owadisp.show_document?p_table=standards&p_id=9735.

Planeau, V. 2005. *Noise hazards associated with the call center industry*. Accessed on 17 of July of 2017, available at: www.higieneocupacional.com.br/download/noise-planeau.doc.

Portela, B.S., Queiroga, M.R., Constantini, A. & Zannin, P.H.T. 2013. Annoyance evaluation and the effect of noise on the health of bus drivers. *Noise & Health* 15 (66): 301–306.

Rodrigues, M.A., Amorim, M., Silva, M.V., Neves, P., Sousa, A. & Inácio, O. 2015. Sound levels and risk perceptions of music students during classes. *Journal of Toxicology and Environmental Health* 78(13–14): 825–839.

Rodrigues, M.A., Freitas, M.A., Neves, M.P. & Silva, M.V. (2014). Evaluation of the noise exposure of symphonic orchestra musicians. *Noise & Health* 16: 40–46.

Rylander, R. 2004. Physiological aspects of noise-induced stress and annoyance. *Journal of Sound and Vibration* 277: 471–478.

Schütte, M., Marks, A., Wenning, E. & Griefahn, B. 2007. The development of the noise sensitivity questionnaire. *Noise & Health* 9: 15–24.

Sjödin, F. 2017. Individual factors and its association with experience noise annoyance in Swedish preschools. *Proceedings of Meetings on Acoustics* 30(1): 015015.

Sjödin, F., Kjellberg, A., Knutsson, A., Landström, U. & Lindberg, L. 2012. Noise and stress effects on preschool personnel. *Noise & Health* 14(59): 166–178.

Skagerstrand, Å., Köbler, S. & Stenfelt, S. 2017. Loudness and annoyance of disturbing sounds – perception by normal hearing subjects. *International Journal of Audiology* 56(10): 775–783.

Stansfeld, S.A. & Matheson, M.P. 2003. Noise pollution: non-auditory effects on health, *British Medical Bulletin* 68(1): 243–257.

Taylor, W., Melloy, B., Dharwada, P., Gramopadhye, A. & Toler, J. 2004. The effects of static multiple sources of noise on the visual search component of human inspection. *International Journal of Industrial Ergonomics* 34(3): 195–207.

van Dijk, F.J.H., Souman, A.M. & Vries, F.F. 1987. Non-auditory effects of noise in industry – A final field study in industry. *International Archives Occupational and Environmental Health* 59: 133–145.

Waye, K.P., Bengtsson, J., Rylander, R., Hucklebridge, F., Evans, P. & Clow, A. 2002. Low frequency noise enhances cortisol among noise sensitive subjects during work performance. *Life Sciences* 70: 745–758.

World Medical Association. 2013. World Medical Association Declaration of Helsinki: ethical principles for medical research involving human subjects. *Journal of American Medical Association* 310(20); 2191–2194.

Assessment and characterization of WMSDs risk in nurses who perform their activity in surgical hospitalization

M. Torres
Surgical Internment Services, Hospital of Braga, Braga, Portugal

P. Carneiro & P. Arezes
ALGORITMI Centre, School of Engineering, University of Minho, Guimarães, Portugal

ABSTRACT: Work-related Musculoskeletal Disorders (WMSDs) are an important problem all over the world, in particular among healthcare workers. The purpose of this study was to identify self-reported musculoskeletal symptoms by nurses that work in surgical services of a hospital unit in the north of Portugal, and to evaluate the risk of developing WMSDs in typical activities of this context. It was used a survey to collect information, which was applied to 146 nurses (84.2% response rate). The results showed a high prevalence of awkward postures and symptoms in different anatomical zones in the last 12 months (88.6%), in particular for lower back region (79.7%), cervical (62.6%) and shoulders (44.7%). The postural analysis technique REBA (Rapid Entire Body Assessment) was used to quantify the WMSDs risk associated with different nursing activities, in real work context. It was observed that the activities which involve moving patients are the ones presenting a higher number of postures with a high or very high risk of developing WMSDs. In this context, everything suggests that high risk postures adopted by nurses may be influenced by the complexity of the working activities involving excessive and repetitive efforts, by the increased pace of work due to excessive activities, lack of work breaks, among others.

Keywords: Nurses, Risk, WMSDs, Self-report, Symptoms, Hospital

1 INTRODUCTION

In the healthcare sector, nursing is one of the jobs most affected by musculoskeletal disorders (AESS, 2000; WHO, 2002). The evidence of this fact is corroborated by the results obtained in various studies, both in national as in international contexts, carried out in hospital environments over the past few years (e.g., BLS, 2008; Torres et al., 2010; Torres et al., 2012).

Nurses usually perform activities that require extreme postures, application of strength, as well as requirements at the level of the spine. It is often observed that they have to face professional other risk factors, namely regular biomechanical and physiological demands beyond the functional capabilities of these professionals, in an organization which does not provide enough recovery time and appropriate rest time (Cail et al., 2000).

Therefore, when performing their activities, nurses are exposed to a variety of risk factors that may contribute to the incidence and development of musculoskeletal pathology, raising the concerns regarding the corresponding incidence of symptoms of WMSDs among these professionals (Hitchings & Smith, 2001).

In hospital contexts, nurses perform their activity in different services, developing a wide variety of tasks, involving patient assistance and treatment, patient motion and transfer (often with high weight) and manual load handling. These interventions are sometimes performed with excessive and repetitive efforts for an extended period of time, as well as by adopting critical postures. In most of the cases, these activities are performed without auxiliary mechanical equipments, contributing as risk factors to the development of WMSDs (Gomes, 2009). The increased pace of work due to excessive activities may lead the nursing personnel to adopt critical postures, therefore being also a risk factor to the development of musculoskeletal complaints (Torres, 2009; Magnago et al., 2010).

The WMSDs associated with the nursing activity are, largely, preventable and, therefore, should not be seen as inevitable. For this purpose, it is relevant that everyone involved can be aware that work can be improved, providing health carers with a healthy and safe work environments.

The high prevalence of musculoskeletal disorders among nursing professionals reinforces the idea that more studies are necessary in order to better understand the problem, and to provide effective approaches considering prevention and promotion of musculoskeletal health in a work context (Faria, 2010).

Taking the above-mentioned considerations into account, the present study had the following aims:

- Identify the self-reported musculoskeletal symptomatology among nurses who perform their activity in the surgical specialties from a hospital unit in northern Portugal
- Assess the risk of developing WMSDs in nursing professionals who develop their activity in this context.

2 MATERIALS AND METHODOLOGY

To achieve the defined research aims, a quantitative exploratory cross-sectional study was carried out. In this study two key tools were used for data gathering: (1) the Nordic musculoskeletal questionnaire (NMQ) (Kuorinka et al., 1987), Portuguese version (Mesquita et al., 2010), complemented by some questions from the questionnaire developed and applied by Torres (2009), which made it possible to collect self-reported data about the musculoskeletal symptomatology associated with nurses' performance; and (2) the REBA method, which allowed the WMSDs' risk quantification associated with several nursing activities.

The target population of this study was all the nurses who, in the months of December 2015 and January 2016, performed their activity in the surgical specialties at a central hospital in the north of Portugal (General Surgery, Plastic and Reconstructive Surgery, Stomatology, Maxillofacial Surgery, Orthopaedics, Neurosurgery (NS), Neurocritical Care Unit (NCCU), Intermediate Care Unit of the Urgency Service (ICUUS)), in a total of 146 professionals. Among these, 123 nurses (84.2% of the population) constituted the sample of this study.

The statistical analysis of data was conducted using the statistical software package SPSS (version 20) to verify the existence of a relation between the variables in study. It was used the Spearman's correlation coefficient, which measures the degree of association or dependence between two variables.

2.1 Data collection

A questionnaire was specifically designed to this study, which contains 3 distinct sections, each of them reporting to a specific dimension, namely: (1) General and demographic data; (2) Identification of the self-reported musculoskeletal symptomatology in the last 7 days, for the last 12 months, including the information about not being able to perform their usual activities due to any problem in the previous year, and lastly a variable to assess the discomfort per body zone; and (3) Working conditions characterization.

Pain intensity was assessed using a Numerical Scale, that consists on a ruler divided into eleven equal parts, numbered successively from 0 (without pain) to 10 (maximum pain).

All the nurses were instructed to complete the questionnaire and deposit it in a properly identified box in the various hospital services. The filled questionnaires were collected at the end of January 2016.

2.2 WMSDs risk assessment

An objective characterization of WMSD risk associated with diverse nursing activities was carried out in different services. This characterization was accomplished by applying the postural analysis REBA methodology (Hignett & McAtamney, 2000). For this purpose, video recording was used in real-life interventions, concerning different nursing activities, performed by different nurses.

351 posture records were analysed in a total of 28 nursing activities performed by 13 nurses, consisted mainly in activities of treating wounds, hygiene and comfort, transfer and patient positioning.

For each analyzed activity, different posture images were selected, in order to be representative of the reality of the work context, in other words, the WMSDs riskier postures were not the only ones analyzed. Since the activities are not cyclical, each was analyzed according to the following procedure:

- Identification of the main postures/movements adopted during the development of the task;
- Estimation of the percentage of working time associated with each of the postures identified above;
- Calculation of the REBA score for each postures/movement category;
- Calculation of the whole REBA score, that is, weighted in relation with the duration of each posture.

3 RESULTS AND DISCUSSION

3.1 Sample characterization

123 nurses (84.2% of the population) who work in the Surgical Specialties of the studied hospital participated in this study. Most of the subjects are female (69.9%), single (51.2%), and have an education level of Nursing degree (87.8%). The remaining 12.2% have a Master degree in several scientific areas. They have a mean age of 34 years old and have more than 10 years of professional experience.

The analysed sample has an average BMI of 23.1 that, according to the Direção Geral de Saúde (2005), shows that these professionals fit a considered healthy pattern, explained by the fact that most of them (56.1%) have physical activity or sports habit apart from their physical labour activity. The average working hours per week in the referred hospital is 40 hours. 15.4% of the sample also work in an additional institution, in which the average working hours per week range between 5 and 40 hours, with an average of 18.2 hours. Respondents were distributed at different services: 18 (14.6%) in the Intermediate Care Unit of the Urgency Service (ICUUS); 20 (16.3%) in the Neurocritical Care Unit (NCCU); 40 (32.5%) in Orthopaedics spread over three wards (B, C and D); 29 (23.6%) in Surgery spread over two wards (B and C); and the others 16 in Neurosurgery (13%). Regarding professional categories, 88.6% of the participants belong to the category 'nurses' and the remaining 11.4% belong to the category of 'specialized nurses'.

3.2 Complaints and musculoskeletal symptomatology per body zone

This study pointed out the high prevalence of self-reported musculoskeletal symptoms (pain, discomfort or paresthesias) in one or more anatomic regions. 88.6% of the respondents mentioned having musculoskeletal complaints related to their job in the last year.

The most affected body regions were the lower back (79.7%), cervical (62.6%), shoulders (44.7%), wrists/hands and knees with the same rate (30.1%) and, at last, ankles/feet (17.1%) and elbows (11.4%). In fact, these are significant values of the frequency of symptoms, which one should bear in mind in any perspective of risk management.

These values clearly express the existence of WMSDs risk among nursing professionals and are even superior to those found in some national and international researches developed in hospital environments (e.g., Trinkoff et al., 2002; Smith et al., 2004; Fonseca, 2005; Barroso et al., 2007; Martins, 2008; Coelho, 2009; Jerónimo, 2013). However, the prevalence of the symptomatology found in the present study is inferior to the one mentioned in a national study characterization of musculoskeletal symptomatology related to work among Portuguese nurses developed by Serranheira et al. (2012), which demonstrated that 98% of the nurses reported some type of symptomatology, at least in an anatomic segment. Thus, results seem to suggest that high prevalence of musculoskeletal symptomatology in the three anatomic regions mentioned before (lower back, cervical and shoulders) may be influenced by a high level of complexity of the developed activities, including excessive and repetitive efforts, by the increased work pace due to an excessive number of activities, and by the lack of work breaks, among others.

It is important to highlight that the most disabling body regions to perform normal daily activities of the respondents (work, household chores or hobbies) in the last 12 months, are the lower back (56.1%), cervical (42.3%), knees (20.3%) and shoulders (17.9%).

3.3 Pain intensity per body zone

The obtained results show that most respondents referred pain complaints with a variable intensity in one or more regions of the body.

Among the nurses who mentioned any type of symptomatology in the last 12 months (n = 109), levels of pain intensity equal or higher than moderate (numerical scale of pain intensity ≥ 4) were identified with emphasis placed on the lower back (n = 80), cervical (n = 55), shoulders (n = 35), knees (n = 26), wrists/hands (n = 21) and dorsal (n = 20). It should be also highlighted the very intense pain level or severe pain that was reported by 55 references to the lower back (51.2%) and 32 references to the cervical (29.3%).

3.4 Working conditions

From the total of respondents (123), 92 (75%) reported that their activity in this environment is a cause of anxiety or stress, the highest levels having been recorded in NS (87%), NCCU (85%), followed by Orthopaedics (3D ward - 81%) and General Surgery (2B ward - 80%). The intensity of that anxiety or stress was classified as moderate by most of the respondents (42%). Most of the nurses who participated in this study (81%) are satisfied with the service where they perform their activity, specially the ones working in the Orthopaedics services – 3C ward (100%), 3D ward (94%) e ICUUS (94%). On the other hand, 21% of the respondents consider themselves anxious or irritable people.

It should also be noted that 75% of the respondents do not take at least a 5 minute break or any other short breaks during their shift work, besides the meal time break. The other 25% who take breaks higher than 5 minutes, take in average 2.2 breaks per work shift.

One hundred and eight respondents (88%) reported the existence of technical aids to patient transfer (patient transfer lifts, transfers and transfer belts, etc) and some nurses pointed out the existence of more than one auxiliary equipment in the service where they perform their activity. From these, only 26 (24%) use these equipment in more than 50% of the situations. The other 82 (76%) nurses reported that the rarely use of auxiliary equipment (in less than 50% of the situations) is mainly due to their reduced number in the workplace (n = 30) and to the lengthy process that its use implies (difficult handling/excessive usage time) (n = 28).

Among the several situations that influence the adopted postures by nurses during the working activities, it should be noted: the increased work pace and intensification of work (88%), handling excessive loads (81%) and patient care (58%). Some of these aspects were also identified in studies previously developed in hospital environments (Barroso et al., 2007; Martins, 2008; Maia, 2002).

Finally, nurses were asked to mention possible interventions which could minimize WMSDs risk in the surgical hospitalization services. From the 123 nurses, 92 answered this question and reported:

- An increase of the ratio nurse/patient for work shift (46.7%);
- The establishment of an occupational gymnastics program (41.3%);
- An ergonomics training program (9.8%).

3.5 Variables correlation

A possible association between the variable 'musculoskeletal complaints over the last 12 months' and the sociodemographic and professional variables were initially tested, and it was noted that:

- There is a weak positive correlation among the musculoskeletal complaints and the variables age ($r_s = 0.336$) and professional seniority ($r_s = 0.327$);
- Based on data analysis, it is evident a statistically significant association ($p < 0,01$) between the musculoskeletal complaints and the variables age and professional seniority, which allow to conclude that musculoskeletal complaints increase with age and professional seniority, being higher on females (70.6%). It is also possible to observe a revealing trend showing that as weekly working hours increase, the number of musculoskeletal complaints intensify. Thus, nurses who work more than 40 hours per week (since 15.4% of the sample work in an additional institution) show a higher rate of musculoskeletal symptomatology (lower back 68.4%, cervical 52.6% and shoulders 52.6%);
- The results also suggest the existence of a statistically significant weak negative correlation ($r_s = -0.309$; $p < 0.01$) between the variables "musculoskeletal complaints" and the use of technical aids to transfer patients, which seems to allow the conclusion that nurses who do not use the auxiliary equipment so often are the ones who reported more musculoskeletal complaints;
- There is a statistically significant weak negative correlation between musculoskeletal complaints and the practice of a physical activity ($r_s = -0.272$; $p = 0.008$). It is noted that nurses who do not practice any physical activity are the ones who tend to report more musculoskeletal complaints.

3.6 WMSDs Risk Assessment—REBA

After the calculation of the whole REBA score for each activity, it was rounded to the unity. The majority of the analyzed activities were classified as having associated a medium global WMSDs risk, that is, they are activities requiring ergonomic intervention, even if not necessary in a short term. However, there is an activity for which the global risk of WMSDs is high, requiring, in this case, a short term ergonomic intervention. This activity is the transfer from a patient from the bed to the armchair, in the Orthopedic service.

Although the global WMSDs risk is classified as average for most of the activities, it is clear that in each activity various postures were analysed with a high or even very high level of WMSDs risk. Activities which involve moving patients (e.g. patient transfer bed/armchair and vice-versa, placing patients in the bed and in the armchair) are the ones that present a higher percentage of postures with a high or very high WMSDs risk. The conjugation of force application and the adoption of critical postures seems to contribute to a substantial increase in the WMSDs risk.

4 CONCLUSIONS

The respondents have demonstrated a high prevalence of work-related musculoskeletal complaints, since nearly 88.6% of them have reported symptomatology, at least in an anatomic segment. The frequency of those symptoms can be seen as an indication of the demanding nature of nursing in the analysed contexts. The most prevalent complaints in the last 12 months are in the lower back (79.7%), cervical region (62.6%) and shoulders (44.7%).

The obtained results clearly suggest the existence of multiple risk factors and their relevant contribution to risk levels rated between moderate and very high.

In this context, the results seem to indicate that risk postures adopted by nurses may be influenced by the complexity of the performed working activities, which regularly involve excessive and repetitive physical efforts, by the increased work pace, and by the lack of work breaks, among other reasons.

In view of the presented data, and the consequences that these problems may have in the nurses' musculoskeletal health and quality of life, and also in patient care quality, it is important to consider a special attention to the analysed work environments conditions.

The presented results also suggest an urgent need to develop new strategies to prevent WMSDs, in order to prevent, on the one hand, absenteeism and, on the other, future complications. Thus, specifically focusing on WMSDs prevention in the surgical services and, therefore, on the effective reduction of risk levels, it is suggested that hospitals should implement some preventive/corrective measures. Some examples of these measures are, for instance, the increase of the ratio nurse/patient per work shift; the establishment of an occupational gymnastics program, and the development of ergonomics training programs. Regardless the adopted measures, they should be carefully analysed and planned but, if appropriately implemented, they may contribute to WMSDs prevention and to an improvement of the nurses' working conditions.

REFERENCES

AESST (2000). Lesões por esforços repetitivos nos Estados-Membros da União Europeia. Síntese de um relatório da Agência, Facts 6.

Barroso, M.; Carneiro, P. & Braga, A. (2007). Characterization of Ergonomic Issues and Musculoskeletal complaints in a Portuguese District Hospital. Proceedings do International Symposium "Risks for Health Care Workers: prevention challenges", ISSA, Atenas, Junho 2007. Manuscrito em CD ROM. 13 p.

BLS (2008). Musculoskeletal disorders and days away from work in 2007. US Bureau of Labor Statistics/ Unites States Departament of Labor. Acedido em 20/03/2012, de: http://www.bls.gov/opub/ted/2008/dec/wk1/art02.htm.

Cail, F.; Aptel, M. & Franchi, P. (2000). Les troubles musculoskelettiques du membre supérieur - guide pour les préventuers. Paris: INRS.

Coelho, M.S.R. (2009). Estudo da frequência das lesões músculo-esqueléticas relacionadas com o trabalho em profissionais de enfermagem – proposta de um programa de ginástica laboral. Porto: Faculdade de Desporto da Universidade do Porto. Acedido em http://repositorio-aberto.up.pt/bitstream/10216/21697/2/16363.pdf.

Direcção-Geral da Saúde (2005). Programa Nacional de Combate à Obesidade. Circular Normativa n.º 3/ DGCG. Lisboa: Ministério da Saúde.

Faria, F. (2010). Condições de Trabalho dos Profissionais de Enfermagem e Atuação Ergonômica. Fisio Clinica de Reabilitação São Joaquim da Barra. Acedido em 20/03/2012, de: http://www.fisiocr.com.br/index.php?pagina=mostra-artigos&id=6.

Fonseca, M. (2005). Contributo para a avaliação da prevalência de sintomatologia músculo-esquelética auto-referida pelos enfermeiros em meio hospitalar. Dissertação de Mestrado em Saúde Pública. Porto: Faculdade de Medicina e Instituto de Ciências Biomédicas Abel Salazar, 125 p.

Gomes, M. (2009). Avaliação da atividade neuromuscular dorsal e lombar em enfermeiros em três posicionamentos de doentes com acidente vascular cerebral. Dissertação de Mestrado em Saúde Ocupacional. Faculdade de Medicina da Universidade de Coimbra. 129 p.

Hignett, S. & McAtamney, L. (2000). Rapid Entire Body Assessment (REBA). Applied Ergonomics, 31:2, 201–205.

Hitchings, G. & Smith, D. (2001). Occupational health and safety issues in contemporary nursing. Safety Science Monitor. 5(1):1–4.

Jerónimo, J.M.A. (2013). Estudo da prevalência e fatores de risco de lesões musculoesqueléticas ligadas ao trabalho em enfermeiros. Dissertação de Mestrado em Enfermagem de Reabilitação. Escola Superior de Enfermagem de Coimbra. 128 p.

Kuorinka, I., Johnsson, B. & Kilborn, A. (1987). Standardized Nordic Questionnaires for Analysis of Musculoskeletal Symptoms. Applied Ergonomics, 18 (3): 233–237.

Magnago, T., Lisboa, M., Griep, R., Kirchhof, A. & Guido, L. (2010). Aspectos psicossociais do trabalho e distúrbio musculoesquelético em trabalhadores de enfermagem. Revista Latino-Americana de Enfermagem, vol.18 n.º3. Ribeirão Preto May/June 2010. ISSN 0104–1169.

Maia, P. (2002). Avaliação da capacidade laboral de enfermeiros em contexto hospitalar. Dissertação de Mestrado em Engenharia Humana. Guimarães: DPS/EE/UM, 173 p.

Martins, J. (2008). Percepção do risco de desenvolvimento de lesões musculoesqueléticas em actividades de enfermagem. Dissertação de Mestrado em Engenharia Humana. Guimarães: DPS/EE/UM, 142 p.

Mesquita, C; Ribeiro, J. & Moreira, P. (2010). Portuguese version of the Standardized Nordic musculoskeletal questionnaire: cross cultural and reliability. Journal of Public Health. ISSN 0943-185318(5), p. 461–466.

Serranheira, F., Cotrim, T., Rodrigues, V., Nunes, C. & Uva, A. (2012). Lesões musculoesqueléticas ligadas ao trabalho em enfermeiros portugueses: «ossos do ofício» ou doenças relacionadas com o trabalho? Revista Portuguesa de Saúde Pública. 30(2): 193–203.

Smith, D., Wei N., Zhao L. & Wang, R. (2004). Musculoskeletal complaints and psychosocial risk factors among Chinese hospital nurses. Occupational Medicine. 54(8): 579–582.

Torres, M. (2009). Percepção do risco de desenvolvimento de lesões musculoesqueléticas em actividades de Enfermagem no contexto de emergência pré-hospitalar. Dissertação de Mestrado em Engenharia Humana. Guimarães: DPS/EE/UM, 211 p.

Torres, M.; Martins, J. & Carneiro, P. (2012). Riscos ergonómicos em atividades de enfermagem no contexto domiciliário. Revista Segurança Comportamental, Ano 3, n.º 5, 1.º Semestre 2012, p. 29–31.

Torres, M.R., Arezes, P.M. & Barroso, M.P. (2010). Caracterização e análise da percepção do risco de LMERT em profissionais de enfermagem em contexto de emergência pré-hospitalar. In Arezes, P., Baptista, J.S., Barroso, M.P., Carneiro, P., Cordeiro, P., Costa, N., Melo, R., Miguel, A.S. & Perestrelo, G.P. (Eds.). Occupational Safety and Hygiene – SHO 2010, ISBN 978-972-99504-6-9, p. 524–528.

Trinkoff, A., Lipscomb, J., Geiger-Brown, J. & Brady, B. (2002). Musculoskeletal Problems of the Neck, Shoulder and Back and Functional Consequences in Nurses. American Journal of Industrial Medicine, 41: 170–178.

WHO (2002). The World Health Report 2002. Reducing Risks, Promoting. Healthy Life. Geneva: World Health Organization, 13 p.

Risk analysis in the execution of the Aguas Santas tunnel

Eliana Carpinteiro
Universidade de Trás-os-Montes e Alto Douro, Vila Real, Portugal

Cristina Reis
Universidade de Trás-os-Montes e Alto Douro, Vila Real, Portugal
Construct & INEGI—FEUP, Portugal

Paula Braga
Universidade de Trás-os-Montes e Alto Douro, Vila Real, Portugal
INEGI—FEUP, Portugal

José António F.O. Correia
INEGI—FEUP, Portugal

J.F. Silva
Instituto Politécnico de Viana do Castelo, Portugal

Carlos Oliveira
Instituto Politécnico de Viana do Castelo, Portugal
INEGI—FEUP, Portugal

ABSTRACT: The construction of underground works carries a high prevention in terms of safety, being considered works of high risk. These works are considered of great importance, being necessary the intervention of knowledge where experience, creativity and theory, play an important role. It is necessary to ensure the safety of all actors involved in performing underground works, and companies are responsible for taking all necessary precautions to ensure that the safety plan is implemented correctly. In order to deepen his knowledge on this subject, the author carried out his study entitled "Risk Analysis in the Execution of the Holy Water Tunnel". With the purpose of develop this work, the researchers were allowed to follow the various phases of the Águas Santas tunnel project. For the risk analysis and hierarchization, the William T. Fine Method was used. It was concluded that In addition to the achievement of a good safety plan, it is the responsibility of all stakeholders to respect all safety rules, from the use of personal and collective protection equipment to signage, in order to reduce or even eliminate the risk of accidents.

1 INTRODUCTION

The term security is susceptible of several interpretations. In general, it can be said that this concept, which derives from the Latin "Securitas", refers to the quality of what is safe, that is, what is protected from any dangers, damages or risks.

The concept of work safety is the set of resources and techniques applied preventively or correctively, for the protection of man, the assets of a company and the environment, aiming the elimination of the risks of accidents arising from the work process or the realization of an assignment.

Accidents at work have high costs for workers and employers. As such, it is important to adopt preventive behaviors and be informed about practices that can reduce and eliminate risks. It is worth noting that, through research on accident costs, Reis, Cristina et al (2005) and Reis C., Oliveira C. (2012), it is possible to prove that it is highly advantageous to prevent accidents at work, as it may be seen from Table 1.

The current data, for the year 2017, shows a decrease in the number of fatal accidents, around

Table 1. Matrix related to the implementation of health and safety plans (Source: Reis, Cristina et al (2005) and Reis C., Oliveira C. (2012)).

From the point of view	Economic advantages (dimensionless)
From the construction company	3
From the insurer	21
Social	5

26%, with the construction sector being the most affected in relation to the other sectors of activity, although the number of accidents has decreased over the last years.

Several causes can be attributed to this type of work-related accidents, the most significant being related to prevention management, work organization, protection and signalling, work space organization or individual factors.

Considering the statistical data provided by the Working Conditions Authority (ACT), construction is one of the priority sectors in establishing preventive measures, such as training, sensitization and evaluation.

Innovation in the sector should also be directed at improvements in occupational risk prevention, by examining the cause of the most common occupational safety and health (OSH) problems and anticipating new risks, by studying the changes that occur in the workplace and in society as a whole.

A greater focus on safety contributes to a broad range of benefits.

2 SAFETY MEASURES IN THE CONSTRUCTION OF TUNNELS

2.1 *Development and specification of the health and safety plan—DEPSS*

The Health and Safety plan, initiated at the project stage, needs to be developed, specified and complemented during the project and respective planning of the work, with a view to its implementation during the execution of the work.

DEPSS develops, in particular, the framework of the relations of all stakeholders in the Shipyard, in particular with regard to its obligations, in order to ensure the prevention of occupational risks and other aspects deemed necessary for the effective organization and operation of the Shipyard.

The DEPSS applies to all stakeholders in the yard, namely the developer and its collaborators; implementing entities and their collaborators; subcontractors and their collaborators; independent workers; suppliers and their collaborators; employer and trade union representatives; visitors, or other persons authorized to access the interior of the yard.

The purposes to be achieved with the implementation of DEPSS in the contract, specifically, are the following:

- Eliminate the accident claims, proposing to complete the work without any accidents, and carry out all activities under adequate safety and health conditions;
- Contribute to the reduction of causes that cause occupational diseases in the construction sector;
- Achieve good levels of productivity due to good working conditions;
- Perform all the works of specific quality, in an organized and environmentally correct space;
- Minimize the social and economic costs resulting from accidents;
- Contribute to the existence of a safety culture, involving all stakeholders in the project;
- Reduce the impact on third parties and safeguard their integrity and that of their assets in the area of site safety management.

2.2 *Specific security plan by activity—PES*

Work involving special risks, as defined in the legislation, should be subject to specific treatment.

In each specific Security Plan—PES, the preventive measures to be adopted for the various identified risks, which will be fulfilled during the work, will be taken into account.

The descriptive memorandum of the Specific Safety Plan will have the following structure:

1. Identification of the activity
2. Situation in space/Scope
3. Purpose
4. Resources
5. Materials with special risks
6. Detailed description of the Constructive Method of Security
7. Identification of the tasks within the activity that constitute risks
8. Identification of constraints
9. Risk analysis and hierarchy
10. Preventive measures to control risk
11. Collective protection equipment
12. Personal protective equipment
13. Before the activity
14. During the activity
15. Monitoring and prevention procedures and records
16. Annexes

All PESs will be the target of training action for all workers, whose activity fits into their content, before the start of work.

2.3 *Risk assessment methodology*

Safety planning integrated into the Specific Safety Plan is based on the identification and evaluation of the risks involved in its implementation and defines the preventive measures to be implemented to eliminate or minimize the likelihood of occupational accidents and/or occupational diseases.

For the risk analysis and hierarchization, the William T. Fine Method was used, the choice of method being a requirement of the developer. The results of the William T. Fine method, semi-quantitative approach, are obtained by means of probability of occurrence of a risk. The methodology is basically summarized in the following steps:

- Identification of potential hazards and risks; assessment of risk or level of risk and definition of actions
- Rectification or preventive measures.

This method allows us to value and order risks according to their hazards, determine the degree of danger, establish the urgency or pertinence of corrective actions, from which we can adequately guide preventive actions and find the economic justification for possible corrective actions.

The formula used to calculate the degree of danger (GP) takes into account the following factors: Severity, Exposure and Probability, presented in Tables 2, 3 and 4 respectively.

Following the valuation of the risks for each factor, according to the predicted indicators, the risk classification is obtained (Table 5), taking into account the degree of calculated hazard, and the respective action measures to be implemented.

Table 2. Risk assessment—Severity (Source: DEPSS).

Factor	Risk assessment	
	Classification	Value
Gravity	very strong catastrophes (many deaths)	100
	Disasters some deaths.	50
	Very serious (a death)	25
	Serious (Extremely serious injuries —low than 15)	15
	Important (accident with temporary incapacity—up to 15 days off)	5
	slight (minor lesions—no lesion)	1

Table 3. Risk assessment—Exposure (Source: DEPSS).

Factor	Classification	Value
Exhibition (E)	Continue	10
	Frequent	6
	Occasional (1 or 2 × per week)	3
	Unusual (1 or 2 × month)	2
	rare	1
	Very rare	0,1

Table 4. Risk assessment—Probability (Source: DEPSS).

Factor	Risk assessment	
	Classification	Value
Probability (P)	Frequent	10
	Likely	6
	Rare but possible to happen	3
	Remote	1
	Extremely Remote	0,5
	practically impossible	0,1

Table 5. Risk classification (Source: DEPSS).

GP	Risk classification	
	Classification	Measures of action
>400	Extreme	Delete urgently
250 to 400	Very high	Requires immediate correction
200 to 250	high	Needs correction
85 to 200	Medium	Need attention
<85	Low	Acceptable

3 CASE STUDY—TUNEL AGUAS SANTAS

The work is located in the municipality of Maia (Figure 1), and consists on "the construction of a new gallery, with an extension of 367 m, north of the two existing downstream (Ermesinde/Porto), including enlargement and improvement to 2 × 4 paths.

The main activities include:

- Excavation (explosives and mechanical means) Provisional excavation support (concrete designed with metal fibers, Swellex type nails, metal crankshafts, micro-cutting tips, fiberglass nails);
- Waterproofing and drainage (geotextile and 2 mm PVC membrane);
- Ultimate Reinforced Coating
- Electromechanical installations (lighting, ventilation, fire system, CCTV).

"This project intends to significantly improve the local traffic, entry and exit of Porto, and because it is a work of great social affection, all the principles of good construction rules are being followed, and the increased care in the monitoring of the existing constructions in the Porto surroundings" according to transcript of the Ramalho Rosas website.

Figure 1. Location (Source: RRC).

3.1 Constructive method

According to Tender, M.L. several criteria are mentioned as relevant for the choice of the method, the most important being the cross-salt section, geological characteristics, cost, time and term.

The constructive method chosen for this underground work was chosen based on the geological analysis of the study area. From a geological point of view, the Águas Santas tunnel is located on the so-called "Granito do Porto", a massif consisting of essentially medium-to-coarse alkaline granite, leucocratic, of two micas.

Three zones ZG1, ZG2 and ZG3 (lower quality mass) were identified, as can be seen in the following figures:

3.1.1 Explosive excavation (ZG1 and ZG2)

The execution of the tunnel is characterized by a sequence of operations aiming at the excavation with mechanical and explosive means, whose phasing is shown in Figure 4:

The sequence of excavations using explosives consists of the following operations:

1. Marking the handle
2. Drilling
3. Loading holes
4. Ventilation of the excavated area
5. Removal of debris into a ditch
6. Sanitation
7. Primary support

3.1.2 Mechanical excavation (ZG3)

The mechanical excavation in ZG3 is carried out with mechanical means, being used for this the rotary equipped with bucket, hammer or brush cutter head and lorries to remove the rubble.

During the excavation, it was observed that the soil characteristics were not those indicated in the geotechnical test and for that reason only 2 advances (9 m) were made by the mechanical excavation method, with the remainder being executed through the excavation with explosives.

Figure 2. Plant (Source: RRC).

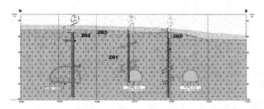

Figure 3. Transverse geological profile—Tunnels (Source: http://www.nonprofitcultivation.org/docs/A48S 172-8_CRP_T6B_102.pdf).

Figure 5. Formwork car (Source: PES and Author).

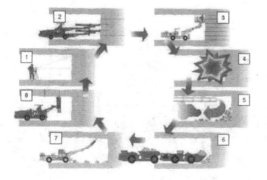

Figure 4. Excavation sequence scheme using explosives (Source: PES).

Figure 6. Union of the Bogie system assemblies Pre-assembly of the vault (Source: Author).

3.2 Mounting formwork carriage

The formwork cart (Figure 5) is used for the execution of the final coating of the Tunnel.

The assembly of the formwork car is divided into the following activities:

1. Pre-assembly of the support system;
2. Pre-assembly of the Bogie system;
3. Union of the Bogie system assemblies (Figure 6)
4. Pre-assembly set of MK supports;
5. Mount MK supports to Bogie system;
6. Pre-assembly of the vault (Figure 6);
7. Mounting of the vault on MK support;
8. Mounting the rods;
9. Assembly of lower catch panels.

3.2.1 Risk analysis and risks hierarchy

The risk assessment was carried out using the method referred to in point 2.3 above.

A risk assessment is carried out at the source, before the implementation of the preventive measures, and the obtained classification of risks is shown in Table 6:

The revaluation of risks with High, Very High classification was carried out after implementation of preventive measures, and the result of the new risk assessment was changed to acceptable values as shown in Table 7:

3.2.2 Preventive measures to risk control

Collective Protective Equipment (Table 8), Individual Protection (Table 9):

Table 6. Risk analysis and hierarchy of coverage car risks.

Activity	Mounting the formwork carriage	Evaluation and hierarchization of risk in origin				
Activity identification	Risk identification	Gravity	Exposition	Probability	Don of danger	Classification
Car mount on scaffold	Trampling	25	5	1	150	Médium
	Injury or crushing by or into objects	25	5	3	450	Extreme
	Blowos and cuts: objects or tools	1	5	3	18	Low
	Objects falls/material manipulation	25	0	3	450	Extreme
	Fall of people in height >2 m	25	0	3	450	Extreme
Application of deciphering oil	Dermatoses	1	3	1	9	Low
	Fall of people in height >2 m	25	0	0	450	Extreme

Table 7. Reassessment of risks of cofra-gem car.

Activity: Assembly of shuttering car		Assessment and hierarchy of risks for the construction phase taking into account the recommended safety measures				
Activity identification	Risk identification	Gravity	Exposition	Probability	Don of danger	Classification
Scaffold car assembly	Trampling	25	6	0,5	75	Low
	Stalking or crushing by or between objects	25	6	0,5	75	Low
	Dropping of objects or tools	1	6	0,5	3	Low
	Fall of objects/materials in handling	25	6	0,5	75	Low
	Fall of people in height >2 m	25	6	0,5	75	Low
	Physical overexertion	5	6	0,5	15	Low
Application of decohering oil	Dermatoses	1	3	0,5	1,5	Low
	Fall of people in height >2 m	25	6	0,5	75	Low

Table 8. Collective protection equipment.

Collective protection	Risks	Places of application
Bodyguards to 90 and 45 cm and baseboard with 15 cm	Fall in height	Peripheral protection of work platforms
High sensitivity differential circuit breaker IΔN less than 30 mA	Electrification	Generators and any electrical installation
Position bays, barriers ET4	Trampling Crushing	Work Area Delimitation
Acoustic signalling (reversing signal) and luminous	Trampling Collision of equipment/vehicles	Self-propelled equipment
Chemical powder extinguisher	Fire	Machines, electric frames and tunnel interior 50 by 50 m

Table 9. Equipment for individual protection of eventual use (E) or permanent use (P).

EPI	Boots with Toe Cap and Steel Insole (S3)	Helmet (EN 397)	Mechanical Protection Gloves	Rubber Gloves (Chemical Protection)	Rubber Gloves (Chemical Protection)	Double sling harness (height over 2 m)
TASKS	P	P	P	P		
Pre-assembling the support system	P	P	P	P		
Pre-assembly of the Bogie system	P	P	P	P		
Union of Bogie System Sets	P	P	P	P		
Pre-assembly set of MK supports	P	P	P	P		
Assembling MK supports to the Bogie system	P	P	P	P		P
Pre-assembly of the vault	P	P	P	P		
Mounting of the vault on MK support	P	P	P	P		P
Assembly of the rods	P	P	P	P		P
Mounting of lower catch panels	P	P	P	P		P
Application of deciphering oil					P	P

4 CONCLUSIONS

The construction of a tunnel is a complex process, encompassing several specialties, where interconnection, between the different phases of design and execution of underground works is of extreme importance.

It is imperative to carry out an in-depth study of all activities, so that specific safety plans can be made as accurately as possible in order that this activity can proceed without accidents and with the best possible safety conditions.

In addition to the achievement of a good safety plan, it is the responsibility of all stakeholders to respect all safety rules, from the use of personal and collective protection equipment to signage, in order to reduce or even eliminate the risk of accidents.

In addition to achieving a good safety plan, all actors must comply with all safety rules, from the use of personal and collective protective equipment to signage, in order to reduce or even eliminate the risk of accidents.

It is essential to reduce accidents, not only in the construction of tunnels, but also in all other engineering works, to demonstrate, at the cost of prolonged absence of fatalities, that the degree of knowledge of the safety rules allows the safe execution of works in the construction.

REFERENCES

Ramalho Rosa Cobetar: DEPSS (Desenvolvimento e especificação do plano de segurança e saúde) e PES (Plano específico de segurança por atividade).

Reis C., C. Oliveria (2012). Análise da relação entre a implementação da diretiva estaleiros e os acidentes na Construção/Analysis of the relation enters the implementation of the directive yard and the accidents in the construction. International Symposium Occupational Safety and Hygiene—SHO 2012. Book in 1 volume, 506 pages. Pág. 375 a 377. ISBN 978-972-99504-8-3. Universidade do Minho. Campus de Azurém, Guimarães, Portugal, 9 e 10 de Fevereiro de 2012.

Reis, Cristina, Soeiro Alfredo – Economia dos Acidentes na construção – Simulação e Análise, Estudo de Alguns Casos, ISHST, Lisboa, 2005.

Tender, M.L., J.P. Couto (2017) – "Safety and health" as a criterion in the choice of tunnelling method.

http://monografias.poli.ufrj.br/monografias/monopoli10020948.pdf.

http://repositorio.ipl.pt/bitstream/10400.21/1570/1/Disserta%C3%A7%C3%A3o.pdf.

http://www.infraestruturasdeportugal.pt/sites/default/files/cet/cap15-09_mar_1998.pdf.

http://www.nonprofitcultivation.org/docs/A48S172-8_CRP_T6B_102.pdf.

http://www.rrc.pt/pt/-/tunel-de-aguas-santas-maia.

Characterization of school furniture in a basic education school

A. Fernandes, P. Carneiro, N. Costa & A.C. Braga
ALGORITMI Centre, University of Minho, Guimarães, Portugal

ABSTRACT: The workplace should be designed to provide adequate physical conditions so that it develops without extra penalization for the worker. Productivity is determine by the quality of the workplace, taking into account early fatigue, musculoskeletal pain and consequent absenteeism. The same can be said for students whose place of work is the classroom. The finding of imbalance of furniture measures in relation to the anthropometric measures of the students, as well as their dimensioning, has been a reason for attention by some researchers. The objective of this work was to characterize school furniture available in a basic education Portuguese public school. Furthermore, it was also intended to verify the compliance of the recommendations on the dimensions of school furniture in the European Standard (EN 1729-1: 2015) as well if the Portuguese government carried out, through the Ministry of Education, the transposition of the Standard taking into account the adjustments resulting from the anthropometric characteristics of the Portuguese population. This study included measuring the height of the table top and of the table structure from 330 tables. Additionally the height, width and depth of 638 chairs were measured. Mismatches were found between the recommendation of EN 1729 and the existing reality. It was verified, in the requirements imposed by the Portuguese government, the inexistence of two intermediate heights for the seat of the chairs as well as for the tables.

1 INTRODUCTION

The school is the extension of the students' home. The majority of young people's time who attend the second or third cycle of schooling in the Portuguese education system is spent in school. On a morning with three ninety minute classes, with two intervals of ten to fifteen minutes, students will be 90% of their time in school, in a classroom, most likely sitting, which means, four hours and thirty minutes seated. It can be said that the adolescent being in a phase of non-linear growth in relation to the body in its totality as well as to the various segments of the body, has a lot of difficulty to pay attention to the class due to the simple fact of continuously trying to find his balance posture while being seated for so long (BBC, 2014). In addition to being seated for a long time there is also the possibility that they might be badly seated. That is, the chairs and tables not being adjusted to the dimensions of their bodies. The truth is that the lack of conjugation between the student and the school furniture, both in terms of dimensions and design, causes discomfort that will require greater muscle strength to maintain stability and balance of the body, resulting in an early fatigue that promotes the lack of concentration (Parcells et al, 1999). Some authors account for more than 80% of the mismatch between the anthropometric measures of the students and the measures of furniture, either at the height of the chair (Gonçalves, 2012) and at the table, or at the level of the seat depth (Agha, 2010), being the girls most penalized by this maladjustment (Parcells et al., 1999). But there are incompatibilities approaching 100%, as is the case of a study done in three schools in Chile, where the gap related to the distance between seat height and table height was above 99% (Castellucci et al., 2010). There are studies that indicate that in children with 6–7 years of age, physical skills in handling objects over work tables with misaligned heights are compromised (Smith-Zuzovsky & Exner, 2004).

Studies based on questionnaires have revealed that there are students' complaints about back and neck pain (Murphy, Buckle & Stubbs, 2007), and it is advisable for students to balance the body in order to compensate for this discomfort (Knight & Noyes, 1999). In addition to the above facts, the frequent observation of incorrect postures adopted by students of the fifth grade in a Portuguese school approximately half of a ninety minutes class on a weekly basis for a duration of a school year is shown in the following images (Figure 1).

The aim of this work was to characterize school furniture available in a basic education Portuguese public school knowing that this type of furniture is transversely used in all schools. Furthermore, it was also intended to verify the compliance of the

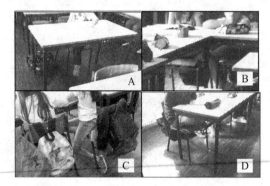
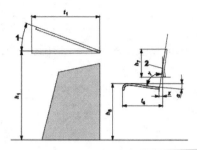

Figure 1. Images of inappropriate postures in the seated position of 5th year of schooling children: A) Student with left leg collected on the seat of the chair and the right leg hanging; B) Student with crossed legs on the seat of the chair; C) Student with right leg collected on the seat of the chair and torso in twist; D) Student with both legs crossed over the seat of the chair.

1 - inclination of the table top (the maximum value for a fixed or inclinable table top is 20°)
2 - Point S
α - inclination of single-sloped seat
γ - angle between seat and backrest
$h1$- height of table top
$h7$ - height of backrest
$h8$ - height of seat
$t1$ - depth of table top
$t4$ - effective depth of seat
x - distance between Point S and back of seat pad

recommendations on the dimensions of school furniture in the European Standard (EN 1729-1: 2015) as well if the Portuguese government carried out, through the Ministry of Education, the transposition of the Standard taking into account the adjustments resulting from the anthropometric characteristics of the Portuguese population. This investigation complements the work, about the same thematic developed in the first cycle, from Gonçalves (2012), augmenting the Portuguese database.

2 METHODOLOGY

To carry out this study, a public school of the 2nd and 3rd cycle of elementary education in the northern area of Portugal was selected. The study was based on the collection of measures of height, width and depth of the seat of the chair, upper height of the backrest of the chair, height of table top and of structure. All dimensions were obtained in accordance with EN 1729-1: 2015 (Figure 2). The study covered 100% of the furniture, specifically 638 chairs and 330 tables. Cumulatively, the age of its users also had to be taken into account in order to verify compliance with the recommendations of the Standard. However the furniture is randomly distributed per classrooms regardless of the student age.

2.1 Characteristics of the school population

The school population under study consists of boys and girls of the 2nd and 3rd cycles, with a total of 373 students (data referring to September 2017). The population ages range from nine to seventeen

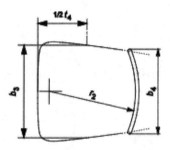

$b3$ - width of seat
$b4$ - width of backrest
$r2$- horizontal radius of backrest
$t4$ - effective depth of seat

Figure 2. Instructions on furniture dimensions (Figure A2 and A5 of EN 1729-1: 2015).

years old. The distribution of students by their ages can be seen in Figure 3.

2.2 Furniture dimensions

Chairs and tables were measured through the use of more than one measuring instrument. Metric tape measure was used to check seats width. To obtain the seat depth measurement, a metal tape measure was also used, aided by a bracket to overcome the limit of the seats collar.

The height of the chairs' seat, the height of the table top and its structure, were also verified with a metal tape measure, aided by a square with a

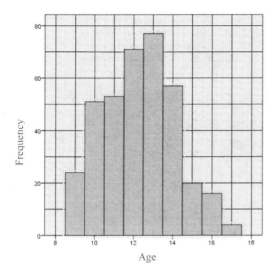

Graphic 1. Frequency of school students' ages (n = 373; Mean = 12.29 years; Standard deviation = 1.857 years).

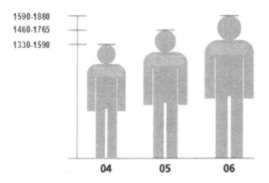

Figure 3. Reference of school furniture in relation to the students' attainment according to the Norm EN 1729-1: 2015.

slide system to guarantee perpendicularity. Given the age of this furniture, some with more than 25 years, it was necessary to double the measurements of the first 60 chairs and 24 tables, corresponding to two classrooms, to contemplate the wear and tear that occurred during their use.

3 RESULTS AND DISCUSSION

3.1 *Chairs' dimensions*

The seats are distributed taking into account the height of the seat. Three different seat height measures were identified. Thus, of the 638 existing seats 18% have a seat height of 410 mm, 25,7% have a seat height of 440 mm and the remaining 56,3% constitute the 450 mm seat height group. Taking into account EN 1729-1: 2015, it can be observed that there is a recommendation for the construction of school furniture with respect to its dimensions. According to the norm, there are 7 categories of chair and table dimensions that cover all users. This classification is related to the dimensions of the users and has a color associated. In the target ages (9–16 years) and based on previous studies (Macedo et al., 2004); Gonçalves, 2012), it can be inferred that the dimensions of the students are between class 4 and 6 of the Standard.

On the other hand, the Portuguese Ministry of Education (ME), through the Shared Services Entity of the Public Administration (eSPap), embodied in the AQ-Mob/Mobiliário 2015 competition, defines as a requirement for manufacturers of school furniture a single seat height size for 1st cycle students and a single size for the remaining cycles (Table 2).

Comparing Table 1 on the European recommendation and Table 2 on the indications of the ME, it is verified that the indication of the ME for the 1st cycle is compatible with reference number 3 of EN 1729-1. In turn, the indication of the ME, relative to the other levels of education are compatible with reference 6 of EN 1729-1. Therefore, references 4 and 5 of EN 1729-1 are not included.

This school, although presenting three measures of seats' height of the chair, are not totally coincident with those of the Standard. With these values, the population aged between 9 and 11 years (N = 153) who is likely to have shorter stature, does not find a compatible seat to sit. This might justify the postures adopted by the 5th graders portrayed in Figure 1.

Taking into account the age and the degradation of the material that equips this school, and assuming that its replacement would be based on a

Table 1. Seat height of chairs (EN 1729-1: 2015).

Seat heigt of chair (mm)				
Reference	3	4	5	6
tolerance (± 10 mm)	350	380	430	460

Table 2. Dimensions of school chairs, according to AQ-Mob/Furniture 2015 (Government of Portugal) W – width; D – depth; H – height.

Dimensions (W × D × H)	Users
360 × 340 × 360	1st cycle
400 × 400 × 450	others

single measurement class (450 mm), one would see a very large number of students without a compatible chair. It is important to add that the school has 115 seats whose seat height measurement (410 mm) does not fit the recommendations of EN 1729-1 (Table 3).

3.2 Table's dimensions

The 330 tables in the school under study are grouped into three classes, a result of the top height of the table top. The height of the table structure must be taken into account to ensure that it does not interfere with the students' thighs (clear space between the seat and the table structure). In this way the tables were grouped in 7 classes (h1, h2, h3, h4, h5, h6, h7), as can be seen in Table 4.

Despite the existence of 3 table's height measurement classes as in EN 1729-1 (Table 5), none of them meet the blue reference measurement (6) of this table. That is, the measure of the highest existing table height is 15 mm below the minimum height of the reference 6.

On the other hand, the Portuguese Ministry of Education, through the Shared Services Entity of the Public Administration (eSPap) based on the AQ-Mob/Mobiliário 2015 competition, requires suppliers to have a single height of 740 mm. Following the same reasoning with the chairs, in a future renovation of furniture, the older students with a lower stature, may have difficulty finding tables for their anthropometric dimensions.

4 CONCLUSION AND FUTURE WORK

As can be seen from the foregoing, the dimensions of school furniture recommended in Standard EN 1729-1: 2015 do not match the existing furniture. Of the 3 dimensions of seat height existing in the school, it was concluded that the 450 mm dimension corresponds to category 6 of the Standard, the dimension of 440 mm corresponds to category 5 of the Standard and the dimension 410 mm does not match the Standard. This situation penalizes the students with lower stature regarding the anthropometric conjugation with the measure of the height of the seat. In relation to the dimensions of the height of the table, we find in the school correspondence for the measurement of 660 mm with category 4 of the Standard and the measurements of 720 mm and 725 mm with category 5 of the Standard. There are no tables matching Category 6 of the Standard. This situation penalizes the older pupils of lower stature, as to the anthropometric combination with the measure of the height of the table. The Portuguese government's guidelines further widen the existing gap.

Given this situation, it will be appropriate to carry out a more detailed study of the situation in order to make more sustained considerations. This means to know if there is no mismatch between the combination of the chair and the table or if on the contrary, and taking into account the study population, the pairs chair 5/Table 4 and chair 6/Table 5 are more favorable than those proposed by the Standard (chair 5/Table 5, chair 6/Table 6).

To do so, this research team will survey the anthropometric measurements of the entire school population (height, seated shoulder height, popliteal height, elbow-seat height, gluteal-popliteal length, and hip width). Based on the survey of the anthropometric measures of the students, the level of adjustment between the students and the available furniture will be verified.

Table 3. Number of existing seats as a function of seat height, seat width and depth.

Seat height	Seat width	Seat depth	Quantity
410 mm			115
440 mm	380 mm	400 mm	164
450 mm			359

Table 4. Existing tables divided by height of the top and height of the structure.

Tables

Top height (mm)	Ref.	Structure height (mm)	Quantity	Total	Percentage	
660	h1	590	52	52	15.8%	15.8%
720	h2	635	26		7.9%	
	h3	645	97	189	29.4%	57.3%
	h4	650	66		20.0%	
	h5	635	11		3.3%	
725	h6	645	17	89	5.2%	27.0%
	h7	650	61		18.5%	

Table 5. Dimension of the top height of the school table top, according to Norm EN 1729-1: 2015.

Seat height of chair (mm)			
H	mín	máx	Reference
380	370	390	4
430	420	440	5
460	450	470	6

ACKNOWLEDGMENTS

For the accomplishment of this work, it was fundamental the collaboration of the school's management, both in the availability of data about the students as well as in the permission to remain in school and logistical support, provided by the operational assistants, for the collection of measures of furniture.

This work has been supported by COMPETE: POCI-01-0145-FEDER-007043 and FCT – Fundação para a Ciência e Tecnologia within the Project Scope: UID/CEC/00319/2013.

REFERENCES

Agha, S.R. (2010). School furniture match to students' anthropometry in the Gaza Strip. *Ergonomics*, *53*(3), 344–354. http://doi.org/10.1080/00140130903398366.

BBC *Homepage – Science: Human Body & Mind*. Consulted on 17/10/2016. Available on: http://www.bbc.co.uk/science/humanbody/body/articles/lifecycle/teenagers/growth.shtml.

Castellucci, H.I., Arezes, P.M., & Viviani, C.A. (2010). Mismatch between classroom furniture and anthropometric measures in Chilean schools. *Applied Ergonomics*, *41*(4), 563–568. http://doi.org/10.1016/j.apergo.2009.12.001.

EN 1729-1:2015. Furniture - Chairs and tables for educational institutions Part 1: Functional dimensions. *European standard norme européenne europäische norm*.

Entidade de Serviços Partilhados da Administração Pública (*eSPap*), *AQ-Mob/Mobiliário 2015*. Consulted on 27/10/2017, available on: https://www.espap.pt/spcp/Paginas/spcp.aspx#maintab5.

Gonçalves, M.A. (2012). Análise das condições ergonómicas das salas de aula do primeiro ciclo do ensino básico. *Doctoral Thesis, Universidade Do Minho, Portugal*.

Knight, G., & Noyes, J. (1999). Children's behaviour and the design of school furniture. *Ergonomics*, *42*(5), 747–760. http://doi.org/10.1080/001401399185423.

Macedo, A.C., Morais, A. V, Martins, H.F., Martins, J.C., Pais, S.M., & Mayan, O.S. (2004). Match Between Classroom Dimensions and Students' Anthropometry : Re-Equipment According to European Educational Furniture Standard. http://doi.org/10.1177/0018720814533991.

Murphy, S., Buckle, P., & Stubbs, D. (2007). A cross-sectional study of self-reported back and neck pain among English schoolchildren and associated physical and psychological risk factors. *Applied Ergonomics*, *38*(6), 797–804. http://doi.org/10.1016/j.apergo.2006.09.003.

Parcells, C., M.S.N., R.N., Stommel, M. & Hubbard, R. (1999). Mismatch of Classroom Furniture and Student Body Dimensions Empirical Findings and Health Implications. *Journal of Adolescent Health*, (98), 265–273.

Smith-Zuzovsky, N., & Exner, C.E. (2004). The effect of seated positioning quality on typical 6- and 7-year-old children's object manipulation skills. *American Journal of Occupational Therapy*, *58*(4), 380–388. http://doi.org/10.5014/ajot.58.4.380.

Gas distribution companies: How can knowledge management promote occupational health and safety?

C.M. Dufour & A. Draghici
Faculty of Management in Production and Transportation, Politehnica University of Timisoara, Timisoara, Romania

ABSTRACT: The article proposes a theoretical framework regarding the use of Knowledge Management (KM) in solving specific Occupational Health and Safety (OHS) issues in the context of Gas Distribution Companies (GDCs). GDCs are subject to the constraints of high-risk industries, but also aging workforce (generational turnover) and rapid technological change (technological turnover). The proposed framework was designed by building on the Australian Standard AS 5037-2005 three-phase cycle for development and implementation of knowledge management. After a brief chapter on the history of gas and on the context of GDCs, a short literature review is presented. In this research context, the following chapter will outline the link between KM and OHS in GDCs. In the core of the paper, the authors expose the theoretical framework for OHS knowledge management for GDCs; finally, some conclusions of the research will be drawn.

1 INTRODUCTION

Gas Distribution Networks (GDN) are a group of interconnected pipes, meters, taps and expansion stations through which gas transits at pressures defined by national norms; these ensure the connection between national pipelines and the end-user for low consumption clients. GDN are characterized by their length and their density (number of end users within a geographical area).

The past decades have brought numerous technological changes for GDN infrastructures (built in grey font, later replaced by ductile iron, steel and finally polyethylene), pipeline installation techniques (development of guided drilling, mechanized drilling and installation, insertion) and metering technology (electronic meter-reading systems). GDCs exploit and maintain a GDN, the topology of which can vary from that of a city to that of a country.

The European Directives of 1996, 1998 and 2003 aimed at a more competitive and healthier energy context and the 2010 and 2012 energy directives on energy consumption reduction in building redefined the ecosystem of GDCs (European Commission, 2017). Therefore, the expectations that lay on the GDCs have been in constant and rapid change in the past decade.

2 OVERVIEW OF THE CONTEXT OF LGDC

GDCs belong to high-risk industries, and as such are submitted to strict regulations and high standards to ensure both operational health and safety and customer and environmental safety.

The organization of GDCs comprises several hierarchical layers defined according to the size of the GDN, its geographical distribution and/or density; these layers include Technical Workers (TW), first-line, middle and upper-managers and a group of experts on technical and OHS issues. TWs can be either specialized in a field of intervention or with several fields of intervention. In his field of intervention and according to defined safety management procedures, a TW or a team of TWs can implement the following technical measures in order to mitigate risks:

1. Preventive, maintenance operations that are inherent to the good working and timely prevention of degradation;
2. Corrective, operations that are made outside routine maintenance checks and which follow the acknowledgement of an early stage malfunction;
3. Intervention, any operation that results from the acknowledgement of a high-risk malfunction which in the immediate future can cause severe damage to people and/or objects.

Where OHS is concerned, TWs receive specific initial training under the form of classroom-training, followed by refreshment-courses. In-depth knowledge is accumulated through dissemination; by working in teams and through field-tutoring knowledge is passed from one TW to another. Behavioral-safety evaluations are the most common method of assessing individual knowledge and application of OHS theoretical and practical know-how, but these assessments remain subject to the knowledge and observation skills of the assessor.

On a day-to-day basis, after receiving the daily workload from the front-line manager, the TW departs to the worksite (individual homes, industrial sites or in the street) and returns at the end of the day for a feedback on what was accomplished and the difficulties encountered. Thus, the work environment is deported (or dislocated each day) and it can be defined as an itinerant work place, the lapse of time during which it is the work environment being correlated to the work tasks to be accomplished; it is characterized by specific OHS risks which require an on-the spot risk analysis. The deported work environment poses two types of risks: the OHS risk related to environment and the OHS risks related to the technical intervention. The OHS risks related to the environment are mitigated through an on-the-spot risk analysis, during which the TW adapts theoretical knowledge to the work environment and may even create new risk-mitigating measures that go beyond those of the theoretical knowledge. The OHS risks relating to the technical intervention derive from the interaction between the human operators and the infrastructure, the mitigation of which relies on both tacit and explicit knowledge.

Another characteristic of GDCs is, what we shall call, generational turnover. For high risk industry workers, specific training and important initial investment in forming the personnel is necessary. Given that it takes a long time to achieve full technical maturity, rupture of contractual relationship is due to two main causes (out of the spectrum of professional fault): retirement and personal circumstances unrelated to the company. In these conditions, a specific type of turnover is born: the generational turnover. This implies that the tacit knowledge remains within the company for many years (decades) and that retirement can cause a knowledge gap.

Generational turnover coexists with technological turnover; new technologies arise and the use of older technologies is limited, knowledge being lost through lack of practice and lack of transmission. The obsolescence of technologies and the inherent decline in number of technical interventions cause an obsolescence spiral of OHS knowledge. The authors have summarized this issue in Figure 1.

Although both turnovers coexist within GDCs, they are managed as opposing KM trends. Technological turnover is a common KM issue as a technological bump causes an input of knowledge with a short-term vision on KM which comprises training, assessment of knowledge and know-how acquisition. Generational turnover is a pending KM issue, visible at certain moments in time and being solved by one-shot interventions but requiring more strategic reflections on KM.

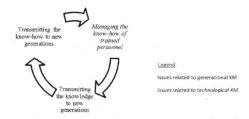

Figure 1. KM and OHS—Obsolescence Spiral.

KM regarding OHS thus becomes a strategic issue that requires a structured approach dealing with the specific context of GDC on one hand, and the technological and generational turnover on the other. Implementing a structured managerial approach for KM regarding OHS gives a competitive edge, increasing the organizations resilience and adaptability.

3 LITERATURE REVIEW

The authors first conducted a literature review concerning the link between knowledge management and gas distribution. The conclusion was that there is a research gap in the literature linking the two fields. In consequence, the authors have focused the literature review firstly on the knowledge management in the context of high-risk industries. This choice was reinforced by the fact that high-risk industries are precursors to understanding how KM can promote OHS. Finally, the authors researched the Australian Standard AS 5037-2005 as a potential foundation for a knowledge management framework tailored for the gas distribution industry.

3.1 *Knowledge management and gas distribution*

Firstly, the authors did a Science Direct review in December 2017, which resulted in the identification of 18 articles. In order to establish this literature base, the expert search engine function was used with the combinations of keywords in Table 1.

After an initial desk-study of keywords and abstract, 8 articles were selected as relating to either the gas industry or knowledge management in similar infrastructures. Of the 8 articles, 1 dealt with urban infrastructure knowledge representation but from an ontological perspective, 5 with methods of enhancing knowledge management and 2 with the gas distribution panorama. El-Diraby and Osman (2011) noted that the infrastructure domains are those with the poorest knowledge management practices.

3.2 *KM, OHS and high risk industries*

Ferguson (2006) accounted for 18 definitions for KM, but here we shall refer to that of the Australian

Table 1. Number of articles found and keywords.

Keywords	Articles
"gas distribution network" AND "knowledge management"	1
"gas distribution" AND "knowledge management"	17

Standard AS 5037-2005 – "a body of understanding and skills that is constructed by people and increased through interaction with other people and with information" (p. 2), whereas knowledge management is "a trans-disciplinary approach to improving organizational outcomes and learning, through maximizing the use of knowledge" (p. 2). KM emerged in the late 20th century as a response to globalization (Podgórski, 2010) and should have brought an answer to issues such as specialization of skills in the context of make or buy, increased turnover and smaller time-lapse competitive advantage—all "relevant knowledge-related aspects that affect the enterprise's viability and success" Wiig (1997, p. 1).

High-risk industries to which GDCs belong are characterized by the need of low reaction time and zero-accident policies, and as such OHS imperatives are numerous. Decision making needs to be rapid, technologically viable and OHS confirmed, meaning that it is not a process that can afford the time lapse of gathering or consulting several hard resources or experts and thus relies extensively on the knowledge of the workers and their speed of interaction (Davenport & Prusak, 1998; Hildreth et al., 1999). Furthermore, it is not a sector open to trial-and-error learning and is characterized by legal and internal procedures as methods of sharing information (Aase & Nybo, 2002). From a KM perspective, these companies struggle to find a practical and tailored-to-needs method for transforming Nonaka's (2007) implicit or tacit knowledge into explicit or formal knowledge. The challenge emerges from the fact that tacit knowledge is acquired through experience and company culture (Podgórski, 2010; Nonaka, 2007) and has a personal component, whereas hard knowledge can be formalized, structured and observed, and then transferred to a written form (Podgórski, 2010).

In this context, the first attempts to link OHS and KM were made in high-risk industries and research has pointed that although desk and training are necessary prerequisites, people will impregnate lessons quicker and more effectively if they hear it in the form of a story (Podgórski, 2010; Davenport & Prusak, 1998). Although the use of narrative disseminates effectively tacit knowledge relating to the local or historical context (Davenport & Prusak, 1998) it has been set aside in favor of more systematic learning methods (Podgórski, 2010) or e-learning programs.

The main focus should be on how to access tacit knowledge, how to evaluate it and how to disseminate it, as it belongs to the individual, is not synthesizable in a formalized form and is closely linked not only to experience in the work field but also to individual and company values, becoming a form of experienced-based intuition (Podgórski, 2010; Nonaka, 2007; Davenport & Prusak, 1998). Furthermore, Nuñez and Villanueva (2011) found that OHS belongs to the broader intellectual capital, "knowledge, procedures and practices that remain in the firm" (p. 61) and not to human capital which is relative to the duration of an employment contract. This relates to the conversion of the tacit knowledge of the individual to the tacit and explicit knowledge of the company and proves that companies are dependent on knowledge, and not individuals or products (Wiig, 1997; Davenport & Prusak, 1998).

KM tools implemented in high-risk industries are mostly IT based tools (e.g., e-learning, knowledge management systems, digital repositories, data mining etc.) but in practice they relate very poorly to the richness and experience of workers. Such tools are suitable for managers who have more desk-time and IT skills to access and process data, information and knowledge, and through (re)interpretation create new knowledge (Davenport & Prusak, 1998; Aase & Nybo, 2002). As concluded by Podgorski (2010), the little research that has been done in coupling OHS and KM points to the need of exploring further methods for acquiring and exchanging tacit knowledge related to OHS (Davenport & Prusak, 1998).

Hildreth et al (1999) point to "legitimate peripheral participation", a specific issue of KM that can also be extended to OHS; it is a process by which new workers learn and become accepted by more experienced workers, as their expertise increases.

Historically, enhancement of OHS has been done through the Behavioral-based Safety (BBS) approach which linked OHS with risky or inappropriate behavior or lack of experience, but only addressed measurable and observable behavior without linking it to relevant knowledge and understanding.

3.3 AS 5037-2005 Australian standards

AS 5037-2005, proposed by Standards Australia, is a hybrid three-phase cycle guideline for managers to enhance tacit flows of knowledge (Burford et al., 2011).

Although several frameworks exist for the implementation of KM in organizations, the authors propose to apply the model presented by AS 5037-2005 as it recognizes the social sources of knowledge, has one of the first national endorsements and proposes a list of 34 practical enablers to help guide managers in the implementation of a KM initiative (Ferguson, 2006).

4 RESEARCH RESULT—THEORETICAL FRAMEWORK

The literature review has pointed out that there is a real need to create managerial tools for the conversion of tacit knowledge into explicit knowledge regarding OHS. This need is particularly poignant in the case of GDCs for which the dual challenge of generational turnover and technological turnover accelerates the knowledge loss rate within the company. In consequence, the authors propose to tailor the Australian Standard AS 5037-2005 for GDCs.

4.1 *Mapping*

The first phase of AS 5037-2005 concerns understanding the context and the organization culture through extensive assessment and mapping of the knowledge ecosystem.

As pointed out in the literature review, the construction of knowledge is a process that comprises acquiring data, acquiring information and giving it a personal shape. In order to achieve this, the authors propose a three point analysis to be conducted within the GDC. The first step consists of formal audits under the form of: i) theoretical tests adapted to the range of activities of the hierarchical level and with regard to OHS in GDCs and ii) skill tests, where applicable. The tests should assess the knowledge of OHS through the knowledge of procedures, data comprehension and contextualization, as well as script-reaction. Furthermore, these audits should be adapted yearly to target "knowledge-gaps" identified in near-miss/accident analysis as well as past records. The second step concerns the analysis of past individual assessments and the establishment of a multi-annual knowledge progression curve. This step also acts as a safeguard, ensuring that the corrective measures resulting from previous audits have been effective.

The third step is related to the analysis of BBS visit records of both line managers and experts. Based on multi-annual records of past tests and/or individual assessments, an individual progression curve should be established with regard to OHS behavior in order to evaluate the effectiveness of corrective actions.

4.2 *Building*

This phase evolves around building experiences and linkages, the outcomes of which being a deeper understanding of the actions that are pertinent, of the organization's culture and its way of innovation (focus on links between people and technology). Research seems to convey that the effectiveness of learning increases with the degree of immersion. As for high-risk industries, trial-and-error methods are not appropriate within the production context, and taking into account that 80% of vital knowledge is tacit knowledge (Podgórski, 2010), the authors have selected the following enablers from the AS 5037-2005:

- Knowledge literacy (basic, technological and information literacy which evolved into skills such as storytelling, mentoring, openness and willingness to share knowledge),
- Share-fairs (special meetings with the purpose of sharing knowledge on a particular subject)
- Narrative management (a means to conveying knowledge and elements of understanding for the transformation of information into knowledge).

Thus, we propose that knowledge transfer, dissemination or sharing be carried out through expertise shaped into hard knowledge through training sessions and disseminated through narratives in hands-on workshops (share-fairs). These workshops link specific technical and OHS issues (identified in the mapping phase) with the purpose of assimilating OHS-knowledge by embedding it into technical knowledge. These workshops should be organized and animated by highly skilled, and therefore experienced operators (possible champions as per AS 5037-2005) that can easily relate and contextualize the day-to-day work environment, and as such, are more apt to identify hazards related to OHS (Podgórski, 2010). Choice of such technical and OHS knowledge disseminators should be based on expertise and recognition by both hierarchy and peers. Furthermore, these workshops should be encouraged at all managerial levels and participants should include managers as OHS literature has shown the necessity of managerial implication.

4.3 *Operationalizing*

The last phase of AS 5037-2005 concerns operationalizing initiatives and capabilities to strengthen and expand the learnings of the previous phase. It is the opinion of the authors that this covers two specters: dissemination of knowledge and individual use of knowledge. The authors propose to achieve reliability and knowledge dissemination (Aase & Nybo, 2002) through organizational redundancy between the managerial line (ML) and the expertise line (EL) on two levels: structural and cultural as per Figure 2.

At a structural level the following knowledge transfers regarding OHS must take place:

- Competence from EL to ML to ensure a clear alignment between observations, feedbacks and corrective actions;

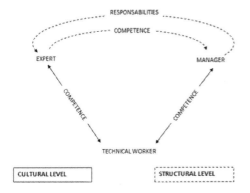

Figure 2. Generation of OHS-related knowledge through organizational redundancy.

- Responsibility from ML to EL to ensure a coherent and unified stand of the enterprise with regard to OHS as well as the effectiveness of any ensuing corrective actions.

At a cultural level the following transfers regarding OHS must take place:

- Competence from ML to EL regarding the day-to-day organization of work and implementation of OHS;
- Competence from the workers to ML and EL through feedback of workarounds (tacit knowledge);
- Competence from ML and EL to workers by organizing training sessions or problem solving sessions based on worker feedback.

5 CONCLUSIONS

The article has a dual research novelty by proposing a hands-on-deck approach to OHS promotion through KM and also by trying to transpose this to the context of gas distribution, an under researched domain in the field of high-risk industries.

The authors have concluded that in the context of GDCs, technological and generational turnover blend together generating specific issues related to the transmission of OHS related knowledge (inherent OHS risks of natural gas, highly technical gestures, aging expertise and fewer technical gestures due to technological change). As, to the knowledge of the authors, there is no framework responding to this dual challenge, this research has focused on creating a theoretical framework based on the general AS 5037-2005 framework. Further fields of research include:

- Validation of this framework through an empirical study;
- Further development of the framework so it is adaptable to the KM maturity level of the LGDC;
- Evaluating the impact of organizational redundancy in terms of cost and full-time equivalent;
- Developing a sustainable knowledge infrastructure relating to the mapping phase;
- Defining and implementing adequate tools for measuring organizational reliability and knowledge dissemination.

REFERENCES

Aase K. and Nybo G. 2002. Organizational Knowledge in High-Risk Industries: What are the alternatives to model-based learning approaches? Third European Conference on Organizational Knowledge, Learning and Capabilities – OKLC 2002, Athens.

AS 2005. Australian Standard AS 5037-2005, Knowledge management – a guide, Retrieved from: http://doc.mbalib.com/view/2f8a2f1840b742c61e2e003b165a9a43.html – consulted 18/09/2017.

Burford S., Ferguson S. 2011. The Adoption of knowledge Management Standards and Frameworks in the Australian Government Sector, Journal of Knowledge Management Practice, 12(1), pp. 1–12.

Davenport, T. and Prusak, L. 1998. Working Knowledge: How Organizations Manage What they Know, Harvard Business Press.

El-Diraby T.E. and Osman H., 2011. A domain ontology for construction concepts in urban infrastructure products, Automation in Construction 20, pp. 1120–1132.

European Commission 2017. Energy, Retrieved from: https://ec.europa.eu/energy/en/topics/energy-efficiency/buildings, consulted May 20th 2017.

Ferguson S. 2006. AS 5037-2005: knowledge management blueprint for Australian organizations? The Australian Library Journal, 55(3), pp. 196–209.

Hildreth P., Wright P. & Kimble C. 1999. Knowledge Man-agement: Are We Missing Something? In Brooks L. and Kimble C. Information Systems – The Next Generation. Proceedings of the 4th UKAIS Conference, York, UK, pp. 347–356.

Hollnagel E., Wears R.L & Braithwaite J. 2015. From Safety-I to Safety-II: a White Paper. The Resilient Health Care Net: Published simultaneously by the University of Southern Denmark, University of Florida, USA and Macquarie University, Australia.

Nonaka I. July-August 2007. The Knowledge Creating Company, Harvard Business Press.

Nuñez, I. & Villanueva, M. 2011. Safety Capital: the management of organizational knowledge on occupational health and safety, Journal of Workplace Learning 23(1), pp. 56–71.

Podgórski D. 2010. The Use of Tacit Knowledge in Occupational Safety and Health Management Systems, International Journal of Occupational Safety and Ergonomics, 16(3), pp. 283–310.

Wiig, K.M. 1997. Knowledge Management: Where Did It Come From and Where Will It Go?, Manuscript for article for the Fall 1997 issue of the Journal of Expert Systems with Applications, 13(1), pp. 1–14.

Cytostatic-drugs handling in hospitals: Impact of contamination at occupational environments

J. Silva, P. Arezes, N. Costa & A.C. Braga
ALGORITMI Research Centre, University of Minho, Guimarães, Portugal

R. Schierl
Institute Outpatient Clinical for Occupational, Social, and Environmental Medicine, University Hospital Munich (LMU), Munich, Germany

ABSTRACT: The manipulation and administration of cytostatics is performed in hospital pharmacies and day hospitals respectively. In each of these sites, the preparation or administration process occurs at different workplaces. In these places spills can occur, which in themselves, cause contamination in the workplace. Of these, the laminar flow hood and the patient's chair are the places where the greatest number of spills occurred. Through the chi-square test of cross-tabulation, an analysis was made to verify if there is a relation of independence or dependence between the different workplaces and work routines. Results show that there is a significant association between the trays, the laminar flow hood, the administration cart and the patient's chair, with the intervals scheduled between the preparation/administration periods.

1 INTRODUCTION

In hospitals, the working environment has a very similar framework towards risks of other activities although the chemical and biological risks are more specific of the health sector. As an example, health care professionals may be exposed to physical risk factors when in contact with or close to radiation equipment, of biological risk due to viruses and bacteria present in these environments, of chemical risk, when handling or administering anesthetic gases, cytostatics, or other drugs with potential for damage, but also of ergonomic risk, such as bad designed workplaces, and also psychosocial risks, such as harassment (Ministério da Saúde, 2010).

Among the listed risks, chemicals contamination stands out because they may cause health problems in professionals who are in contact with dangerous chemical substances, and some of these are the cause of neoplastic diseases (Turci, et al., 2003) (NIOSH, 2004).

Even if surface contamination is at a low level, it is essential to keep collaborators aware so that occupational exposure remains as low as desirable (Merger, Tanguay, Langlois, Lefebvre, & Bussières, 2014) (Berruyer et al., 2015). Berruyer et al. (2015) proposed that an annual prevention program be considered. This program should include the monitoring of surface contamination, since it seems to be an instrument of great importance for the promotion of good working practices and contributes to the reduction of occupational exposure of health professionals to cytostatic drugs (Odraska et al., 2014).

The employers should ensure that workers receive training regularly, that safe handling procedures comply with current national guidelines, that they support their implementation and the availability of personal protective equipment to employees and that they know how to use it. They should also provide medical surveillance, exposure monitoring, and other administrative controls (Boiano et al., 2015).

Another measure to be implemented is a periodic environmental monitoring, as a way to educate employees of health units' in the interpretation of results and their respective corrective measures (Merger et al., 2014).

The development of this research implied a set of previous actions for its accomplishment. First, a literature review was carried out, followed of, observations, routine records, the development of the questionnaire and its application were performed.

The objective of this study was to analyze the relationship between workplaces where the spills occur and work routines.

This work can contribute to the implementation of strategies and corrective measures to reduce the risk of occupational exposure to cytostatics.

2 METHODOLOGY

The health units were designated by hospital center A, hospital center B and hospital center C, respectively, the first, second and third hospitals that accepted to participate in the current study.

Meetings were held with service providers and employees, i.e. hospital pharmacies, day care hospitals and vascular surgery (control group). During these meetings, a presentation of the research project was made and requested the involvement of all the collaborators. The project was disseminated to the collaborators by the heads of the services, through a message sent to their respective institutional addresses.

2.1 Observation and recording of routines

The project was followed by three sessions of observation and recording of the procedures and practices through the verification checklist of the procedures and the consultation of the procedures manual, in the three hospital centers, in the hospital pharmacies and in the medical oncology Day-care hospitals.

2.2 Questionnaire development

The questionnaire was developed and used considering that it is an elementary and auxiliary approach to data research (Amorim, 1995). The questionnaire was designed to allow the data collection about the side effects due to exposure to cytostatics by pharmacists, preparation technicians and nurses.

The questionnaire is a research technique that contains a quantitative approach, allowing to transform the information provided by the respondents through their answers into numbers.

2.2.1 Questionnaire description

The survey consisted of 4 pages, being the first the cover page for informed consent, and the other 3 included 18 multiple choice questions. These aimed at evaluating the maximum information in the most objective way possible. Some questions are sociodemographic and others refer to scientific knowledge. In the construction of the survey, the published literature and the pertinence of the questions were taken into account in order to obtain relevant information regarding the queries and aims of the study, and in order to obtain information that correspond to the reality.

The sample was characterized at the sociodemographic level by the following elements: gender, place of service, qualifications, age, descendants, length of service in the health sector, time of service in the manipulation/administration task, perception of the risk, use of individual protective equipment, separation of waste, practices and frequency of training actions.

The factors considered of greater relevance for the questionnaire were:

- Sociodemographic elements such as age, gender, qualifications, etc;
- Elements related to the factors that can promote errors and cause spillage;
- Individual commitment, such as reading information on cytostatics, and presentation of proposals for improving practices.

The sample is represented by a population of 154 health professionals who work in hospital pharmacies, medical oncology Day-care hospital, and at the vascular surgery service of hospital centers A, B and C.

The questionnaire was applied to two groups: the first, those that manipulate and administer cytostatics, an exposed group, consisting of 98 professionals (pharmacists, pharmacy technicians, nurses and nursing auxiliaries), who perform their tasks in the hospital pharmacy, in oncology and drug transport, respectively. The second, the control group, who consists of 56 nurses, who are not exposed to cytostatics and develop their activity at the vascular surgery service in three hospitals.

The analysis of the places where spills occurred is addressed in this article through the application of the statistical test Chi-square (χ^2) cross-tabulation with different questions. In previous article only the place where more spills occurred was mentioned (Silva, et al. 2017).

2.2.2 Exposed/control group

This research was based on health professionals who perform their tasks in different workplaces and, in the current case, the differentiating element between the two groups (exposed/control group) consists in the exposure or not to cytostatics.

Thus, the exposed group, which was composed of professionals that develop their activity in hospital pharmacies (pharmacists, preparation technicians and auxiliaries) and those who develop the activity in the medical oncology Day-care hospital, (nurses and nursing assistants). All of them manipulate/administer cytostatics.

On the other hand, the control group consisted of nurses who perform their activity in the vascular surgery service. In this service, no cytostatics are prepared or administered.

2.3 Statistical analysis

The data collected were treated and analyzed using IBM SPSS Statistics 24.0 software (SPSS Inc, Chicago, IL).

The chi-square test (χ^2) was used to verify the independence of spill sites. These data were crossed with the following questions: time of service in the task of manipulation of cytostatics? Do you have scheduled break intervals between preparation/administration periods? In the last 12 months have you attended training courses on cytostatic manipulation/administration? By your own initiative, do you often read articles on cytostatics? Have you ever made any proposals for improving your practice?

3 RESULTS AND DISCUSSION

The results under analysis refer to the sites where the spills occur due to the manipulation and administration of cytostatics.

Regarding the occurrence of spillages during cytostatics transport, manipulation and/or administration, 53 of the respondents answered affirmatively, 29 of hospital A, 12 of hospital B; and 12 of hospital C. The locations where those spillages occurred are diverse and vary significantly, as can be seen in (Figure 1). Thus, for the storage, the affirmative answers were 7, in a percentage of 4.5 percent. For the shelf, the affirmative answers were 6, which correspond to the percentage of 3.9%.

On the tray, the number of spillages was 10, corresponding to 6.5%. On the other hand, in the laminar flow hood, the number of spillages was 19, corresponding to 12.3%. On the packaging table, the number of spillages was 6 (3.9%). In the transport box, the number of spillages was 5, equivalent to 3.2%. In the administration trolley, there were 7 spillages, which correspond to 4.5% of the professionals surveyed. Finally, the patient's chair represents the place where more spillages occurred, 25 in total, equivalent to 16.2%.

The application of the cross-statistical test with the "storage and transport box", for the questions, time of service in the task of manipulation of cytostatics?

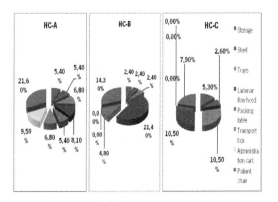

Figure 1. Locations where spillages occurred.

By your own initiative, do you often read articles on cytostatics? Have you ever made any proposals for improving your practice? The statistical test show that there is a significant association between these issues and the warehouse and transportation box. The values of "p" are respectively p = 0.008, 0.004 and 0.000 and p = 0.011, 0.018 and 0.034, with p < 0.05, so they are dependent. However, the issues: Do you have scheduled breaks between preparation/administration periods? In the last 12 months have you attended training courses on cytostatic manipulation/administration? The statistical test allowed to verify that the questions are independent p > 0.05 of the places where spills occur.

Regarding the "shelf" for the questions: Time of service in the task of manipulating cytostatics? Do you have scheduled break intervals between preparation/administration periods? In the last 12 months have you attended training courses on cytostatic manipulation/administration? The statistical test shows independence, p > 0.05. However, for the questions: By your own initiative, do you often read articles on cytostatics? Have you ever made any proposals for improving your practice? The statistical test shows that there is a significant association between these questions and the "shelf". The values of "p" are respectively p = 0.008 and 0.001, so p < 0.05, so they are related.

Locations such as "trays and laminar flow hood" to the questions: Service time in the task of manipulating cytostatics? In the last 12 months have you attended training courses on cytostatic manipulation/administration? The statistical test confirms its independence p > 0.05. However, for questions, do you have scheduled break intervals between preparation/administration periods? By your own initiative, do you often read articles on cytostatics? Have you ever made any proposals for improving your practice? Statistical testing shows that there is a significant association between these issues and the trays and the laminar flow hood. The values of "p" are respectively p = 0.047, 0.004 and ~0.000, and p = 0.021, 0.002 and 0.002 so p < 0.05, so they are related.

Regarding the "packing table" for the questions, time of service in the task of manipulating cytostatics? Do you have scheduled break intervals between preparation/administration periods? In the last 12 months have you attended training courses on cytostatic manipulation/administration? By your own initiative, you often read articles on cytostatics? The statistical test confirms its independence p > 0.05. However, to the question: Have you ever proposed any improvement to your practices? The statistical test shows that there is a significant association between these issues and the packaging table. The value of p = 0.012, and therefore p < 0.05, so are dependent.

Concerning the "cart of administration" for questions, time of service in the task of manipulating cytostatics? In the last 12 months have you attended training courses on cytostatic manipulation/administration? By your own initiative, you often read articles on cytostatics? Have you ever made any proposals for improving your practice? The statistical test confirms its independence $p > 0.05$. However, for the question: Do you have scheduled pause intervals between preparation/administration periods? The statistical test shows that there is a significant association between these questions and the administration cart. The value of $p = 0.006$, and therefore $p < 0.05$, so they are dependent.

Finally, the "chair of the patient" for the questions: Time of service in the task of handling cytostatics? Do you have scheduled break intervals between preparation/administration periods? The statistical test shows that there is a significant association between these questions and the administration cart. The "p" values are $p = 0.003$ and 0.001, so $p < 0.05$, so they are related. For questions: In the last 12 months have you attended training courses on manipulation/administration of cytostatics? By your own initiative, do you often read articles on cytostatics? Have you ever made any proposals for improving your practice? The statistical test confirms its independence $p > 0.05$.

4 CONCLUSIONS

In the places where the spills occurred, it was verified the existence of situations of dependence and independence. The dependency condition has been found in almost all places, mainly in the questions (Do you often read articles on cytostatics on your own initiative? Have you ever made any proposals to improve your practices?). However, in the trays, in the laminar flow hood, in the administration cart and in the patient's chair, dependence is associated with the question (Does it have scheduled break intervals between preparation/administration periods?). In the storage, transport box and patient's chair there is an association of dependence in the question (Time of service in the task of manipulation/administration of cytostatics). In the remaining tabulations in the result of the statistical test $p > 0.05$, that is to say, being independent.

ACKNOWLEDGMENT

This work has been supported by COMPETE: POCI-01-0145-FEDER-007043 and FCT – Fundação para a Ciência e Tecnologia within the Project Scope: UID/CEC/00319/2013.

REFERENCES

Amorim, A. (1995) Introdução às Ciências Sociais. Aveiro: Estante Editora.
ARS (2010) Manual Gestão Risco Profissional. Retrieved from https://www.dgs.pt/delegado-de-saude-regional-de.../gestao-do-risco-1-pdf.aspx.
Berruyer, M., Tanguay, C., Caron, N.J., Lefebvre, M., & Bussières, J.F. (2015) Multicenter study of environmental contamination with antineoplastic drugs in 36 Canadian hospitals: a 2013 follow-up study. Journal of Occupational and Environmental Hygiene, 12(2), 87–94. http://doi.org/10.1080/15459624.2014.949725.
Boiano, J.M., Steege, A.L., & Sweeney, M.H. (2015) Adherence to Precautionary Guidelines for Compounding Antineoplastic Drugs: A Survey of Nurses and Pharmacy Practitioners. Journal of Occupational and Environmental Hygiene, 12(9), 588–602. http://doi.org/10.1080/15459624.2015.1029610.
Merger, D., Tanguay, C., Langlois, É., Lefebvre, M., & Bussières, J.F. (2014) Multicenter study of environmental contamination with antineoplastic drugs in 33 Canadian hospitals. International Archives of Occupational and Environmental Health, 87(3), 307–313. http://doi.org/10.1007/s00420-013-0862-0.
NIOSH (2004) NIOSH Alert: preventing occupational exposures to antineoplastic and other hazardous drugs in health care settings. DHHS (NIOSH) Publ No. 2004-165. Retrieved from http://www.cdc.gov/niosh/docs/2004-165/pdfs/2004-165.pdf%5Cn.
Odraska, P., Dolezalova, L., Kuta, J., Oravec, M., Piler, P., Synek, S., & Blaha, L. (2014) Association of surface contamination by antineoplastic drugs with different working conditions in hospital pharmacies. Archives of Environmental & Occupational Health, 69(3), 148–58. http://doi.org/10.1080/19338244.2013.763757.
Silva, J., Arezes, P., Schierl, R. & Costa, N. (2017) Risk of occupational exposure to cytotoxic drugs: the role of handling procedures in hospital workers. International Journal of Medical and Health Sciences, vol 11(10), pp. 522–530.
Turci, R., Sottani, C., Spagnoli, G., & Minoia, C. (2003) Biological and environmental monitoring of hospital personnel exposed to antineoplastic agents: A review of analytical methods. Journal of Chromatography B: Analytical Technologies in the Biomedical and Life Sciences, 789(2), 169–209. http://doi.org/10.1016/S1570-0232(03)00100-4.

Author index

Abreu, M.J. 541
Adamides, Em. 183
Aia, S. 313
Albuquerque Neto, H.C. 555, 577
Almeida, M.N. 555, 577
Alves, L.A. 325
Alves, L.V. 367
Alves, O. 29
Antunes, F. 235
Araújo Mieiro, M.A. 513
Araújo, E.R. 371
Arcas, G. 447
Arezes, P. 23, 189, 279, 447, 589, 613
Atosuo, J. 341
Azevedo, R. 231

Baltazar, A. 261
Barkokébas Jr., B. 137, 143, 279, 377
Barreira, D. 155
Barros, B.L. 389
Beja, J. 235
Bernardo, F.G. 67, 73
Blazik-Borowa, E. 335
Bonifácio, D.S. 35
Bozilov, A. 219
Bracci, M. 443
Braga, A.C. 601, 613
Braga, P. 513, 551, 595
Bragança, S. 23
Brito, P. 29
Broday, E.E. 13

Calvo-Cerrada, B. 149
Candeias, S.M. 91
Capitão, F. 195
Carneiro, P. 437, 589, 601
Carpinteiro, E. 595
Carrillo-Castrillo, J.A. 121, 517
Carvalho, B.C.C.B. 555, 577
Carvalho, F. 307, 459
Carvalho, F.P. 125
Carvalho, J.M. 61

Caseiro, A. 319
Castellucci, I. 23
Castelo Branco, J. 383
Cerqueira, D. 459
Chaves, T.F. 367
Chichorro Gonçalves, M. 401
Clemente, M. 235
Coelho, D.A. 201
Correia, C. 495
Correia, J.A.F.O. 513, 551, 595
Costa, D.M.B. 325, 331
Costa, L.S. 483
Costa, M. 91
Costa, N. 413, 437, 447, 601, 613
Costa, P. 161
Costa, S. 267
Costa, S. 437
Cotrim, T.P. 91
Couto, J.P. 177
Cruz, F.M. 279
Cunha, L. 239
Cunha, M. 109
Cunha, T. 177
Cvetkovic, M. 219
Cvetković, M. 523
Czarnocka, E. 335
Czarnocka, K. 335
Czarnocki, K. 335
Czernecka, W. 419

da Conceição Peixoto Teixeira, D. 297
da Cruz, F.M. 137, 143
da Fonseca, P.G. 307
da Silva, I. 377
Dahlke, G. 395, 477
de F. Cavalcanti, L.L. 225
de França, T.C.M. 143
de la Hoz-Torres, M.L. 273
de P. Xavier, A.A. 13, 453
Delerue-Matos, C. 251, 267
Dias, R.B.B. 213
Dinis, M.L. 407, 425, 523
Dionísio, F.D. 413

do Couto, A. 377
Domingues, P. 189
Dores, A.R. 389
Draghici, A. 19, 607
Duarte, J. 383
Dufour, C.M. 607

Falcão, G.A.M. 367
Fernandes, A. 267
Fernandes, A. 601
Fernandes, C. 561
Ferreira, A. 109, 155, 165, 195, 245, 257, 261, 285, 319
Ferreira, I. 239
Ferreira, R.V. 325
Figueiredo, J.P. 109, 155, 165, 195, 261, 285, 291, 319
Fiuza, A.M. 523
Fotopoulou, K. 183

Gabriel, A. 319
Galante, E.B.F. 35, 325
Gaspar da Silva, A. 47
Gibson, M. 529
Glisovic, S. 219
Golubovic, T. 219
Gomes, V. 551
Gonçalves, F.S.M. 361
Gonçalves, G. 85
Gonçalves, M. 29
Górny, A. 301, 419
Guedes, J.C. 67, 73, 347, 355
Guerreiro, M.J. 91

Haddad, A.N. 35
Hankiewicz, K. 471
Hartwig, M. 97
Hippert, M.A.S. 79
Hola, B. 335

Jacinto, C. 29
Janicas, A.S. 91
Job, A. 431

Katsakiori, P. 183
Kivistö-Rahnasto, J. 131
Kumara Jayathilaka, K.R. 471

Lago, E.M.G. 137, 143
Lança, A. 155, 245, 257, 261
Laranjeira, P. 335
Leão, C.P. 413
Lilius, E.-M. 341
Lima, T.G. 361, 507
Liston, P.M. 443
Longo, O.C. 79
Lopes, C. 177
López, J.M. 447
López-Guillen, A. 149
López-Guillén, A. 149
Lourenço, M.L. 201

Machado, C.F. 207
Machado, J.I.F. 555
Machado, O. 231
Maheronnaghsh, S. 547
Malta, M. 125
Mamrot, E. 97
Manta, R. 137
Marchigiani, E. 443
Marinho, R.M.F. 361
Marques, M. 351
Martí-Armengual, G. 149
Martínez-Aires, M.D. 273
Martín-Morales, M. 273
Martins, A.R.B. 137, 143
Martins, E.B. 279
Martins, L.B. 225, 371
Martins, R.P. 67, 73
Matos, A.C. 401
Matos, M.L. 161, 383
Melo, M.B.F.V. 367
Melo, R.B. 307, 459
Mendes, C. 245, 257
Miguel, A.S. 561
Mocan, A. 19
Monteiro, A. 351
Monteiro, E. 29
Monteiro, R. 231
Monteiro, R. 583
Morais, S. 251, 267
Moreira, F. 155, 195
Moreira, P.H. 347
Morgado, C.V. 325, 331
Mrugalska, B. 529

Naves, L.B. 541
Neves, R. 235
Neves, T. 261
Nóbrega, J.S. 207

Nóbrega, J.S.W. 325
Nunes, F.O. 47
Nunes, I.L. 489

Oliveira, A.D.S. 577
Oliveira, C. 513, 551, 595
Oliveira, G. 285
Oliveira, J.M. 125
Oliveira, M. 103
Oliveira, M. 267
Oliveira, P.A.A. 297, 495, 567
Oostingh, G. 319
Osório, N. 319

Pais, R. 567
Paixão, S. 245, 257
Palmitesta, P. 443
Pardo-Ferreira, M.C. 121, 517
Parlangeli, O. 443
Paulos, I. 351
Pavón, I. 447
Pechmann, A. 489
Pedro, L. 235
Pedrosa, L. 161
Pedrosa, L.R. 143
Peixoto, C. 251
Pereira, A. 351
Pereira, C. 239
Pereira, J. 535
Pereira, M. 85
Pereira, M.C. 251, 267
Pérez, J.A. 355
Phatrabuddha, N. 171
Pinho, R. 501
Pinto, D. 115
Pinto, E. 85
Pisco Almeida, A.M. 235
Porto, S. 137
Ptak, T. 395

Quintela, D.A. 103

Rabello, M.V.T. 331
Raimundo, A.M. 103
Raposo, A. 501
Rebelo, M. 335
Reinvee, M. 313
Reis, A.C. 53, 61
Reis, A.M. 573
Reis, C. 513, 551, 595
Reis, R. 177
Reis, V. 91
Ribeiro, C.A. 91
Ribeiro, E. 351
Rigatou, V. 183
Rodrigues, C. 1, 371

Rodrigues, F. 115, 235
Rodrigues, M.A. 389, 583
Rodrigues, N. 561
Roxo, I.N. 165
Rubio-Romero, J.C. 121, 517
Ruiz Padillo, D.P. 273

Sá, M.M. 231
Sampaio, P. 189
Sa-ngiamsak, T. 171
Santos Baptista, J. 115, 347, 377, 383, 495, 501, 561
Santos, C. 165
Santos, J. 291, 547
Santos, M. 239, 483
Santos, V. 291
Sanz-Gallen, P. 149
Saraiva, C. 189
Schierl, R. 613
Sgourou, E. 183
Sigcha, L. 447
Silva, A.A.R. 213
Silva, A.J.A. 507
Silva, A.R.M.V. 555, 577
Silva, A.S. 407, 425
Silva, D. 239
Silva, F. 109, 285
Silva, I.S. 431
Silva, J. 613
Silva, J.E.M.R. 1
Silva, J.F. 513, 551, 595
Silva, L.B. 213
Silva, M. 465
Silva, R.M. 213
Silveira, F. 335
Simões, H. 285, 535
Slezakova, K. 251, 267
Sousa, A. 85
Sousa, C. 85
Sousa, D. 231
Sousa, M. 245, 257
Souza, R. 535
Stojiljković, E. 219, 523
Suominen, E. 341
Swart, T. 351
Szer, J. 335

Talaia, M. 465
Tappura, S. 131
Tavares, J. 231
Teixeira, A. 319
Teixeira, A.F. 161
Teixeira, J.P. 267
Teixeira, L. 465
Teixeira, S. 561
Tender, M. 177

Tomé, D. 583
Torres Costa, J. 501
Torres, M. 589
Turkiewicz, K. 477

Valado, A. 319
Vasconcelos Pinto, M. 291
Vasconcelos, B.M. 137, 143

Vaz, M. 53, 547
Vergara, L.L.G. 573
Viegas, C. 7
Viegas, S. 41
Vieira, C. 351
Vieira, R. 235
Vilén, L. 341
Virgílio, A. 103

Wictor, I.C. 453

Yingratanasuk, T. 171

Zambrana-Ruíz, A. 517
Zarte, M. 489
Zeglin, B.V. 573
Zlatar, T. 137, 143